W9-BEC-170

Algebra 1

An Incremental Development

Second Edition

Algebra 1

An Incremental Development

Second Edition

JOHN H. SAXON, JR.

SAXON PUBLISHERS, INC.

Algebra 1: An Incremental Development
Second Edition
Teachers Edition

Printed in the United States of America

ISBN: 0-939798-43-3

Editor and production supervisor: Nancy Warren
Compositor: Black Dot Graphics

Second printing: January 1991

Printed on recycled paper

Saxon Publishers, Inc.
1320 W. Lindsey
Norman, Oklahoma 73069

Contents

Preface

This is the second edition of the first book in a three-book series designed to prepare students for calculus. In addition to the standard topics of algebra I, this book begins the study of geometry by concentrating on the concepts of area and volume. The study of probability is begun with problems that teach average, basic probability, and probability of a series of independent events.

This series of books is designed to teach students to be successful mathematical problem solvers. Students who have completed the series make high scores on tests of problem solving. Mathematical problem solving is simply the use of mathematical concepts in new situations. Teaching concepts is the first task. The best way to teach the concepts is to let the student work carefully designed problems that lead to use of productive thought patterns which utilize the concepts. Students do not grasp concepts quickly, and long-term practice is required at each hierarchical level. Whereas students understand slowly, they seem to forget quickly. The research of Benjamin Bloom has indicated that long-term practice beyond mastery is required to permit students to achieve what Bloom calls "automaticity." Bloom tells us that it is necessary to "overlearn" to achieve automaticity. When automaticity has been achieved, the student will be able to read a new problem, and the concept or concepts that are required to solve the problem will come automatically to mind. Long-term practice with the skills necessary to apply the concepts has allowed the student to automate these skills. Because concept recognitions and skill applications have been automated, the student's mind is freed from the lower-level mechanics of the problem, and the student can consider the problem at a higher cognitive level. When automaticity is achieved, the concepts are emblazoned in the long-term memory of the student and can be recalled when needed.

philosophy We have found that it is difficult to design problems that teach concepts and applications at the same time. In this book many word problems are designed so that the concept is the only real thing in the problem. For example, one ratio problem considers "the ratio of the erudite to the unlettered." If students protest that they don't know what the words mean, tell them it makes no difference. If these words bother students, they should use *eaters* and *uncles* or any two words that begin with *e* and *u*. If students learn to work word problems that contain unfamiliar words, they will not have trouble with chemistry problems containing words such as *sulfur dioxide* and *trinitrotoluene.* **If students learn the concepts, they can apply the concepts in any situation.** Books that try to teach applications and concepts at the same time often fail. **The teaching of the concepts must come first.**

Thus, the philosophy of this book is that students learn by doing and that students cannot learn a concept on the day it is introduced. This is the reason that the problem sets contain only three or four problems of the new kind and contain twenty-six or twenty-seven review problems. This emphasis on review allows students to practice every concept previously presented in every problem set. Some teachers who use the book for the first time remark that there are not enough problems of the new kind for students to "get it." They seem to forget that the students did not "get it" when they had 20 problems of the new kind and no review. Teachers have historically handed out review problem sets because the students were not "getting it." This book simply goes one step further and totally reverses the emphasis. We introduce the new concept with three or four problems with the realization that understanding will take time and that constant practice over a long period of time is the key ingredient of success.

We do not try for total understanding on the first day. **Since students learn by doing the problems repetitively, the teacher's overriding responsibility is to ensure that every student does every problem in every problem set.** If students work the problems, they will learn and they will understand. If students do not work the problems, they will not learn. It's that simple. The best way to ensure that students work the problems is to devote most of the class period to doing the homework.

After students have practiced a concept at one level for a period of time, another lesson on the concept is presented in which the concept is discussed at a higher cognitive level. This is the reason we say that the **incremental development** is not just distributed practice. The incremental development builds instead of just reviewing.

teaching procedure

The following recommended procedure has been used successfully by many teachers. Begin the class by projecting the answers to the previous problem set on the overhead so that students can mark their incorrect answers. Then collect the homework. **Do not answer questions at this time.** Every problem on last night's homework will be in the next homework problem set. Tell students to hold their questions until they begin to do the homework. Then explain the new topic thoroughly but briefly. Remember that the students will not grasp the concept totally on the first day. Then say, "Let's begin the homework now while we still have 40 minutes of class time. Begin with the problem that is the hardest for you. If you need help, hold up your hand and I will help you." Suppose seven hands go up. You have five bright students. You must use them. Say "Mary help Frank and Roger help Susan. I will help you, Jimmy." The students who give the help will benefit almost as much as the students who receive help. Then walk around the room for the rest of the period, giving help one-on-one and directing help as needed. We have found that this method permits every student to complete at least half of the homework problems in class and permits students to complete every hard problem in class. Thus they only have a few easy problems to do at home. If this procedure is followed, there will be no questions about tonight's homework at the beginning of class tomorrow, because all the hard problems will have been completed in class today. This method gives you time to move through the classroom unhurriedly and to have the one-on-one contact with students that the lecture method of teaching does not permit.

Do not attempt to grade the homework. Thirty students with 30 problems each for five classes equals 4500 problems a day and 90,000 problems a month. It is impossible for a teacher to check 90,000 problems a month. However, the homework must be spot-checked to ensure that all problems have been completed and that all steps have been shown. We have found that 30 papers can be spot-checked and marked satisfactory or unsatisfactory in less than 4 minutes. This includes entries in the grade book. If one point is awarded for each satisfactory homework assignment, there will be four homework points available between two tests. Weight the test problems to total 96 points. Giving too much credit for homework is not a good practice. This is one way to do it. There are other ways to handle homework. Use what works for you.

innovative techniques The inclusion of geometry in this book and other considerations required that several topics normally taught in the *Algebra 1* book be moved to the *Algebra 2* book. Among these are factoring trinomials that have a nonunity lead coefficient, chemical mixture word problems, and "boat-in-the-river" problems. This was done so that these important topics can be taught effectively to all students by giving them the necessary emphasis over a considerable period of time. Geometry topics include perimeter, area, volume, degree measure, volume and area conversions. This incorporation of geometry provides an integrated as well as incremental approach.

Algebra books have historically taught word problems by using special procedures that have limited applicability. For example, nickel and dime problems may be taught by using the format $(32 - x)$ for the number of nickels if the total number of coins is 32. This is done because coin problems have been attempted before two equations in two unknowns have been introduced and before students have been taught to solve these systems by using substitution or elimination. The basic equation is that the number of nickels plus the number of dimes equals 32. The $(32 - x)$ is a trick that requires the substitution to be made mentally. When the student encounters problems about nickels, dimes and quarters, the $(32 - x)$ trick cannot be used and the student has not been taught the skills necessary to handle this problem. In this book, coin problems are introduced after systems of equations have been mastered. Students have no difficulty with the problems and are prepared for the more advanced problems when they are encountered in algebra 2.

The quadratic formula is introduced in the latter part of the text. The student, however, will have been using parts of the quadratic formula over a period of months in evaluation practice. When the student then is introduced to the quadratic formula in its entirety, most of it will already be familiar.

Another innovation in the book is the use of meaningful diagrams to teach percent problems. These diagrams are a must! A student may say, "I can work the problem but I cannot draw the diagram." We believe that this student is really saying, "I have a procedure that will get an answer that is often wrong and I do not understand the problem well enough to draw the diagram." We believe that the diagram is very important and should be required.

grade placement This book is not a grade-level book. It is designed to teach fundamental concepts to students of all ability levels. Bright students use this book in the eighth grade. Good students will use this book in the ninth grade. Average students will use this book in the tenth grade, and very slow students will use this book in the eleventh or twelfth grade. The Saxon *Algebra 2* book is designed to be used the year after this book and should be followed by the Saxon *Advanced Mathematics* book. We do not believe that it is fair to the slower or less motivated students to place them in a watered-down book. Almost all students can master the concepts if a gentle, patient approach is used and if students are allowed the time to practice the concepts.

Placement is very important in using the Saxon series. After three to four weeks a teacher can identify several students that are not prepared for algebra 1 and are doomed to fail. These students should be moved to an algebra $\frac{1}{2}$ class where they can succeed. Students who successfully complete *Algebra* $\frac{1}{2}$ will be ready for *Algebra 1* the following year and will succeed. Algebra 1 students who do not earn a good solid C the first semester should be awarded an incomplete (I) for the semester and should be allowed to repeat the first semester of algebra 1 in the spring. Giving the student an F or a D and letting the student continue leads to another F or D in June. We have found that "looping" the students is successful, as about 50 percent make an A or a B in the spring and about 80 percent make at least a good solid C. Then they can complete the *Algebra 1* book in the summer or in an out-of-sequence class in the fall. "Looping" works and is highly recommended.

Almost every school that has used the entire Saxon series has doubled senior math enrollment, increased college board scores 20 percent, quadrupled calculus enrollment, doubled physics enrollment, increased chemistry enrollment significantly, and, best of all, reduced the number of students who take nonacademic math courses by over 50 percent.

acknowledgments I am indebted to Tom Brodsky for his help in revising this book. His help in selecting and ordering the geometry problems in the problem sets is especially appreciated. I thank Gary Cavender for his suggestions. I thank Joan Coleman for supervising the preparation of the manuscript, and David Pond, Scott Kirby, Chad Threet, John Chitwood, Julie Webster, Smith Richardson, Tony Carl, Gary Skidmore, and Kevin McKeown for their artwork.

John Saxon
Norman, Oklahoma

Basic

3 Review Lessons
114 Algebra Lessons

REVIEW LESSON A Addition and subtraction of fractions · Lines and segments

A.A addition and subtraction of fractions

To add or subtract fractions that have the same denominators, we add or subtract the numerators as indicated below, and the result is recorded over the same denominator.

$$\frac{5}{11} + \frac{2}{11} = \mathbf{\frac{7}{11}} \qquad \frac{5}{11} - \frac{2}{11} = \mathbf{\frac{3}{11}}$$

If the denominators are not the same, it is necessary to rewrite the fractions so that they have the same denominators.

	PROBLEM	REWRITTEN WITH EQUAL DENOMINATORS	ANSWER
(a)	$\frac{1}{3} + \frac{2}{5}$	$\frac{5}{15} + \frac{6}{15}$	$\mathbf{\frac{11}{15}}$
(b)	$\frac{2}{3} - \frac{1}{8}$	$\frac{16}{24} - \frac{3}{24}$	$\mathbf{\frac{13}{24}}$

A mixed number is the sum of a whole number and a fraction. Thus the notation

$$13\frac{3}{5}$$

does not mean 13 multiplied by $\frac{3}{5}$ but instead 13 plus $\frac{3}{5}$.

$$13 + \frac{3}{5}$$

When we add and subtract mixed numbers, we handle the fractions and the whole numbers separately. In some subtraction problems it is necessary to borrow, as shown in (e).

	PROBLEM	REWRITTEN WITH EQUAL DENOMINATORS	ANSWER
(c)	$13\frac{3}{5} + 2\frac{1}{8}$	$13\frac{24}{40} + 2\frac{5}{40}$	$\mathbf{15\frac{29}{40}}$
(d)	$13\frac{3}{5} - 2\frac{1}{8}$	$13\frac{24}{40} - 2\frac{5}{40}$	$\mathbf{11\frac{19}{40}}$
		BORROWING	
(e)	$13\frac{3}{5} - 2\frac{7}{8}$	$13\frac{24}{40} - 2\frac{35}{40} = 12\frac{64}{40} - 2\frac{35}{40}$	$\mathbf{10\frac{29}{40}}$

A.B lines and segments

It is impossible to draw a mathematical line because a mathematical line is a **straight line** that has no **width** and has **no ends.** To show the location of a mathematical line, we draw a pencil line and put arrowheads on both ends to emphasize that the mathematical line goes on and on in both directions.

A X C

We can name a line by naming any two points on the line and using an overbar with two arrowheads. We can designate the line shown by writing $\overleftrightarrow{CX}$, $\overleftrightarrow{XC}$, $\overleftrightarrow{AX}$, $\overleftrightarrow{XA}$, $\overleftrightarrow{AC}$, or $\overleftrightarrow{CA}$. A part of a line is called a **line segment.** To show the location of a line segment, we use a pencil line with no arrowheads. We name a segment by naming the endpoints of the segment.

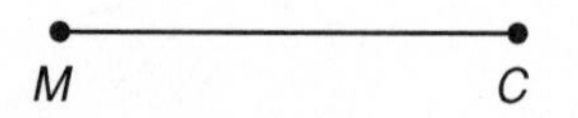

This is segment MC or segment CM. We can indicate that two letters name a segment by using an overbar with no arrowheads. Thus $\overline{MC}$ means segment MC. If we use two letters without the overbar, we designate the length of the segment. Thus MC is the length of $\overline{MC}$.

example A.1 Segment AC measures $10\frac{1}{4}$ units. Segment AB measures $4\frac{3}{7}$ units. Find BC.

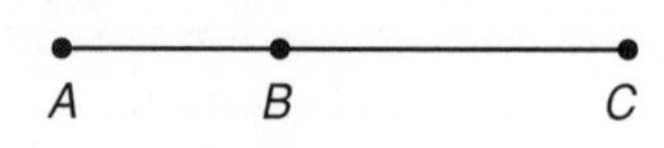

solution We need to know the length of segment BC. We know AC and AB. We subtract to find BC.

$$\begin{aligned} BC &= AC - AB \\ &= 10\frac{1}{4} - 4\frac{3}{7} \\ &= 10\frac{7}{28} - 4\frac{12}{28} && \text{common denominators} \\ &= 9\frac{35}{28} - 4\frac{12}{28} && \text{borrowed} \\ &= \mathbf{5\frac{23}{28}\ units} && \text{subtracted} \end{aligned}$$

problem set A Add or subtract as indicated. Write answers as proper fractions reduced to lowest terms or as mixed numbers.

1. $\frac{1}{5} + \frac{2}{5}$ **2.** $\frac{3}{8} - \frac{2}{8}$ **3.** $\frac{4}{3} - \frac{1}{3} + \frac{8}{3}$

Different denominators:

4. $\frac{1}{3} + \frac{1}{5}$ **5.** $\frac{3}{8} - \frac{1}{5}$ **6.** $\frac{2}{3} - \frac{1}{8}$

7. $\frac{1}{13} + \frac{1}{5}$ **8.** $\frac{17}{15} - \frac{2}{3}$ **9.** $\frac{5}{9} + \frac{2}{5}$

10. $\frac{14}{17} - \frac{6}{34}$ **11.** $\frac{5}{13} + \frac{1}{26}$ **12.** $\frac{4}{7} - \frac{2}{5}$

13. $\frac{4}{7} + \frac{1}{8} + \frac{1}{2}$ **14.** $\frac{3}{5} + \frac{1}{8} + \frac{1}{8}$ **15.** $\frac{5}{11} - \frac{1}{6} + \frac{2}{3}$

Addition of mixed numbers:

16. $2\frac{1}{2} + 3\frac{1}{5}$ **17.** $7\frac{3}{8} + 4\frac{7}{3}$ **18.** $1\frac{1}{8} + 7\frac{2}{5}$

Subtraction with borrowing:

19. $15\frac{1}{3} - 7\frac{4}{5}$ **20.** $42\frac{3}{8} - 21\frac{3}{4}$ **21.** $22\frac{2}{5} - 13\frac{7}{15}$

22. $421\frac{1}{11} - 17\frac{4}{3}$ **23.** $78\frac{2}{5} - 14\frac{7}{10}$ **24.** $43\frac{1}{13} - 6\frac{5}{8}$

25. $21\frac{1}{5} - 15\frac{7}{13}$ **26.** $21\frac{2}{19} - 7\frac{7}{10}$ **27.** $43\frac{3}{17} - 21\frac{9}{10}$

28. The measure of $\overline{AB}$ is $7\frac{1}{8}$ units. The measure of $\overline{BC}$ is $5\frac{2}{7}$ units. Find AC.

A B C

29. XL is $42\frac{1}{7}$ units. LC is $24\frac{2}{11}$ units. Find CX.

X C L

30. KX is $74\frac{1}{11}$ units. YZ is $22\frac{1}{3}$ units. KZ is $44\frac{2}{3}$ units. Find YX.

X Y Z K

REVIEW LESSON B *Geometry review · Perimeter · Area*

B.A perimeter

If two lines cross, we say that the lines **intersect.** The place where the lines cross is called the **point of intersection.** Two lines in the same plane either intersect or do not intersect. If two lines in the same plane do not intersect, we say that the lines are **parallel lines.** The perpendicular distance between two parallel lines is everywhere the same.

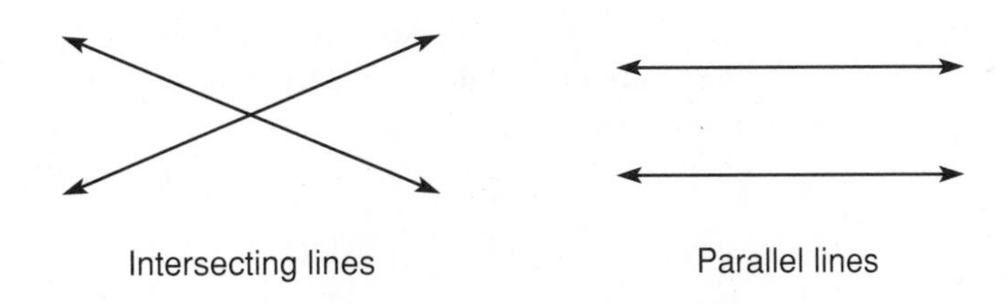

Intersecting lines Parallel lines

If two lines make square corners at the point of intersection, we say that the lines are **perpendicular.** The angles made by perpendicular lines are called **right angles.** We can draw a little square at the point of intersection to indicate that all four angles formed are right angles. Two right angles form a straight angle. An angle smaller than a right angle is called an **acute angle.** An angle greater than a right angle but less than a straight angle is called an **obtuse angle.**

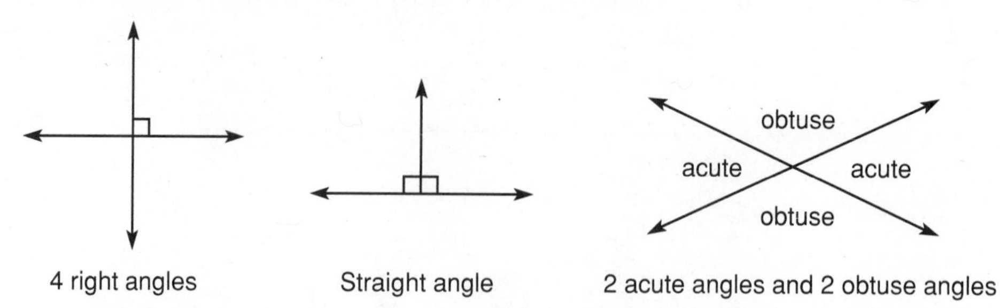

4 right angles　　Straight angle　　2 acute angles and 2 obtuse angles

Polygons are simple, closed, planar (flat) geometric figures whose sides are line segments.

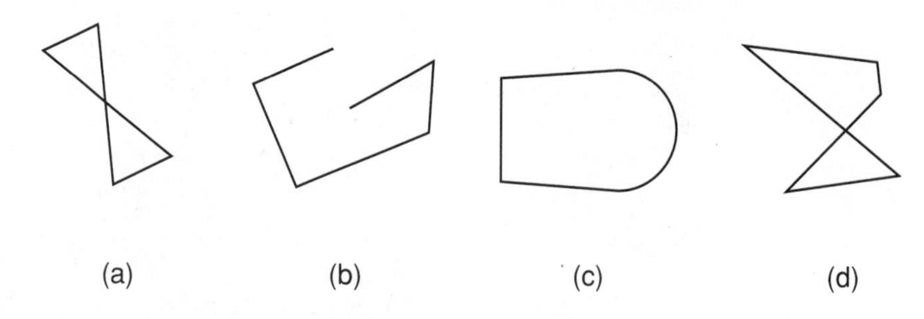

(a)　(b)　(c)　(d)

Figures (a) and (d) are not polygons because the segments cross and the figures are not simple closed figures. Figure (b) is not a polygon because it is not closed. Figure (c) is not a polygon because one of the "sides" is curved.

If a polygon has an indentation (a cave), the polygon is a **concave** polygon. If there is no indentation, the polygon is a **convex** polygon.

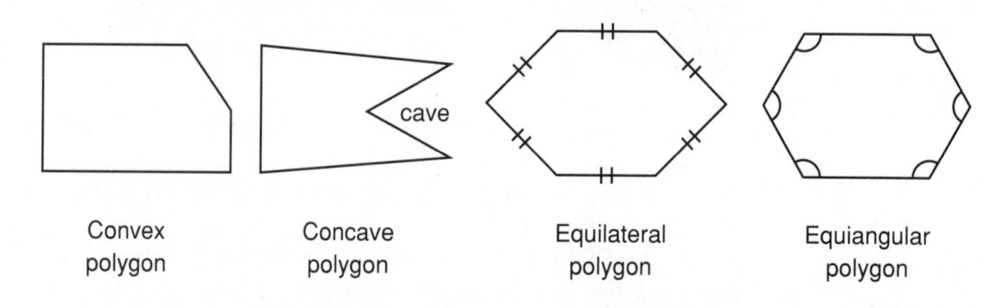

Convex polygon　Concave polygon　Equilateral polygon　Equiangular polygon

A "corner" of a polygon is called a **vertex.** The plural of vertex is **vertices.** We note that if a polygon has four sides it also has four vertices. If a polygon has five sides, it has five vertices. The number of vertices always equals the number of sides. If all the sides of a polygon have the same length, the polygon is an **equilateral polygon.** If all the angles have the same measure, the polygon is an **equiangular polygon.**

Polygons are named according to the number of sides.

3 sides　4 sides　5 sides　6 sides　7 sides　8 sides

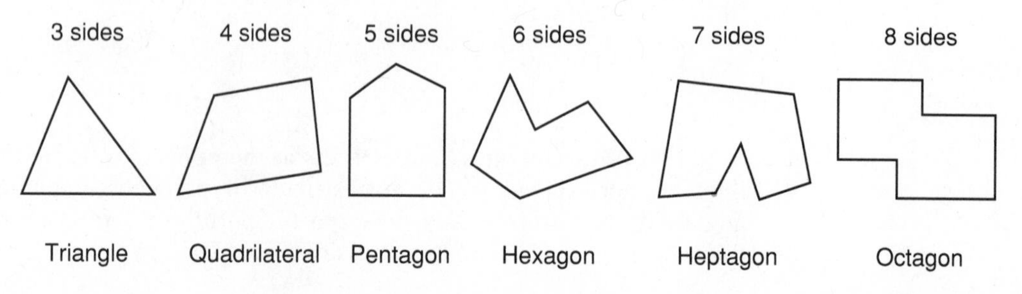

Triangle　Quadrilateral　Pentagon　Hexagon　Heptagon　Octagon

A polygon of 9 sides is a nonagon. A polygon of 10 sides is a decagon. A polygon of 11 sides is an undecagon. A polygon of 12 sides is a dodecagon. Polygons of more than 12 sides do not have special names. To name these polygons, we use a number and the suffix *-gon.* Thus a polygon of 42 sides would be a 42-gon.

A **trapezoid** is a quadrilateral that has exactly two parallel sides. A **parallelogram** is a quadrilateral with two pair of parallel sides. A **rectangle** is a parallelogram with four right angles. A **rhombus** is an equilateral parallelogram. A **square** is a rhombus with four right angles.

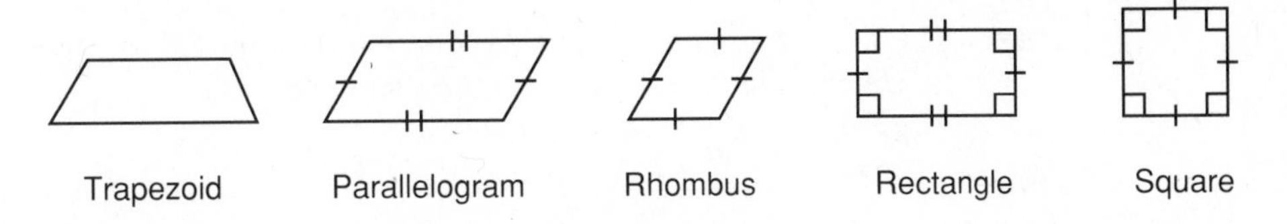

Trapezoid Parallelogram Rhombus Rectangle Square

In these figures we use equal tick marks to denote sides whose lengths are equal. Polygons in which all sides have the same length and all angles have the same measure are called **regular polygons.**

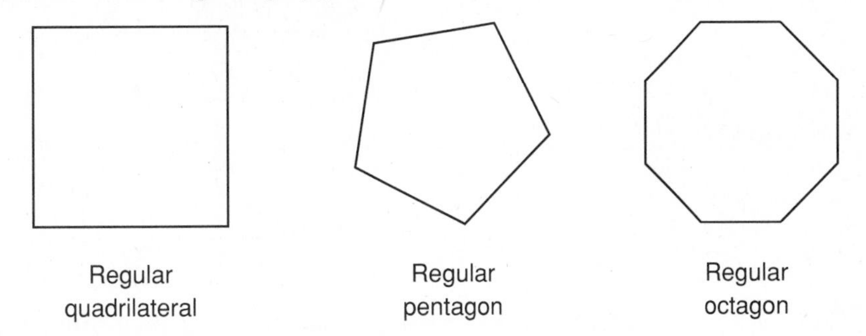

Regular quadrilateral Regular pentagon Regular octagon

The Greek prefix *peri-* means around. The Greek word *metron* means measure. Thus the word **perimeter** means the measure around.

example B.1 Find the perimeter of this figure. Dimensions are in feet. All angles are right angles.

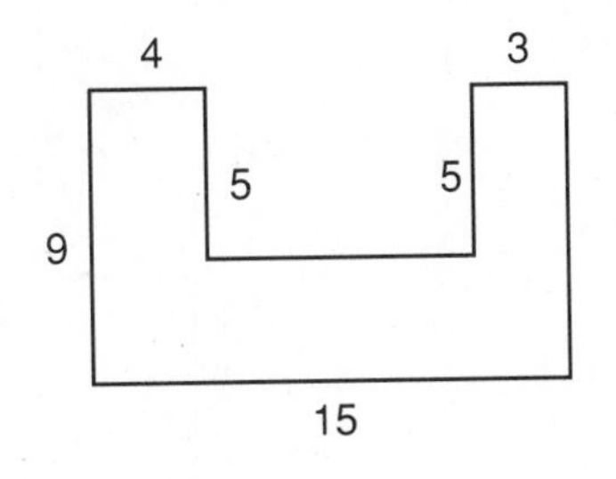

solution All the angles are right angles, so we begin by finding the missing lengths.

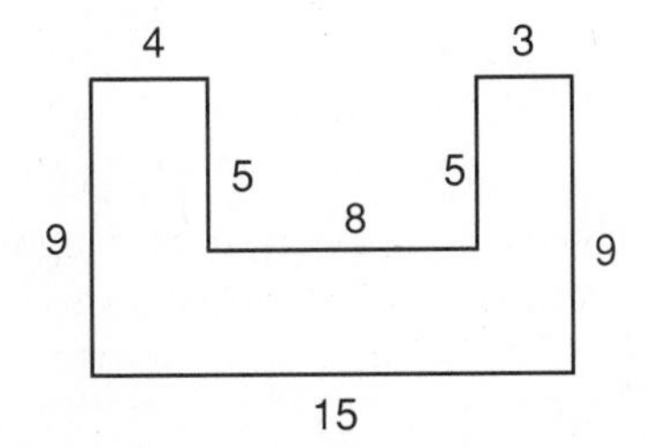

Since it is 15 feet across the bottom, it must be 15 feet across the top. The missing length is 8 feet because $4 + 3 + 8 = 15$. The height of the left-hand side is 9 feet, so the height of the right-hand side is 9 feet. Thus

$$\text{Perimeter} = (9 + 4 + 5 + 8 + 5 + 3 + 9 + 15) \text{ feet}$$
$$= \mathbf{58 \text{ feet}}$$

B.B area

The area of a figure tells us the number of square floor tiles it will take to cover the figure completely. Here are three rectangles.

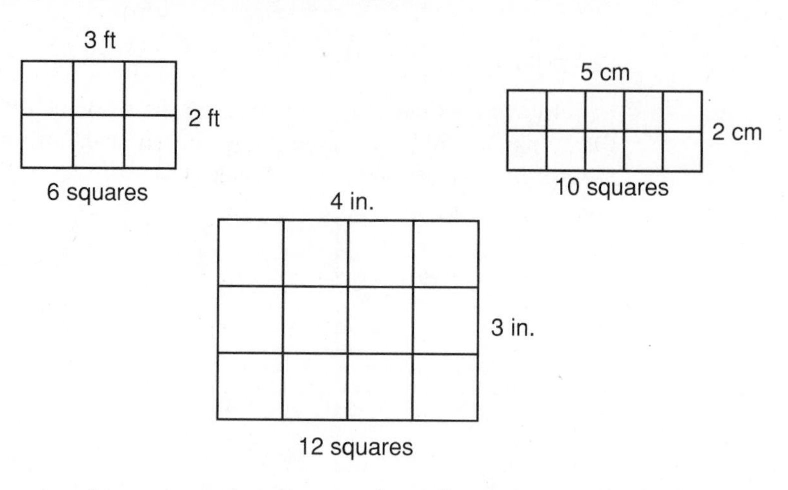

In each rectangle, we note that the number of squares equals the length times the width. This is true for any rectangle.

Area of a rectangle = length × width

example B.2 Find the area of this figure. Dimensions are in centimeters. All angles are right angles.

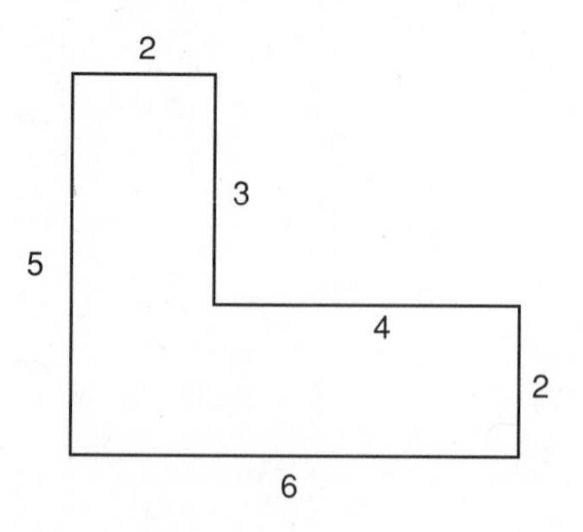

solution To find the area, we divide the figure into rectangles. Then we find the area of each of the rectangles, and add. We show two different ways to work this problem.

Area $A = 5\text{ cm} \times 2\text{ cm} = 10\text{ cm}^2$
Area $B = 4\text{ cm} \times 2\text{ cm} = \underline{\ 8\text{ cm}^2}$
18 cm²

Area $C = 3\text{ cm} \times 2\text{ cm} = 6\text{ cm}^2$
Area $D = 6\text{ cm} \times 2\text{ cm} = \underline{12\text{ cm}^2}$
18 cm²

B.C triangles

To find the area of a triangle, we choose one side and call it the **base.** The perpendicular distance from the base to the other vertex of the triangle is called the **height** of the triangle or the **altitude** of the triangle. The area of a triangle equals one-half of the product of the base and the height.

$$\text{Area of a triangle} = \frac{\text{base} \times \text{height}}{2}$$

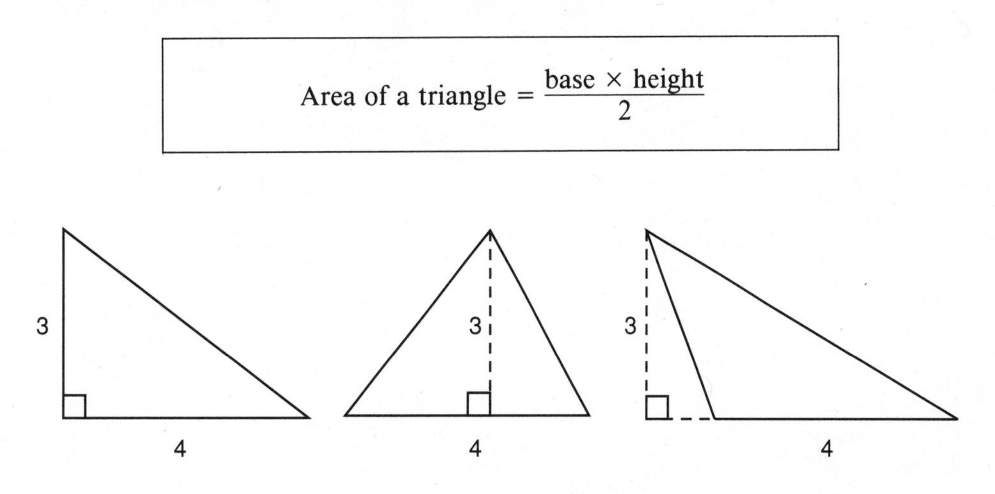

Each of these triangles has an area of 6 square units because each has a base of 4 and an altitude of 3.

$$\text{Area} = \frac{b \times h}{2} = \frac{4 \times 3}{2} = \textbf{6 square units}$$

example B.3 Find the area of this figure. Corners that look square are square.

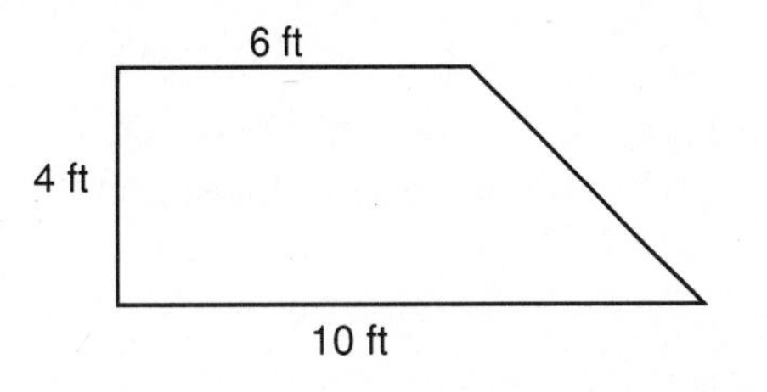

solution First we divide the figure into a rectangle and a triangle.

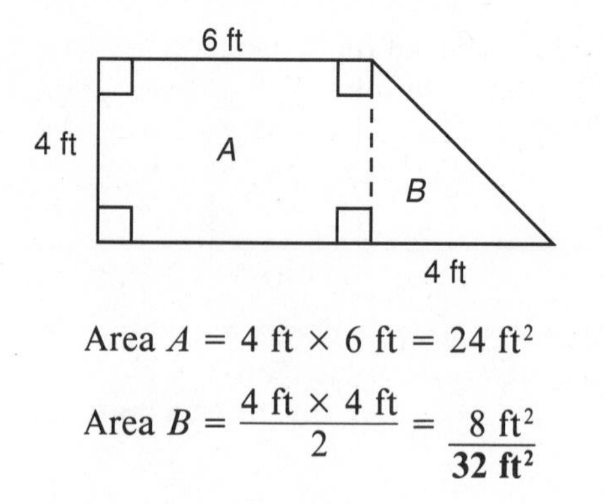

$$\text{Area } A = 4 \text{ ft} \times 6 \text{ ft} = 24 \text{ ft}^2$$

$$\text{Area } B = \frac{4 \text{ ft} \times 4 \text{ ft}}{2} = \underline{8 \text{ ft}^2}$$

$$\mathbf{32 \text{ ft}^2}$$

example B.4 Find the area of the shaded portion of this figure. All angles are right angles. Dimensions are in meters.

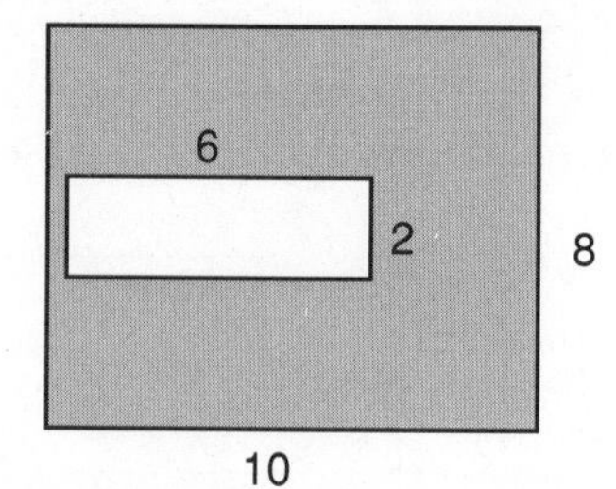

solution The area of the figure in the preceding example was found by adding smaller areas. We can find the shaded area in this problem by subtracting the area of the small rectangle from the area of the large rectangle.

$$\begin{aligned} \text{Large area} &= 8 \text{ m} \times 10 \text{ m} = 80 \text{ m}^2 \\ -\ \text{Small area} &= 2 \text{ m} \times 6 \text{ m} = \underline{-12 \text{ m}^2} \\ & \qquad\qquad\qquad\quad \mathbf{68 \text{ m}^2} \quad \text{subtracted} \end{aligned}$$

B.D circles

Every point on a circle is the same distance from the center of the circle. This distance is called the **radius** of the circle. The diameter of a circle is twice the length of the radius of the circle.

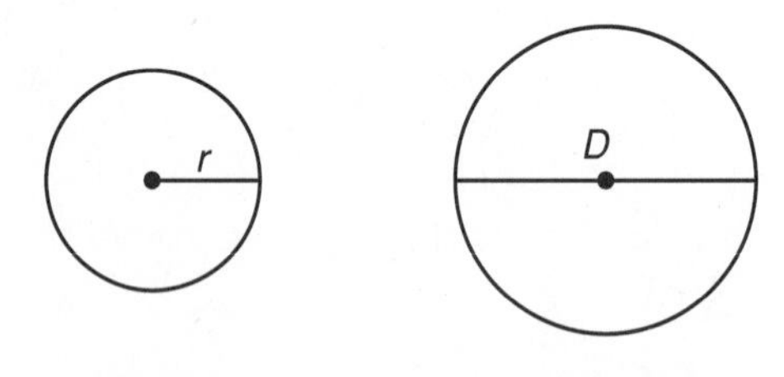

Radius Diameter

It takes more than 3 diameters to go around a circle. It takes fewer than 4 diameters to go around a circle. It takes approximately 3.14 diameters to go around a circle.

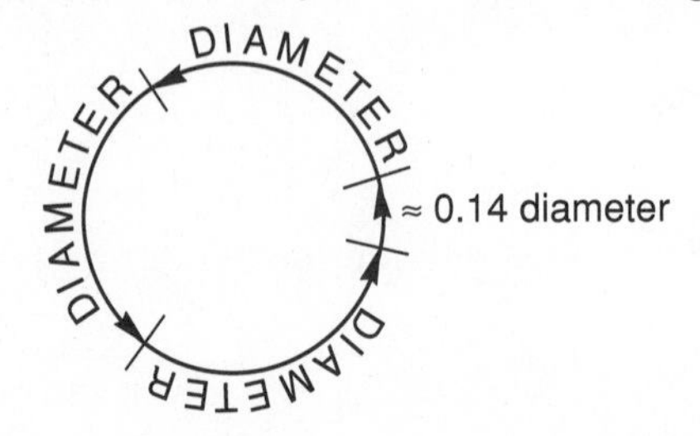

The number of diameters it takes to go around a circle is the same for every circle. We call this number **pi** (pie) and use the symbol π to represent this number. To write the number exactly would take an infinite number of digits. If you depress the π key on your calculator, you will get

$$3.141592654$$

We call the perimeter of a circle the **circumference** of the circle. If we know the diameter, we can find the circumference. If the diameter is 12 inches, the distance around the circle is 12 times π, which is approximately (12×3.14) inches. Since the radius is one-half the diameter, it takes 2π radii to go around the circle.

$$\text{Circumference} = \pi D = \pi(2r) = 2\pi r$$

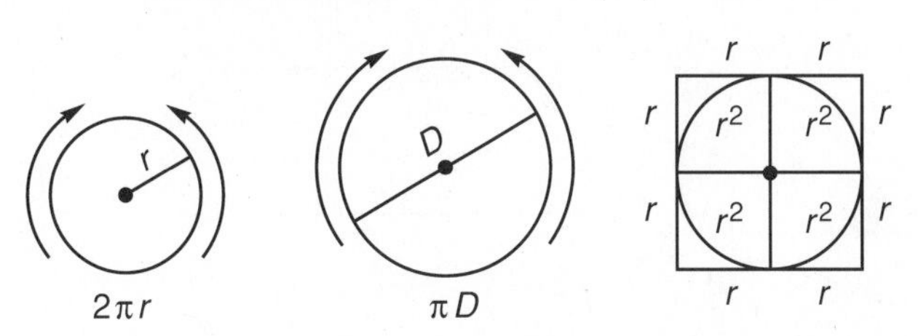

In the figure on the right we show a circle inside a big square. Each side of each little square has the length r. So the area of each little square is r times r, or r^2. There are four little squares, and their total area is $4r^2$.

$$\text{Area of the big square} = 4r^2$$

In the figure, we see that the area of the circle is less than the area of the big square, which equals the area of the four little squares. The area of the circle is greater than the area of three little squares. The area of the circle is exactly π little squares. Isn't that a coincidence? It takes π diameters to go all the way around, and the area of the circle equals π times the area of one of the little squares!

$$\text{Area of a circle} = \pi r^2$$

example B.5 Find the area of this figure. The dimensions are in meters. Lines that look parallel are parallel.

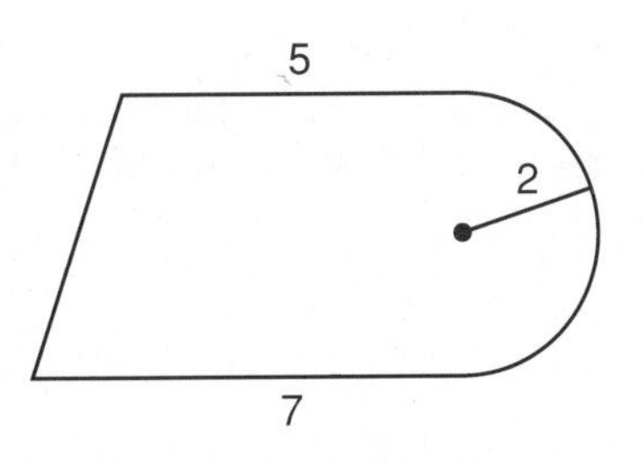

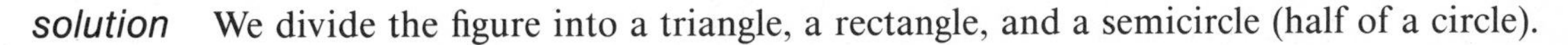

solution We divide the figure into a triangle, a rectangle, and a semicircle (half of a circle).

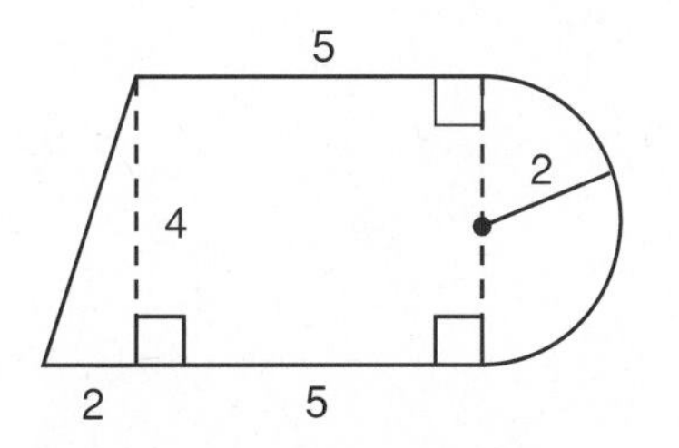

$$\text{Total area} = \text{area of triangle} + \text{area of rectangle} + \text{area of semicircle}$$

$$= \frac{2 \times 4}{2} + (5 \times 4) + \frac{\pi(2)^2}{2}$$

$$= 4 + 20 + 6.28 = \mathbf{30.28\ m^2}$$

These problems are arithmetic problems. A calculator can be used to help with the arithmetic. Use the π key on the calculator, or use 3.14 as an approximation for π as we did here.

example B.6 (a) Find the area of the trapezoid.
(b) Find the area of the parallelogram. Dimensions are in meters.

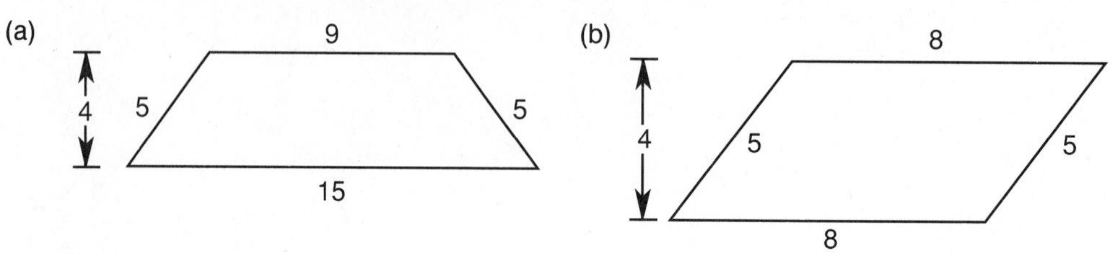

solution There is a formula for the area of a trapezoid. There is a formula for the area of a parallelogram. These formulas are hard to remember because they are used so seldom. The easiest way to find the areas is to divide the figures into two triangles.

(a) The altitude of both triangles is 4. The base of one triangle is 9, and the base of the other triangle is 15.

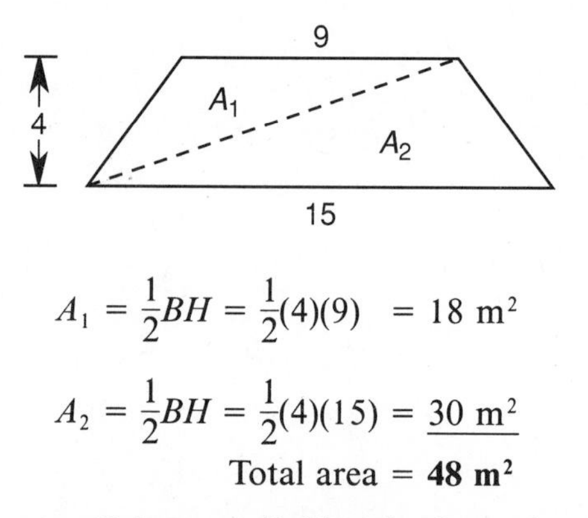

$$A_1 = \frac{1}{2}BH = \frac{1}{2}(4)(9) = 18\ \text{m}^2$$

$$A_2 = \frac{1}{2}BH = \frac{1}{2}(4)(15) = \underline{30\ \text{m}^2}$$

$$\text{Total area} = \mathbf{48\ m^2}$$

(b) The diagonal of a parallelogram divides the figure into two triangles whose areas are equal. Both bases are 8 and both altitudes are 4.

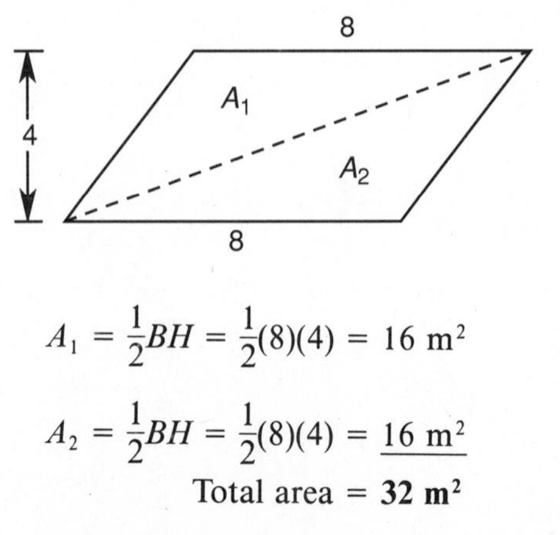

$$A_1 = \frac{1}{2}BH = \frac{1}{2}(8)(4) = 16\ \text{m}^2$$

$$A_2 = \frac{1}{2}BH = \frac{1}{2}(8)(4) = \underline{16\ \text{m}^2}$$

$$\text{Total area} = \mathbf{32\ m^2}$$

practice a. Find the perimeter of this figure. Dimensions are in centimeters. Lines that look parallel are parallel.

b. Find the area of the same figure.

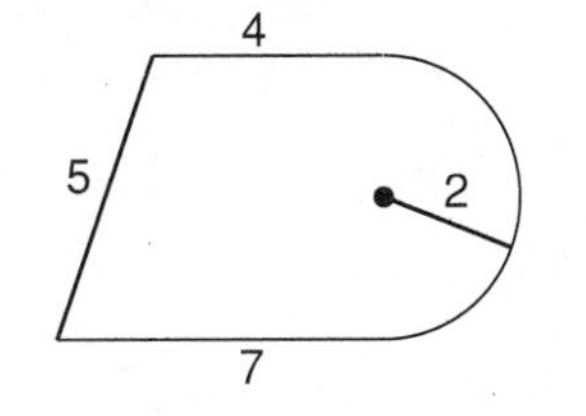

c. The area of a rectangle is 42 square meters. The length of one side is 6 meters. What is the length of the other side?

d. The circumference of a circle is 14 inches. What is the approximate radius of the circle? Use a calculator as necessary.

e. The area of a rectangle is 42 square feet. The length of one side is 6 feet. What is the perimeter of the rectangle?

f. Find the area of this trapezoid. Dimensions are in meters.

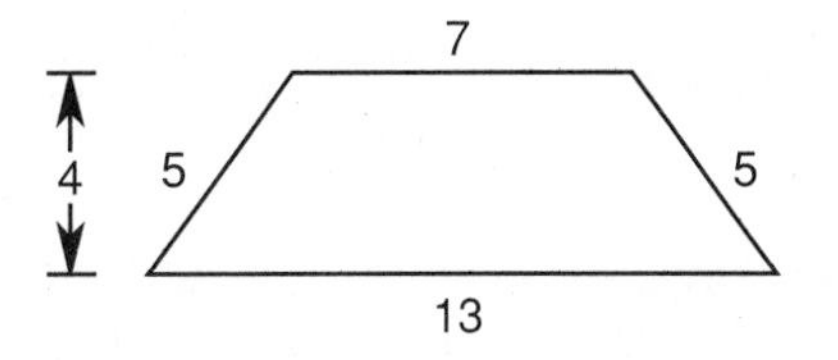

g. Find the area of this parallelogram. Dimensions are in feet.

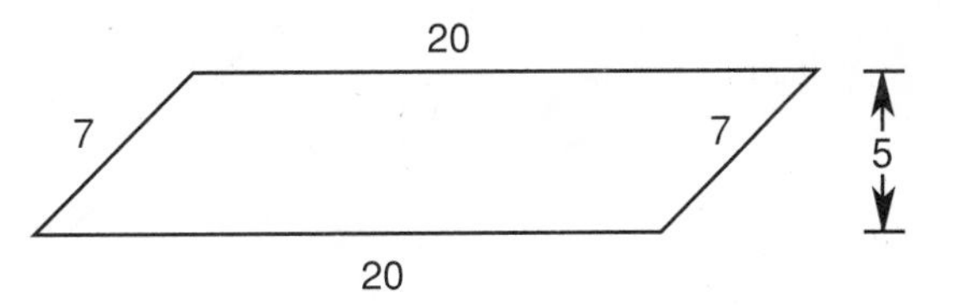

problem set B Add or subtract as indicated:

1. $\frac{3}{5} + \frac{2}{7}$
2. $\frac{7}{2} + \frac{13}{4}$
3. $\frac{8}{3} + \frac{1}{15}$
4. $5\frac{2}{5} + 7\frac{11}{10}$
5. $8\frac{1}{3} + 3\frac{7}{5}$
6. $9\frac{3}{5} + 5\frac{8}{3}$
7. $23\frac{7}{10} - 14\frac{6}{7}$
8. $42\frac{3}{8} - 14\frac{3}{4}$
9. $22\frac{2}{5} - 14\frac{8}{15}$
10. $426\frac{1}{11} - 16\frac{5}{3}$
11. $8\frac{2}{5} - 3\frac{7}{3}$
12. $42\frac{3}{13} - 5\frac{2}{5}$
13. *BD* is $10\frac{3}{7}$ inches, and *CD* is $3\frac{1}{5}$ inches. Find *CB*.

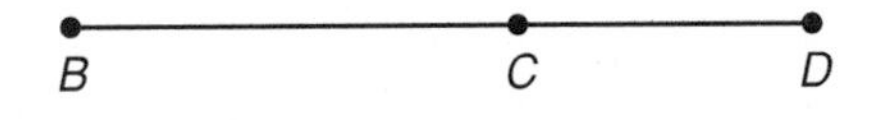

14. The diameter of a circle is 4 centimeters. What is the area of the circle?
15. The area of a square is 36 square inches. What is the length of one side of the square?
16. The circumference of a circle is 42 meters. What is the approximate radius of the circle?

Find the perimeter of each figure. Dimensions are in centimeters. All angles are right angles.

17. **18.**

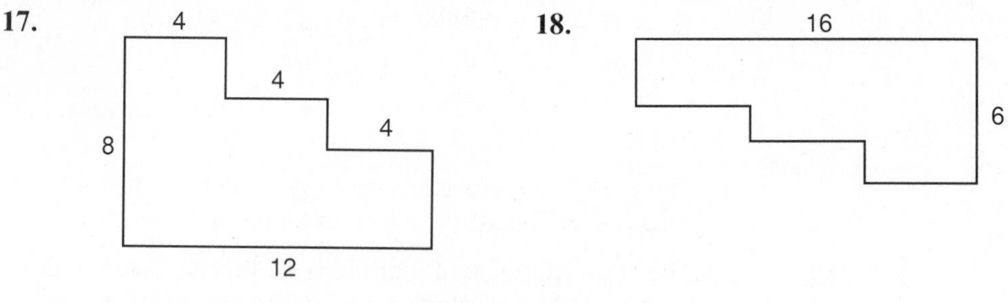

Find the area of each figure. Dimensions are in meters.

19. **20.**

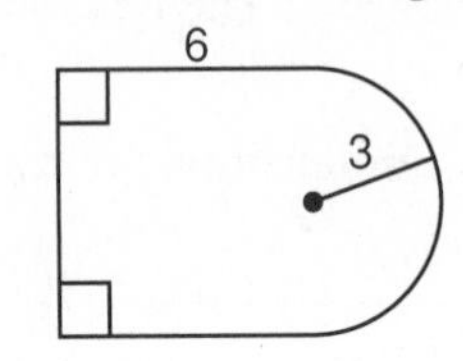

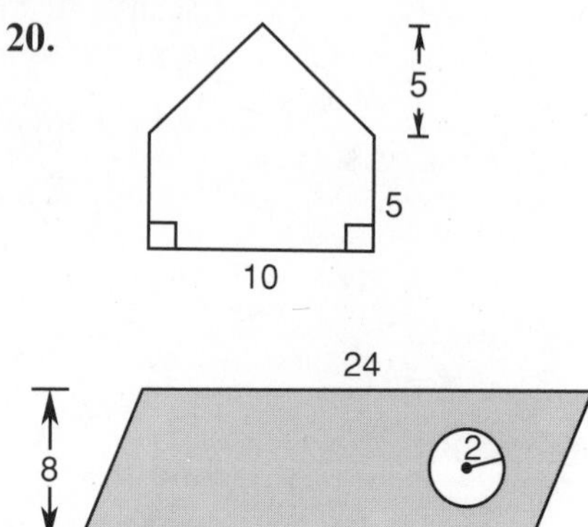

21. Find the area of the shaded portion of this parallelogram. Dimensions are in centimeters.

Subtract:

22. $4\frac{2}{3} - 1\frac{8}{3}$ **23.** $16\frac{1}{4} - 5\frac{17}{2}$ **24.** $14\frac{9}{2} - 12\frac{4}{3}$

25. $121\frac{5}{8} - 6\frac{21}{3}$ **26.** $26\frac{5}{7} - 4\frac{8}{5}$ **27.** $93\frac{2}{7} - 12\frac{14}{5}$

28. $14\frac{7}{3} - 7\frac{14}{5}$ **29.** $15\frac{2}{11} - 3\frac{11}{2}$ **30.** $93\frac{1}{5} - 6\frac{8}{3}$

REVIEW LESSON C *Geometric shapes · Volume · Degree measure*

C.A geometric shapes

In mathematics we call a flat surface a **plane.** Planar geometric figures are figures that can be drawn on a flat surface. Polygons are closed planar geometric figures whose sides are line segments. The polygon with the fewest number of sides is the triangle. If a triangle has one right angle, the triangle is called a **right triangle.** If one angle in the triangle is an obtuse angle, the triangle is called an **obtuse triangle.** If all the angles in a triangle are acute angles, the triangle is called an **acute triangle.**

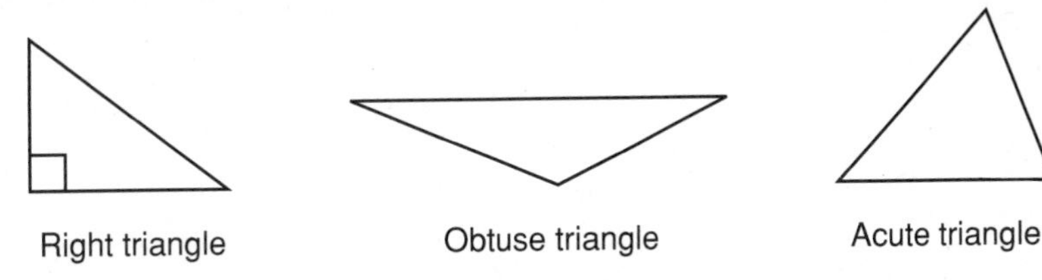

An **equilateral triangle** has three sides whose lengths are equal. An **isosceles triangle** has two sides whose lengths are equal. A **scalene triangle** has no sides whose lengths are equal. We use tick marks to designate equal lengths and to designate angles whose measures are equal.

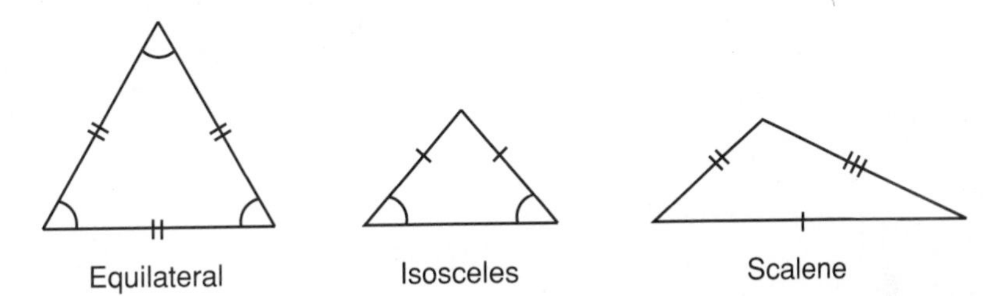

All three sides in an equilateral triangle have the same length. The three angles also have equal measures. Two sides in an isosceles triangle have equal lengths. The angles opposite these sides are also equal (have equal measures). **In any triangle the angles opposite equal sides are equal angles (have equal measures). Also, the sides opposite equal angles have equal lengths.** The scalene triangle has no equal sides, so no two angles have equal measures.

C.B volume

We remember that the area of a surface tells us the number of square floor tiles it will take to cover the surface. An area of 46 square feet (ft^2) can be covered with 46 square floor tiles whose sides are 1 foot long. An area of 46 square miles (mi^2) can be covered with 46 square floor tiles whose sides are 1 mile long. An area of 46 square meters (m^2) can be covered with 46 square floor tiles whose sides are 1 meter long.

We use the word *volume* to tell us the number of cubes it takes to completely occupy a particular three-dimensional space. Geometric figures that have three dimensions are called **geometric solids.** A **cube** is a six-sided geometric solid whose faces are all squares of the same dimensions.

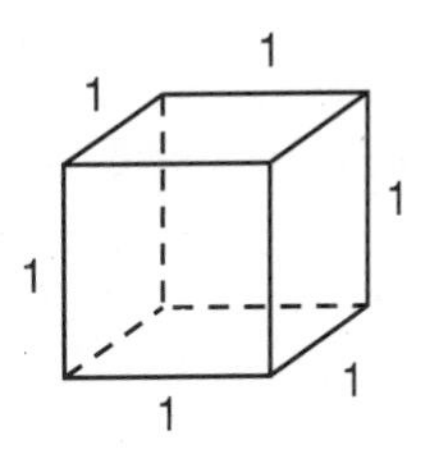

If each edge is 1 meter long, the cube has a volume of 1 cubic meter (1 m^3). If each edge is 1 mile long, the cube has a volume of 1 cubic mile (1 mi^3). If each edge is 1 foot long, the cube has a volume of 1 cubic foot (1 ft^3).

When we discuss volume, it is helpful to think of sugar cubes. We can visualize volume by mentally stacking sugar cubes. On the left we have a rectangle that measures 2 feet by 3 feet. It has an area of 6 square feet.

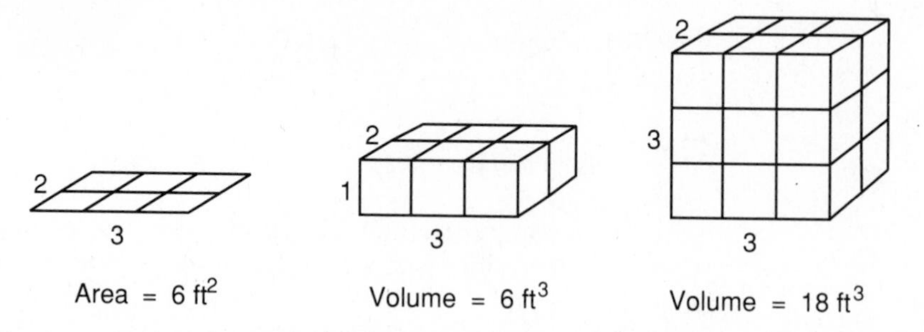

In the center figure we have placed a 1-cubic-foot sugar cube on each square. There are 6 cubes, so we say the volume is 6 ft^3. In the right-hand figure we have stacked the cubes 3 deep. There are 18 cubes, so we say that the volume is 18 ft^3. From this we can induce that the volume equals the number of cubes on the bottom layer times the number of layers. We can extend this idea to any geometric solid whose sides are perpendicular to the base. Because the sides of a solid are perpendicular to the base, we call the solid a **right solid.** The "top" and the "bottom" of a right solid are called the **bases** of the solid and are identical geometric figures. If the bases of a right solid are polygons, the solid is called a **prism.** The volume of a right solid equals the number of cubes that we can place on the bottom layer (area of the base) times the number of layers (height of the solid).

> VOLUME OF A RIGHT SOLID
>
> The volume of a right solid equals the product of the area of the base and the height of the solid.

example C.1 The base of a right solid with vertical sides is the isosceles triangle shown. The height of the solid is 6 meters. What is the volume of the solid?

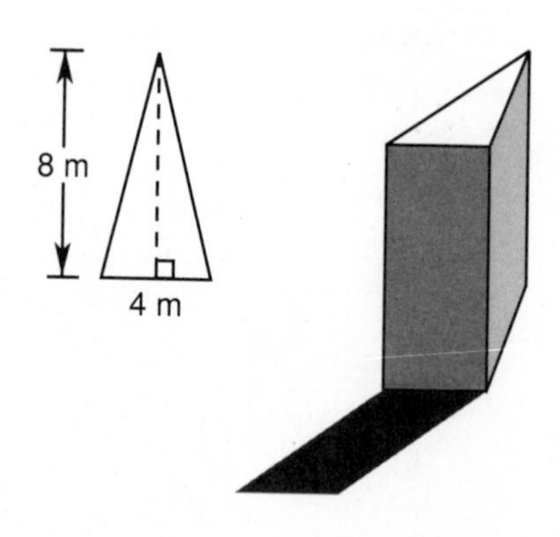

solution The area of the base is 4×8 divided by 2, or 16 m^2. This means that 16 one-meter sugar cubes (crushed, of course) would cover the base completely to a depth of one meter. The height of the solid is 6 meters. If we stacked the sugar cubes 6 deep, we would have 6 layers of 16 cubes in a layer, or 96 sugar cubes in all.

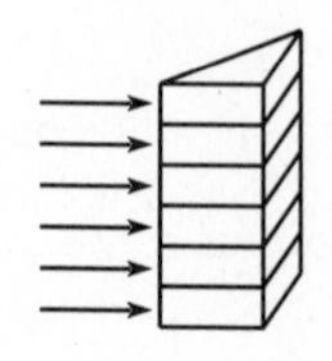

$$\text{Volume} = (\text{area of base}) \times (\text{height})$$
$$= (16 \text{ m}^2)(6 \text{ m}) = \mathbf{96\ m^3}$$

example C.2 A right circular cylinder has a radius of 6 cm and a height of 20 cm as shown. What is the volume?

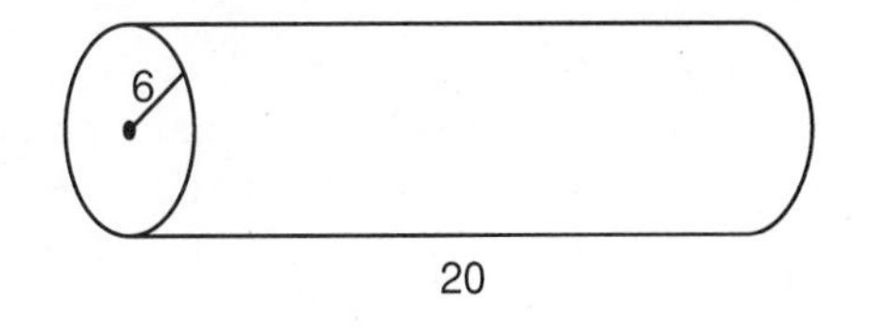

solution We multiply the area of the base times the height.

$$\text{Volume} = \text{area} \times \text{height}$$
$$= \pi(6^2)(20) = \mathbf{720\pi\ cm^3}$$

example C.3 The base of an irregular solid has an area of 426.3 square feet. The sides are perpendicular to the base and the height is 20 feet. What is the volume of the solid?

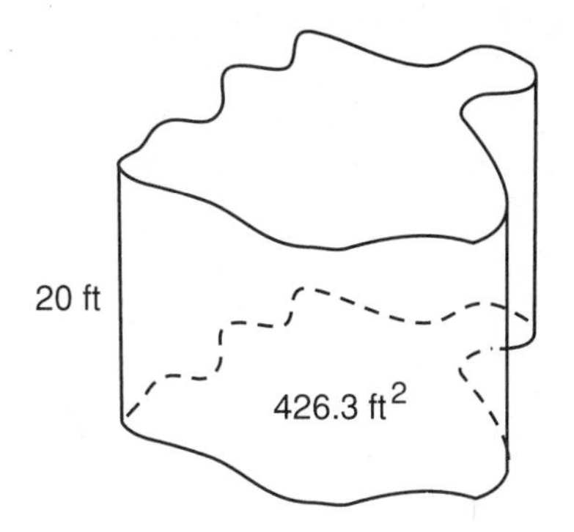

solution We can put 426.3 one-cubic-foot sugar cubes in the bottom layer. If we stack the cubes 20 deep, we get

$$\text{Volume} = (426.3 \text{ ft}^2)(20 \text{ ft}) = \mathbf{8526\ ft^3}$$

C.C degree measure

If the central angle of a circle is divided into 360 equal parts, each of the parts has a measure of 1 degree. One-fourth of a central angle is a right angle. A right angle has a measure of 90 degrees, which we can write as 90°. Two right angles form a straight angle. A straight angle has a measure of 180°.

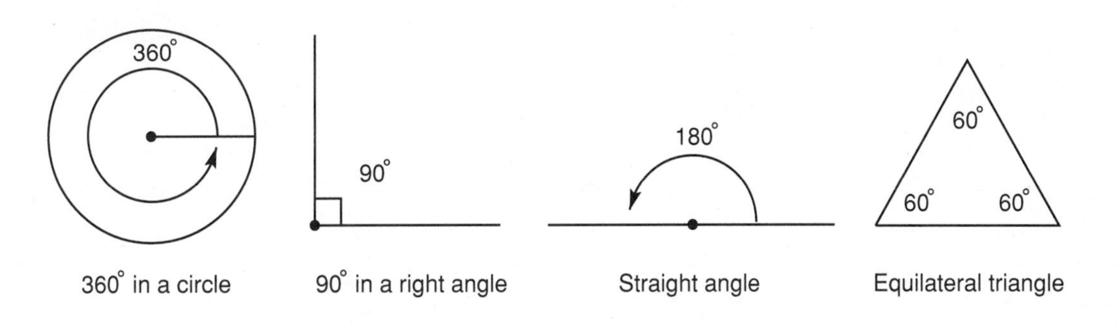

360° in a circle | 90° in a right angle | Straight angle | Equilateral triangle

The sum of the measurements of the three angles of any triangle is 180°. An equilateral triangle has three equal angles, and each angle has a measure of 60°.

practice

a. Find the area of this shape. Dimensions are in feet.

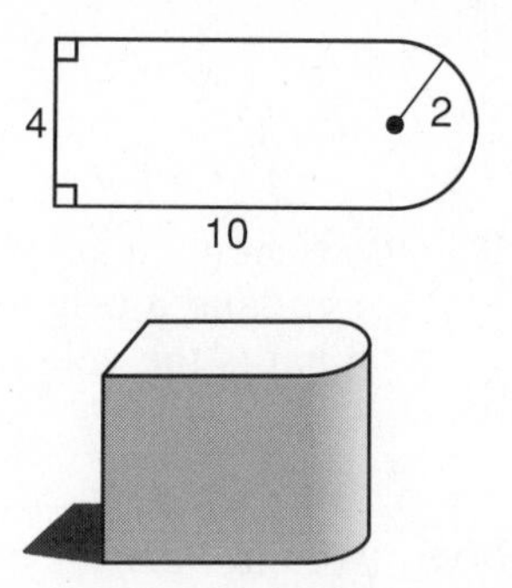

b. The shape shown is the base of a right solid whose sides are 10 feet high. What is the volume of the solid?

c. What kind of triangle has three sides whose lengths are equal?

d. What kind of triangle has only two sides whose lengths are equal?

e. What do we call a quadrilateral that has exactly two parallel sides?

f. What is the degree measure of a straight angle? What is the degree measure of a right angle?

problem set C

Add or subtract as indicated:

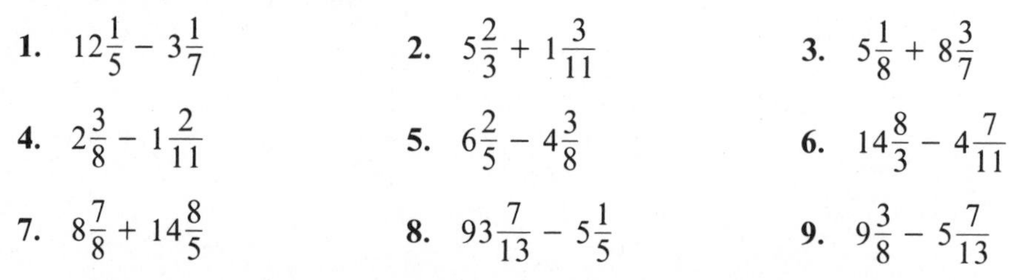

1. $12\frac{1}{5} - 3\frac{1}{7}$ **2.** $5\frac{2}{3} + 1\frac{3}{11}$ **3.** $5\frac{1}{8} + 8\frac{3}{7}$

4. $2\frac{3}{8} - 1\frac{2}{11}$ **5.** $6\frac{2}{5} - 4\frac{3}{8}$ **6.** $14\frac{8}{3} - 4\frac{7}{11}$

7. $8\frac{7}{8} + 14\frac{8}{5}$ **8.** $93\frac{7}{13} - 5\frac{1}{5}$ **9.** $9\frac{3}{8} - 5\frac{7}{13}$

10. The base of a right circular cylinder has a diameter of 10 feet. The cylinder is 10 feet long. What is the volume of the cylinder?

11. The base of a right solid is a triangle whose altitude is 14 in. and whose base is 4 in. The solid is 10 in. high. What is the volume of the solid?

12. Define a rectangle. Define a square. Is every square also a rectangle?

13. KA is $14\frac{1}{3}$ cm. MK is $12\frac{2}{5}$ cm. What is AM?

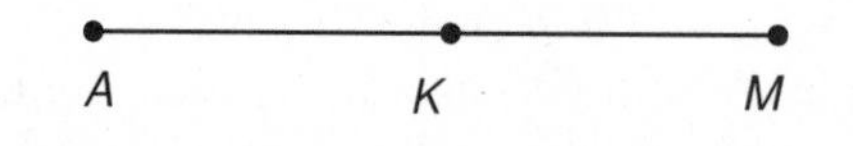

14. The circumference of a circle is 10 centimeters. What is the approximate diameter of the circle?

15. The perimeter of a rectangle is 140 ft. Two of the sides are 10 ft long. What is the length of each of the other two sides?

Find the perimeter of each figure. Dimensions are in miles. All angles are right angles.

16. **17.**

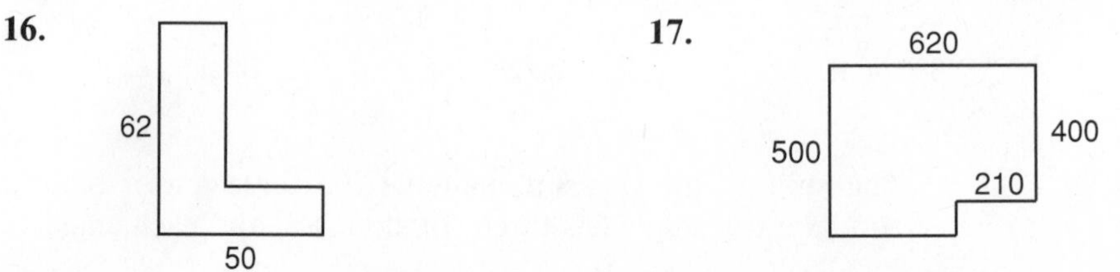

Find the area of each figure. Dimensions are in meters. Lines that look parallel are parallel.

18. **19.**

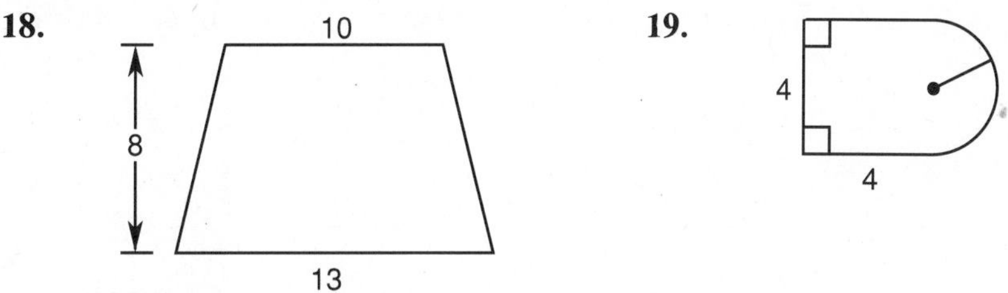

20. Find the area of the shaded portion of this figure. Dimensions are in kilometers.

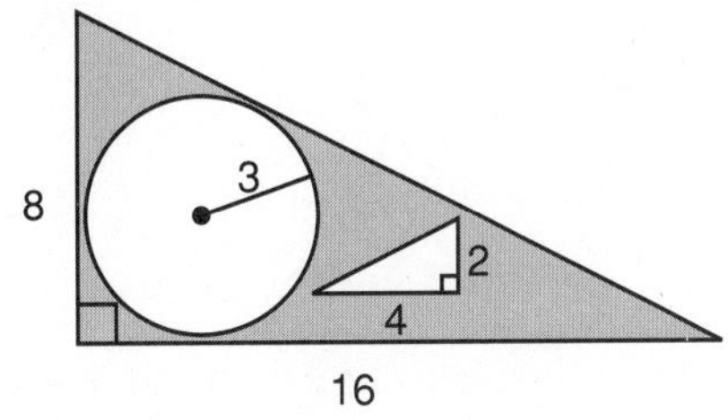

Add or subtract as indicated:

21. $5\frac{7}{8} - 4\frac{11}{12}$ **22.** $6\frac{8}{5} + 14\frac{9}{10}$ **23.** $33\frac{5}{8} - 7\frac{2}{5}$

24. $5\frac{11}{12} - 4\frac{12}{13}$ **25.** $7\frac{1}{8} + 2\frac{3}{11}$ **26.** $93\frac{2}{5} - 1\frac{11}{12}$

27. $4\frac{9}{13} - 2\frac{1}{11}$ **28.** $9\frac{2}{5} - 7\frac{1}{8}$ **29.** $35\frac{1}{7} - 6\frac{2}{3}$

30. Find the sum without using a calculator:

$$41 + 163 + 97.5 + 0.072 + 94.32 + 0.05$$

LESSON 1 *Real numbers and the number line · Multiplication and division of fractions · Unit multipliers*

1.A numerals and numbers

A number is an idea. A numeral is a single symbol or a collection of symbols that we use to express the idea of a particular number.

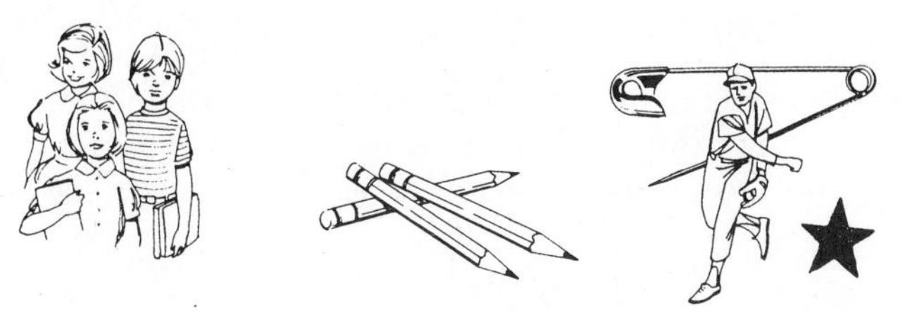

The three drawings above all have the quality of *threeness.* The three children and the three pencils both bring to mind the idea of *three.* The drawing at the right also brings to mind the idea of *three*, although all the things in the drawing are not of the same kind.

If we wish to use a symbol to designate the idea of three, we could write any of the following:

$$\text{III}, \quad 3, \quad \frac{30}{10}, \quad \frac{27}{9}, \quad \frac{33}{11}, \quad 2 + 1, \quad 6 \div 2, \quad 11 - 8$$

Each of these is a symbolic representation of the idea of 3. Throughout the book when we use the word *number*, we are describing the idea. We will use numerals to designate the numbers. But we will remember that none of the marks we make on paper are numbers because

A number is an idea!

Since the symbols

$$3 \quad \text{and} \quad \frac{30}{10}$$

are both numerals that represent the same number, we say that they have the same value. **Thus, the value of a numeral is the number represented by the numeral, and we see that the words *value* and *number* have the same meaning.**

1.B natural or counting numbers

The system of numeration that we use to designate numbers is called the **decimal system.** It was invented by the Hindus of India, passed to their Arab neighbors, and finally transmitted to Europe circa 1200 A.D. The decimal system uses 10 symbols that we call **digits.** These digits are

$$0, \quad 1, \quad 2, \quad 3, \quad 4, \quad 5, \quad 6, \quad 7, \quad 8, \quad 9$$

We use these digits by themselves or in combination with one another to form the numerals that we use to designate decimal numbers.

We call the numbers that we use to count objects or things the *natural numbers* or the *counting numbers*. When we begin counting, we always begin with the number 1 and follow it with the number 2, etc.

$$1, \; 2, \; 3, \; 4, \; 5, \; 6, \; 7, \; 8, \; 9, \; 10, \; 11, \; 12, \; 13, \ldots$$

It would not be natural to try to count by using numbers such as $\frac{1}{2}$ or 0 or $\frac{3}{4}$, so these numbers are not called natural or counting numbers. We designate the natural or counting numbers with the listing above. The three dots after the number 13 indicate that this listing continues without end.

1.C real numbers

The numbers of arithmetic are zero and the positive real numbers. **We say that a positive real number is any number that can be used to describe a physical distance greater than zero.** Thus, all the numbers shown here

$$\frac{3}{4} \qquad 0.000163 \qquad 363 \qquad 3\frac{3}{8} \qquad 46 \qquad \frac{11}{7} \qquad 400.1623232323$$

are positive real numbers, for all of them can be used to describe physical distances when used with descriptive units such as feet, yards, etc.

$\frac{3}{4}$ mile   0.000163 yard   363 feet $3\frac{3}{8}$ meters

46 inches $\frac{11}{7}$ kilometers   400.1623232323 centimeters

The number zero is not a positive number, but it can be used to describe a physical distance of no magnitude. Thus we say that zero is a real number. In addition to the positive numbers and zero, in algebra we use numbers that we call **negative numbers,** and these numbers are also called real numbers. The ancients did not understand or use negative numbers. A man could not own negative 10 sheep. If he owned any sheep at all, the number of sheep had to be designated by a number greater than zero. The ancients could subtract 4 from 6 and get 2, but they felt that it was impossible to subtract 6 from 4 because that would result in a number that was less than zero itself. To their way of thinking, this was clearly impossible.

While some might tend to agree with the ancients, to the modern mathematician, physicist, or chemist, the idea or concept of negative numbers does exist, and it is a useful concept. **We say that every positive real number has a negative counterpart, and we call these numbers the negative real numbers.** We must always use a minus sign when we designate a negative number, as we see here by writing negative seven.

$$-7$$

We may use a plus sign to designate a positive number, as we see by writing positive seven.

$$+7$$

Or we may leave off the plus sign as we did in arithmetic and just write the numerical part with no sign.

$$7$$

We must remember that when we write a numeral with no sign, we designate a positive number. When we are talking about negative numbers as well as positive numbers, we say that we are talking about **signed numbers.** As we shall see later, the use of signed numbers will enable us to lump the operations of addition and subtraction into a single operation which we will call algebraic addition.

1.D number lines

In the 1950s the so-called new math appeared, and among other things it introduced the **number line** at the elementary algebra level. The number line can be used as a graphic aid when discussing signed numbers, and it is especially useful when discussing the addition of signed numbers.

To construct a number line, we first draw a line and divide it into equal units of length. The units may be any length as long as they are all the same length.

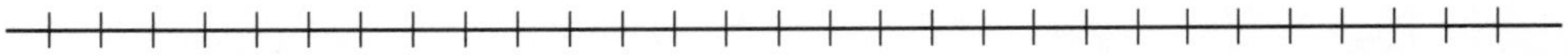

Many books show small arrowheads on the ends of number lines to emphasize that the lines continue without end in both directions, as we show below. The arrowheads are not necessary and may be omitted. Now we choose a point on the line as our base point. We call this base point the **origin,** and we associate the number zero with this point.

0

Then we associate the positive real numbers with the points to the right of the origin and the negative real numbers with the points to the left of the origin.

−7 −6 −5 −4 −3 −2 −1 0 1 2 3 4 5 6 7

On the number line above we have indicated the location of zero, the counting numbers, and the negative counterpart of each counting number. As required, we can indicate the position of any real number by locating it in relation to the numbers shown. For example, on the number line below we indicate the position of $+\frac{3}{4}$, $-1\frac{1}{2}$, and $+2.6$ by placing a dot at the approximate location of these numbers.

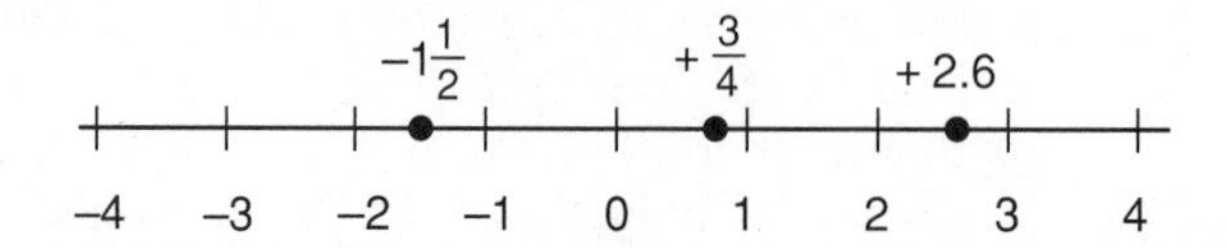

When we place a dot on the number line to indicate the location of a number, we say that we have *graphed* the number and that the dot is the graph of the number. Conversely, the number is said to be the coordinate of the point that we have graphed. We use the number line to tell if one number is greater than another number by saying that a number is *greater* than another number if its graph lies to the right of the graph of the other number. Thus $\frac{3}{4}$ is greater than $-1\frac{1}{2}$ because the graph of $\frac{3}{4}$ lies to the right of the graph $-1\frac{1}{2}$. This topic will be discussed in considerable detail in later lessons.

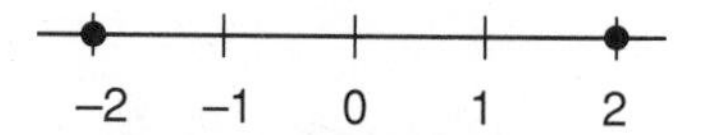

On the line above we have graphed $+2$ and -2. The number $+2$ (usually the + sign is omitted) lies 2 units to the right of the origin, and the number -2 lies 2 units to the left of the origin. Since the graphs of these numbers are equidistant from the origin but in opposite directions, it is sometimes helpful to think of each of these numbers as being the **opposite** of the other number. In this example, **we say that -2 is the opposite of 2 and that 2 is the opposite of -2.**

1.E multiplication and division of fractions

Fractions are multiplied by multiplying the numerators to get the new numerator, and by multiplying the denominators to get the new denominator.

	PROBLEM	SOLUTION
(a)	$\frac{4}{3} \times \frac{7}{5}$	$\frac{4 \times 7}{3 \times 5} = \frac{28}{15} = \mathbf{1\frac{13}{15}}$

We divide fractions by inverting the divisor and then multiplying.

	PROBLEM	INVERTING	SOLUTION
(b)	$\frac{4}{3} \div \frac{15}{8}$	$\frac{4}{3} \times \frac{8}{15}$	$\frac{4 \times 8}{3 \times 15} = \mathbf{\frac{32}{45}}$

If cancellation is possible, it is easier if we cancel before we multiply.

	PROBLEM	CANCELLATION	SOLUTION
(c)	$\frac{7}{3} \times \frac{30}{9}$	$\frac{7}{\cancel{3}_{1}} \times \frac{\cancel{30}^{10}}{9}$	$\frac{70}{9} = \mathbf{7\frac{7}{9}}$

(d) $\frac{3}{5} \times \frac{5}{6} \times \frac{21}{23}$	$\frac{\overset{1}{\cancel{3}}}{\underset{1}{\cancel{5}}} \times \frac{\overset{1}{\cancel{5}}}{\underset{2}{\cancel{6}}} \times \frac{21}{23}$	$\mathbf{\frac{21}{46}}$

We change mixed numbers to improper fractions first and then multiply or divide as indicated.

PROBLEM	IMPROPER FRACTION	SOLUTION
(e) $2\frac{1}{2} \times 5\frac{1}{3}$	$\frac{5}{\underset{1}{\cancel{2}}} \times \frac{\overset{8}{\cancel{16}}}{3}$	$\frac{40}{3} = \mathbf{13\frac{1}{3}}$
(f) $12\frac{1}{3} \div 2\frac{1}{6}$	$\frac{37}{\underset{1}{\cancel{3}}} \times \frac{\overset{2}{\cancel{6}}}{13}$	$\frac{74}{13} = \mathbf{5\frac{9}{13}}$
(g) $\dfrac{3\frac{1}{3}}{2\frac{1}{5}}$	$\dfrac{\frac{10}{3}}{\frac{11}{5}}$	$\frac{10}{3} \cdot \frac{5}{11} = \frac{50}{33} = \mathbf{1\frac{17}{33}}$

1.F symbols of equality and inequality

We use the equals sign (=) to designate that two quantities are equal. Thus we can write

$$5 + 2 = 7$$

because the number represented by the notation 5 + 2 is the same number as that represented by the numeral 7. In the same way we use the symbol ≠ to designate that two quantities are not equal. Thus we can write that

$$5 + 2 \neq 11$$

because 7 is not equal to 11.

1.G basic operations

The four basic operations of arithmetic are also the basic operations of algebra. The operations are addition, subtraction, multiplication, and division. We will review these operations here and will restrict our discussion to the numbers of arithmetic, which are the positive real numbers and zero.

addition

When we wish to add two numbers to get a result, we use the plus sign (+) to indicate the operation of addition. We call each of the numbers an **addend,** and we call the result a **sum.**

$$2 + 3 = 5$$

In this example, we use the plus sign to indicate addition; we say that the numbers 2 and 3 are addends, and we say that 5 is the sum.

We note that the sum of zero and any particular real number is the particular real number itself.

$$4 + 0 = \mathbf{4} \qquad \text{and} \qquad 15 + 0 = \mathbf{15}$$

subtraction When we wish to subtract one number from another number, we use the minus sign (−) to indicate the operation of subtraction. We call the first number the **minuend;** the second number, the **subtrahend;** and the result, the **difference.**

$$9 - 5 = \mathbf{4}$$

In this example, 9 is the minuend, 5 is the subtrahend, and 4 is the difference.

multiplication If two numbers are to be multiplied to achieve a result, each of the numbers is called a **factor** and the result is called a **product.** There are several ways to indicate the operation of multiplication.

$$4 \cdot 3 = \mathbf{12} \qquad 4(3) = \mathbf{12} \qquad (4) \cdot (3) = \mathbf{12} \qquad (4)(3) = \mathbf{12} \qquad 4 \times 3 = \mathbf{12}$$

In each of the five examples shown here, the notation indicates that 4 is to be multiplied by 3 and the result is 12. In algebra, we will avoid the last notation because the cross can be confused with the letter x, a symbol we will use for other purposes. In each of the above, we say that 4 and 3 are factors, and we say that 12 is the product.

We note that the product of a particular real number and the number 1 is the particular real number itself.

$$4 \cdot 1 = \mathbf{4} \qquad \text{and} \qquad 15 \cdot 1 = \mathbf{15}$$

The number zero also has a unique multiplicative property. **The product of any real number and the number zero is the number zero.**

$$4 \cdot 0 = \mathbf{0} \qquad \text{and} \qquad 15 \cdot 0 = \mathbf{0}$$

division If one number is to be divided by another number to achieve a result, the first number is called the **dividend,** the second number is called the **divisor,** and the result is called the **quotient.**

$$\frac{10}{5} = \mathbf{2} \qquad 10 \div 5 = \mathbf{2}$$

Both of the notations shown here indicate that 10 is to be divided by 5 and that the result is 2. We call 10 the dividend, call 5 the divisor, and say that the quotient is 2. When the indicated division is expressed in the form of a fraction such as $\frac{10}{5}$, we say that 10 is the **numerator** of the fraction and that 5 is the **denominator** of the fraction.

1.H review of operations with decimal numbers

We must align the decimal points vertically when we add and subtract decimal numbers, as we show here.

$$\begin{array}{r} 1.005 \\ +300.012 \\ \hline \mathbf{301.017} \end{array}$$

example 1.1 Add 4.0016 and 0.02163.

solution We remember to place the numbers so that the decimal points are aligned.

$$\begin{array}{r} 4.0016 \\ +0.02163 \\ \hline \mathbf{4.02323} \end{array}$$

example 1.2 Subtract 0.02163 from 4.0016.

solution Again we align the decimal points.

$$\begin{array}{r} 4.0016 \\ -0.02163 \\ \hline \mathbf{3.97997} \end{array}$$

example 1.3 Multiply 4.06 × 0.016.

solution We do not align the decimal points when we multiply.

$$\begin{array}{r} 4.06 \\ \times 0.016 \\ \hline 2436 \\ 406 \\ \hline \mathbf{0.06496} \end{array}$$

example 1.4 Divide 6.039 by 0.03.

solution As the first step, we adjust the decimal points as necessary. Then we divide.

$$0.03\overline{)6.039} \qquad \begin{array}{r} \mathbf{201.3} \\ 3\overline{)603.9} \\ 6 \\ \hline 3 \\ 3 \\ \hline 9 \\ 9 \\ \hline \end{array}$$

1.I unit multipliers

If we multiply a number by a fraction that has a value of 1, we do not change the value of the number. We just change the numeral we use to represent the number. To write 5 with a denominator of 7, we multiply 5 by 7 over 7.

$$\frac{5}{1} \times \frac{7}{7} = \frac{35}{7}$$

The fraction 7 over 7 has a value of 1, so we have just multiplied 5 by 1. Thirty-five over 7 has a value of 5 and is just another way to write 5. The fractions

$$\frac{3 \text{ ft}}{1 \text{ yd}} \quad \text{and} \quad \frac{1 \text{ yd}}{3 \text{ ft}}$$

have units and are equal to 1 because 3 ft is another name for 1 yd. We call these fractions **unit multipliers.** We can use unit multipliers to change the units of a number.

example 1.5 Use unit multipliers to change 32 feet to inches.

solution There are two unit multipliers that we can consider.

$$\frac{1 \text{ ft}}{12 \text{ in.}} \quad \text{and} \quad \frac{12 \text{ in.}}{1 \text{ ft}}$$

Let's try the first unit multiplier and see what happens.

$$\frac{32 \text{ ft}}{1} \times \frac{1 \text{ ft}}{12 \text{ in.}} = \frac{32 \text{ ft} \times 1 \text{ ft}}{12 \text{ in.}} = \frac{32 \text{ ft}^2}{12 \text{ in.}}$$

This answer is correct but is not what we want. Let's try the other unit multiplier.

$$\frac{32\,\cancel{ft}}{1} \times \frac{12\text{ in.}}{1\,\cancel{ft}} = \mathbf{(32)(12)\ in.}$$

We can cancel the ft on the bottom with the ft on top because ft over ft has a value of 1. Since we are not interested in a numerical answer, we will not do the multiplication.

example 1.6 Use unit multipliers to convert 36 ft to miles (5280 ft = 1 mi).

solution There are two unit multipliers that we may consider.

$$\frac{5280\text{ ft}}{1\text{ mi}} \quad \text{and} \quad \frac{1\text{ mi}}{5280\text{ ft}}$$

We will use the second multiplier because the ft on the bottom will cancel the ft on the top. The symbol ≈ means approximately equal to.

$$\frac{36\,\cancel{ft}}{1} \times \frac{1\text{ mi}}{5280\,\cancel{ft}} = \frac{36}{5280}\text{ mi} \approx \mathbf{0.0068\ mi}$$

example 1.7 Use unit multipliers to convert 47.25 inches to centimeters.

solution The inch is **defined** to be **exactly** 2.54 cm, so there are two unit multipliers that we may consider.

$$\frac{1\text{ in.}}{2.54\text{ cm}} \quad \text{and} \quad \frac{2.54\text{ cm}}{1\text{ in.}}$$

Since we have 47.25 in., we will use the second unit multiplier because it has in. on the bottom, which will cancel in. on the top.

$$47.25\,\cancel{in.} \times \frac{2.54\text{ cm}}{1\,\cancel{in.}} = \mathbf{(47.25)(2.54)\ cm}$$

A numerical answer is not necessary, so we will leave the answer as it is.

example 1.8 Use unit multipliers to convert 42 m to centimeters.

solution There are 100 cm in 1 m. The two possible unit multipliers are

$$\frac{100\text{ cm}}{1\text{ m}} \quad \text{and} \quad \frac{1\text{ m}}{100\text{ cm}}$$

We will use the first one because it has meters on the bottom.

$$\frac{42\,\cancel{m}}{1} \times \frac{100\text{ cm}}{1\,\cancel{m}} = \mathbf{4200\ cm}$$

example 1.9 Convert 42 ft to centimeters.

solution Many people in the United States still use feet, inches, and miles to make measurements. The rest of the world uses centimeters, meters, and kilometers. Thus U.S. engineers often find it necessary to convert from one system to another. The crossover point is the **exact relationship** 1 in. = 2.54 cm. We will convert feet to inches and then convert inches to centimeters.

$$42\,\cancel{ft} \times \frac{12\,\cancel{in.}}{1\,\cancel{ft}} \times \frac{2.54\text{ cm}}{1\,\cancel{in.}} = \mathbf{42(12)(2.54)\ cm}$$

We will not do the multiplication because a decimal answer is not required. We are interested in the method, not in an exact numerical answer.

example 1.10 Convert 4 miles to inches.

solution We will convert from miles to feet and then convert feet to inches.

$$4 \cancel{\text{miles}} \times \frac{5280 \cancel{\text{ft}}}{1 \cancel{\text{mile}}} \times \frac{12 \text{ in.}}{1 \cancel{\text{ft}}} = \mathbf{4(5280)(12) \text{ in.}}$$

The following table provides the basic equivalent measures.

TABLE OF EQUIVALENT MEASURES

1 ft=12 in.	3 ft=1 yd	5280 ft=1 mi
1 m=100 cm	10 mm=1 cm	1000 m=1 km
	1 in=2.54 cm	

Problems to provide practice in operations with decimal numbers will appear in the problem sets. Do not use a calculator when working these problems.

practice

Perform operations as indicated. Do not use a calculator.

a. 47.123 + 8.416 + 705.4 **b.** 800.62 − 75.88

c. 47.05 × 6.42 **d.** 4.006 ÷ 0.032 **e.** 5.412 ÷ 0.123

f. Use two unit multipliers to convert 75 feet to centimeters.

g. Use two unit multipliers to convert 450 inches to miles. (Go from inches to feet to miles.)

problem set 1

1. What is the difference between a number and a numeral?

2. What do we call our system of numeration?

3. Who invented this system?

4. List the digits that we use in this system.

5. What numbers are called the counting numbers?

6. What numbers are called natural numbers?

7. The numbers of arithmetic are zero and the positive real numbers. How do we define positive real numbers?

8. What do we call the point on the number line with which we associate the number zero?

9. The radius of a circle is 4 cm.
(a) What is the circumference of the circle?
(b) What is the area of the circle?

10. The height of a triangle is 10 in. The area of the triangle is 40 in.^{2}. What is the length of the base of the triangle?

11. The circumference of a circle is 628 centimeters. What is the approximate radius of the circle?

12. The circumference of a circle is 314 ft. What is the approximate radius of the circle?

13. Find the perimeter of this figure. All angles are right angles.

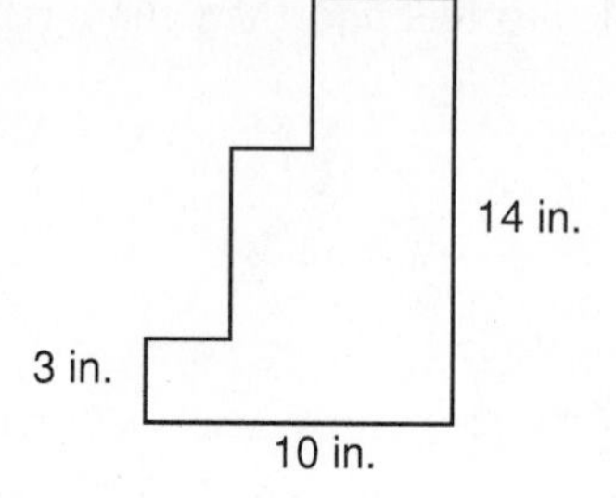

14. Find the area of the shaded portion of this parallelogram. Dimensions are in centimeters.

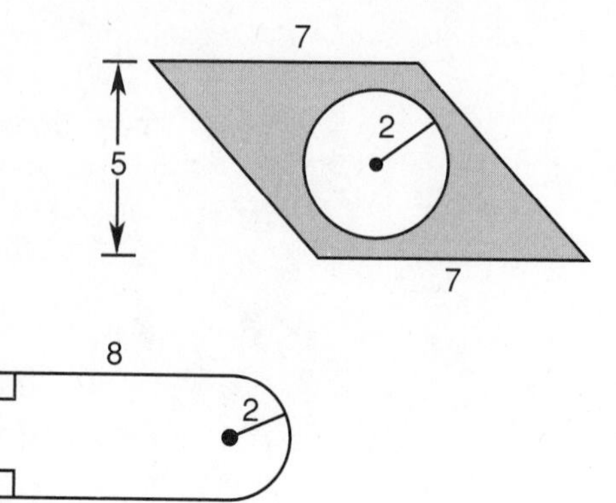

15. Find the area of this figure. Dimensions are in inches.

Add, subtract, multiply, or divide as indicated:

16. $4\frac{1}{2} \times 6\frac{2}{3}$

17. $4\frac{1}{2} \div 6\frac{2}{3}$

18. $\dfrac{14\frac{2}{3}}{3\frac{1}{4}}$

19. $\frac{8}{5} \times \frac{25}{2}$

20. $\frac{4}{5} \div 2\frac{1}{3}$

21. $7\frac{1}{8} \times 5\frac{1}{16}$

22. $\frac{4}{7} \times 5\frac{2}{5}$

23. $\frac{4}{3} \times \frac{7}{2} \times \frac{9}{5}$

24. $4\frac{1}{2} \times 5\frac{1}{16}$

25. $146\frac{1}{2} - 13\frac{2}{3}$

26. $7\frac{4}{5} + 19\frac{5}{3}$

27. $18\frac{3}{5} - 6\frac{4}{9}$

28. XM is $10\frac{1}{5}$ miles. CX is $4\frac{2}{3}$ miles. Find MC.

X C M

29. Use two unit multipliers to convert 800 ft to centimeters. (Go from feet to inches to centimeters.)

30. The base of a right solid is shown. The area of the base is 426 cm^2. The sides are 10 cm high. What is the volume of the solid?

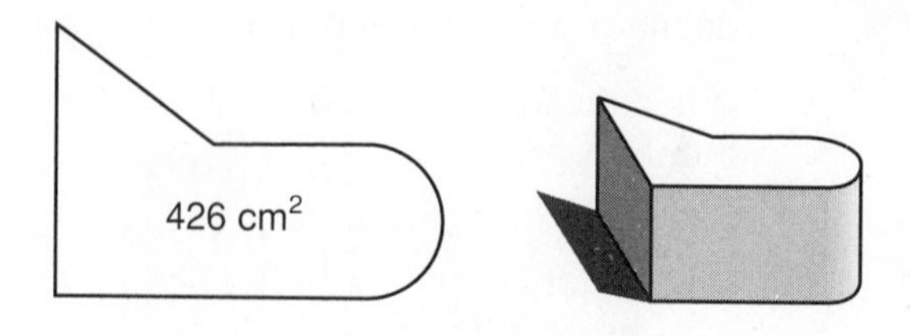

LESSON 2 Sets · Surface area

2.A sets

We use the word **set** to designate a well-defined collection of numbers, objects, or things. We say that the individual objects or things that make up the set are the elements of the set or the members of the set. It is customary to designate a set by enclosing the members of the set within braces.

$$A = \{1, 2, 3, 4, 5\}$$

We have designated set A as the set whose members are the counting numbers 1 through 5 inclusive. We could also designate this set by placing a verbal phrase within the braces as

$$A = \{\text{the counting numbers 1 through 5 inclusive}\}$$

or we could designate the members of this set by writing a sentence:

Set A is the set whose members are the counting numbers 1 through 5 inclusive.

Since we can designate which numbers are natural or counting numbers so that there is no doubt as to whether a number is or is not a natural or counting number, we say that these numbers constitute a set. Both *natural* and *counting* are names for the same set, and we normally use one name or the other. Thus we say

$$\text{Natural numbers} = \{1, 2, 3, 4, 5, \ldots\}$$

The three dots indicate that the listing of numbers continues without end.

If we include the number zero with the set of natural numbers, we say that we have designated the set of *whole numbers*.

$$\text{Whole numbers} = \{0, 1, 2, 3, 4, 5, \ldots\}$$

We designate that the listing of the members of the set of whole numbers continues without end by using the three dots after the last digit recorded. **Now if we include in our list the negative of every member of the set of counting numbers, we have designated the set that we call the *integers*.**

$$\text{Integers} = \{\ldots, -3, -2, -1, 0, 1, 2, 3, \ldots\}$$

The three dots on each end indicate that the listing continues without end in both directions.

2.B surface area

The surface area of a solid is the total area of all the exposed surfaces of the solid. A rectangular box has six surfaces. It has a top and a bottom, a front and a back, and two sides. The surface area is the sum of these six areas.

example 2.1 Find the surface area of this box. All dimensions are in meters.

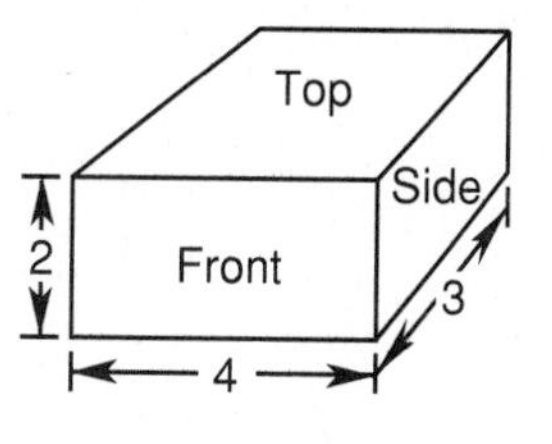

solution Since the box is rectangular, the areas of the top and the bottom are equal. The area of the front equals the area of the back, and the areas of the two sides are equal.

$$\begin{aligned}
\text{Area of front} &= 4\text{ m} \times 2\text{ m} = 8\text{ m}^2\\
\text{Area of back} &= 4\text{ m} \times 2\text{ m} = 8\text{ m}^2\\
\text{Area of top} &= 4\text{ m} \times 3\text{ m} = 12\text{ m}^2\\
\text{Area of bottom} &= 4\text{ m} \times 3\text{ m} = 12\text{ m}^2\\
\text{Area of side} &= 3\text{ m} \times 2\text{ m} = 6\text{ m}^2\\
\text{Area of side} &= 3\text{ m} \times 2\text{ m} = \underline{6\text{ m}^2}\\
\text{Surface area} &= \quad \text{total} \quad = \mathbf{52\text{ m}^2}
\end{aligned}$$

example 2.2 Find the surface area of this prism. All dimensions are in centimeters.

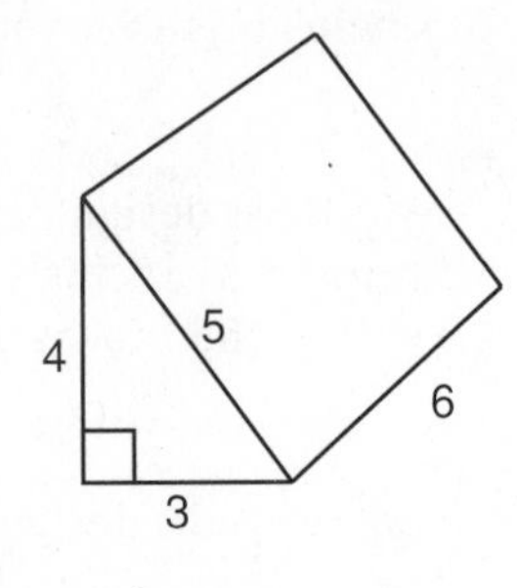

solution The prism has two ends that are triangles. It has three faces that are rectangles.

$$\begin{aligned}
\text{Area of one end} &= \frac{4\text{ cm} \times 3\text{ cm}}{2} = 6\text{ cm}^2\\
\text{Area of one end} &= \frac{4\text{ cm} \times 3\text{ cm}}{2} = 6\text{ cm}^2\\
\text{Area of bottom} &= 3\text{ cm} \times 6\text{ cm} = 18\text{ cm}^2\\
\text{Area of back} &= 4\text{ cm} \times 6\text{ cm} = 24\text{ cm}^2\\
\text{Area of front} &= 5\text{ cm} \times 6\text{ cm} = \underline{30\text{ cm}^2}\\
\text{Surface area} &= \quad \text{total} \quad = \mathbf{84\text{ cm}^2}
\end{aligned}$$

example 2.3 Find the surface area of the right circular cylinder shown. Dimensions are in meters.

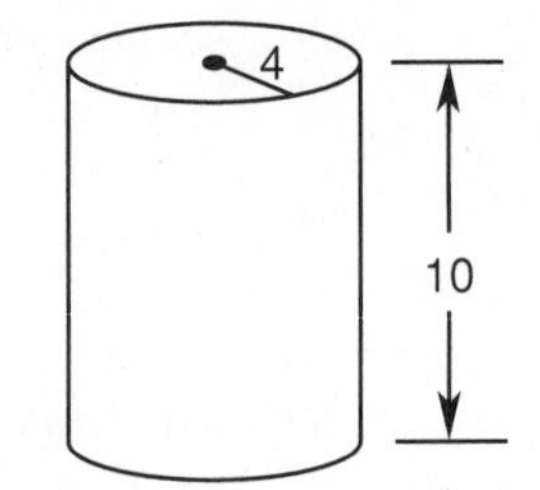

solution The cylinder has two ends that are circles. The area of one end is πr^2, so the area of both ends is

$$\begin{aligned}
\pi r^2 + \pi r^2 &\approx (3.14)(4\text{ m})^2 + (3.14)(4\text{ m})^2\\
&= 100.48\text{ m}^2
\end{aligned}$$

We can easily calculate the lateral surface area if we think of the cylinder as a tin can which we can cut down the dotted line and then press flat.

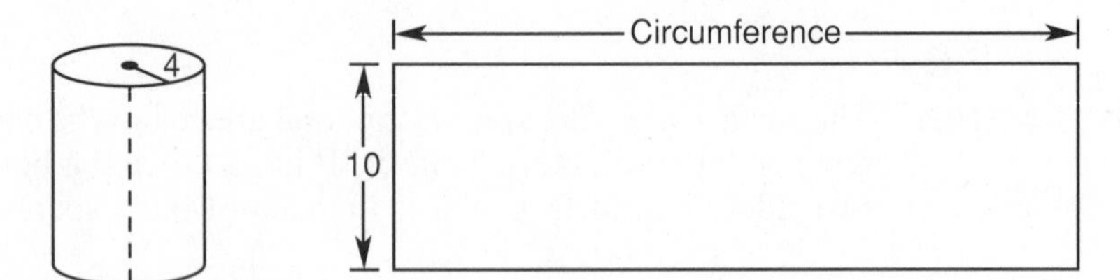

We note that the height of the rectangle is 10 m, and that the length is the circumference of the circle, which is π times the diameter. The radius of this cylinder is 4 m, so the diameter is 8 m.

$$\text{Circumference} = \pi D = (3.14)(8\text{ m}) = 25.12\text{ m}$$

Thus the area is 10 m × 25.12 m = 251.2 m^2. Therefore

$$\text{Total surface area} = 100.48\text{ m}^2 + 251.2\text{ m}^2 = \mathbf{351.68\text{ m}^2}$$

example 2.4 The base of a right solid 10 ft high is shown. Find the surface area of the solid. Dimensions are in feet.

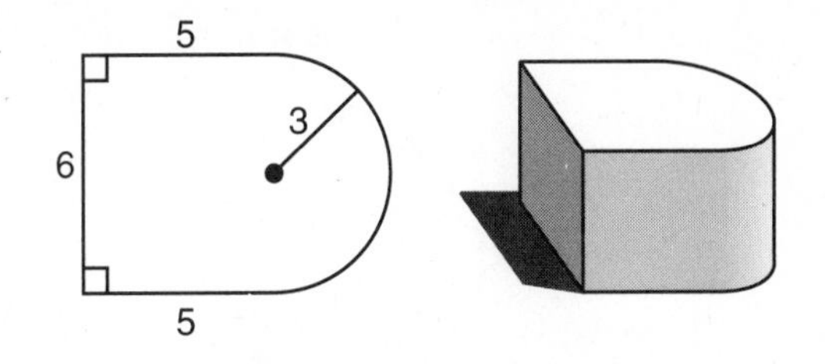

solution The surface area consists of the areas of the two equal bases and the lateral surface area.

$$\text{Area of one base} = \square + D$$

$$= 5 \times 6 + \frac{\pi(3)^2}{2} = 30 + 4.5\,\pi \approx 44.14 \text{ ft}^2$$

The lateral surface area of any right solid equals the perimeter times the height. We can see this if we cut our solid and mash it flat:

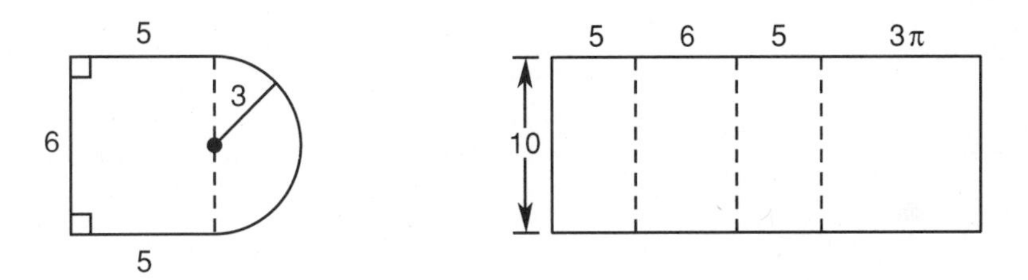

The length of the curved side equals the circumference of a whole circled divided by 2.

$$\text{Length of curve} = \frac{\pi D}{2} = \frac{\pi \cancel{2} r}{\cancel{2}} = \pi r = 3\pi$$

The perimeter of the figure is

$$\text{Perimeter} = 5 + 6 + 5 + 3\pi \approx 16 + 9.42 = 25.42 \text{ ft}$$

Thus the lateral surface area is the area of the rectangle.

$$\text{Lateral surface area} = 10(25.42) = 254.2 \text{ ft}^2$$

We add this to the surface area of both bases to get the total surface area.

Base area	44.14 ft^2
Base area	44.14 ft^2
Lateral surface area	254.2 ft^2
Total surface area	**342.48 ft^2**

practice

a. Use braces and digits to designate the set of integers.

b. Use braces and digits to designate the set of whole numbers.

c. Use braces and digits to designate the set of natural numbers.

d. Find the surface area of this rectangular prism.

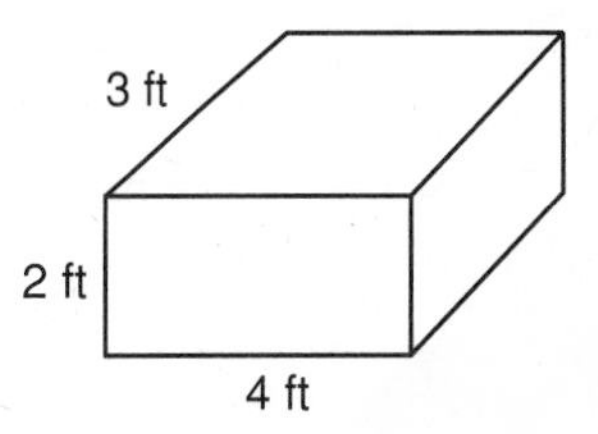

e. Find the surface area of this right circular cylinder. Dimensions are in centimeters.

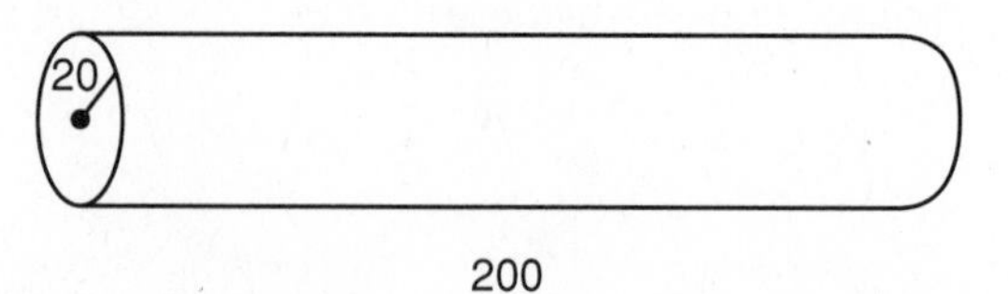

f. This is the base of a right solid whose sides are 10 cm high. Find the lateral surface area of the solid. Dimensions are in centimeters.

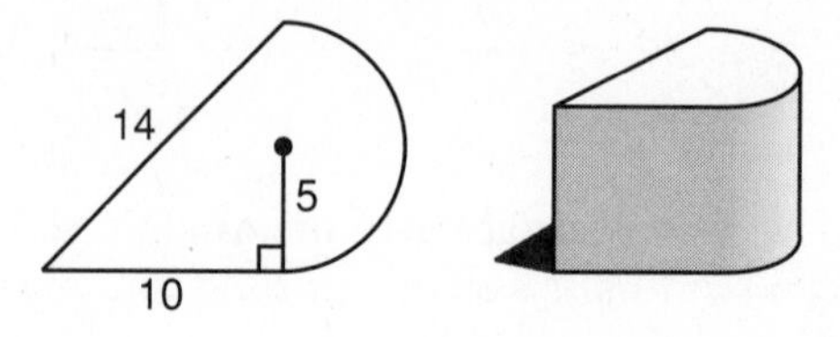

problem set 2

1. What do we call the total area of all exposed sides of a solid?

Divide or multiply as indicated:

2. $1.509 \div 0.02$ **3.** 64.09×1.3

4. Designate the set of counting numbers.

5. A right angle has how many degrees?

6. Elements of a set are also called what?

7. When two numbers are added to get an answer, what do we call the numbers and what do we call the answers?

8. Define a positive real number.

9. A straight angle has how many degrees?

10. How can we tell if one number is greater than another number?

11. Use two unit multipliers to convert 4000 ft to centimeters.

12. What do we call the answer to a division problem?

13. Use the numbers 4 and 0 to illustrate the two special properties of the number 0.

14. The circumference of a circle is 40 cm.
(a) What is the approximate radius of the circle?
(b) What is the approximate diameter of the circle?

15. Find the perimeter of this figure. Dimensions are in feet. Corners that look square are square.

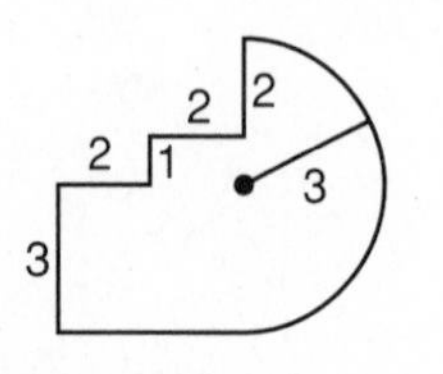

16. Find the area of this figure. Dimensions are in centimeters. The lines that look parallel are parallel.

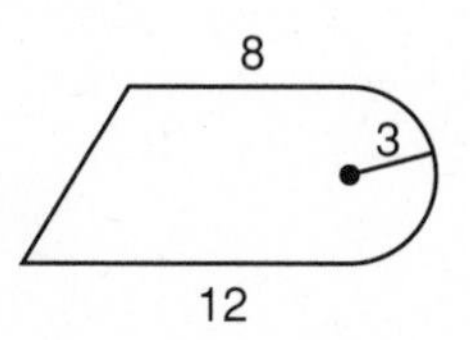

17. Find the area of the shaded portion of this figure. Dimensions are in inches. The radius of the circle is 4 in.

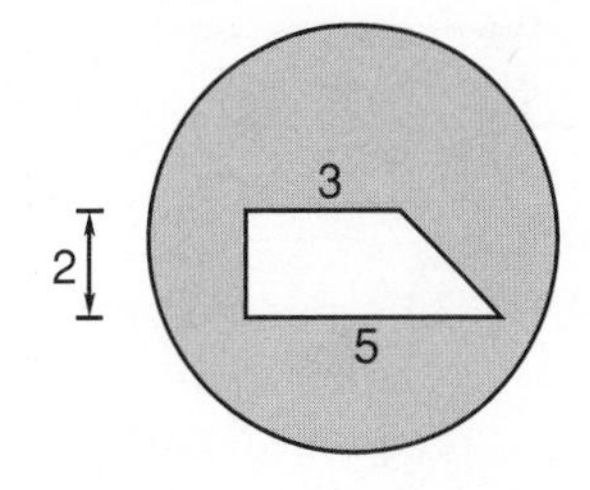

18. Use one unit multiplier to convert 42.6 inches to centimeters (2.54 cm = 1 in.).

19. Use two unit multipliers to convert 5.6 miles to inches (5280 ft = 1 mi).

20. A right circular cylinder has a radius of 6 ft and a length of 18 ft. What is the volume of the cylinder?

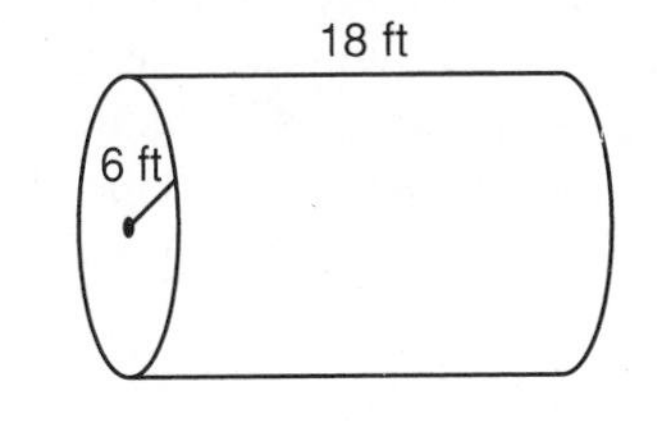

Add, subtract, multiply, or divide as indicated:

21. $4\frac{1}{2} \times 7\frac{3}{3}$ **22.** $4\frac{1}{2} \div 7\frac{2}{3}$ **23.** $\dfrac{7\frac{1}{8}}{4\frac{2}{5}}$

24. $47\frac{3}{4} - 14\frac{7}{8}$ **25.** $95\frac{1}{8} - 4\frac{13}{16}$ **26.** $94\frac{2}{5} - 7\frac{3}{6}$

27. XK is $7\frac{2}{3}$. KM is $4\frac{1}{2}$. Find MX.

X M K

28. Find the area of this rectangle in square feet. Do not use a calculator.

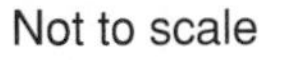

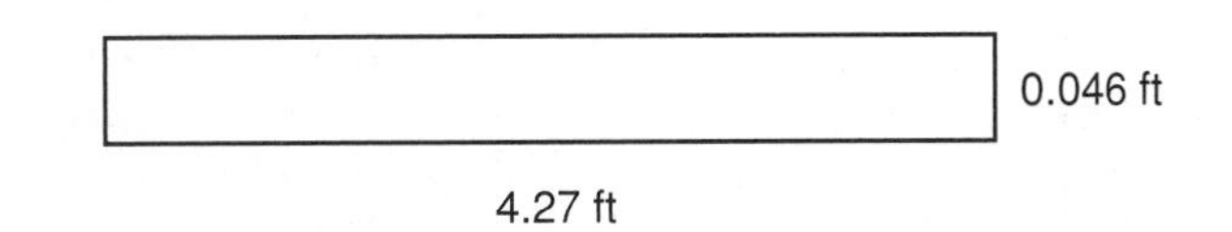

Add, subtract, multiply, or divide as indicated:

29. $15\frac{3}{8} - 2\frac{1}{7}$ **30.** $15\frac{3}{8} \times 2\frac{1}{7}$ **31.** $15\frac{3}{8} \div 2\frac{1}{7}$

32. Find the surface area of this rectangular prism. Dimensions are in meters.

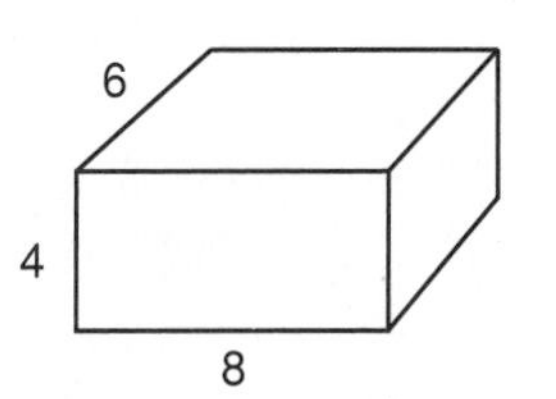

33. Find the surface area of this right prism. Dimensions are in inches.

3
5
8
4

LESSON 3 Absolute value · Addition on the number line

3.A absolute value

The number zero is neither positive nor negative and can be designated with the single symbol 0. Every other real number is either positive or negative and thus requires a two-part numeral. One of the parts is the plus or the minus sign, and the other part is the numerical part. If we look at the two numerals

$$+7 \quad \text{and} \quad -7$$

we note that the numerical part of each one is the same and that the numerals differ only in their signs. We can think of the numerical part as designating the quality of "bigness" of the number, and we use the words **absolute value** to describe this quality. **However, when we try to write the absolute value of one of these numbers by just writing the numerical part**

$$7$$

we find that we have written a positive number because we have agreed that a numeral written with no sign designates a positive number. Because of this agreement, we are forced to define the absolute value of any nonzero real number to be a positive number. We define the absolute value of zero to be zero. If we enclose a number[†] within vertical lines, we are designating the absolute value of the number. We will demonstrate this notation by designating the absolute value of zero, the absolute value of positive 7, and the absolute value of negative 7.

$|0| = 0$ read "the absolute value of zero equals zero"

$|+7| = +7$ read "the absolute value of 7 equals 7"

$|-7| = +7$ read "the absolute value of -7 equals 7"

Since the plus sign in front of a positive number is customarily omitted, the above can be written without recording the plus signs:

$|7| = 7$ read "the absolute value of 7 equals 7"

$|-7| = 7$ read "the absolute value of -7 equals 7"

Thus we have two rules for stating the absolute value of a real number.

1. **The absolute value of zero is zero.**
2. **The absolute value of any nonzero real number is a positive number.**

Here we designate the absolute value of zero and several other real numbers.

(a) $|0| = \mathbf{0}$ (b) $|-7.12| = \mathbf{7.12}$ (c) $|7.12| = \mathbf{7.12}$

(d) $|-5| = \mathbf{5}$ (e) $|5| = \mathbf{5}$ (f) $\left|\frac{3}{4}\right| = \mathbf{\frac{3}{4}}$

example 3.1 Simplify: (a) $|-5|$ (b) $|11 - 2|$ (c) $-|20 - 2|$

[†] We really should use the word *numeral* here, but from now on we will often use the words *number* and *numeral* interchangeably because excessive attention to the difference between these words is counterproductive.

solution (a) The absolute value of -5 is 5.

$$|-5| = \mathbf{5}$$

(b) First we simplify within the vertical lines:

$$|11 - 2| = |9|$$

The absolute value of 9 is 9.

$$|9| = \mathbf{9}$$

(c) Again we simplify within the vertical lines:

$$-|20 - 2| = -|18|$$

The absolute value of 18 is 18, but we want the opposite of this, so our answer is -18.

$$-|18| = \mathbf{-18}$$

3.B addition of signed numbers

In arithmetic the minus sign always means to subtract, but in algebra we also use the minus sign to designate that a number is a negative number. This can be confusing at first, as we see if we look at the expression

$$3 - 2$$

and ask if the minus sign means to subtract or if it means that -2 is a negative number. It turns out that we will find the same answer with either thought process, but in algebra we normally prefer the second process in which we think of the negative sign as designating that -2 is a negative number. If we do this, we say that we are using algebraic addition.

To help explain the rules for algebraic addition, we will use diagrams drawn on a number line.

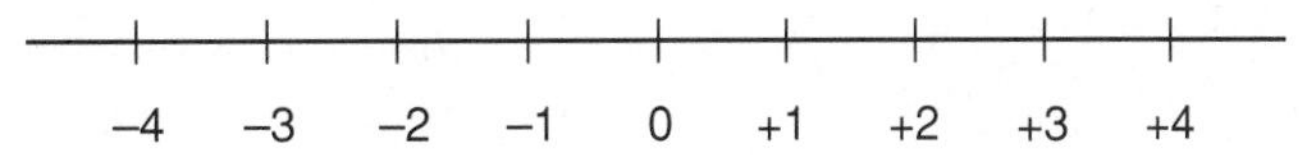

We will represent signed numbers with arrows and say that the arrows indicate directed numbers. We represent positive numbers with arrows that point to the right and negative numbers with arrows that point to the left. The length of each arrow corresponds to the absolute value of the number represented. For instance, $+3$ and -2 can be represented with the following arrows.

To use these arrows to add $+3$ and -2, we begin at the origin and draw the $+3$ arrow pointing to the right.

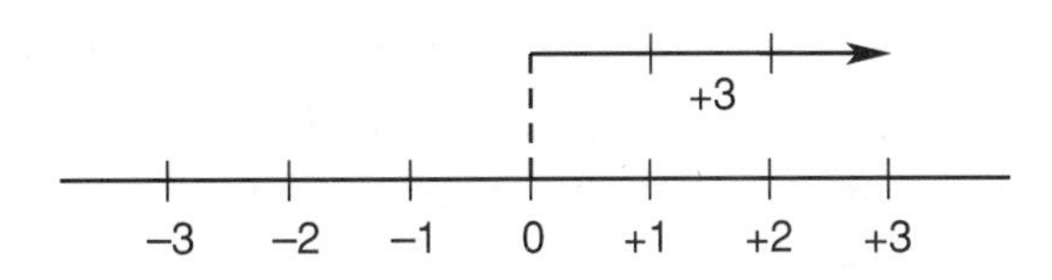

Then from the head of this arrow we draw the -2 arrow, which points to the left.

The head of the −2 arrow is over +1 on the number line.

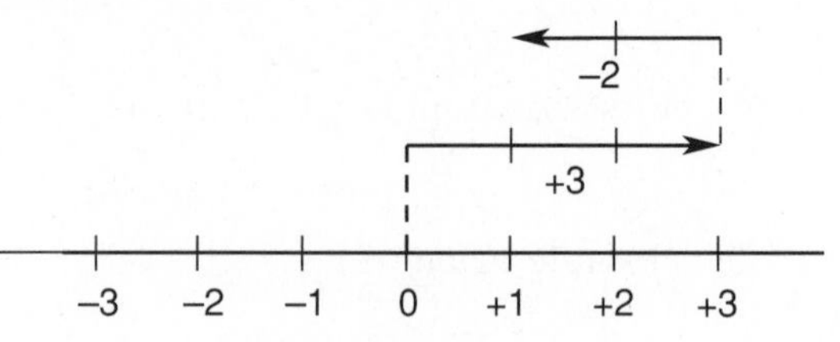

This is a graphical solution to the problem

$$(+3) + (-2) = \mathbf{+1}$$

Note that we obtain the same answer when we add signed numbers algebraically as we obtain when we use only the positive numbers of arithmetic and subtract!

$$3 - 2 = \mathbf{1}$$

It may seem that we are trying to turn an easy problem into a difficult problem, but such is not the case. **In algebra the operations of addition and subtraction are lumped together in the one operation of algebraic addition, and this enables a straightforward solution to problems that would be very confusing if the concepts of signed numbers and algebraic addition were not used.**

example 3.2 Use directed numbers and the number line to add +3 and −5.

solution

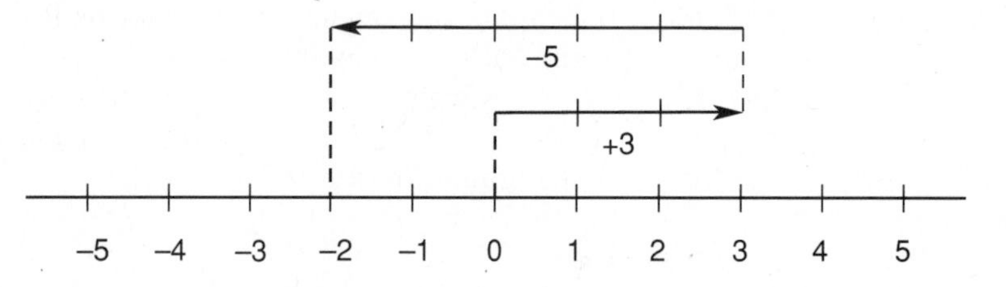

We begin at the origin and draw an arrow 3 units long that points to the right to represent the number +3. From the head of this arrow we draw an arrow 5 units long that points to the left to represent the number −5. The head of the second arrow is just above the number −2 on the number line. Thus we see that

$$(+3) + (-5) = \mathbf{-2}$$

example 3.3 Use directed numbers and the number line to add −5 and +3.

solution

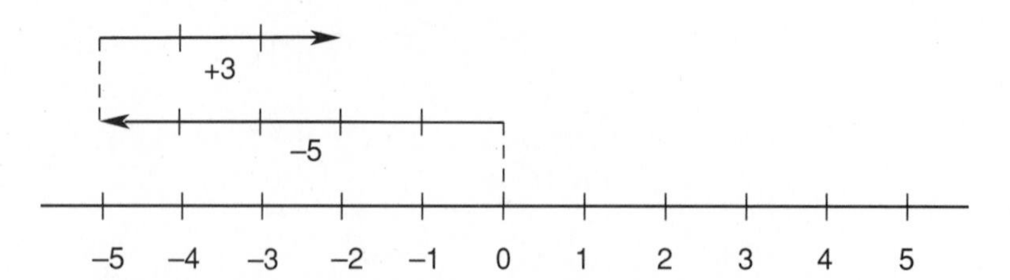

We will use the same arrows, but this time we will draw the −5 arrow first and then draw the +3 arrow. We note that again we get an answer of −2.

This example demonstrates that we may exchange the order in which we add two numbers without changing the answer we get. The Latin word for exchange is *commutare*, so we call this peculiarity or property of real numbers the **commutative property for addition.**

COMMUTATIVE PROPERTY FOR ADDITION

The order in which two real numbers are added does not affect the sum. For example

$$4 + 3 = 7 \qquad \text{and} \qquad 3 + 4 = 7$$

This property can be used to show that any number of numbers can be added in any order and the answer will be the same every time.

When the signed numbers to be added have the same signs, the arrows will point in the same direction, as we see in the next two examples.

example 3.4 Use directed numbers and the number line to add +2 and +1.

solution

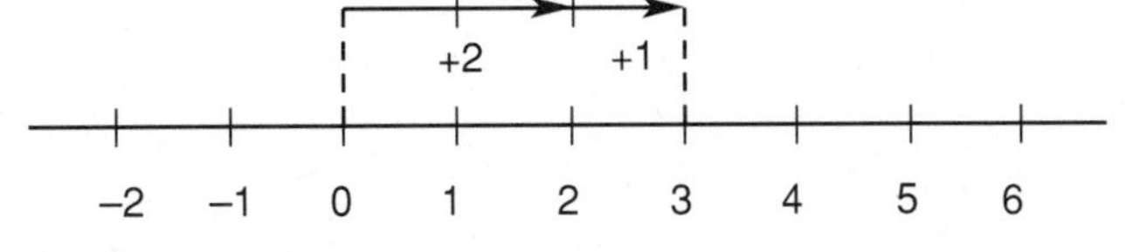

We see from the graph that the solution is +3.

$$(+2) + (+1) = \mathbf{+3}$$

example 3.5 Use directed numbers and the number line to add −2 and −1.

solution

−1 −2

−5 −4 −3 −2 −1 0 1 2

We see from the graph that the solution is −3.

$$(-2) + (-1) = \mathbf{-3}$$

The numbers to be added may also be exchanged when three or more numbers are being added. To demonstrate this we will add four signed numbers, and then exchange the order of the numbers and work the problem again. The sum will be the same.

example 3.6 Use directed numbers and the number line to add these numbers:

$$(-4) + (+2) + (-1) + (+5)$$

solution We will use arrows and add the numbers in the order they are written.

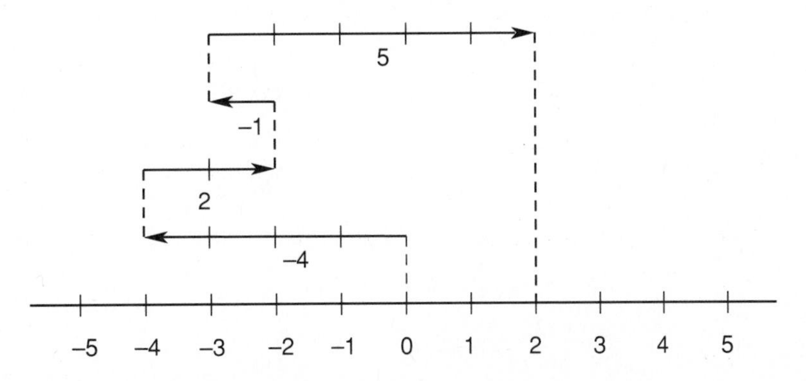

We began at the origin and moved 4 units to the left for -4, then 2 units to the right for $+2$, then 1 unit to the left for -1, and finally 5 units to the right for $+5$. We find that we end up directly above the number $+2$ on the number line. Thus

$$(-4) + (+2) + (-1) + (+5) = \mathbf{2}$$

the answer will be the same regardless of the order in which we draw the arrows. To show this, we will work the problem again with the order of the numbers changed.

$$(-1) + (+2) + (-4) + (+5)$$

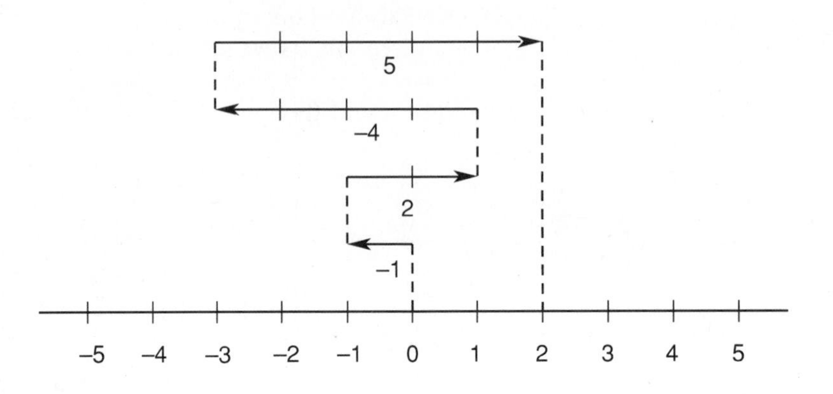

Again we find that the sum of the numbers is **2.**

practice Simplify:

a. $|-4|$ **b.** $|4.2|$ **c.** $-|10 - 6|$

d. Use a number line to add: $(-5) + (2) + (-3) + (3)$

problem set 3 Simplify:

1. $|-8|$ **2.** $|+8|$ **3.** $|-12|$

4. $-|30 - 12|$ **5.** $-|15 - 5|$

Draw a number line for each of the following problems and use directed numbers (arrows) to add the signed numbers.

6. $(+3) + (-8)$ **7.** $(-1) + (+2)$

8. $(+4) + (+3)$ **9.** $(-4) + (+2) + (-4) + (+8)$

10. $(+3) + (-5) + (+7) + (-1)$ **11.** $(+7) + (-5) + (-3) + (-4)$

12. Designate the set of natural numbers. **13.** Designate the set of integers.

14. What do we call the answer to a division problem?

15. What is the coordinate of a point on the number line?

16. What is the graph of a number?

17. What do we call the answer to a multiplication problem?

18. What is a factor?

19. When we divide 10 by 5 and get an answer of 2, we call 2 the quotient. What do we call 10 and 5?

20. What real number cannot be graphed on the number line?

Add, subtract, multiply, or divide as indicated:

21. $1472\frac{1}{2} - 1432\frac{15}{16}$ 22. $\frac{1}{2} + \frac{7}{4} + \frac{9}{8} - \frac{1}{16}$ 23. $\frac{14}{32} \times \frac{8}{21}$

24. $5\frac{1}{3} + 7\frac{3}{8} - 1\frac{1}{4}$ 25. $8.48636 \div 2.12$

26. Find the perimeter of this figure. All angles are right angles. Dimensions are in inches.

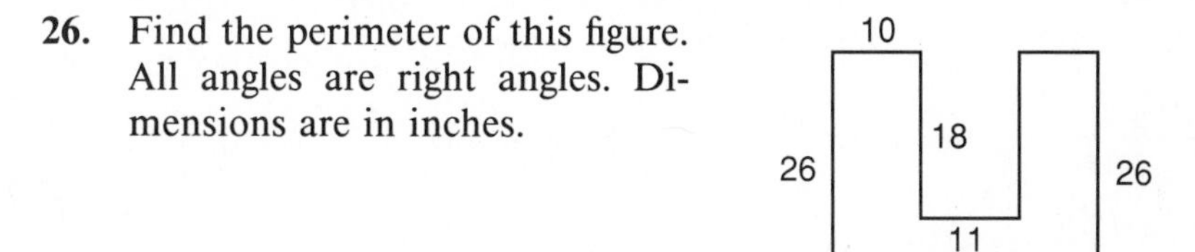

27. This figure is in the base of a right prism whose sides are 6 cm high. Find the volume and the surface area of the prism. Dimensions are in centimeters. All angles are right angles. (Remember that the lateral surface area of a right prism equals the perimeter times the height.)

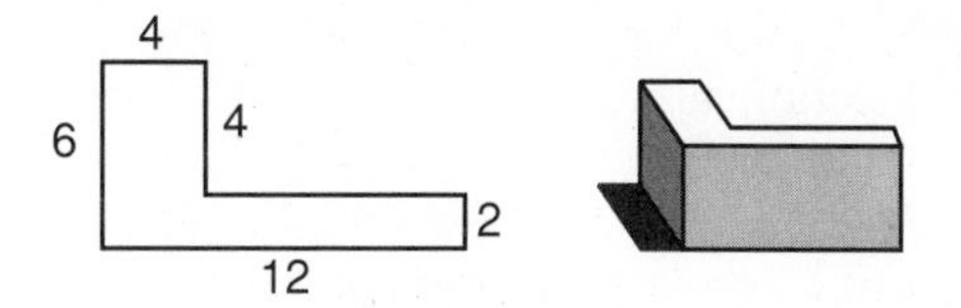

Add, subtract, multiply, or divide as indicated:

28. $2\frac{1}{4} \div 3\frac{1}{8}$ 29. $7\frac{2}{5} \times 3\frac{5}{7}$ 30. $7\frac{3}{8} + 7\frac{3}{5} - 3\frac{3}{10}$

31. $12.16608 \div 3.04$ 32. $0.00143 + 0.012 + 443.6 + 0.0007$

33. 3.628×0.0404 34. $4\frac{1}{4} \div 3\frac{2}{5}$ 35. $\dfrac{2\frac{1}{8}}{3\frac{4}{3}}$

36. Use two unit multipliers to convert 100 feet to centimeters (2.54 cm = 1 in.).

LESSON 4 Rules for addition · Definition of subtraction

4.A rules for addition

In the preceding lesson we learned to add signed numbers by using a number line and arrows to represent the numbers. This procedure allows us to have a graphical picture of what we are doing. Unfortunately this method is slow and time-consuming. We do not have time to go through the entire algebra course drawing number lines and arrows, so we must develop rules that will allow us to do algebraic addition quickly. We need two rules—one to use when the numbers to be added have the same signs and one to use when the numbers have different signs. In the following example we will draw two diagrams that will help us state the first rule.

example 4.1 Use directed numbers and the number line to add +1 and +3 algebraically, and use directed numbers and the number line to add −1 and −3 algebraically.

solution

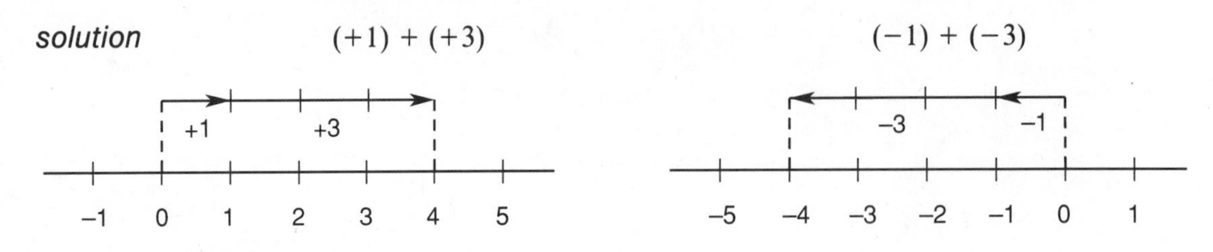

We see from these figures that

$$(+1) + (+3) = \mathbf{+4} \qquad \text{and} \qquad (-1) + (-3) = \mathbf{-4}$$

Now we will generalize. **To add algebraically two signed numbers that have the same sign, we add the absolute values of the numbers and give the result the same sign as the sign of the numbers.**

Now we will use two examples in which numbers with different signs are added algebraically to help us state the second rule.

example 4.2 Use directed numbers and the number line to add −2 and +5 algebraically, and use directed numbers and the number line to add +2 and −5 algebraically.

solution

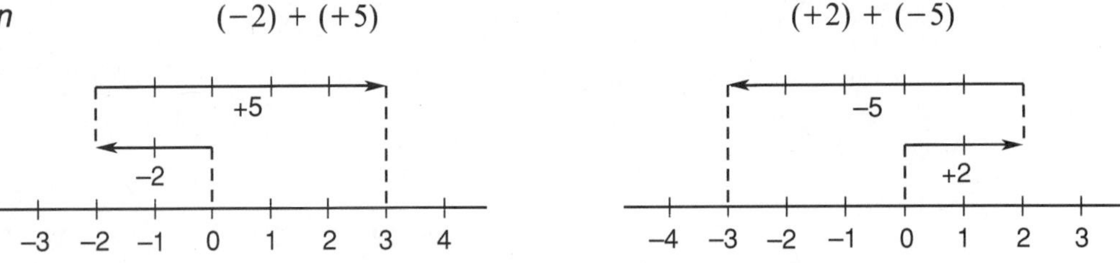

From the figure we see that the absolute value of each answer is 3 but that one of the answers is +3 and that one of the answers is −3.

$$(-2) + (+5) = \mathbf{+3} \qquad (+2) + (-5) = \mathbf{-3}$$

In the first case, the number +5 had the larger absolute value and thus the sign of the result was positive. In the second case, the number −5 had the larger absolute value and thus the sign of the result was negative. In both cases, the absolute value of the answer was the difference in the absolute values of the numbers.

Now we will generalize. **To add algebraically two signed numbers that have opposite signs, we take the difference in the absolute values of the numbers and give to this result the sign of the original number whose absolute value is the greatest.**

When two numbers have the same absolute value but different signs, their sum is zero. For instance, the sum of (−5) and (+5) is zero.

$$(-5) + (+5) = \mathbf{0}$$

We say that −5 is the opposite of +5 and that +5 is the opposite of −5. **Every real number except zero has an opposite, and the sum of any number and its opposite is zero. Another name for the opposite of a number is the additive inverse of the number, so we can also say that the sum of any number and its additive inverse is zero.**

Test your understanding of the rules by covering the answers to the following problems and seeing if your answers are the same.

(a) $(+7) + (-3) = \mathbf{+4}$
(b) $(-7) + (-3) = \mathbf{-10}$
(c) $(-7) + (+3) = \mathbf{-4}$
(d) $(-4) + (-1) = \mathbf{-5}$
(e) $(+2) + (+6) = \mathbf{+8}$
(f) $(-2) + (-8) = \mathbf{-10}$
(g) $(-2) + (+8) = \mathbf{+6}$
(h) $(+2) + (-8) = \mathbf{-6}$

4.B adding more than two numbers

We have noted that signed numbers may be added in any order and the answer will not change. Some people add from left to right, and others begin by first adding numbers that have the same sign.

example 4.3 Add: $(-5) + (4) + (-3) + (+2)$

solution This time we will add from left to right.

$(-5) + (4) + (-3) + (+2)$	original problem
$(-1) + (-3) + (+2)$	added (-5) and (4)
$(-4) + (+2)$	added (-1) and (-3)
$\mathbf{-2}$	added (-4) and $(+2)$

example 4.4 Add: $(-3) + (+2) + (-2) + (+4)$

solution We see that we have two negative numbers and two positive numbers. As the first step, we will add (-3) to (-2) and $(+2)$ to $(+4)$ and then add these sums.

$(-3) + (+2) + (-2) + (+4)$	original problem
$(-5) + (+6)$	added (-3) to (-2) and $(+2)$ to $(+4)$
$\mathbf{1}$	added (-5) to $(+6)$

4.C inserting parentheses mentally

Most signed number problems are written without parentheses enclosing the signed numbers. We must insert the parentheses mentally before we can add. **We will let the sign preceding the number designate whether the number is a positive number or a negative number, and we will mentally insert a plus sign in front of each number to indicate algebraic addition.** If we use this process,

$$4 - 3 + 2 \qquad \text{can be read as} \qquad (+4) + (-3) + (+2)$$

and

$$-6 - 3 - 2 + 5 \qquad \text{can be read as} \qquad (-6) + (-3) + (-2) + (+5)$$

Thus, to simplify an expression such as

$$-4 + 2 - 3 - 3 - 2 + 6$$

we mentally enclose each of the numbers in parentheses, insert the extra plus signs, and then add.

$$(-4) + (+2) + (-3) + (-3) + (-2) + (+6) = \mathbf{-4}$$

Care must be used to avoid associating the signs with the wrong numbers. If the mental parentheses are not used, some would incorrectly read the original expression from right to left as "6 plus 2 minus 3," etc. Guard against this.

example 4.5 Simplify: $-4 - 3 + 2 - 4 - 3 - 2$

solution We mentally enclose each number in parentheses and use plus signs so that we can read the problem as

$$(-4) + (-3) + (+2) + (-4) + (-3) + (-2)$$

Now we add the numbers and get a sum of -14.

$$(-4) + (-3) + (+2) + (-4) + (-3) + (-2) = \mathbf{-14}$$

example 4.6 Simplify: $-2 + 11 - 4 + 3 - 2$

solution We mentally enclose the numbers in parentheses and add algebraically to get a sum of $+6$.

$$(-2) + (+11) + (-4) + (+3) + (-2) = \mathbf{+6}$$

example 4.7 Simplify: $(-4) + |-2| + 3 - 7 - 2$

solution We mentally insert parentheses so that the problem reads as follows:

$$(-4) + (|-2|) + (3) + (-7) + (-2)$$

Now we simplify and get an answer of -8.

$$(-4) + (2) + (3) + (-7) + (-2) = \mathbf{-8}$$

4.D algebraic subtraction

As we have seen, if we use algebraic addition, we can handle minus signs without using the word *subtraction.* We let the signs tell whether the numbers are positive or negative, and we mentally insert parentheses and extra plus signs as necessary. Thus the subtraction problem on the left

$$7 - 4 = 3 \qquad 7 + (-4) = 3$$

can be turned into the algebraic addition problem on the right. A definition of algebraic subtraction does exist, however, and some people prefer to use it rather than using mental parentheses. The result is exactly the same, but the definition uses the word *subtraction.* To subtract algebraically, we change the sign of the subtrahend and add.

$$7 - 4 = 3 . \qquad 7 + (-4) = 3$$

The formal definition of the operation of algebraic subtraction is as follows.

ALGEBRAIC SUBTRACTION

If a and b are real numbers, then

$$a - b = a + (-b)$$

where $-b$ is the opposite of b.

Thus there are two thought processes that may be used to simplify expressions that contain minus signs such as

$$7 - 4$$

Since we prefer to consider that the minus sign designates a negative number, we will emphasize algebraic addition in this book and will avoid the use of the word *subtraction*.

practice Simplify:

a. $-5 - 2 + 7 - 6$

b. $-4 - |-2| - 6 + (-5)$

c. $-|-8| - 3 + 5 - 11$

d. $-8 + |-6| - |5| - 7$

problem set 4

1. State the rule for adding two numbers whose signs are alike.

2. State the rule for adding two numbers whose signs are different.

3. What is (a) a factor? (b) a quotient? (c) a product?

4. Find the surface area of this rectangular prism. Dimensions are in meters.

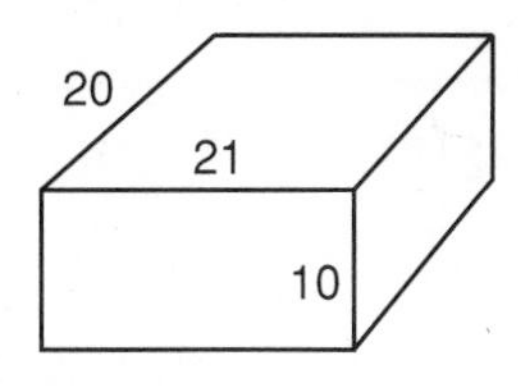

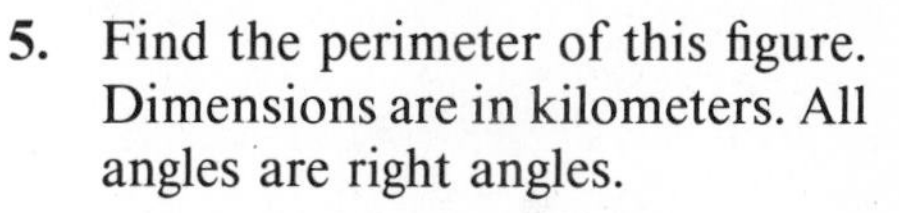

5. Find the perimeter of this figure. Dimensions are in kilometers. All angles are right angles.

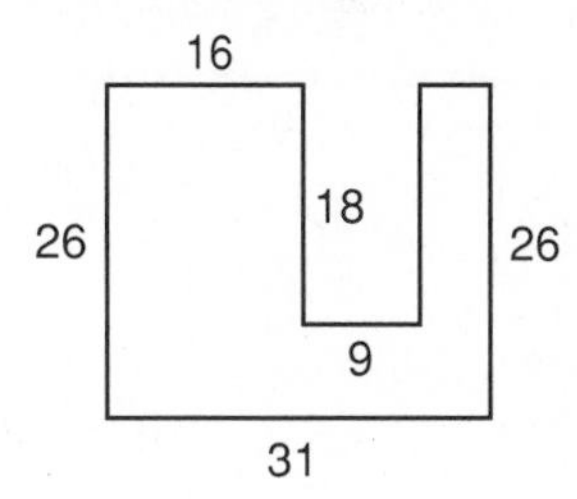

6. Find the volume of the rectangular prism in Problem 4.

7. Find the surface area of this right circular cylinder. Dimensions are in centimeters.

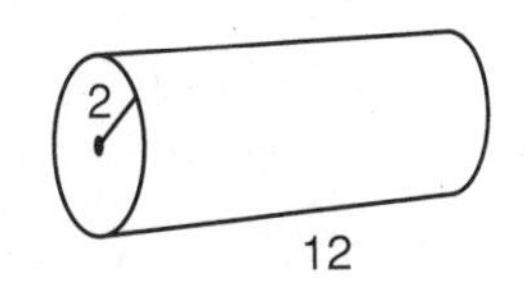

8. Use two unit multipliers to convert 3000 inches to miles.

Simplify:

9. $(+3) + (-14)$

10. $(-3) + (-14)$

11. $(-14) + (-21)$

12. $(-32) + (+4)$

13. $(-7) + (-24)$

14. $(-5) + (4) + (-3) + (+8)$

15. $(-3) + (+2) + (-2) + |-2|$

16. $(-2) + (-5) + (3) + (-5)$

17. $(+2) + (-5) + (-3) + (-7)$

18. $(-5) + (-3) + (11) + (-2)$

19. $(-14) + (-3) + (-7) + (-14)$

Insert parentheses mentally and simplify:

20. $-4 - 3 + 2 - 4 - 3 - 8$

21. $-2 + 11 - 4 + 3 - 8$

22. $-11 - 3 + 14 - 2 - 5 + 7$

23. $-5 - 11 + 20 - 14 + 5$

24. $-2 - 8 + 3 - 2 + 5 - 7$

25. $7 - 3 - 2 - 11 + 4 - 5 + 3$

26. $-7 - 4 - 13 + 4 - 2 + 7$

27. $-8 + 13 - 4 + 13 - 2 - 5 - 7$

Both the preceding notations and absolute value are included in the following problems:

28. $-7 + (-8) + 3$

29. $+|-2| + (-5)$

30. $|-2 - 3| + (-5)$

31. $-7 + (-3) + 4 - 3 + (-2)$

32. $-6 + |-2| + (-3) - 1$

33. $(-8) + (5) + (-10) - 4 - |-2|$

34. $|-2 - 3| -2$

35. $-4 - 2 + (-8) + |-5|$

36. $+|-2 - 3| - 4 + (-8)$

37. $|-2| + (-2) - 2$

38. $XY = 9\frac{4}{5}$ in. $XZ = 17\frac{3}{8}$ in. $YZ = ?$

X Y Z

39. Divide: $0.037 \div 0.0004$

LESSON 5 *Opposites and multiple signs*

5.A the opposite of a number

We can use the thought "opposite of a number" to help us understand and simplify expressions such as $-(-2)$, $-[-(-2)]$, $-\{-[-(2)]\}$, etc. We begin by graphing the number 2 and the number -2.

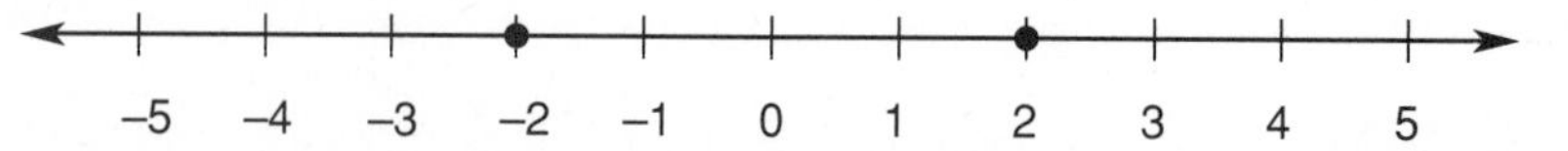

From the figure we see that the number -2 lies on the **opposite** side of the origin from 2 and is exactly the same distance from the origin. Thus we can think of $+2$ as being the **opposite of** -2 and -2 as being the opposite of $+2$. Often it is helpful to read a negative sign as "the opposite of." If we use this wording, it is easy to locate $-(-2)$, for we read this as the opposite of the opposite of 2. Well, the opposite of 2 is -2, so the opposite of that must be 2 itself.

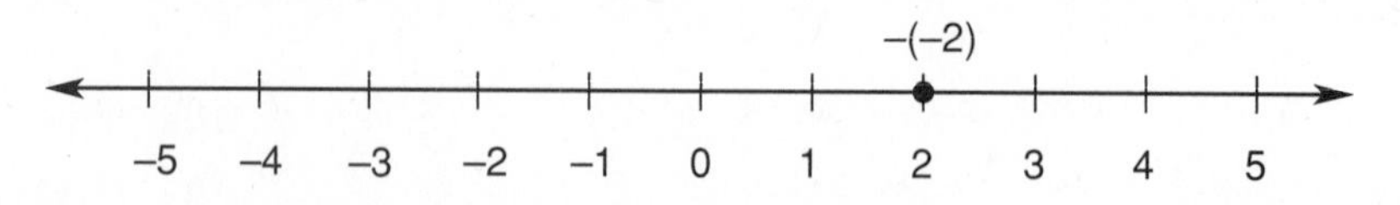

Thus 2 and $-(-2)$ are different numerals or symbols for the same number. If this is true, then where does $-[-(-2)]$ lie? We can read this as the opposite of the opposite of the opposite of 2.

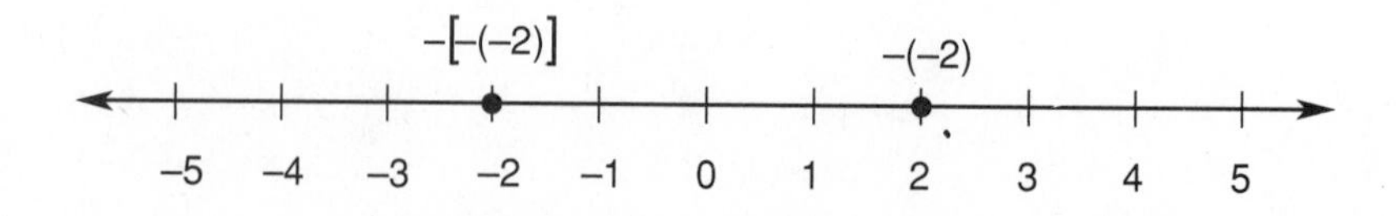

If we begin at 2, we can find the opposite of 2 at -2. Then the opposite of the opposite of 2 is back on the right side, so the opposite of the opposite of the opposite of 2 is on the left and is another way to write -2. Of course, we could go on forever with this process—but we won't.

5.B simplifying more difficult notations

Complicated expressions such as

$$-(-4) + (-2) + [-(-6)]$$

can be simplified by using algebraic addition and the concept of opposites. We begin by noting that algebraic addition of three numbers is indicated. We emphasize this by enclosing the numbers that are to be added and writing plus signs between the enclosures.

$$\boxed{-(-4)} + \boxed{+(-2)} + \boxed{+[-(-6)]}$$

The number in the first enclosure is $+4$, in the second is -2, and in the third is $+6$. So we can write

$$(4) + (-2) + (6) = \mathbf{8}$$

example 5.1 Simplify: $-(+4) - (-5) + 5 - (-3) + (-6)$

solution This problem indicates addition of five numbers.

$$\boxed{-(+4)} + \boxed{-(-5)} + \boxed{(+5)} + \boxed{-(-3)} + \boxed{+(-6)}$$

We simplify within each enclosure and add algebraically.

$$(-4) + (+5) + (+5) + (+3) + (-6) = \mathbf{3}$$

example 5.2 Simplify: $-(-3) - [-(-2)] + [-(-3)]$

solution We see three numbers are to be added. We begin by enclosing each number and inserting the necessary plus signs.

$$\boxed{-(-3)} + \boxed{-[-(-2)]} + \boxed{+[-(-3)]}$$

Now we simplify within each enclosure and then add.

$$3 + (-2) + (3) = \mathbf{4}$$

example 5.3 Simplify: $-(-4) + (-2) - [-(-6)]$

solution This time we will picture the enclosures mentally but won't write them down. If we do this, we can simplify the given expression as

$$(4) + (-2) + (-6) = \mathbf{-4}$$

example 5.4 Simplify: $-(+4) - (-5) + 5 - (-3) + (-6)$

solution This time we won't even use parentheses but will write the simplification directly as

$$-4 + 5 + 5 + 3 - 6 = \mathbf{3}$$

It will take a lot of practice to become adept in doing simplifications such as this one. Don't get discouraged if you find these problems troublesome.

practice Simplify:

a. $-(-3) - (-4)$

b. $+(-5) + [-(-6)]$

c. $-(+6) - (-8) + 7 - (-3) + (-5)$

d. $-(-3) - [-(-4)] + [-(-6)]$

problem set 5

1. The opposite of 4 is -4 and the opposite of -4 is 4. What is the opposite of $-45{,}654$?

2. What is the sum of a number and its opposite?

Use the concept of opposites to simplify:

3. $-(+4)$

4. $-(-4)$

5. $-[-(-4)]$

6. $-\{-[-(-4)]\}$

7. Find the perimeter of the following figure. Dimensions are in feet. All angles are right angles.

30 20 11 30

8. Find the area of this figure. Dimensions are in meters.

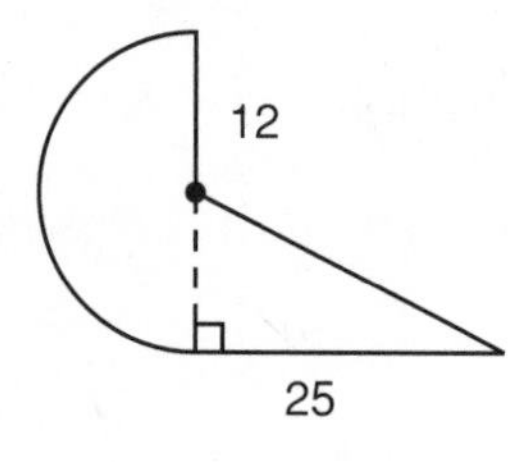

9. Use two unit multipliers to convert 444 feet to centimeters.

Use the concept of opposites and algebraic addition to simplify the following. Use additional plus signs and brackets as required.

10. $+7 - (-3) + (-2)$

11. $-3 + (-2) - (-3)$

12. $4 - (-3) - 7 + (-2)$

13. $3 - (+4) - (-2)$

14. $-6 - (-8) - (-6)$

15. $-2 - |-2|$

16. $6 + |-2|$

17. $-3 - (-3) + |-3|$

18. $-2 - (-3) - \{-[-(-4)]\}$

19. $-2 + 5 - (-3) + |-3|$

20. $-|-10| - (-10)$

21. $-2 - (-(-6)) + |-5|$

22. $-7 + (-5) - (+5) - |2|$

23. $|-2 - 5 - 7| - |-4|$

24. $-8 - 3 - 4 - (-10) + |12|$

25. $|7 - 3| - (-2) + 7 - 4 + |-11|$

26. $-4 - (-3) - 7 + (-3)$

27. $-(-5) + (-2) - 3 - |-14|$

28. $-(-2) - (+2) - 3 - (-3)$

29. $-|-3 - 2| - (-3) - 2 - 5$

30. $-(-3) - [-(-4)] - 2 + 7$

Simplify:

31. $31\frac{3}{8} - 4\frac{7}{15}$

32. $3\frac{2}{5} \div 3\frac{1}{4}$

33. $\dfrac{7\frac{1}{3}}{3\frac{9}{7}}$

34. $0.416 + 5.007$ **35.** $0.00402 \div 0.01$ **36.** 0.3004×21.02

37. $\dfrac{0.0612}{1.02}$

LESSON 6 Rules for multiplication of signed numbers

6.A multiplication of signed numbers

The sum of three 2s is 6. Also the sum of two 3s is 6.

$$2 + 2 + 2 = \mathbf{6} \qquad 3 + 3 = \mathbf{6}$$

We can get the same results from multiplication by writing

$$3 \cdot 2 = \mathbf{6} \qquad \text{or} \qquad 2 \cdot 3 = \mathbf{6}$$

because multiplication is just a shorthand notation for repeated addition of the same numeral. Thus, if we wish to use the number line to explain the multiplication of $3 \cdot 2$, we can do it two ways. We can show the sum of two 3s or the sum of three 2s.

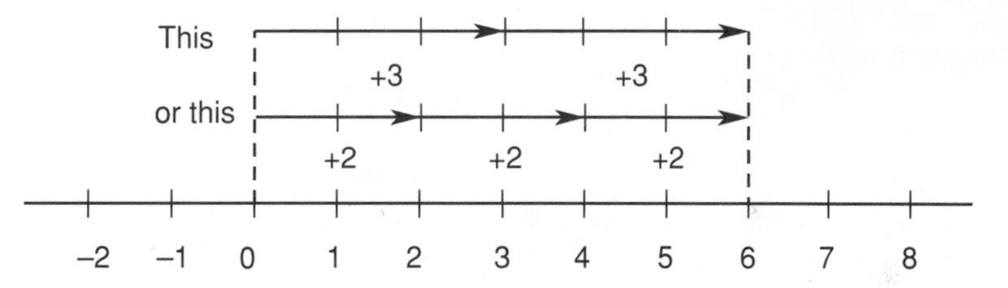

Now let's find the product of 2 and -3 on the number line. We can obtain the same answer by adding two -3s.

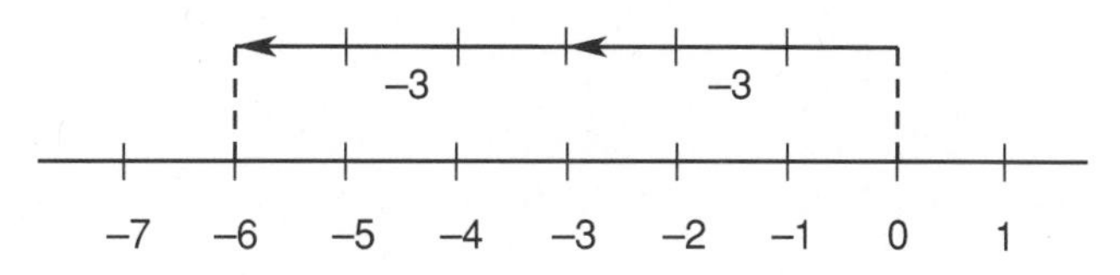

Thus we see that

$$(-3) + (-3) = \mathbf{-6} \qquad \text{so} \qquad 2 \text{ times } -3 = \mathbf{-6}$$

But now if we attempt to use the number line to show -3 times 2 by trying to draw -3 arrows that are $+2$ units long, we find that the task is impossible because we don't know how to draw -3 arrows, for any number of arrows we draw will be a number equal to or greater than 1.

The number line was a useful graphic aid in understanding the concept of signed numbers and the algebraic addition of signed numbers but is of less help when discussing the multiplication of signed numbers (and also the division of signed numbers), so we will not try to use it further for this purpose.

We could now just give the rules for multiplication and be done with it, but let's try to get some feeling for why the rules are as they are.

We need three rules for multiplication:

1. A rule for use when both factors are positive.
2. A rule for use when one factor is negative and the other factor is positive.
3. A rule for use when both factors are negative.

6.B multiplication with both factors positive

We use the first figure in this lesson to justify the following rule.

The product of two positive real numbers is a positive real number whose absolute value is the product of the absolute values of the two numbers.

examples (a) $(+4)(+5) = \mathbf{20}$ (b) $4(3) = \mathbf{12}$ (c) $2 \cdot 9 = \mathbf{18}$

6.C multiplication with one negative factor and one positive factor

Now we use the same figure again and note that 3 times 2 equals 6.

$$3(2) = \mathbf{6}$$

So it seems reasonable that the product of the opposite of 3 times 2 would be the opposite of 6, or -6. It is!

$$(-3)(2) = \mathbf{-6} \quad \text{and also} \quad (3)(-2) = \mathbf{-6}$$

The product of two signed real numbers that have opposite signs is a negative real number whose absolute value is the product of the absolute values of the numbers.

examples (a) $(-3)(5) = \mathbf{-15}$ (b) $4(-2) = \mathbf{-8}$ (c) $(-2)(6) = \mathbf{-12}$

6.D multiplication with both factors negative

We know that 3 times 2 equals 6,

$$3 \cdot 2 = \mathbf{6}$$

and in the preceding section we said that 3 times the opposite of 2 equals the opposite of 6.

$$3(-2) = \mathbf{-6}$$

Now it seems reasonable to guess that the product of the opposite of 3 times the opposite of 2 would be the opposite of the opposite of 6, which is 6 itself. It is!

$$(-3)(-2) = \mathbf{+6}$$

The product of two negative real numbers is a positive real number whose absolute value is the product of the absolute values of the two numbers.

examples (a) $-3(-2) = \mathbf{6}$ (b) $-5(-4) = \mathbf{20}$ (c) $-5(-3) = \mathbf{15}$

practice Simplify:

(a) $(-5)(-4)$ (b) $(-5)(8)$ (c) $3(-6)$

(d) $(-4)(+3)$ (e) $(-8)(-5)$ (f) $-(5)(-8)$

problem set 6

1. What is the coordinate of a point on the number line?

2. What is the additive inverse of -2?
3. How do we tell if one number is greater than another number?
4. When we multiply numbers, what do we call the answer?
5. What do we call the answer to a division problem?
6. Indicate the set of integers.
7. Find the perimeter of this figure. Dimensions are in meters. All angles are right angles.

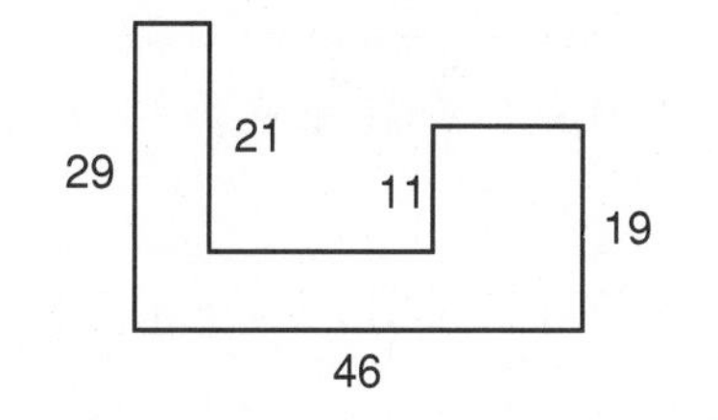

8. Find the area of the shaded portion of this rectangle. Dimensions are in centimeters. The figure in the center is a trapezoid with a height of 2 centimeters.

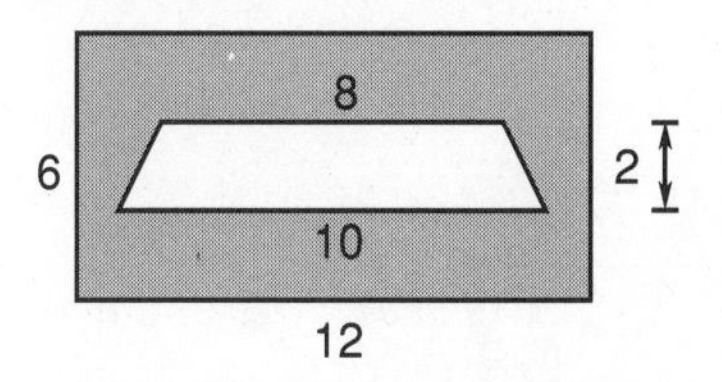

9. Find the surface area of this rectangular solid. All dimensions are in meters.

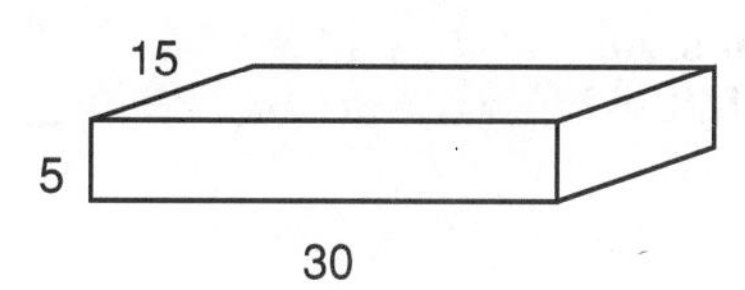

10. Find the volume and surface area of this right prism. Dimensions are in feet.

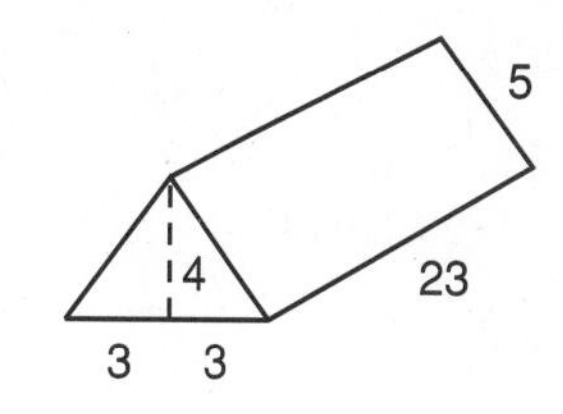

Simplify:

11. $(-3)(-5)$
12. $5(-2)$
13. $-5(2)$
14. $-[-(-4)]$
15. $-|-2| - (-2)$
16. $4 + (-2) + (-4)$
17. $-(-4) + (-2) - (-3)$
18. $|-4| + 5 - 6 - |-2 - 4|$
19. $-3 + 7 - 8 - 5 - (4)$
20. $-|-2| + |2| - (-2)$
21. $-7 + 3 - 2 - 5 + (-6)$
22. $-3 + (-3) + (-6) - 2$
23. $-5 + 3 - 2 - 5 - (-2)$
24. $5 - 3 - (-2) - (-(-3))$
25. $-3 - (-3) + (-2) - (3)$
26. $|-4 - 3| - 2 + 7 - (-3)$
27. $7 - 4 - 5 + 12 - 2 - |-2|$
28. $|-3 - 2| - (-3) - 4 - 6$
29. $5 - |-2 + 5| - (-3) + 2$
30. $-8 + 5 - 3 - (-2) + (-3)$
31. $-3 + 7 - (-2) + (-3) - 2$
32. $4 - 3 - (-2) - |12 - 3 + 4|$
33. $41.263 + 0.002$
34. $21\frac{2}{5} - 7\frac{7}{8}$
35. $4.05 \div 0.0005$
36. $\dfrac{5\frac{1}{2}}{6\frac{2}{3}}$

37. $-3\frac{1}{5} + 2\frac{1}{8}$

38. $\frac{15}{7} \times \frac{21}{5} \times \frac{2}{49}$

39. Use two unit multipliers to convert 620 centimeters to feet.

LESSON 7 Inverse operations · Rules for multiplication and division

7.A inverse operations

If one operation will *undo* another operation, the two operations are called *inverse operations*. If we take a particular number and then add and subtract the same number, the result is the particular number itself. For example, if we begin with the number 7 and add 3 and then subtract 3, the result is 7.[†]

$$7 + 3 - 3 = 7$$

Thus addition and subtraction are inverse operations.

Multiplication and division are also inverse operations. If we multiply 7 by 2 and then divide by 2, the result is 7.

$$\frac{7 \cdot 2}{2} = 7$$

There has been no change since dividing by 2 *undoes* the effect of multiplying by 2.

Since multiplication and division are inverse operations, our rules for the multiplication and division of signed numbers must be stated in such a way that these operations are inverse operations.

7.B division of one positive number by another positive number

In Lesson 6 we said that the product of two positive real numbers is a positive real number. For example, 4 times 3 equals 12.

$$4 \cdot 3 = \mathbf{12}$$

Since division undoes multiplication, we can divide the product by one of the factors to give the other factor:

$$\frac{12}{3} = \mathbf{4} \qquad \text{and} \qquad \frac{12}{4} = \mathbf{3}$$

If one positive real number is divided by another positive real number, the quotient is a positive real number whose absolute value is the quotient of the absolute values of the original numbers.

[†] Now we use algebraic addition instead of subtraction.

7.C division of negative numbers

In Lesson 6 we said that the product of two real numbers with opposite signs is a negative real number. Therefore, we must define the division of two numbers of opposite signs so that the division will be an inverse operation. Hence, 2 times the opposite of 3 is the opposite of 6.

$$2(-3) = \mathbf{-6}$$

Division must be defined in such a way that it will *undo* **the multiplication.**

$$\frac{-6}{-3} = \mathbf{2} \qquad \text{and} \qquad \frac{-6}{2} = \mathbf{-3}$$

Thus we see that we will have to define division so that the quotient of two negative real numbers is a positive real number and that the quotient of a negative real number divided by a positive real number is a negative real number.

If one negative real number is divided by another negative real number, the quotient is a positive real number whose absolute value is the quotient of the absolute values of the original numbers.

If a negative real number is divided by a positive real number, the quotient is a negative real number whose absolute value is the quotient of the absolute values of the original numbers.

We need to cover one more possibility. In Lesson 2 we found that the product of the opposite of 2 times the opposite of 3 equals 6.

$$-2(-3) = \mathbf{6}$$

Thus, since division is the inverse operation of multiplication and must *undo* the multiplication, the following must be true.

$$\frac{6}{-2} = \mathbf{-3} \qquad \text{and also} \qquad \frac{6}{-3} = \mathbf{-2}$$

This gives us the last case. The quotient of a positive real number divided by a negative real number is a negative real number.

If a positive real number is divided by a negative real number, the quotient is a negative real number whose absolute value is the quotient of the absolute values of the original numbers.

The following examples demonstrate all the rules for the division of signed numbers.

(a) $\frac{-12}{4} = \mathbf{-3}$ (b) $\frac{-10}{-2} = \mathbf{5}$ (c) $\frac{20}{-2} = \mathbf{-10}$ (d) $\frac{20}{2} = \mathbf{+10}$

7.D rules for division and multiplication of signed real numbers

We can consolidate all we have learned about the multiplication and division of signed numbers into two rules:

1. ***Like signs.*** **The product or the quotient of two signed numbers that have the same sign is a positive number whose absolute value is the absolute value of the product or the quotient of the absolute values of the original numbers.**
2. ***Unlike signs.*** **The product or the quotient of two signed numbers that have opposite signs is a negative number whose absolute value is the absolute value of the product or the quotient of the absolute values of the original numbers.**

We can state the rules above in a less rigorous but more easily remembered way if we say

> IN BOTH MULTIPLICATION AND DIVISION
>
> 1. *Like* signs $\xrightarrow{\text{yield}}$ a positive number.
> 2. *Unlike* signs $\xrightarrow{\text{yield}}$ a negative number.

practice Simplify:

a. $\frac{4}{2}$ **b.** $-4(2)$ **c.** $\frac{-4}{2}$ **d.** $(-4)(-2)$

e. $-4(-2)$ **f.** $\frac{4}{-2}$ **g.** $-2(4)$ **h.** $\frac{-4}{-2}$

problem set 7

1. What operation is the inverse operation of multiplication?

2. What operation is the inverse operation of division?

3. Use two unit multipliers to convert 4000 cm to feet.

Simplify:

4. $-4(-3)$

5. $4(-12)$

6. $-3(8)$

7. $\frac{-16}{-2}$

8. Find the volume of this rectangular solid. Dimensions are in feet.

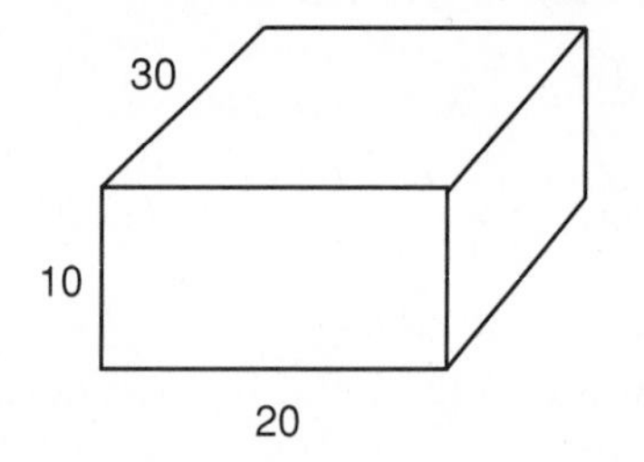

9. Find the perimeter of this figure. Dimensions are in meters. All angles are right angles.

10. Find the surface area of the rectangular prism shown. Dimensions are in centimeters.

11. XM is $7\frac{6}{15}$. MB is $3\frac{2}{5}$. Find BX.

Simplify:

12. $-|-5 + 3 - 2| + 2$

13. $-4 + 5 - 7 + (-3) - 2 + 7$

14. $-[-3(-2)] + (-5)$

15. $-3 - (+6) + (-6) - 2$

16. $-|-6| - [-(-2)] + 5$

17. $-5 - (3) - 2 + (-3)$

18. $-7 - 4 - (-3) + |-3|$

19. $-2 + (-3) - (-4) + 2$

20. $-|-3 - 3| - 2$

21. $-3 + (-3) - (-5) - |7|$

22. $-4 - (6) - (-3) - 2$

23. $-|-2 - 5 + 7| - (-3)$

24. $7 - |-3 - 5 + 1| - (-2)$

25. $-6 - 4 - (3) - (-3) + 3$

26. $5 - |-2 - 3| + |+7 - 3|$

27. $-6 + (-3) - [-(-2)] + 7$

28. $|-3| + (-2) - 4 + 7$

29. $-6 - (-6) - 4 + (-2) - 1$

30. $-5 + (-2) - (-7) - 4$

31. $8 - 7 + (-6) - 3 + (-5) - |-2|$

32. $\dfrac{4\frac{2}{3}}{3\frac{1}{9}}$

33. $\dfrac{3\frac{2}{5}}{7\frac{6}{15}}$

34. Divide (do not use a calculator): $\dfrac{0.0832}{4.16}$

LESSON 8 *Division by zero · Exchange of factors in multiplication · Conversions of area and volume*

8.A division by zero

The operation of division is the inverse operation of the operation of multiplication, for division is defined as the process that will *undo* multiplication. Thus if $3 \times 2 = 6$, it is necessary that

$$\frac{6}{3} = 2 \quad \text{and} \quad \frac{6}{2} = 3$$

We will now use the same thought process to try to decide what the result will be if we divide a nonzero number by zero. We will use the example of 6 divided by 0.

$$\frac{6}{0} = ?$$

Since we say that division *undoes* multiplication, the multiplication that is to be undone by the above division must be

$$6 = ? \cdot 0$$

But the product of zero and any real number is zero—it is not 6. There is no number that we can substitute for ? so that the product of ? and 0 equals 6. Therefore, we say that since the multiplication does not exist that is to be undone by the division, the expression

$$\frac{6}{0}$$

has no meaning, or is undefined. A similar reasoning process is used to show that we can't divide zero by zero, and we say that zero divided by zero is indeterminate rather than saying it is undefined. Thus indicated divisions such as

$$\frac{0}{0} \qquad \frac{142}{0} \qquad \frac{6}{0} \qquad \frac{-5}{0}$$

have no meaningful simplifications.

example 8.1 Evaluate: (a) $\dfrac{4 - 2 - 2}{13}$ (b) $\dfrac{13}{4 - 2 - 2}$

solution (a) First we simplify the numerator.

$$\frac{4 - 2 - 2}{13} = \frac{0}{13} = 0$$

Zero over 13 is read as "zero thirteenths" and has a value of **zero. If the bottom of a fraction is not zero and the top is zero, the fraction equals zero.**

(b) $$\frac{13}{4 - 2 - 2} = \frac{13}{0} = ?????$$

The expression 13 over 0 has no meaning and thus has no value. **It does not have a value of infinity. It does not have a value of zero. It is a meaningless expression.**

8.B exchange of factors in multiplication

In Lesson 3, we noted that we can change the order in which signed numbers are added without changing the answer. This is called the commutative property of addition. Now we note that the order of multiplying signed numbers does not affect the answer. This is called the **commutative property for multiplication.**

> COMMUTATIVE PROPERTY FOR MULTIPLICATION
>
> The order in which two real numbers are multiplied does not affect the product. For example
>
> $4 \cdot 3 = 12$ and $3 \cdot 4 = 12$

The order in which signed numbers are multiplied does not affect the value of the product! This property can be used to show that any number of numbers can be multiplied in any order and the answer will be the same every time.

example 8.2 Find the product: $-4(3)(-6)(-2)$

solution

$-4(3)(-6)(-2)$	given
$-12(-6)(-2)$	multiplied -4 by 3
$72(-2)$	multiplied -12 by -6
$\mathbf{-144}$	multiplied 72 by -2

In the first step, we multiplied -4 by 3 and got -12. In the second step, we multiplied -12 by -6 and got $+72$, which we multiplied by -2 to get the final result of -144.

example 8.3 Find the product: $-6(-2)(3)(-4)$

solution

$-6(-2)(3)(-4)$	given
$12(3)(-4)$	multiplied -6 by -2
$36(-4)$	multiplied 12 by 3
$\mathbf{-144}$	multiplied 36 and -4

example 8.4 Find the product: $-6(-4)(3)(-2)$

solution

$-6(-4)(3)(-2)$	given
$24(3)(-2)$	multiplied -6 by -4
$72(-2)$	multiplied 24 by 3
$\mathbf{-144}$	multiplied 72 by -2

In each of the three preceding examples the same factors were multiplied, but the order of multiplication was different. The product was -144, however, regardless of the order of the factors.

8.C conversions of area and volume

Two unit multipliers are required to convert area measurements, and three unit multipliers are required to convert volume measurements.

example 8.5 Change 44 in.^{2} to square centimeters.

solution We will write 44 in.^{2} as 44 in. · in. Two unit multipliers are necessary.

$$44\,\cancel{\text{in.}}\cdot\cancel{\text{in.}} \times \frac{2.54\text{ cm}}{1\,\cancel{\text{in.}}} \times \frac{2.54\text{ cm}}{1\,\cancel{\text{in.}}} = \mathbf{44(2.54)(2.54)\ cm^2}$$

example 8.6 Change 180 ft^3 to cubic inches.

solution We will write 180 ft^3 as 180 ft · ft · ft. This shows why we need three unit multipliers.

$$180\,\cancel{\text{ft}}\cdot\cancel{\text{ft}}\cdot\cancel{\text{ft}} \times \frac{12\text{ in.}}{1\,\cancel{\text{ft}}} \times \frac{12\text{ in.}}{1\,\cancel{\text{ft}}} \times \frac{12\text{ in.}}{1\,\cancel{\text{ft}}} = \mathbf{180(12)(12)(12)\ in.^3}$$

practice Simplify:

a. $\dfrac{-3 - 2}{-2 + 8 - 6}$

b. $\dfrac{-8 + 6 + 2}{8 - 4 - 4}$

c. $-(-4)(-1)(-4)$

d. $2(-6)(10)(-2)$

e. Use two unit multipliers to convert 44 square miles to square feet.

f. Use three unit multipliers to convert 4400 cubic feet to cubic yards.

problem set 8

1. Is the product of two negative numbers always a positive number?

2. Is the sum of a positive number and a negative number always a negative number?

3. Why is division by zero said to be undefined or indeterminate?

Simplify:

4. $-4(3)(-2)$

5. $4(-3)(-4)$

6. $-2(3)(4)$

7. $\dfrac{-15}{-3}$

8. $\dfrac{-4}{2}$

9. $\dfrac{4}{-2}$

10. $-3(2)(-1)(3)$

11. $-2(-3)(-2)(2)$

12. $-4(-2)(-3)$

13. $4(-2)(-3)(2)$

14. $-3 - 6 + 5 - 2 + 4 - 3$

15. $-3 + (-2) + 3 - (-4)$

16. $|-2| + |-4 - 5| + 2$

17. $-2 - (-3) + (-4) - |-3|$

18. $-|-2 + 3 - 5| - |-3 - 6|$

19. $-2 - (-3) - |-4 - 3| + 2$

20. $-\{-[-(-2)]\} - |-4 - 2|$

21. $-5 + (-3) - (-2) + 2$

22. $-|-3 - 2| + (-5)$

23. $7 + 5 - 3 - 2 + (-5)$

24. $-[-(-4)] - (-3) + 2$

25. $|-6| + |-3| - 5 + (-3)$

26. $-4 - 3 - (+3) + (-3)$

27. $|-2 - 5 + 7| - (-3) + 2$

28. $3 - (-4) + (-3) - (-4)$

29. $3 - |-2 - 3| + (-6) - (-3)$

30. Use three unit multipliers to convert 147 m^3 to cubic centimeters.

31. Find the surface area of this right circular cylinder. Dimensions are in centimeters.

32. Find the number of 1-centimeter cubes this right prism will hold. Find the surface area of the prism. Dimensions are in centimeters.

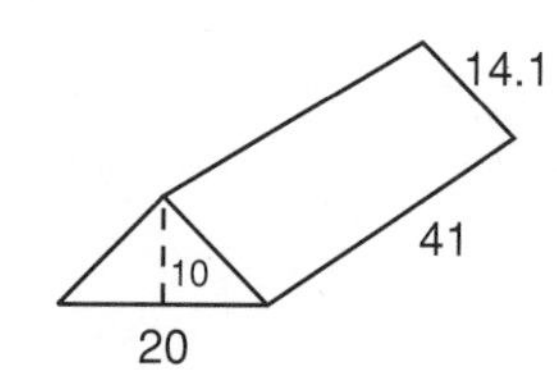

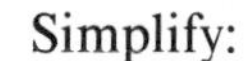

Simplify:

33. $\dfrac{3\frac{2}{5}}{7\frac{1}{3}}$

34. $7\frac{1}{3} \div 3\frac{2}{5}$

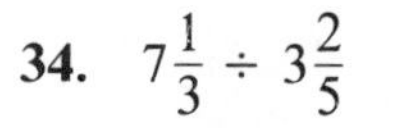

35. $4.00165 - 1.00072$

36. $0.008484 \div 0.0028$

37. $-2\frac{3}{5} + 1\frac{2}{3}$

38. $-7\frac{4}{11} + 2\frac{7}{8}$

LESSON 9 *Reciprocal and multiplicative inverse · Order of operations*

9.A reciprocal or multiplicative inverse

If one fraction is the inverted form of another fraction, each of the fractions is said to be the **reciprocal** of the other fraction.

$$\frac{2}{3} \quad \text{is the reciprocal of} \quad \frac{3}{2}$$

$$\frac{3}{2} \quad \text{is the reciprocal of} \quad \frac{2}{3}$$

$$-\frac{4}{11} \quad \text{is the reciprocal of} \quad -\frac{11}{4}$$

$$-\frac{11}{4} \quad \text{is the reciprocal of} \quad -\frac{4}{11}$$

Since numbers such as 4 can also be written in a form such as $\frac{4}{1}$, these numbers also have reciprocals.

$$\frac{1}{4} \quad \text{is the reciprocal of} \quad 4$$

$$4 \quad \text{is the reciprocal of} \quad \frac{1}{4}$$

$$-5 \quad \text{is the reciprocal of} \quad -\frac{1}{5}$$

$$-\frac{1}{5} \quad \text{is the reciprocal of} \quad -5$$

The number zero does not have a reciprocal because if we try to write the reciprocal of zero we get

$$\frac{1}{0} \quad \text{(meaningless)}$$

which we say is a meaningless notation because division by zero is undefined. **Zero is the only real number that does not have a reciprocal.** The **reciprocal** of a number is often called the **multiplicative inverse** of the number.

DEFINITION OF RECIPROCAL OR MULTIPLICATIVE INVERSE

For any nonzero real number a, the reciprocal, or multiplicative inverse, of the number is $\frac{1}{a}$.

If a number is multiplied by its multiplicative inverse (its reciprocal), the product is the number 1. Thus

$$4 \cdot \frac{1}{4} = \mathbf{1} \qquad -5 \cdot \frac{1}{-5} = \mathbf{1} \qquad \text{and} \qquad -\frac{1}{13} \cdot (-13) = \mathbf{1}$$

This simple fact is of great importance and will be very useful in the solutions of equations, a topic that will be discussed in later lessons.

9.B order of operations

If we wish to compute the value of

$$4 + 3 \cdot 2$$

we have a problem. It appears that there are two possible solutions.

(a) $4 + 3 \cdot 2$

Here we will first multiply 3 by 2 to get 6

$$4 + 6$$

and then add to get 10.

$$4 + 6 = \mathbf{10}$$

(b) $4 + 3 \cdot 2$

Here we will first add 4 and 3 to get 7

$$7 \cdot 2$$

and then multiply to get 14.

$$7 \cdot 2 = \mathbf{14}$$

We worked the problem two ways and got two different answers. **Neither way is necessarily more correct than the other, but since there are two possible ways to work the problem, mathematicians have found it necessary to agree on one way so that everyone will get the same answer. They have agreed to do the multiplications first and then to do the additions.** Thus, to simplify an expression such as

$$4 \cdot 3 + 5 - 6 + 4 - 3 \cdot 5 + 6 - 4 \cdot 2$$

we will use a two-step process. First we will perform all the multiplications and get

$$12 + 5 - 6 + 4 - 15 + 6 - 8$$

Now we will do the algebraic additions.

$17 - 6 + 4 - 15 + 6 - 8$	added 12 and 5
$11 + 4 - 15 + 6 - 8$	added 17 and -6
$15 - 15 + 6 - 8$	added 11 and 4
$0 + 6 - 8$	added 15 and -15
$\mathbf{-2}$	added 6 and -8

example 9.1 Simplify: $4 \cdot 3 + 2 + (-3)5$

solution We perform the multiplications first and get

$$12 + 2 - 15$$

which we now add algebraically.

$$12 + 2 - 15 = \mathbf{-1}$$

example 9.2 Simplify: $-5(2) - 3 + 6(3)$

solution We perform the two multiplications first

$$-10 - 3 + 18$$

and now add to get

$$\mathbf{5}$$

example 9.3 Simplify: $4 \cdot 3 - 2 \cdot 5 + 6 - 5 \cdot 2$

solution

$12 - 10 + 6 - 10$	performed multiplications
$\mathbf{-2}$	added algebraically

9.C identifying multiplication and addition

When confronted with an expression such as

$$4 - 3(5) - 7(-6) - (4)(-5)$$

the beginner often has difficulty telling whether quantities are to be added or multiplied. There is an easy way to identify indicated multiplication. **If there is no + or − sign between symbols, multiplication is indicated.** Let's simplify the expression above from left to right by first performing the indicated multiplications. The first place where there is no sign between the symbols is between the 3 and the parentheses enclosing the 5. The second place is between the 7 and the parentheses enclosing the −6, and the third is between the parentheses enclosing both the 4 and the −5.

$$4 - 3\underset{\uparrow}{}(5) - 7\underset{\uparrow}{}(-6) - (4)\underset{\uparrow}{}(-5)$$

The places where multiplication is indicated are designated by arrows. If we perform the indicated multiplications, we have

$$4 - 15 - (-42) - (-20)$$

Now we can simplify this expression and add.

$$4 - 15 + 42 + 20 = \mathbf{51}$$

practice

Simplify:

a. $6 \cdot 3 - 4(5)(6)$

b. $13 - 4(-5) - 3(10)$

c. $\dfrac{-4 - (5) - (6)}{10 - (-5)(-2)}$

d. $\dfrac{30 - 6 + (-8)(-3)}{6(-5) + (10) + (-5)(4)}$

problem set 9

1. Which real number does not have a reciprocal and why?

2. (a) What is the product of any nonzero real number and its reciprocal?
(b) Is this product sometimes a negative number?

3. (a) What is a quotient?
(b) What is a product?

4. What is the additive inverse of −8?

5. Use three unit multipliers to convert 420 ft^3 to cubic inches.

6. Use two unit multipliers to convert 465 meters to inches.

7. Find the volume of this right circular cylinder. Dimensions are in meters.

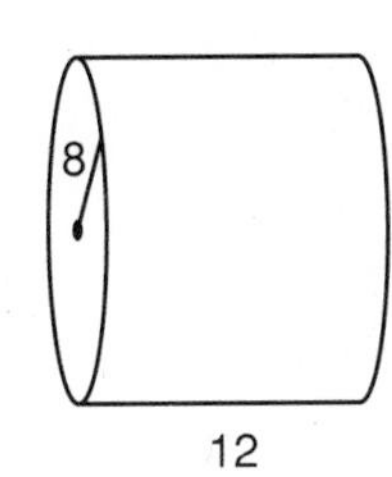

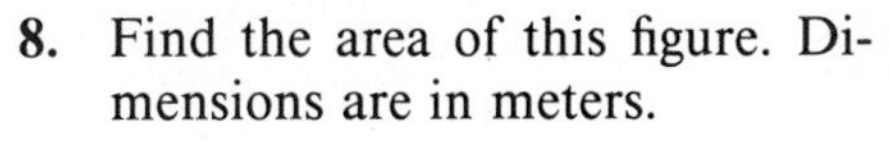

8. Find the area of this figure. Dimensions are in meters.

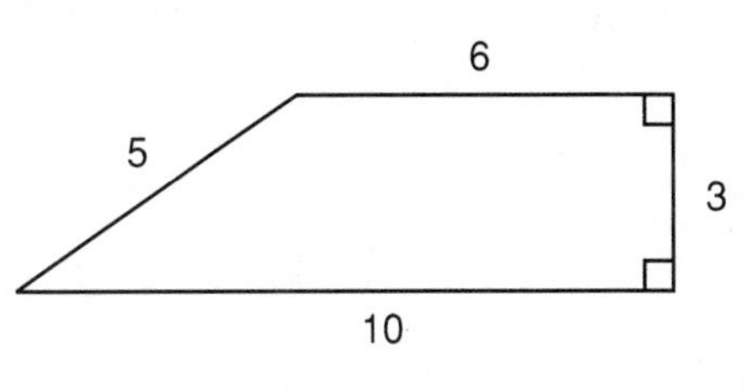

Simplify. Remember that multiplication is done before addition.

9. $6 - 8 + 2(3)$

10. $3 - 2 \cdot 4 + 3 \cdot 2$

11. $-5 - 7 - 3 \cdot 2$

12. $4 - 5(-5) + 3$

13. $-2(-2) - 2 - 2$

14. $-3(-2)(-3) - 2$

15. $-3 - 6 + 2 \cdot 5$

16. $-6(-2) - 3(-2)$

17. $-4(-3) + (-2)(-5)$

18. $-2 - 3(+6)$

19. $\dfrac{-18}{-9}$

20. $(-2)(-2) - 2$

21. $-3 + (-2) - (+2)(2)$

22. $-2 - 2(-2) + (-2)(-2)$

23. $(-2)(-2)(-2) - |-8|$

24. $-(-5) + (-2) + (-5)|-3|$

25. $(-7)(2) - 2(-3) + 6$

26. $(-5) - (-5) + 2(-2) + 4$

27. $-3 - (-2) + (-3) - 2(-2)$

28. $3(-2) - |(-3)(+3)| + 9 - 7(-2)$

29. $(-5)(-5) - 5(2) + 3$

30. $(-3)(-3) - 3 - 2|(-3)(2) + 5|$

31. $3\frac{2}{7} - 7\frac{6}{15}$

32. $\dfrac{3\frac{1}{5}}{7\frac{6}{15}}$

33. $\dfrac{4.16}{0.52}$

34. $2\frac{3}{8} \cdot 5\frac{3}{5}$

LESSON 10 Symbols of inclusion · Order of operations

10.A symbols of inclusion

In Lesson 9 we found that the simplification of

$$4 + 3 \cdot 2$$

is 10 because we have agreed to do the multiplication first and then do the addition. So,

$$4 + 3 \cdot 2 = 4 + 6 = \mathbf{10}$$

Parentheses, brackets, braces, and bars are called **symbols of inclusion,** and can be used to help us emphasize the meaning of our notation. Using these symbols, the notation above could be written in any of the following ways:

(a) $4 + (3 \cdot 2)$ (b) $4 + [3 \cdot 2]$ (c) $4 + \{3 \cdot 2\}$ (d) $4 + \overline{3 \cdot 2}$

Each of the notations emphasizes that 3 is to be multiplied by 2 and that 4 is to be added to this product. A further benefit of the use of symbols of inclusion is that a nonstandard order of operations can be indicated. For example, we can use parentheses to indicate that 4 is to be added to 3 and the result multiplied by 2 by writing

$$2(3 + 4) \qquad \text{or} \qquad (3 + 4)2$$

While bars and braces can be used as indicated above, we normally reserve the use of braces to indicate a set, and bars are most often used as fraction lines as shown here.

$$\frac{4 + (3 \cdot 2)}{5(2 - 3)} = \frac{4 + 6}{-5} = \frac{10}{-5} = \mathbf{-2}$$

The parentheses in the numerator are used to emphasize that 4 is to be added to the product of 3 and 2, and the parentheses in the denominator are used to designate that 5 is to be multiplied by the algebraic sum of 2 and -3.

10.B order of operations

To simplify numerical expressions that contain symbols of inclusion, we begin by simplifying within the symbols of inclusion. Then we simplify the resulting expression, remembering that multiplication is performed before addition.

example 10.1 Simplify: $4(3 + 2) - 5(6 - 3)$

solution First we will simplify within the parentheses.

$4(5) - 5(3)$	simplified within parentheses
$20 - 15$	multiplied
$\mathbf{5}$	added algebraically

example 10.2 Simplify: $-3(2 - 3 + 5) - 6(4 + 2) - 3$

solution First we simplify within the parentheses, then multiply and finish by adding.

$-3(4) - 6(6) - 3$	simplified within parentheses
$-12 - 36 - 3$	multiplied
$\mathbf{-51}$	added algebraically

example 10.3 Simplify: $-2(-3 - 3)(-2 - 4) - (-3 - 2) + 3(4 - 2)$

solution We begin by simplifying within the parentheses.

$-2(-6)(-6) - (-5) + 3(2)$	simplified within parentheses
$-72 + 5 + 6$	multiplied
$\mathbf{-61}$	added algebraically

When the expression is in the form of a fraction, we begin by simplifying both the numerator and the denominator. Then we have our choice of dividing or leaving the result in the form of a fraction.

example 10.4 Simplify: $\dfrac{5(-5 + 3) + 7(-5 + 9) + 2}{(4 - 2) + 3 + 5}$

solution First we will simplify the numerator and the denominator.

(a) $\dfrac{5(-2) + 7(4) + 2}{2 + 3 + 5}$ simplified within parentheses

(b) $\dfrac{-10 + 28 + 2}{2 + 3 + 5}$ multiplied

(c) $\dfrac{20}{10}$ added algebraically

(d) **2** divided

example 10.5 Simplify: $\dfrac{-3(4 - 2) - (-5)}{4 - (3)(-3)}$

solution First we simplify above and below.

(a) $\dfrac{-3(2) + 5}{4 - (3)(-3)}$ simplified within parentheses

(b) $\dfrac{-6 + 5}{4 + 9}$ multiplied

(c) $\dfrac{-1}{13}$ added algebraically

(d) $\mathbf{-\dfrac{1}{13}}$ this doesn't divide evenly, and we will leave it in fractional form

practice Simplify:

a. $\dfrac{5(-6 + 4) + 7(-3 + 9) + 3}{(5 - 3) + 3 + 5}$ b. $\dfrac{-3(10 - 8) - (-4)}{4 - 3(-3) - 13}$

problem set 10

1. Designate the set of whole numbers.
2. Designate the set of integers.
3. (a) What is a factor? (b) What is a quotient? (c) What is a sum?

Simplify:

4. $(-2 - 2)(-3 - 4)$
5. $-3(-6 - 2) + 3(-2 + 5)$
6. $(-3 - 2) - (-6 + 2)$
7. $(-4 + 7) + (-3 - 2)$
8. $5(9 + 2) - (-4)(5 + 1)$
9. $(-3 - 2)(-2)(-2 - 2)$
10. Use two unit multipliers to convert 420 centimeters to feet.
11. Use three unit multipliers to convert 420 cubic inches to cubic feet.
12. Find the area of this figure. Dimensions are in meters.

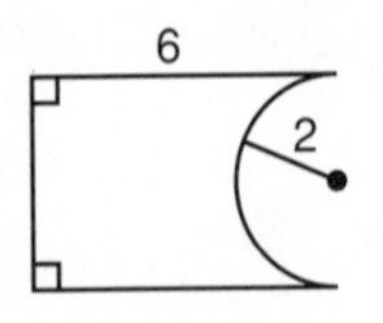

13. Find the surface area of this rectangular prism. Dimensions are in inches.

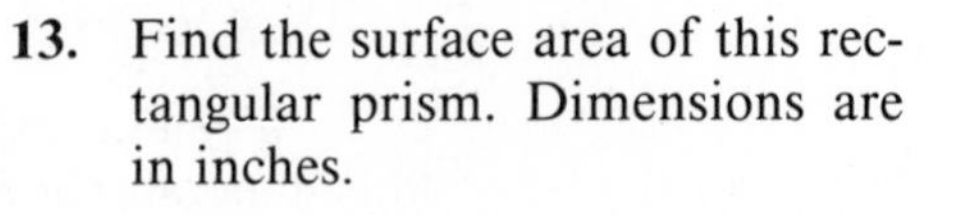

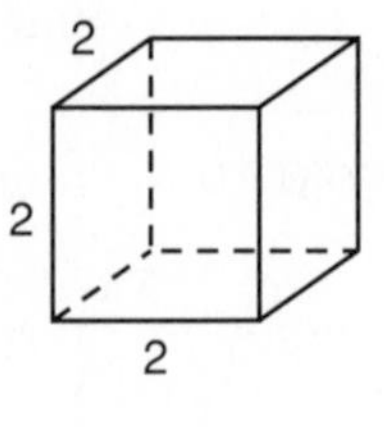

Simplify:

14. $\dfrac{-150}{-25}$
15. $\dfrac{75}{-3}$

16. $-2(-5 - 7) - 3(-8 + 2)$

17. $(-2 - 7 + 4) - (-3 - 2)$

18. $(2 - 3)(-8 + 2) + |-3 + 5|$

19. $-4 - 6 - (-3) - (-3 - 8)$

20. $\frac{1}{4}(8 - 4) - 5(8 - 2) - 2$

21. $(6 - 2)(-3 - 5) - (-5)$

22. $-|-2 - 5 + 3|(5 - 2)$

23. $2(2 - 4) - 8 - 6(7 + 3) - |-2|$

24. $4(8 + 4) + 7(10 - 8)$

25. $-8 - 4 - (-2) - (+2)(-3)$

26. $6(10 + 3) + 2(-3 - 2)(-2 - 2)$

27. $5(12 + 2) - 6(-3 + 8) - (2 + 3)$

28. $-6 - (+3) + (-3) - 5(4 - 3)$

29. $4 - 6 - 2(-3) - 5(6) + 7$

30. $7(14 - 7) - 6(-12 - 4)$

31. $2 - 4 - 5(-2) + 5(-2) - 4$

32. $-8(+2) + 3(-2)(4 - 3)$

33. $\frac{0.01608}{-0.004}$

34. $\frac{3\frac{2}{3}}{-5\frac{1}{6}}$

35. $-4\frac{1}{5} + 2\frac{1}{3}$

LESSON 11 **Multiple symbols of inclusion**

Often we encounter expressions such as

$$-3[(-2 - 4) - 3] - 2$$

where symbols of inclusion are within other symbols of inclusion. We simplify these expressions by beginning with the innermost symbol of inclusion and working our way out. Here we will simplify within the parentheses and then within the brackets.

$-3[(-6) - 3] - 2$	simplified within parentheses
$-3[-9] - 2$	simplified within brackets
$27 - 2$	multiplied
25	added

example 11.1 Simplify: $4\{2[(-3 - 2)(-7 + 4) - 5]\} - 2$

solution We will begin on the inside with the parentheses and work our way out.

$4\{2[(-5)(-3) - 5]\} - 2$	simplified within parentheses
$4\{2[10]\} - 2$	simplified within brackets
$4\{20\} - 2$	simplified within braces
$80 - 2$	multiplied
78	added

example 11.2 Simplify: $\dfrac{-3\{[(-2 - 3)][-2]\}}{-3(4 - 2)}$

solution First we will simplify the numerator and the denominator. Then we will divide as the last step.

$$\frac{-3\{[(-5)][-2]\}}{-3(2)} \qquad \text{simplified within parentheses}$$

$$\frac{-3\{10\}}{-3(2)} \qquad \text{simplified within braces}$$

$$\frac{-30}{-6} \qquad \text{multiplied}$$

$$5 \qquad \text{divided}$$

practice Simplify:

a. $3\{2[(-4 - 3)(-8 - 2) - 4]\}$

b. $\dfrac{-3\{[(-4 - 1)3] - 5\}}{2(4 - 7)}$

problem set 11

1. Designate the set of whole numbers.
2. How do we define real numbers?
3. (a) What is a factor? (b) A product? (c) A quotient?

Simplify:

4. $-2(-6 - 3) + \dfrac{0}{5}$
5. $(-7) - [-(-2)]5$
6. $-2 + 3|-4|$
7. $(-3 + 2)(-6)$
8. $(-3 - 2) - (5 + 2)$
9. $(-3 + 5)(2 - 3)$
10. $(5 - 3)(-3) + (-2)7$
11. $-2 - 3 - (-2) + (-3)$
12. $-2 - (-2) - |-2|(2)$
13. $-|-4 - 2|(-2) + 3$
14. $-3(-3)(-2 - 5 + |-11|)$
15. $-2 + 3 - 2(-2)3$
16. $-|-11| + (-3)|-3 + 5|$
17. $-3\{[(-5 - 2)](-1)\}$
18. Find (a) the perimeter and (b) the area of this figure. Dimensions are in meters.

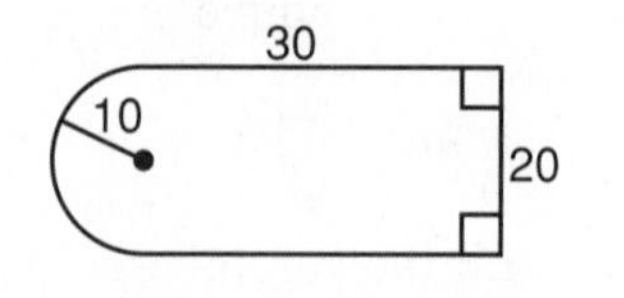

19. Find the area of the shaded portion of this parallelogram. Dimensions are in inches.

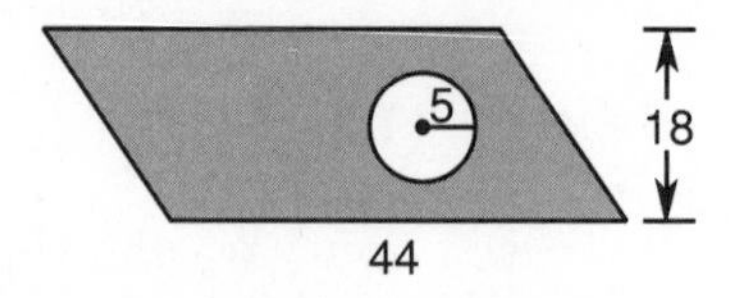

Simplify:

20. $\dfrac{-3(4 - 2) - (-5)}{4 - (3)(-3)}$
21. $\dfrac{-(-2 - 6)}{(-2)(-1 - 3) + 8}$
22. $\dfrac{-(-3 + 5) + 7}{4 - (-3)}$
23. $-3 - (-2) + (3 - 5)(-2) - 5$
24. $\dfrac{3(-4 - 2)}{2(-3)(-4)}$
25. $-3 - [-(-2)] + (-3)(5)$

26. $-6 - 2 - (-3) + 2 - 6$

27. $\dfrac{-6(2)(-2)}{-(-5 - 3)}$

28. $\dfrac{-4(-2 - 2)}{-3 -(-2)}$

29. $\dfrac{-3(-2 + 5)}{-5(-6 + 4)}$

30. $\dfrac{3.1563}{3.006}$

31. Use two unit multipliers to convert 485 inches to meters.

32. Use two unit multipliers to convert 476 square miles to square feet.

LESSON 12 More on order of operations · Products of signed numbers

12.A more on order of operations

In the discussion of the order of operations in Lesson 9 we said that mathematicians have agreed that when they write

$$4 \cdot 3 + 2$$

the **multiplication should be done first** and then the addition.

$$12 + 2 \qquad \text{multiplied}$$

$$\mathbf{14} \qquad \text{added}$$

We did not discuss division because if symbols of inclusion are properly used, the order in which division is to be performed is apparent.

If we write

$$\frac{4 \cdot 3 + 2}{-7 + 5}$$

we find the value of the numerator and the value of the denominator and then divide.

$$\frac{14}{-2} = \mathbf{-7}$$

If the following problem is encountered, however,

$$4 + \frac{14}{2} - 3 \cdot 6$$

the notation clearly indicates that only 14 is to be divided by 2, and if we do this first, we get

$$4 + 7 - 3 \cdot 6$$

Now we do the multiplication and conclude with algebraic addition.

$$4 + 7 - 18$$

$$11 - 18$$

$$\mathbf{-7}$$

At this point in an algebra book, however, it is customary to give a rule for finding the number represented by

$$6 + 3 \cdot 6 \div 2 - 6 \cdot 2$$

The rule is to perform the operations from left to right in the following order.

1. Multiplication and division
2. Algebraic addition

First we will go through the problem from **left to right,** performing the multiplications and divisions **in the order in which they are encountered.**

$6 + 3 \cdot 6 \div 2 - 6 \cdot 2$	original problem
$6 + 18 \div 2 - 6 \cdot 2$	multiplied 3 times 6
$6 + 9 - 6 \cdot 2$	divided 18 by 2
$6 + 9 - 12$	multiplied 6 times 2

Now we go through the problem again from left to right, performing the algebraic additions as they are encountered.

$6 + 9 - 12$	from above
$15 - 12$	added 6 and 9
$\mathbf{3}$	added 15 and -12

If symbols of inclusion had been properly used, instead of stating the problem as

$$6 + 3 \cdot 6 \div 2 - 6 \cdot 2$$

the problem would have been written as follows

$$6 + \frac{(3 \cdot 6)}{2} - (6 \cdot 2)$$

and here the method of solution is clearly indicated. We simplify within the parentheses as the first step.

$6 + \frac{18}{2} - 12$	simplified within parentheses
$6 + 9 - 12$	divided 18 by 2
$15 - 12$	added 6 and 9
$\mathbf{3}$	added 15 and -12

We will use symbols of inclusion to include the use of a bar as a fraction line when stating problems. Thus problems such as the one just discussed will not be encountered again in this book.

12.B products of signed numbers

Let's review the concept of the opposite of a number by watching the pattern that develops here.

	READ AS	WHICH IS
2	2	2
-2	the opposite of 2	-2
$-(-2)$	the opposite of the opposite of 2	2
$-[-(-2)]$	the opposite of the opposite of the opposite of 2	-2
$-\{-[-(-2)]\}$	the opposite of the opposite of the opposite of the opposite of 2	2

The expressions in the left-hand column are all equivalent expressions for 2 or for -2. If we look at the right-hand column, we see that every time an additional $(-)$ is included in the left-hand expression, the right-hand expression changes sign.

A similar alternation in sign occurs whenever a particular number is multiplied by a negative number. For instance,

$$-2 = -2$$

$$(-2)(-2) = +4$$

$$(-2)(-2)(-2) = -8$$

$$(-2)(-2)(-2)(-2) = +16$$

$$(-2)(-2)(-2)(-2)(-2) = -32$$

The numbers on the right have different absolute values, but they *alternate in sign.* We note that

The product of **two** negative factors is **positive.**

The product of **three** negative factors is **negative.**

The product of **four** negative factors is **positive.**

The product of **five** negative factors is **negative.**

Without proof we will generalize these observations.

1. **The product of an even number of negative real numbers is a positive real number.**
2. **The product of an odd number of negative real numbers is a negative real number.**

We can use these observations to determine the sign of the product of several signed numbers. Let's consider

$$(4)(-3)(-4)(-2)(11) = ?$$

Here we have $+4$ and $+11$ as two of the five factors. Since multiplication by a positive number does not affect the sign of the product, we will not consider these numbers. The other three factors are negative. We can look at the rules stated above and see that the sign of the product of three negative numbers is negative. Thus our answer can be expressed as

$$-(4 \cdot 3 \cdot 4 \cdot 2 \cdot 11) = \mathbf{-1056}$$

example 12.1 Determine the signs of the following products and give the reasons. Do *not* do the multiplications.

solution

		SIGN	REASON
(a)	$(-4)(-3)(2)(+5)(+6)$	positive	Even number of negative factors
(b)	$(3)(+2)(6)$	positive	No negative factors
(c)	$(-3)(-2)(6)(4)(-2)$	negative	Odd numbers of negative factors
(d)	$(-3)(-2)(-5)(-7)(-2)$	negative	Odd number of negative factors
(e)	$(-3)(-4)(-2) + 2(-3)$	?	Rule does not apply, as this is an indicated *sum*. We will do the problem in three steps.
(f)	$(-3)(-4)(-2)$	negative	Odd number of negative factors
(g)	$2(-3)$	negative	Odd number of negative factors
(h)	$-(3)(4)(2) - (2)(3)$	negative	Algebraic sum of two negative numbers is a negative number

practice Simplify:

a. $(-2)(-2)(-3)(-3)$

b. $4 \cdot 2(-3)(-2)$

c. Is the following product a positive number or negative number? Do not multiply. Give the reason.

$$4(-3)(5)(-4)(-3)(-7)(-21)(5)(14)(-5)(-8)$$

Simplify:

d. $-[-(-3)]$

e. $-\{-[-(-2)]\}$

problem set 12

1. Is the product of 33 negative numbers and 2 positive numbers a positive number or a negative number?

2. Is the product of 33 negative numbers and 3 positive numbers a positive number or a negative number?

3. What is the reciprocal of -5?

4. What is another name for the reciprocal of a number?

5. The figure on the left is the base of a right solid that is 100 cm high. What is the volume of the solid? Dimensions are in centimeters.

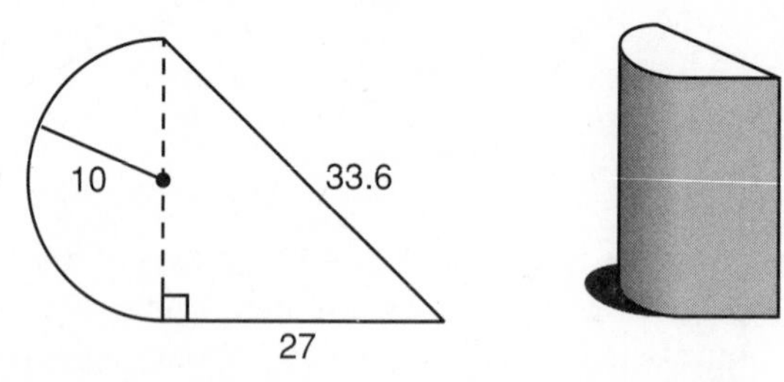

6. Find the surface area of this rectangular solid. Dimensions are in inches.

7. Use two unit multipliers to convert 60 meters to inches.

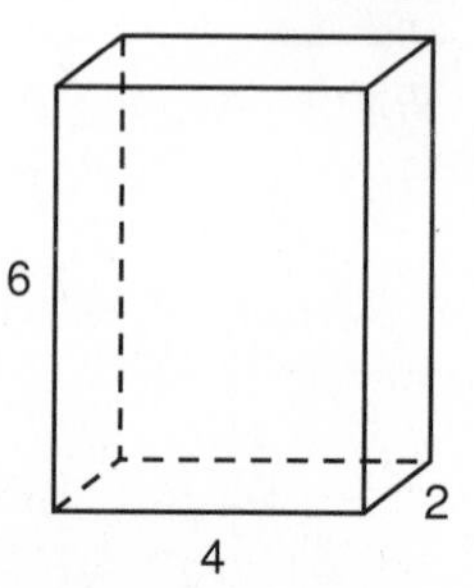

8. Use three unit multipliers to convert 48,700 cubic feet to cubic miles.

Simplify:

9. $4 - \frac{(+12)}{(-3)} + 2$

10. $\frac{-6}{-1} + (-3)(-2) + 3|-4 - 2|$

11. $-3(-2 - 5)4$

12. $-5(-3 - 2) + (-2) - (-3 - 4)$

13. $\frac{-6 - 3}{-4 + (4 - 3)}$

14. $\frac{-5(-2) - 4}{(-2 - 1)(-1)}$

15. $\frac{-2(-6) - 2}{-3 + (-7 + 2)}$

16. $\frac{-2 - 4(-3 - 2)}{3 + (-2)(+7)}$

17. $-3 - (2) + (-2) - (-3)(-2)$

18. $\frac{(-3)|-1 - 4|}{-3 - |-2|}$

19. $\frac{(-8 - 2)(-2)}{-2 - 6(2)}$

20. $-2 + (-2) - (-4)5$

21. $-2(-6 - 1 - 2) - (-2 + 7)$

22. $-3 - (-2 - 6)(-2 + 4)$

23. $(-5 - 6) - 2(3 - 6) + |-4|$

24. $-2 - 3 - (2)(-2) + (-1)(-3)$

25. $-2[3 - 2 - (-3)] - [(3 - 2)(-2)]$

26. $(-2)(-3)(-4 + 2) - (3 + 1)$

27. $-7 - (2) + (-2) - 3|-4|$

28. $5 - 6 - 4 + 3 - (-2)(-3 - 2)$

29. $-3(-2|-11|) + 5(-3 + 2)$

30. $3 - 6 - 2 - (-3)(-4) + 2$

31. $-3\frac{1}{5} + 2\frac{1}{6}$

32. $-0.1386 \div 0.063$

33. $-3\frac{1}{5} \div 2\frac{1}{6}$

LESSON 13 Evaluation of algebraic expressions

In Lesson 1 we said that a **number** is an **idea** and that when we wish to write down something to represent this **idea,** we use a **numeral.** If we wish to bring to mind the number 7, we could write any of the following:

$$7 \qquad \frac{14}{2} \qquad 4 + 3 \qquad \frac{-21}{-3} \qquad 2 + 2 + 2 + 1$$

We call each of these notations a **numerical expression** or just a **numeral.** Every numerical expression represents only one number and we call this number the **value** of the expression. Each of the numerical expressions shown above has a **value** of 7.

In algebra we often use letters to represent numbers. When letters as well as numbers are used in an expression, we don't call the expression a numerical expression but we call it by the more general name of **algebraic expression or mathematical expression.** These words are used to describe expressions that contain only numbers or only letters or contain both numbers and letters.

If we write the algebraic expression

$$4 + x$$

the expression has a **value** that depends on the value that we assign to x. If we give x a value of 5, then the expression has a value of 9 because

$$4 + 5 = \mathbf{9}$$

If we give x a value of 11, then the expression has a value of 15 because

$$4 + 11 = \mathbf{15}$$

Because the value assigned to x can be changed or varied, we call letters such as x **variables.**

We also call the letters **unknowns** since they represent unknown or unspecified numbers. The numeral 4 in this example does not change value and has a constant value of 4. For this reason the symbol that we use to denote a number is called a **constant.**

When we use variables in algebraic expressions, the notations that we use to indicate the operations of division and algebraic addition are the same as the notations that we use to indicate the division and algebraic addition of real numbers. The notation for the multiplication of variables is sometimes slightly different. We can denote that we wish to multiply 4 by the variable x by writing any of the following:

(a) $4x$ (b) $4(x)$ (c) $(4)(x)$ (d) $4 \cdot x$ (e) $(4) \cdot (x)$

The notations (b) through (e) are the same as the notations that we use for real number multiplication, but the notation shown in (a) is different from the real number notation.

$4x$ indicates that 4 is to be multiplied by the value of x

whereas

45 does not indicate that 4 is to be multiplied by 5
but instead is a numeral that represents the number 45

Thus the expression xym indicates that the values of x, y, and m are to be multiplied. If we give x a value of 1, y a value of 2, and m a value of 3, the value of the expression xym can be found.

$$1 \cdot 2 \cdot 3 = \mathbf{6}$$

If we write the algebraic expression

$$4x + mx$$

we indicate that 4 is to be multiplied by the value of x and that the value of m is to be multiplied by the value of x and that the two products are to be added. If we give x a value of 3 and m a value of 5, then we can find the value of the expression.

$$4 \cdot 3 + 5 \cdot 3$$
$$12 + 15 \quad = \mathbf{27}$$

Thus the value of the expression when x equals 3 and m equals 5 is 27.

If we give x the value of 2 and m the value of 6, then the expression will have a different value.

$$4x + mx = 4 \cdot 2 + 6 \cdot 2 = 8 + 12 = \mathbf{20}$$

In this case the value of the expression is 20. It is of **utmost importance** to note that in the first case when we gave x a value of 3, the value of x everywhere in the expression was 3. When we gave x a value of 2, the value of x everywhere in the expression was 2. **While the values assigned to variables may change or be changed, under any set of conditions the value assigned to a particular variable in an expression is the same value throughout the expression. Also, when we begin solving equations and working problems, we must remember that the value assigned to any particular variable under any set of conditions must be the same value regardless of where the particular variable appears in the equation or the problem.**

example 13.1 Find the value of: xmp if $x = 4$, $m = 5$, and $p = 2$

solution We replace x with 4, m with 5, and p with 2.

$$xmp = 4 \cdot 5 \cdot 2 = 20 \cdot 2 = \mathbf{40}$$

example 13.2 Evaluate (find the value of): $4yz - 5$ if $y = 2$ and $z = 10$

solution We replace y with 2 and z with 10 and then simplify.

$$4yz - 5 = 4(2)(10) - 5 = 80 - 5 = \mathbf{75}$$

example 13.3 Evaluate: $y - z$ if $y = -2$ and $z = -6$

solution We replace y with -2 and $-z$ with $+6$ since $-z$ represents the opposite of z.

$$y - z = -2 + 6 = \mathbf{+4}$$

example 13.4 Evaluate: $-a - b - ab$ if $a = -3$ and $b = -4$

solution The value of a is -3, so the opposite of a is 3. The value of b is -4, so the opposite of b is $+4$. Finally, $ab = 12$ and the opposite of this is -12. Thus, we get -5 for an answer.

$$3 + 4 - 12 = \mathbf{-5}$$

example 13.5 Evaluate: $-x - (-a + b)$ if $x = 2$, $a = -4$, and $b = -6$

solution Some people find that it is helpful to replace each variable with parentheses. Then the proper number is written inside the parentheses.

$-(\ \ ) - [-(\ \ ) + (\ \ )]$	replaced variables with parentheses
$-(2) - [-(-4) + (-6)]$	numbers inserted

The first entry can be read as the opposite of 2, or -2. Inside the brackets we have $-(-4)$, read the opposite of the opposite of 4, which is 4 itself. The last entry inside the brackets is $+(-6)$, read plus the opposite of 6, which is the same as -6. Thus we have

$$-2 - (4 - 6) = -2 - (-2) = -2 + 2 = \mathbf{0}$$

example 13.6 Evaluate: $x - y(-a + x)$ if $x = -2$, $y = +3$, and $a = -4$

solution We will replace each variable with parentheses.

$(\ \ ) - (\ \ )[-(\ \ ) + (\ \ )]$	replaced variables with parentheses
$(-2) - (+3)[-(-4) + (-2)]$	numbers inserted
$= -2 - (3)(4 - 2) = -2 - (3)(2)$	simplified
$= -2 - 6 = \mathbf{-8}$	simplified

example 13.7 Evaluate: $-(m + x)(-a + mx)$ if $m = 2$, $x = -3$, and $a = -4$

solution We will replace each variable with parentheses.

$-[(\ \ ) + (\ \ )][-(\ \ ) + (\ \ )(\ \ )]$	replaced variables with parentheses
$-[(2) + (-3)][-(-4) + (2)(-3)]$	numbers inserted
$= -(2 - 3)[4 + (-6)] = -(-1)(4 - 6)$	simplified
$= -(-1)(-2) = -(2) = \mathbf{-2}$	simplified

example 13.8 Evaluate: $-xa(x - a) + a$ if $a = -2$ and $x = 4$

solution We will replace each variable with parentheses.

$-(\ \ )(\ \ )[(\ \ ) - (\ \ )] + (\ \ )$	replaced variables with parentheses
$-(4)(-2)[(4) - (-2)] + (-2)$	numbers inserted
$= -(-8)(4 + 2) - 2 = 8(6) - 2$	simplified
$= 48 - 2 = \mathbf{46}$	simplified

practice Evaluate:

a. $x - xy$ if $x = -2$ and $y = 3$

b. $a - (ab - a)$ if $a = -4$ and $b = -2$

c. $x - ab(a - b)$ if $x = -3$, $a = -2$, and $b = -4$

d. $-xa(a + x) + x$ if $x = -4$ and $a = -2$

problem set 13

1. What is the difference between a numerical expression and an algebraic expression?
2. What do we mean by the value of an expression?
3. What is (a) a variable? (b) an unknown?
4. Use two unit multipliers to convert 75 feet to centimeters.
5. Use two unit multipliers to convert 7000 square miles to square feet.
6. Find the volume of this right circular cylinder. Dimensions are in meters.

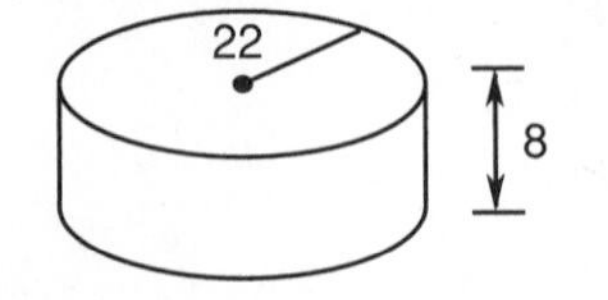

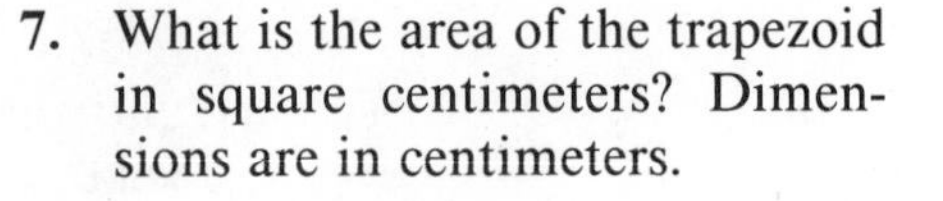

7. What is the area of the trapezoid in square centimeters? Dimensions are in centimeters.

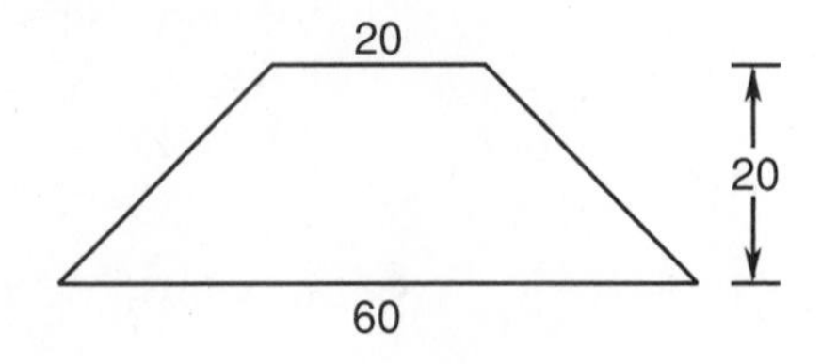

Evaluate:

8. $xm - 2m$ if $x = -2$ and $m = -3$
9. $ma - m - a$ if $m = -2$ and $a = -4$
10. $2abc - 3ab$ if $a = 2$, $b = -3$, and $c = 4$

11. $xy - 3y$ if $x = 2$ and $y = 4$
12. $-x(a + b)$ if $x = 4$, $a = -3$, and $b = -5$
13. $-a + b + ab$ if $a = -5$ and $b = -2$
14. $x - y(a - x)$ if $x = -3$, $y = 4$, and $a = 4$
15. $-(m - x)(a - mx)$ if $m = 3$, $x = -4$, and $a = -2$
16. $-xa(x + a) - a$ if $a = -4$ and $x = 2$
17. $|-b - a| + a$ if $a = -4$ and $b = 2$
18. $-a + (-a + b)$ if $a = -3$ and $b = -5$
19. $-xy - (-x + y)$ if $x = -3$ and $y = -4$
20. $-c - (p - c)$ if $p = -5$ and $c = 2$
21. $-xy - x(x - y)$ if $x = -4$ and $y = -1$

Simplify:

22. $-2[-3(-2 - 5)(3)]$
23. $-3 - (-2) - 3(-2 + 5) + 2|-3|$
24. $-|-3|(2 - 5) - [-(-3)]$
25. $-4 - 2(3 - 2) - (-2 - 5)$
26. $-3(-2 - 3)(5 - 7) - 2$
27. $5 - 3(-2 + 6) - (5 - 7) - 2$
28. $-5(-3 + 7)(-2)(-3 + 2)$
29. $-3(5 - 3) - (-2)(-6 - 1)$
30. $\dfrac{(-5 - 2) + (-3 - 2)}{-3 - (-2)}$
31. $\dfrac{-2[-(-3)]}{(-2)(-4 + 3)}$
32. $-3\frac{1}{4} + 2\frac{3}{11}$
33. $-5\frac{1}{3} \div 6\frac{2}{3}$

LESSON 14 More complicated evaluations

The procedures discussed in Lesson 13 are also used to evaluate more complicated expressions. The use of parentheses, brackets, and braces is often helpful in preventing mistakes. We will use all of these symbols of inclusion in the following examples.

example 14.1 Evaluate: $-a[-a(p - a)]$ if $p = -2$ and $a = -4$

solution We use parentheses, brackets, and braces as required.

$$-(\ \ )\{-(\ \ )[(\ \ ) - (\ \ )]\}$$

Now we will insert the numbers inside the parentheses.

$$-(-4)\{-(-4)[(-2) - (-4)]\}$$

Lastly, we simplify:

$$4\{4[2]\} = 4\{8\} = \mathbf{32}$$

example 14.2 Evaluate: $ax[-a(a - x)]$ if $a = -2$ and $x = -6$

solution This time we will not use parentheses. We will replace a with -2, $-a$ with 2, x with -6, and $-x$ with 6.

$$12[2(-2 + 6)]$$

Now we simplify, remembering to begin with the innermost symbol of inclusion.

$$12[2(-2 + 6)] = 12[2(4)] = 12[8] = \mathbf{96}$$

example 14.3 Evaluate: $-b[-b(b - c) - (c - b)]$ if $b = -4$ and $c = -6$

solution We replace b with -4, $-b$ with 4, c with -6, and $-c$ with 6.

$$4[4(-4 + 6) - (-6 + 4)]$$

Now we simplify, remembering to begin within the innermost symbols of inclusion and to multiply before adding.

$$4[4(2) - (-2)]$$
$$4[8 + 2]$$
$$4[10]$$
$$\mathbf{40}$$

practice Evaluate:

a. $-a[-a(p - a)]$ if $p = -4$ and $a = 2$

b. $pa[-p(-a)]$ if $p = -2$ and $a = -4$

c. $-x[-x(x - a) - (a - x)]$ if $x = -2$ and $a = -5$

problem set 14

1. What is (a) a factor? (b) a quotient? (c) a sum?

Evaluate:

2. $x - xy$ if $x = -2$ and $y = -3$

3. $x(x - y)$ if $x = -2$ and $y = -3$

4. $(x - y)(y - x)$ if $x = 2$ and $y = -3$

5. $(x - y) - (x - y)$ if $x = -2$ and $y = 3$

6. $(-x) + (-y)$ if $x = -2$ and $y = 3$

7. $-xa(x - a)$ if $a = -2$ and $x = 4$

8. $(-x + a) - (x - a)$ if $x = -4$ and $a = 5$

9. $-x(a - xa)$ if $x = 3$ and $a = -5$

10. $-mp(p - m)$ if $m = -5$ and $p = 2$

11. $(p - x)(a - px)$ if $a = -3$, $p = 2$, and $x = -4$

12. $(p - px) + (a + p)$ if $a = -3$, $p = 2$, and $x = -4$

13. $(p - px) + (a + p)$ if $a = -5$, $p = -3$, and $x = 4$

14. $-a[(-x - a) - (x - y)]$ if $a = -3$, $x = 4$, and $y = -5$

15. Find the volume and the surface area of this right prism. Dimensions are in meters.

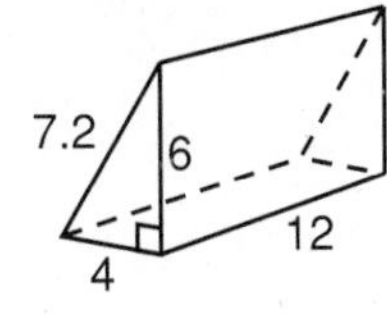

16. Find the perimeter of this figure. All angles are right angles. Dimensions are in feet.

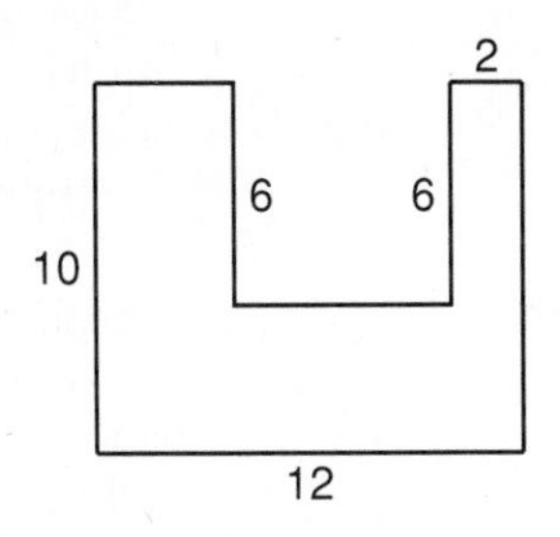

17. Use three unit multipliers to convert 4700 cubic feet to cubic yards.

Simplify:

18. $-3[-2 - 5(3 - 7)]$

19. $-3 - (-2) - \{-[-(5)]\}$

20. $-2 + (-3) - |-5 + 2|3$

21. $-8 - 6(-2 - 1) + (-5)$

22. $-3[(2 - 5) - (3 - 1)]$

23. $\dfrac{-3(-6 - 2) + 5}{-3(-2 + 1)}$

24. $\dfrac{3(-2 - 1)}{-7(2 - 4)}$

25. $\dfrac{3(-2) - 5}{-3(-2)}$

26. $-2(-4) - \{-[-(-6)]\}$

27. $-3 - 2[(5 - 3)2 - (2 - 3)]$

28. $-2(-5 - 2) + (-3)(-6) - 2$

29. $5(-2 - 3) - 3(-2 + 5)$

30. $-3 - 2 - 5(-2 - 1) + (-3)$

31. $4[2(3 - 2) - (6 - 4)]$

32. $-2 - |-2 - 5| + (-3)(-6 - 2)$

LESSON 15 *Terms and the distributive property*

15.A factors and coefficients

If the form in which variables and constants are written in an expression indicates that the variables and constants are to be multiplied, we say that the expression is an **indicated product.** If we write

$$4xy$$

we indicate that 4 is to be multiplied by the product of x and y. Each of the symbols is said to be a factor of the expression. Any one factor of an expression or any product of factors of an expression can also be called the **coefficient** of the rest of the expression. Thus in the expression $4xy$ we can say that

(a) 4 is the coefficient of xy $\quad 4(xy)$

(b) x is the coefficient of $4y$ $\quad x(4y)$

(c) y is the coefficient of $4x$ $\quad y(4x)$

(d) xy is the coefficient of 4 $\quad xy(4)$

(e) $4y$ is the coefficient of x $\quad 4y(x)$

(f) $4x$ is the coefficient of y $\quad 4x(y)$

As mentioned earlier, the value of a product is not affected by the order in which the multiplication is performed, so we may arrange the factors in any order without affecting the value of the expression. You note that we change the order at will in (a) through (f) above.

If the coefficient is a number as in (a) above, we call it a **numerical coefficient,** and if the coefficient consists entirely of variables or letters as in (b), (c), and (d) above, we call it a **literal coefficient.** We need to speak of numerical coefficients so often that we usually drop the adjective *numerical* and use the single word *coefficient.* Thus in the following expressions

$$4xy \qquad -15pq \qquad 81xmz$$

4 is the coefficient of xy, -15 is the coefficient of pq, and 81 is the coefficient of xmz.

15.B terms

A **term** is an algebraic expression that

1. Consists of a single variable or constant.
2. Is the indicated product or quotient of variables and/or constants.
3. Is the indicated product or quotient of expressions that contain variables and/or constants.

$$4 \qquad x \qquad 4x \qquad \frac{4xy(a + b)}{p} \qquad \frac{3x + 2y}{m}$$

All the expressions above can be called **terms.** The first two are terms because they consist of a single symbol. The third is a term because it is an indicated product of symbols. The fourth and fifth are terms because they are considered to be indicated quotients even though the numerator of the fourth term is an indicated product and the numerator of the fifth term is an indicated sum. **A term is thought of as a single entity that represents or has the value of one particular number.** The word *term* is very useful in allowing us to identify or talk about the parts of a larger expression. For instance, the expression

$$x + 4xym - \frac{6p}{y + 2} - 8$$

is an expression that has four terms. We can speak of a particular term of this expression, say the third term, without having to write out the term in question. The terms of an expression are numbered from left to right beginning with the number 1. Thus, for the expression above:

The first term is $+x$. The third term is $-\dfrac{6p}{y + 2}$.

The second term is $+4xym$. The fourth term is -8.

If we consider that the sign preceding a term indicates addition or subtraction, then the sign is not a part of the term. In this book we prefer to use the thought of algebraic addition, and thus most of the time we will consider the sign preceding a term to be a part of the term. But we must be careful.

Let's look at the third term in the expression we are considering.

$$-\frac{6p}{y+2}$$

If p and y are given values such that $\frac{6p}{y+2}$ is a negative number, then $-\frac{6p}{y+2}$ will be positive. For example, if p is equal to -4 and y is equal to 1, then the expression has a value of $+8$.

$$-\frac{6p}{y+2} = -\frac{6(-4)}{1+2} = -(-8) = \mathbf{+8}$$

15.C the distributive property

We have noted that the order of adding two real numbers does not change the answer. Also, the order of multiplying two real numbers does not change the answer. We call these two properties or peculiarities of real numbers the **commutative property for addition** and the **commutative property for multiplication.**

Now we will discuss another property of real numbers that is of considerable importance, the **distributive property of real numbers.** If we write

$$4(5 - 3)$$

we indicate that we are to multiply 4 by the algebraic sum of the numbers 5 and -3. A property (or peculiarity) of real numbers permits the value of this product to be found two different ways.

$4(5 - 3)$	$4(5 - 3)$
$4(2)$	$4 \cdot 5 + 4(-3)$
8	$20 - 12$
	8

On the left we first added 5 and -3 to get 2, and then multiplied by 4 to get 8. On the right we first multiplied 4 by both 5 and -3, and then added the products 20 and -12 to get 8. Both methods of simplifying the expression led to the same result. **We call this property or peculiarity of real numbers the *distributive property* because we get the same result if we distribute the multiplication over the algebraic addition.**[†]

DISTRIBUTIVE PROPERTY

For any real numbers a, b, c,

$$a(b + c) = ab + ac$$

It is possible to extend the distributive property so that the extension is applicable to the indicated product of a number or a variable and the algebraic sum of any number of real numbers or variables.

[†] Note that while multiplication can be distributed over addition, the reverse is not true, for addition cannot be distributed over multiplication. For example,

$$2 + (3 \cdot 5) \neq (2 + 3) \cdot (2 + 5)$$

because $17 \neq 35$

EXTENSION OF THE DISTRIBUTIVE PROPERTY

For any real numbers $a, b, c, d, \ldots,$

$$a(b + c + d + \cdots) = ab + ac + ad + \cdots$$

example 15.1 Use the distributive property to find the value of $4(6 - 2 + 5 - 7)$.

solution We begin by multiplying 4 by each of the terms within the parentheses, and then we add the resultant products.

$$\begin{aligned} 4(6 - 2 + 5 - 7) &= 4(6) + 4(-2) + 4(5) + 4(-7) \\ &= 24 - 8 + 20 - 28 = \mathbf{8} \end{aligned}$$

example 15.2 Use the distributive property to expand $mn(x + y + 2p)$.

solution We will multiply mn by each of the terms within the parentheses.

$$mn(x + y + 2p) = \boldsymbol{mnx + mny + 2mnp}$$

example 15.3 Use the distributive property to expand $(x - 3y + xz)mp$.

solution The order is different, but we use the same procedure. Thus, mp is multiplied by each term within the parentheses.

$$(x - 3y + xz)mp = \boldsymbol{mpx - 3ymp + mpxz}$$

example 15.4 Use the distributive property to expand $-3(2x - 4)$.

solution **This can be read two ways! The first way is to read it as the opposite of** $3(2x - 4)$. Since $3(2x - 4)$ can be expanded as

$$3(2x - 4) = 6x - 12$$

we can write the opposite of this as

$$\boldsymbol{-6x + 12}$$

The second way is to multiply −3 by both 2*x* and −4 and write the result as an algebraic sum. If we do this, we get the same answer.

$$-3(2x - 4) = \boldsymbol{-6x + 12}$$

practice

a. Use the letters a, b, and c, and parentheses to write the distributive property.

b. Evaluate $4(5 - 3)$ by adding and then multiplying.

c. Evaluate $4(5 - 3)$ by multiplying and then adding.

d. When we multiply first, we say we are using the distributive property. Use the distributive property to evaluate $4(6 - 2 + 5 - 7)$.

Expand by using the distributive property:

e. $xy(a + b - 2c)$ **f.** $(mn - 3)(4x) - 3(2x-4)$

problem set 15

1. What is a coefficient?

2. What is a literal coefficient?

3. Find the area of the shaded portion of this trapezoid. Dimensions are in feet.

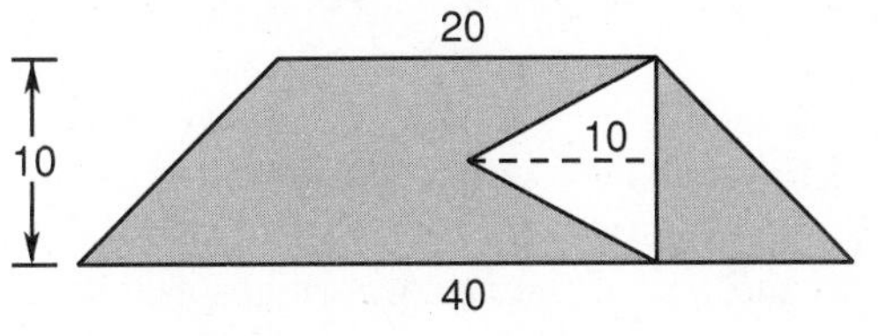

4. Find the volume of this right circular cylinder. Dimensions are in centimeters.

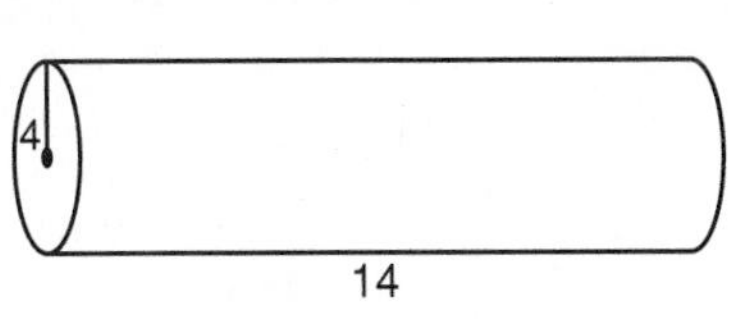

5. Use three unit multipliers to convert 280 cubic inches to cubic feet.

Evaluate by using the distributive property:

6. $-7(-8 + 3)$

7. $5(-3 - 6)$

Expand by using the distributive property:

8. $mx(ab - b)$

9. $-4y(d + cx)$

10. $(a + bc)2x$

11. $3a(x + 2y)$

Evaluate:

12. $-a(a - b)$ if $a = -2$ and $b = -7$

13. $(-a + b) + (-a)$ if $a = -2$ and $b = 5$

14. $(a - x)(-x)$ if $a = 2$ and $x = -5$

15. $(x - y) - (y - x)$ if $x = -2$ and $y = -4$

16. $x - 2a(-a)$ if $x = 4$ and $a = -3$

17. $-x(a - xa)$ if $x = -4$ and $a = -3$

18. $-y[-ay - (xy)]$ if $a = -2$, $x = 2$, and $y = -3$

Simplify:

19. $-|-2| + (-3) - 3 - (-4 - 2)$

20. $4[(2 - 4) - (6 - 3)]$

21. $5[(3 - 2)(-5 - 3)]$

22. $-3 - 5(-2) - 4 + (-6)(3)$

23. $-2[-5 - 3(-2)][(-4) + 2]$

24. $-5(-2)(-2 - 3) - (-|-2|)$

25. $-[-(-3)] - 2(-2) + (-3)$

26. $3(-2)(-3 - 2) - (-4 - 2)$

27. $\dfrac{3 - (-2)(4)}{5 - (-3)}$

28. $-|-3| - 2(-3) + (-3) - 5 - 2$

29. $\dfrac{3 + 7(-3)}{-6 - 2(-4)}$

30. $(-2 - 5 + 3)(-2) - [-6 + 3(-2)]$

31. $\dfrac{-0.06561}{4.05}$

32. $\dfrac{-3\frac{1}{3}}{2\frac{1}{5}}$

33. $-3\frac{1}{5} + 1\frac{3}{8}$

LESSON 16 Like terms · Addition of like terms

16.A like terms

Like terms are terms that have the same variables in the same form or in equivalent forms so that the terms (excluding numerical coefficients) **represent the same number regardless of the nonzero values assigned to the variables.** Let's look at the indicated sum of terms

$$4xmp - 2pmx + 6mxp$$

Now whether terms are **like terms** or not does not depend on the signs of the terms or on the values of the numerical coefficients. So we won't consider the + and − signs or the numbers 4, 2, and 6. We just need to know if

$$xmp, \qquad pmx, \qquad \text{and} \qquad mxp$$

are in the same form or equivalent forms and if each expression represents the same number regardless of the nonzero values that are assigned to the variables. We state the following:

1. They are in equivalent forms, for they have the same variables in the form of an indicated product, and the order of multiplication of the factors does not affect the value of the product.
2. They represent the same number regardless of the values assigned to the variables.

We will not attempt to prove statement #2 but will demonstrate it with one set of values for the variables. If we let $x = 4$, $p = 6$, and $m = 2$, we have

xpm	pmx	mxp
$4 \cdot 6 \cdot 2$	$6 \cdot 2 \cdot 4$	$2 \cdot 4 \cdot 6$
48	**48**	**48**

Thus $4xmp$, $2pmx$, and $6mpx$ are like terms because the variables represent the same number regardless of the real number replacements used for the variables.

16.B addition of like terms

The extension of the distributive property of Lesson 15 can be rewritten as

$$ba + ca + da + \cdots = (b + c + d + \cdots)a$$

We note that a is a common factor of each of the terms on the left and is written outside the parentheses on the right. If we look at the indicated sum of terms

$$4xmp - 2pmx + 6mxp$$

we see that the factor xpm is a factor of all three terms and can be treated in the same manner as the a factor on the left side of the statement of the distributive property.

Thus we can write the sum of three terms as a product of $(4 - 2 + 6)$ and pmx as shown here.

$$4xpm - 2pmx + 6mxp = (4 - 2 + 6)pmx = \mathbf{8pmx}$$

The factors of the three variables in the expression $8pmx$ can be written in any order without changing the *value* of the expression. Thus any of the following would be equally acceptable.

$$8pmx \quad 8pxm \quad 8xmp \quad 8xpm \quad 8mpx \quad 8mxp$$

The above is a rather detailed approach to justify the following statement:

To add like terms, we algebraically add the numerical coefficients. Thus to add

$$4xpm - 2pmx + 6mxp$$

we simply add the numerical coefficients 4, −2, and +6 to get $8pxm$.

$$4xpm - 2pmx + 6mxp = \mathbf{8pxm}$$

If the expression contains signed numbers, these are added separately, as shown in the following examples.

example 16.1 Simplify by adding like terms: $3x + 5 - xy + 2yx - 5x$

solution The first term and the fifth term are like terms, and the third term and the fourth term are like terms. If we add these terms, we get

$$\mathbf{-2x + xy + 5}$$

example 16.2 Simplify by adding like terms: $3xy + 2xyz - 10yx - 5yzx$

solution The first term and the third term are like terms and may be added. Also, the second and fourth terms are like terms and may be added. If we add these terms, we get

$$\mathbf{-7yx - 3xyz}$$

example 16.3 Simplify by adding like terms: $4 + 7mxy + 5 + 3yxm - 15$

solution We add like terms and get

$$\mathbf{-6 + 10mxy}$$

example 16.4 Simplify by adding like terms:

$$3x - x - y + 5 - 2y - 3x - 10 - 8y$$

solution We add the x terms, the y terms, and the numbers and get

$$\mathbf{-x - 11y - 5}$$

example 16.5 Simplify by adding like terms:

$$-3 + xmy - y - 5 + 8ymx - 3y - 14$$

solution We add the y terms, the ymx terms, and the numbers in any order and get

$$\mathbf{-22 - 4y + 9myx}$$

Of course, the letters myx could be in any order.

practice Simplify by adding like terms:

a. $-2xy + 3x + 4 - 4yx - 2x$ **b.** $2xyz + 3xy - 5zyx$

c. $3yac - 2ac + 6acy$ **d.** $4 - x - 2xy + 3x - 7yx$

problem set 16

1. What is a term?
2. What kind of terms may be added?

3. What is (a) a factor? (b) a product? (c) a quotient?

4. Find the area of this figure. Dimensions are in feet.

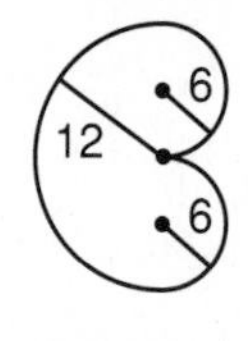

5. Find the surface area of this rectangular prism. Dimensions are in inches.

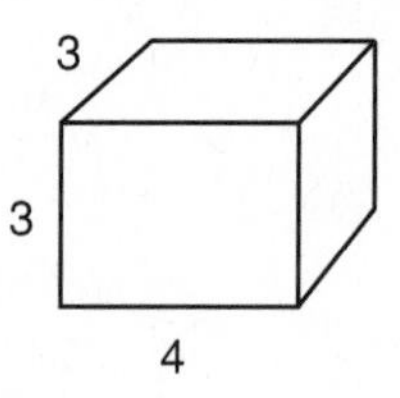

6. Use three unit multipliers to convert 6 feet to meters. Go from feet to inches, to centimeters, to meters.

7. Use four unit multipliers to convert 42,000 square inches to square miles. Use two unit multipliers to go from square inches to square feet and two unit multipliers to go from square feet to square miles.

Simplify by adding like terms:

8. $3xyz + 2zxy - 7zyx + 2xy$

9. $4x + 3 - 2xy - 5x - 7 + 4yx$

Expand by using the distributive property:

10. $(4 + 2y)x$

11. $3x(y - 2m)$

12. $2p(xy - 3k)$

Evaluate:

13. $-a(x - a)$ if $a = -3$ and $x = 6$

14. $-x - (-a)(a - x)$ if $x = -2$ and $a = 4$

15. $(m - p)p$ if $m = 3$ and $p = -2$

16. $-p(-x) - px$ if $p = -3$ and $x = 4$

17. $(x - y) - (y - x)$ if $x = -3$ and $y = -2$

18. $-x(-y) - xy$ if $x = 3$ and $y = -2$

19. $-px(x - p)$ if $x = -4$ and $p = 5$

20. $(-a)(b)(-a + b)$ if $a = 6$ and $b = -3$

Simplify:

21. $-3 - 2(-4 + 7) - 5 - |-2 - 5|$

22. $-\{3(-2)(-4 + 2) - [3 - (-2)]\}$

23. $-4 - (-2) - [-(-2)] - |-3|$

24. $-6 - 2(-3)(-1) - 5(3 - 2 - 2)$

25. $3 - (-6 + 8)2 - 4(-3) + (-3)$

26. $-5 - 2 - 6(-3 + 7)2 - 2(-3)$

27. $-2[(-3 + 5)(-2) - (3 - 2)]$

28. $\dfrac{-2(-3 + 7)}{(-2)(-3)}$

29. $\dfrac{-7(-2 + 3)}{-2(-3)}$

30. $\dfrac{-3\frac{1}{5}}{5\frac{7}{10}}$

31. $-5\frac{1}{5} + 7\frac{2}{3}$

32. $\dfrac{0.09338}{-0.046}$

LESSON 17 Exponents · Powers of negative numbers · Roots

17.A exponents

Often we find it necessary to indicate that a number is to be used as a factor a given number of times. For instance, if we wish to indicate that 5 is to be used as a factor seven times, we could write $5 \cdot 5 \cdot 5 \cdot 5 \cdot 5 \cdot 5 \cdot 5$. This is a cumbersome expression and mathematicians have developed a sort of mathematical shorthand called **exponential notation** that allows the expression to be written more concisely. The exponential notation for 5 used as a factor seven times is 5^7, read **"five to the seventh."** The general form of the expression is x^n, which indicates that x is to be used as a factor n times and is read **"x to the n."**

DEFINITION OF EXPONENTIAL NOTATION

$$\underbrace{x \cdot x \cdot x \cdot \ldots \cdot x}_{n \text{ factors}} = x^n$$

In this definition the letter x represents a real number and is called the **base** of the expression. The letter n represents a positive integer and is called the **exponent.**

For example,

$$x^4 = x \cdot x \cdot x \cdot x \qquad \text{The base is } x \text{ and the exponent is 4.}$$

$$(-4)^3 = (-4)(-4)(-4) \qquad \text{The base is } (-4) \text{ and the exponent is 3.}$$

$$\left(\frac{1}{3}\right)^4 = \left(\frac{1}{3}\right)\left(\frac{1}{3}\right)\left(\frac{1}{3}\right)\left(\frac{1}{3}\right) \qquad \text{The base is } \frac{1}{3} \text{ and the exponent is 4.}$$

The value of an exponential expression is called a **power** of the base.

$$2^4 = (2)(2)(2)(2) = 16$$

The value of 2 used as a factor four times is 16. **We say that the fourth power of 2 is 16.** We find it convenient to sometimes use the word *power* to name the exponent. To do this, we would say that the expression

$$2^4$$

represents 2 raised to the **fourth power.**

17.B powers

When a positive number is raised to a positive power, the result is always a positive number.

example 17.1 Simplify: (a) 3^2 (b) 3^3 (c) 3^4 (d) -3^4

solution Each of these notations tells us that $+3$ is the base. The exponents 2, 3, and 4 tell us to use 3 as a factor twice, three times, and four times.

(a) $3^2 = (3)(3) = \mathbf{9}$ (b) $3^3 = (3)(3)(3) = \mathbf{27}$

(c) $3^4 = (3)(3)(3)(3) = \mathbf{81}$

(d) We must be careful here because -3^4 means **the opposite of** 3^4 and not $(-3)^4$.

$$-3^4 = -(3)(3)(3)(3) = \mathbf{-81}$$

When a negative number is raised to an even power, the result is always positive; and when a negative number is raised to an odd power, the result is always negative, as demonstrated in the next example.

example 17.2 Simplify: (a) $(-3)^2$ (b) $(-3)^3$ (c) $(-3)^4$ (d) $-(-3)^4$

solution The first three are straightforward.

$$\text{(a)}\ (-3)^2 = (-3)(-3) = \mathbf{9} \qquad \text{(b)}\ (-3)^3 = (-3)(-3)(-3) = \mathbf{-27}$$

$$\text{(c)}\ (-3)^4 = (-3)(-3)(-3)(-3) = \mathbf{81}$$

(d) Be careful here. We want the opposite of $(-3)^4$.

$$-(-3)^4 = -(-3)(-3)(-3)(-3) = \mathbf{-81}$$

example 17.3 Simplify: $-3^3 - (-3)^2 + (-2)^2$

solution Be careful with the first term.

$$-(3)(3)(3) - (-3)(-3) + (-2)(-2) = -27 - 9 + 4 = \mathbf{-32}$$

example 17.4 Simplify: $-2^2 - 4(-3)^3 - 2(-2)^2 - 2$

solution -2^2 is -4, $(-3)^3$ is -27, and $(-2)^2$ is 4, so we get

$$-4 - 4(-27) - 2(4) - 2 = -4 + 108 - 8 - 2 = \mathbf{94}$$

17.C roots

If we use 3 as a factor twice, the result is 9. Thus 3 is the positive square root of 9. We use a **radical sign** to indicate the root of a number.

$$(3)(3) = 9 \qquad \text{so} \qquad \sqrt{9} = 3$$

If we use 3 as a factor three times, the result is 27. Thus 3 is the cube root of 27.

$$(3)(3)(3) = 27 \qquad \text{so} \qquad \sqrt[3]{27} = 3$$

If we use 3 as a factor four times, the result is 81. Thus 3 is the positive fourth root of 81.

$$(3)(3)(3)(3) = 81 \qquad \text{so} \qquad \sqrt[4]{81} = 3$$

If we use 3 as a factor five times, the result is 243. Thus 3 is the fifth root of 243.

$$(3)(3)(3)(3)(3) = 243 \qquad \text{so} \qquad \sqrt[5]{243} = 3$$

Because $(-3)(-3)$ equals $+9$, we say that -3 is the negative real root of 9. Because $(-3)(-3)(-3)(-3)$ equals $+81$, we say that -3 is the negative fourth root of 81. **If *n* is an even number, every positive real number has a positive *n*th root and a negative *n*th root. We use the radical sign to designate the positive even root.** To designate the negative even root, we also use a minus sign.

$$\sqrt{9} = 3 \qquad -\sqrt{9} = -3 \qquad \sqrt[4]{81} = 3 \qquad -\sqrt[4]{81} = -3$$

The number under the radical sign is called the **radicand,** and the little number that designates the root is called the **index.** If n is an odd number, every positive number has exactly one real nth root. Also, every negative number has exactly one real odd nth root.

$$\sqrt[3]{27} = 3 \qquad \sqrt[3]{-27} = -3$$

In this book we will consider even roots of positive numbers and odd roots of both positive and negative numbers. Even roots of negative numbers will be discussed in the next book in this series.

example 17.5 Simplify: (a) $\sqrt{64}$ (b) $\sqrt[4]{16}$ (c) $\sqrt[3]{-27}$ (d) $-\sqrt{81}$

solution (a) The notation $\sqrt{64}$ designates the positive number which used as a factor twice has a product of 64. The answer is **8** because 8 times 8 equals 64.The notations $\sqrt[2]{64}$ and $\sqrt{64}$ both designate the positive square root of 64. If the index is not written, it is understood to be 2.

$$\sqrt{64} = \mathbf{8}$$

(b) The fourth root of 16 is 2 because

$$\sqrt[4]{16} = \mathbf{2} \qquad \text{because} \qquad (2)(2)(2)(2) = 16$$

(c) Every real number has exactly one nth root, where n is odd.

$$\sqrt[3]{-27} = \mathbf{-3} \qquad \text{because} \qquad (-3)(-3)(-3) = -27$$

(d) The square root of 81 is 9. We want the opposite of this.

$$-\sqrt{81} = \mathbf{-9}$$

practice Simplify:

a. $(-2)^2$ **b.** -2^2 **c.** $(-3)^3 + \sqrt[3]{-27}$

d. $-3^3 - (-2)^2 - 2^2$ **e.** $-3^2 - \sqrt[3]{-8} - \sqrt{16}$

problem set 17

1. Use the numbers 2, 3, and 4 to demonstrate the distributive property.

2. Which integer is not a real number?

3. How many 1-inch-square floor tiles would it take to cover this figure? Dimensions are in inches.

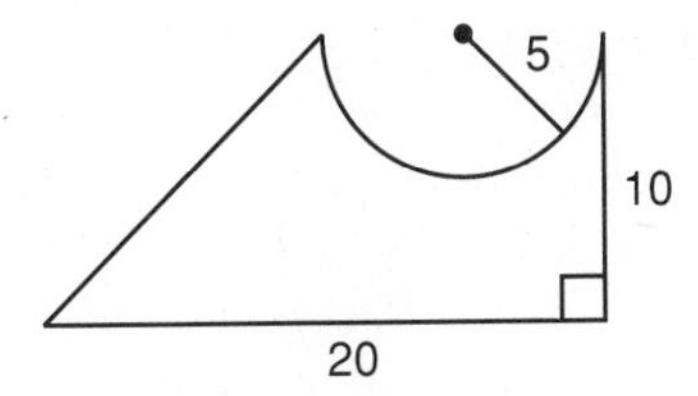

4. Find the volume and surface area of this right prism. Dimensions are in feet.

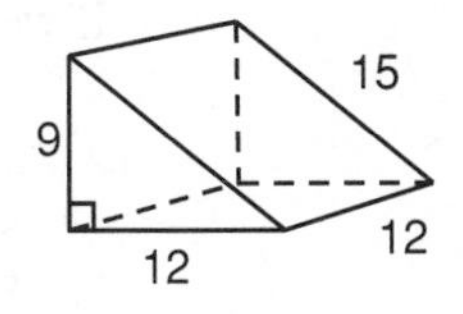

5. Use three unit multipliers to convert 42 feet to meters. Convert feet to inches to centimeters to meters.

6. Use four unit multipliers to convert 170 square inches to square yards. Use two unit multipliers to go from square inches to square feet and two unit multipliers to go from square feet to square yards.

7. Use three unit multipliers to convert 42 cubic meters to cubic centimeters.

Simplify by adding like terms:

8. $xym - 3ymx - 4xmy - 3my + 2ym$ **9.** $-3pxk + pkx - 3kpx - kp - 3kx$

10. $m + 4 + 3m - 6 - 2m + mc - 4mc$

11. $a - 3 - 7a + 2a - 6ax + 4xa - 5$

12. $-p - 5 - 3p - 6 - 2p + 7 - ax + 3xa$

Expand by using the distributive property to multiply:

13. $x(4 - ap)$
14. $(5p - 2c)4xy$
15. $4k(2c - a + 3m)$

Evaluate:

16. $|x - a| - a(-x)$ if $a = -3$ and $x = 4$
17. $(-x - a) - a(x - a)$ if $a = -3$ and $x = -4$
18. $-a(b - a)$ if $a = -4$ and $b = -3$
19. $-(a - x)(x - a)$ if $a = -5$ and $x = 3$
20. $(-p) - a(p - a)$ if $a = -4$ and $p = 5$
21. $-a[(x - a) + (2x + a)]$ if $a = -4$ and $x = 3$
22. $-(x + xy)$ if $x = -3$ and $y = 2$
23. $m[(x + 2xm) - (3x - mx)]$ if $x = -4$ and $m = 2$

Simplify:

24. $-3(4 - 3) - 3 - |-3|$
25. $-2^2 + (-3 - 5) - (-2)$
26. $-4(-3 + 7) - (-2) - 3$
27. $-2(-5 - 2)(-2)(-2 - 3)$
28. $-5 + (-5) - (3) + (2)$
29. $\dfrac{-4(2 - 4)}{(-2)(-4)}$
30. $-2(-4) - 3^2 + 2 - 5$
31. $(-2)^3 - 2(2) - 3(-5)$
32. $-7 + (6) - (-3) + (-2)^2$

LESSON 18 Evaluation of powers

Evaluation of expressions with exponents is straightforward when the replacements of the variables are all positive numbers. To evaluate

$$yx^2m^3$$

with $y = 3$, $x = 4$, and $m = 2$, we proceed as follows:

$$(3)(4)^2(2)^3 = (3)(16)(8) = \mathbf{384}$$

We must be careful, however, when the expressions contain minus signs or when some replacement values of the variables are negative numbers.

example 18.1 If $a = -2$, what is the value of each of the following:

(a) a^2 (b) $-a^2$ (c) $-a^3$ (d) $(-a)^3$

solution (a) a^2 means a times a, or $(-2)(-2) = \mathbf{+4}$

(b) $-a^2$ asks for the opposite of a^2, or $-(-2)(-2) = \mathbf{-4}$

(c) $-a^3$ means the opposite of a times a times a, or

$$-(-2)(-2)(-2) = \mathbf{+8}$$

(d) $(-a)^3$ means $(-a)(-a)(-a) = (2)(2)(2) = \mathbf{+8}$

example 18.2 Evaluate: x^2z^3y if $x = 2$, $z = 3$, and $y = -2$

solution We replace x with 2, z with 3, and y with -2.

$$(2)^2(3)^3(-2) = (4)(27)(-2) = \mathbf{-216}$$

example 18.3 Evaluate: $pm^2 - z^3$ if $p = 1$, $m = -4$, and $z = -2$

solution We replace p with 1, m with -4, and z with -2.

$$(1)(-4)^2 - (-2)^3 = (1)(16) - (-8) = 16 + 8 = \mathbf{24}$$

practice Evaluate:

a. x^2z^3y if $x = -3$, $z = -2$, and $y = -2$

b. $b^2 - 4ac$ if $b = -4$, $a = -3$, and $c = -5$

c. $m^2 - pq^2$ if $m = -2$, $p = 4$, and $q = -3$

problem set 18

1. Designate (a) the set of whole numbers; (b) the set of integers.

2. What is another name for the multiplicative inverse?

3. Is the product of 142 negative numbers and 3 positive numbers a positive number or a negative number?

Simplify:

4. $3^2 + (-3)^2$ **5.** $-2^3 + (-2)^3$ **6.** $-2^2 + (-4)^2$

7. $-(-3)^2 - (-2)^3$ **8.** $-3^2(-2)^2 - 2$

9. $-3 - 2^3 - (-3)^3 + \sqrt[6]{64}$

10. If the figure on the left is the base of a right prism that is 8 feet high, how many sugar cubes that measure 1 foot on a side will the solid hold? What is the surface area of the prism? Dimensions are in feet. All angles are right angles.

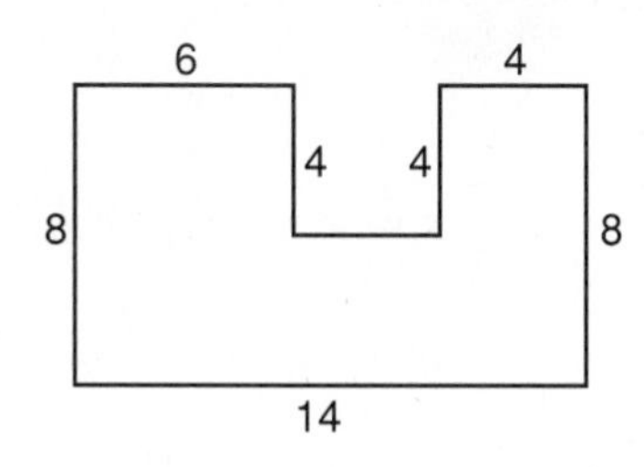

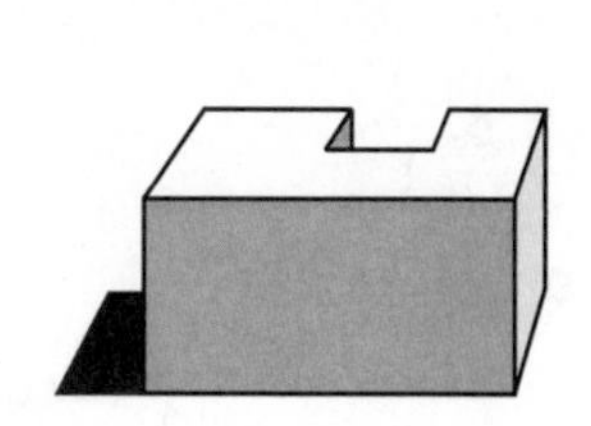

11. How many 1-inch-square floor tiles would it take to cover the trapezoid? Dimensions are in inches.

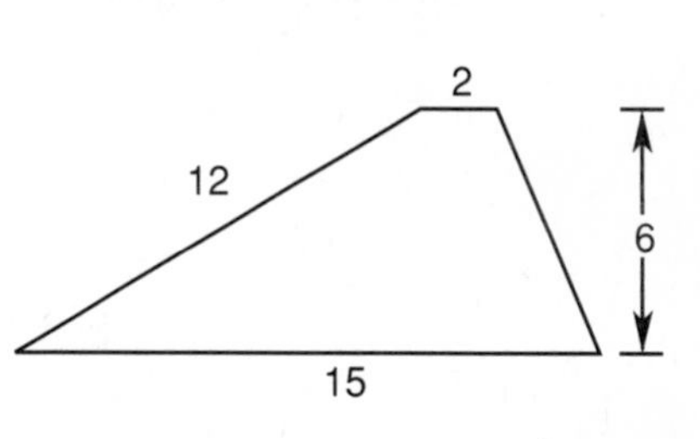

12. Use three unit multipliers to convert 49 meters to feet. Go from meters to centimeters to inches to feet.

Evaluate:

13. $-x^2 - y^3$ if $x = -3$ and $y = -2$

14. $a^2 - b^2a$ if $a = -2$ and $b = 3$

Simplify:

15. $-3^2 - 2(-3 - 4)$

16. $-2^2 - 4(-3)$

17. $-2(-2)(-2 - 3)$

18. $-|-2| - 3 + (-3 - 2)$

19. $5(3 - 4)(-2) + (-5 - 2)$

20. $2[-3(-2 - 4)(3 - 2)]$

Simplify by adding like terms:

21. $5 - x + xy - 3yx - 2 + 2x$

22. $-3bpx - 3bp - 3 + 5pb$

23. $7 - 3k - 2k + 2kx - xk + 8$

24. $-8 - py + 2yp + 4 - y$

Expand by using the distributive property:

25. $(4 - 2p)4x$

26. $-3(-x - 4)$

27. $-2x(a - 3p)$

Evaluate:

28. $-x(a - 3x) + x$ if $a = 3$ and $x = 4$

29. $-p(-a + 2p) + p$ if $p = -3$ and $a = 2$

30. $k(ak - 4a) + k$ if $k = -3$ and $a = 2$

Simplify:

31. $\dfrac{2\frac{3}{5}}{-4\frac{7}{8}}$

32. $-1\frac{1}{8} + 2\frac{7}{16}$

33. $-0.000012 \div 0.003$

LESSON 19 Product rule for exponents · Like terms with exponents

19.A product rule for exponents

The product theorem for exponents can be deduced from the definition of exponential notation. We know that

	(a) 3^5	means	$3 \cdot 3 \cdot 3 \cdot 3 \cdot 3$
and thus	(b) $3^2 \cdot 3^3$	means	$(3 \cdot 3) \cdot (3 \cdot 3 \cdot 3)$, or 3^5
and also	(c) $3 \cdot 3^4$	means	$(3) \cdot (3 \cdot 3 \cdot 3 \cdot 3)$, or 3^5

We see that when we multiply exponentials whose bases are the same, the exponent of the product is obtained by adding the exponents of the factors. Thus,

(a) $x^5 \cdot x^7 \cdot x^2 = \mathbf{x^{14}}$ (b) $5^2 \cdot 5^3 \cdot 5^2 = \mathbf{5^7}$

(c) $p^5 \cdot p^{12} = \mathbf{p^{17}}$ (d) $4^2 \cdot 4^3 \cdot 4^{25} = \mathbf{4^{30}}$

We call this rule **the product rule for exponents** and give the formal definition as follows:

PRODUCT RULE FOR EXPONENTS

If m and n are real numbers and $x \neq 0$,

$$x^m \cdot x^n = x^{m+n}$$

We can use this theorem to help simplify expressions that contain exponents.

example 19.1 Simplify: $x^2y^2x^5y^3$

solution Since rearranging the order of the factors of a product does not change the value of the product, we may write $x^2y^2x^5y^3$ as

$$x^2x^5y^2y^3 = \boldsymbol{x^7y^5}$$

example 19.2 Simplify: $x^2y^3m^5x^3y^2$

solution First we rearrange the factors and then we simplify.

$$x^2x^3y^2y^3m^5 = \boldsymbol{x^5y^5m^5}$$

Now we define the notation x^1.

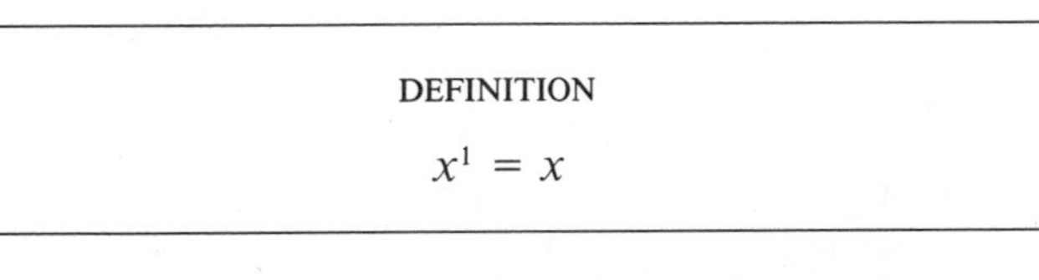

DEFINITION

$$x^1 = x$$

This says that x means the same thing as x^1 and that x^1 means the same thing as x. **If any variable or constant is written without an exponent, it is understood to have an exponent of 1. Thus,**

5 equals 5^1, $\quad x^1$ equals x, $\quad$ and $\quad 7^1$ equals 7

example 19.3 Simplify: xyy^2x^3

solution First we rearrange the letters. Then we simplify, remembering that x means x^1 and y means y^1.

$$xx^3yy^2 = \boldsymbol{x^4y^3}$$

example 19.4 Simplify: $m^3pmxm^2x^3p^5$

solution We rearrange and simplify to get the result.

$$m^3mm^2xx^3pp^5 = \boldsymbol{m^6x^4p^6}$$

We note that there are six ways that this result can be written:

$$m^6x^4p^6 \qquad m^6p^6x^4 \qquad p^6x^4m^6 \qquad p^6m^6x^4 \qquad x^4m^6p^6 \qquad x^4p^6m^6$$

Because the order of factors does not affect the product, all six of these answers are equally correct and none is preferred. In mathematics preference is reserved to the individual. There is no requirement to alphabetize the variables.

19.B adding like terms that contain exponents

Above we noted that we multiply exponential expressions with like bases by adding the exponents. The task of adding like terms that contain exponents appears similar, but the rule is different. When we add like terms that contain exponents, we do not add the exponents. Thus

$$3x^2 + 2x^2 = \mathbf{5x^2}$$

and does not equal $5x^4$. Addition and multiplication are often confused, so we discuss them in the same lesson so that we can point out the difference.

When we add, we can only add like terms. We recall that letters stand for unspecified numbers and that the order of multiplication of real numbers can be changed. This means that

$$x^2yp^5 \quad \text{and} \quad p^5x^2y$$

are like terms, and that

$$xy^2p^5 \quad \text{and} \quad y^2xp^5$$

are like terms because the literal factors of the terms are the same.

example 19.5 Simplify by adding like terms:

$$x^2yp^5 + 2xy^2p^5 + 3p^5x^2y - 7y^2xp^5$$

solution The first and third terms are like terms, and the second and fourth terms are also like terms. We add the numerical coefficients of these terms and get

$$\mathbf{4x^2yp^5 - 5xy^2p^5}$$

example 19.6 Simplify by adding like terms:

$$2m^3xy^2p + 3pxy^2m^3 - 10xy^2m^3p + yx^2m^3p$$

solution The first three terms are like terms and may be added.

$$\mathbf{-5xy^2m^3p + x^2ym^3p}$$

example 19.7 Simplify by adding like terms:

$$2x^2y + 3yx^2 + x^2y^2 - x^2y - 4x^2y^2$$

solution The first, second, and fourth terms are like terms and so are the third and fifth terms. Thus we add the first, second, and fourth terms and the third and fifth terms to obtain

$$\mathbf{4x^2y - 3x^2y^2}$$

practice Simplify:

a. $x^3xy^2y^5x^7mm$ **b.** $xyx^4x^3y^5$

Simplify by adding like terms:

c. $2x^2y^3 + xy - 8y^3x^2 - 5yx$ **d.** $x^6y + yx^6 + 3xy - 5xy^6$

problem set 19

1. What is a variable?
2. When can terms be called like terms?

3. How many 1-inch-square floor tiles would cover this figure? Dimensions are in inches. All angles are right angles.

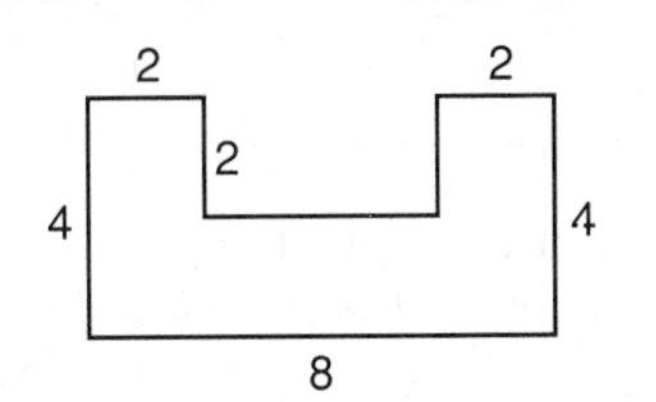

4. What is the area of the shaded portion of the parallelogram? Dimensions are in centimeters.

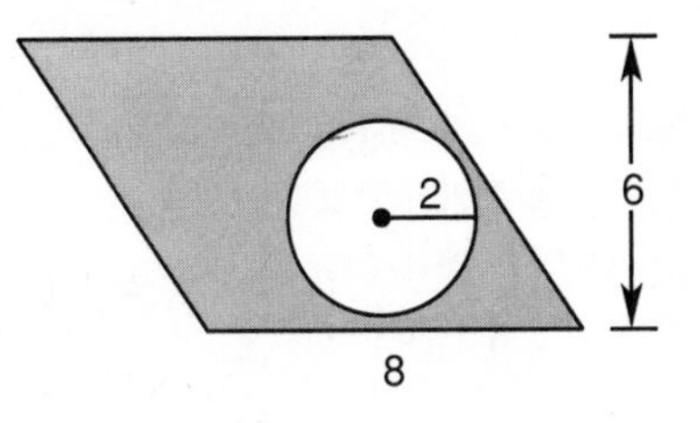

Simplify:

5. x^2yyyx^3yx

6. $xm^2xm^3x^3m$

7. $ky^2k^3k^2y^5$

8. $a^2ba^2b^3ab^4$

9. Use three unit multipliers to convert 40 meters to feet.

10. Use three unit multipliers to convert 400 cubic meters to cubic kilometers.

Simplify by adding like terms:

11. $3ab^2 - 2ab + 5b^2a - ba$

12. $x^2 - 3yx + 2yx^2 - 2xy + yx$

13. $xym^2p - 3m^2yxp + 7pm^2xy - 3y^2mxp$

Expand by using the distributive property:

14. $x(3p - 2y)$

15. $(3 - 2b)a^2$

16. $5(2 - 4p)$

Evaluate:

17. $x^2 - y^2$ if $x = -3$ and $y = -2$

18. $x(y^2 - x^2)$ if $x = -3$ and $y = 2$

19. $a^3 - y^3$ if $a = -3$ and $y = 2$

20. $a(b^3 - a)$ if $a = -2$ and $b = -4$

21. $(a - x)(x - a)$ if $a = -3$ and $x = 4$

22. $x^2(a - x)$ if $a = -5$ and $x = 3$

23. $m(x - m) + x$ if $m = -3$ and $x = -4$

24. $|-p + a| - a^2$ if $p = 4$ and $a = -2$

Simplify:

25. $-4 - (-2)(-2 + 5) - 3 + \sqrt[3]{-27}$

26. $\dfrac{-3 - [-(-3)]}{-(-2)}$

27. $-6 - [-3 + 5(-2)] - 2$

28. $-3^2 - 2^2 - (-3)^3 - \sqrt[3]{-8}$

29. $-5 - (-5)^2 - 3 + (-2)$

30. $-2[(-3 - 5) + (-7 + 2)]$

31. $\dfrac{1\frac{3}{4}}{-2\frac{1}{3}}$

32. $5\frac{2}{3} - 7\frac{9}{10}$

33. $(0.004)(0.012)$

LESSON 20 Statements and sentences · Conditional equations

20.A numerical and algebraic expressions (again)

Each of the following expressions

(a) 4 (b) $6 + 3$ (c) $4(2 + 3)$ (d) $\dfrac{7(8 + 4)}{5 + 2}$

is called a **numerical expression** because it consists of **a meaningful arrangement of numerals and symbols that designate specific operations.** Every numerical expression represents a particular number, and we say that this number is the **value** of the expression. The **values** of expressions (a), (b), (c), and (d) are 4, 9, 20, and 12, respectively.

We use the words **algebraic expression** to describe **numerical expressions** and also to describe expressions that contain variables.

(e) 6 (f) $x + 4$ (g) $x^2 - 6$ (h) $x(x + 4)$

Each of the expressions shown in (e) through (f) is an algebraic expression. The value of expression (e) is 6, but the values of expressions (f), (g), and (h) depend on the value that we assign to the variable x. If we assign to x the value of 3, then the values of expressions (f), (g), and (h) are 7, 3, and 21, respectively.

20.B statements and sentences

If we wish to make a statement that certain quantities are equal or are not equal, we can do so by writing a grammatical **sentence** in English.

(a) The number of peaches equals the number of apples.
(b) The number of peaches does not equal the number of apples.
(c) The number of peaches is greater than the number of apples.
(d) The number of peaches is less than the number of apples.

If we use the variables

$$N_p \quad \text{and} \quad N_a$$

to represent the number of peaches and the number of apples, and if we use the symbols

$=$	to mean	equal
$\neq$	to mean	not equal
$>$	to mean	greater than (read left to right)
$<$	to mean	less than (read left to right)

we can make the same statements by writing algebraic sentences.

(a) $N_p = N_a$

(b) $N_p \neq N_a$

(c) $N_p > N_a$

(d) $N_p < N_a$

We see that all four are called statements or sentences but only the first one, (a), uses the equals sign. This algebraic statement is called an **equation.** The other three statements, (b), (c), and (d), do not use the equals sign and are called **inequalities.** We will discuss equations here and hold the topic of inequalities for a later lesson.

20.C equations

An *equation* is an algebraic statement consisting of two algebraic expressions connected by an equals sign. Thus all the following are statements and all are also equations.

(a) $4 = 3 + 1$ (b) $4 + x = 2 + 2 + x$ (c) $4 = 6$

(d) $4 + x = 6 + x$ (e) $x + 4 = 8$

Equations are not always true equations as we see here. Two of these equations are true equations and two are false equations, and the truth or falsity of one of them depends on the number we use as a replacement for x in the equation.

(a) This is a true equation because 4 does equal the sum of 3 and 1.

(b) This is a true equation regardless of the number we use as a replacement for x. We will demonstrate this by replacing x with -3 and then by replacing x with $+7$.

WITH (-3)		WITH $(+7)$	
(b) $4 + (-3) = 2 + 2 + (-3)$		(b) $4 + (+7) = 2 + 2 + (+7)$	
$4 - 3 = 4 - 3$		$4 + 7 = 4 + 7$	
$1 = 1$	True	$11 = 11$	True

(c) This is a false equation because 4 is not equal to 6.

(d) This is a false equation regardless of the number we use as a replacement for x. We will demonstrate this by replacing x with -3 and then by replacing x with $+7$.

WITH (-3)		WITH $(+7)$	
(d) $4 + (-3) = 6 + (-3)$		(d) $4 + (+7) = 6 + (+7)$	
$4 - 3 = 6 - 3$		$4 + 7 = 6 + 7$	
$1 = 3$	False	$11 = 13$	False

(e) We call this equation a **conditional equation** because its truth or falsity is conditioned by the number used as a replacement for x. If we use -2 as the replacement for x, we get a false equation; but if we use 4 as the replacement for x, we get a true equation.

WITH (-2)		WITH (4)	
(e) $(-2) + 4 = 8$		(e) $(4) + 4 = 8$	
$-2 + 4 = 8$		$4 + 4 = 8$	
$2 = 8$	False	$8 = 8$	True

Replacement values of the variable that turn the equation into a true equation are called *solutions* of the equation or *roots* of the equation and are said to satisfy the equation. Thus in the equation

$$x + 4 = 8$$

we say that the number 4 is a **solution** or **root** of the equation, and we also say that the number 4 **satisfies** the equation.

example 20.1 Does -2 or $+5$ satisfy the equation $x^2 = -5x - 6$?

solution First we try -2.

$$(-2)^2 = -5(-2) - 6 \qquad \text{replaced } x \text{ with } -2$$
$$4 = 10 - 6 \qquad \text{simplified}$$
$$4 = 4 \qquad \text{True}$$

Now we try $+5$.

$$(5)^2 = -5(5) - 6 \qquad \text{replaced } x \text{ with } 5$$
$$25 = -25 - 6 \qquad \text{simplified}$$
$$25 = -31 \qquad \text{False}$$

Thus **-2 is a solution** and **$+5$ is not a solution.**

example 20.2 Is -2 or $+5$ a root of the equation $x^2 - 3x = 10$?

solution First we try -2.

$$(-2)^2 - 3(-2) = 10$$
$$4 + 6 = 10 \qquad \text{True}$$

Now we try $+5$.

$$(5)^2 - 3(5) = 10$$
$$25 - 15 = 10 \qquad \text{True}$$

Thus, both -2 and $+5$ are solutions or roots to the given equation, and we say that **both -2 and $+5$ satisfy this equation.**

practice Is either -2 or -5 a root of these equations?

a. $x - 2 = 0$

b. $x^2 + 7x = -10$

problem set 20

1. Give an example of (a) a true equation, (b) a false equation, (c) a conditional equation.

2. Find the perimeter of the following figure. Dimensions are in feet. All angles are right angles.

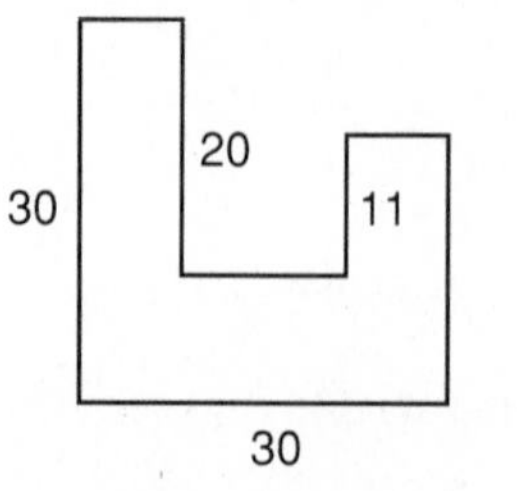

3. Find the surface area of the right circular cylinder shown. Dimensions are in inches.

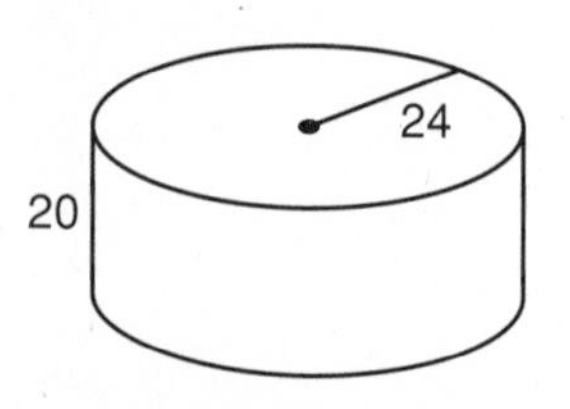

4. Use three unit multipliers to convert 12 cubic feet to cubic yards.

Simplify:

5. $x^2xxy^2xy^3$

6. $p^2m^5ypp^3my^2$

7. $8k^5nn^2kn^3k^5k$

8. $a^2aba^3b^2a^5$

9. $m^2pap^2ma^2aa^3$

10. $4p^2x^2kpx^3k^2k$

Simplify by adding like terms:

11. $3x + 2 - x^2 + 2x^2 - 4$ **12.** $xy - 3xy^2 + 5y^2x - 4xy$

13. $-3x^2ym + 7x - 5ymx^2 + 16x$ **14.** $5mp^2y - 6myp^2 + 3ymp^2 - 2p^2my$

15. $x + 2x^2 - 3 + 5x - 6x^2 - 10$ **16.** $m^2y - 6ym^2 + 2y - 3m^2y + 4y$

17. $5 - 3x + 7 - 4 + 4x^2 - 2x - x^2$

Expand by using the distributive property:

18. $a(3x - 2)$ **19.** $4xy(5 - 2a)$ **20.** $(3a - 4)6x$

Simplify:

21. $(-3)^2 - 2^3$ **22.** $-3^2 - (-2)^2 + \sqrt[3]{-27}$

23. $(-3)^3 + (-2)^3 - |-2|$ **24.** $-3[(-3 + 5)(-2 - 6)] - 3$

25. $-2[(5 - 3) - (5 - 8)]$

Evaluate:

26. $a^3 - b^3$ if $a = -2$ and $b = 3$

27. $a - b(a^2 - b)$ if $a = -2$ and $b = 3$

28. $cy[(cx - y)]$ if $x = -3$, $y = 3$, and $c = -2$

29. $-b^2a(a - b)$ if $a = -3$ and $b = 2$

30. $b(b^2) - a^2$ if $a = 3$ and $b = -2$

Simplify:

31. $\dfrac{3\frac{1}{4}}{-2\frac{5}{7}}$ **32.** $-3\frac{1}{4} + 2\frac{5}{7}$ **33.** $\dfrac{0.003636}{0.0303}$

LESSON 21 *Equivalent equations · Additive property of equality*

21.A equivalent equations

Two equations are said to be equivalent equations if *every* solution of either one of the equations is also a solution of the other equation.

$$\text{(a)} \quad x + 6 = 9 \qquad \text{(b)} \quad x + 10 = 13$$

The two equations shown are equivalent equations, for the number 3 will satisfy both equations and 3 is the only number that will satisfy either equation.

21.B additive property of equality

If we begin with the true statement that

$$2 = 2$$

and add +4 to both sides of the equality, we get the true statement that 6 equals 6.

$$2 + 4 = 2 + 4 \qquad 4 + 2 = 4 + 2$$
$$6 = 6 \qquad 6 = 6$$

On the left we placed the 4s after the 2s and on the right we placed the 4s before the 2s. We note that both procedures yield the same result. We emphasize this fact in the formal definition given in the box by writing the definition twice, the second time with the order of the addends reversed.

> ADDITIVE PROPERTY OF EQUALITY
>
> If a, b, and c are any real numbers and if $a = b$, then
>
> $$a + c = b + c \quad \text{and also} \quad c + a = c + b$$

The additive property of equality can be used to find the solution of conditional equations such as

$$x + 4 = 6$$

This equation is a conditional equation and is neither true nor false because no value has been assigned to x, so the additive property of equality does not apply. Thus we must hedge a little. We assume that some real number exists that when substituted for x will make the equation a true equation. We further assume that x in the equation represents this number. Now we have assumed that

$$x + 4 \text{ equals } 6$$

is a true statement, and thus we can use the additive property of equality. We will use the additive property to eliminate the $+4$ that is now on the left side with x. We will add -4 to both sides of the equation.

$$\begin{array}{rcrl} x + 4 &=& 6 & \\ -4 & & -4 & \quad \text{add } -4 \text{ to both sides} \\ \hline x + 0 &=& 2 & \\ x &=& 2 & \end{array}$$

Now since we made an assumption to permit the use of the additive property of equality we must check our solution in the original equation.

$$\begin{array}{rcll} x + 4 &=& 6 & \quad \text{original equation} \\ (2) + 4 &=& 6 & \quad \text{substitute 2 for } x \\ 6 &=& 6 & \quad \text{True} \end{array}$$

Since using the number 2 for x in the equation makes the equation a true equation, we say that the number 2 is a root or solution of the equation and that the number 2 satisfies the equation. It can be shown that the use of the additive property of equality will not change the numbers that satisfy the equation, so we say that the use of the additive property of equality results in an equation that is an **equivalent equation** to the original equation.

> If the same quantity is added to both sides of an equation, the resulting equation will be an **equivalent equation** to the original equation, and thus every solution of one of these equations will also be a solution of the other equation.

The rule says that we may add the same quantity to both sides of an equation, but it does not specify a particular format to be used, and one format is usually just as acceptable as another. Below are shown three possible formats for the problem worked above.

(a)
$$\begin{array}{rcr} x + 4 & = & 6 \\ -4 & = & -4 \\ \hline x & = & \mathbf{2} \end{array}$$

(b)
$$\begin{array}{rcr} x + 4 & = & 6 \\ -4 & & -4 \\ \hline x & = & \mathbf{2} \end{array}$$

(c)
$$\begin{aligned} x + 4 + (-4) &= 6 + (-4) \\ x + 0 &= 2 \\ x &= \mathbf{2} \end{aligned}$$

In (a) we added -4 to both sides of the equation and placed an equals sign between the -4s to *emphasize* that they are equal. In (b) we added -4 to both sides but omitted the equals sign since it really isn't necessary. In (c) we added the -4s on the same line with the rest of the numbers and variables. This form is adequate for very simple problems such as this one but is less desirable for more complicated problems. By the end of the book one should be able to perform this calculation mentally, without writing anything down, so that none of these formats will be necessary.

example 21.1 Solve: $x - 3 = 12$

solution To solve the equation, we want to isolate x on one side of the equals sign. We can do this if we eliminate the -3. Thus we will add $+3$ to both sides of the equation.

$$\begin{array}{rcr} x - 3 & = & 12 \\ +3 & & +3 \\ \hline x & = & \mathbf{15} \end{array}$$

This same procedure can be used when the equation contains fractions or mixed numbers, as we see in the next two examples.

example 21.2 Solve: $x + \frac{1}{4} = -\frac{3}{8}$

solution To isolate x we must eliminate the $\frac{1}{4}$. Thus we add $-\frac{1}{4}$ to both sides of the equation.

$$\begin{array}{rcl} x + \frac{1}{4} & = & -\frac{3}{8} \\ -\frac{1}{4} & & -\frac{1}{4} \\ \hline x & = & -\frac{3}{8} - \frac{1}{4} \longrightarrow x = \mathbf{-\frac{5}{8}} \end{array}$$

example 21.3 Solve: $k + 2\frac{1}{3} = \frac{2}{9}$

solution This time we add $-2\frac{1}{3}$ to both sides and then simplify the result.

$$\begin{array}{rcl} k + 2\frac{1}{3} & = & \frac{2}{9} \\ -2\frac{1}{3} & & -2\frac{1}{3} \\ \hline k & = & \frac{2}{9} - 2\frac{1}{3} \longrightarrow k = \frac{2}{9} - \frac{21}{9} \longrightarrow k = \mathbf{-\frac{19}{9}} \end{array}$$

practice Solve:

a. $x - \frac{1}{2} = \frac{3}{8}$

b. $k - 27 = -38$

c. $d + 4\frac{1}{7} = 3\frac{1}{6}$

d. $p + 5 = 17$

problem set 21

1. What does "solve an equation" mean?
2. What are equivalent equations?
3. Use four unit multipliers to convert 30 square meters to square inches.

Solve:

4. $x - 15 = 30$
5. $y - 13 = 23$
6. $x + 1\frac{1}{4} = -\frac{9}{10}$
7. $k + 7 = 93$
8. $m - 2 = 17$
9. $4 + k = -7$
10. Find the area of the shaded portion of the parallelogram. Dimensions are in centimeters.

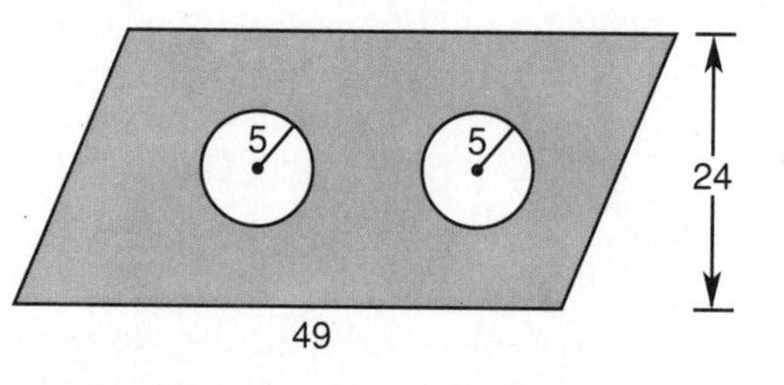

11. Find the perimeter of this figure. Dimensions are in yards.

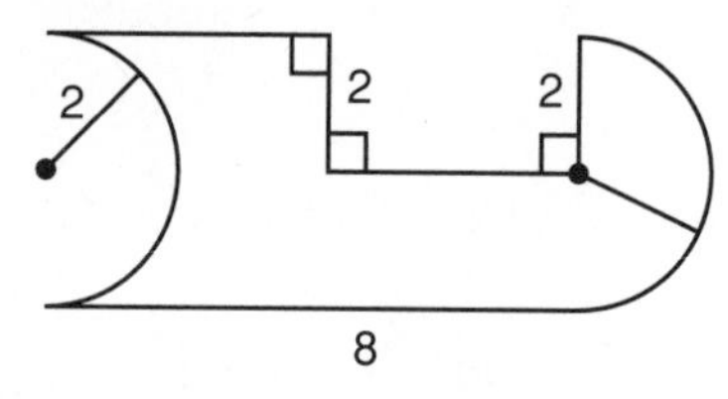

Simplify:

12. $x^2ym^5x^2y^4$
13. x^3y^2myxm
14. xxx^2yyy^3xy

Simplify by adding like terms:

15. $x^2y + 3yx - 2y^2x - 4yx^2$
16. $3x - 3 - x^2 - 2x^3 + 7 - 2x + 6x^2$
17. $p^2xy - 3yp^2x + 2xp^2y - 5$
18. $-4 + 7x - 3x - 5 + 2x - 4x^2$

Expand by using the distributive property:

19. $4x(a + 2b)$
20. $(2x + 4)3$
21. $4px(my - 3ab^2)$

Evaluate:

22. $a^3 - b^3$ if $a = -2$ and $b = 3$
23. $(a - b)(b - x)$ if $x = 2$, $a = 3$, and $b = -3$
24. $x^3 - a(a - b)$ if $x = -2$, $a = 3$, and $b = -3$
25. $x(x - y) - y$ if $x = -2$ and $y = 3$

Simplify:

26. $(-2)^3 - 2^3$
27. $-3^2 - (-3)^2 + \sqrt[5]{32}$

28. $-4(-3 - 2) - 5(-2) - 2|-4|$

29. $\frac{5(-2^2 - 4)}{-4 - 6(-2)}$

30. $\frac{-3[5(-2 - 1) - (6 - 2)]}{2(-3 - 2)}$

LESSON 22 Multiplicative property of equality

To demonstrate the **multiplicative property of equality,** we will begin with the true equation

$$2 = 2 \qquad \text{true}$$

and multiply both sides of the equation by 3.

$$3 \cdot 2 = 3 \cdot 2$$

The result is the true equation that 6 equals 6.

$$6 = 6 \qquad \text{still true}$$

The formal statement of the multiplicative property of equality in the box below is made twice to emphasize the fact that the order of the factors does not affect the product.

MULTIPLICATIVE PROPERTY OF EQUALITY

If a, b, and c are any real numbers and if

$$a = b,$$

then

$$ca = cb \qquad \text{and also} \qquad ac = bc$$

It is possible to use the multiplicative property of equality to prove that multiplying or dividing every term on both sides of an equation by the same nonzero number does not change the solution(s) to the equation. This means that the new equation is an equivalent equation to the original equation.

If **every term** on both sides of an equation is multiplied or divided by the same nonzero quantity, the resulting equation will be an equivalent equation to the original equation, and thus every solution of one of these equations will be a solution of the other equation.

We can use this rule with either of two thought processes to solve equations such as

$$4x = 12$$

The first way is to remember that division by 4 will undo multiplication by 4 because division and multiplication are inverse operations. Thus, we solve by dividing both sides of the equation by 4.

$$4x = 12 \longrightarrow \frac{4x}{4} = \frac{12}{4} \longrightarrow \mathbf{x = 3}$$

The second way is to remember that the product of 4 and its reciprocal is 1. To solve by using this thought, we will multiply both sides of the equation by $\frac{1}{4}$, which is the reciprocal of 4:

$$4x = 12 \longrightarrow \frac{1}{4} \cdot 4x = \frac{1}{4} \cdot 12 \longrightarrow \mathbf{x = 3}$$

Both of the preceding thought processes are correct and either one can be used at any time.

example 22.1 Solve: $5x = 20$

solution We can solve by (a) multiplying both sides by $\frac{1}{5}$, or by (b) dividing both sides by 5.

$$\text{(a)}\ \frac{1}{5} \cdot 5x = \frac{1}{5} \cdot 20 \qquad \text{(b)}\ \frac{5x}{5} = \frac{20}{5}$$

$$\mathbf{x = 4} \qquad\qquad \mathbf{x = 4}$$

example 22.2 Solve: $\frac{2}{5}x = 7$

solution We can solve by (a) multiplying both sides by $\frac{5}{2}$, or by (b) dividing both sides by $\frac{2}{5}$.

$$\text{(a)}\ \frac{5}{2} \cdot \frac{2}{5}x = \frac{5}{2} \cdot 7 \qquad \text{(b)}\ \frac{\frac{\cancel{2}}{\cancel{5}}x}{\frac{\cancel{2}}{\cancel{5}}} = \frac{7}{\frac{2}{5}}$$

$$\mathbf{x = \frac{35}{2}} \qquad\qquad x = 7 \cdot \frac{5}{2} \longrightarrow \mathbf{x = \frac{35}{2}}$$

Many beginning algebra students would prefer to write the preceding answer as the mixed number $17\frac{1}{2}$. Both forms of the answer are equally correct, but we prefer the improper fraction $\frac{35}{2}$ to the mixed number $17\frac{1}{2}$ because the improper fraction is easier to use. Suppose the instructions in this problem had been to, say, solve and then multiply the answer by $\frac{11}{4}$. If we had written the answer as the mixed number $17\frac{1}{2}$, we would have to change it back to the improper fraction $\frac{35}{2}$ before the multiplication could be performed. Instructors at this level and in higher courses usually prefer the improper fraction to the mixed number.

example 22.3 Solve: $2\frac{1}{4}x = 3$

solution We can undo multiplication by $2\frac{1}{4}$ by dividing by $2\frac{1}{4}$,

$$2\frac{1}{4}x = 3 \longrightarrow \frac{\cancel{2\frac{1}{4}}x}{\cancel{2\frac{1}{4}}} = \frac{3}{2\frac{1}{4}} \longrightarrow x = \frac{3}{\frac{9}{4}} \longrightarrow x = \frac{12}{9} = \mathbf{\frac{4}{3}}$$

or by rewriting $2\frac{1}{4}$ as the improper fraction $\frac{9}{4}$ and then multiplying both sides by $\frac{4}{9}$, the reciprocal of $\frac{9}{4}$.

$$\frac{9}{4}x = 3 \longrightarrow \frac{4}{9} \cdot \frac{9}{4}x = \frac{4}{9} \cdot 3 \longrightarrow \boldsymbol{x = \frac{4}{3}}$$

example 22.4 Solve: $\frac{x}{3} = 9$

solution We can undo division by 3 by multiplying by 3. Thus, we multiply both sides of the equation by 3 and cancel.

$$\cancel{3} \cdot \frac{x}{\cancel{3}} = 9 \cdot 3 \qquad \text{and thus} \qquad \boldsymbol{x = 27}$$

We can also use the concept of inverse operations when x is divided by a fraction or a mixed number.

example 22.5 Solve: (a) $\dfrac{x}{2\frac{1}{2}} = 7$ (b) $\dfrac{p}{\frac{3}{2}} = 4\frac{1}{3}$

solution (a) We can undo division by $2\frac{1}{2}$ by multiplying by $2\frac{1}{2}$.

$$\frac{x}{2\frac{1}{2}} = 7 \longrightarrow \frac{2\frac{\cancel{1}}{\cancel{2}}x}{2\frac{\cancel{1}}{\cancel{2}}} = 7 \cdot 2\frac{1}{2} \longrightarrow x = 7 \cdot \frac{5}{2} \longrightarrow \boldsymbol{x = \frac{35}{2}}$$

(b) We can undo division by $\frac{3}{2}$ by multiplying by $\frac{3}{2}$.

$$\frac{p}{\frac{3}{2}} = 4\frac{1}{3} \longrightarrow \frac{\frac{\cancel{3}}{\cancel{2}}p}{\frac{\cancel{3}}{\cancel{2}}} = 4\frac{1}{3} \cdot \frac{3}{2} \longrightarrow p = \frac{13}{3} \cdot \frac{3}{2} = \boldsymbol{\frac{13}{2}}$$

practice Solve:

a. $\dfrac{x}{\frac{1}{4}} = 20$ **b.** $\dfrac{x}{2\frac{2}{8}} = 5$ **c.** $\frac{3}{5}x = 27$ **d.** $3\frac{1}{5}z = 32$

problem set 22

1. What is another name for multiplicative inverse?

2. What is the product of any number and its reciprocal?

Solve:

3. $1\frac{5}{8}y = 26$ **4.** $3\frac{4}{5}x = -38$ **5.** $7x = 49$

6. $\frac{x}{7} = 5$ **7.** $2x = 20$ **8.** $\frac{x}{3} = 5$

9. Use three unit multipliers to convert 16,000 centimeters to miles.

10. Find the area of the shaded portion of this trapezoid. Dimensions are in meters.

11. Find the surface area of this right circular cylinder. Dimensions are in centimeters.

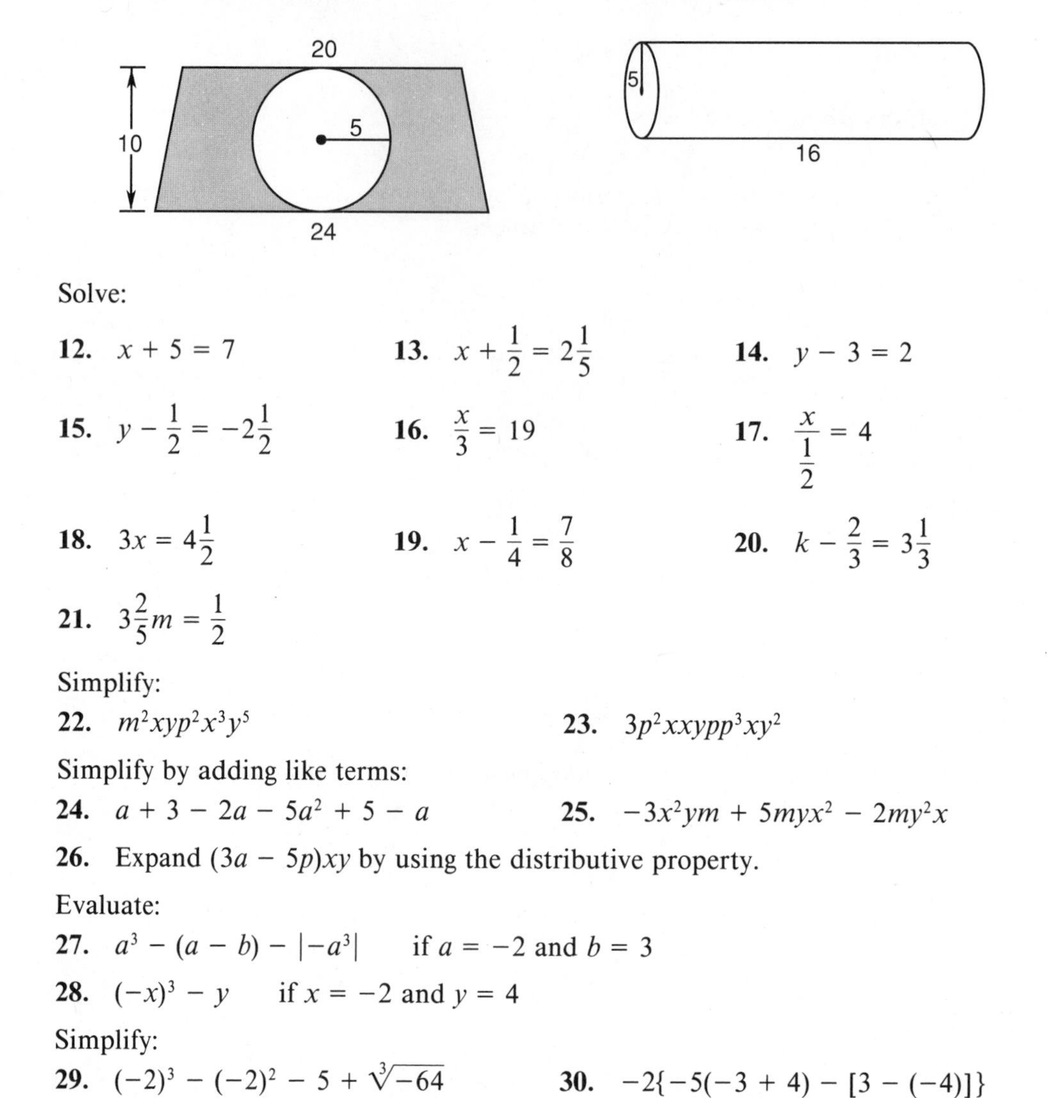

Solve:

12. $x + 5 = 7$

13. $x + \frac{1}{2} = 2\frac{1}{5}$

14. $y - 3 = 2$

15. $y - \frac{1}{2} = -2\frac{1}{2}$

16. $\frac{x}{3} = 19$

17. $\frac{x}{\frac{1}{2}} = 4$

18. $3x = 4\frac{1}{2}$

19. $x - \frac{1}{4} = \frac{7}{8}$

20. $k - \frac{2}{3} = 3\frac{1}{3}$

21. $3\frac{2}{5}m = \frac{1}{2}$

Simplify:

22. $m^2xyp^2x^3y^5$

23. $3p^2xxypp^3xy^2$

Simplify by adding like terms:

24. $a + 3 - 2a - 5a^2 + 5 - a$

25. $-3x^2ym + 5myx^2 - 2my^2x$

26. Expand $(3a - 5p)xy$ by using the distributive property.

Evaluate:

27. $a^3 - (a - b) - |-a^3|$ if $a = -2$ and $b = 3$

28. $(-x)^3 - y$ if $x = -2$ and $y = 4$

Simplify:

29. $(-2)^3 - (-2)^2 - 5 + \sqrt[3]{-64}$

30. $-2\{-5(-3 + 4) - [3 - (-4)]\}$

LESSON 23 *Solution of equations*

In Lessons 21 and 22, we were introduced to the two rules for solving equations. These are two of the most important rules in algebra. In order to emphasize this fact, they are repeated here in boldface type.

1. **The same quantity can be added to both sides of an equation without changing the answers[†] to the equation.**
2. **Every term on both sides of an equation can be multiplied or divided by the same nonzero quantity without changing the answers to the equation.**

In many equations, it is necessary to use both of these rules to find the solution. We will always use the addition rule first and then the multiplication/division rule. We do this because we are undoing a normal order of operations problem. To demonstrate we will begin with the number 4, multiply by 3, and then add -2 for a result of 10.

$$3(4) - 2 = 10$$

Now, to undo what we have done and get back to 4, we must undo the addition of -2 first and then undo the multiplication. **This is the reason that in solving equations, we reverse the normal order of operations and undo addition first and then undo multiplication or division.** We demonstrate this procedure by replacing 4 with x in the above expression and getting the equation $3x - 2 = 10$. Now we solve to find that x equals 4.

$$\begin{array}{rcl} 3x - 2 &=& 10 \\ +2 && +2 \\ \hline 3x &=& 12 \end{array} \qquad \begin{array}{l} \text{replaced 4 with } x \\ \text{add } +2 \text{ to both sides} \\ \; \end{array}$$

$$\frac{\cancel{3}x}{\cancel{3}} = \frac{12}{3} \longrightarrow \boldsymbol{x = 4} \qquad \text{divided both sides by 3}$$

example 23.1 Solve: $4x + 5 = 17$

solution We must use the addition rule to eliminate the $+5$ and then use the multiplication/division rule to eliminate the 4. We always use the addition rule first.

$$\begin{array}{rcl} 4x + 5 &=& 17 \\ -5 && -5 \\ \hline 4x &=& 12 \end{array} \qquad \begin{array}{l} \text{original equation} \\ \text{add } -5 \text{ to both sides} \\ \; \end{array}$$

$$\frac{4x}{4} = \frac{12}{4} \longrightarrow \boldsymbol{x = 3} \qquad \text{divided both sides by 4}$$

example 23.2 Solve: $-5m + 6 = 8$

solution To isolate m, we must first eliminate the 6 and then eliminate the -5.

$$\begin{array}{rrcl} & -5m + 6 &=& 8 \\ & -6 && -6 \\ \hline \text{(a)} & -5m &=& 2 \end{array} \qquad \begin{array}{l} \text{original equation} \\ \text{add } - 6 \text{ to both sides} \\ \; \end{array}$$

Now we complete the solution by dividing both sides by -5.

$$\frac{-5m}{-5} = \frac{2}{-5} \longrightarrow \boldsymbol{m = -\frac{2}{5}}$$

Dividing by a negative number sometimes leads to errors. In this problem, division by

[†] We remember that the answers to an equation are formally called the roots or solutions of the equation.

a negative number can be avoided by mentally multiplying both sides of equation (a) by −1. This changes the signs on both sides.

$$-5m = 2 \xrightarrow{\text{mentally multiplying both sides by } -1} 5m = -2$$

Now we can finish by dividing both sides by +5.

$$\frac{5m}{5} = \frac{-2}{5} \longrightarrow \boldsymbol{m = -\frac{2}{5}} \qquad \text{divided by 5}$$

example 23.3 Solve: $-7k - 4 = -21$

solution We begin by adding +4 to both sides.

$$\begin{array}{rcll} -7k - 4 &=& -21 & \text{original equation} \\ +4 && +4 & \text{add } +4 \text{ to both sides} \\ \hline -7k &=& -17 & \end{array}$$

Now we change the signs by mentally multiplying both sides of the equation by −1 and get

$$7k = 17$$

We finish by dividing both sides by +7.

$$\frac{7k}{7} = \frac{17}{7} \longrightarrow \boldsymbol{k = \frac{17}{7}} \qquad \text{divided by 7}$$

example 23.4 Solve: $-11p + 5 = 17$

solution We first eliminate the +5 and then the −11.

$$\begin{array}{rcll} -11p + 5 &=& 17 & \text{original equation} \\ -5 && -5 & \text{add } -5 \text{ to both sides} \\ \hline -11p &=& 12 & \\ 11p &=& -12 & \text{multiplied by } -1 \\ \dfrac{11p}{11} &=& \dfrac{-12}{11} \longrightarrow \boldsymbol{p = -\dfrac{12}{11}} & \text{divided by 11} \end{array}$$

example 23.5 Solve: $\frac{1}{5}m - \frac{1}{2} = \frac{3}{4}$

solution We first eliminate the $-\frac{1}{2}$.

$$\begin{array}{rcll} \dfrac{1}{5}m - \dfrac{1}{2} &=& \dfrac{3}{4} & \text{original equation} \\ +\dfrac{1}{2} && +\dfrac{1}{2} & \text{add } \dfrac{1}{2} \text{ to both sides} \\ \hline \dfrac{1}{5}m &=& \dfrac{5}{4} & \end{array}$$

Now we finish by multiplying both sides by $\frac{5}{1}$.

$$\frac{5}{1} \cdot \frac{1}{5}m = \frac{5}{4} \cdot \frac{5}{1} \qquad \text{multiplied by } \frac{5}{1}$$

$$\boldsymbol{m = \frac{25}{4}}$$

example 23.6 Solve: $0.4x - 0.2 = -0.16$

solution We first add +0.2 and then we divide by 0.4.

$$\begin{array}{rcll} 0.4x - 0.2 &=& -0.16 & \text{original equation} \\ +0.2 && +0.2 & \text{add 0.2 to both sides} \\ \hline 0.4x &=& 0.04 & \end{array}$$

Now we divide both sides by 0.4.

$$\frac{\cancel{0.4}x}{\cancel{0.4}} = \frac{0.04}{0.4} \longrightarrow \mathbf{x = 0.1} \qquad \text{divided by 0.4}$$

practice

Solve:

a. $\frac{2}{5}x - \frac{3}{10} = \frac{1}{2}$ **b.** $2\frac{1}{4}x + \frac{3}{7} = \frac{5}{14}$

c. $1.2 = -1.4 + 20x$ **d.** $0.7x - 0.4 = 0.16$

problem set 23

Solve:

1. $3x = 4\frac{1}{2}$ **2.** $x + \frac{1}{2} = \frac{2}{3}$ **3.** $3x - 4 = 10$

4. $5k - 4 = -30$ **5.** $-2p + 3 = -29$ **6.** $\frac{2}{3}y = 5\frac{1}{2}$

7. $-2x - 2 = 5$ **8.** $\frac{x}{3\frac{1}{2}} = 5$ **9.** $\frac{1}{8}m - \frac{1}{3} = \frac{3}{4}$

10. $x + 3\frac{1}{3} = 5$ **11.** $y - \frac{1}{4} = \frac{5}{7}$ **12.** $-3y + \frac{1}{2} = \frac{5}{7}$

13. Use three unit multipliers to convert 300 kilometers to inches.

14. This figure is the base of a right prism that is 10 inches high. Find the volume and the surface area of the prism. Dimensions are in inches.

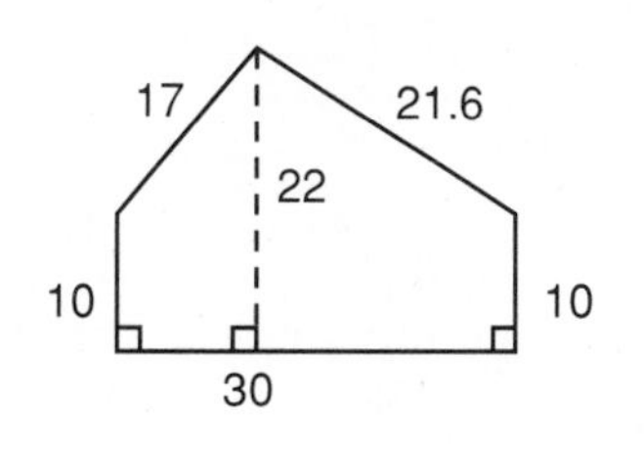

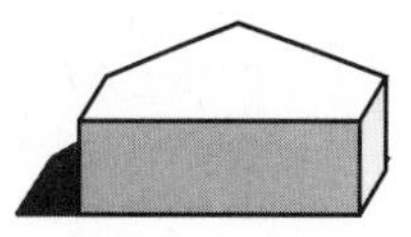

Simplify:

15. $x^2kxk^2x^2ykx^2$ **16.** $aaa^3bxa^2b^3abx^4$

Simplify by adding like terms:

17. $3a^2xy + 5xa^2y - 2ya^2x$ **18.** $6c - 6 - 2c - 5 - 3c + 7$

19. $a^2xx + a^2x^2 - 3x^2aa$

Expand by using the distributive property:

20. $4x(2y - 3 + 2a)$ **21.** $(x - 4)3$

Evaluate:

22. $a(a^3 - b) - b$ if $a = -2$ and $b = 3$

23. $x(y - a) + a(y - x)$ if $x = -3$, $y = 2$, and $a = -1$

24. $-b(b - a)$ if $a = -2$ and $b = 1$

25. $(-d)(d - c)$ if $c = -3$ and $d = -5$

Simplify:

26. $-4[(-5 + 2) - 3(-2 - 1)]$

27. $-3^2 + (-3)^3 - 4 - \sqrt[3]{27}$

28. $\dfrac{2^2 - 3^3 - 4^2}{(-2)^3}$

29. $(-2)^2 - 3^3 - (-4)^2 - |-2 - 3|$

LESSON 24 More complicated equations

Often we will encounter equations that have variables on both sides of the equation. When this occurs, we will begin the solution by using the addition rule to eliminate the variable on one side or the other. It makes no difference which side we choose, as we will demonstrate by working the first problem both ways.

example 24.1 Solve: $3x - 4 = 5x + 7$

solution We begin by eliminating the variable on the right side by adding $-5x$ to both sides.

$$\begin{array}{rcrl} 3x - 4 &=& 5x + 7 & \text{original equation} \\ -5x & & -5x & \text{add } -5x \text{ to both sides} \\ \hline -2x - 4 &=& 7 & \end{array}$$

We finish by adding +4 and then dividing by −2.

$$\begin{array}{rcrl} -2x - 4 &=& 7 & \\ +4 & & +4 & \text{add } +4 \text{ to both sides} \\ \hline \end{array}$$

$$-2x = 11 \longrightarrow \frac{-2x}{-2} = \frac{11}{-2} \longrightarrow \boldsymbol{x = -\frac{11}{2}} \qquad \text{divided by } -2$$

example 24.2 Solve: $3x - 4 = 5x + 7$

solution This time we begin by eliminating the variable on the left side by adding $-3x$ to both sides.

$$\begin{array}{rcrl} 3x - 4 &=& 5x + 7 & \text{original equation} \\ -3x & & -3x & \text{add } -3x \text{ to both sides} \\ \hline -4 &=& 2x + 7 & \end{array}$$

Now we finish by adding −7 and dividing by 2. We will get $-\frac{11}{2}$, the same answer that we got in the last example.

$$\begin{array}{rcrl} -4 &=& 2x + 7 & \\ -7 & & -7 & \text{add } -7 \text{ to both sides} \\ \hline \end{array}$$

$$-11 = 2x \longrightarrow \frac{-11}{2} = \frac{2x}{2} \longrightarrow \boldsymbol{-\frac{11}{2} = x} \qquad \text{divided by } 2$$

In many problems we must begin by simplifying on both sides by adding like terms.

example 24.3 Solve: $3x + 2 - x + 4 = -5 - x - 4$

solution We begin by adding like terms on both sides of the equation to get

$$2x + 6 = -9 - x$$

This time we decide to eliminate the x term on the right side so we add $+x$ to both sides.

$$\begin{array}{rcl} 2x + 6 &=& -9 - x \\ +x & & +x \\ \hline 3x + 6 &=& -9 \end{array} \qquad \text{add } +x \text{ to both sides}$$

Now we can finish by adding -6 to both sides and then dividing both sides by 3.

$$\begin{array}{rcl} 3x + 6 &=& -9 \\ -6 & & -6 \\ \hline \end{array} \qquad \text{add } -6 \text{ to both sides}$$

$$3x = -15 \longrightarrow \frac{3x}{3} = \frac{-15}{3} \longrightarrow \boldsymbol{x = -5} \qquad \text{divided by 3}$$

example 24.4 Solve: $k + 3 - 4k + 7 = 2k - 5$

solution We will begin by adding like terms on both sides to get

$$-3k + 10 = 2k - 5$$

Next, we eliminate the $-3k$ on the left by adding $+3k$ to both sides and then finish the solution.

$$\begin{array}{rcll} -3k + 10 &=& 2k - 5 & \text{added like terms} \\ +3k & & +3k & \text{add } 3k \text{ to both sides} \\ \hline 10 &=& 5k - 5 & \\ +5 & & +5 & \text{add } +5 \text{ to both sides} \\ \hline 15 &=& 5k & \\ \frac{15}{5} &=& \frac{5k}{5} & \text{divide both sides by 5} \\ \boldsymbol{3} &=& \boldsymbol{k} & \end{array}$$

example 24.5 Solve: $-7n + 3 + 2n = 4n - 5 + n$

solution We begin by simplifying and then eliminating the $5n$ term on the right side.

$$\begin{array}{rcll} -5n + 3 &=& 5n - 5 & \text{simplified both sides} \\ -5n & & -5n & \text{add } -5n \text{ to both sides} \\ \hline -10n + 3 &=& -5 & \\ -3 &=& -3 & \text{add } - \text{ to both sides} \\ \hline -10n &=& -8 & \\ 10n &=& 8 & \text{multiplied both sides by } (-1) \end{array}$$

$$\frac{10n}{10} = \frac{8}{10} \longrightarrow \boldsymbol{n = \frac{4}{5}} \qquad \text{divided each side by 10 and simplified answer}$$

Note that we used four steps in solving this equation. The steps were:

1. **Simplify by adding like terms.**
2. **Eliminate x on one side.**
3. **Eliminate the constant term.**
4. **Eliminate the coefficient of x.**

Check the order of these steps in the next problem.

example 24.6 Solve: $2x - 5 + 7x = 5 + 3x + 10$

solution After we simplify, we will eliminate the $3x$ term on the right side.

$$\begin{array}{rcll} 9x - 5 &=& 3x + 15 & \text{simplified} \\ -3x & & -3x & \text{add } -3x \text{ to both sides} \\ \hline 6x - 5 &=& 15 & \\ +5 & & +5 & \text{add } +5 \text{ to both sides} \\ \hline 6x &=& 20 & \\ \frac{6x}{6} &=& \frac{20}{6} & \text{divide both sides by 6} \\ x &=& \mathbf{\frac{10}{3}} & \text{simplify answer} \end{array}$$

practice Solve:

a. $3m - 7m = 8m - 6$

b. $5 - 6p = 9p - 7 + 8p - 3 + 2p$

c. $2x + 3x + 4x - 5 = 2 + 3 + 4x$

d. $3p + 7 - (-3) = p + (-2)$

problem set 24

1. What are equivalent equations?

2. Are $x + 2 = 6$ and $3x - 14 + x - 5 = -3$ equivalent equations? Why?

3. Use four unit multipliers to convert 120 square miles to square inches.

4. Use three unit multipliers to convert 200 cubic meters to cubic centimeters.

5. How many 1-yard-square floor tiles would it take to cover the area shown? Dimensions are in yards. Corners that look square are square.

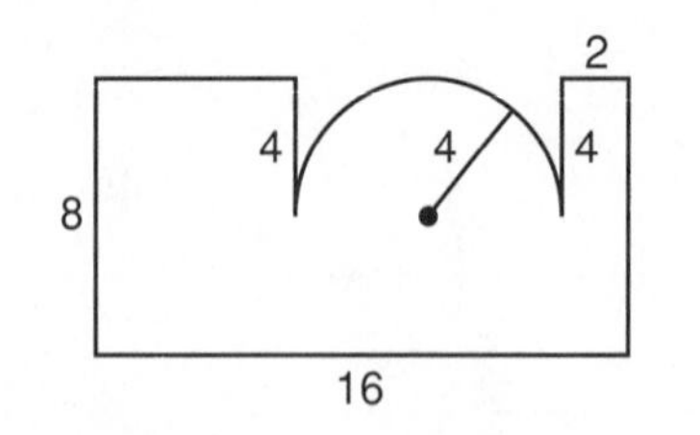

6. Find the volume of this right circular cylinder. Dimensions are in centimeters.

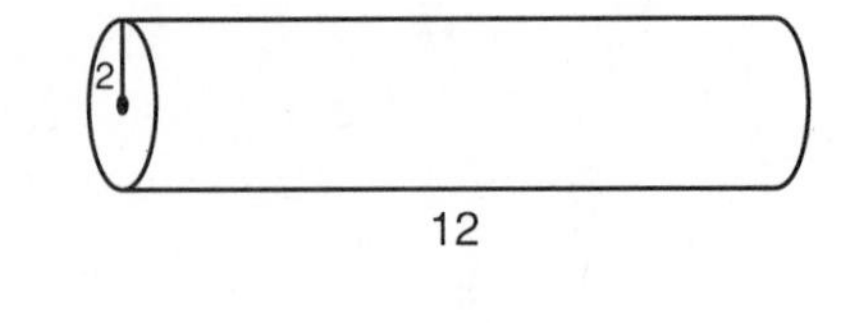

Solve:

7. $2\frac{1}{2}x = \frac{3}{7}$

8. $-x = 3$

9. $3x - 4 = 7$

10. $0.02p - 2.4 = 0.006$

11. $2x + x + 3 = x + 2 - 5$

12. $5x - 3 - 2 = 3x - 2 + x$

13. $m + 4m - 2 - 2m = 2m + 2 - 3$ **14.** $-m - 6m + 4 = -2m - 5$

15. $y + 2y - 4 - y = 3y - 2 + y$

Simplify:

16. $m^2y^5myy^3m^3$ **17.** $k^5mmm^2k^2m^2k^3aa^2$

Simplify by adding like terms:

18. $xym^2z - 3m^2zxy + 2xm^2yz - 5xym^2z$

19. $a^2bc + 2bc - bca^2 + 5ca^2b - 3cb$ **20.** $a - ax + 2xa - 3a - 3$

Expand by using the distributive property:

21. $4(7 - 3x)$ **22.** $(m + 2p)3axy$

Evaluate:

23. $a(-a^2 + b)$ if $a = -2$ and $b = 4$ **24.** $x(x - y)$ if $x = -3$ and $y = 5$

25. $p(a - 2p^2)$ if $a = -2$ and $p = -4$

26. $(y - m^2) - m$ if $m = 2$ and $y = -3$

Simplify:

27. $-3^2 - (-3)^3 + (-2) - \sqrt[3]{-125}$

28. $-4(-3 + 2) - 3 - (-4) - |-3 + 2|$

29. $\dfrac{-3(-2 - 3 - 4)}{-(-3 + 2)}$ **30.** $\dfrac{-2(-3) + (-6)}{-4(-2)}$

LESSON 25 More on the distributive property · Simplifying decimal equations

25.A the distributive property

Remember that we can simplify expressions such as

$$4(2 + 7)$$

by adding first or by using the distributive property and multiplying first.

ADDING FIRST	MULTIPLYING FIRST
$4(2 + 7)$	$4(2 + 7)$
$4(9)$	$8 + 28$
36	**36**

Thus far, we have restricted our use of this property to expanding simple expressions such as $4p(x + 3y)$.

$$4p(x + 3y) = \mathbf{4px + 12py}$$

In the following examples, we will use the distributive property to expand expressions that are more complicated. We remember that in each case the expression on the outside is multiplied by every term inside the parentheses.

example 25.1 Expand: $xy(y^2 - x^2z)$

solution The xy is multiplied by y^2 and also by $-x^2z$.

$$xy(y^2 - x^2z) = (xy)(y^2) + (xy)(-x^2z) = \mathbf{xy^3 - x^3yz}$$

example 25.2 Expand: $4xy^3(x^4y - 5x)$

solution $4xy^3$ is to be multiplied by both x^4y and $-5x$.

$$4xy^3(x^4y - 5x) = (4xy^3)(x^4y) + (4xy^3)(-5x) = \mathbf{4x^5y^4 - 20x^2y^3}$$

example 25.3 Expand: $(ay - 4y^5)2x^2y$

solution It is not necessary to write down two steps. We can do the multiplication in our head if we are careful.

$$(ay - 4y^5)2x^2y = \mathbf{2ax^2y^2 - 8x^2y^6}$$

example 25.4 Expand: $8m^2x(5m^3x - 3x^5 + 2x)$

solution This time $8m^2x$ must be multiplied by all three terms inside the parentheses.

$$(8m^2x)(5m^3x) + (8m^2x)(-3x^5) + (8m^2x)(2x) = \mathbf{40m^5x^2 - 24m^2x^6 + 16m^2x^2}$$

25.B simplifying decimal equations

Finding the solutions of equations such as

$$0.4 + 0.02m = 4.6 \qquad \text{and} \qquad 0.002k + 0.02 = 4.02$$

can be facilitated if we begin by multiplying every term on both sides of the equation by the power of 10 that will make every decimal coefficient an integer. The value of the smallest decimal number in the problem often determines whether we multiply by 10 or 100 or 1000 or 10,000, etc.

example 25.5 Solve: $0.4 + 0.02m = 4.6$

solution The smallest decimal number in the problem is 0.02. We can convert 0.02 to 2 if we multiply by 100. Thus, we will multiply every term on both sides of the equation by 100 and then solve.

$0.4 + 0.02m = 4.6$	original equation
$40 + 2m = 460$	multiplied every term by 100
$2m = 420$	added -40 to both sides
$\mathbf{m = 210}$	divided both sides by 2

example 25.6 Solve: $0.002k + 0.02 = 4.02$

solution This time the smallest number is 0.002, so we will use 1000 as our multiplier.

$0.002k + 0.02 = 4.02$	original equation
$2k + 20 = 4020$	multiplied every term by 1000
$2k = 4000$	added -20 to both sides
$\mathbf{k = 2000}$	divided both sides by 2

practice Expand:

a. $xy^2(y^2p - p)$

b. $(xy - x)2xy$

c. $3xp^3(p^5 - x^2p^8)$

d. $2x^2m^2(m^2 - 4m)$

Solve:

e. $0.7m + 0.6m = 3.4$

f. $0.08x - 0.1 = 16.7$

problem set 25

1. Use letters a, b, and c to state the distributive property.

2. What is the reciprocal of $-\frac{1}{4}$?

3. Use three unit multipliers to convert 430 feet to meters.

4. Use two unit multipliers to convert 6500 square feet to square yards.

Solve:

5. $3x + 2 = 7$

6. $\frac{1}{2}x + \frac{3}{4} = -\frac{3}{7}$

7. $3\frac{1}{3}x - \frac{1}{2} = 5$

8. $-\frac{1}{7} + \frac{4}{3}x = \frac{1}{5}$

9. $x + 3 - 5 - 2x = x - 3 - 7x$

10. $4x - 5 + 2x = 3x - 4 + x$

11. $-2y - 4 - y = -y + 2 + 3y$

12. $-5 - x - 2 + 7x = 3 - 4x$

13. $-a - 2a + 4 - 4a = 7 - 3a$

14. $p - 2p - 5 + 7p = 3 - 2p$

Simplify:

15. $p^2xyy^2x^2yx^2x$

16. $3p^2x^4yp^5xxyy^2$

Simplify by adding like terms:

17. $-3x^2y + 5 - yx^2 - 13 + xy$

18. $-4x + x^2 - 3x - 5 + 7x^2$

19. $xyp^2 - 4p^2xy + 5xp^2y - 7yxp^2$

Expand by using the distributive property:

20. $x^2y(x^3 - xyz^3)$

21. $4x^2(ax - 2)$

22. How many 1-cm-square floor tiles will it take to completely cover the shaded area of the big circle? Dimensions are in centimeters.

16

6

23. This figure is the base of a right solid that is 3 inches high. How many 1-inch sugar cubes will the solid hold? What is the surface area of the solid? Dimensions are in inches. The long sides are parallel.

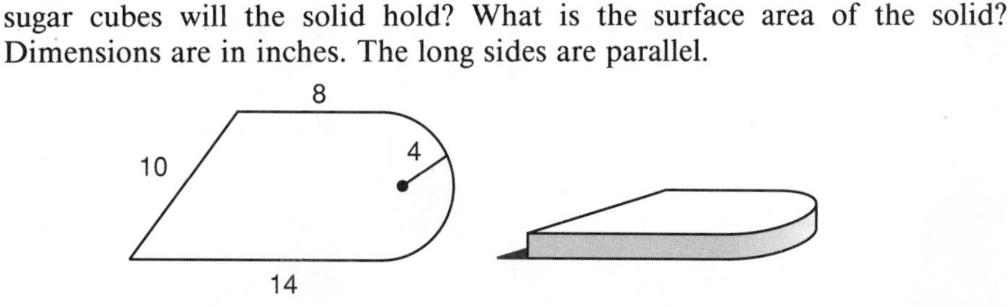

Evaluate:

24. $(ay - 4y^5)2x^2y$ if $a = -2$, $y = -1$, and $x = 2$

25. $-(-a - x) - x^2$ if $x = -3$ and $a = 4$

26. $-y^2(y - 2x)$ if $y = -2$ and $x = 3$

27. $-(-p)^2 + (p - x)$ if $p = -2$ and $x = 5$

28. $(a - b) + (-a)^2$ if $a = -3$ and $b = 6$

Simplify:

29. $-3^2 - 3(3^2 - 4) - \sqrt[4]{16} - |-7 + 2|$ **30.** $\dfrac{-6 - (-2 - 3)}{4 - (-3)}$

LESSON 26 Fractional parts of numbers

When we multiply a number by a fraction, we say that the result is a fractional part of the number. If we multiply $\frac{7}{8}$ by 48, we get 42. We say this mathematically by writing

$$\frac{7}{8} \times 48 = 42$$

and if we use words we say that

(seven-eighths) (of 48) (is 42)

We can generalize this problem into an equation that has three parts.

$$(F) \times (\text{of}) = (\text{is})$$

The letter F stands for fraction, and the words *of* and *is* associate the parts of the statement as we note in the following problems. We will use the variable WN to represent what number and WF to represent what fraction. We will avoid the use of the meaningless variable x.

example 26.1 $\frac{3}{4}$ of what number is 69?

solution In this problem, the fraction is $\frac{3}{4}$, the word *of* associates with what number (WN), and the word *is* associates with 69. We make these replacements and get.

$$(F) \times (\text{of}) = (\text{is}) \longrightarrow \left(\frac{3}{4}\right) \times (WN) = 69$$

We can undo multiplication by $\frac{3}{4}$ by multiplying by $\frac{4}{3}$. Thus we solve by multiplying both sides of the equation by $\frac{4}{3}$.

$$\frac{4}{3} \cdot \frac{3}{4} WN = 69 \cdot \frac{4}{3} \longrightarrow \mathbf{WN = 92}$$

example 26.2 What fraction of 40 is 24?

solution This time the fraction is unknown, *of* associates with 40, and *is* associates with 24.

We make these replacements. Then to solve, we divide both sides of the equation by 40.

$$(F) \times (\text{of}) = (\text{is}) \longrightarrow (WF)(40) = (24) \longrightarrow \frac{WF \cdot \cancel{40}}{\cancel{40}} = \frac{24}{40} \longrightarrow \boldsymbol{WF = \frac{3}{5}}$$

example 26.3 $2\frac{1}{2}$ of 240 is what number?

solution This time the fraction is written as the mixed number $2\frac{1}{2}$. We see that *of* associates with 240 and *is* with what number (WN). We make these substitutions and solve by multiplying $2\frac{1}{2}$ and 240.

$$(F) \times (\text{of}) = (\text{is}) \longrightarrow (2\tfrac{1}{2})(240) = WN \longrightarrow \boldsymbol{600 = WN}$$

practice

a. $\frac{4}{3}$ of what number is 62?

b. What fraction of 60 is 28?

c. $4\frac{1}{2}$ of 824 is what number?

d. $3\frac{1}{5}$ of what number is 40?

problem set 26

1. Use three unit multipliers to convert 50 feet to meters.

2. Use three unit multipliers to convert 50 cubic meters to cubic centimeters.

3. Find the surface area of this right solid. Dimensions are in centimeters.

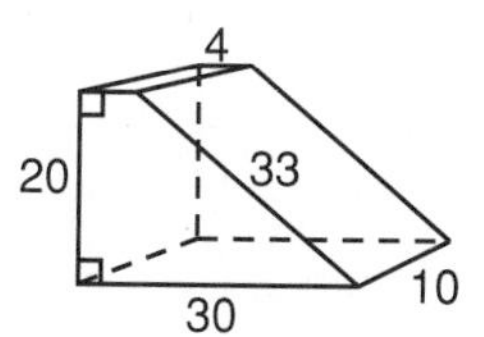

4. Find the area of this figure. Dimensions are in feet.

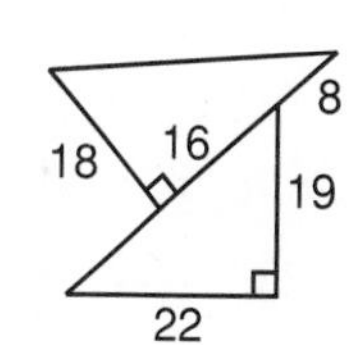

Solve:

5. What fraction of 324 is 270?

6. $\frac{7}{3}$ of 42 is what number?

Solve:

7. $\frac{1}{2}x + \frac{1}{2} = 2\frac{1}{5}$

8. $-\frac{7}{8} + \frac{1}{2}x = \frac{3}{4}$

9. $0.005p + 1.4 = 0.005$

10. $x - 2 - 2x = 3 - x + 4x$

11. $3y - y + 2y - 5 = 7 - 2y + 5$

12. $p - 2p + 4 - 7 = p + 2$

13. $k - 4 - 2k = 7 + 3k - k + 5$

14. $x - 5 - (-2) + 2x = 7$

15. $3x - 4x + 7 = 5 + x - 6$

Simplify:

16. $y^5x^2y^3yxy^2$

17. $m^2myy^2m^3ym$

Simplify by adding like terms:

18. $xym^2 + 3xy^2m - 4m^2xy + 5mxy^2$

19. $pc - 4cp + c - p + 7pc - 7c$

20. $a^2xy + 4xa^2y - yxa^2 + 3yx^2a$

Expand by using the distributive property:

21. $x^2y^3(3xy - 5y)$ **22.** $3x^4a(x^3 - 2x^4a^3)$ **23.** $(xyp - 3xp)p^2xy$

Evaluate:

24. $x^3y(x - y)$ if $x = -3$ and $y = 1$

25. $p^2 - a^2(p - a)$ if $p = 3$ and $a = 5$

26. $ka(-a) - k + a$ if $a = -3$ and $k = 4$

27. $p(a) - xp(-a)$ if $p = -2$, $a = 3$, and $x = 4$

Simplify:

28. $-3^2 + (-2)^3 - \sqrt[4]{81}$ **29.** $-4(-7 + 5)(-2) - |-2 - 5|$

30. $\dfrac{-5(-3 + 7)}{-5 + (-2)}$

LESSON 27 Negative exponents

It is convenient to have an alternative way to write the reciprocal of an exponential expression. Here we show an alternative way to write 1 over 5^2 and 1 over 5^{-2}.

$$\frac{1}{5^2} = 5^{-2} \qquad \frac{1}{5^{-2}} = 5^2$$

In the formal definition we will use x and n to represent the base and the exponent.

> DEFINITION OF x^{-n}
>
> If n is any real number and x is any real number that is not zero,
>
> $$\frac{1}{x^n} = x^{-n}$$

(a) $\dfrac{1}{3^4} = 3^{-4}$ (b) $7^{-2} = \dfrac{1}{7^2}$ (c) $\dfrac{1}{5^{-8}} = 5^8$ (d) $6^{-3} = \dfrac{1}{6^3}$

In (a) we moved 3^4 from the denominator to the numerator and **changed the sign of the exponent** from plus to minus. In (b) we moved 7^{-2} from the numerator to the denominator and **changed the sign of the exponent** from minus to plus. In (c) we moved 5^{-8} from the denominator to the numerator and **changed the sign of the exponent** from minus to plus. In (d) we moved 6^{-3} to the denominator and **changed the sign of the exponent.** The formal definition of x^{-n} is stated in the box above. We will now state the definition informally.

> A number or a variable that is written as an exponential expression can be written in reciprocal form if the sign of the exponent is changed.

If the exponent is positive, the exponent is negative in the reciprocal form. If the exponent is negative, the exponent is positive in the reciprocal form.

example 27.1 Simplify: 3^{-2}

solution The negative exponent is meaningless as an operation indicator. Thus the first step in the solution is to write 3^{-2} in reciprocal form and change the negative exponent to a positive exponent.

$$3^{-2} = \frac{1}{3^2}$$

Now we can complete the simplification because a positive exponent is an operation indicator because 3^2 means $3 \cdot 3$.

$$3^{-2} = \frac{1}{3^2} = \frac{1}{3 \cdot 3} = \mathbf{\frac{1}{9}}$$

example 27.2 Simplify: $\frac{1}{3^{-3}}$

solution Again, as the first step we write the expression in reciprocal form so that the negative exponent can be changed to a positive exponent.

$$\frac{1}{3^{-3}} = 3^3$$

Now 3^3 is meaningful as $3 \cdot 3 \cdot 3$, and 3^3 equals 27, so

$$\frac{1}{3^{-3}} = 3^3 = 3 \cdot 3 \cdot 3 = \mathbf{27}$$

example 27.3 Simplify: $(-3)^{-2}$

solution We first change the negative exponent to a positive exponent by writing the exponential expression in reciprocal form.

$$(-3)^{-2} = \frac{1}{(-3)^2} = \mathbf{\frac{1}{9}}$$

We have defined negative exponents so that their use will not conflict with the use of the product rule, which is repeated here.

PRODUCT RULE FOR EXPONENTS

If m and n are real numbers and $x \neq 0$,

$$x^m \cdot x^n = x^{m+n}$$

When the bases are the same, we multiply exponential expressions by adding the exponents. This is true even if some of the exponents are negative numbers.

(a) $x^{-5}x^2 = \mathbf{x^{-3}}$ (b) $y^7y^5y^{-2} = \mathbf{y^{10}}$ (c) $p^{10}p^{-15} = \mathbf{p^{-5}}$

example 27.4 Simplify: $x^4m^2x^{-2}m^{-5}$

solution We first change the order of multiplication

$$x^4x^{-2}m^2m^{-5}$$

and then add the exponents of the exponential expressions whose bases are the same.

$$\mathbf{x^2m^{-3}}$$

example 27.5 Simplify: $x^{-2}y^{-6}y^5x^4zxz^5$

solution We change the orders of the factors and add the exponents of the exponential expressions that have the same bases to get

$$x^{-2}x^4xy^{-6}y^5zz^5 = \mathbf{x^3y^{-1}z^6}$$

practice Simplify:

a. 4^{-2} b. $\frac{1}{2^{-3}}$ c. $(-4)^{-2}$

d. $\frac{1}{(-3)^{-2}}$ e. $x^5m^3x^{-2}m^{-8}$ f. $x^{-3}y^{-8}y^5x^4zx^2z^5$

problem set 27

1. Designate (a) the set of integers, (b) the set of whole numbers.

2. $2\frac{1}{4}$ of what number is 750?

3. What fraction of 72 is 63?

4. $3\frac{1}{8}$ of 72 is what number?

Solve:

5. $-5 + 2\frac{1}{2}x = 17$

6. $3\frac{1}{2}x + 2 = 9$

7. $\frac{1}{4}x + \frac{1}{2} = \frac{7}{8}$

8. $3x + 5 - x = x + 5$

9. $3y - 5 = 7 - 2y + 8$

10. $7p - 14 = 4p - 5 + p$

11. $0.0025k + 0.06 = 4.003$

12. $3m - 2 - m = -2 + m - 5$

13. $x - 3x - 5 - 2x = 7x + 3 - 5 + 2x$

14. Use two unit multipliers to convert 7200 inches to meters.

15. This figure is the base of a right solid that is 4 feet high. Find the volume and the surface area of the solid. Dimensions are in inches.

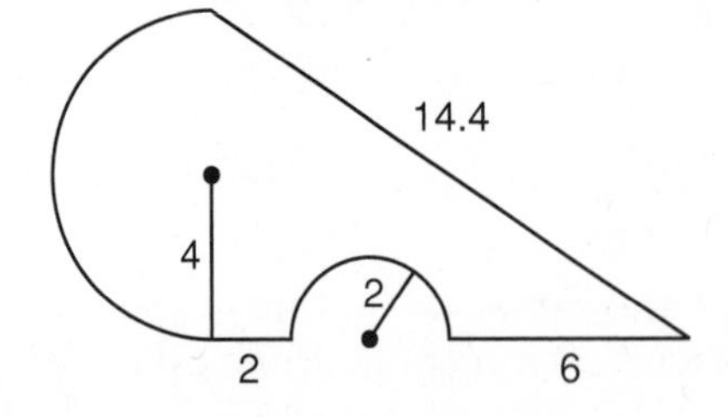

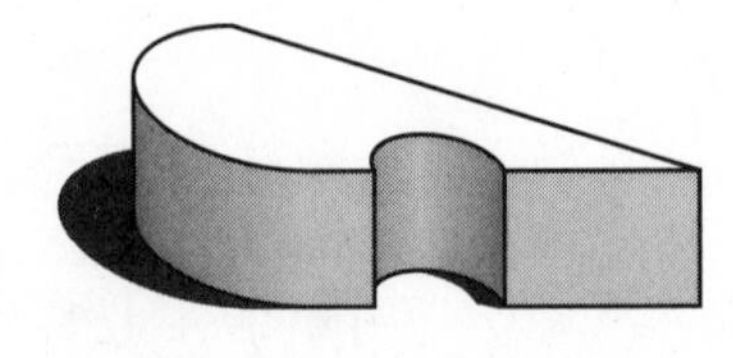

16. What is the surface area of this right prism? Dimensions are in feet.

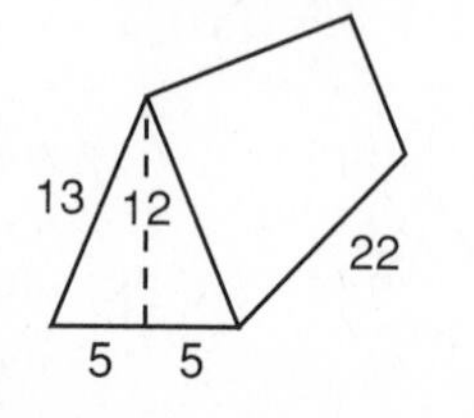

Simplify:

17. $\dfrac{1}{(-2)^{-2}}$ 18. 2^{-4} 19. $\dfrac{1}{2^{-2}} - \sqrt[5]{243}$

20. $x^{-5}y^5axy^5y^{-8}a^{-4}$ 21. $m^2p^{-4}m^{-2}p^6m^4m^{-5}$

Simplify by adding like terms:

22. $4x^2yp - px^2y + 3ypx^2 - 4$ 23. $5m^2x^2y - 2x^2m^2y + 8m^2y^2x$

Expand:

24. $(6x^2yp - 4p + 2)x^2y^3p$ 25. $4mz(m^3cz^2 - 5mz^5)$

Evaluate:

26. $-xa^2(a + x)$ if $x = -2$ and $a = 3$

27. $m^2p - p(m - p)$ if $m = -3$ and $p = 5$

28. $4x(a + x)(-x)$ if $x = -3$ and $a = 2$

29. $m(k^2 - m^3)$ if $k = -2$ and $m = -1$

Simplify:

30. $\dfrac{4(2 - 5)}{-2(4 - 2) - (-2)}$

LESSON 28 *Zero exponents · Decimal parts of a number · Volume conversions*

28.A zero exponents

We know that a nonzero number divided by itself equals 1. For instance,

$$1 = \frac{4^2}{4^2}$$

We can simplify this expression by moving the 4^2 on the bottom to the top and changing the exponent from 2 to -2. Then we multiply 4^2 by 4^{-2} and get 4^0.

$$1 = \frac{4^2}{4^2} = 4^2 \cdot 4^{-2} = 4^0$$

Now, 4^0 must equal 1 because 4^2 divided by 4^2 equals 1. In the same way, we see that any nonzero quantity raised to the zero power must have a value of 1.

$$(x + y + z^2)^0 = 1 \qquad (pm)^0 = 1 \qquad (px^{-4})^0 = 1$$

Each of the above has a value of 1 if the expression in parentheses does not have a value of zero. Zero raised to the zero power has no meaning.

DEFINITION

If x is any real number that is not zero,

$$x^0 = 1$$

example 28.1 Simplify the following expressions: (a) $x^2y^5y^{-2}x^{-2}$ (b) $m^5b^2mb^{-2}$

solution (a) $x^2y^5y^{-2}x^{-2} = y^3x^0 = \boldsymbol{y^3}$ (because $x^0 = 1$ if $x \neq 0$)

(b) $m^5b^2mb^{-2} = m^6b^0 = \boldsymbol{m^6}$ (because $b^0 = 1$ if $b \neq 0$)

Since we must not use the expression 0^0, it is necessary in problems such as (a) and (b) to assume that the variable with the zero exponent is not zero. Further, in the problem sets, we will assume a nonzero value for any variable that has zero for its exponent.

example 28.2 Expand: $x^5y^0z(p^{-3}z^0 - 4x^{-5}z^{-1})$

solution We choose to begin by simplifying x^5y^0z and $p^{-3}z^0$, remembering that $y^0 = 1$ and that $z^0 = 1$. Now we have

$$x^5z(p^{-3} - 4x^{-5}z^{-1})$$

We finish by doing the two multiplications and get

$$x^5zp^{-3} - 4x^0z^0 = \boldsymbol{x^5zp^{-3} - 4}$$

example 28.3 Expand: $x^{-2}y^{-2}(x^2y^2 + 4x^4y^2)$

solution We do the two multiplications and get

$$x^{-2}y^{-2}x^2y^2 + x^{-2}y^{-2}4x^4y^2$$

solution Now we simplify by remembering that both x^0 and y^0 equal 1.

$$x^0y^0 + 4x^2y^0 = (1)(1) + 4x^2(1) = \boldsymbol{1 + 4x^2}$$

example 28.4 Simplify: (a) 2^0 (b) -2^0 (c) $(-2)^0$

solution Any expression (except 0) raised to the zero power has a value of 1. Thus the answer to both (a) and (c) is **1**. The expression in (b) asks for the value of the opposite of 2^0. The answer to (b) is **−1.**

28.B decimal parts of a number

Many people call decimal numbers decimal fractions because terminating decimal numbers can be written as fractions. For example,

$$\text{(a)}\quad 28.6132 = \frac{286{,}132}{10{,}000} \qquad \text{(b)}\quad 0.000463 = \frac{463}{1{,}000{,}000}$$

We have been working problems concerning fractional parts of a number by using the relationship

$$(F) \times (\text{of}) = (\text{is})$$

We can solve statements about decimal parts of a number by using the slightly different relationship

$$(D) \times (\text{of}) = (\text{is})$$

where D stands for the decimal (decimal fraction) part of the number and *of* and *is* have the same meanings as before.

example 28.5 0.32 of what number is 24.32?

solution We will use

$$(D) \times (\text{of}) = (\text{is})$$

We replace D with 0.32, *of* with WN, and *is* with 24.32 and then solve.

$$0.32WN = 24.32 \longrightarrow \frac{0.32WN}{0.32} = \frac{24.32}{0.32} \longrightarrow \mathbf{WN = 76}$$

example 28.6 What decimal part of 42 is 26.04?

solution In $(D) \times (\text{of}) = (\text{is})$, we replace D with WD, *of* with 42, and *is* with 26.04. Then we solve.

$$WD(42) = 26.04 \longrightarrow \frac{WD(42)}{42} = \frac{26.04}{42} \longrightarrow \mathbf{WD = 0.62}$$

example 28.7 0.42 of 86 is what number?

solution This time 0.42 replaces D and 86 replaces *of*. Then we multiply to find WN.

$$(0.42)(86) = WN \longrightarrow \mathbf{36.12 = WN}$$

28.C volume conversions

Three unit multipliers are necessary for each step in a volume conversion.

example 28.8 Use six unit multipliers to convert 800 cm^3 to cubic feet.

solution We will use six unit multipliers to go from cubic centimeters to cubic feet.

$$800\ \cancel{\text{cm}}^3 \times \frac{1\ \cancel{\text{in.}}}{2.54\ \cancel{\text{cm}}} \times \frac{1\ \cancel{\text{in.}}}{2.54\ \cancel{\text{cm}}} \times \frac{1\ \cancel{\text{in.}}}{2.54\ \cancel{\text{cm}}} \times \frac{1\ \text{ft}}{12\ \cancel{\text{in.}}} \times \frac{1\ \text{ft}}{12\ \cancel{\text{in.}}} \times \frac{1\ \text{ft}}{12\ \cancel{\text{in.}}}$$

$$= \frac{\mathbf{800}}{\mathbf{(2.54)(2.54)(2.54)(12)(12)(12)}}\ \mathbf{ft^3}$$

practice

a. What decimal part of 60 is 80?

b. 0.16 of what number is 48.16?

c. 0.48 of 8 is what number?

Simplify:

d. 8^0 **e.** -8^0 **f.** $(-3)^0$ **g.** $(k + j + k^2)^0$

h. Use six unit multipliers to convert 42 cubic centimeters to cubic feet.

problem set 28

1. Use six unit multipliers to convert 80 cubic yards to cubic inches.

2. Use three unit multipliers to convert 42 cubic centimeters to cubic inches.

3. 0.8 of what number is 7.68?

Solve:

4. $3x - 7 = 42$ **5.** $2\frac{1}{2}x - 5 = 17$

6. $\frac{3}{4} + \frac{1}{2}x + 2 = 0$

7. $0.4k + 0.4k - 0.02 = 4.02$

8. $-2x - 5 - x - 8 - 5x - 3 = 0$

9. $5m - m - 2m + 5 = -3m - 2$

10. $-3x - 2 = 3x - 5 + 8$

11. $-p - 2p - 4 + 7p = 5 + 2p - 6$

12. $-2k + k - 3k = 7 + 2k - 2$

Simplify:

13. $(-3)^{-2}$

14. $\frac{1}{4^{-2}} - \sqrt[3]{-27}$

15. $\frac{1}{(-4)^{-3}}$

16. $\frac{1}{2^{-3}}$

17. $(-5)^{-3}$

18. Find the area of this trapezoid. The dimensions are in centimeters.

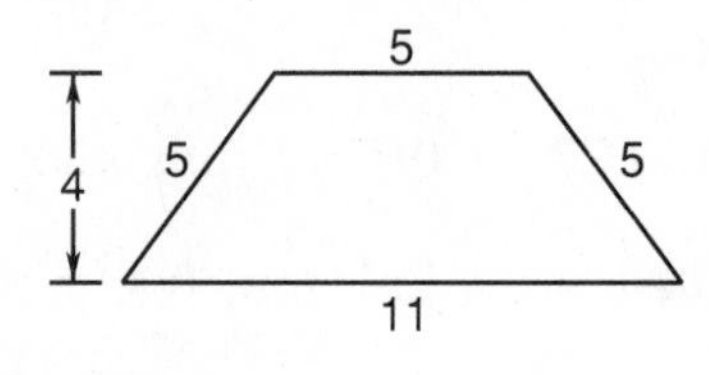

19. This figure is the base of a right solid 6 ft high. Find the volume and the surface area of the solid. Dimensions are in feet.

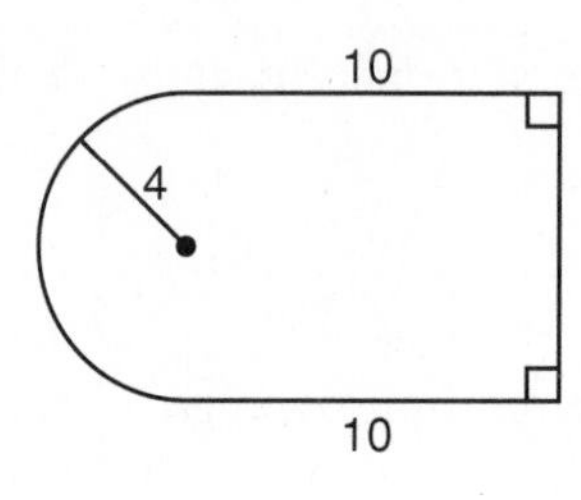

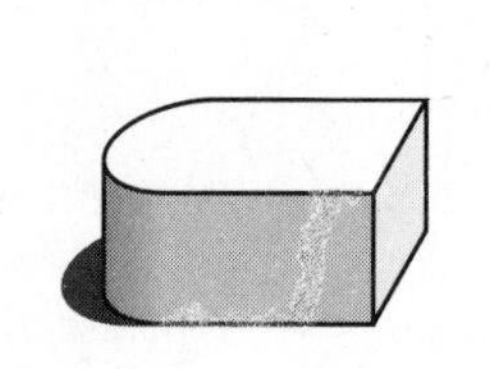

Expand:

20. $(x^2 - 4x^5y^{-5})3p^0x^{-2}$

21. $2x^{-2}y^0(x^2y^0 - 4x^{-6}y^4)$

22. $x^{-4}y^0(x^4 - 3y^2x^5p^0)$

23. $m^0x(x^{-1}y^0 - y^2m^0)$

Simplify by adding like terms:

24. $x^2ym^3 - 3x^2y + 6ym^3x^2 + yx^2$

25. $abc^2 - 2ab^2c + 6c^2ab - 4b^2ac$

Evaluate:

26. $-a^3(a^0 - b)$ if $a = -2$, and $b = 4$

27. $-c(c - b)$ if $c = -2$ and $b = 4$

28. $x(x^0 - y)(y - 2x)$ if $x = -3$ and $y = 5$

Simplify:

29. $-3^2 + (-3)^3 - 3^0 - |-3 - 3|$

30. $\frac{-4[2 - (-2)]}{-7(5 - 2)}$

LESSON 29 Algebraic phrases

In algebra we learn to answer verbal questions by turning these questions into algebraic equations. Then we solve the equations to get the desired answers. The equations that we write contain algebraic phrases that have the same meanings as the verbal phrases used in the questions. There are several keys to writing these phrases. The word *sum* means that things are added, and the word *product* means that things are multiplied. Seven more than, or increased by 7, means to add 7; while 7 less than, or decreased by 7, means to subtract 7. If we use N to represent an unknown number, then we will use $-N$ to represent the opposite of the unknown number. In the same way, twice a number would be represented by $2N$, and 5 times the opposite of a number would be represented by $5(-N)$. If we write the sum of twice a number and negative 10 as $2N - 10$, we could write 7 times this sum by writing $7(2N - 10)$. Cover the answers in the right-hand column below and see if you can write the algebraic phrase that is indicated.

The sum of a number and 7	$N + 7$
Seven less than a number	$N - 7$
The opposite of a number decreased by 5	$-N - 5$
The sum of the opposite of a number and -5	$-N - 5$
The product of twice a number and 8	$8(2N)$
The sum of twice a number and -5	$2N - 5$
Five times the sum of twice a number and -5	$5(2N - 5)$
Six times the sum of twice the opposite of a number and -8	$6[2(-N) - 8]$
The product of 7 and the sum of a number and 10	$7(N + 10)$
The sum of 3 times a number and -4, multiplied by 5	$5(3N - 4)$
The sum of -10 and 6 times the opposite of a number	$6(-N) - 10$

practice Write the algebraic phrases that correspond to these word phrases.

a. Five times the sum of 3 times a number and -5.

b. The product of 3 and a number decreased by 50.

c. The sum of 5 times a number and -13.

d. Three times the sum of the opposite of a number and negative 7.

problem set 29

1. 1.6 of what number is 3200?

2. What decimal part of 80 is 8400?

3. $4\frac{5}{6}$ of 4596 is what number?

4. Use six unit multipliers to convert 4200 cubic inches to cubic meters.

5. Find the area of this figure. Dimensions are in meters.

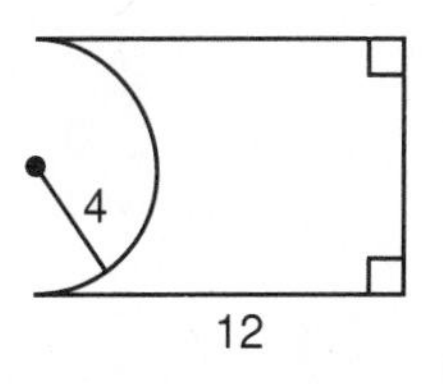

Write the algebraic phrases that correspond to these word phrases.

6. The sum of 5 times a number and -8.

7. Three times the sum of the opposite of a number and -7.

Solve:

8. $\frac{1}{2}x + 2 = -\frac{3}{4}$

9. $0.3k + 0.85k - 2 = 2.6$

10. $\frac{1}{7}k - \frac{4}{7} = -7$

11. $2k - 5 + k - 3 = 2 + 2k + 5$

12. $-2 - 6p + p - 2 = 7 - p$

13. $3m - m = 5 - 4 + 2m + 5 - 5m$

Simplify:

14. $(-6)^{-2}$

15. 3^{-3}

16. $\frac{1}{(-4)^{-2}}$

17. $\frac{1}{5^{-3}} - \sqrt[3]{-125}$

Expand:

18. $p^0x^{-1}(x - 2x^0)$

19. $4x^2p^0(3xp^5 - 2x^{-2})$

20. $(y^{-2}x^{-1} + 3p^2y^{-2}k^0)xp^0y^2$

21. $4m^2x^{-5}(2m^{-2}x^{-5} - 3m^{-2}k^0)$

22. $2p^{-4}x^2y^{-2}(p^4x^2y^2k^0 - 3p^2x)$

23. $4x^2yp^{-3}(x^{-2}y^4p^6 - 5x^4yp^{-3})$

Simplify by adding like terms:

24. $xmp^{-2} - 4p^{-2}xm + 6p^{-2}mx - 5mx$

25. $k^2p^{-4}y - 5k^2yp^{-4} + 2yk^2p^{-4} - 5k^2yp^{-4}$

Evaluate:

26. $a^3 - (b^0 - a)$ if $a = -3$ and $b = 4$

27. $p - a(p - ap)$ if $a = -2$ and $p = -3$

28. $-k^0 - (-km)$ if $m = 3$ and $k = -5$

Simplify:

29. $-2(-3) - (-4)^0(-3)|-5 - 2|$

30. $\frac{-7(-4 - 6)}{-(-4) - [-(-6)]}$

LESSON 30 *Equations with parentheses*

When equations contain parentheses, we begin by eliminating the parentheses. If the parentheses are preceded by a number, we use the distributive property. We multiply the number by every term inside the parentheses and discard the parentheses.

example 30.1 Solve: $2(3 - b) = b - 5$

solution As the first step, we will use the distributive property on the left side to expand $2(3 - b)$. Then we will complete the solution.

$$
\begin{array}{rcll}
2(3 - b) &=& b - 5 & \text{original equation} \\
6 - 2b &=& b - 5 & \text{multiplied} \\
+ 2b & & + 2b & \text{add } 2b \text{ to both sides} \\
\hline
6 &=& 3b - 5 & \\
+ 5 & & + 5 & \text{add 5 to both sides} \\
\hline
11 &=& 3b \longrightarrow \dfrac{11}{3} = b & \text{divided by 3}
\end{array}
$$

example 30.2 Solve: $3(1 + 2x) + 7 = -4(x + 2)$

solution This equation has parentheses on both sides. Thus we begin by using the distributive property on the left side and again on the right side to eliminate both sets of parentheses.

$$
\begin{array}{rcll}
3(1 + 2x) + 7 &=& -4(x + 2) & \text{original equation} \\
3 + 6x + 7 &=& -4x - 8 & \text{used distributive property} \\
10 + 6x &=& -4x - 8 & \text{added like terms} \\
+ 4x & & +4x & \text{add } 4x \text{ to both sides} \\
\hline
10 + 10x &=& -8 & \\
-10 & & -10 & \text{add } -10 \text{ to both sides} \\
\hline
10x &=& -18 & \\
x &=& -\dfrac{18}{10} \longrightarrow x = -\dfrac{9}{5} & \text{divided by 10 and simplified}
\end{array}
$$

In this problem, we used all of the five steps that we will use to solve equations. Sometimes one of the steps is not necessary, as in Example 30.1 above where addition of like terms was not required. If the variable is x, the five steps are:

1. **Eliminate parentheses.**
2. **Add like terms on both sides.**
3. **Eliminate x on one side or the other.**
4. **Eliminate the constant term.**
5. **Eliminate the coefficient of x.**

example 30.3 Solve: $15(4 - 5x) = 16(4 - 6x) + 10$

solution As the first step, we will use the distributive property as required on both sides of the equation.

$$
\begin{array}{rcll}
60 - 75x &=& 64 - 96x + 10 & \text{used distributive property} \\
60 - 75x &=& 74 - 96x & \text{added like terms} \\
+ 96x & & + 96x & \text{add } 96x \text{ to both sides} \\
\hline
60 + 21x &=& 74 & \\
-60 & & -60 & \text{add } -60 \text{ to both sides} \\
\hline
21x &=& 14 & \\
x &=& \dfrac{14}{21} \longrightarrow x = \dfrac{2}{3} & \text{divided both sides by 21 and simplified}
\end{array}
$$

In the preceding three examples we began by using the distributive property. To solve the next two problems, we need to have two rules for eliminating parentheses preceded by a plus sign or a minus sign. The rules are:

1. **When parentheses are preceded by a plus sign, both the parentheses and the sign may be discarded as demonstrated here.**

$$+(-4 + 3x) = \mathbf{-4 + 3x}$$

2. **When parentheses are preceded by a minus sign, both the minus sign and the parentheses may be discarded if the signs of all terms within the parentheses are changed. This rule is used because the minus sign indicates the negative of, or the opposite of, the quantity within the parentheses.**

$$-(x - 3y + 6 - k) = \mathbf{-x + 3y - 6 + k}$$

example 30.4 Solve: $12 - (2x + 5) = -2 + (x - 3)$

solution As the first step we drop the parentheses, remembering that if the parentheses are preceded by a minus sign, we must change all signs inside the parentheses.

$$12 - 2x - 5 = -2 + x - 3$$

Now we simplify on both sides of the equation

$$7 - 2x = x - 5$$

and solve for x:

$$\begin{array}{rcl} 7 - 2x = & x - 5 & \\ +5 + 2x & +2x + 5 & \text{add } +5 + 2x \text{ to both sides} \\ \hline 12 \quad = & 3x & \longrightarrow \frac{12}{3} = \frac{3x}{3} \longrightarrow \mathbf{4 = x} \quad \text{divided by 3} \end{array}$$

example 30.5 Solve: $-(4y - 17) + (-y) = (2y - 1) - (-y)$

solution Again we remember that when we discard parentheses preceded by a minus sign, the signs of all terms within the parentheses are changed.

$$-4y + 17 - y = 2y - 1 + y$$

First we add like terms and then we solve.

$$\begin{array}{rcl} -5y + 17 = & 3y - 1 & \\ +5y + \ \ 1 & +5y + 1 & \text{add } +5y + 1 \text{ to both sides} \\ \hline 18 = & 8y & \longrightarrow \frac{18}{8} = \frac{8y}{8} \longrightarrow \mathbf{\frac{9}{4} = y} \quad \text{divided by 8} \end{array}$$

practice Solve:

a. $-3(2 - c) = c - 2$

b. $-(6c - 5) = 4(7c - 8) + 3$

c. $-(7 - 9)z - 6z = 8(-6 + 2)$

problem set 30 Write the algebraic phrases that correspond to these word phrases.

1. Seven times the sum of a number and -5.

2. Seven less than twice the opposite of a number.

3. The sum of 7 times a number and -51.

4. A number is multiplied by 4 and this product decreased by 15.

5. 3.25 of what number is 585?

6. What fraction of $6\frac{7}{8}$ is $\frac{1}{4}$?

7. $2\frac{5}{8}$ of 21 is what number?

8. Use six unit multipliers to convert 70 square feet to square meters.

9. Use four unit multipliers to convert 10,000 kilometers to feet. (Go from kilometers to meters to centimeters to inches to feet.)

10. Find the surface area of this right circular cylinder. Dimensions are in centimeters.

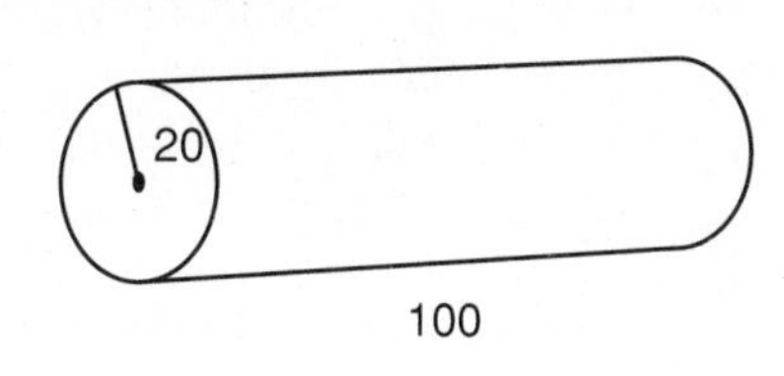

11. Find the area of the figure shown. Dimensions are in feet.

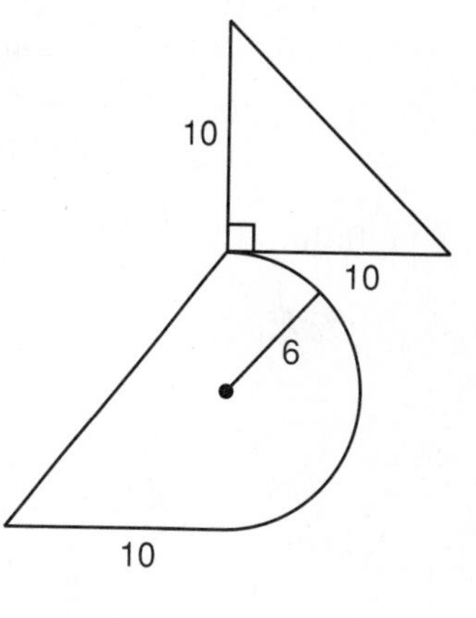

Solve:

12. $0.1p - 0.2p + 2 = -4.6$

13. $\frac{1}{4} + 4\frac{1}{2}k = \frac{1}{8}$

14. $12 - 2x + 5 = -2 + (x - 3)$

15. $-4(4y - 17) + (-y) = (2y - 1) - (-y)$

Simplify:

16. $\frac{1}{(-4)^{-3}} - \sqrt{4}$

17. 4^{-3}

Expand:

18. $y^0x^{-4}(x^4 - 5y^4x^4)$

19. $(y^{-5} + 3x^5y^2)x^0y^5$

20. $-2x^2(3x^4 - 6x^{-2}y^0p)$

21. $5x^0y^2(y^4x^6 - 5x^0y^{-4})$

22. $3m^2n^2(p^0m^4n - m^{-2}n^{-2})$

23. $(x^0p^5 - 3x^0p^{-5})2p^0x^5$

Simplify by adding like terms:

24. $3xym^2z^3 + 2x^2xy^2y^{-1}m^2z^2 - xym^5m^{-3}z^2x^2$

25. $3xy - 2x^2yx^{-1} + 5x^3x^{-2}y^3y^{-2} + 5xxxx^{-2}y$

Evaluate:

26. $-a^2 - 3a(a - b)$ if $a = -2$ and $b = -1$

27. $-c(ac - a^0)$ if $a = -3$ and $c = 4$

28. $-n(n^0 - m) - |m^2|$ if $n = -4$ and $m = 6$

Simplify:

29. $-2^0 - 2^2 - (-2)^3$

30. $-3 + (-3)(-3)^2 + (-3)^3$

LESSON 31 Word problems

To solve word problems, we look for statements in the problems that describe equal quantities. Then we use algebraic phrases and equals signs to write equations that make the same statements of equality. We will begin by solving problems that contain only one statement of equality. These problems require that we write only one equation. Later, we will encounter problems that contain more than one statement of equality. These problems will require more than one equation for their solution.

We will avoid the use of the letters x and y in writing these equations. We will try to use variables whose meaning is easy to remember. The problems in this lesson discuss some unknown number. We will use the letter N to represent the unknown number.

example 31.1 The sum of twice a number and 13 is 75. Find the number.

solution **The word *is* means equal to. Thus, the sum of twice a number and 13 equals 75.**

$$2N + 13 = 75 \qquad \text{equation}$$

We can solve this equation by adding -13 to both sides and then dividing by 2.

$$\begin{array}{rcrl} 2N + 13 & = & 75 & \quad\text{equation} \\ -13 & & -13 & \quad\text{add } -13 \text{ to both sides} \\ \hline 2N \quad & = & 62 & \\ N & = & \mathbf{31} & \quad\text{divided both sides by 2} \end{array}$$

Solutions to word problems should always be checked to see if they really do solve the problem.

Check: $2(31) + 13 = 75 \longrightarrow 62 + 13 = 75 \longrightarrow 75 = 75$ Check

example 31.2 Find a number such that 13 less than twice the number is 137.

solution We will use N to represent the unknown number. Then twice the unknown number is $2N$ and 13 less than that is $2N - 13$.

$$\begin{array}{rcrl} 2N - 13 & = & 137 & \quad\text{equation} \\ +13 & & +13 & \quad\text{add } +13 \text{ to both sides} \\ \hline 2N \quad & = & 150 & \\ N & = & \mathbf{75} & \quad\text{divided both sides by 2} \end{array}$$

Check: $2(75) - 13 = 137 \longrightarrow 150 - 13 = 137 \longrightarrow 137 = 137$ Check

example 31.3 Find a number such that if 5 times the number is decreased by 14, the result is twice the opposite of the number.

solution If we use N for the number, then $2(-N)$ will represent twice the opposite of the number.

$$5N - 14 = 2(-N) \qquad \text{equation}$$

$$\begin{array}{rcll} 5N - 14 & = & -2N & \quad\text{multiplied} \\ 2N + 14 & & 2N + 14 & \quad\text{add } 2N + 14 \text{ to both sides} \\ \hline 7N \quad & = & 14 & \end{array}$$

$$N = 2 \qquad \text{divided both sides by 7}$$

Check: $5(2) - 14 = 2(-2) \longrightarrow 10 - 14 = -4 \longrightarrow -4 = -4$ Check

example 31.4 Find a number which decreased by 18 equals 5 times its opposite.

solution Again we use N for the number and $-N$ for its opposite.

$$\begin{array}{rclr} N - 18 &=& 5(-N) & \text{equation} \\ N - 18 &=& -5N & \text{multiplied} \\ \underline{5N + 18} && \underline{5N + 18} & \text{add } 5N + 18 \text{ to both sides} \\ 6N &=& 18 & \end{array}$$

$$N = 3 \qquad \text{divided both sides by 6}$$

Check: $3 - 18 = 5(-3) \longrightarrow 3 - 18 = -15 \longrightarrow -15 = -15$ Check

example 31.5 We get the same result if we multiply a number by 3 *or* if we multiply the number by 5 and then add 2. Find the number.

solution The statement of the problem leads to the following equation.

$$\begin{array}{rclr} 3N &=& 5N + 2 & \text{equation} \\ \underline{-5N} && \underline{-5N} & \text{add } -5N \text{ to both sides} \\ -2N &=& 2 & \end{array}$$

$$N = -1 \qquad \text{divided both sides by } -2$$

Check: $3(-1) = 5(-1) + 2 \longrightarrow -3 = -5 + 2 \longrightarrow -3 = -3$ Check

practice

a. Four times a number decreased by 8 equals 92. Find the number. Check your answer.

b. If the product of 4 and a number is decreased by 12, the result is twice the opposite of the number. Find the number. Check your answer.

problem set 31

1. Use six unit multipliers to convert 800 cubic centimeters to cubic feet.

2. Find a number which decreased by 21 equals twice the opposite of the number.

3. If the product of 3 and a number is increased by 7, the result is 23 greater than the number. Find the number.

4. This trapezoid is the base of a right prism 10 in. high. Find the volume and the surface area. Dimensions are in inches.

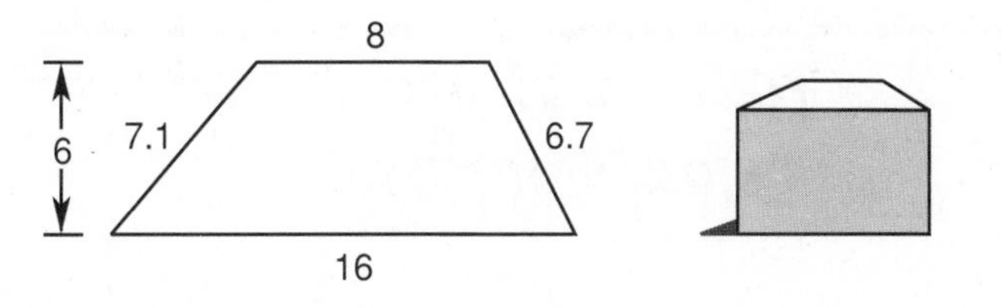

5. Find the area of this figure in square inches. The tick marks show that the altitude is 20 in. and that the base measures 4 ft.

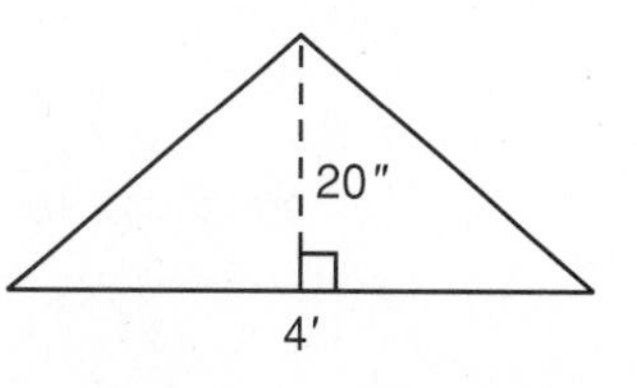

6. $2\frac{1}{9}$ of what number $= 3\frac{4}{5}$?

7. What decimal part of 0.05 is 1.25?

Solve:

8. $\frac{p}{7} - 2 = \frac{15}{3}$

9. $2\frac{1}{4}k + \frac{1}{4} = \frac{1}{8}$

10. $1.3p + 0.3p - 2 = 1.2$

11. $3(p - 2) = p + 7$

12. $+2(3x - 5) = 7x + 2$

13. $-(x - 3) - 2(x - 4) = 7$

14. $-5(p - 4) - 3(-2 - p) = p - 2$

15. $2(3p - 2) - (p + 4) = 3p$

Simplify:

16. 6^{-2}

17. $\frac{1}{(-3)^{-3}}$

18. $(-2)^{-2} - 2^2$

19. $\frac{1}{(-4)^{-3}}$

Expand:

20. $-3x^0p(-5xp^0 - 2p^{-1})$

21. $2xp^{-4}(x^{-4}p - 3x^2p^{-2})$

22. $2p^4x^0y(p^5m^4x - 5x^{-2}y^{-4})$

23. $(x^4 - 2p^2)3x^0p^{-4}$

Simplify by adding like terms:

24. $-3x^2x^0xy^2 + 2x^3y^{-3}y^5 + 5x^{-3}x^{-6}yy^2y^{-5}$

25. $2xym^2 + 3x^2ym - 4y^2my^{-1}m^0x^4x^{-2}$

Evaluate:

26. $a - a(b^0 - a)$ if $a = -2$ and $b = 5$

27. $x^2 - xy^2(x - y)$ if $x = -3$ and $y = 4$

28. $m - m^2 - (m - n)$ if $m = -3$ and $n = -5$

Simplify:

29. $-3^3 - 3^2 - (-3)^4 - |-3^2 - 3|$

30. $\frac{-5(-5 - 4)}{-2^0(-8 - 1)}$

LESSON 32 *Products of prime factors · Statements about unequal quantities*

32.A prime numbers

The number 6 can be composed by multiplying the two counting numbers 3 and 2.

$$3 \cdot 2 = 6$$

Because 6 can be composed by multiplying two counting numbers that are both greater than 1, we say that 6 is a composite number. The number 35 is also a composite number because it can be composed as the product of the counting numbers 5 and 7.

$$5 \cdot 7 = 35$$

The number 1 must be one of the factors if we wish to compose 17 by multiplying.

$$17 \cdot 1 = 17$$

The number 1 must also be a factor if we wish to compose either 3 or 11 or 23.

$$1 \cdot 3 = 3 \qquad 1 \cdot 11 = 11 \qquad 1 \cdot 23 = 23$$

Since these numbers can be composed only if 1 is one of the factors, we do not call these numbers composite numbers. We call them **prime numbers.**

A *prime number* is a counting number greater than 1 whose only counting number factors are 1 and the number itself.

The number 12 can be written as a product of integral factors in four different ways.

(a) $12 \cdot 1$ (b) $4 \cdot 3$ (c) $2 \cdot 6$ (d) $2 \cdot 2 \cdot 3$

In (a), (b), and (c), one of the factors is not a prime number, but in (d) all three of the factors are prime numbers. **A prime factor is a factor that is a prime number.** To find the prime factors of a counting number, we divide by prime numbers, as we see in the following examples.

example 32.1 Express 80 as a product of prime factors.

solution We will divide by prime numbers.

$$\frac{80}{2} = 40 \qquad \frac{40}{2} = 20 \qquad \frac{20}{2} = 10 \qquad \frac{10}{2} = 5$$

Using the five factors we have found, we can express 80 as a product of prime factors as $\mathbf{2 \cdot 2 \cdot 2 \cdot 2 \cdot 5}$.

example 32.2 Express 147 as a product of prime factors.

solution 147 is not divisible by 2 or by 5, so let's try 3.

$$\frac{147}{3} = 49 \qquad \text{and} \qquad \frac{49}{7} = 7$$

So 147 expressed as a product of prime factors is $\mathbf{3 \cdot 7 \cdot 7}$.

32.B statements about unequal quantities

Often a word problem makes a statement about quantities that differ by a specified amount. Thus, the statement tells us that the quantities are not equal, and our task is to write an equation about quantities that are equal. To perform this task, we must add as required so that both sides of the equation represent equal quantities.

example 32.3 Twice a number is 42 less than -102. Find the number.

solution **We must be careful because the problem tells us about things that are not equal. We begin by writing an equation that we know is incorrect.**

$$2N = -102 \qquad \text{incorrect}$$

The problem said that $2n$ was 42 less than -102, so we must add 42 to $2N$ or we must add -42 to -102.

ADDING 42 TO $2N$ or ADDING -42 TO -102

$$\begin{array}{ll} 2N + 42 = -102 & \text{correct} \\ \quad -42 \quad\; -42 & \\ \hline 2N \quad = -144 & \\ N = -72 & \end{array} \qquad \begin{array}{ll} 2N = -102 - 42 & \text{correct} \\ 2N = -144 & \\ N = -72 & \end{array}$$

Check: $2(-72) + 42 = -102 \longrightarrow -144 + 42 = -102$
$\longrightarrow -102 = -102$ Check

example 32.4 Five times a number is 72 greater than the opposite of the number. Find the number.

solution **This statement is tricky because it describes quantities that are unequal. As the first step in writing the desired equation, we will write an equation that we know is incorrect.**

$$5N = -N \qquad \text{incorrect}$$

This equation is incorrect because $5N$ is really 72 greater. We can make the equation correct by adding -72 to $5N$ or by adding $+72$ to $-N$.

ADDING -72 TO $5N$ or ADDING $+72$ TO $-N$

$$\begin{array}{l} 5N - 72 = -N \\ +N + 72 \quad +N + 72 \\ \hline 6N \quad = \quad 72 \\ N = 12 \end{array} \qquad \begin{array}{l} 5N = -N + 72 \\ +N \quad +N \\ \hline 6N = \quad 72 \\ N = 12 \end{array}$$

Check: $5(12) - 72 = -12 \longrightarrow 60 - 72 = -12 \longrightarrow -12 = -12$ Check

example 32.5 If the sum of twice a number and -14 is multiplied by 2, the result is 12 greater than the opposite of the number. Find the number.

solution **Again we begin by writing an equation that we know is incorrect.**

$$2(2N - 14) = -N \qquad \text{incorrect}$$

We know that the left side is greater by 12. We can write a correct equation by adding -12 to the left side or by adding $+12$ to the right side.

ADDING -12 TO THE LEFT SIDE or ADDING $+12$ TO THE RIGHT SIDE

$$\begin{array}{r} 2(2N - 14) - 12 = -N \\ 4N - 28 - 12 = -N \\ 5N = 40 \\ N = 8 \end{array} \qquad \begin{array}{r} 2(2N - 14) = -N + 12 \\ 4N - 28 = -N + 12 \\ 5N = 40 \\ N = 8 \end{array}$$

Check: $2(2 \cdot 8 - 14) - 12 = -8 \longrightarrow 2(2) - 12 = -8$
$\longrightarrow -8 = -8$ Check

example 32.6 Five times a number is 21 less than twice the opposite of the number. What is the number?

solution **We must be careful because 5 times the number is 21 less. Thus we will add 21 so that it will be equal.**

$$5N + 21 = 2(-N)$$

$$\begin{array}{rcl} 5N + 21 & = & -2N \\ \underline{2N - 21} & & \underline{2N - 21} \\ 7N & = & -21 \end{array}$$

$$N = -3$$

Check: $5(-3) + 21 = 2(3) \longrightarrow -15 + 21 = 6 \longrightarrow 6 = 6$ Check

practice

a. Find a number such that 8 times the number is 36 greater than the opposite of the number.

b. If the sum of a number and 6 is multiplied by 2, the result is 10 greater than the number. Find the number.

Write these numbers as products of prime factors:

c. 400 **d.** 108

problem set 32

1. Use 6 unit multipliers to convert 120 cubic feet to cubic centimeters.

2. Find the surface area of this rectangular prism in square centimeters. Dimensions are in centimeters.

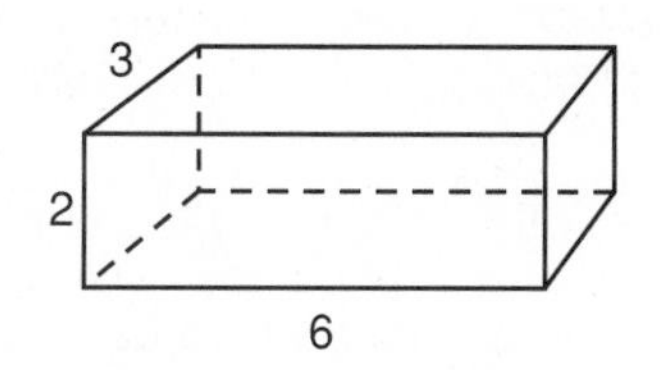

3. Find the area of the triangle in square inches. The tick marks show that the altitude is in feet and the base is in inches.

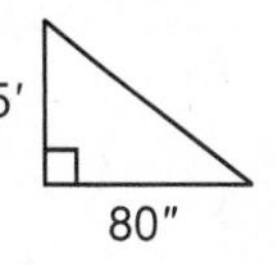

4. $\frac{3}{7}$ of what number is $2\frac{1}{5}$?

5. What fraction of 40 is 90?

6. 1.025 of 50 is what number?

Solve:

7. $\frac{5}{8}x - 3 = \frac{1}{2}$

8. $\frac{1}{7}y + 10 = 14\frac{1}{4}$

9. $0.3 + 0.06p + 0.02 - 0.02p = 4$

10. $3p - 4 - 6 = -2(p - 5)$

11. $k + 4 - 5(k + 2) = 3k - 2$

12. $x - 4(x - 3) + 7 = 6 - (x - 4)$

13. $p - 3(p + 4) = 2(p + 1)$

Write the following numbers as products of prime factors:

14. 160 **15.** 294 **16.** 250 **17.** 450

Simplify:

18. $(-3)^{-3}$

19. 2^{-3}

Expand:

20. $2x^{-2}(x^{-2}y^0 + x^2y^5p^0)$

21. $x^{-3}p^0(x^6p^5 - 3x^3p^0)$

22. $4x^2y^0(x^0y^2 - 3x^2y^{-2})$

23. $(4p^{-2} - 3x^{-3}p^5)p^2x^0$

Simplify by adding like terms:

24. $-3x^{-2}y^2x^5 + 6x^3y^{-2}y^4 - 3x^3y^2 + 5x^2y^3$

25. $-xyz^5z^{-4} + 5xy^{-4}y^5z - 3zxy^7y^{-6}$

Evaluate:

26. $m - (-m)(m^0 - a)$ if $m = -2$ and $a = 3$

27. $k^2 - k^3(km^0)$ if $k = -3$ and $m = 2$

28. $a^3x - x^3$ if $a = -3$ and $x = -2$

Simplify:

29. $-3^3 - 2^2 - 4^3 - |-2^2 - 2|$

30. $\dfrac{-(-3 + 7) - 4^0}{(-2)(-3 + 5)}$

LESSON 33 Greatest common factor

The number 210 has four prime number factors as shown here:

$$2 \cdot 3 \cdot 5 \cdot 7 = 210$$

We call factors that are numbers **numerical factors.** Some expressions have factors that are letters, and some expressions have both numbers and letters as factors, as does $210xy^2z^3$.

$$210xy^2z^3 = 2 \cdot 3 \cdot 5 \cdot 7 \cdot x \cdot y \cdot y \cdot z \cdot z \cdot z$$

We call the letter factors **literal factors,** and we use the words **algebraic factor** as a general term to describe factors that are either numbers or letters or both numbers and letters.

> DEFINITION
>
> The greatest common factor (GCF) of two or more terms is the product of all prime algebraic factors common to every term, each to the highest power that it occurs in all of the terms.

The expression $6x^2y^2m^2 + 3xy^3m^2 + 3x^3y^2$ can be written as

$$2 \cdot 3 \cdot x \cdot x \cdot y \cdot y \cdot m \cdot m + 3 \cdot x \cdot y \cdot y \cdot y \cdot m \cdot m + 3 \cdot x \cdot x \cdot x \cdot y \cdot y$$

Now only the first term has 2 as a factor, so 2 is not of the greatest common factor. Each term has 3 as a factor at least once, so 3 is a factor of the greatest common factor of all the terms.

$$3$$

Each term has x as a factor at least once in every term, so x is a factor of the GCF.

$$3x$$

In the same way, y is used as a factor at least twice in every term, so the greatest common factor of the three given terms is

$$3xy^2$$

The variable m is not included because it is not a factor of the third term of the original expression.

example 33.1 Find the GCF of $8z^4m^2p - 12z^3m^4p^2$.

solution The greatest common factor of the term is $\mathbf{4z^3m^2p}$.

example 33.2 Find the greatest common factor of $4x^2y^3z - 8y^2xz^3$.

solution The GCF is $\mathbf{4xy^2z}$.

example 33.3 Find the GCF of $16x^2yp^3 - 4x^3y^2p + 2x^2y^2p^{15}$.

solution The GCF is $\mathbf{2x^2yp}$.

practice Find the greatest common factor of:

a. $6x^2y^3m - 14xy^5m^2 + 24x^{10}y^6m^4$

b. $5a^2b^2c^2 + 60a^3b^3c^3 - 30a^4b^4c^4$

c. $12xy^5p^3 - 16x^6y^2p^{16} + 28x^3yp^5$

problem set 33

1. Zollie had a secret number. She found that the sum of 3 times her number and 60 equaled -50. What was her number?

2. Twice the sum of 3 times a number and 60 is 155 greater than the opposite of the number. Find the number.

3. 0.125 of what number is 5.25?

4. What fraction of 4 is $\frac{1}{4}$?

5. $\frac{3}{5}$ of $6\frac{2}{3}$ is what number?

Solve:

6. $3\frac{1}{2} + 2\frac{1}{4}x = \frac{5}{4}$

7. $3(x - 2) + (2x + 5) = x + 7$

8. $-4.2 + 0.02x - 0.4 = 0.03x$

9. $-p - 4 - (2p - 5) = 4 + 2(p + 3)$

10. $8 - k + 2(4 - 2k) = k + 2k$

11. Use three unit multipliers to convert 500 cubic inches to cubic centimeters.

12. Use six unit multipliers to convert 500 cubic inches to cubic meters.

13. Find the area of the rectangle in square inches. Note that the length is 32 inches and the width is 1 foot.

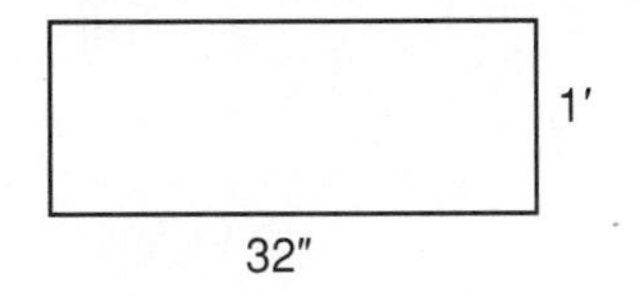

Find the greatest common factor of:

14. $4ab^2c^4 - 2a^2b^3c^2 + 6a^3b^4c$

15. $5x^2y^5m^2 - 10xy^2m^2 + 15x^2y^4m^2$

Write as products of prime factors:

16. 630

17. 600

Simplify:

18. 2^{-4}

19. $\frac{1}{4^{-3}} - \sqrt[5]{-32}$

Expand:

20. $3x^2y^0(x^{-2} - 3y^2x^4)$

21. $2p^{-5}(p^2x^5 - 3x^0p^5)$

22. $4x^{-3}y^2(x^{-3}y^{-2} - 2x^4y^4)$

23. $(y^{-5} - 2y^7x^5)x^0y^5$

Simplify by adding like terms:

24. $3xyz^2 - 4z^2xy + 7yx^2z - 5zx^2y$

25. $3x^2xyy^3y^{-1} + 2x^2xyyy - 4x^{-2}yx^5y^2 + 7x^2$

Evaluate:

26. $-x^0 - a(x - 2a)$ if $x = -5$ and $a = 3$

27. $p^3 - a^2 + ap$ if $p = -3$ and $a = 2$

28. $a^2 - a^3 - a^4$ if $a = -2$

Simplify:

29. $-3^3 - 3^2 - (-3)^2 - |-2^2|$

30. $\frac{-4(3^0 - 6)(-2)}{-4 - (-3)(-2) - 3}$

LESSON 34 Factoring the greatest common factor

When we use the distributive property, we change an expression from a product to a sum. The expression $2a(x + c)$ tells us to multiply $2a$ by $x + c$. If we do this multiplication, we get the algebraic sum $2ax + 2ac$:

$$2a(x + c) = 2ax + 2ac$$

If we reverse the process and write $2ax + 2ac$ as the product of the two factors $2a$ and $(x + c)$, we say that we are factoring.

***Factoring* is the process of writing an indicated sum as a product of factors.**

example 34.1 Factor $2ax + 2ac$.

solution We will factor in three steps. The first step is to write two empty parentheses to indicate a product.

$$(\quad)(\quad)$$

The second step is to write the greatest common factor of the terms in the first parentheses.

$$(2a)(\quad)$$

The third step is to write the proper terms in the second parentheses so that $2a$ times these terms gives us $2ax + 2ac$.

$$(2a)(x + c)$$

And since the first parentheses are not necessary, the answer can be written as

$$\mathbf{2a(x + c)}$$

example 34.2 Factor $a^3x^2m^2 + a^2xm^3 - a^4x^3m^2$.

solution We want to write this sum as a product. We begin by writing two sets of parentheses.

$$(\quad)(\quad)$$

In the first parentheses we want to write the greatest common factor of all three terms. To find this GCF, we will write the three terms as products of individual factors.

$$a^3x^2m^2 + a^2xm^3 - a^4x^3m^2$$

$$a \cdot a \cdot a \cdot x \cdot x \cdot m \cdot m + a \cdot a \cdot x \cdot m \cdot m \cdot m - a \cdot a \cdot a \cdot a \cdot x \cdot x \cdot x \cdot m \cdot m$$

Look at the a's. Each term has at least two a's, so a^2 is part of the greatest common factor.

$$(a^2 \quad)(\quad)$$

Each term has at least one x, so x is a part of the greatest common factor.

$$(a^2x \quad)(\quad)$$

Finally, each term has at least two m's, so m^2 is a part of the greatest common factor. No other factors are common to all three terms.

$$(a^2xm^2)(\quad)$$

Now, the first entry in the second parentheses must be ax because a^2xm^2ax equals $a^3x^2m^2$, the first term of the original expression.

$$(a^2xm^2)(ax \quad)$$

The second entry in the second parentheses must be m because $a^2xm^2(m) = a^2xm^3$, the second term of the original expression.

$$(a^2xm^2)(ax + m \quad)$$

The third entry must be $-a^2x^2$ because $a^2xm^2(-a^2x^2) = -a^4x^3m^2$, the last entry in the original expression. The desired factored expression is $a^2xm^2(ax + m - a^2x^2)$ because

$$\mathbf{a^2xm^2(ax + m - a^2x^2)} = a^3x^2m^2 + a^2xm^3 - a^4x^3m^2$$

example 34.3 Factor $4a^3b^4z^3 + 2a^2bz^4$.

solution

$$2 \cdot 2 \cdot a \cdot a \cdot a \cdot b \cdot b \cdot b \cdot b \cdot z \cdot z \cdot z + 2 \cdot a \cdot a \cdot b \cdot z \cdot z \cdot z \cdot z$$

Each term has, at least one 2, two a's, one b, and three z's as factors. Thus the greatest common factor is $2a^2bz^3$, so we write

$$(2a^2bz^3)(\quad)$$

The first term in the second parentheses is $2ab^3$ because $2a^2bz^3(2ab^3) = 4a^2b^4z^3$, the first term in the original expression.

$$(2a^2bz^3)(2ab^3 \quad)$$

The second term in the second parentheses is z because $2a^2bz^3(z) = 2a^2bz^4$, the second term in the original expression. Now

$$\mathbf{(2a^2bz^3)(2ab^3 + z)}$$

is our answer because

$$(2a^2bz^3)(2ab^3 + z) = 4a^3b^4z^3 + 2a^2bz^4$$

example 34.4 Factor $6a^2x^2 + 2a^3x^3 + 4a^4x^3$.

solution The greatest common factor is $2a^2x^2$.

$$(2a^2x^2)(\qquad)$$

The entry in the second parentheses is

$$(3 + ax + 2a^2x)$$

because

$$\mathbf{(2a^2x^2)(3 + ax + 2a^2x)} = 6a^2x^2 + 2a^3x^3 + 4a^4x^3$$

example 34.5 Factor $3m^3xy^2 + m^2y$.

solution $\mathbf{m^2y(3mxy + 1)}$ is the answer because

$$\mathbf{m^2y(3mxy + 1)} = 3m^3xy^2 + m^2y$$

practice Factor:

a. $2a^2b^2 + 2a^3b^4 - 2a^2b^6$

b. $15a^2z^8 - 35z^5a$

c. $28xmz^{10} - 7x^2m^3z^4$

problem set 34

1. If the product of 5 and a number is increased by 7, the result is -42. What is the number?

2. If the product of 5 and a number is increased by 7 and this sum multiplied by 3, the result is 11 greater than the opposite of the number. Find the number.

3. $4\frac{1}{3}$ of what number is $3\frac{5}{8}$?

4. What decimal part of 0.42 is 0.00504?

5. $3\frac{2}{5}$ of $3\frac{1}{8}$ is what number?

Solve:

6. $3\frac{1}{2}n - \frac{1}{2} = \frac{4}{3}$

7. $x - 4 - 2x + 5 = 3(2x - 4)$

8. $0.2m + 4.34 - m = 2.3$

9. $3(-k - 4) + 6 = k + 7$

10. The average of a group of numbers is the sum of the numbers divided by the number of numbers. To find the average of 6, 7, 11, and 15, we would divide the sum by 4.

$$\text{Average} = \frac{6 + 7 + 11 + 15}{4} = \frac{39}{4} = 9.75$$

Find the average of 473.11, 742.8, and 947.61. Round the answer to two decimal places.

11. Use six unit multipliers to convert 20 square miles to square centimeters.

12. Express the perimeter of this figure in meters. All angles are right angles. Begin by changing centimeters to meters (10 cm = 0.1 m).

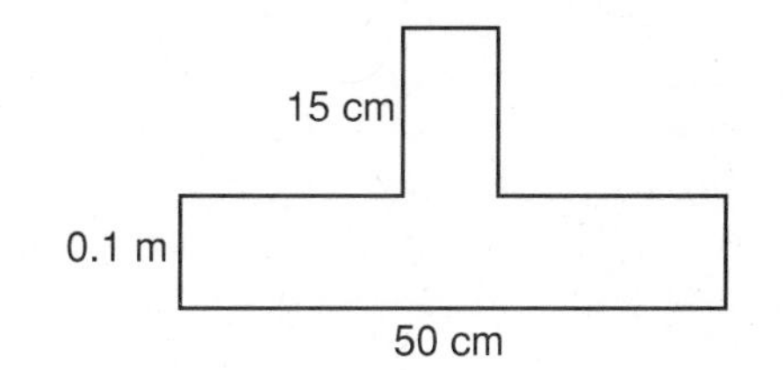

13. Find the area of the shaded portion of this parallelogram in square meters. Dimensions are in meters.

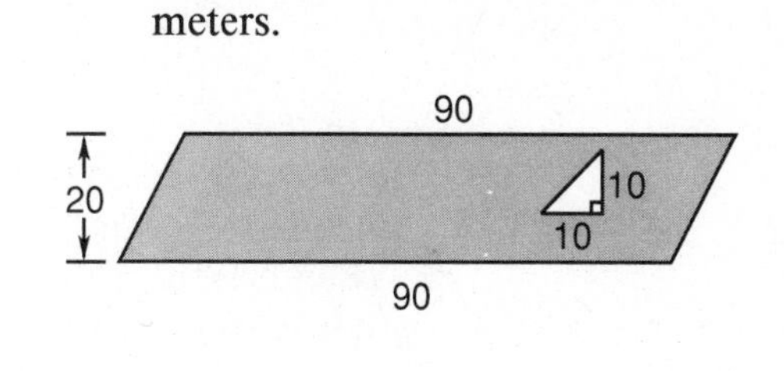

Factor the greatest common factor:

14. $3x^4y^2p - 6x^2y^5p^4$

15. $6a^3x^2m^5 + 2a^4x^5m^5 + 4a^2x^2m$

Write as products of prime factors:

16. 250

17. 360

Simplify:

18. 3^{-4}

19. $\frac{1}{4^{-3}}$

Expand:

20. $y^0x^{-2}(x^2 - 4x^4y^6)$

21. $(p^5y^5 - y^{-5})p^0y^5$

22. $p^0x^2y(x^3y^{-1} - 3x^5y^{-2})$

23. $x^2(2x^{-2} - 4x^0p^5y^5)$

Simplify by adding like terms:

24. $4x^2y^{-2}p^4 - 3y^{-2}x^2p^4 + 7yyx^2p^4$

25. $3xxy^2x^{-2} - 2x^0yy + 5y^2 - 6x^2 - 4x^3x^{-1}$

Evaluate:

26. $-p(x - px)$ if $p = -3$ and $x = 4$

27. $x^3 - x^2 + 2x$ if $x = -2$

28. $x(y - xy^0)$ if $x = -2$ and $y = 4$

Simplify:

29. $3^2 - 3^3 - 3^0 + |-3^0|$

30. $\frac{4(-2 - 3) - 4^0}{-2(-4 + 6) - 3}$

LESSON 35 Canceling

We have been solving equations by using the fact that multiplication and division are inverse operations because they "undo" one another. If we want to solve the equation

$$4x = 20$$

we see that x is multiplied by 4. To undo multiplication by 4, we must divide by 4. If we divide one side of an equation by 4, we must also divide the other side of the equation by 4.

$$\frac{\cancel{4}x}{\cancel{4}} = \frac{20}{4} \longrightarrow x = 5$$

On the left, we say that we have canceled the 4s. Some people prefer to say that 4 over 4 "reduced to 1" instead of saying "'canceled" because the use of these words helps to prevent canceling when canceling is not permissible. For instance, the 4s cannot be canceled in the following expression because addition and division are not inverse operations and do not undo one another.

$$\frac{x + \cancel{4}}{\cancel{4}} = x + 1 \qquad \text{incorrect}$$

In this problem nothing "reduces to 1." However, the following expression can be simplified by canceling

$$\frac{\cancel{4}(x + 1)}{\cancel{4}}$$

because multiplication by 4 and division by 4 do undo each other. We can see that 4 over 4 "reduces to 1."

$$\frac{4(x + 1)}{4} = \boldsymbol{x + 1}$$

Cancellation or reduction to 1 is possible when the numerator and the denominator contain one or more common factors.†

example 35.1 Simplify: (a) $\frac{4(a - 3)}{4}$ (b) $\frac{3(x - 2)}{x - 2}$

solution (a) Here the common factor is 4, and 4 over 4 equals 1.

$$\frac{\cancel{4}(a - 3)}{\cancel{4}} = \boldsymbol{a - 3}$$

(b) Here the common factor is $x - 2$, and $x - 2$ reduces to 1.

We will assume in all problems of this type that the denominator does not equal zero.

$$\frac{3\cancel{(x - 2)}}{\cancel{x - 2}} = \mathbf{3}$$

---

† In the expression being discussed we remember that the 4 in the denominator can be written as $4 \cdot 1$. Thus both the numerator and denominator have 4 as a factor.

example 35.2 Simplify: $\dfrac{3p + 3}{3}$

solution We cannot simplify in this form because the numerator is not a product. However, if we factor $3p + 3$, we see that we can cancel because both the numerator and the denominator will have 3 as a factor.

$$\frac{\cancel{3}(p + 1)}{\cancel{3}} = \mathbf{p + 1}$$

example 35.3 Simplify: (a) $\dfrac{3x - 9x^2}{3x}$ (b) $\dfrac{5x - 25x^2}{5xy}$

solution In (a) we can factor out a $3x$ and in (b) a $5x$.[†]

$$\text{(a)}\quad \frac{3x(1 - 3x)}{3x} = \mathbf{1 - 3x} \qquad \text{(b)}\quad \frac{5x(1 - 5x)}{5x(y)} = \frac{\mathbf{1 - 5x}}{\mathbf{y}}$$

practice Simplify:

a. $\dfrac{4 - 4x}{4}$ **b.** $\dfrac{4x + 4}{4}$ **c.** $\dfrac{7x - 49x^2}{7x}$

problem set 35

1. Jay and Bill found that 4 times the sum of a number and -6 equaled 20. What was the number?

2. If the sum of 4 times a number and 6 is multiplied by 3, the result is 5 greater than the opposite of the number. Find the number.

3. $5\frac{7}{10}$ of what number is $9\frac{1}{2}$?

4. What fraction of $2\frac{1}{4}$ is $7\frac{1}{8}$?

5. 1.05 of 0.043 is what number?

Solve:

6. $-5\frac{1}{2} + 2\frac{2}{5}p = 6\frac{1}{4}$

7. $-n + 0.4n + 1.8 = -3$

8. $x - (3x - 2) + 5 = 2x + 4$

9. $5(x - 2) - (-x + 3) = 7$

Factor the greatest common factor:

10. $4a^2xy^4p - 6a^2x^4$

11. $3a^2x^4y^6 + 9ax^2y^4 - 6x^4a^2y^5z$

12. Use six unit multipliers to convert 3000 cubic meters to cubic inches.

13. Find the area of this figure in square centimeters. Begin by changing 0.4 m to centimeters.

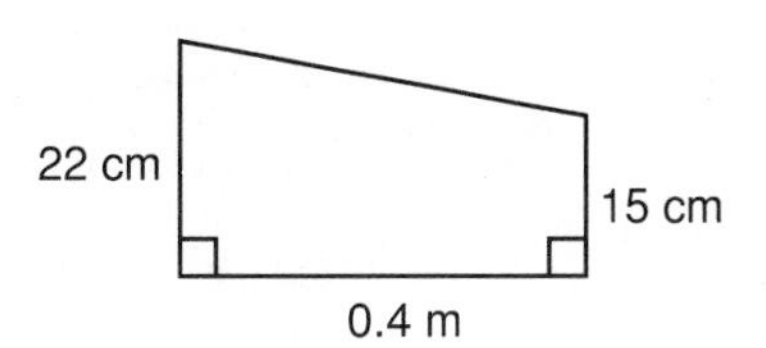

† Note the use of the words *factor out*. This phrase is meaningful even though some authorities insist that it is redundant and that the single word *factor* will suffice. However, this slight redundancy is not harmful, especially since it is a natural redundancy.

14. Express in cubic feet the volume of a right prism whose base is shown and whose sides are 2 feet high. What is the surface area of the prism? Dimensions are in feet. Corners that look square are square.

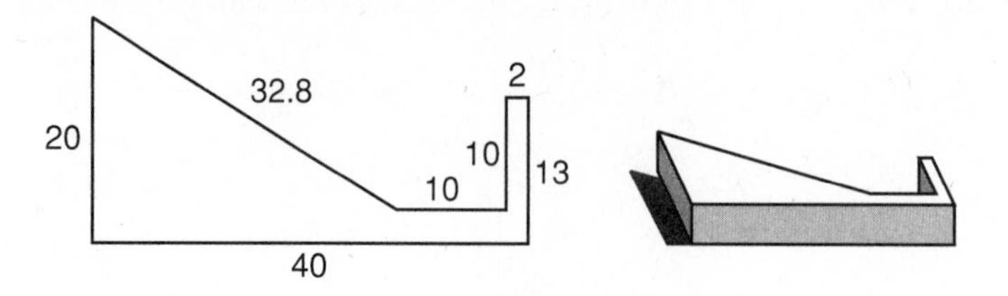

Simplify. Factor if necessary:

15. $\dfrac{2x + 6}{2}$ 16. $\dfrac{k^2x - k^3x}{k^2xy}$

17. Find the average of 648.32, 475.61, 983.56, and 811.4 to two decimal places.

18. Write 270 as a product of prime factors.

19. Simplify: $\dfrac{1}{4^{-4}} - \sqrt[3]{125}$

Expand:

20. $(x^3y^0 - p^0x^2y^4)x^{-3}$ 21. $3x^0y^{-3}(4y^3z - 7x^2)$

22. $3x^4y^2(xy^{-4} - 3x^{-4}y^5)$ 23. $2x^0y^{-5}(4xyy^5 - 3x^5y^4)$

Simplify by adding like terms:

24. $3x^2ym^5 - 2xym^5 + 4m^5yx^2 - 6m^5yx$

25. $2x^4y^{-3} - 3x^2x^2y^{-7}y^4 + 6x^3xy^{-1}y^{-3} + xxy^{-3}$

Evaluate:

26. $a^2 - a^0(a - ab)$ if $a = -3$ and $b = 5$

27. $b - ab(b - a)$ if $a = -3$ and $b = 5$

28. $-k - kp^0 - (-pk^2)$ if $k = -3$ and $p = 2$

Simplify:

29. $2^2 - 2^3 - (-3)^2 + \sqrt[4]{81}$ 30. $\dfrac{-3^2 + 4^2 + 3^3}{2(-5 + 2) - 3^0}$

LESSON 36 *Multiplying fractions · Minus signs and negative exponents*

36.A multiplication of fractions

Two fractions are multiplied by multiplying the numerators to form the new numerator and by multiplying the denominators to form the new denominator. For example,

$$\frac{3}{2} \cdot \frac{5}{7} = \mathbf{\frac{15}{14}}$$

Since variables stand for unspecified real numbers, all the rules for real numbers also apply to variables. Thus fractions that contain variables are multiplied by using the same rule.

$$\frac{mx}{4y} \cdot \frac{ax}{2y} = \frac{amx^2}{8y^2}$$

The distributive property of real numbers is also applicable to expressions that contain fractions. Expressions that contain fractions are often called **rational expressions.** We can expand the following rational expression by multiplying x over y by both of the terms inside the parentheses.

$$\frac{x}{y}\left(\frac{a}{y} - b\right) = \frac{xa}{y^2} - \frac{bx}{y}$$

example 36.1 Use the distributive property to expand: $\frac{x^2}{y^2}\left(\frac{x^2}{y} - \frac{3y^3}{m}\right)$

solution Two multiplications are indicated. We multiply $\frac{x^2}{y^2}$ by $\frac{x^2}{y}$ and then multiply $\frac{x^2}{y^2}$ by $\frac{-3y^3}{m}$. This gives

$$\frac{x^2x^2}{y^2y} - \frac{x^2 3y^3}{y^2 m}$$

Lastly, we simplify both expressions and get

$$\frac{x^4}{y^3} - \frac{3yx^2}{m}$$

example 36.2 Expand: $\frac{m}{z}\left(\frac{axp}{mk} - 2m^4p^4\right)$

solution Again we will use two steps. First, we multiply and then we simplify.

$$\frac{maxp}{zmk} + \frac{m}{z}(-2m^4p^4) = \frac{axp}{zk} - \frac{2m^5p^4}{z}$$

example 36.3 Expand: $\frac{ab}{c^2}\left(\frac{xab}{c} + 2bx - \frac{4}{c^2}\right)$

solution $\frac{ab}{c^2}$ must be multiplied by all three terms inside the parentheses. We do this and get

$$\frac{xa^2b^2}{c^3} + \frac{2ab^2x}{c^2} - \frac{4ab}{c^4}$$

36.B minus signs and negative exponents

Expressions that contain both minus signs and negative exponents can be troublesome. **A minus sign in front of an expression indicates the opposite of the expression, whereas a negative exponent has a meaning that is entirely different.**

(a) 4^2 (b) 4^{-2} (c) -4^2

(d) $(-4)^2$ (e) -4^{-2} (f) $(-4)^{-2}$

The notations in (a) and (b) are easy to simplify because in each of these we have a positive number raised to a power.

(a) $4^2 = \mathbf{16}$ (b) $4^{-2} = \frac{1}{4^2} = \mathbf{\frac{1}{16}}$

The notations in (c) and (d) often give difficulty because of the problem caused by the minus sign in front of the 4. When the minus sign is not enclosed in parentheses as in (c),

$$-4^2$$

it is helpful to cover up the minus sign with a fingertip. If we cover the minus sign in -4^2, we get

$$4^2$$

Now we raise 4 to the second power and get 16.

$$16$$

Now we remove our fingertip and uncover the minus sign. We see that the result is negative 16.

$$\mathbf{-16}$$

From this we see that -4^2 is read as the "opposite of" 4 squared and is not read as the "opposite of 4" squared. The notation in (d) is "the opposite of 4" squared.

$$(-4)^2$$

If we try to cover up the minus sign with a fingertip, we cannot because the minus sign is "protected" by the parentheses.

$$(-4)^2$$

This reminds us that (-4) is to be used as a factor twice.

$$(-4)^2 = (-4)(-4) = \mathbf{+16}$$

The notations in (e) and (f) are similar to the two we have just discussed. They are

$$\text{(e)} \quad -4^{-2} \qquad \text{and} \qquad \text{(f)} \quad (-4)^{-2}$$

The minus sign in (e) is "unprotected," so we can cover it with a fingertip.

$$4^{-2}$$

Now we simplify 4^{-2} as

$$4^{-2} = \frac{1}{4^2} = \frac{1}{16}$$

We finish by removing the fingertip and finding that the answer is negative one-sixteenth.

$$\mathbf{-\frac{1}{16}}$$

The minus sign in (f) is "protected" by the parentheses. Thus, the simplification is

$$(-4)^{-2} = \frac{1}{(-4)^2} = \frac{1}{(-4)(-4)} = \mathbf{\frac{1}{16}}$$

practice Expand:

a. $\frac{x^2}{y}\left(\frac{x^2}{y} - \frac{3y^3}{m}\right)$ **b.** $\frac{x}{z}\left(\frac{15y}{x} - \frac{4x}{y}\right)$

Simplify:

c. 3^{-2} **d.** $(-3)^{-2}$ **e.** $(-3)^2$ **f.** -3^{-2}

problem set 36

1. If the sum of twice a number and -3 is multiplied by 4, the answer is 28. Find the number.

2. If the product of a number and -3 is reduced by 5, the result is 25 less than twice the opposite of the number. Find the number.

3. $2\frac{5}{8}$ of what number is 14?

4. What fraction of $3\frac{3}{4}$ is $22\frac{1}{2}$?

5. 2.625 of what number is 8.00625?

Solve:

6. $3\frac{1}{4}n - \frac{2}{5} = 3$

7. $x - 3(x - 2) = 7x - (2x + 5)$

8. $-3m - 3 + 5m - 2 = -(2m + 3)$

9. $0.2k - 4.21 - 0.8k = 2(-k + 0.1)$

Factor the greatest common factor:

10. $12a^2x^5y^7 - 3ax^2y^2$

11. $15a^5x^4y^6 + 3a^4x^3y^7 - 9a^2x^6y$

Simplify (factor if necessary):

12. $\dfrac{4x^2 - 4x}{4x}$

13. $\dfrac{2 - 6x}{2}$

14. $\dfrac{x^2ym + xym}{xym}$

15. $\dfrac{x^2y - xy}{xym}$

16. Write 750 as a product of prime factors.

Simplify:

17. -3^{-2}

18. $(-3)^{-2}$

19. $\dfrac{1}{-3^{-2}} - \sqrt[3]{8}$

20. $\dfrac{1}{(-3)^2}$

21. Use six unit multipliers to convert 500 cubic inches to cubic meters.

22. Find the surface area of this rectangular solid in square feet.

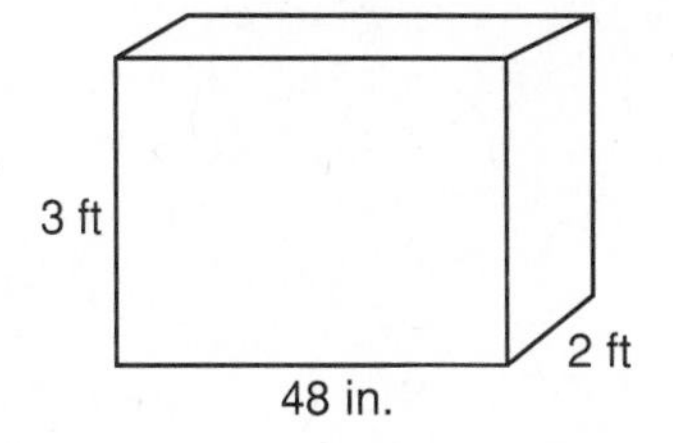

Expand:

23. $\dfrac{ax}{c^2}\left(\dfrac{b^4}{xk} - 2b^2\right)$

24. $(p^0x^2 - 4p^{-6}xy^5)x^{-2}$

Simplify by adding like terms:

25. $-xym^2 + 6ym^2x - 3x^2ym^2 - 9yx^2m^2$

26. $3x^4x^{-3}y^0 + xy^0y^{-2}y^2 - 7x^4x^{-3}p^0$

Evaluate:

27. $m(a^0 - ma)(-m) + |m^2 - 2|$ if $m = 2$ and $a = -4$

28. $k^3 - k(a)^2$ if $k = -3$ and $a = 2$

29. $-mx(a - x) - a$ if $m = -3$, $x = 2$, and $a = 2$

30. Simplify: $\dfrac{-3^2 + 4^2 - 5(4 - 2)}{3^0(5 - 2)}$

LESSON 37 Graphing inequalities

37.A inequalities

We use the symbols

(a) $\neq$ (b) $>$ (c) $<$

to designate that quantities are not equal, and we say that these symbols are symbols of **inequality.** They can be read from left to right or from right to left. We read

(d) $4 \neq 5$

from left to right as "4 is not equal to 5" or from right to left as "5 is not equal to 4." The symbols $>$ and $<$ are inequality symbols and are also called greater than/less than symbols. The small or pointed end is read as "less than" and the big or open end is read as "greater than." When we read, we read only one end of the symbol, the end that we come to first. Thus we read

(e) $4 > 2$

from left to right as "4 is greater than 2" or from right to left as "2 is less than 4." If the sign is combined with an equals sign, only one of the conditions must be met. We read

(h) $4 \geq 2 + 2$

from right to left as "2 plus 2 is less than or equal to 4," or from left to right as "4 is greater than or equal to 2 plus 2." This combination symbol is also called an inequality symbol although half of it is an equals sign.

Inequalities can be false inequalities, true inequalities, or conditional inequalities.

(a) $4 + 2 \leq 3$ (b) $x + 2 \geq x$ (c) $x < 4$

Inequality (a) is false, (b) is true, and the truth or falsity of (c) depends on the replacement value used for the variable. If a number that we use as a replacement for the variable makes the inequality a true inequality, we say that the number is a **solution of the inequality** and say that the number **satisfies the inequality.** If the number that we use as a replacement for the variable makes the inequality a false inequality, then the number is not a solution of the inequality and does not satisfy the inequality. Since more than one number will often satisfy a given inequality, there is often more than one solution to the inequality.

> We call the **set of numbers** that will satisfy a given equation or inequality the **solution set** of the equation or inequality.

37.B greater than and less than

Zero is a real number and can be used to describe a distance of zero. Any other number that can be used to describe a physical distance is a positive real number, and the opposite of each of these numbers is a negative real number.

We use the number line to help us picture the way real numbers are ordered (are arranged in order) and to help us define what we mean by *greater than.* On this number line we have graphed 2 and 4.

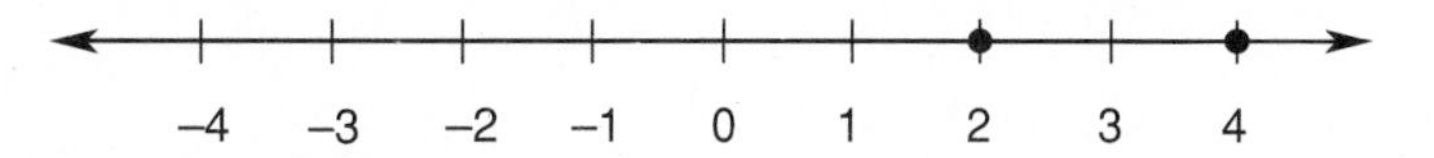

We remember that one number is greater than another number if its graph is to the right of the graph of the other number. Since the graph of 4 is to the right of the graph of 2, we say that 4 is greater than 2.

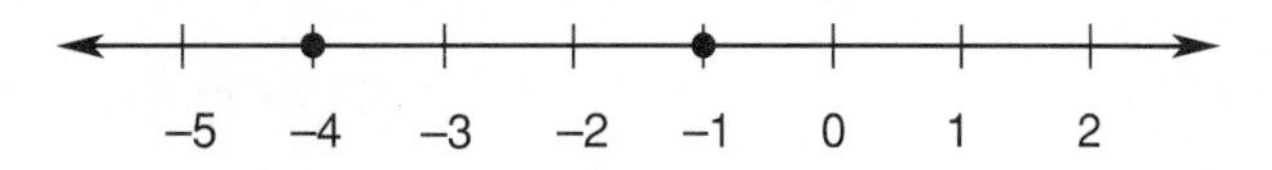

Using the same definition, we can say that -1 is greater than -4 because the graph of -1 is to the right of the graph of -4.

In the following section the small arrows will not be drawn on the ends of the number line because these arrows can be confused with the arrows drawn to indicate the solutions to the problems.

37.C graphical solutions of inequalities

We can use the number line to display the graph or the picture of the solution to many problems.

example 37.1 Graph: $x > 2$

solution This problem asks that we graph all numbers that are greater than 2. We draw an arrow to designate these numbers.

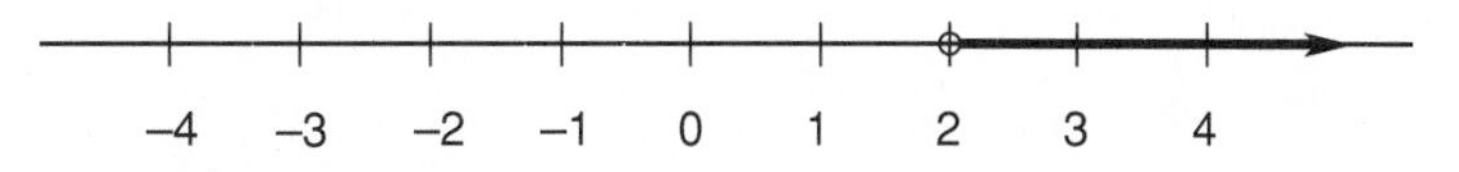

The open circle at 2 indicates that 2 is not a part of the solution because 2 is not greater than 2.

example 37.2 Graph the solution of $x \leq 2$.

solution This inequality is read from left to right as x **is less than or equal to 2.** Thus we are asked to show the location on the number line of all numbers that are equal to 2 or are less than 2.

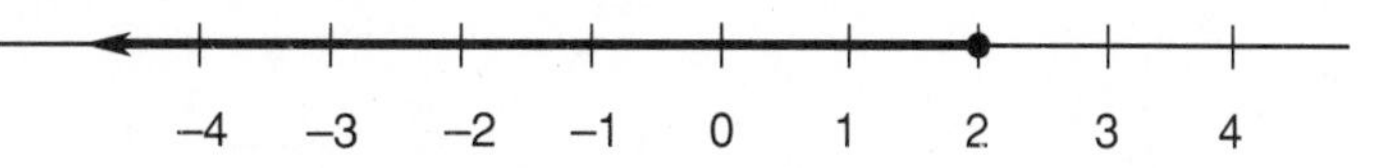

The locations of the numbers that satisfy the condition are indicated by the heavy line. The solid circle at 2 indicates that the number 2 is a part of the solution of $x \leq 2$.

example 37.3 Write the inequality whose graph is shown here.

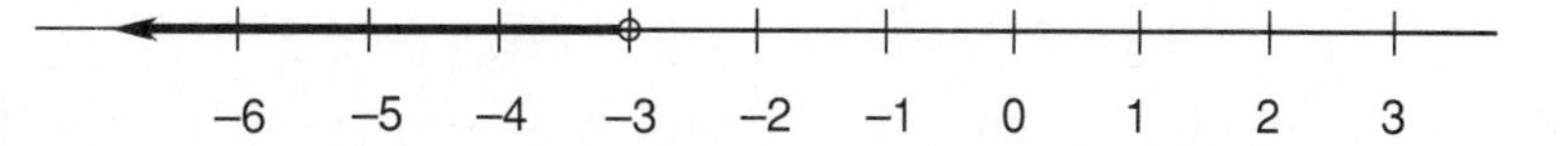

solution The graph indicates all real numbers that are less than -3, so the inequality is

$$x < -3$$

practice a. Graph $x \leq 5$ on a number line.

b. Write the inequality indicated by this graph.

problem set 37

1. If a number is multiplied by 3 and this product is reduced by 5, the result is 40. What is the number?

2. If a number is multiplied by 7 and this product is increased by 7, the result is 1 less than 9 times the number. What is the number?

3. $\frac{1}{4}$ of what number is $\frac{3}{8}$?

4. What decimal part of 41.25 is 2.475?

5. $\frac{1}{3}$ of $7\frac{1}{6}$ is what number?

Solve:

6. $-\frac{1}{5} + 2\frac{1}{2}p = 2$

7. $3(x - 2) - (2x + 5) = -2x + 10$

8. $4x - 3(x + 2) = 2x - 5$

9. $0.004m - 0.001m + 0.002 = -0.004$

Factor the greatest common factor:

10. $4a^2x^3y^5 - 8a^4x^2y^4$

11. $6a^2xm^5 + 2ax^4m^6 - 18a^5x^3m^5$

Simplify (factor if necessary):

12. $\frac{3x - 9}{3}$

13. $\frac{4px^2 - 8px}{px^2}$

14. Use five unit multipliers to convert 100 kilometers to miles. (Go from kilometers to meters to centimeters to inches to feet to miles.)

15. This figure is the base of a right solid whose sides are 50 cm high. Find the volume and the surface area of the solid. Dimensions are in centimeters. The sides that look parallel are parallel.

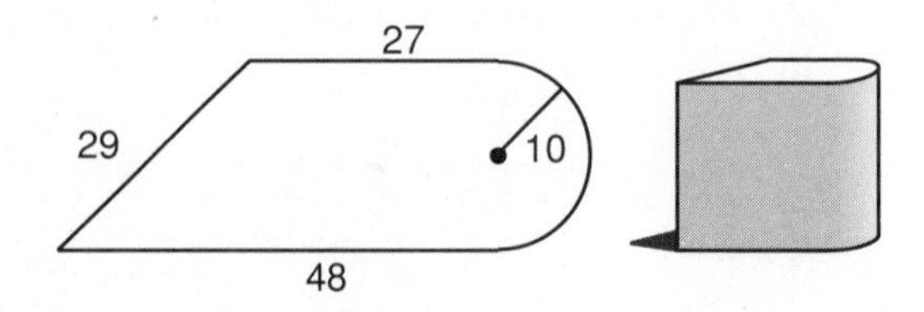

16. Draw a number line and graph $x < 2$.

17. Write the inequality indicated by this graph.

Simplify:

18. -3^{-3}

19. $(-3)^{-3}$

20. $\frac{1}{-5^{-2}} - \sqrt[3]{125}$

Expand:

21. $\frac{p^2}{x}\left(\frac{k^2p}{x^2} - \frac{p^2}{x}\right)$

22. $\left(\frac{a^2}{x} - \frac{2x}{a}\right)\frac{4x^2}{a}$

23. $\frac{ax}{c^2}\left(\frac{ax^2}{c} - \frac{3c^2a}{x^3}\right)$

24. $x^{-2}(x^2p^0 - 5x^4p^7)$

Simplify by adding like terms:

25. $x^3y^3p + px^3y^3 - 4x^2xyy^2p^2p^{-1}$

26. $5a^2x + 7xa^2 + aax^2x^{-1} - 2ax^3x^{-1}$

Evaluate:

27. $m^2 - (m - p)$ if $m = 2$ and $p = -2$

28. $a^2 - y^3(y - a)$ if $a = -2$ and $y = -3$

29. $k(x - ka)$ if $a = -2$, $k = 3$ and $x = -3$

30. Simplify: $\frac{-3^2 - (-3)^3 - 3}{-3(-3)(+3)}$

LESSON 38 ***Ratio***

When we write the numbers 3 and 4 separated by a fraction line as

$$\frac{3}{4}$$

we say that we have written the fraction three-fourths. Another name for a fraction is **ratio,** and we can also say that we have written the ratio of 3 to 4. All of the following ratios designate the same number and thus are equal ratios.

$$\frac{3}{4} \quad \frac{6}{8} \quad \frac{300}{400} \quad \frac{15}{20} \quad \frac{27}{36} \quad \frac{111}{148}$$

An equation or statement in which two ratios are equal is called a **proportion.** Thus, we can say that

$$\frac{3}{4} = \frac{15}{20}$$

is a proportion. We note that the cross products of equal ratios are equal as shown here.

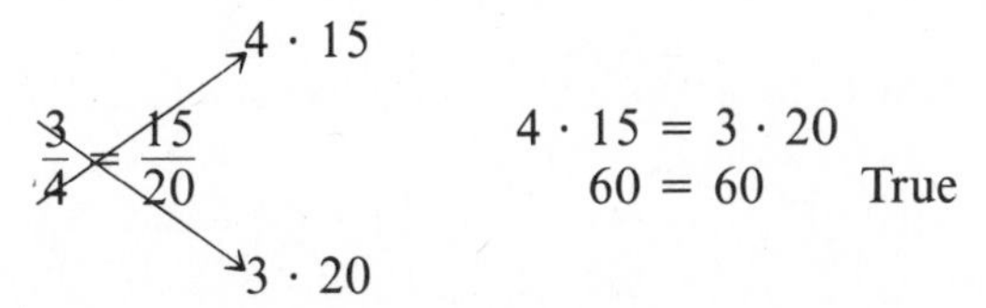

We can solve proportions that contain an unknown by setting the cross products equal and then dividing to complete the solution. To solve

$$\frac{7}{5} = \frac{91}{g}$$

we first set the cross products equal

$$7g = 5 \cdot 91$$

and then finish by dividing both sides by 7.

$$\frac{7g}{7} = \frac{5 \cdot 91}{7} \longrightarrow g = \frac{455}{7} \longrightarrow \mathbf{g = 65}$$

When we set the cross products equal, we say that we have *cross multiplied.*

example 38.1 Solve: $\frac{4}{m} = \frac{21}{5}$

solution We begin by setting the cross products equal.

$$4 \cdot 5 = 21m$$

Now we finish by dividing both sides by 21.

$$\frac{4 \cdot 5}{21} = \frac{\cancel{21}m}{\cancel{21}} \longrightarrow \mathbf{\frac{20}{21} = m}$$

We use proportions and cross multiplication to solve ratio word problems. In these problems, we wish to maintain a constant ratio between two things. We will use meaningful variables to represent the things and avoid the meaningless variables x, y, and z.

example 38.2 The ratio of pigs to goats in the barnyard was 7 to 5. If there were 91 pigs, how many goats were there?

solution We first note that we are comparing pigs and goats. Either one may be on top. If it is on top on one side, it must also be on top on the other side. We will demonstrate this by working the problem two ways.

$$\text{(a)}\ \frac{P}{G} = \frac{P}{G} \qquad \text{(b)}\ \frac{G}{P} = \frac{G}{P}$$

Now we read the problem and find that the ratio of pigs to goats was 7 to 5. So on the left side of both equations, we replace P with 7 and G with 5.

$$\text{(a)}\ \frac{7}{5} = \frac{P}{G} \qquad \text{(b)}\ \frac{5}{7} = \frac{G}{P}$$

Now we read the problem again and find that there were 91 pigs. Thus, we can replace P in both equations with 91.

$$\text{(a)}\ \frac{7}{5} = \frac{91}{G} \qquad \text{(b)}\ \frac{5}{7} = \frac{G}{91}$$

We solve both of these the same way—by cross multiplying and then dividing by 7.

$$7G = 5 \cdot 91 \longrightarrow 7G = 455 \longrightarrow \frac{7G}{7} = \frac{455}{7} \longrightarrow \mathbf{G = 65}$$

example 38.3 In the same barnyard, the ratio of chickens to ducks was 9 to 4, and there were 108 chickens. How many ducks were there?

solution Either chickens or ducks may go on top. We decide to put the ducks on top, so we write

$$\frac{D}{C} = \frac{D}{C}$$

The ratio of chickens to ducks was 9 to 4, so on the left we replace C with 9 and D with 4.

$$\frac{4}{9} = \frac{D}{C}$$

Finally, we replace C on the right with 108, and finish by cross multiplying and then dividing.

$$\frac{4}{9} = \frac{D}{108} \longrightarrow 9D = 4 \cdot 108 \longrightarrow \frac{9D}{9} = \frac{4 \cdot 108}{9} \longrightarrow \mathbf{D = 48}$$

practice

a. The ratio of neophytes to masters at the tryout was 7 to 2. If there were 714 neophytes, how many masters were there?

b. The crowd in the Belgrade town square was made up of Croatians and Serbs in the ratio of 5 to 9. If there were 18,000 Serbs, how many Croatians were there?

problem set 38

1. Use three unit multipliers to convert 17 miles to centimeters.

2. Find the surface area of the right circular cylinder in square centimeters. Dimensions are in centimeters.

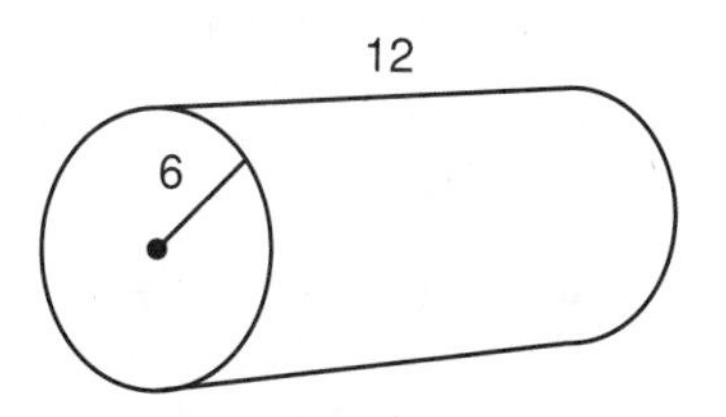

3. In a picaresque novel about the Spanish Main, the ratio of rascals to good guys was 13 to 5. If 600 were good guys, how many rascals were in the novel?

4. If the sum of twice a number and -7 is increased by 8, the result is 16 greater than the opposite of the number. What is the number?

5. 2.125 of what number equals 0.1275?

6. What fraction of $\frac{7}{8}$ is $2\frac{5}{11}$?

Solve:

7. $0.06 + 0.06x = -0.042$

8. $3\frac{1}{2}k + \frac{3}{4} = -\frac{7}{8}$

9. $2(5 - x) - (-2)(x - 3) = -(3x - 4)$

10. $3(-2x - 2 - 3) - (-x + 2) = -2(x + 1)$

11. $-x - 2(-x - 3) = -4 - x$

Factor the greatest common factor:

12. $3x^2y^3z^5 - 9xy^6z^6$

13. $4x^2y - 12xy^2 + 24x^3y^3$

Simplify (factor if necessary):

14. $\dfrac{7x + 7}{7}$

15. $\dfrac{k^4p - 2k^5p^2}{k^4p}$

16. Draw a number line and graph $x \geq -2$.

17. Write an inequality that describes this graph.

(number line: 3, 4; open circle at 4, arrow to the right)

Simplify:

18. $(-2)^{-2}$ **19.** -2^{-2} **20.** $(-2)^{-3}$

Expand:

21. $\frac{p^2}{x}\left(\frac{4x}{p^2} - \frac{p^4}{p^2x}\right)$ **22.** $\frac{4a^2}{x}\left(\frac{ax}{x} - \frac{3x}{a}\right)$

23. $\frac{mp}{k}\left(\frac{mp}{k} - \frac{k}{mp}\right)$ **24.** $x^{-3}(y^{-2}k^0 - 3xk^5)$

Simplify by adding like terms:

25. $3p^3x^{-4}xp^4 - 2pp^2p^3x^{-1}x^{-2} - 4x^2x^{-2}x^{-3}pp^6$

26. $xy - 3yx + 7x^3y^2x^{-2}y^{-1} - 2x^2yy^5y^{-4}y^{-1}x^{-1}$

Evaluate:

27. $(m - x^2)x - (-m)$ if $x = -2$ and $m = -3$

28. $(a^3 - y^2)(a - y)$ if $a = -3$ and $y = 4$

29. $x(x - y)(3 - 2xy)$ if $x = -2$ and $y = 5$

30. Simplify: $-2^2 + (-2)^3 - 2(-2 - 2) - 2$

LESSON 39 Trichotomy axiom · Negated inequalities · Advanced ratio problems

39.A trichotomy axiom

Johnny wrote a number on a piece of paper. Then he turned the paper over and wrote a number on the other side. There are exactly three possibilities.

1. The second number is the same number as the first number.
2. The second number is less than the first number.
3. The second number is greater than the first number.

While this is seemingly self-evident, it is not trivial. Mathematicians recognize that this property of real numbers reveals that the real numbers are an ordered set. Since this property has three parts, we give it the name **trichotomy.**†

> TRICHOTOMY AXIOM
>
> For any two real numbers a and b, exactly one of the following is true:
>
> $a = b \qquad a > b \qquad a < b$

† From the Greek *trikha* meaning "in three parts."

39.B negated inequalities

Symbols of inequality can be negated by drawing a slash through the symbol. We read

$$\text{(a)} \quad x \ngtr 10$$

from left to right as "x is not greater than 10." There are only three possibilities under the trichotomy axiom, and if x is not greater than 10, then it must be less than or equal to 10. So we can say the same thing by writing

$$\text{(b)} \quad x \leq 10$$

In the same way, if we write

$$\text{(c)} \quad x \ngeq 6$$

that x is not greater than or equal to 6, then x must be less than 6 because that's the only other possibility. Thus, both (c) and (d) make the same statement.

$$\text{(c)} \quad x \ngeq 6 \qquad \text{(d)} \quad x < 6$$

example 39.1 Graph the solution of $x \nless 2$.

solution If x is not less than 2, then x has to be equal to or greater than 2. Thus, the solution is the graph of $x \geq 2$.

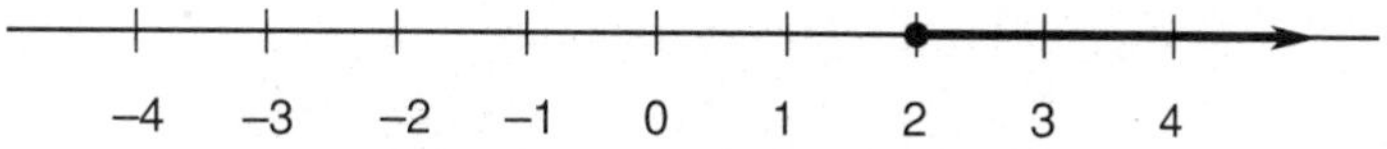

−4 −3 −2 −1 0 1 2 3 4

The solid circle at 2 indicates that 2 is a part of the solution of this inequality.

example 39.2 Write both an inequality and a negated inequality that describe this graph.

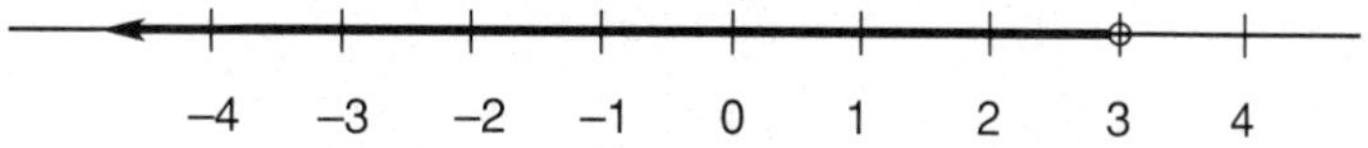

−4 −3 −2 −1 0 1 2 3 4

solution The graph indicates the real numbers that are less than 3. These are the real numbers that are not greater than or equal to 3.

$\boldsymbol{x \ngeq 3}$ means the same thing as $\boldsymbol{x < 3}$

39.C advanced ratio problems

Some ratio problems are difficult because key information is hidden by the way the problem is worded. If we are told that the ratio of red marbles to blue marbles is 5 to 7, we would write

$$\frac{R}{B} = \frac{5}{7}$$

Now if we are told that we have a total of 156 marbles and are asked for the number of marbles that are red, we would have difficulty because there is no place for total in the equation we have written. If we use the following four-step procedure, we can work any ratio problem with ease because this method will produce three useful equations.

1. Write the information given to include the total.
2. Use the cover-up method to write three equations.
3. Reread the problem to determine which equation to use.
4. Substitute in the selected equation and solve the problem.

example 39.3 The ratio of red marbles to blue marbles is 5 to 7. If there are 156 marbles in the bag, how many marbles are red?

solution **Step 1.** Write all the information to include the total. If 5 are red and 7 are blue, the total is 12.

$$R = 5$$

$$B = 7$$

$$T = 12$$

Step 2. Write all the implied equations. In this problem there are three implied equations. We can recognize the equations if we cover up part of the information with a finger, as we show here.

$B = 7$
$T = 12$ (a) $\frac{B}{T} = \frac{7}{12}$

$R = 5$
$T = 12$ (b) $\frac{R}{T} = \frac{5}{12}$

$R = 5$
$B = 7$ (c) $\frac{R}{B} = \frac{5}{7}$

Step 3. Now we reread the question. It says we have 156 total and asks for the number that are red. This tells us to use equation (b) because the variables in this equation are T and R.

$$\text{(b)} \quad \frac{R}{T} = \frac{5}{12}$$

Step 4. Now we substitute 156 for T and solve for R.

$$\frac{R}{156} = \frac{5}{12} \longrightarrow 12R = 5 \cdot 156 \longrightarrow \frac{12R}{12} = \frac{5 \cdot 156}{12} \longrightarrow \mathbf{R = 65}$$

example 39.4 The ratio of fish to crabs in the sea cave was 13 to 4. If there were 119 fish and crabs in the cave, how many were fish?

solution **The first step is very important.** If we record the information to include the total, the equations can be written by inspection.

$$F = 13$$

$$C = 4$$

$$T = 17$$

From this we can write the three equations

$$\text{(a)} \quad \frac{C}{T} = \frac{4}{17} \qquad \text{(b)} \quad \frac{F}{T} = \frac{13}{17} \qquad \text{(c)} \quad \frac{F}{C} = \frac{13}{4}$$

Since we are given the total and asked for fish, we will use equation (b), substitute 119 for total, and solve for F.

$$\frac{F}{119} = \frac{13}{17} \longrightarrow 17F = 13 \cdot 119 \longrightarrow \frac{17F}{17} = \frac{13 \cdot 119}{17} \longrightarrow \mathbf{F = 91}$$

practice

a. The team played 65 games. If the ratio of wins to losses was 3 to 2, how many games did the team win?

b. The ratio of hard problems to easy problems was 23 to 7. How many easy problems were there if the total number of problems was 930?

Draw two number lines and graph the solutions to these inequalities.

c. $x \not< -5$

d. $x \not\leq -2$

problem set 39

1. Courtney and Kristofer thought of the same number. They multiplied the number by 3 and then increased the product by 5 for a final result of -55. What number were they thinking of?

2. Twice the opposite of a number is increased by 5, and this sum is multiplied by 2. The result is 22 greater than the number. Find the number.

3. Use four unit multipliers to change 28,000 square inches to square miles.

4. The first four cows down the chute weighed 740 kg, 832 kg, 804 kg, and 760 kg. What was the average weight of the four cows?

5. $\frac{2}{5}$ of $23\frac{1}{3}$ is what number?

Solve:

6. $2\frac{1}{3}k - 4 = 7$

7. $p - 3(p - 4) = 2 + (2p + 5)$

8. $5x - 4(2x - 2) = 5 - x$

9. $0.02x - 4 - 0.01x - 2 = -6.3$

Factor the greatest common factor:

10. $12a^2x + 4ax^4$

11. $2x^2a^3y - xay + 4ay^2$

Simplify (factor first if necessary):

12. $\dfrac{3x - 9}{3}$

13. $\dfrac{5xy + 20xy^2}{5xy}$

14. The figure is the base of a right solid 5 feet high. How many 1-inch sugar cubes will the solid hold? What is the surface area of the solid? Dimensions are in inches.

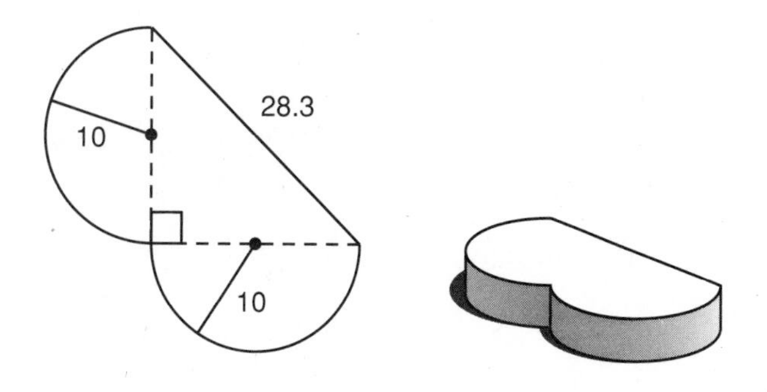

15. Express the perimeter in centimeters. All angles are right angles. Begin by changing meters to centimeters.

1 m

300 cm

16. Write a negated inequality that describes this graph.

1 3 5

17. Draw a number line and graph this inequality: $y \geq 3$

Simplify:

18. -2^{-2} **19.** $(-2)^{-2}$ **20.** $-(-2)^{-2}$

Expand:

21. $\left(\frac{p}{a} - \frac{p}{xa}\right)\frac{p^2}{a}$ **22.** $\frac{4x^2y}{m}\left(\frac{x^3}{m^2} - \frac{y}{m}\right)$

23. $\frac{mp}{x}\left(\frac{a}{x} - \frac{1}{x^2}\right)$ **24.** $p^{-2}x^0(p^2 - 3p^5x^5)$

Simplify by adding like terms:

25. $5yxp^2 - p^2yx + 2ppyx - 3p^2y^2y^{-1}x$ **26.** $3ay - 5ya - 6y^{-2}y^3a - 4a^3ay$

Evaluate:

27. $kp^0 - (k - p)$ if $k = -4$ and $p = 1$

28. $m^2 - m^{-3}(p)$ if $m = -2$ and $p = 3$

29. $a^2b - (a - b)$ if $a = -2$ and $b = -1$

30. Simplify: $\frac{-4(3 - 5) - 2^0}{(-2)^3 - 2(-2)}$

LESSON 40 Quotient rule for exponents

Let's review our rules and definitions for exponents.

DEFINITION:	$\underbrace{x \cdot x \cdot x \cdots x}_{n \text{ factors}} = x^n$
DEFINITION:	If x is not zero, $x^0 = 1$
DEFINITION:	$x^1 = x$
PRODUCT RULE:	If x is not zero, $x^m \cdot x^n = x^{m+n}$
DEFINITION:	If x is not zero, $x^{-n} = \frac{1}{x^n}$

The **quotient rule for exponents** is really an extension of the last definition above that says $x^{-n} = \frac{1}{x^n}$. If we wish to multiply

$$x^5 \qquad \text{times} \qquad \frac{1}{x^2}$$

we can use the definition of x^{-n} to write $\frac{1}{x^2}$ as x^{-2} and then multiply by using the product rule.

$$x^5 \cdot \frac{1}{x^2} = x^5 \cdot x^{-2} = x^{5-2} = \mathbf{x^3}$$

The quotient rule permits the same procedure in just one step.

QUOTIENT RULE FOR EXPONENTS

If m and n are real numbers and $x \neq 0$,

$$\frac{x^m}{x^n} = x^{m-n} = \frac{1}{x^{n-m}}$$

example 40.1 Simplify: $\frac{x^6}{x^4}$

solution We know that we can move the x^4 from the denominator to the numerator if we change the sign of the 4 from plus to minus.

$$\frac{x^6}{x^4} = x^6x^{-4} = x^{6-4} = \mathbf{x^2}$$

If we use the quotient rule, we can omit the first step and write

$$\frac{x^6}{x^4} = x^{6-4} = \mathbf{x^2}$$

example 40.2 Simplify $\frac{x^6}{x^4}$, but this time write the x in the denominator.

solution

$$\frac{x^6}{x^4} = \frac{1}{x^{4-6}} = \mathbf{\frac{1}{x^{-2}}}$$

example 40.3 Simplify: $\frac{x^6}{x^{-4}}$

solution We will work the problem twice. The first time we will use the quotient rule so that the x is in the numerator.

(a)
$$\frac{x^6}{x^{-4}} = x^{6+4} = \mathbf{x^{10}}$$

(b) This time we will put the x in the denominator.

$$\frac{x^6}{x^{-4}} = \frac{1}{x^{-4-6}} = \mathbf{\frac{1}{x^{-10}}}$$

example 40.4 Simplify: $\frac{x^{-a}}{x^b}$

solution We will use the quotient rule to simplify so that the x is in the numerator.

(a)
$$\frac{x^{-a}}{x^b} = x^{-a-b}$$

(b) And this time we will put the x in the denominator.

$$\frac{x^{-a}}{x^b} = \frac{\mathbf{1}}{\boldsymbol{x^{b+a}}}$$

The two answers shown in (a) and (b) are **equivalent expressions,** which means that they have the same value no matter which nonzero real number is used as a replacement for x. Since they are equivalent expressions, neither expression can be designated as the preferred answer because the preference of one person will not necessarily be the same as the preference of another person.

example 40.5 Simplify: $\dfrac{x^{-5}y^6z}{z^{-3}y^2x}$

solution We will find four equivalent expressions for this expression.

(a) $\boldsymbol{x^{-6}y^4z^4}$ (b) $\boldsymbol{\dfrac{1}{x^6y^{-4}z^{-4}}}$ (c) $\boldsymbol{\dfrac{y^4z^4}{x^6}}$ (d) $\boldsymbol{\dfrac{x^{-6}}{y^{-4}z^{-4}}}$

Answer (a) is written with all variables in the numerator; (b) has all variables in the denominator; (c) has all exponents positive; and (d) has all exponents negative. No one of these forms is more correct than another. We will emphasize this by using different forms for the answers in the back of the book.

practice Simplify. Write all variables in the numerator.

a. $\dfrac{p^{-4}y^5z}{z^{-5}y^3p^2}$ **b.** $\dfrac{x^{-6}y^4z^5}{x^{-6}y^5z^2}$ **c.** $\dfrac{m^4p^3z^{10}d}{m^{-2}p^4z^{-6}d^{-2}}$

problem set 40

1. Seven times a number is decreased by 4 for a result of -25. What is the number?

2. Three times the opposite of a number is increased by 16, and this result is 26 greater than the number. Find the number.

3. The ratio of gaudy scarves to tawdry scarves was 7 to 11. If there were 2520 scarves in the pile, how many were merely gaudy?

4. What fraction of $2\frac{1}{8}$ is 6?

5. 1.205 of 3.2 is what number?

Solve:

6. $\frac{2}{3}k + 5 = 12$

7. $x - 5x + 4(x - 2) = 3x - 8$

8. $2p - 5(p - 4) = 2p + 12$

9. $0.4x - 0.02x + 1.396 = 0.598$

Factor the greatest common factor:

10. $3x^2y^5p^6 - 9x^2y^4p^3 + 12x^2yp^4$

11. $2x^2y^2 - 6y^2x^4 - 12xy^5$

Simplify (factor if necessary):

12. $\dfrac{5x^2 - 25x}{5x}$

13. $\dfrac{4xy + 16x^2y^2}{4xy}$

14. Graph: $x \ngeq -2$

15. Write 360 as a product of prime factors.

Simplify:

16. -2^{-4}

17. $\dfrac{1}{-(-3)^{-3}} - \sqrt[3]{-27}$

18. Use four unit multipliers to convert 200 square miles to square inches.

19. The figure shown is the base of a right solid that is 1 foot high. How many 1-inch sugar cubes will this solid hold? What is the surface area? Dimensions are in inches. Sides that look parallel are parallel.

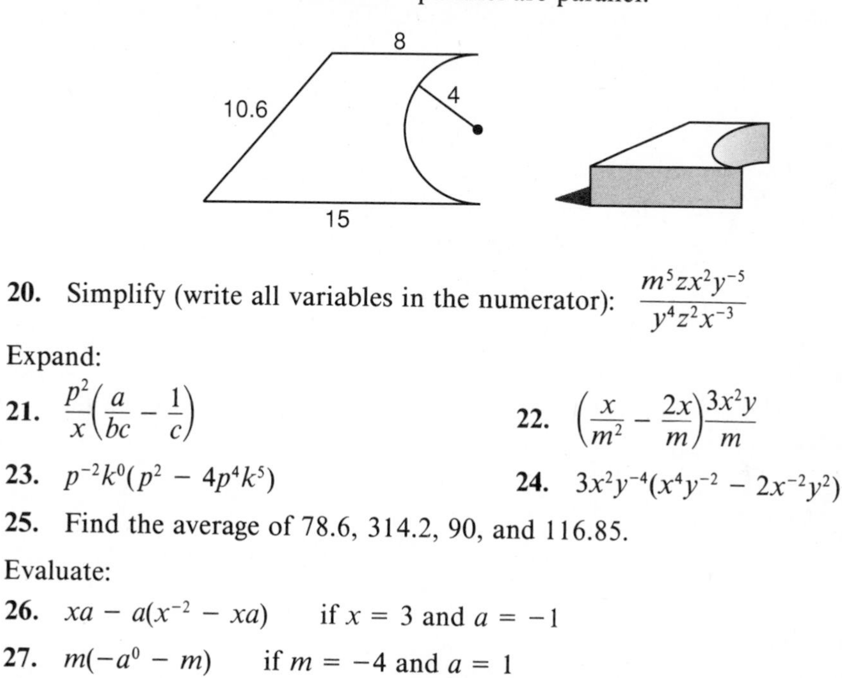

20. Simplify (write all variables in the numerator): $\dfrac{m^5zx^2y^{-5}}{y^4z^2x^{-3}}$

Expand:

21. $\dfrac{p^2}{x}\left(\dfrac{a}{bc} - \dfrac{1}{c}\right)$

22. $\left(\dfrac{x}{m^2} - \dfrac{2x}{m}\right)\dfrac{3x^2y}{m}$

23. $p^{-2}k^0(p^2 - 4p^4k^5)$

24. $3x^2y^{-4}(x^4y^{-2} - 2x^{-2}y^2)$

25. Find the average of 78.6, 314.2, 90, and 116.85.

Evaluate:

26. $xa - a(x^{-2} - xa)$ if $x = 3$ and $a = -1$

27. $m(-a^0 - m)$ if $m = -4$ and $a = 1$

28. $p - (m - pm)$ if $p = -3$ and $m = 4$

29. Simplify: $\dfrac{-2^0(-5 - 7)(-3) - |-4|}{-2(-(-6))}$

LESSON 41 *Distributive property and negative exponents*

In the preceding five problem sets, we have used the distributive property to expand expressions that contain fractions, such as

$$\frac{4x^2}{y^4}(xy + 2y^2x^3)$$

In this lesson, we will do the same expansions, but now we will also consider expressions that contain negative exponents.

example 41.1 Expand and write the answer with all exponents positive: $\frac{4x^{-2}}{y^4}\left(y^4x^2 - \frac{3x^4}{y^{-2}}\right)$

solution We will use two steps. First we multiply to get

$$\frac{4x^{-2}y^4x^2}{y^4} - \frac{12x^{-2}x^4}{y^4y^{-2}}$$

Now we simplify and write the answer with all exponents positive.

$$\mathbf{4 - \frac{12x^2}{y^2}}$$

example 41.2 Multiply and write the product with all variables in the numerator:

$$x^{-2}y\left(\frac{y^4}{x^2} - \frac{x^4}{y^2}\right)$$

solution We begin by using the distributive property to multiply.

$$x^{-2}y\left(\frac{y^4}{x^2} - \frac{x^4}{y^2}\right) = \frac{x^{-2}yy^4}{x^2} - \frac{x^{-2}yx^4}{y^2} = \frac{y^5}{x^4} - \frac{x^2}{y}$$

Now we can write this result with all variables in the numerator as

$$\mathbf{x^{-4}y^5 - x^2y^{-1}}$$

example 41.3 Multiply and write the product with all variables in the denominator:

$$\frac{k^2b}{p^{-2}}\left(\frac{ab^{-1}}{k^2} - \frac{4p^2}{b}\right)$$

solution Again we begin by using the distributive property to multiply.

$$\frac{k^2bab^{-1}}{p^{-2}k^2} - \frac{4p^2k^2b}{p^{-2}b}$$

Now we simplify and write the product with all variables in the denominator and get

$$\mathbf{\frac{1}{a^{-1}p^{-2}} - \frac{4}{p^{-4}k^{-2}}}$$

practice Expand. Write all variables in the numerator.

a. $\frac{z^{-3}}{m}\left(\frac{x^4}{m^2} - \frac{3z}{m}\right)$ **b.** $\left(\frac{m^{-5}}{w^4x} - \frac{cw^2}{x^3m}\right)\frac{m^{-2}}{3cm}$

problem set 41

1. Three more than 5 times a number is −27. What is the number?
2. If 7 times a number is decreased by 5 and this difference is doubled, the result is 14 less than twice the number. What is the number?
3. War Eagle spied 1428 antelope and wildebeests grazing on the savannah. If they were in the ratio of 9 to 5, how many antelope were there?
4. What fraction of 72 is 16?

5. Write 130 as a product of prime factors.

Solve:

6. $\frac{3}{5}p + 7 = 22$

7. $3x - (x - 2) + 5 = 4x + 6$

8. $3p - 2(p - 4) = 7p + 6$

9. $0.004k - 0.002 + 0.002k = 4$

Factor the greatest common factor:

10. $4x^2m^5y - 2x^4m^3y^3$

11. $4m^2x^5 - 2m^2x^2 + 6m^5x^2$

Simplify:

12. $\dfrac{4 - 4x}{4}$

13. $\dfrac{9x - 3x^2}{3x}$

14. Graph: $x \leq -5$

15. Write two inequalities that describe this graph.

−3 −2 −1

Simplify:

16. -3^{-2}

17. $\dfrac{1}{(-3)^2}$

Simplify. Write all variables with positive exponents.

18. $\dfrac{x^3y^2}{xy^4}$

19. $\dfrac{x^3y^{-3}z}{z^5x^2y}$

20. $\dfrac{x^{-4}y^{-3}p^2}{x^{-5}yp^4}$

21. Use five unit multipliers to convert 390 miles to kilometers.

22. Find the area of the figure in square centimeters. Dimensions are in centimeters.

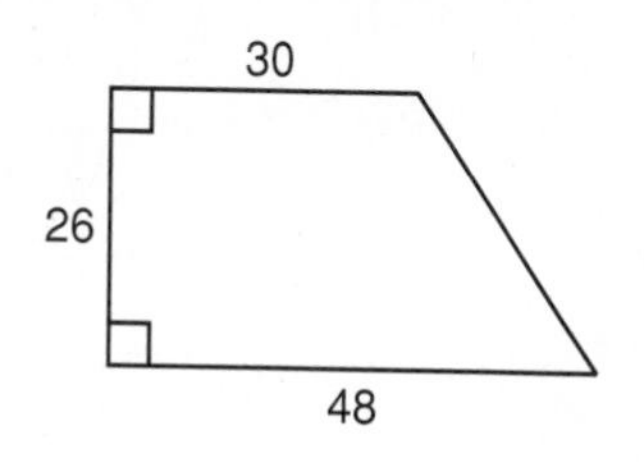

Expand. Write the answers with positive exponents.

23. $\left(\dfrac{x^{-4}}{a^3} - \dfrac{a^3}{x}\right)\dfrac{a^{-3}}{x}$

24. $\dfrac{m^{-2}}{b}\left(\dfrac{b^2}{m^3} - \dfrac{4am^2}{b^4}\right)$

Simplify by adding like terms:

25. $x^2yp - 4xxyp - 3x^2py$

26. $3x^2ym - 2m^2x^2y - 5x^2my + 4ym^2x^2$

Evaluate:

27. $m - 3m^{-3}$ if $m = -3$

28. $a(b^0 - ab)$ if $a = 3$ and $b = -5$

29. $x - (-y) - y^2$ if $x = -2$ and $y = 3$

30. Simplify: $-3 - 2^0|-4 - 3| - 2(-2)$

LESSON 42 Like terms and negative exponents · Two-step problems

42.A adding like terms

Sometimes it is difficult to determine if terms in an expression are like terms. For instance, if we look at the expression

$$\frac{bx}{x^3y^{-2}} - 3by^2x^{-2} + \frac{4y^2}{bx^2}$$

it is rather difficult to see that two of the terms are like terms and thus may be added. If each of the terms is written in the same form, however, it is easy to identify like terms. Let's write each of the three terms with all exponents positive.

$$\frac{by^2}{x^2} - \frac{3by^2}{x^2} + \frac{4y^2}{bx^2}$$

We see that the first two terms are like terms and may be added. The last term is different and thus cannot be added.

$$\frac{by^2}{x^2} - \frac{3by^2}{x^2} + \frac{4y^2}{bx^2} = \mathbf{-\frac{2by^2}{x^2} + \frac{4y^2}{bx^2}}$$

Now we will work the problem again by first writing the original terms so that all exponents are negative.

$$\frac{x^{-2}}{b^{-1}y^{-2}} - \frac{3x^{-2}}{b^{-1}y^{-2}} + \frac{4b^{-1}x^{-2}}{y^{-2}} = \mathbf{-\frac{2x^{-2}}{b^{-1}y^{-2}} + \frac{4b^{-1}x^{-2}}{y^{-2}}}$$

Again we see that the first two terms are like terms and may be added. To see if terms are like terms, we can put them in any form we wish as long as we use the same form for every term. However, many people feel more comfortable with all exponents positive.

example 42.1 Add like terms: $\frac{bx}{x^3y^{-2}} - 3by^2x^{-2} + \frac{4y^2}{b^{-1}x^2}$

solution To help us identify like terms, we will rewrite each term so that all exponents are positive.

$$\frac{by^2}{x^2} - 3\frac{by^2}{x^2} + \frac{4by^2}{x^2}$$

We see that all three terms are like terms and can be added by adding the numerical coefficients.

$$\frac{by^2}{x^2} - \frac{3by^2}{x^2} + \frac{4by^2}{x^2} = \mathbf{\frac{2by^2}{x^2}}$$

example 42.2 Add like terms: $\frac{a^{-3}b}{b^{-3}} + \frac{2b^4}{a^3}$

solution We begin by writing each term with all exponents positive.

$$\frac{b^4}{a^3} + \frac{2b^4}{a^3}$$

Now we see that the terms are like terms and may be added.

$$\frac{b^4}{a^3} + \frac{2b^4}{a^3} = \mathbf{\frac{3b^4}{a^3}}$$

example 42.3 Add like terms: $\dfrac{7a^{-3}b^2}{c^{-1}} - \dfrac{5b^2}{a^3c^{-1}} + \dfrac{3b^2}{a^3c}$

solution We begin by writing each term with all exponents positive.

$$7\frac{cb^2}{a^3} - 5\frac{cb^2}{a^3} + 3\frac{b^2}{a^3c}$$

Now we see that the first two terms are like terms and may be added but that the third term is different. Thus the simplification is

$$\mathbf{\frac{2cb^2}{a^3} + \frac{3b^2}{a^3c}}$$

42.B two-step problems

Some problems require two steps for their solution. The answer for the first step is used in the second step.

example 42.4 If $x + 3 = 7$, what is the value of $x - 8$?

solution First we solve $x + 3 = 7$ to find x.

$$\begin{array}{rcl} x + 3 & = & 7 \\ -3 & & -3 \\ \hline x & = & 4 \end{array} \quad \text{add } -3 \text{ to both sides}$$

Now we use 4 for x to find the value of $x - 8$.

$x - 8$	expression
$4 - 8$	substitution
-4	simplified

example 42.5 The average of four numbers is 8. Three of the numbers are 2, 4, and 7. What is the fourth number?

solution If the average of 4 numbers is 8, the sum of the numbers is 4 times 8, or 32. The sum of the given numbers is only 13, so the fourth number has to be **19** because

$$(2 + 4 + 7) + 19 = 32$$

practice Simplify by adding like terms:

a. $x^{-2}y + \dfrac{3y}{x^2} - 5xy$ **b.** $\dfrac{a^{-8}b^2}{b^{-9}} - \dfrac{4b^{11}}{a^8} + \dfrac{6b^{11}}{a^8b^5}$

c. If $x - 5 = 7$, what is the value of $x + 4$?

d. The average of three numbers is 10. Two of the numbers are 4 and 7. What is the other number?

problem set 42

1. Three less than 7 times a number is -31. What is the number?

2. The sum of 7 times a number and -3 is multiplied by 2. This result is 114 greater than the opposite of the number. Find the number.

3. The ratio of poseurs to outright frauds was 14 to 3. If they totaled 2244, how many were poseurs?

4. What decimal part of 7 is 14.14?

5. $2\frac{1}{4}$ of $3\frac{5}{8}$ is what number?

Solve:

6. $\frac{1}{2} + \frac{1}{8}x - 5 = 10\frac{1}{2}$

7. $4(x - 2) - 4x = -(3x + 2)$

8. $-5x + 2 = -2(x - 5)$

9. $0.3z - 0.02z + 0.2 = 1.18$

Factor the greatest common factor:

10. $6k^5m^2 - 2mk^3 - mk$

11. $x^4y^2m - x^3y^3m^2 + 5x^6y^2m^2$

Simplify:

12. $\dfrac{3x - 9x^2}{3x}$

13. $\dfrac{4xy - 4x}{4x^2}$

14. Graph: $x \not> 5$

15. Write the inequality indicated by this graph.

2 4

16. Write 280 as a product of prime factors.

Simplify. Write all variables in the numerator.

17. $\dfrac{x^2y^5}{x^4y^{-3}m^2}$

18. $\dfrac{x^5y^5mm^{-2}}{xx^3y^{-3}m^4}$

19. $\dfrac{x^2xyp^{-5}}{p^{-3}p^{-4}y^{-4}}$

Expand. Write answers with positive exponents.

20. $\dfrac{x^{-2}}{y}\left(\dfrac{xz}{y} - \dfrac{1}{y^{-4}}\right)$

21. $\left(\dfrac{a}{b} - \dfrac{2b}{a}\right)\dfrac{a^{-2}}{b^{-2}}$

22. Use three unit multipliers to convert 80 kilometers to inches.

23. If $x - 9 = 4$, what is the value of $x - 19$?

24. Find the volume and the surface area of this right prism. Dimensions are in meters.

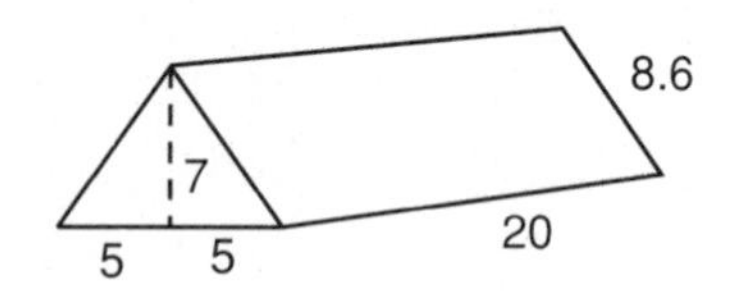

25. Simplify by adding like terms: 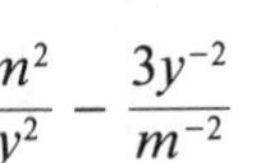 $\dfrac{m^2}{y^2} - \dfrac{3y^{-2}}{m^{-2}}$

Evaluate:

26. $-a^{-3}(a - a^2x)$ if $a = -2$ and $x = 4$

27. $b - (-c^0)$ if $b = -2$ and $c = 4$

28. $k^3 - (k - c)$ if $k = -2$ and $c + 2 = 6$

Simplify:

29. $|-3^{-3}|$

30. $\dfrac{-(-2 - 5) - (-3 - 6)}{-2^0(-4)(-2)}$

LESSON 43 Solving multivariable equations

When we are asked to solve an equation in one unknown, such as

$$12 + 4x - 3 + 4 - 2x = 6x - 3 - 2 + 5x$$

we are asked to simplify both sides and to finally write the equation with x all by itself on one side and a number on the other side. When we do this, we say that we have **isolated x** on one side of the equation. When we have isolated x in this problem, we get

$$x = 2$$

If an equation contains more than one variable and we are asked to solve the equation for one of the variables, our task is the same as that described above. **We are asked to rearrange the equation so that the designated variable is the sole member of one side of the equation (either side).** In the following problems, however, the other side of the equation will contain variables as well as numbers.

example 43.1 Solve for y: $6y - x + z = 4$

solution We will begin the process of **isolating** y by eliminating $-x$ and $+z$ from the left-hand side of the equation.

$$\begin{array}{rl} 6y - x + z = 4 & \text{original equation} \\ +x - z \quad\quad +x - z & \text{add } +x - z \text{ to both sides} \\ \hline 6y \quad\quad = 4 + x - z & \end{array}$$

Now we complete the isolation by dividing every term by 6.

$$\frac{6y}{6} = \frac{4}{6} + \frac{x}{6} - \frac{z}{6} \longrightarrow \mathbf{y = \frac{2}{3} + \frac{x}{6} - \frac{z}{6}}$$

example 43.2 Solve for y: $4x - 2y + 2 = y - 4$

solution The first step is to eliminate the y term on one side or the other. We choose to eliminate the $-2y$, so we add $+2y$ to both sides.

$$\begin{array}{rl} 4x - 2y + 2 = \quad y - 4 & \text{original equation} \\ +2y \quad\quad +2y & \text{add } +2y \text{ to both sides} \\ \hline 4x \quad\quad + 2 = \quad 3y - 4 & \end{array}$$

Now we have all the y's on the right-hand side. To **isolate** y on the right, we must eliminate the -4 and the 3 that are on the right-hand side. To eliminate the -4, we add $+4$ to both sides.

$$\begin{array}{rl} 4x + 2 = 3y - 4 & \\ +4 \quad\quad +4 & \text{add } +4 \text{ to both sides} \\ \hline 4x + 6 = 3y & \end{array}$$

Now we complete the isolation of y by dividing every term by 3.

$$\frac{4x}{3} + \frac{6}{3} = \frac{3y}{3} \qquad \text{divide by 3}$$

$$\mathbf{\frac{4x}{3} + 2 = y} \qquad \text{simplified}$$

example 43.3 Solve for p: $4p + 2a - 5 = 6a + p$

solution We begin by eliminating the p on the right-hand side of the equation.

$$\begin{array}{rl} 4p + 2a - 5 = 6a + p & \text{original equation} \\ \underline{-p \qquad\qquad\qquad -p} & \text{add } -p \text{ to both sides} \\ 3p + 2a - 5 = 6a \quad & \end{array}$$

Now we eliminate the $+2a$ and -5 on the left side.

$$\begin{array}{rl} 3p + 2a - 5 = \quad 6a & \\ \underline{-\ 2a + 5 \quad -2a + 5} & \text{add } -2a + 5 \text{ to both sides} \\ 3p \qquad\qquad = \quad 4a + 5 & \end{array}$$

As the final step, we divide every term by 3 and get

$$\boldsymbol{p = \frac{4}{3}a + \frac{5}{3}}$$

example 43.4 Solve for x: $5y + x - 2y - 4 + 3x = 0$

solution We will begin by adding like terms. Then we will eliminate the $3y$ and the -4 by adding $-3y$ and $+4$ to both sides.

$$\begin{array}{rl} 3y + 4x - 4 = \quad 0 & \text{added like terms} \\ \underline{-3y \qquad + 4 = -3y + 4} & \text{add } -3y + 4 \text{ to both sides} \\ 4x \qquad = -3y + 4 & \end{array}$$

Now we complete the process of isolating y by dividing every term by 4.

$$\frac{4x}{4} = \frac{-3y}{4} + \frac{4}{4} \qquad \text{divide by 4}$$

$$\boldsymbol{x = -\frac{3}{4}y + 1} \qquad \text{simplified}$$

example 43.5 Solve for y: $4y + 6x - 4 = 2$

solution Since only one term contains a y, we begin by moving all other terms to the right-hand side.

$$\begin{array}{rl} 4y + 6x - 4 = \quad 2 & \text{original equation} \\ \underline{-\ 6x + 4 \quad +4 - 6x} & \text{add } +4 - 6x \text{ to both sides} \\ 4y \qquad\qquad = \quad 6 - 6x & \end{array}$$

$$\frac{4y}{4} = \frac{6}{4} - \frac{6x}{4} \qquad \text{divide by 4}$$

$$\boldsymbol{y = \frac{3}{2} - \frac{3}{2}x} \qquad \text{simplified}$$

practice

a. Solve for y: $8y - 13x - 8 = 4$

b. Solve for p: $8p + 3w = w - 15 - 2p$

problem set 43

1. If the sum of twice a number and -10 is multiplied by -4, the result is 61 greater than the opposite of the number. What is the number?

2. The defense budget was spent on halberds and other armor in the ratio of 2 to 19. If the total budget was 84,000 farthings, how much went for halberds?

3. What fraction of 30 is 18?

Solve:

4. $2\frac{1}{3}x + 5 = 19$

5. $3(-x - 4) = 2x + 3(x - 5)$

6. $-(5 - 2x) + x = 7(x - 2)$

7. $-(0.2 - 0.4z) - 0.4 = z - 1.47$

8. Use five unit multipliers to convert 400 kilometers to miles.

9. Find the volume in cubic feet of a right solid whose base is shown and whose sides are 2 yards high. Find the surface area. Dimensions are in feet.

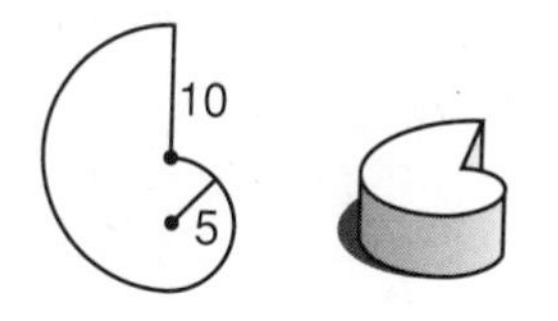

10. The four numbers averaged 58.8. Find the fourth number if the first three numbers were 60.6, 90.08, and 56.92.

11. If $x - 4 + 2x - 5 = 6$, find the value of $3x - 2$.

Solve for y:

12. $3x + 2y = 5 - y$

13. $-2y + 6y - x - 4 = 0$

Factor the greatest common factor:

14. $6x^4y^2 - 4zx^2y^2$

15. $8x^5y^2z - 16x^2y^2z^2 - xyz$

Simplify:

16. $\dfrac{4x - 8xy}{4x}$

17. $\dfrac{5x^2y^2 - 25x^3y^3}{x^2y^2}$

18. Graph: $x \not\leq 2$

19. Write 1125 as a product of prime factors.

Simplify. Write all variables with positive exponents.

20. $\dfrac{x^5yx^{-7}y^2}{x^4yy^3x^3}$

21. $\dfrac{4x^{-2}y^{-6}m}{x^5y^5m^{-4}}$

22. Expand: $x^2z^{-2}\left(\dfrac{x^4z^{-4}}{x} - \dfrac{3z^2}{x^2}\right)$

Simplify by adding like terms:

23. $\dfrac{3x^{-2}x^3y}{y^{-4}} - 2xy^5$

24. $\dfrac{3x^{-2}y^2}{m^{-2}} - \dfrac{5m^2y^2}{x^2} + \dfrac{2my^2}{m^{-1}x^2}$

25. $xy^2 - \dfrac{3xy}{y^{-1}} + \dfrac{2x^0y^2}{x^{-1}} - \dfrac{4x^2}{y^2} + 2x^2y^{-2}$

Evaluate:

26. $|x| - x(y)(-x)$ if $x = -2$ and $y + 4 = 8$

27. $cx - c^3x$ if $c = -3$ and $x = 5$

28. $ab^0(a^{-3} - b^{-2})$ if $a = -2$ and $b = \sqrt[3]{27}$

Simplify:

29. $-3 - \dfrac{1}{3^{-2}}$

30. $\dfrac{-4[(-2 + 5) - (-3 + 8)]}{-2^0|5 - 1|}$

LESSON 44 Least common multiple

If we are given the numbers

$$4, \qquad 5, \qquad \text{and} \qquad 8$$

and are asked to find the **smallest number that is evenly divisible by each of the numbers,** a reasonable guess would be the product of the numbers, which is 160, because we know that each of the numbers will divide 160 evenly.

$$\frac{160}{4} = 40 \qquad \frac{160}{5} = 32 \qquad \frac{160}{8} = 20$$

But 160 is not the smallest number that is evenly divisible by the three numbers. The number 40 is.

$$\frac{40}{4} = 10 \qquad \frac{40}{5} = 8 \qquad \frac{40}{8} = 5$$

We call the smallest number that can be divided evenly by each of a group of specified numbers the *least common multiple* of the specified numbers.

We can give a two-step procedure for finding the least common multiple.

1. Write each of the given numbers as a product of prime factors.
2. Write the least common multiple as a product of factors in which **each factor of the numbers appears as many times as it is used as a factor in any one of the original numbers.**

Thus to find the least common multiple of 4, 5, and 8, we first write each of the numbers as a product of prime factors.

$$\overbrace{2 \cdot 2}^{4} \qquad \overbrace{5}^{5} \qquad \overbrace{2 \cdot 2 \cdot 2}^{8}$$

Now, 2 is a factor of two of the numbers, and it appears three times as a factor of 8. Thus it must appear three times as a factor of the least common multiple.

$$2 \cdot 2 \cdot 2$$

Every factor of the original numbers must be a factor of the least common multiple, so 5 must be a factor. Thus the least common multiple of 4, 5, and 8 is 40 because

$$5 \cdot 2 \cdot 2 \cdot 2 = \mathbf{40}$$

example 44.1 Find the least common multiple (LCM) of 100, 15, and 8.

solution We begin by writing each of the numbers as a product of prime numbers.

$$\begin{array}{ccc} 100 & 15 & 8 \\ 2 \cdot 2 \cdot \underbrace{5 \cdot 5} & 3 \cdot 5 & \underbrace{2 \cdot 2 \cdot 2} \end{array}$$

We note that 5 is a factor twice, 3 is a factor once, and 2 is a factor three times. Thus the LCM is

$$\underbrace{5 \cdot 5} \cdot \underbrace{3} \cdot \underbrace{2 \cdot 2 \cdot 2} = \mathbf{600}$$

So 600 is the smallest number that is evenly divisible by each of the three numbers, 100, 15, and 8.

example 44.2 Find the LCM of 80, 75, and 30.

solution We begin by writing each of the numbers as a product of prime numbers.

$$\begin{array}{ccc} 80 & 75 & 30 \\ \underbrace{2 \cdot 2 \cdot 2 \cdot 2} \cdot 5 & \underbrace{3} \cdot \underbrace{5 \cdot 5} & 2 \cdot 3 \cdot 5 \end{array}$$

So the LCM is

$$\underbrace{2 \cdot 2 \cdot 2 \cdot 2} \cdot \underbrace{3} \cdot \underbrace{5 \cdot 5} = \mathbf{1200}$$

practice

Find the least common multiple of:

a. 20, 30, and 14 **b.** 18, 8, and 27 **c.** 40, 50, and 3

problem set 44

1. If twice a number is increased by 5 and this sum is multiplied by -3, the result is -57. What is the number?
2. The village was polyglot. If the ratio of bilingual denizens to trilingual denizens was 14 to 3 and the denizens totaled 3400, how many were trilingual?
3. What fraction of $2\frac{1}{8}$ is $\frac{1}{5}$?

Solve:

4. $-4\frac{3}{4} + 8\frac{1}{3}x = 13\frac{1}{4}$
5. $-2 - |-3| - 2^2 - (3 - x) = -(-3)^3$
6. $-3x - 2(5 - 7x) = 14$
7. $5p - 6(2p + 1) = -4p - 2$

Solve each equation for y:

8. $x + 3y - 4 = 0$
9. $4y - x = 7$
10. $2y + 2k + 4x - 4 = 0$
11. $3y - 2x - 7 = 0$

Factor the greatest common factor:

12. $3a^2b^4c^5 - 6a^2b^6c^6$
13. $8x^2a - 4x^2a^2 + 2xa^2$

Simplify:

14. $\dfrac{3xyz - 3xy}{3xy}$
15. $\dfrac{2xy - 2x}{y - 1}$
16. Write two different inequalities that describe this graph.

−2 −1

17. The average of the five numbers was 6.8. If the first four numbers were 4.3, 5.2, 7, and 6.8, what was the fifth number?
18. Use three unit multipliers to convert 80 yards to centimeters.
19. Find the least common multiple of: 16, 12, and 50

Simplify. Write all variables in the numerator.

20. $\dfrac{p^5p^{-4}z^2}{z^{-5}zp^3}$
21. $\dfrac{akp^2p^4}{a^{-3}a^5p^5k^4}$
22. $\dfrac{mm^{-4}pp^5}{m^{-3}pp^6}$

23. Expand: $m^{-2}z^4\left(m^2z^{-4} - \frac{3m^6z}{z^4}\right)$

Simplify by adding like terms:

24. $aaxxy^{-3} + \frac{2a^2x^2}{y^3} - \frac{4axxx}{y^3}$

25. $m^2xy^{-2} - 3mmxy^{-2} + \frac{4m^2x}{y^2} - 3mmx^{-1}y^2$

Evaluate:

26. $a^2 - b^{-3}c$ if $a = 3$, $b = -2$, and $c = -1$

27. $ab(b^0 - bc)$ if $a = 3$, $b = -1$, and $c + 1 = -3$

28. $m - (m - x)$ if $m = -5$, and $x + 1 = 4$

Simplify:

29. $-2^4 + \frac{1}{-(-2)^3} + \sqrt[3]{64}$

30. $\frac{-2[(-3 + 2)(-3 + 5^0)]}{-2 - |-3 - 1|}$

LESSON 45 Least common multiples of algebraic expressions

The least common multiple is most often encountered when it is used as the least common denominator. If we are asked to add the fractions

$$\frac{1}{4} + \frac{5}{8} + \frac{7}{12}$$

we rewrite each of these fractions as a fraction whose denominator is 24, which is the least common multiple of 4, 8, and 12.

$$\frac{6}{24} + \frac{15}{24} + \frac{14}{24} = \mathbf{\frac{35}{24}}$$

In Lesson 47 we will discuss the method of adding algebraic fractions such as

$$\frac{b}{15a^2b} + \frac{c}{10ab^3}$$

To prepare for this lesson, we will practice finding the least common multiple of algebraic expressions.

example 45.1 Find the LCM of $15a^2b$ and $10ab^3$.

solution We begin by writing the expressions as products of factors whose exponents are 1.

$$15a^2b \qquad\qquad 10ab^3$$

$$3 \cdot 5 \cdot \underbrace{a \cdot a} \cdot b \qquad 2 \cdot 5 \cdot a \cdot \underbrace{b \cdot b \cdot b}$$

The LCM is

$$2 \cdot 3 \cdot 5 \cdot a \cdot a \cdot b \cdot b \cdot b = \mathbf{30a^2b^3}$$

example 45.2 Find the LCM of $4x^2m$ and $6x^3m$.

solution We begin the same way.

$$4x^2m \qquad\qquad 6x^3m$$

$$\underbrace{2 \cdot 2} \cdot x \cdot x \cdot m \qquad \underbrace{3} \cdot 2 \cdot \underbrace{x \cdot x \cdot x} \cdot m$$

The LCM is

$$2 \cdot 2 \cdot 3 \cdot x \cdot x \cdot x \cdot m \quad \text{or} \quad \mathbf{12x^3m}$$

example 45.3 Find the LCM of $12x^2am^2$ and $14x^3am^4$.

solution The LCM of 12 and 14 is 84. The most that x, a, and m are used as factors is x^3am^4. Thus, the LCM is

$$\mathbf{84x^3am^4}$$

practice Find the least common multiple of:

a. $6y^2w$ and $4y^3w^2$

b. $15x^4y^2m^3$ and x^6ym^3

c. $12a^3b^4$ and $20a^{10}b^2$

problem set 45

1. If the sum of twice a number and -10 is multiplied by 4, the result is 2 less than the number. What is the number?

2. At the Mardi Gras ball the guests roistered and rollicked until the wee hours. If the ratio of roisterers to rollickers was 7 to 5 and 1080 were in attendance, how many were rollickers?

3. What decimal part of 2.25 is 1.3995?

Solve:

4. $3\frac{1}{4}x - 4 = 21$

5. $7(x - 3) - 6x + 4 = 2 - (x + 3)$

6. $-2x + 3(-5 - x) = x$

7. $0.04x + 0.2 - 0.4x = 0.38$

Solve each equation for y:

8. $2x - 5y + 4 = 0$

9. $4 + 2x + 2y - 3 = 5$

Factor the greatest common factor:

10. $5x^2y^5m^2 - 10x^4y^2m^3$

11. $3x^2yz - 4zyx^2 + 2xyz^2$

Simplify:

12. $\dfrac{2x + 2}{2}$

13. $\dfrac{4xy + 4x}{4x}$

14. Graph: $x \not\leq 2$

Find the least common multiple of:

15. 75, 8, and 30

16. 18, 27, and 45

17. This figure is the base of a right prism 10 inches high. Find the volume and the surface area of the prism.

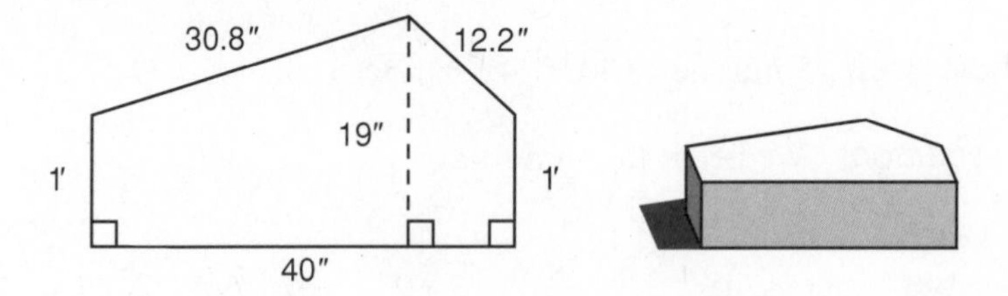

18. Use six unit multipliers to convert 200 square centimeters to square yards.

19. If $x + 3 = 4$, evaluate $x^2 - 19$.

Simplify. Write all variables with positive exponents.

20. $\dfrac{x^{-4}}{y^2p^{-4}}$ **21.** $\dfrac{k^5m^2}{k^7m^{-5}}$ **22.** $\dfrac{a^2bc^{-2}c^5}{a^2b^{-3}a^2c^3}$

23. Expand: $\left(\dfrac{m^2}{y^{-1}} + 4m^5y^6\right)m^{-2}y$

Simplify by adding like terms:

24. $axy^2 + \dfrac{2ax}{y^{-2}} - \dfrac{3ay^2}{x^{-1}} + 5ay^2x^{-1}$

25. $3m^2k^5 - \dfrac{2m^3k^6}{mk} + 4mmk^6k^{-1} - 3mk^5$

Evaluate:

26. $-|-a|(a - x)$ if $a = -2$ and $x = 4$

27. $-xy(y - x^0)$ if $x = 3$ and $y = -2$

28. $p^{-2}(a^{-5} - y)$ if $p = 2$, $y = -4$, and $a = \sqrt[3]{-1}$

Simplify:

29. -3^{-3} **30.** $\dfrac{4(-3 + 2) - |-5 + 2^0|}{3(-2)^2}$

LESSON 46 Addition of rational expressions

If we add one-eleventh to two-elevenths, we get three-elevenths.

$$\frac{1}{11} + \frac{2}{11} = \frac{1 + 2}{11} = \mathbf{\frac{3}{11}}$$

This is a demonstration of the rule for adding fractions whose denominators are the same.

RULE FOR ADDING FRACTIONS

Fractions with equal denominators are added by adding the numerators algebraically and recording the sum over a single denominator.

$$\frac{4}{11} - \frac{14}{11} + \frac{2}{11} - \frac{5}{11} = \frac{4 - 14 + 2 - 5}{11} = \mathbf{-\frac{13}{11}}$$

This rule also applies if the denominators are algebraic expressions.

$$\frac{5}{a + 6} + \frac{a + b}{a + 6} + \frac{2}{a + 6}$$

We see that the denominators are the same, so we can add the numerators and record the sum over a single denominator.

$$\frac{5}{a + 6} + \frac{a + b}{a + 6} + \frac{2}{a + 6} = \frac{5 + a + b + 2}{a + 6} = \mathbf{\frac{7 + a + b}{a + 6}}$$

example 46.1 Add: $\frac{4}{2x^2 + y} - \frac{6ax}{2x^2 + y}$

solution The denominators are the same so we can add the numerators.

$$\frac{4}{2x^2 + y} - \frac{6ax}{2x^2 + y} = \mathbf{\frac{4 - 6ax}{2x^2 + y}}$$

example 46.2 Add: $\frac{5}{a^2 + 7y} - \frac{3}{a^2 + 7y} + \frac{z}{a^2 + 7y}$

solution The denominators are the same so we can add the numerators.

$$\frac{5}{a^2 + 7y} - \frac{3}{a^2 + 7y} + \frac{z}{a^2 + 7y} = \frac{5 - 3 + z}{a^2 + 7y} = \mathbf{\frac{2 + z}{a^2 + 7y}}$$

example 46.3 Add: $\frac{5x + 7}{5a^2x} - \frac{3x - 2}{5a^2x}$

solution The denominators are the same so we add the numerators and get

$$\frac{5x + 7 - 3x + 2}{5a^2x} = \mathbf{\frac{2x + 9}{5a^2x}}$$

practice Add:

a. $\frac{1}{11} + \frac{5}{11}$

b. $\frac{4m - 2}{3m + 2} + \frac{6m - 4}{3m + 2}$

c. $\frac{2x + m}{3x^2m} + \frac{2x + m + 1}{3x^2m}$

d. $\frac{9}{xy^3 + m} - \frac{7ap}{xy^3 + m}$

problem set 46

1. The sum of twice a number and -5 is 35 greater than the opposite of the number.

2. When the defense gathered, the ratio of pikemen to archers was 3 to 10. If the defense totaled 27,989 soldiers, how many were archers?

Solve:

3. $-2x - 4(x - 2) = 3x + 5$

4. $-2(x - 4) + 8 = 4 - (x + 2)$

5. $(-2)^3(-x - 4) - |-2| - 3^2 = -2(x - 4) - x$

Solve for y:

6. $3x + 2y = 5$

7. $x - 3y + 7 = 0$

8. Use four unit multipliers to convert 50.8 square centimeters to square feet.

9. If $x + 9 = 3$, evaluate $-x^2 + 4$.

10. The average of five numbers is 60.12. Find the fifth number if the first four numbers are 58.8, 11.4, 73, and 62.2.

11. Find the surface area of this right prism in square centimeters. Find the volume in cubic centimeters. Dimensions are in meters.

12. Add: $\frac{a}{b} + \frac{c^2 - a}{b} + \frac{4}{b}$

13. Write two inequalities that describe this graph.

Find the least common multiple of:

14. 125, 75, and 45

15. c^3, c^2, and 2

16. $4c^3$, c^2, and $3c^4$

17. b^3, b^2c, and b^2c^2

18. Factor the greatest common factor of: $4x^2y^5p^2 - 3x^5y^4p^2$

19. Simplify: $\frac{4x^2 - 4x}{4x}$

Simplify: Write all exponents in the denominator.

20. $\frac{x^5y^2}{x^3y^5}$

21. $\frac{m^4p^5}{p^{-3}m^6}$

22. $\frac{xxx^3y^5y^{-2}}{x^{-3}yy^{-6}}$

23. Expand: $\frac{x^{-2}p^0}{y^4}\left(\frac{x^2}{p^4} - x^4p^6\right)$

Simplify by adding like terms:

24. $\frac{x^2}{y^2} - 3x^2y^{-2} + 4x^{-3}x^5y^{-2} - \frac{8y^{-2}}{x^{-2}}$

25. $xa^2y - 2a^2xy + 4ya^2x - \frac{6ax}{ay}$

Evaluate:

26. $-a - |a - x|$ if $a = -2$ and $x = -3$

27. $xy^0 - (x - y)$ if $x = -3$ and $y = 5$

28. $p^{-3}n^2 - n(p)$ if $n = -2$ and $p + 5 = 3$

Simplify:

29. $-\frac{1}{(-4)^{-2}} - \sqrt[3]{-27}$

30. $\frac{2[(-4 - 6^0)(5 - 2)]}{6 - [-(-2)]}$

LESSON 47 *Addition of abstract fractions*

There are three rules of algebra that some people believe are more important than all the rest of the rules put together. Two of them are the addition rule for equations and the multiplication rule for equations. We have used these, and they are restated very informally here:

1. **The same quantity can be added to both sides of an equation.**
2. **Every term on both sides of an equation can be multiplied or divided by the same quantity.**[†]

The other important rule is that

3. **The denominator and numerator of a fraction can be multiplied by the same quantity.**[†]

This theorem is usually called the *fundamental theorem of fractions* or the *fundamental theorem of rational expressions.* We will call it the **denominator-numerator same-quantity rule** because this name is more meaningful.

> DENOMINATOR-NUMERATOR SAME-QUANTITY RULE
>
> The denominator and the numerator of a fraction may be multiplied by the same nonzero quantity without changing the value of the fraction.

We cannot find the sum of

$$\frac{1}{4} + \frac{1}{2}$$

in this form because the denominators are not the same. But if we use the **denominator-numerator same-quantity rule** and multiply both the numerator and the denominator of $\frac{1}{2}$ by 2, we get $\frac{2}{4}$, which is an equivalent expression for $\frac{1}{2}$.

$$\frac{1}{4} + \frac{1}{2} = \frac{1}{4} + \frac{1(2)}{2(2)}$$

Now the fractions may be added, for they both have a denominator of 4.

$$\frac{1}{4} + \frac{2}{4} = \mathbf{\frac{3}{4}}$$

If the fractions to be added have different denominators, the procedure shown here can be used to rewrite the fractions as equivalent fractions that have the same denominators.

example 47.1 Add: $\frac{3}{4} + \frac{2}{b}$

[†] Except zero.

solution We will use the **denominator-numerator same-quantity rule** and a three-step procedure to rewrite the fractions as equivalent fractions that have the same denominators.

(a) As the first step we write the fraction lines with the proper sign between them:

$$\frac{}{} + \frac{}{}$$

(b) Now we write the least common multiple of the denominators as the new denominators. When the least common multiple is used in this fashion, we call it the **least common denominator.**

$$\frac{}{4b} + \frac{}{4b}$$

(c) The first two steps were automatic. We used no theorems or rules. Now we use the **denominator-numerator same-quantity rule.** We have multiplied the denominator, 4, or the first fraction by b to get $4b$, so we must also multiply the numerator, 3, by b and get $3b$.

$$\frac{3b}{4b} + \frac{}{4b}$$

We have multiplied the denominator, b, of the second fraction by 4 to get $4b$, so we must also multiply the numerator, 2, by 4 to get the new numerator of 8.

$$\frac{3b}{4b} + \frac{8}{4b}$$

Now the fractions have the same denominators and can be added.

$$\frac{3b}{4b} + \frac{8}{4b} = \mathbf{\frac{3b + 8}{4b}}$$

example 47.2 Add: $\frac{4}{b} + \frac{1}{c} + \frac{1}{2}$

solution

(a) $\frac{}{} + \frac{}{} + \frac{}{}$ — write the fraction lines

(b) $\frac{}{2bc} + \frac{}{2bc} + \frac{}{2bc}$ — use the LCM as the new denominator of every term

(c) $\frac{8c}{2bc} + \frac{2b}{2bc} + \frac{bc}{2bc} = \mathbf{\frac{8c + 2b + bc}{2bc}}$ — find the new numerators and add

example 47.3 Add: $\frac{5}{x} + \frac{x}{b} + \frac{a}{c}$

solution

(a) $\frac{}{} + \frac{}{} + \frac{}{}$ — write the fraction lines

(b) $\frac{}{xbc} + \frac{}{xbc} + \frac{}{xbc}$ — use the LCM as the new denominator of every term

(c) $\frac{5bc}{xbc} + \frac{x^2c}{xbc} + \frac{axb}{xbc} = \mathbf{\frac{5bc + x^2c + axb}{xbc}}$ — find the new numerators and add

example 47.4 Add: $\frac{m}{c^3} + \frac{4}{c^2} - 6$

solution When a term does not have a denominator, a good first step is to write a denominator of 1.

$$\frac{m}{c^3} + \frac{4}{c^2} - \frac{6}{1}$$ insert denominator

Now we use three steps to complete the process.

(a) $\text{—} + \text{—} - \text{—}$ write the fraction lines

(b) $\frac{}{c^3} + \frac{}{c^3} - \frac{}{c^3}$ use the LCM as the new denominator of every term

(c) $\frac{m}{c^3} + \frac{4c}{c^3} - \frac{6c^3}{c^3} = \boldsymbol{\frac{m + 4c - 6c^3}{c^3}}$ find the new numerators and add

example 47.5 Add: $\frac{p}{4} - \frac{a}{2} + \frac{c}{b}$

(a) $\text{—} - \text{—} + \text{—}$ write the fraction lines

(b) $\frac{}{4b} - \frac{}{4b} + \frac{}{4b}$ use the LCM as the new denominator of every term

(c) $\frac{pb}{4b} - \frac{2ab}{4b} + \frac{4c}{4b} = \boldsymbol{\frac{pb - 2ab + 4c}{4b}}$ find the new numerators and add

example 47.6 Add: $\frac{a}{c^2} + \frac{3}{4c^3} + \frac{m}{3c^4}$

solution (a) $\text{—} + \text{—} + \text{—}$ write the fraction lines

(b) $\frac{}{12c^4} + \frac{}{12c^4} + \frac{}{12c^4}$ use the LCM as the new denominator of every term

(c) $\frac{12ac^2}{12c^4} + \frac{9c}{12c^4} + \frac{4m}{12c^4} = \boldsymbol{\frac{12ac^2 + 9c + 4m}{12c^4}}$ find the new numerators and add

practice Add:

a. $\frac{5}{x} + \frac{1}{y} + \frac{1}{4}$

b. $\frac{c}{4} + \frac{5}{2} + x$

c. $\frac{x}{m^3} + \frac{1}{c^3} + \frac{a}{m^4}$

d. $\frac{m}{p^2} + \frac{3}{p^3} - 4$

problem set 47

1. The product of 8 and a number is 10 less than 3 times the number. What is the number?

2. When Oberon and Titania assembled the little people, they found that the pixies and leprechauns were in a ratio of 3 to 13. If there were 6816 in all, how many were pixies?

3. What decimal part of 0.46 is 0.01058?

Solve:

4. $2\frac{1}{5}x + 5 = -15$

5. $-4x - 3(x - 3) = x + 2$

6. $-4x + (-2x + 5) = -2x$

7. $0.2p + 2.2 + 2.2p = 4.36$

Solve for y:

8. $5x + 4 = 3y$

9. $2y - 5 = x$

10. Use eight unit multipliers to convert 50 square miles to square meters.

11. If $3x - 4 + x - 6 = 10$, evaluate $4 - 2x$.

12. If $5x - 9 + x - 3 = 6$, evaluate $9 - 3x$.

13. Write an inequality that describes this graph.

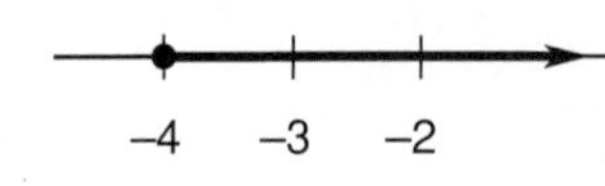

14. Find the surface area of this right circular cylinder in square centimeters. Dimensions are in meters.

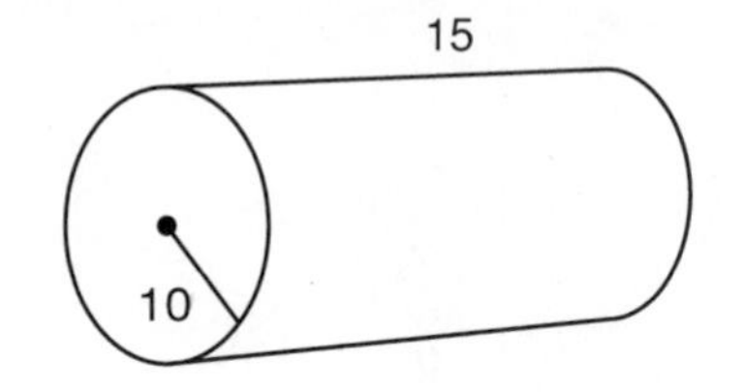

15. If this figure is the base of a right solid 10 inches high, how many 1-inch sugar cubes would it hold? What is the surface area? Dimensions are in inches.

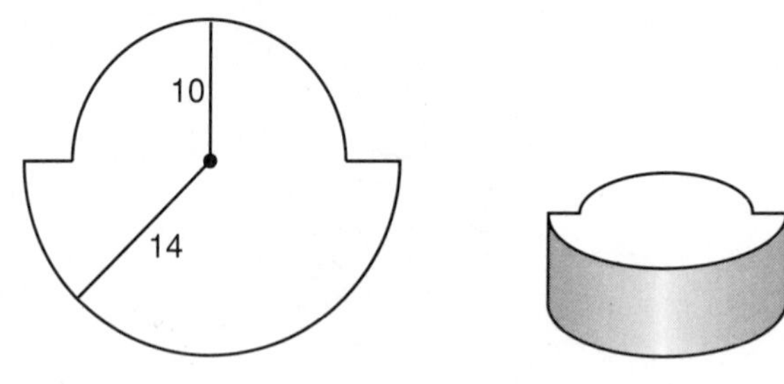

Add:

16. $\frac{1}{2} + \frac{3}{4} + \frac{6}{7}$

17. $\frac{x}{y} + \frac{b}{4y} + c$

18. $\frac{4}{a} + \frac{c}{4a} + 5$

19. $\frac{ad}{4d^3} + \frac{8}{d} + \frac{mx}{d^4}$

20. Find the least common multiple of: 40, 35, and 18

21. Factor: $x^3y^2m^5 - 3x^2ym^6$

22. Expand: $\left(\frac{a^2}{x^{-1}} - 4a^6x^4\right)\frac{a^{-2}}{x}$

Simplify:

23. $\frac{4ax - axy}{ax}$

24. $\frac{x^4y^2}{x^{-2}y^{-3}}$

25. $\frac{x^2xyy^{-4}}{x^4y^{-5}}$

26. Simplify by adding like terms: $\frac{a^2x^3}{y} - \frac{2x^3a^2}{y} + \frac{4xx^2a^2}{y} - 3a^2x^3$

Evaluate:

27. $a^{-2}(2a - a^{-3})$ if $a = -3$

28. $x - y(x^0 - y)$ if $x = -2$ and $y + 1 = 4$

Simplify:

29. $\frac{1}{-3^{-3}} + \sqrt[5]{32}$

30. $\frac{-2[(-4 - 2) - (5^0 - 3)]}{-2 - |2|}$

LESSON 48 Conjunctions

If we wish to designate the numbers that are greater than 5, we can write

$$x > 5$$

If we wish to designate the numbers that are less than 10, we can write

$$x < 10$$

If we wish to designate the numbers that are greater than 5 and that are also less than 10, we can write either of the following.

$$x > 5 \text{ and } x < 10 \qquad \text{or} \qquad 5 < x < 10$$

Both of these notations mean the same thing and designate the numbers that are between 5 and 10. We use the word **conjunction** to describe a statement of two conditions, both of which must be met. Thus, both of the above statements are conjunctions. In the concise notation on the right, we note that the symbols point in the same direction. They always do, as we see when we reverse both the symbols and the numbers to make the same statement another way.

$$10 > x > 5$$

But, we must be careful when we write conjunctions because

$$10 < x < 5$$

designates the numbers that are greater than 10 and are also less than 5. Of course, there are no numbers that fall into this category.

example 48.1 Graph the solution to $5 < x < 10$.

solution This conjunction designates the numbers between 5 and 10.

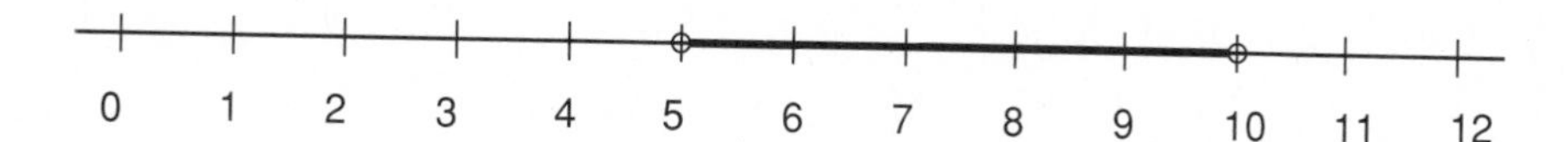

example 48.2 Graph the solution to $-2 < x \leq 4$.

solution This conjunction asks for the graph of the numbers between -2 and 4. Note that the symbol $<$ excludes -2 and that the symbol ≤ 4 includes 4 in the solution set.

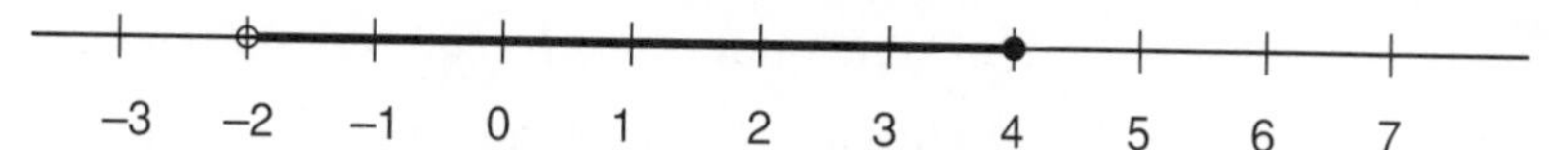

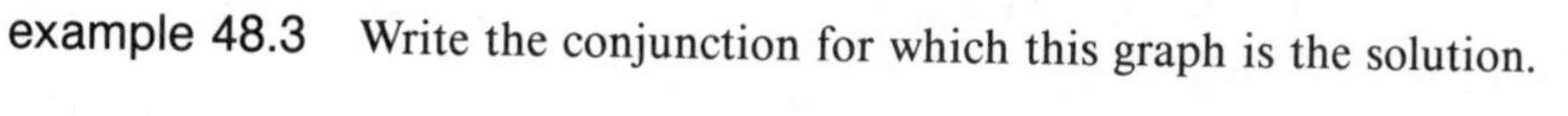

example 48.3 Write the conjunction for which this graph is the solution.

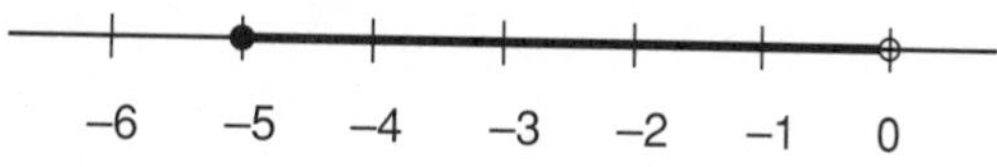

solution The graph shows the numbers that are less than 0 and greater than or equal to -5. We write this conjunction as

$$\mathbf{-5 \leq x < 0}$$

example 48.4 Write the conjunction that designates the numbers that are greater than -1 and less than or equal to 5.

solution The conjunction is $\mathbf{-1 < x \le 5}$.

practice

Graph on a number line:

a. $-5 < x \le 4$ **b.** $-5 \le x < 4$ **c.** $-7 < x < -2$

d. Write the conjunction that describes the numbers that are greater than -4 and that are also less than or equal to 10.

problem set 48

1. If the sum of 4 times a number and 5 is multiplied by 3, the result is 24 less than the opposite of the number. What is the number?

2. At the prestidigitator's banquet, the ratio of real magicians to charlatans was 7 to 2. If there were 324 at the banquet, how many were real magicians?

3. $4\frac{1}{5}$ of what number equals 28?

Solve:

4. $3\frac{1}{3}x + 7 = -2$

5. $5p - 4p - (p - 2) = 3(p + 4)$

6. $(-2)^3(-k - |-3|) - (-2) - 2k = k - 3^2$

Solve for y:

7. $2x + 4y = 6$

8. $3y - 4 = 2x$

Find the least common multiple of:

9. 8, 36, and 75

10. x, c^2x^2, and cdx

Add:

11. $\frac{1}{3} + \frac{2}{5} + \frac{3}{10}$

12. $\frac{3}{7} + \frac{8}{9} - \frac{1}{3}$

13. $\frac{a}{x} + \frac{b}{c^2x^2} + d$

14. $\frac{4}{x^2} + \frac{6}{2x^3} - \frac{3}{4x^4}$

15. $\frac{4}{x^2} + \frac{c}{4x^3} + m$

16. $\frac{a}{b} + \frac{c}{4b^2} + \frac{a^2}{8b^3}$

17. $\frac{m}{a^5} + \frac{k}{2a^4} - \frac{3}{4a^3}$

18. $\frac{1}{2a^3} + \frac{3}{4ab^2} + \frac{c}{8a^2b^2}$

19. Use four unit multipliers to convert 1000 square inches to square miles.

20. Find the number of 1-inch-square tiles necessary to cover this figure. Dimensions are in feet.

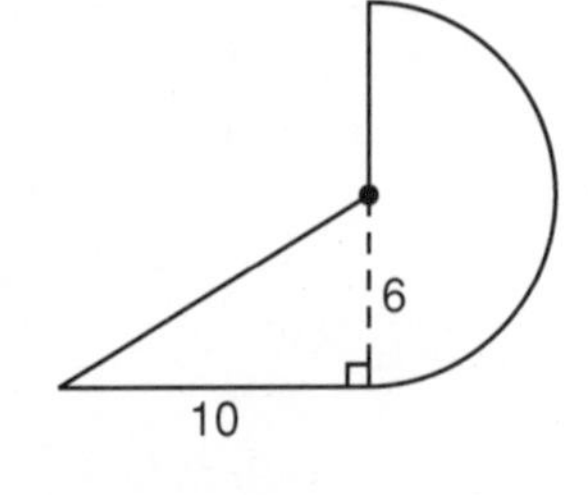

21. Factor: $8m^3x^2y^4p - 4m^2xpm$

22. Expand: $\frac{x^{-2}}{y^{-3}}\left(\frac{x^2}{y^3} - \frac{ax^3}{y^{-4}}\right)$

Simplify:

23. $\dfrac{4x^4 - 4}{4}$ 24. $\dfrac{mm^2p^3y^{-3}}{m^{-3}m^{-2}p^{-3}y^4}$ 25. $\dfrac{xx^{-3}y^5x^0}{x^2y^{-5}xy^2}$

Simplify by adding like terms:

26. $\dfrac{m}{y} - \dfrac{3m^2y}{my^2} - \dfrac{5m^{-3}m^4}{y^{-3}y^4} + \dfrac{2ym}{ym^2}$

27. Evaluate: $-x - |xa|(x^0 - a)$ if $x = -2$ and $a = \sqrt[3]{-27}$

28. If $x + 4 = 2$ and $y = 3$, evaluate $-x^2 - y^2(xy)$.

Simplify:

29. $(-3)^{-2}$ 30. $-4[(-3 - 2^0) - (5 - 2) + |3|]$

LESSON 49 Percents less than 100

We have been working problems about fractional and decimal parts of numbers by using one of the following equations.

(a) $(F) \times (\text{of}) = \text{is}$ or (b) $(D) \times (\text{of}) = \text{is}$

The percent equation is exactly the same as (a) except that the fraction has a denominator of 100. *Centum* is the Latin word for 100, and thus percent literally means "by the 100." We often use the symbol % to represent the word percent. The percent equation is

(c) $\dfrac{P}{100} \times (\text{of}) = \text{is}$ which can also be written as (d) $\dfrac{P}{100} = \dfrac{\text{is}}{\text{of}}$

The part identified by the word *of* is often called the **base,** and the part identified by the word *is* is called the **percentage.** If we use these words, we get equation (e). In equation (f) $\dfrac{P}{100}$ is called the **rate.**

(e) $\dfrac{P}{100} \times \text{base} = \text{percentage}$ (f) $\text{rate} \times \text{base} = \text{percentage}$

All four equations produce the same result. We prefer equation (c) because it is just like equation (a), which we have used for fractional parts of numbers. However, your teacher may prefer one of the other forms. Many teachers like (d) because this form can be explained using the concepts and vocabulary of ratios. Others prefer (e) or (f) because these forms are used almost exclusively in the business world. The four equations (c), (d), (e), and (f) are not different equations but different forms of the same equation. Each form has advantages and disadvantages. None is perfect. Some people find form (c) to be the most difficult to solve. The ratio form, form (d), is almost never used except in beginning algebra classes. Form (f) does not use percent as such but uses rate, which is percent divided by 100. We find that it is best to pick one form and stick with it. If you don't like form (c), use another.

To solve word problems about percent, it is necessary to be able to visualize the problem. We will begin to work on achieving this visualization by drawing diagrams of percent problems after we work the problems. **Learning to draw these diagrams is very important.**

example 49.1 Twenty percent of what number is 15? Work the problem and then draw a diagram of the problem.

solution We will use equation (c) and use 20 for *percent, WN* for *what number*, and 15 for *is*.

$$\frac{P}{100} \times \text{of} = \text{is} \longrightarrow \frac{20}{100} \cdot WN = 15$$

We will solve by multiplying both sides by $\frac{100}{20}$.

$$\frac{\cancel{100}}{\cancel{20}} \cdot \frac{\cancel{20}}{\cancel{100}} WN = 15 \cdot \frac{100}{20} \longrightarrow WN = \frac{1500}{20} \longrightarrow WN = \mathbf{75}$$

The "before" diagram is 75, which represents 100 percent. The "after" diagram shows that 15 is 20 percent. Thus the other part must be 60, which is 80 percent.

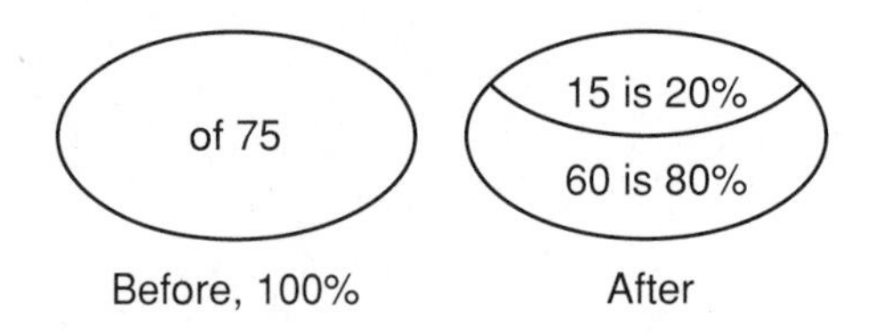

example 49.2 What percent of 140 is 98? Work the problem and then draw a diagram of the problem.

solution We use *WP* for *what percent*, 140 for *of*, and 98 for *is*.

$$\frac{P}{100} \times \text{of} = \text{is} \longrightarrow \frac{WP}{100} \cdot 140 = 98$$

We will solve by multiplying both sides by $\frac{100}{140}$.

$$\frac{\cancel{100}}{\cancel{140}} \cdot \frac{WP}{\cancel{100}} \cdot \cancel{140} = 98 \cdot \frac{100}{140} \longrightarrow WP = \frac{9800}{140} \longrightarrow \mathbf{WP = 70\%}$$

The diagrams show that 140 or 100 percent was divided into two parts: 98, which is 70 percent, and 42 which is 30 percent.

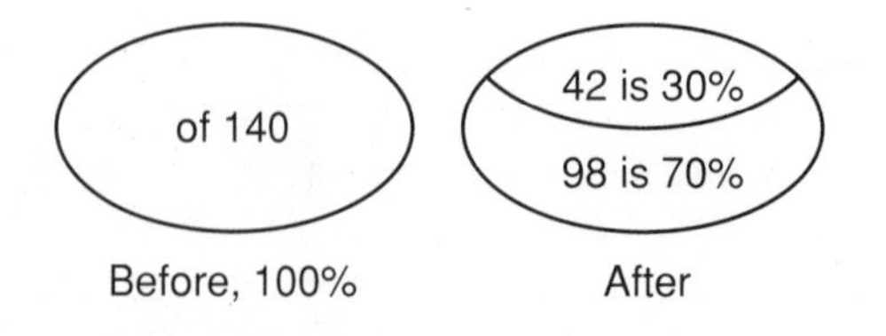

example 49.3 Fifteen percent of 300 is what number? Work the problem and then draw a diagram of the problem.

solution We use 15 for percent, 300 for *of*, and *WN* for *is*.

$$\frac{P}{100} \times \text{of} = \text{is} \longrightarrow \frac{15}{100} \cdot (300) = WN$$

We multiply to solve and get

$$\frac{4500}{100} = WN \longrightarrow \mathbf{45 = WN}$$

The diagrams show that 300 is divided into two parts. One part is 45 or 15 percent. Thus the other part must be 255 or 85 percent.

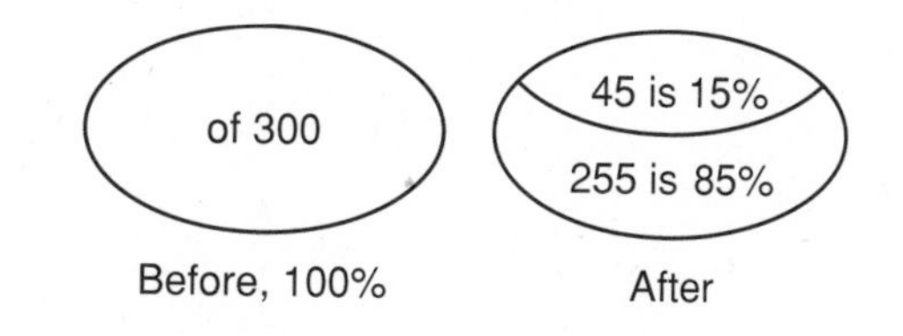

practice Draw a diagram after completing each problem:

a. Eighty-two percent of 400 is what number?

b. Twenty percent of what number is 800?

c. What number is 18% of 360?

problem set 49

1. The sum of 3 times a number and 25 is 5 greater than the opposite of the number. What is the number?

2. Leonardo and Michelangelo turned out paintings whose areas were in the ratio of 14 to 13. During the period in question, the total area of their paintings was 1080 square units. How many square units were painted by Leonardo?

3. What fraction of $2\frac{1}{8}$ is $\frac{1}{4}$?

4. Firenze, Venezia, and Milano played one game each. The average score for the three games was 16. If Firenze scored 14 and Venezia scored 11, what was Milano's score?

5. Use six unit multipliers to convert 16 cubic feet to cubic centimeters.

6. This figure is the base of a right prism whose sides are 100 inches high. How many 1-inch sugar cubes will this prism hold? What is the total surface area? The dimensions are in feet.

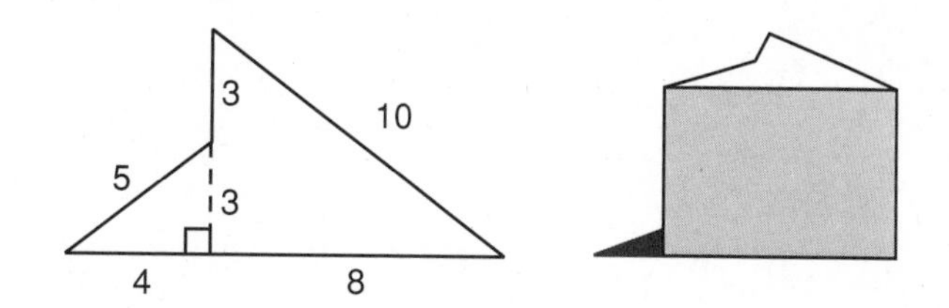

Draw the diagram after completing each problem.

7. What percent of 8300 is 996?

8. Eighty percent of what number is 1120?

Solve:

9. $4x - 5(x + 2) = -(2x - 4)$

10. $0.02 + 0.02x - 0.4 - 0.4x = 3.116$

Solve for y:

11. $3x - 4y = 7$

12. $-2y + 5 + 3x = 0$

Find the least common multiple of:

13. 21, 24, and 60

14. $4x^2, yx^2, 8m^3x^2$

Add:

15. $\frac{3}{4} + \frac{2}{5} - \frac{3}{20}$

16. $\frac{a}{x} + \frac{b}{c} + d$

17. $\frac{m}{xc} + \frac{d^2}{xkc^3} - \frac{3p}{xk^2c^3}$

18. $\frac{4}{a^2b^2} - \frac{c}{ad} - \frac{m}{a^3b}$

Graph:

19. $x \not\leq 2$

20. $2 \leq x < 5$

21. Factor: $18x^5y^2m - 9x^3ym^5$

22. Expand: $\frac{x^{-1}}{y}\left(\frac{y}{x} - \frac{3xy^{-5}}{p^6}\right)$

Simplify:

23. $\frac{4x^2 - 4x}{4x}$

24. $\frac{kp^2k^{-1}p^{-3}p^{-4}}{k^2pp^2k^{-5}}$

25. $\frac{m^2xym^3x^{-5}}{yy^{-4}m^{-3}x^2}$

26. Simplify by adding like terms: $m^2x^2 - \frac{3m^{-2}x^{-2}}{m^4x^{-4}} + \frac{2m^2}{xm} - \frac{5x^{-1}}{m^{-1}}$

Evaluate:

27. $-p^{-2} - (p^2 - x)$ if $p = -3$ and $x = 5$

28. $-p^2 - p^0x^2$ if $p = -3$ and $x + 2 = 4$

Simplify:

29. $\frac{1}{-(-3)^{-2}} - \sqrt[3]{-64}$

30. $-5[(-2 + 3)(-2 - 4^0) - |-5|]$

LESSON 50 Polynomials · Addition of polynomials

50.A polynomials

Thus far we have encountered expressions such as

(a) $\frac{x^{-3} + m}{x}$ (b) $\frac{2y}{x^3}$ (c) $4a^{-2}x + m^3$

(d) rt^{n-2} (e) 4 (f) $-4a^2$

(g) $7x^2 + 2$ (h) $4x^2y$ (i) $7y^3 + 3y + 2$

All of these expressions are called **algebraic expressions** or **mathematical expressions** and are individual terms or indicated sums of terms. The more complicated expressions as shown in (a) through (d) have no special names and are just called **terms** or **expressions.** The simple expressions as shown in (e) through (i) occur so often and are so useful that we give these expressions a special name: **polynomial.** It is unfortunate that we use such an intimidating word to describe the simplest kind of algebraic expression. We should think "simplenomial" when we see the word polynomial.

A polynomial in one variable is one term or a sum of individual terms each of which has the form

$$ax^n$$

where **a is a real number** and **n is a whole number,** such as the following:

(a) $4x^2$ (b) $-x^3$ (c) $-1.414x^{32}$

(d) $2x^4$ (e) $-7x$ (f) -7

Each of the six expressions meets all three of the requirements for being called a polynomial:

1. **Each expression is in the form ax^n.**
2. **The numerical coefficient of each expression, a, is a real number.**
3. **The exponent of the variable, n, is a positive integer or is the number zero.**

The last polynomial shown, -7, can be thought of as being $-7x^0$, which is the same as -7 if x has a value other than zero because **any nonzero quantity raised to the zero power has a value of 1!**

We refresh our memory with the following examples.

$(-15)^0 = 1$ $\left(\frac{75}{14}\right)^0 = 1$ $x^0 = 1$ $(x + y)^0 = 1$ $(x, x + y \neq 0)$

Thus, since x^0 equals 1 if x is not zero, we can write

$$-7x^0 = -7(1) = -7$$

A **polynomial of one term** is called a **monomial,** so each of the six expressions in (a) through (f) above can be called a polynomial or may be described by using the more restrictive name of monomial.

A **polynomial of two terms** is called a **binomial,** and a **polynomial of three terms** is called a **trinomial.** Thus each of the following

(a) $4 + x$ (b) $p^{15} - 4p$ (c) $-y^{10} + 3y$

can be described by using either the word polynomial or the word binomial. The following expressions

(a) $x^2 + 2x + 4$ (b) $y^{14} - 1.6y^2 + 4$ (c) $m^4 + 2m - 1.6$

can be called either polynomials or trinomials. Indicated sums of more than three monomial terms have no special names and are just called polynomials. Thus we can call any of the following expressions a polynomial.

(a) -14.2 (b) $\frac{7}{2}$ (c) $4x^2$ (d) $6x^2 + 4$

(e) $x^4 - 3x^2 + 2x$ (f) $-7x^{15} + 2x^3 - 5x^2 + 6x + 4$

Expressions (a) and (b) are polynomials because the exponents of the understood variable is the number zero.

(a) $-14.2m^0 = -14.2$ (b) $\frac{7}{2}y^0 = \frac{7}{2}$

Although expressions (c), (d), and (e) are all polynomials, they can also be called a monomial, a binomial, and a trinomial in that order. The last expression, (f), is a polynomial because each term in the indicated sum is a monomial. This expression does not have a more restrictive name.

> DEFINITION OF A POLYNOMIAL IN ONE UNKNOWN
>
> A polynomial in one unknown is an algebraic expression of the form
>
> $$4x^{15} + 2x^{14} - 3x^{10} + \cdots + 2x + 2$$
>
> where the coefficients are real numbers, x is a variable, and the exponents are whole numbers.

Thus none of the following expressions is a polynomial.

(a) $4x^{-3}$ (b) $\dfrac{-6x + y}{z}$ (c) $-15y^{-5}$

The expressions (a) and (c) have real number coefficients, but the exponent of the variable is not a whole number. Expression (b) is not a polynomial because it is not in the required form of ax^n.

50.B polynomials, general

> DEFINITION OF A POLYNOMIAL
>
> A polynomial in one or more unknowns is an algebraic expression having only terms of the form $ax^n y^m z^p \cdots$, where the coefficient a is a real number and the exponents $n, m, p, \ldots,$ are whole numbers.

Thus the general definition of a polynomial has the same restrictions that are given by the definition of a polynomial in one unknown, namely:

1. The numerical coefficients of the individual terms must be **real numbers.**
2. The exponents of the variables must be **whole numbers.**

The following can therefore be called polynomials in more than one variable.

(a) xyz^2m (b) $4x^{15}ym^3 + pq^5$ (c) $-11x^2p^4 + 2$

50.C degree

The **degree of a term** of a polynomial is the sum of the exponents of the variables in the term. Thus

$4x^3, 6xym, 2x^2y$ are third-degree terms

$4x^2m^3, 3y^5, 2xypmz$ are fifth-degree terms

The **degree of a polynomial** is the same as the degree of its highest-degree term.

$3x^2 + xyz + m$ is a third-degree polynomial because the degree of its highest degree term (xyz) is 3

$4x^5 + yx^3 + 2x^2 + 2$ is a fifth-degree polynomial because the degree of its highest degree term ($4x^5$) is 5

Polynomials are usually written in descending powers of one of the variables. The polynomials

$$x^5 - 3x^4 + 2x^2 - x + 5$$
$$x^4m + x^3m^2 - 2xm^5 - 6$$
$$-2xm^5 + x^3m^2 + x^4m - 6$$

are written in descending powers of a particular variable. The first two polynomials are written in descending powers of x. The third polynomial is the same polynomial as the second but is written in descending powers of m instead of descending powers of x.

50.D addition of polynomials

Since polynomials are composed of individual terms, the rule for adding polynomials is the same rule that we use for adding terms—**like terms may be added.**

example 50.1 Add: $(x^3 + 3x^2 + 2) + (2x^3 + 4)$

solution We remember that we can discard parentheses preceded by a plus sign without changing the signs of the terms therein.

$$(x^3 + 3x^2 + 2) + (2x^3 + 4) = x^3 + 3x^2 + 2 + 2x^3 + 4$$
$$= \mathbf{3x^3 + 3x^2 + 6}$$

example 50.2 Add: $(3x^4 - 2x^2 + 3) - (x^4 - 2x^3 + x^2)$

solution When we discard the parentheses in this problem, we remember to **change the sign of every term in the second parentheses** because the second parentheses are preceded by a minus sign.

$$3x^4 - 2x^2 + 3 - x^4 + 2x^3 - x^2 = \mathbf{2x^4 + 2x^3 - 3x^2 + 3}$$

example 50.3 Add: $(3x^3 + 2x^2 - x + 4) - (x^2 - 7x - 5)$

solution The first parentheses can be removed with no change to the terms inside, but because the second parentheses are preceded by a minus sign, when we remove these parentheses, we must **change the sign of all terms therein.**

$$(3x^3 + 2x^2 - x + 4) - (x^2 - 7x - 5) = 3x^3 + 2x^2 - x + 4 - x^2 + 7x + 5$$
$$= \mathbf{3x^3 + x^2 + 6x + 9}$$

practice Add. Write the answers in descending order of the variable.

a. $-(3x^5 + 6x^4 - 7x^3 - 5) + (x^5 - x^4 + 3x^2 - 2x - 8)$

b. $(3x^2 + x^4 - 6x + 2) - (15x^4 + 2x^3 - 6x^2 + 5x - 3)$

c. Which of these are polynomials?

(1) $4x^{-3}$ (2) $-2x^0$ (3) 5 (4) $-0.04x^2$

problem set 50

1. Glicken found that 4 times the sum of twice a number and -5 was 92 less than the opposite of the number. What was the number?

2. In the remuda, the ratio of piebalds to mottled mares was 7 to 11. If there were 756 remounts in the remuda, how many were piebalds?

3. What fraction of $\frac{1}{3}$ is $\frac{2}{27}$?

Draw the diagrams after working these problems.

4. Harding and Jack found that 8 percent of their number was 72. What was their number?

5. What percent of 860 is 43?

6. Sixteen percent of 4200 is what number?

7. Forty-three percent of what number is 2150?

8. What percent of 5400 is 108?

Solve:

9. $3x - 5(-2x - 8) + 4 = 2 - (x - 4)$

10. $-(-3)^3 - |-2| - 2^2 - (-k - 3) = -4^2 - (3k - 4)$

11. Solve for y: $3x - 4y + 7 = 0$

12. Find the least common multiple of: 250, 75, and 20

Add:

13. $\frac{1}{7} + \frac{5}{21} + \frac{3}{5}$

14. $\frac{m}{x} + \frac{b}{xy} + \frac{ac}{x^2ym}$

15. $\frac{3}{cd} + \frac{5}{4c^2d} + \frac{7}{8cd^2}$

16. $\frac{p}{xa} + \frac{5}{xam} + x$

17. Use two unit multipliers to convert 65,000 square feet to square miles.

18. Find the perimeter of this figure in meters. All angles are right angles.

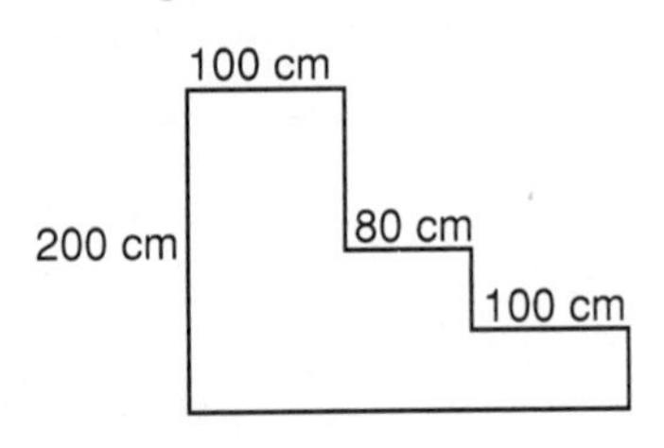

19. If $x + 1 = 4$ and $y - 2 = 3$, evaluate $x^2 - y^2$.

20. Add. Write the answers in descending order of the variable:

$$(-x^3 - 2x - 3x^2 + 5) - 2(x^3 - x + 2x^2 - 3)$$

21. Graph: $-4 < x \leq 1$

22. Factor: $x^2ym - 4x^2ym^3 + 2x^4y^3m^6$

Simplify:

23. $\frac{4x^2 - 4x^4}{4x^2}$

24. $\frac{x^2y^{-2}m^{-5}y^0}{xxy^2y^{-5}x^{-3}}$

25. Expand: $\left(\frac{x^2}{yp^{-4}} - \frac{3x^2y}{p^{-4}}\right)\frac{x^{-2}}{y^4p}$

26. Simplify by adding like terms: $x^2y^{-2}p + \frac{3xxp}{y^2} - \frac{4x}{y^{-2}} + 6xy^2$

Evaluate:

27. $-xy(x - y)$ if $x = -3$ and $y = 5$

28. $-p^2 - p^{-3}(xp^0)$ if $p = -2$ and $x + 2 = -2$

Simplify:

29. -3^{-3}

30. $-2[(-4 - 3^0)(5 - 2) - (-6)] - \sqrt[3]{-125}$

LESSON 51 Multiplication of polynomials

We remember that we use the distributive property

$$a(b + c) = ab + ac$$

when we multiply a monomial by a binomial. The expression on the outside of the parentheses is multiplied by both terms inside the parentheses.

example 51.1 Multiply: $4x(x^2 - 2)$

solution We must multiply $4x$ by x^2 and by -2. Then we sum the products.

$$4x(x^2 - 2) = 4x(x^2) + 4x(-2) = \mathbf{4x^3 - 8x}$$

To develop a general procedure for multiplying two polynomials, we will use the distributive property to multiply

$$(a + b)(c + d)$$

The notation $(a + b)(c + d)$ tells us that $(a + b)$ is to be multiplied by c and that $(a + b)$ is also to be multiplied by d and that the two products are to be summed.

$$(a + b)(c + d) = (a + b)c + (a + b)d$$

Now we use the distributive property again to multiply $(a + b)$ by c

$$(a + b)c = ac + bc$$

and to multiply $(a + b)$ by d,

$$(a + b)d = ad + bd$$

and the products are summed.

$$\mathbf{ac + bc + ad + bd}$$

This has been a rather involved development to illustrate the following rule.

RULE FOR MULTIPLYING POLYNOMIALS

To multiply one polynomial by a second polynomial, each term of the first polynomial is multiplied by each term of the second polynomial and then the products are summed.

example 51.2 Multiply: $(4x + 5)(3x - 2)$

solution The notation indicates that $4x$ is to be multiplied by both $3x$ and -2

$$4x(3x - 2) = 12x^2 - 8x$$

and that $+5$ is to be multiplied by both $3x$ and -2

$$+5(3x - 2) = 15x - 10$$

and that the products are to be added alegbraically.

$$(4x + 5)(3x - 2) = 12x^2 - 8x + 15x - 10 = \mathbf{12x^2 + 7x - 10}$$

We can also do the multiplication if the binomials are written one above the other. Either one may be on top.

example 51.3 Multiply: $(4x + 2)(x - 5)$

solution We begin writing the binomials one above the other.

$$\begin{array}{r} 4x + 2 \\ \underline{x - 5} \end{array}$$

Now, the x of $x - 5$ is multiplied by both terms of $4x + 2$, and the products are recorded.

$$\begin{array}{rl} 4x \;+ 2 & \\ \underline{x \;- 5} & \\ 4x^2 + 2x & \quad \text{product of } x \text{ and } 4x + 2 \end{array}$$

Now, the -5 of $x - 5$ is multiplied by both terms of $4x + 2$, and the products are recorded so that like terms (if any) are recorded below like terms to facilitate addition.

$$\begin{array}{rrrl} 4x & + & 2 & \\ x & - & 5 & \\ \hline 4x^2 & + & 2x & \quad \text{product of } x \text{ and } 4x + 2 \\ & - & 20x - 10 & \quad \text{product of } -5 \text{ and } 4x + 2 \\ \hline \mathbf{4x^2} & \mathbf{-} & \mathbf{18x - 10} & \quad \text{sum of the products} \end{array}$$

The product $-20x$ was recorded below the term $+2x$. There was no constant in the first product for -10 to be recorded below, so -10 was written out to the right.

example 51.4 Multiply: $(4x + 2)(3x - 5)$

solution We will use the vertical format to multiply.

$$\begin{array}{rrrl} 4x & + & 2 & \\ 3x & - & 5 & \\ \hline 12x^2 & + & 6x & \quad \text{product of } 3x \text{ and } 4x + 2 \\ & - & 20x - 10 & \quad \text{product of } -5 \text{ and } 4x + 2 \\ \hline \mathbf{12x^2} & \mathbf{-} & \mathbf{14x - 10} & \quad \text{sum of the products} \end{array}$$

example 51.5 Expand: $(3x + 2)^2$

solution When we write x^2, we mean x times x. Thus, when we write $(3x + 2)^2$, we mean $3x + 2$ times $3x + 2$. We will use a vertical format.

$$\begin{array}{ll} 3x + 2 & \\ 3x + 2 & \\ \hline 9x^2 + 6x & \text{product of } 3x \text{ and } 3x + 2 \\ \quad + 6x + 4 & \text{product of 2 and } 3x + 2 \\ \hline \mathbf{9x^2 + 12x + 4} & \text{sum of the products} \end{array}$$

This same procedure can be used if one or both expressions have three or more terms. Each term in one expression is multiplied by every term in the other expression and the products are then summed algebraically. The next example illustrates the procedure for multiplying a binomial by a trinomial.

example 51.6 Multiply: $(4x - 2)(x^2 + x + 4)$

solution We will multiply both $4x$ and -2 by all three terms in the second parentheses. This will give us six products. Then we simplify by adding the like terms.

$$4x^3 + 4x^2 + 16x - 2x^2 - 2x - 8 = \mathbf{4x^3 + 2x^2 + 14x - 8}$$

practice Multiply:

a. $(a + 6)(a - 6)$ **b.** $(5x + 3)(2x - 4)$ **c.** $(5x - 6)^2$

problem set 51

1. Jo Ellen found that if the product of a number and 5 is increased by 20, the result is 28 less than the opposite of the number. What is the number?
2. The ratio of expressions frowned on to expressions proscribed was 3 to 17. If 1620 expressions were considered, how many were proscribed?
3. What fraction of $7\frac{1}{8}$ is $3\frac{2}{7}$?

Draw the diagrams after working the next three problems.

4. Seventeen percent of what number is 952?
5. What percent of 300 is 60?
6. Thirty-eight percent of 700 is what number?

Solve:

7. $-4x - x - 3(x - 2) = 4 - (x - 2)$
8. $1.591 + 0.003k - 0.002 + 0.002k = -(0.003 - k)$
9. Add. Write the answer in descending order of the variable:

$$(4x^2 - 2x + 7x^5) - 2(x - 4 + 2x^2 - 3x^4)$$

10. Multiply: $(5x^2 + 12x + 7)(x + 1)$
11. Use six unit multipliers to convert 28,000 square inches to square kilometers.
12. Write a conjunction that describes this graph.

−3 −2 −1 0 1 2 3 4 5

13. Berne (the Bear) is one of the four largest cities in Switzerland and is the capital. The average population of the four largest cities is about 210,000. If the population of Geneva is about 155,000 and if the population of Basel and Zurich are approximately 180,000 and 365,000, respectively, what is the population of Berne?

14. How many 1-inch-square floor tiles would it take to cover the area shown? Dimensions are in inches. Corners that look square are square.

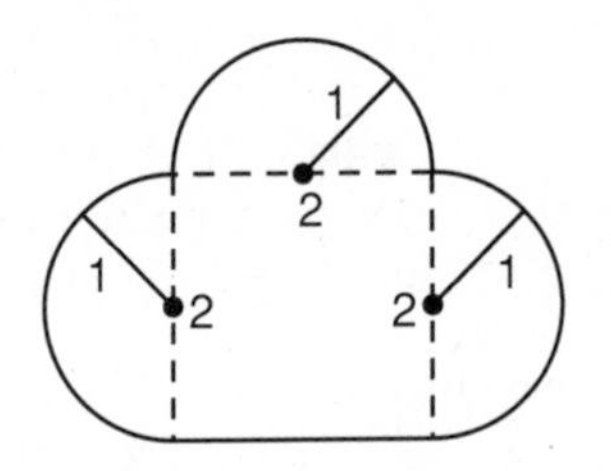

15. Find the volume in cubic inches of the right prism whose base is shown and whose height is 1 foot. Find the surface area. Dimensions are in inches. Corners that look square are square.

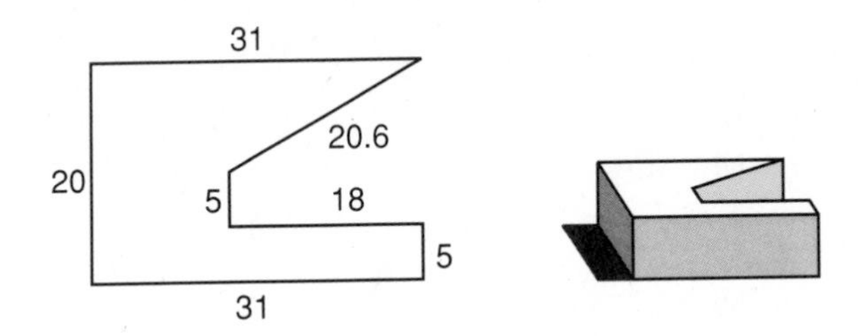

16. Find the least common multiple of: 15, 175, and 225

Add:

17. $\frac{1}{2a} + \frac{k}{4a^2} + \frac{x}{8m^2a}$

18. $\frac{m}{x^2y} + \frac{4}{yx^2} - \frac{3y}{x^4}$

19. $\frac{1}{c} + \frac{x}{mc^2} + d$

20. $\frac{x}{ya} + \frac{b}{xa^2} - k$

21. Graph: $x \not< 5$

22. Factor: $xy - 4x^2y^2m - 3x^2y$

Simplify:

23. $\frac{3x^4 - 3x^2}{3x^2}$

24. $\frac{x^{-4}yy^{-3}x^0x^2}{x^{-3}y^3y^2x^{-4}}$

25. Expand: $\frac{x}{y^{-1}}\left(\frac{x}{y} - \frac{3x^2}{xy}\right)$

26. Simplify by adding like terms: $\frac{x^2y}{p^{-3}} - \frac{4x^2p^3}{y^{-1}} - \frac{2xp}{y^{-1}p^2} - \frac{5y}{p^{-3}x^{-2}}$

Evaluate:

27. $-x - x^{-2} - xy^{-2}$ if $x + 1 = -1$ and $y = 3$

28. $-x(x - y^0)|y|$ if $x = -1$ and $y = -4$

Simplify:

29. $\frac{1}{-4^{-2}} - \sqrt[5]{-243}$

30. $-3[(-2^0 + 5) - (-3 - 7) - |-2|]$

LESSON 52 Percents greater than 100

When a problem discusses a quantity that increases, the final quantity is greater than the initial quantity. If we let the initial quantity represent 100 percent, the final percent will be greater than 100. This means that the "after" diagram representing the final quantity will be larger than the "before" diagram. **The "after" diagrams in this book will not be drawn to scale.**

To demonstrate, we will work three problems of this type. We will finish each problem by drawing diagrams that give a visual representation of the problem.

example 52.1 What number is 160 percent of 60? Find the number and then draw diagrams that depict the problem.

solution We substitute *WN* for *is*, 160 for *p*, and 60 for *of*.

$$\frac{P}{100} \times \text{of} = \text{is} \longrightarrow \frac{160}{100} \times 60 = WN \longrightarrow \frac{9600}{100} = WN \longrightarrow \mathbf{96 = WN}$$

Thus, our diagrams show a before of 60 and an after with 96, which is 160 percent.

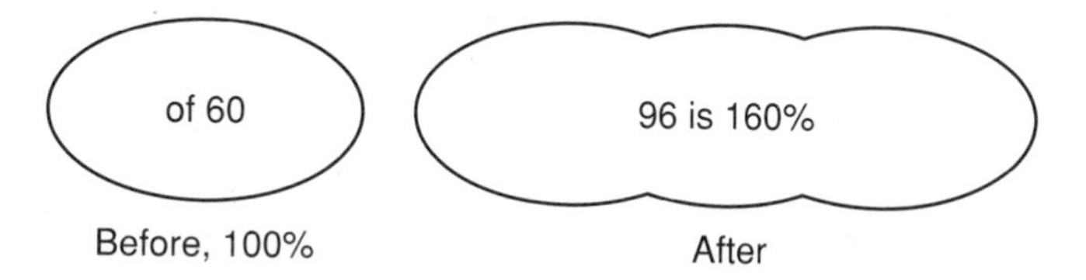

example 52.2 If 75 is increased by 150 percent, what is the result? Work the problem and then draw diagrams that depict the problem.

solution **We must be careful here. Seventy-five is the original number and is 100 percent. If we increase the percentage by 150 percent, the final percentage will be 250 percent.** We can restate the problem as follows: 250 percent of 75 is what number? We will use 250 for percent, 75 for *of*, and *WN* for *is*.

$$\frac{250}{100} \times 75 = WN \longrightarrow 2.5 \times 75 = WN \longrightarrow \mathbf{187.5 = WN}$$

Thus, our diagrams show a before of 75 and an after with 187.5, which is 250 percent.

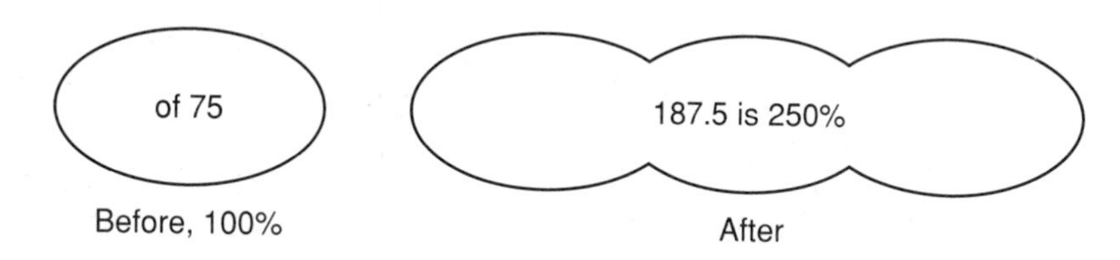

example 52.3 What percent of 90 is 306? Work the problem and then draw diagrams that depict the problem.

solution We use *WP* for *P*, 90 for *of*, and 306 for *is*.

$$\frac{P}{100} \times \text{of} = \text{is} \longrightarrow \frac{WP}{100} \times 90 = 306 \longrightarrow \frac{\cancel{100}}{\cancel{90}} \cdot \frac{WP}{\cancel{100}} \times \cancel{90} = 306 \cdot \frac{100}{90}$$

$$\mathbf{WP = 340\%}$$

Thus the diagram of the problem should show a before of 90 and an after with 306, which is 340 percent of 90.

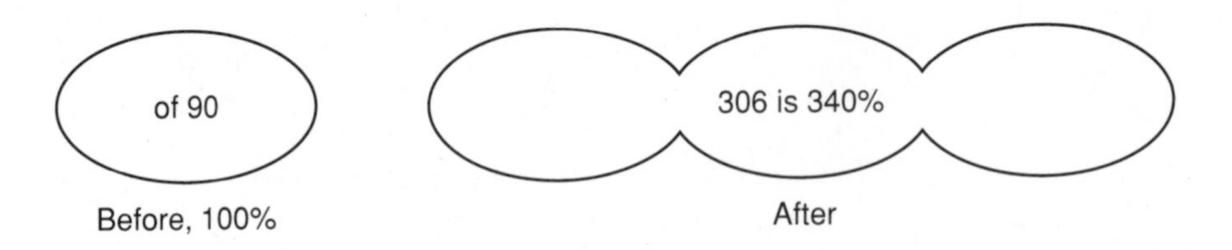

practice Work each problem and then draw the diagram.

a. If 20 is increased by 90 percent, what is the result?

b. What number is 270 percent of 80?

problem set 52

1. Lancelot and Guinevere found that the sum of twice a number and 5 is 13 less than the opposite of the number. Find the number.

2. The ratio of quotidian chores to exotic chores was 7 to 2. If 3780 chores were considered, how many were quotidian?

3. What decimal part of 1.07 is 2.1721?

4. If $-x - 11 = -9$ and $a + 1 = 1$, evaluate $x^2 - 2a$.

5. Find the area of this figure. Dimensions are in centimeters. Corners that look square are square.

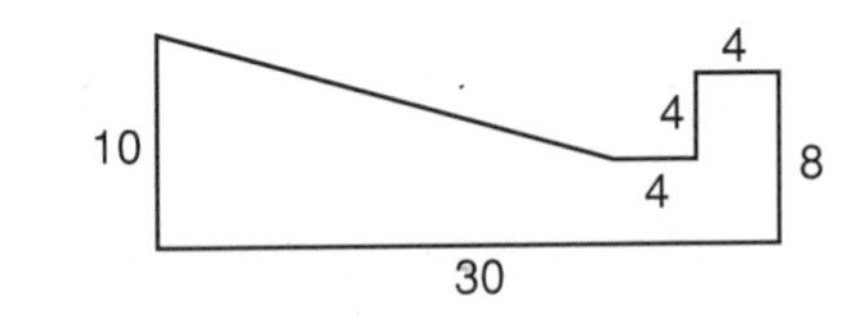

6. What percent of 50 is 700? Draw the diagram.

Solve:

7. $5 + 3\frac{1}{2}x = 2\frac{1}{4}$

8. $3x - 5(x - 4) = 2x + 7$

9. $-2[(-k - 3)(-2) - 3] = (-3 - 3k)(-2)^3 - 3^2$

10. Add. Write the answer in descending order of the variable:

$$4(x^2 - 3x + 5) - 2(x^3 + 2x^2 - 4) - (2x^4 - 3x^3 + x^2 + 3)$$

Multiply:

11. $(2x + 4)(5x - 3)$

12. $(x + 3)^2$

13. $(x + 3)(3x - 4)$

14. $(2x + 7)(2x - 7)$

15. Find the least common multiple of: 24, 60, and 450

Add:

16. $\frac{a}{x} + \frac{b}{cx^2} + d$

17. $\frac{m}{p^2k} - \frac{4a}{3pk} + \frac{6}{5pk^2}$

18. $a + \frac{bc}{m} - \frac{4mc}{x^2}$

19. $\frac{x}{mc} + \frac{b}{c} - \frac{4}{2kc^2}$

20. Write a conjunction that describes this graph.

21. Factor: $4x^2ym - 6xym + 2x^2y^2m^2$

Simplify:

22. $\dfrac{5x - 5}{5}$

23. $\dfrac{x^3y^{-4}p^0y^4p^2}{x^4xx^{-7}y^2p^4}$

24. Expand: $\left(\dfrac{ax^{-5}}{y^{-2}} + \dfrac{4x^3}{ay^2}\right)\dfrac{x^5}{ay^2}$

25. Simplify by adding like terms: $\dfrac{m^2x}{y^{-1}} - \dfrac{3m^2y}{x^{-1}} + 5mmyx - \dfrac{4x^2ym^2}{x}$

Evaluate:

26. $a^{-2}(a - a^3)$ if $a = -3$

27. $(p - a)(a - 2pa)$ if $a = -2$ and $p + 2 = 7$

Simplify:

28. -3^{-4}

29. $\dfrac{1}{-2^{-4}} - \sqrt[3]{125}$

30. $4 - [5(6 - 5^0) - (-5 - 2) - |-7|]$

LESSON 53 *Rectangular coordinates*

53.A first-degree equations in two unknowns

In Lesson 50, we noted that the **degree of a term** of a polynomial is the sum of the exponents of the variables of the term. Thus

$2x^3$	is a third-degree term
x	is a first-degree term
xyz	is a third-degree term

Also we said that the **degree of a polynomial** is the same as the degree of its highest-degree term. Thus

$x^3 + xy + m$	is a third-degree polynomial
$2x + 4y$	is a first-degree polynomial
$2x$	is a first-degree polynomial

If two polynomial expressions are connected by an equals sign, we call the equation a **polynomial equation. The degree of a polynomial equation is the same as the degree of its highest-degree term.** Thus

$x^4 - 3x^2 + 2 = 0$	is a fourth-degree polynomial equation
$xyz + y = 4$	is a third-degree polynomial equation
$y = 2x + 4$	is a first-degree polynomial equation

There is an infinite number of pairs of values of x and y that are solutions to any first-degree polynomial equation in two variables. We will use the equation

$$y = 2x + 4$$

to investigate. If we assign a value to x, the equation will then indicate the value of y that is paired with the assigned value of x. For instance, if we assign to x a value of 2, then

$$y = 2(2) + 4 \longrightarrow y = 8$$

the paired value of y is 8. If we give x a value of 2 and y a value of 8 in the original equation, we find that these values of x and y satisfy the equation and are solutions to the equation because the replacement of the variables by these numbers makes the equation a true statement.

$$\begin{aligned} y &= 2x + 4 \\ 8 &= 2(2) + 4 \\ 8 &= 4 + 4 \\ 8 &= 8 \qquad \text{True} \end{aligned}$$

If, in the original equation, we give x a value of -5, then

$$y = 2(-5) + 4 \longrightarrow y = -10 + 4 \longrightarrow \mathbf{y = -6}$$

we find that the paired value of y is -6. Thus the pair of values $x = -5$ and $y = -6$ will also satisfy the equation, for the use of both of these numbers in the original equation in place of x and y will cause the equation to become a true equation, as shown here.

$$\begin{aligned} y &= 2x + 4 \\ -6 &= 2(-5) + 4 \\ -6 &= -10 + 4 \\ -6 &= -6 \qquad \text{True} \end{aligned}$$

We can replace x with any real number and use the equation to find the value of y that the equation pairs with this value of x.

In both of the foregoing examples we **assigned** a value to the variable x. The variable x, to which we assign a value, is called the **independent variable.** We see that, in each case, the value of y **depends** on the value that we assigned to x. Therefore, in our examples, we call the variable y the **dependent variable.** We could have assigned a value to y and then used the equation to find the corresponding value of x, in which case y would be the independent variable and x would be the dependent variable. **It is customary, however, to avoid confusion, to use the letter x to designate the independent variable and to use the letter y to designate the dependent variable. We will follow this custom in this book.**

53.B ordered pairs

In the preceding section, we found that, given the equation

$$y = 2x + 4$$

if we let $x = 2$ and $y = 8$, this pair of values of x and y will make the equation a true equation. Also, the pair of values $x = -5$ and $y = -6$ will make the equation a true equation. Since writing $x = 2$ and $y = 8$ and writing $x = -5$ and $y = -6$ is rather cumbersome, it is customary to write just (2, 8) and (−5, −6), **with the x value always designated by the first number in the parentheses and the y value always designated by the second number.** Since the numbers are written in order with x first and y second, we designate this notation as an **ordered pair** of x and y. The general form of an ordered pair of x and y is (x, y). If two other variables are used instead of

x and y, it is necessary to designate which variable will be represented by each of the entries in the parentheses. If the variables m and p are to be used and we wish to write the m value first, we could designate this at the outset by making a statement about the order pair (m, p).

It is important to remember that in ordered pairs of x and y, the first number will always designate the value of x and the second number will always designate the value of y.

53.C Cartesian coordinate system

In Lesson 37 we learned that the solution of an equation or inequality in one variable can be presented in graphical form on a single number line as we show here.

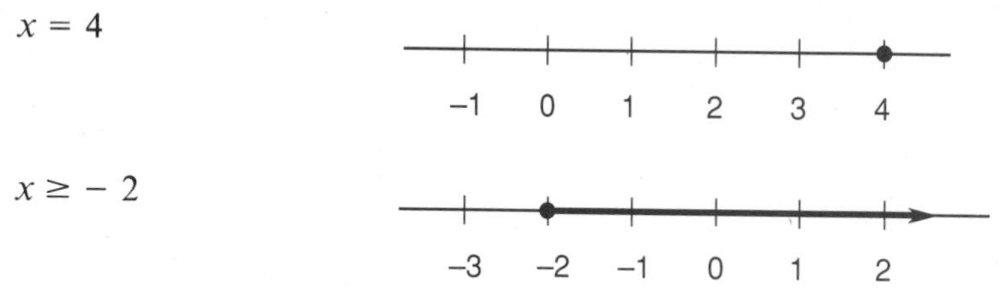

The graphical solution to equations or inequalities that contain two variables cannot be displayed on a single number line. We must have one number line for one of the variables and another number line for the other variable. It is customary to use the variables x and y and draw the x number line horizontally, and the y number line vertically, and to let the number lines intersect at the origin of both number lines. The positive values of x are located to the right of the origin on the horizontal or x number line, and the positive values of y are located above the origin on the vertical or y number line. The x number line is called the **x axis** or **horizontal axis** and the y number line is called the **y axis** or **vertical axis.**

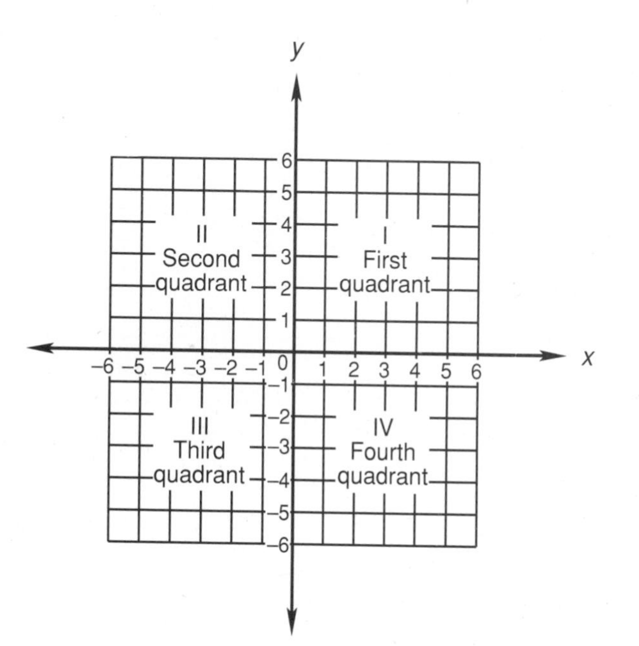

The figure shows that the two number lines divide the plane into four *quarters,* or *quadrants.* The quadrants are named the **first quadrant, second quadrant, third quadrant,** and **fourth quadrant** as shown. The figure in its entirety is called a system of **Cartesian coordinates** or a **Cartesian coordinate system** after the famous seventeenth-century French philosopher and mathematician René Descartes. It is also called a **rectangular coordinate system** and is sometimes called a **coordinate plane.**

When we use a single number line to graph the solution to an equation in one variable such as $x = 2$ (below), we call the mark we make on the number line the **graph** of the point, and conversely we call the number the **coordinate** of the point on the line designated by the graph. The point is defined to be without size, and thus the graph is not the point but denotes the location of the point.

−3 −2 −1 0 1 2 3

Since a rectangular coordinate system has two number lines, **it is necessary to associate with every point on a rectangular coordinate system two numbers or coordinates.** These numbers designate the location of the point. The following figure shows the graphs of four points. Written by each point is the ordered pair of values of x and y that we associate with the point; these numbers are the x and y coordinates of the point. **The number written first is always the x coordinate of the point and is called the abscissa of the point.** This number denotes the measure of the distance of the point to the right (+) or left (−) of the vertical axis. **The number written second is always the y coordinate of the point and is called the ordinate of the point.** This number denotes the measure of the distance of the point above (+) or below (−) the horizontal axis.

example 53.1 Graph the points whose coordinates are given:

(a) (3, 3) (b) (−4, 0) (c) (0, −3) (d) (4, −2)

solution In the figure below we place four dots, one to mark the location of each point.

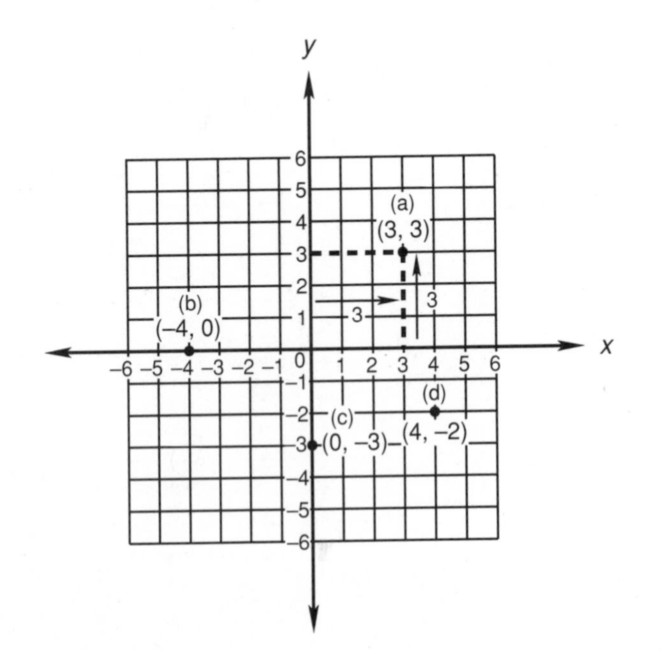

In the first quadrant we show the graph of the point (3, 3), which is **3 units to the right** of the y axis and thus has an **x coordinate of +3.** The point is also **3 units above** the x axis and thus has a **y coordinate of + 3.** The next point, (−4, 0), is located 4 units to the left of the y axis and thus has an x coordinate of −4. The point is no units above or below the x axis (it is on the x axis) and thus has a y coordinate of 0. The exact positions of the other two points shown are designated by the ordered pairs $(0, -3)$ and $(4, -2)$. Thus we see that the location of every point on the coordinate plane can be designated by stating the x and y coordinates of the point.

We have seen that a point on a number line is without size and that the graph of a point is not the point but only denotes the location of the point. In the same way, a point in a rectangular coordinate system is without size, and the graph of the point is not the point but designates the location of the point.

practice Graph the points whose coordinates are:

a. $(-6, -7)$ **b.** $(4, -5)$

problem set 53

1. Gwen and Norma found that the sum of 4 times the opposite of a number and -3 is 27 larger than the number. What is the number?

2. Pragmatism was on the rise and thus the ratio of pragmatists to quixotics was 7 to 4. If 14,740 delegates attended the convention, how many were pragmatists?

3. What fraction of $2\frac{1}{8}$ is $\frac{1}{4}$?

Work each problem and then draw the diagrams.

4. One hundred thirty percent of what number is 78?

5. Fifteen percent of what number is 10.5?

6. What percent of 450 is 288?

Solve:

7. $3\frac{2}{5}x - 3 = 4\frac{1}{8}$ **8.** $0.4m - 2 - 0.2m = 1.4 + m$

Multiply:

9. $(3x - 2)(x + 4)$ **10.** $(5x - 3)^2$

11. $(2x - 5)(3x - 2)$ **12.** $(5x - 7)(6x - 1)$

13. Find the volume and the surface area of the right prism whose base is shown and whose sides are 2 yards high. Dimensions are in yards.

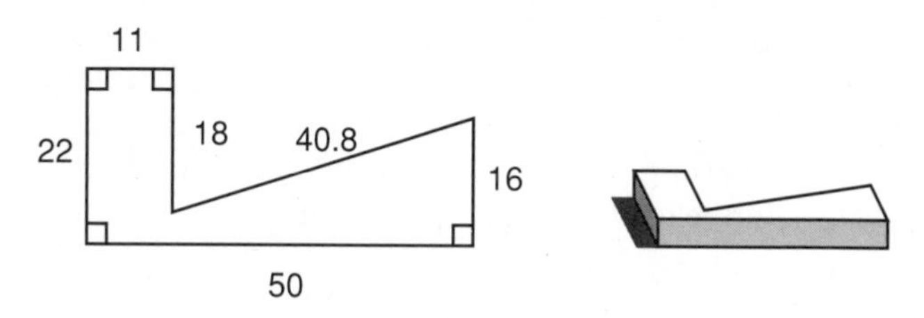

14. Use eight unit multipliers to convert 400 square feet to square kilometers.

Graph these ordered pairs of x and y on a rectangular coordinate system:

15. $(-3, 4)$ **16.** $(-1, -3)$

Add:

17. $4 + \frac{2}{x^2} - \frac{5}{a^2x}$ **18.** $\frac{p}{a^2m} - \frac{4}{a} - k$

19. $\frac{3ax}{m} + \frac{4x}{am^2} + \frac{2}{mx}$

20. $\frac{2a}{x} - \frac{5}{p^2x} - 3m$

21. Graph $x \not\leq 4$ on a number line.

Simplify:

22. $\frac{3ax - 3a}{3a^2}$

23. $\frac{myy^{-3}m^{-4}y^{-2}}{x^0y^2y^{-4}y^2m^{-7}}$

24. Expand: $\frac{3x^{-4}}{y^4}\left(\frac{2x^{-4}}{y^4} - \frac{3x^2}{y^2a}\right)$

25. Simplify by adding like terms: $\frac{3x^2y^{-2}}{m^5} - \frac{3x^2y^2}{m^5} - \frac{4xx^3m^{-5}}{x^2y^2} + \frac{6m^{-5}}{x^{-2}y^{-2}}$

Evaluate:

26. $-xa(a - xa^0)$ if $a = -3$ and $x - 1 = 6$

27. $(xa - a)(-a^{-4}x)$ if $a = -3$ and $x = 7$

Simplify:

28. $-(-2)^{-2}$

29. $\frac{1}{(-2)^{-3}} - \sqrt[5]{243}$

30. $4[(6 - 2^0) - (-3 + 5)2]$

LESSON 54 Graphs of linear equations

If we use a rectangular coordinate system to graph the ordered pairs that satisfy a first-degree polynomial equation in two variables, we find that the graph is the graph of a straight line. For this reason, we call a first-degree polynomial equation in one or more variables a **linear equation.** The following equations are examples of linear equations in two variables,

$$y - 2x + 1 = 0 \qquad x - 4 = -2y \qquad x + y = 0$$

for the graph of each of the equations is a straight line.

To graph a linear equation in two variables, we need to know only two ordered pairs of x and y that satisfy the equation since only two points are needed to determine the graph of a line. But since the topic is a new one, in the examples we will learn how to find three or more ordered pairs of x and y that lie on the line.

example 54.1 Graph: $y = 2x - 1$

solution We begin by making a table and choosing convenient values for x. The values chosen for x should not be too close together. Also they should not be so large that they will graph off of our coordinate system. Numbers such as 0, 3, −3, 2, and −2 usually work well.

x	0	3	−3	2	−2
y					

Now we will use the numbers 0, 3, −3, 2, and −2 one at a time as replacement values for x in the equation $y = 2x - 1$ to find the paired values of y.

WHEN $x = 0$	WHEN $x = 3$	WHEN $x = -3$	WHEN $x = 2$	WHEN $x = -2$
$y = 2(0) - 1$	$y = 2(3) - 1$	$y = 2(-3) - 1$	$y = 2(2) - 1$	$y = 2(-2) - 1$
$y = -1$	$y = 5$	$y = -7$	$y = 3$	$y = -5$

Next we complete the table by entering the values of y that the equation has paired with the chosen values of x.

x	0	3	−3	2	−2
y	−1	5	−7	3	−5

Thus we have found five ordered pairs of x and y that satisfy the equation and therefore lie on the graph of the equation. These ordered pairs are (0, −1), (3, 5), (−3, −7), (2, 3), and (−2, −5). In the next figure we have graphed the points designated by these ordered pairs and have connected them with a straight line. The point (−3, −7) was not graphed since this point fell outside the borders of our coordinate system.

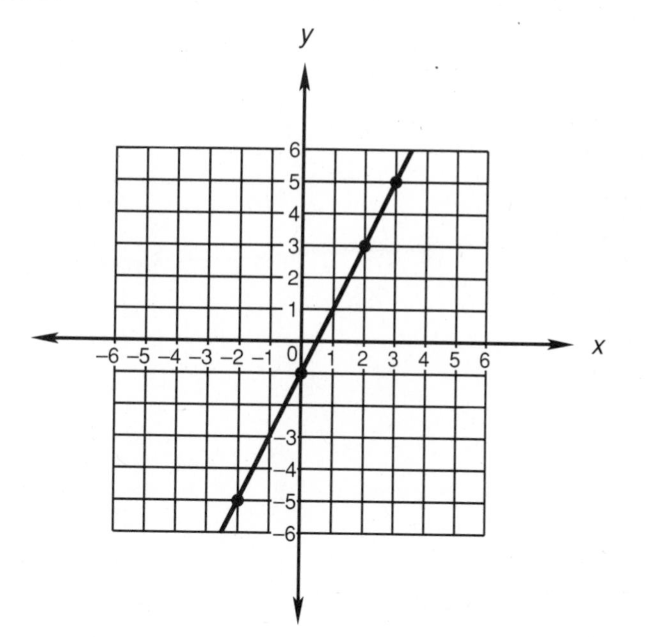

Since a line (straight line) in mathematics is defined to be infinite in length, we have graphed only a segment of the line, or a line segment. Also, we remember that a mathematical line has no width, and since the line we have drawn has a width that can be measured, it is not a mathematical line but is a **graph of a mathematical line** and indicates the location of the mathematical line in question.

example 54.2 Graph the equation: $y = -\frac{1}{2}x + 2$

solution We begin by making a table and choosing convenient values for x.

x	0	2	−2
y			

Now we find the values of y that the equation $y = -\frac{1}{2}x + 2$ pairs with the chosen values of x.

WHEN $x = 0$	WHEN $x = 2$	WHEN $x = -2$
$y = -\dfrac{1}{2}(0) + 2$	$y = -\dfrac{1}{2}(2) + 2$	$y = -\dfrac{1}{2}(-2) + 2$
$y = 2$	$y = 1$	$y = 3$

We insert 2, 1, and 3 to complete the table.

x	0	2	−2
y	2	1	3

We finish by graphing the points and drawing the line through the points.

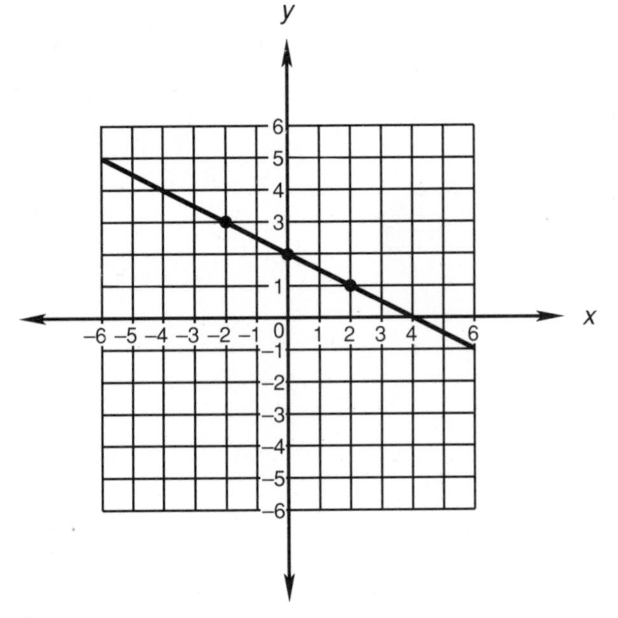

example 54.3 Graph the equation: $y = -x$

solution (a) Make a table.

x	0	3	−3
y			

(b) Find the value of y for each value of x.

WHEN $x = 0$	WHEN $x = 3$	WHEN $x = -3$
$y = -(0)$	$y = -(3)$	$y = -(-3)$
$y = 0$	$y = -3$	$y = 3$

(c) Complete the table.

x	0	3	−3
y	0	−3	3

(d) Graph the points and draw the line.

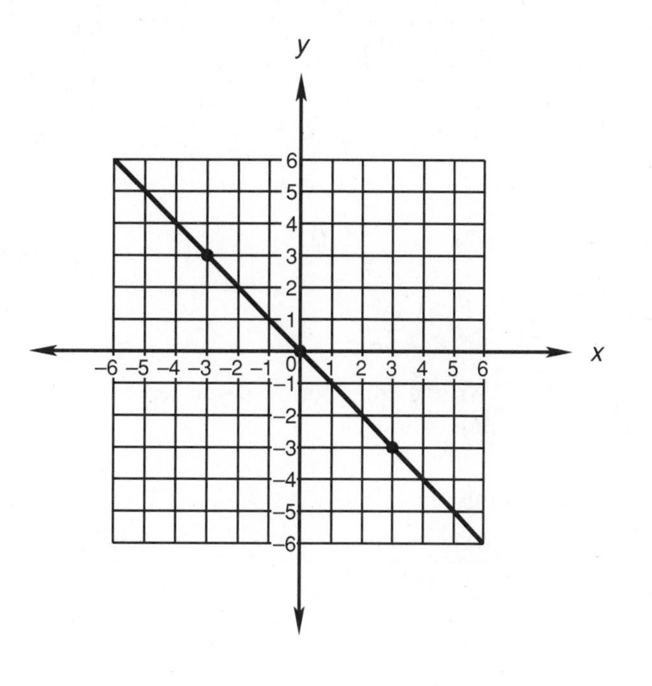

practice Graph the equations:

a. $y = -x + 1$

b. $y = -\frac{1}{3}x + 2$

problem set 54

1. If the product of 5 and a number is decreased by 10 and this difference is multiplied by 3, the result is 22 less than the opposite of the number. What is the number?

2. The mourners wailed and ululated. If the ratio of the former to the latter was 2 to 21 and 805 mourners were in attendance, how many merely wailed?

3. What fraction of $3\frac{1}{5}$ is $2\frac{3}{4}$?

Work the problem and then draw the diagrams:

4. Three hundred seventy-five percent of 1300 is what number?

5. Sixty-five percent of what number is 260?

6. What percent of 18 is 27?

Solve:

7. $-1\frac{2}{9} + 2\frac{1}{5}p = -\frac{1}{3}$

8. $3x - [-(-2)]x + (-3)(x + 2) = 5x + (-7)$

Multiply:

9. $(5x - 4)(2x + 3)$ **10.** $(-5x - 2)(-x + 4)$ **11.** $(7x - 5)^2$

12. Find the volume of this right prism in **cubic feet.** Find the surface area in **square feet.** Dimensions are in **yards.**

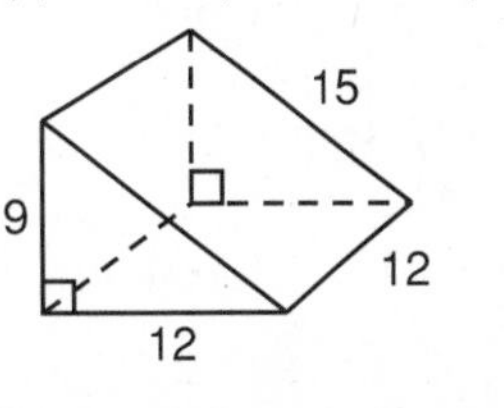

13. If $x - 3^2 = -9$, evaluate x^4.

14. The average of the first six scores was 20. The first five scores were 5, 7, 12, 14, and 22. What was the sixth score?

Graph on a rectangular coordinate system:

15. $y = -2x + 4$

16. $y = x - 3$

Add:

17. $4 + \frac{x}{y^2} - \frac{2a}{xy^3}$

18. $\frac{a}{x^2c} - \frac{3a}{x^3c^2} - \frac{4}{xc} + \frac{3a + 2}{x^3c}$

19. $\frac{b}{m^2p} - \frac{1}{c^2m^3} + 4$

20. $x + \frac{4}{2x^2p^5} + \frac{xy}{4xp}$

21. Write a conjunction described by this graph.

(Number line: open circle at −2, closed circle at 4, segment between.)

Simplify:

22. $\frac{4xy^2 - xy^2}{xy}$

23. $\frac{m^2p^4x^{-2}x^2x^0p^6}{m^2p^{-4}x^0p^0x^2}$

24. $\frac{x^3y^0x^{-1}p^2y^{-2}}{ppx^{-4}x^6y^{-2}}$

25. Expand: $\frac{2x^{-4}}{y^2}\left(\frac{x^2}{2y^{-2}} - \frac{6x^4}{y^7p}\right)$

26. Simplify by adding like terms: $\frac{x}{y} - \frac{3x^2x^{-1}y^2}{y^3} + \frac{2x^2}{xy^2} - \frac{4xxy^{-1}}{xy}$

Evaluate:

27. $p - x(p^{-2} - x^0)$ if $p = -2$ and $x = 5$

28. $x - ax(x - ax)$ if $x = -4$ and $a - 3 = -4$

Simplify:

29. $\frac{1}{-3^{-2}} - [3 - (-3)^3]$

30. $-3[-2(-2^0 - 5) - (3 - 2) - |3|]$

LESSON 55 Vertical and horizontal lines · Overall average

55.A vertical and horizontal lines

In Lesson 54 we graphed the following equations.

$$y = 2x - 1 \qquad y = -\frac{1}{2}x + 2 \qquad y = -x$$

We note that each of these equations has both an x term and a y term. We also note that none of the graphs was a vertical line or a horizontal line. **Some equations of a straight line contain either an x term or a y term but not both an x term and a y term, and the graph of these equations is either a vertical line or a horizontal line.**

example 55.1 Graph: $y = 4$

solution This equation tells us that the y coordinate of every point on the line is 4. On the line, we indicate three points. Note that each of these points has a y coordinate of 4.

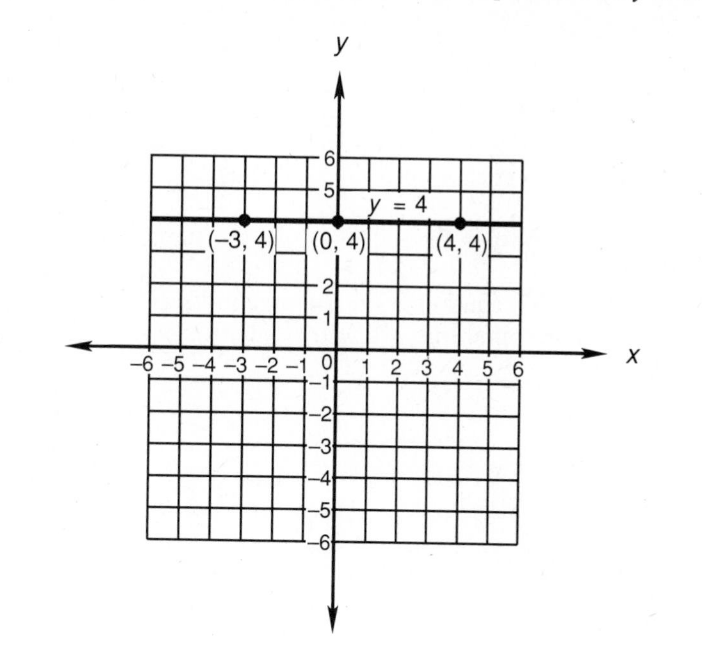

example 55.2 Graph: $x = -2$

solution This equation places no restriction on the y coordinates of the points on the line. It tells us that the x coordinate of every point on the line is -2 regardless of the y coordinate of the point. Note that each point graphed has an x coordinate of -2.

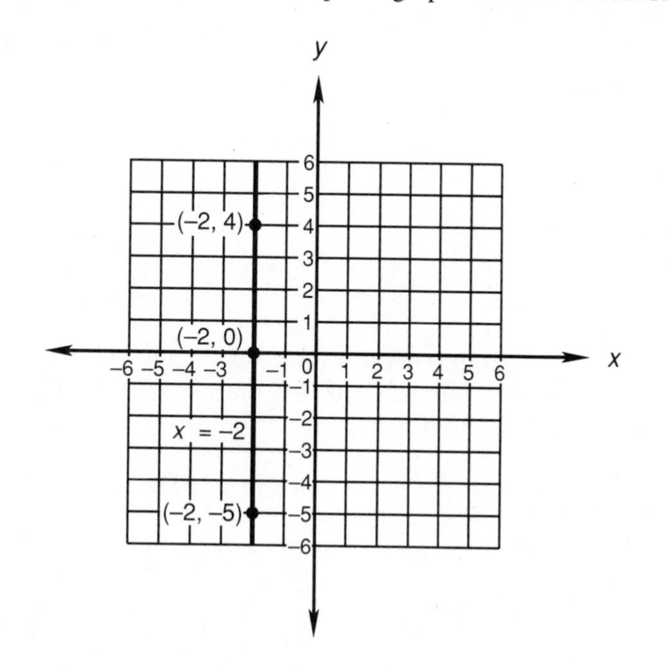

We will remember that equations of the form

$$y = \pm k \qquad \text{or} \qquad x = \pm k$$

such as

$$y = 2 \qquad y = -4 \qquad x = 5 \qquad x = -4$$

are special forms of the equation of a straight line. Equations of the form

$$y = \pm k$$

are horizontal lines, while equations of the form

$$x = \pm k$$

are vertical lines.

example 55.3 Graph $y = 2x$ and $y = 2$ on the same coordinate system.

solution These equations are often confused. The equation $y = 2$ is the equation of a horizontal line, while the equation $y = 2x$ is not.

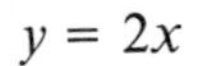

x	0	2	−2
y	0	4	−4

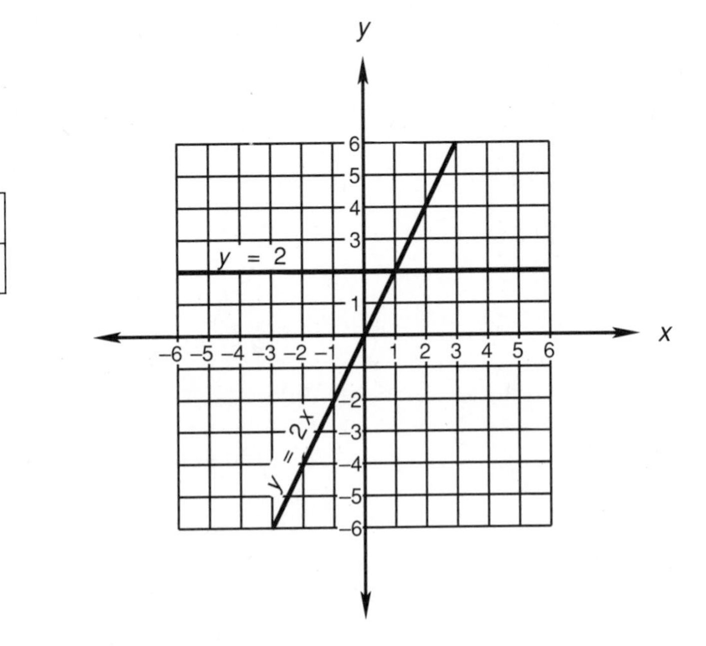

55.B overall average

The average of the averages is seldom the overall average. Suppose the average of the first six numbers is 8 and the average of the next four numbers is 10. If we find the average of the averages, we add 8 to 10 and divide by 2.

$$\text{Average of the averages} = \frac{8 + 10}{2} = 9$$

This is the average of the averages but is not the overall average. The overall average is

$$\text{Overall average} = \frac{\text{sum of all the numbers}}{\text{number of numbers}}$$

The average of the first six numbers is 8, so the sum of the first six numbers is 6×8, or 48. The average of the next four numbers is 10, so the sum of the next four numbers is $4 \times 10 = 40$. The number of numbers is $6 + 4 = 10$, so the correct overall average is

$$\text{Overall average} = \frac{(6 \times 8) + (4 \times 10)}{6 + 4} = \frac{48 + 40}{10} = \mathbf{8.8}$$

example 55.4 The average of the first 90 numbers was 4. The average of the next 10 numbers was 6. What was the overall average?

solution **The overall average is the sum of all the numbers divided by the number of numbers.**

$$\text{Overall average} = \frac{(4 \times 90) + (10 \times 6)}{90 + 10} = \frac{360 + 60}{100} = \mathbf{4.2}$$

The overall average is much closer to the average of the first group than to the average of the second group because the first group had 90 numbers and the second group had only 10 numbers.

practice Graph these lines on a rectangular coordinate system:

a. $y = -2x$ **b.** $y = -2$

c. The average of the first 5 weights was 10. The average of the next 10 weights was 25. What was the average of all the weights?

problem set 55

1. If the product of a number and -5 is increased by 6, the result is 2 less than 3 times the opposite number. Find the number.

2. Unfortunately, the ratio of the erudite to the unlettered was 2 to 7. If there were 3717 under consideration, how many were erudite?

3. What fraction of $\frac{1}{5}$ is $2\frac{7}{8}$?

Work the problems and then draw the diagrams:

4. Four hundred-sixty percent of 700 is what number?

5. Ninety-three percent of what number is 651,000?

6. What percent of 2000 is 10?

Solve:

7. $3\frac{1}{6}k + \frac{3}{4} = 2\frac{1}{5}$

8. $-(-x) - 3(-2x) + 3(-2 + x) = -(x - 5)$

Multiply:

9. $(3x + 5)(2x + 7)$ **10.** $(4x - 3)^2$

11. Find the least common multiple of: 1575, 25, and 14

Graph these lines on a rectangular coordinate system:

12. $y = \frac{1}{3}x - 3$ **13.** $y = -3$

14. Find the surface area and the volume of the right prism whose base is shown and whose sides are 2 yards high. Dimensions are in yards. Corners that look square are square.

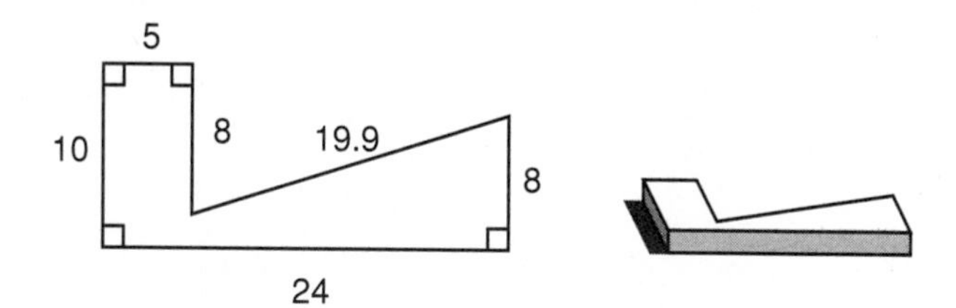

15. There were 17, 27, and 56 indigenous plants in the first three gardens. If the four gardens had an average of 50 indigenous plants, how many indigenous plants were in the fourth garden?

16. Use five unit multipliers to convert 24,000 kilometers to miles.

Add:

17. $x + \frac{2x}{m^2} - \frac{3}{m^2x^3}$

18. $\frac{x}{k^2p} - \frac{3x}{k^2p} + 7$

19. $-3x + \frac{2}{xp^2} - \frac{5x}{x^3p}$

20. $\frac{k}{4m^2} + \frac{k^2}{8} - 3$

21. Graph $-3 \not\leq x$ on a number line.

22. Factor: $4k^2pz - 6k^3p^2z^5 - 2k^2p^2z^2 - 4kp$

Simplify:

23. $\frac{6xay - 24xay^2}{6xay}$

24. $\frac{k^2m^{-2}kp^{-2}k^5p^0}{p^5kp^{-5}k^{-4}m^4}$

25. Expand: $\left(\frac{x^2m^2}{3} - \frac{5x^5p^0}{m^{-4}}\right)\frac{3x^{-2}y^0}{m^2}$

26. Simplify by adding like terms: $a^2k^2y^{-1} - \frac{4k^2}{a^{-2}y} + \frac{2k^2a}{a^{-1}y} - \frac{6k^{-4}}{k^2y}$

Evaluate:

27. $a^3(a - a^{-2})$ if $a = -2$

28. $p - pm(m - pm)$ if $m = -2$ and $p = 4$

Simplify:

29. $27(-3)^{-3} - 5^2 - \sqrt[3]{-125}$

30. $-2(-1) - 3[(2 - 4^0) - 2(-3 - 5)] + |2|$

LESSON 56 Addition of rational expressions

If the denominator of a term in an addition problem is in the form of a sum, this sum must be a factor of the least common multiple of the denominators.

example 56.1 Add: $\frac{a}{x} + \frac{b}{x + y}$

solution The least common multiple of the denominators is $x(x + y)$, and we will use this expression as the new denominator. We will use the **denominator-numerator same-quantity theorem** to find the new numerators.

$$\frac{\quad}{x(x + y)} + \frac{\quad}{x(x + y)}$$

The original denominator of the first term was x, and it has been multiplied by

$(x + y)$, so the original numerator, a, must also be multiplied by $x + y$.

$$\frac{a(x + y)}{x(x + y)} + \frac{\quad}{x(x + y)}$$

The original denominator of the second term was $x + y$, and it has been multiplied by x, so the original numerator, b, must also be multiplied by x. Then we add the numerators.

$$\frac{a(x + y)}{x(x + y)} + \frac{xb}{x(x + y)} = \frac{\mathbf{a(x + y) + xb}}{\mathbf{x(x + y)}}$$

There are many equally correct forms of the answer. We may multiply out in either the numerator or the denominator or both. Thus all three of the following forms are correct.

$$\frac{\mathbf{ax + ay + xb}}{\mathbf{x(x + y)}} \qquad \frac{\mathbf{a(x + y) + xb}}{\mathbf{x^2 + xy}} \qquad \frac{\mathbf{ax + ay + xb}}{\mathbf{x^2 + xy}}$$

Since all the forms are correct, no one form is preferred. In mathematics, preference is reserved to the individual.

example 56.2 Add: $\frac{4x + a}{a + b} + \frac{c}{x}$

solution We begin by writing the fraction lines and the new denominators. Then we find the new numerators and add.

$$\frac{\quad}{x(a + b)} + \frac{\quad}{x(a + b)}$$

$$\frac{x(4x + a)}{x(a + b)} + \frac{c(a + b)}{x(a + b)} = \frac{\mathbf{4x^2 + ax + ca + cb}}{\mathbf{x(a + b)}}$$

We could multiply $x(a + b)$ in the denominator, but we decide to leave it in the more concise form.

example 56.3 Add: $\frac{a + b}{x} + \frac{c}{m} + d$

solution We begin by writing the fraction lines with the new denominators, and then we find the new numerators and add.

$$\frac{\quad}{xm} + \frac{\quad}{xm} + \frac{\quad}{xm}$$

$$\frac{m(a + b)}{mx} + \frac{cx}{mx} + \frac{dxm}{mx} = \frac{\mathbf{m(a + b) + cx + dxm}}{\mathbf{xm}}$$

We could multiply $m(a + b)$ in the numerator, but we decide to leave it as it is.

example 56.4 Add: $\frac{a}{b + c} - x + \frac{d + m}{k}$

solution The first step is to write the new denominators. Then we find the new numerators and add.

$$\frac{\quad}{k(b + c)} - \frac{\quad}{k(b + c)} + \frac{\quad}{k(b + c)}$$

$$\frac{ka}{k(b + c)} - \frac{xk(b + c)}{k(b + c)} + \frac{(d + m)(b + c)}{k(b + c)} = \frac{\mathbf{ka - xk(b + c) + (d + m)(b + c)}}{\mathbf{k(b + c)}}$$

Again we choose not to use the distributive property, and we leave the answer in the more concise form.

practice Add:

a. $\frac{x}{y} - b + \frac{c + d}{m}$

b. $\frac{5b + c}{a + b} - \frac{x}{b} + c$

problem set 56

1. Virginia and Weir found that if the product of 3 and a number was increased by 13, the result was 12 less than twice the opposite of the number. What was their number?

2. The apes were either prognathous or platyrrhine. If the ratio of the former to the latter was 2 to 23 and 3500 were observed, how many were prognathous?

3. 2.14 times what number is 0.00642?

Solve and then draw the diagrams:

4. Three hundred forty-seven percent of what number equals 2429?

5. Eighty-seven percent of 300 is what number?

6. What percent of 460 is 1150?

Solve:

7. $3\frac{1}{8}p + 2\frac{1}{4} = \frac{1}{6}$

8. $-[-|-3|(-2 - m) - 4] = -2[(-3 - m) - 2m]$

Multiply:

9. $(4x - 2)(3x + 5)$ 10. $(5x - 1)^2$ 11. $(5x - 7)(7x - 5)$

Graph on a rectangular coordinate system:

12. $y = -3$ 13. $x = 4$

14. $y = -2x + 1$ 15. $y = 3x - 4$

Add:

16. $\frac{3}{2x^2y} - \frac{ab}{4x^3y} - c$

17. $\frac{a}{b + c} - x\frac{4}{b^2}$

18. What is the volume and the surface area of the right prism whose base is the figure shown and whose height is 2 meters? Dimensions are in meters. Corners that look square are square.

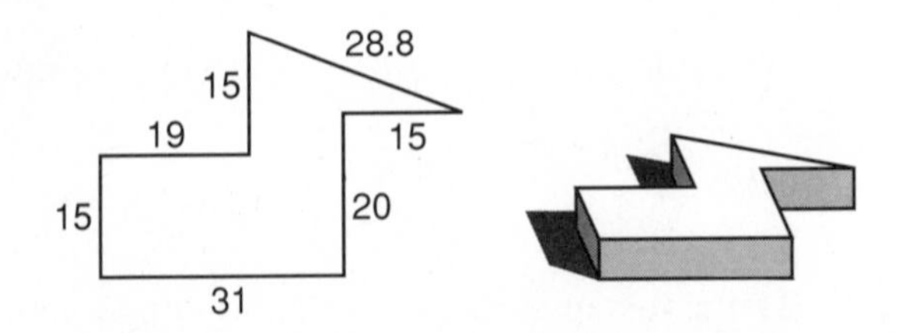

19. Use five unit multipliers to convert 60,000 miles to kilometers.

20. The average weight of the first 3 students was 135 pounds. The average weight of the next 97 students was 163 pounds. What was the average weight of all 100 students?

21. Write a conjunction that describes this graph.

(Number line: closed dot at 4, open dot at 7, segment between them shaded.)

22. Factor: $9k^2bm^4 - 3kb^4m^2 + 12kb^3m^3$

Simplify:

23. $\dfrac{3ap^2m - 6ap^2m^2}{3ap^2m}$

24. $\dfrac{m^{-2}mmm^{-3}xx^{-4}}{xm^3x^{-3}m^{-3}x^4}$

25. Expand: $\dfrac{x^2y^0p}{m^{-2}}\left(\dfrac{p^{-3}m^2}{k} - \dfrac{p^0pm^2}{x^2}\right)$

26. Simplify by adding like terms: $mx - \dfrac{3}{m^{-1}x^{-1}} + \dfrac{4mx}{m^2x^2} + \dfrac{5m^2x^{-1}}{mx^{-2}}$

Evaluate:

27. $k^2 - kp - p(-k)$ if $k = -3$ and $p - 4 = 1$

28. $x - x(x^{-2} - x^0)$ if $x = -3$

Simplify:

29. $\dfrac{1}{(-2)^{-3}} + \sqrt[3]{27}$

30. $-\{[(-2) - 3^0] - [(2 - 3)(-2) + 3]\}$

LESSON 57 Power rule for exponents

The last eight problem sets have contained problems whose solutions have required the use of one or more of the following definitions and theorems.

DEFINITION:	$x \cdot x \cdot x \cdot x \cdots x = x^m$	(m is a whole number)
DEFINITION:	$x^0 = 1$	$(x \neq 0)$
DEFINITION:	$x^1 = x$	
DEFINITION:	$x^{-n} = \dfrac{1}{x^n}$	$(x \neq 0)$
PRODUCT RULE:	$x^m \cdot x^n = x^{m+n}$	$(x \neq 0)$
QUOTIENT RULE:	$\dfrac{x^m}{x^n} = x^{m-n} = \dfrac{1}{x^{n-m}}$	$(x \neq 0)$

To complete the list of rules for exponents, we will introduce the **power rule for exponents,** which is a logical extension of the first definition listed above. By using this definition, we can show that x is to be used as a factor three times by writing x^3 as

$$x \cdot x \cdot x = x^3 \qquad \text{or} \qquad (x)^3$$

If we wish to show that x^5 should be used as a factor three times, we could write $(x^5)^3$:

$$x^5 \cdot x^5 \cdot x^5 = (x^5)^3$$

Since $x^5 \cdot x^5 \cdot x^5$ equals x^{15}, then $(x^5)^3$ must also equal x^{15}.

$$(x^5)^3 = x^{15}$$

Thus, when an exponential expression is raised to a power, we simplify by multiplying the exponents. We call this rule the *power rule for exponents.*

POWER RULE FOR EXPONENTS

If m and n and x are real numbers and $x \neq 0$,

$$(x^m)^n = x^{mn}$$

examples (a) $(a^5)^2 = \mathbf{a^{10}}$ (b) $(x^{-4})^{-2} = \mathbf{x^8}$ (c) $(m^{-2})^4 = \mathbf{m^{-8}}$

(d) $(p^{-7})^{-7} = \mathbf{p^{49}}$ (e) $(x^3)^5 = \mathbf{x^{15}}$ (f) $(m^k)^2 = \mathbf{m^{2k}}$

The use of the power rule in expressions such as those above usually gives little trouble, but the use of this rule to simplify expressions such as $(2x^5y^2z)^3$ is more complicated. We know that the notation indicates that $2x^5y^2z$ is to be used as a factor three times because 3 is the exponent of the whole expression.

$$(2x^5y^2z)(2x^5y^2z)(2x^5y^2z)$$

Since the order of multiplication of factors of a product does not affect the value of the product, we will rearrange the order of multiplication to get

$$2 \cdot 2 \cdot 2 \cdot x^5 \cdot x^5 \cdot x^5 \cdot y^2 \cdot y^2 \cdot y^2 \cdot z \cdot z \cdot z$$

which could be written by the use of the product rule as

$$2^3x^{15}y^6z^3 \qquad \text{which equals} \qquad \mathbf{8x^{15}y^6z^3}$$

To obtain the same result by using the power rule,

$$(2x^5y^2z)^3$$

we would have to multiply the exponent of each factor of the given term by 3 or raise each factor of the given term to the third power.

$$(2x^5y^2z)^3 = (2)^3(x^5)^3(y^2)^3(z)^3 = \mathbf{8x^{15}y^6z^3}$$

This is clear to the mathematician from the notation that $(x^m)^n = x^{mn}$, but it is sometimes not clear to the student at this level. Therefore, we will state it as follows: **To raise a term that contains no indicated additions to a given power, the exponent indicating the power is multiplied by the exponent of every factor of the numerator and the denominator** (if there is one) **of the term.** For example,

(a) $\left(\dfrac{3x^{-2}y^5}{z^4}\right)^{-2} = \dfrac{\mathbf{x^4y^{-10}}}{\mathbf{9z^{-8}}}$ (b) $(2a^{-2}b^2z^{-10})^{-5} = \dfrac{\mathbf{a^{10}b^{-10}z^{50}}}{\mathbf{32}}$

(c) $\left(\dfrac{4xy}{m^{-2}}\right)^2 = \dfrac{4^2x^2y^2}{m^{-4}} = \dfrac{\mathbf{16x^2y^2}}{\mathbf{m^{-4}}}$ (d) $\left(\dfrac{3xy^{-2}}{p^5}\right)^{-2} = \dfrac{3^{-2}x^{-2}y^4}{p^{-10}} = \dfrac{\mathbf{x^{-2}y^4}}{\mathbf{9p^{-10}}}$

(e) $\left(\dfrac{3x^0y}{k^4}\right)^3 = \dfrac{\mathbf{27y^3}}{\mathbf{k^{12}}}$ (f) $\left(\dfrac{2^0x^{-2}y^4}{z}\right)^{15} = \dfrac{\mathbf{x^{-30}y^{60}}}{\mathbf{z^{15}}}$

practice Simplify:

a. $\left(\frac{3x^0}{y^2}\right)^{-2}$ b. $\left(\frac{3y^{-2}z^5}{x^3}\right)^{-3}$ c. $(3m^{-3}c^2y^{-9})^{-4}$

problem set 57

1. If the sum of a number and -5 is multiplied by -3, the result is 2 greater than 9 times the opposite of the number. What is the number?

2. The proboscidians led the parade of pachyderms. If the ratio of proboscidians to nonproboscidians was 13 to 5 and there were 756 pachyderms in the parade, how many were proboscidians?

3. What fraction of $3\frac{7}{8}$ is $\frac{1}{4}$?

Solve and then draw the diagrams:

4. Three hundred-twenty percent of what number equals 192?

5. What percent of 98 is 3.92?

6. If 72 is increased by 130 percent, what is the result?

Solve:

7. $\frac{1}{4} + 2\frac{1}{5}k + 3\frac{2}{9} = 0$

8. $-[-(-k)] - (-2)(-2 + k) = -k - (4k + 3)$

Multiply:

9. $(2x - 4)(x - 3)$ 10. $(3x + 5)^2$

Graph on a rectangular coordinate system:

11. $x = -1\frac{1}{2}$ 12. $y = 2x + 2$ 13. $y = -\frac{1}{3}x - 2$

Add:

14. $\frac{x + y}{x^2y} + \frac{y}{x^4}$ 15. $\frac{4}{x - y} - \frac{3}{y}$ 16. $\frac{9 + 2b}{x} - 3 + \frac{6}{x^2y}$

17. Graph $4 \not\leq x$ on a number line.

18. Factor: $12x^4yp^3 - 4x^3y^2pz - 8x^2p^2y^2$

19. Use four unit multipliers to convert 2000 square miles to square inches.

20. Find the area of the following figure in **square meters.** All angles are right angles.

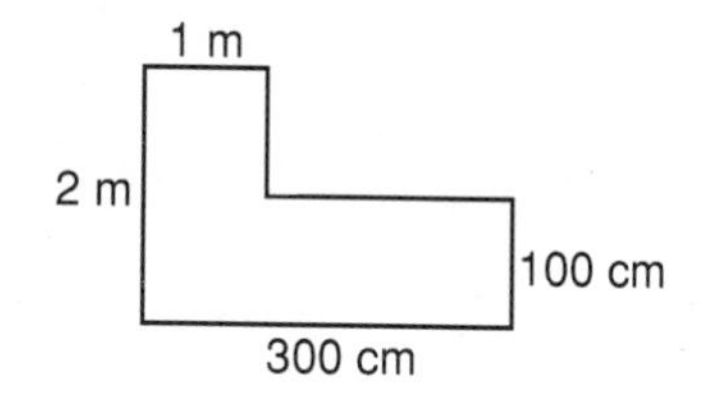

Simplify:

21. $\frac{mx - 5mx}{mx}$ 22. $(2x^2y^{-2}z)^{-2}(xy)^4$

23. $(5x^{-3}y^2)^2(x^0y)^5xy$ 24. $(xy)x(x^{-4}y)^2(x)^3$

25. Expand: $\frac{3x^{-2}}{y^{-4}}\left(\frac{ax^2}{3y^4} - \frac{3x^5}{y^{-2}}\right)$

26. Simplify by adding like terms:

$$2xy - 3yx + x^2y - y^2x + 4y^2xy^{-1} - \frac{2x^4x^{-2}}{y^{-1}}$$

Evaluate:

27. $xy - y^2(x - y)$ if $x = -2$ and $y + 11 = 14$

28. $x^{-3} - x^2(-x)$ if $x = -2$

Simplify:

29. $\frac{1}{-2^{-3}} - 2 - \sqrt[4]{16}$

30. $-2\{(2 - 4) - (3 - 6^0)(-2) - 2[(-3)(-2 - 1)]\}$

LESSON 58 Substitution axiom · Simultaneous equations

58.A substitution axiom

Since Lesson 13 we have been finding the value of a particular expression that contains variables by assigning a value to each variable. Thus the value of

$$\frac{x^2y}{p} \quad \text{when } x = 2,\ y = 4 \text{ and } p = -1 \text{ is} \quad \frac{(2)^2(4)}{(-1)} = \frac{16}{-1} = \mathbf{-16}$$

To do this we have been using what is usually called the **substitution axiom.**

The substitution axiom is stated in different ways by different authors. Three statements of this axiom that are frequently used follow:

SUBSTITUTION AXIOM

1. Changing the numeral by which a number is named in an expression does not change the value of the expression.
2. For any numbers a and b, if $a = b$, then a and b may be substituted for each other.
3. If $a = b$, then a may replace b or b replace a in any statement without changing the truth or falsity of the statement.

Definition 1 seems to apply only to individual expressions. Definition 2 is general enough but not sufficiently specific. Definition 3 seems to apply only to statements and not to individual expressions. **We will use the definition below to state formally and exactly the thought that if two expressions have equal value, it is permissible to use either expression.**

SUBSTITUTION AXIOM

If two expressions a and b are of equal value, $a = b$, then a may replace b or b may replace a in any expression without changing the value of the expression. Also, a may replace b or b may replace a in any statement without changing the truth or falsity of the statement. Also, a may replace b or b replace a in any equation or inequality without changing the solution set of the equation or inequality.

Thus the substitution axiom applies to expressions, equations, and inequalities. We have already been using this axiom to evaluate expressions as shown in the problem worked out at the beginning of this section. Now we will use the axiom to solve a system of first-degree linear equations in two unknowns.

58.B simultaneous equations

If we consider the two equations

$$\text{(a)}\quad 2x - y = 1 \qquad \text{and} \qquad \text{(b)}\quad x = -3y + 11$$

we see that (3, 5) is a solution to equation (a) and that (5, 2) is a solution to equation (b).

$$\begin{array}{rr} \text{(a)} & 2x - y = 1 \\ & 2(3) - (5) = 1 \\ & 6 - 5 = 1 \\ & 1 = 1 \quad \text{True} \end{array} \qquad \begin{array}{rr} \text{(b)} & x = -3y + 11 \\ & 5 = -3(2) + 11 \\ & 5 = -6 + 11 \\ & 5 = 5 \quad \text{True} \end{array}$$

But neither (3, 5) nor (5, 2) is a solution to both equations at the same time or simultaneously. If we need one solution that will satisfy two or more equations, we call the equations a **system of simultaneous equations.** We can designate that two or more equations form a system of equations by so stating in words or by using a brace as shown here.

$$\begin{cases} 2x - y = 1 \\ x = -3y + 11 \end{cases}$$

58.C solution of simultaneous equations by substitution

We can find the common solution to a system of two first-degree simultaneous equations in two unknowns by using the substitution axiom.

example 58.1 Solve: $\begin{array}{l} \text{(a)} \\ \text{(b)} \end{array} \begin{cases} 2x - y = 1 \\ x = -3y + 11 \end{cases}$

solution **First we must assume that an ordered pair of real numbers exists which will satisfy both of these equations and that x and y in the equations represent these real numbers.** If our assumption is correct, we can find the value of the members of this ordered pair. If our assumption is incorrect, the attempted solution will degenerate into an expression involving real numbers that may be true or false. Examples are $1 = 2$ or $4 + 2 = 6$ or $0 = 0$.†

Now, since both equations are assumed to be true equations and also since x in both equations stands for the number that will satisfy both equations, we can **replace the variable x in equation (a) with the equivalent expression for x given by equation (b).**

† To be discussed in a later lesson.

$$
\begin{aligned}
2x - y &= 1 && \text{equation (a)} \\
2(-3y + 11) - y &= 1 && \text{replaced } x \text{ with } -3y + 11 \\
-6y + 22 - y &= 1 && \text{multiplied} \\
-7y + 22 &= 1 && \text{added like terms} \\
-7y &= -21 && \text{added } -22 \text{ to both sides} \\
\mathbf{y} &= \mathbf{3} && \text{divided both sides by } -7
\end{aligned}
$$

We have found that the y value of the desired ordered pair is 3. Now we may find the value of x by substituting the number 3 for y in either of the original equations.

IN EQUATION (a)	IN EQUATION (b)
$2x - y = 1$	$x = -3y + 11$
$2x - (3) = 1$	$x = -3(3) + 11$
$2x - 3 = 1$	$x = -9 + 11$
$2x = 4$	$\mathbf{x = 2}$
$\mathbf{x = 2}$	

Thus the ordered pair of x and y that will satisfy both equations is **(2, 3).**

example 58.2 Solve: $\begin{cases} 2x + 3y = -13 \\ y = x - 6 \end{cases}$

solution The bottom equation states that y equals $x - 6$. Therefore, in the top equation we will replace y with $x - 6$ and solve for x.

$$
\begin{aligned}
2x + 3y &= -13 && \text{top equation} \\
2x + 3(x - 6) &= -13 && \text{substituted } x - 6 \text{ for } y \\
2x + 3x - 18 &= -13 && \text{multiplied} \\
5x &= 5 && \text{added like terms} \\
\mathbf{x} &= \mathbf{1} && \text{divided both sides by 5}
\end{aligned}
$$

Now the paired value for y may be found by substituting 1 for x in either of the original equations.

TOP EQUATION	BOTTOM EQUATION
$2x + 3y = -13$	$y = x - 6$
$2(1) + 3y = -13$	$y = (1) - 6$
$2 + 3y = -13$	$y = 1 - 6$
$3y = -15$	$\mathbf{y = -5}$
$\mathbf{y = -5}$	

Thus the solution is the ordered pair **(1, −5).**

example 58.3 Solve: $\begin{cases} 3x + 2y = 3 \\ x = 3y - 10 \end{cases}$

solution We will replace x in the top equation with its equivalent $(3y - 10)$ from the bottom equation.

$$
\begin{aligned}
3x + 2y &= 3 && \text{top equation} \\
3(3y - 10) + 2y &= 3 && \text{replaced } x \text{ with } (3y - 10)
\end{aligned}
$$

$$9y - 30 + 2y = 3 \qquad \text{multiplied}$$
$$11y = 33 \qquad \text{simplified}$$
$$\mathbf{y = 3} \qquad \text{divided by 11}$$

Now the paired value for x may be found by replacing y with 3 in either of the original equations.

TOP EQUATION	BOTTOM EQUATION
$3x + 2(3) = 3$	$x = 3(3) - 10$
$3x + 6 = 3$	$x = 9 - 10$
$3x = -3$	$\mathbf{x = -1}$
$\mathbf{x = -1}$	

Thus the solution is the ordered pair **(−1, 3).**

example 58.4 Solve: $\begin{cases} -x - 2y = 4 \\ y = -3x + 8 \end{cases}$

solution We will replace y in the top equation by its equivalent $(-3x + 8)$ from the bottom equation.

$$-x - 2y = 4 \qquad \text{top equation}$$
$$-x - 2(-3x + 8) = 4 \qquad \text{replaced } y \text{ with } (-3x + 8)$$
$$-x + 6x - 16 = 4 \qquad \text{multiplied}$$
$$5x - 16 = 4 \qquad \text{simplified}$$
$$5x = 20 \qquad \text{added 16 to both sides}$$
$$x = 4 \qquad \text{divided by 5}$$

Now we can replace x with 4 in either of the original equations to find the value of y.

TOP EQUATION	BOTTOM EQUATION
$-x - 2y = 4$	$y = -3x + 8$
$-(4) - 2y = 4$	$y = -3(4) + 8$
$-4 - 2y = 4$	$y = -12 + 8$
$-2y = 8$	$\mathbf{y = -4}$
$\mathbf{y = -4}$	

Thus the solution is the ordered pair **(4, −4).**

example 58.5 Solve: $\begin{cases} 2x + 3y = 5 \\ x = y \end{cases}$

solution In the top equation we replace x with its equivalent (y) from the bottom equation.

$$2x + 3y = 5 \qquad \text{original equation}$$
$$2(y) + 3y = 5 \qquad \text{substituted } y \text{ for } x$$
$$5y = 5 \qquad \text{added}$$
$$\mathbf{y = 1} \qquad \text{divided by 5}$$

and since $x = y$ $\qquad \mathbf{x = 1}$

Thus the solution is the ordered pair **(1, 1).**

practice Use substitution to solve for x and y:

a. $\begin{cases} 2x + 3y = 24 \\ x = 4y - 10 \end{cases}$ b. $\begin{cases} y = x + 7 \\ x + 2y = -16 \end{cases}$

problem set 58 Solve and then draw the diagrams:

1. Forty-seven percent of what number equals 188?
2. What percent of 68 is 95.2?
3. If 80 is increased by 210 percent, what is the result?
4. Solve: $-2[(-3 - 2) - 2(-2 + m)] = -3m - 4^2 - |-2|$

Use substitution to solve for x and y:

5. $\begin{cases} 2x + y = 7 \\ x = -3y + 11 \end{cases}$ 6. $\begin{cases} 2x + 3y = -14 \\ y = 2x - 7 \end{cases}$

7. Use six unit multipliers to convert 1000 square meters to square feet.
8. The average of Becky Jo's first two tests was 90. The average of her next eight tests was 95. What was her overall average?
9. Find the volume and the surface area of the right prism whose base is shown and whose height is 2 feet. Dimensions are in inches. Corners that look square are square.

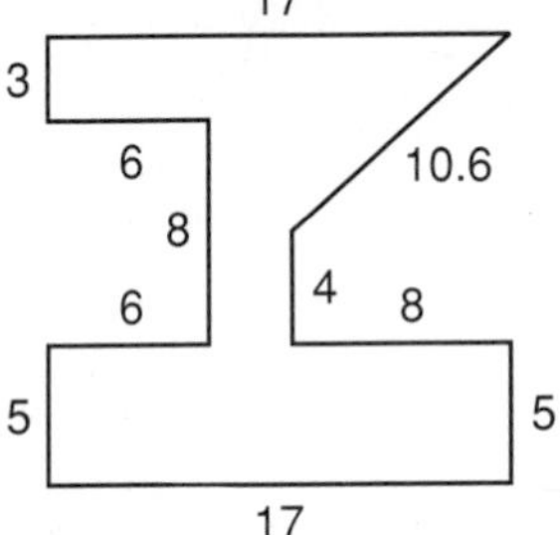

10. Find the least common multiple of: 175, 147, and 45
11. Expand: $(3 - x)^2$

Graph on a rectangular coordinate system:

12. $y = -3\frac{1}{2}$ 13. $y = 2x - 5$ 14. $y = -\frac{1}{2}x + 4$

Add:

15. $\dfrac{-x}{a^2b} + \dfrac{a - b}{b}$ 16. $\dfrac{a + b}{a} - \dfrac{3}{ax}$ 17. $4 - \dfrac{7}{a} + \dfrac{a + b}{a^2}$

18. Factor: $25x^2y^5p^2 - 15x^5y^4p^4 + 10x^4y^4p^4z$

Simplify:

19. $\dfrac{5pq - 5p^2q^2}{5pq}$ 20. $x(x^2)(x^3)^{-3}x^{-2}y^2$

21. $(x^2y^0)(x^3y^2)xy$ 22. $xx(x^2)^{-2}(x^2)(x^{-3}y^0)^5$

23. $\left(\dfrac{3x^{-2}y^5}{p^{-3}}\right)^{-2}\left(\dfrac{2x^{-2}}{y}\right)^2$ 24. $\dfrac{x(x^2y^0)^{-2}xy^{-2}}{(xy^{-3})^2xy^{-3}}$

25. Expand: $\left(\dfrac{x^2}{2y^{-1}} - \dfrac{4x^{-2}}{y}\right)\dfrac{2x^{-2}}{y}$

26. Simplify by adding like terms: $x^2(p^5)^2y - \frac{3x^2p^{10}}{y^{-1}} + \frac{4x^3py}{x^2p^{-2}} - 2xp^{10}y$

Evaluate:

27. $x^2 - xy^{-2} - (-y)^{-2}$ if $x = 3$ and $2y - 5 = -11$

28. $p^2x - xp^0(x - p)$ if $x = -3$ and $p = 3$

Simplify:

29. $\frac{1}{-2^{-3}} + 8 + \sqrt{64}$

30. $-3\{(-4 - 2) - (-2 - 1^0) - [(-2) - (-2 - 1)]\}$

LESSON 59 Dividing fractions

59.A complex fractions

To review the **denominator-numerator same-quantity rule,** we will begin with the number 5. Then if we multiply 5 by 2 and divide by 2 the answer is 5,

$$5 = \frac{5 \cdot 2}{2} = \frac{10}{2}$$

because $\frac{10}{2}$ is another way to write 5. We have changed the numeral, but the number it represents is unchanged because 2 over 2 has a value of 1. This is why we can multiply the denominator and the numerator of a fraction by any nonzero number without changing the value of the fraction.

Fractions of fractions are called **complex fractions.** We will simplify complex fractions by multiplying the numerator and the denominator by the same quantity.

$$\frac{\frac{a}{b}}{\frac{c}{d}} \qquad \begin{pmatrix} b \neq 0 \\ c \neq 0 \\ d \neq 0 \end{pmatrix}^{\dagger}$$

If we multiply the denominator of this complex fraction by the reciprocal of the denominator, we obtain a product of 1 because we remember that the product of a number and the reciprocal of the same number is the number 1.

$$\frac{c}{d} \cdot \frac{d}{c} = \frac{cd}{dc} = 1$$

But if we wish to multiply the denominator of the fraction by its reciprocal, which is $\frac{d}{c}$, we must also multiply the numerator by $\frac{d}{c}$ so that the value of the original expression will not be changed.

$$\frac{\frac{a}{b}}{\frac{c}{d}} = \frac{\frac{a}{b} \cdot \frac{d}{c}}{\frac{c}{d} \cdot \frac{d}{c}} = \frac{\frac{ad}{bc}}{\frac{cd}{dc}} = \frac{\frac{ad}{bc}}{1} = \boldsymbol{\frac{ad}{bc}}$$

† If neither b, c, nor d equals zero, the denominator cannot equal zero. We provide this notation to ensure the reader that we are not implying that division by zero is permissible.

We have simplified the original fraction by multiplying both the denominator and the numerator by the **reciprocal of the denominator.**

example 59.1 Simplify: $\dfrac{\frac{a}{b}}{c}$ $(b, c, \neq 0)$

solution We multiply both the denominator and the numerator by $\frac{1}{c}$, which is the reciprocal of the denominator c.

$$\frac{\frac{a}{b}}{c} = \frac{\frac{a}{b} \cdot \frac{1}{c}}{\frac{c}{1} \cdot \frac{1}{c}} = \frac{\frac{a}{bc}}{\frac{c}{c}} = \frac{\frac{a}{bc}}{1} = \boldsymbol{\frac{a}{bc}}$$

example 59.2 Simplify: $\dfrac{\frac{1}{c}}{\frac{1}{b}}$ $(b, c \neq 0)$

solution We multiply both the denominator and the numerator by b, which is the reciprocal of $\frac{1}{b}$.

$$\frac{\frac{1}{c}}{\frac{1}{b}} = \frac{\frac{1}{c} \cdot \frac{b}{1}}{\frac{1}{b} \cdot \frac{b}{1}} = \frac{\frac{b}{c}}{1} = \boldsymbol{\frac{b}{c}}$$

example 59.3 Simplify: $\dfrac{a}{\frac{1}{c}}$ $(c \neq 0)$

solution We multiply both the numerator and the denominator by c, which is the reciprocal of $\frac{1}{c}$.

$$\frac{a}{\frac{1}{c}} = \frac{\frac{a}{1} \cdot \frac{c}{1}}{\frac{1}{c} \cdot \frac{c}{1}} = \frac{\frac{ac}{1}}{1} = \boldsymbol{ac}$$

example 59.4 Simplify: $\dfrac{\frac{a}{x}}{\frac{b}{a + x}}$ $(x, b \neq 0)$

solution We will multiply both the denominator and the numerator by $\frac{a + x}{b}$, which is the reciprocal of $\frac{b}{a + x}$.

$$\frac{\frac{a}{b}}{\frac{b}{a + x}} = \frac{\frac{a}{x} \cdot \frac{a + x}{b}}{\frac{b}{a + x} \cdot \frac{a + x}{b}} = \boldsymbol{\frac{a(a + x)}{bx}}$$

59.B division rule

The rule for dividing fractions is sometimes stated as follows: **To divide one fraction by another fraction, invert the fraction in the denominator and multiply.** If we use this rule to simplify

$$\frac{\frac{m}{n}}{\frac{x}{y}} \qquad (x, n, y \neq 0)$$

the solution is

$$\frac{m}{n} \cdot \frac{y}{x} = \frac{my}{nx}$$

If we use the denominator-numerator same-quantity rule to perform the same simplification, we obtain

$$\frac{\frac{m}{n}}{\frac{x}{y}} = \frac{\frac{m}{n} \cdot \frac{y}{x}}{\frac{x}{y} \cdot \frac{y}{x}} = \boldsymbol{\frac{my}{nx}}$$

This procedure yields the same result as that obtained by using the rule for dividing fractions, but hopefully we have some understanding of what we did and can justify our procedure.

Many algebraic manipulations can be justified by one of the following:

1. **The denominator-numerator same-quantity rule.**
2. **The multiplicative property of equality.**
3. **The additive property of equality.**

We will remember to justify our manipulations by one of these three rules whenever possible.

practice Simplify:

a. $\dfrac{\frac{x}{m}}{d}$ b. $\dfrac{\frac{1}{r}}{\frac{1}{z}}$ c. $\dfrac{\frac{n}{a}}{\frac{b}{d}}$ d. $\dfrac{w}{\frac{1}{w + c}}$

problem set 59

1. If the product of a number and 2 is increased by 7, the result is 2 less than the opposite of the number. What is the number?
2. The prognosticators outscored the retrospectors in the ratio of 5 to 2. If 98 total points were scored, how many belonged to the prognosticators?
3. What fraction of $\frac{1}{4}$ is $3\frac{7}{8}$?

Solve and then draw the diagrams:

4. Seventy-two percent of what number equals 216?
5. One hundred thirty-five percent of what number equals 405?
6. If 48 is increased by 250 percent, what is the result?

Solve:

7. $2\frac{1}{4} + 3\frac{1}{5}k + \frac{1}{8} = 0$

8. $-[-(-p)] - (-4)(-2 - p) = -(4 - 2p)$

Use substitution to solve for both x and y:

9. $\begin{cases} 3x + 2y = 7 \\ x = 7 - 3y \end{cases}$ **10.** $\begin{cases} x + 2y = -6 \\ y = 3x + 4 \end{cases}$ **11.** $\begin{cases} x + y = 6 \\ x = 9 - 2y \end{cases}$

Multiply:

12. $(3x - 4)(2 - x)$ **13.** $(x + 1)(x^2 + 2x + 2)$

Graph on a rectangular coordinate system:

14. $y = -3$ **15.** $y = -3x - 2$ **16.** $y = \frac{1}{2}x - 2$

Simplify:

17. $\dfrac{\frac{m}{n}}{z}$ **18.** $\dfrac{m + 1}{\frac{n}{d}}$ **19.** $\dfrac{\frac{am}{n}}{\frac{x}{dc}}$

20. Find the area of this figure in square centimeters.

4 m
5 m

Add:

21. $\dfrac{4}{x + y} - \dfrac{3}{y^2}$ **22.** $\dfrac{a}{x^2y} + 4a - \dfrac{m}{x + y}$ **23.** $\dfrac{x + y}{a^2} + \dfrac{y^2}{a}$

Simplify:

24. $\dfrac{(x^2)^{-3}(yx)^2x^0}{x^2y^{-2}(xy^{-2})^3}$ **25.** $\left(\dfrac{x^{-2}}{y^4}\right)^2\left(\dfrac{x}{y}\right)$

26. $\dfrac{p^9(3x)^2y}{x^0(x^{-1}y^{-2})^3}$ **27.** $\dfrac{4x^2 - 4}{4}$

28. Expand: $\dfrac{x^{-4}y}{p^2}\left(\dfrac{y^{-1}}{x^4} + \dfrac{2x^{-4}p^{-2}}{y^{-1}}\right)$

29. Simplify by adding like terms: $\dfrac{xy}{z} - \dfrac{3x^2y^2}{xyz} + \dfrac{2x^3x^{-2}yz^{-2}}{z^{-1}} + \dfrac{5xy}{zy^{-2}}$

30. Simplify: $-\{3(-3 \cdot 2^0)[-(3 - 2)(-2)] - |-4|\} + \sqrt[3]{64}$

LESSON 60 *Set notation · Rearranging before graphing*

60.A finite and infinite sets

The words **finite** and **infinite** are basic words and are difficult to define. The word *finite* implies the thought of bounded or limited, while the word *infinite* implies the thought of unbounded or without limit. Thus when we say that a set has a finite number of members, we are describing a set such as

$$\{6, 7, 8, 9, 10\}$$

in which the listing of the members has an end. **A set with a finite number of members is called a finite set.**

When we say that a set has an infinite number of members, we are describing a set such as

$$\{1, 2, 3, 4, 5, \ldots\}$$

in which the listing of the members of the set continues without end. **Sets that have an infinite number of members are called infinite sets.**

Some authors define a finite set as a set whose members could be counted if we could live the number of lifetimes necessary to do the counting. Thus the set that has

$$63{,}072{,}000{,}000{,}000{,}000$$

members is a finite set because we could count the members of this set if we counted for 1 billion years at two counts per second (not considering leap years).

If we use the same definition, the set

$$\{1, 2, 3, 4, \ldots\}$$

is an infinite set because we could never count the number of members of this set since the listing has no end.

60.B membership in a set

If we have the numbers

$$0, 0, 0, 0, 0, 0, 1, 1, 1, 1, 2, 2, 2, 2$$

and wish to designate these numbers as members of a set B, we would write

$$B = \{0, 1, 2\}$$

which is read "B is the set whose members are the numbers 0, 1, and 2." **Note that each of the numbers is listed only once.** Thus, if we have the set

$$D = \{0, 1, 2, 3, 4\}$$

we have said that the set consists of 0, or any number of 0s; also the set consists of 1s, 2s, 3s, and 4s, but not necessarily just one 1 and one 2 and one 3 and one 4.

example 60.1 Represent the following numbers as being members of set K:

$$1, 0, 2, 1, 0, 5, 7, 4, 5, 7$$

solution We list each number only once. The order in which the numbers are listed is unimportant.

$$\mathbf{K = \{0, 1, 2, 4, 5, 7\}}$$

example 60.2 Represent the following numbers as constituting set L: 7, 15, 0, 1, 15, 0, 8, −13, 42

solution We list each number only one time. The order in which the numbers are listed is unimportant.

$$L = \{7, 15, 0, 1, 8, -13, 42\}$$

We use the symbols

$$\in \quad \text{and} \quad \notin$$

to designate that a particular symbol or number is a member of a given set or is not a member of the given set.

$$0 \in L \qquad 23 \notin L$$

We would read the above as "zero is a member of set L" and "23 is not a member of set L."

example 60.3 Given the sets $A = \{0, 1, 3, 5\}$, $B = \{0, 4, 6, 7\}$, and $C = \{1,2, 3, 5, 7\}$, are the following statements true or false?

(a) $5 \in A$ (b) $4 \in C$ (c) $5 \notin B$

solution (a) $5 \in A$ True, because 5 is a member of set A.

(b) $4 \in C$ False, because 4 is not a member of set C.

(c) $5 \notin B$ True, because 5 is not a member of set B.

example 60.4 Given the sets $M = \{0, 1, 2, 3\}$, $L = \{5, 6, 7\}$, and $N = \{0, 1\}$, are the following statements true or false?

(a) $6 \in M$ (b) $0 \in N$

solution (a) False, because 6 is not a member of set M.

(b) True, because 0 is a member of set N.

60.C rearranging before graphing

Often we encounter linear equations that have not been solved for y. Graphing these equations is easier if we first rearrange them by solving for y.

example 60.5 Graph: $3x + 2y = 4$

solution We first solve for y by adding $-3x$ to both sides and then dividing by 2.

$$\begin{array}{rl} 3x + 2y = 4 & \text{original equation} \\ -3x \qquad\quad - 3x & \text{add } -3x \text{ to both sides} \\ \hline 2y = 4 - 3x & \\ y = 2 - \frac{3}{2}x & \text{divided each term by 2} \\ y = -\frac{3}{2}x + 2 & \text{order of terms changed} \end{array}$$

Now we make a table and choose convenient values for x.

x	0	2	-2
y			

Now we find the values of y that the equation $y = -\frac{3}{2}x + 2$ pairs with 0, 2, and -2.

WHEN $x = 0$

$y = -\frac{3}{2}(0) + 2$

$y = 2$

WHEN $x = 2$

$y = -\frac{3}{2}(2) + 2$

$y = -1$

WHEN $x = -2$

$y = -\frac{3}{2}(-2) + 2$

$y = 5$

Now we complete the table, graph the points, and draw the line.

x	0	2	-2
y	2	-1	5

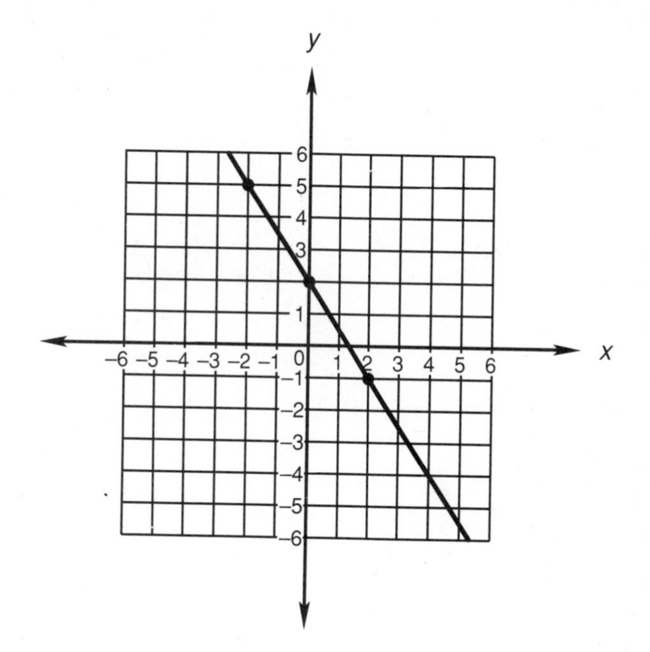

example 60.6 Graph: $y - x = 0$

solution We (a) solve for y, (b) complete the table, and then (c) draw the graph.

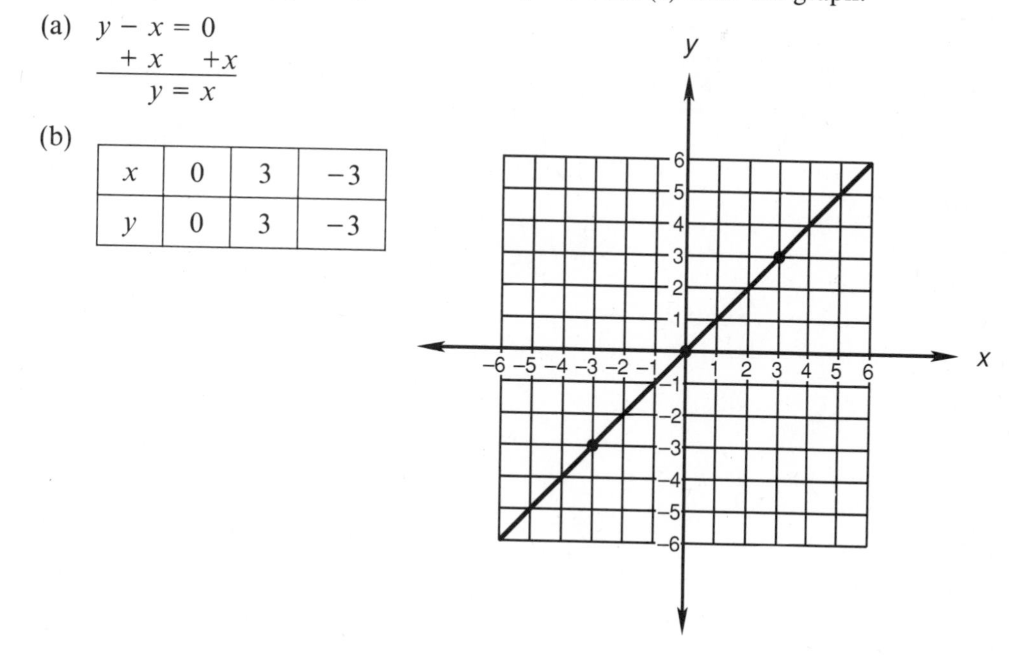

(a)
$$\begin{array}{rcl} y - x &=& 0 \\ +\,x && +x \\ \hline y &=& x \end{array}$$

(b)

x	0	3	-3
y	0	3	-3

practice **a.** Represent 3, 1, 0, 5, 0, 9, and 5 as making up the set K.

b. Given the sets $A = \{0, 2, 4\}$, $B = \{0, 1, 5, 6\}$, and $C = \{1, 3, 4, 6\}$. Which of the following is true?

(a) $5 \in B$ (b) $5 \notin C$ (c) $0 \in A$ (d) $3 \notin C$

c. Graph on a rectangular coordinate system: $2x - 3y = 6$

problem set 60

Work Problems 1–4 and then draw the diagrams:

1. Lisa and Dennis found that 190 percent of their number equaled 76. What was their number?

2. What percent of 180 is 36?

3. Ninety percent of 0.4 is what number?

4. Tricia and Heather increased 170 by 60 percent. What number did they get?

5. The average of the first 5 numbers was 200. The average of the second 20 numbers was 400. What was the overall average?

6. Find the volume and the surface area of the right prism whose base is shown and whose sides are 10 centimeters high. Dimensions are in meters.

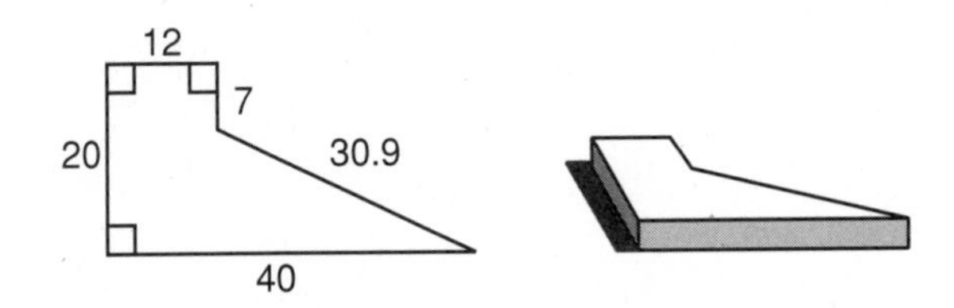

Use substitution to solve for x and y:

7. $\begin{cases} 2x - 2y = 18 \\ x = 6 - 2y \end{cases}$ 8. $\begin{cases} 3x - y = 4 \\ y = 6 - 2x \end{cases}$ 9. $\begin{cases} 5x - 3y = 6 \\ y = 2x + 3 \end{cases}$

Multiply:

10. $(x + 12)^2$ 11. $(2x - 3)(2x^2 - 3x + 4)$

12. Given the sets $A = \{0, 1, 3, 5, 7\}$, $B = \{2, 4, 6, 8\}$, and $C = \{0, 1, 2, 3, 4, 5\}$, which of the following is true?

(a) $0 \in B$ (b) $0 \in C$ (c) $2 \notin B$ (d) $2 \notin A$

13. Graph on a rectangular coordinate system: $3y + 2x = 3$

Simplify:

14. $\dfrac{\frac{1}{a}}{x}$ 15. $\dfrac{\frac{b}{c}}{\frac{1}{a + b}}$ 16. $\dfrac{\frac{x}{c}}{x + y}$ 17. $\dfrac{m}{\frac{a}{mc^2}}$

Add:

18. $\dfrac{4}{2x^2} - \dfrac{3}{4x^2y} + \dfrac{2a}{8x^3p}$ 19. $\dfrac{m}{b(b + c)} + \dfrac{k}{b}$ 20. $\dfrac{3p}{xm^2} - \dfrac{a}{2m^3} + \dfrac{4k}{m^4a}$

21. Factor: $4k^2ax - 8ka^2x^2 + 12k^3a^4x^2$

Simplify:

22. $\dfrac{x^2y - x^2yz}{xyz}$ 23. $(4x^0y^2m)^{-2}(2y^{-4}m^0x)^4$

24. $\dfrac{(k^3p^0)^{-2}k^2p^5}{p^{-5}p^0k^{-1}}$ 25. $\left(\dfrac{x^2y^{-2}}{p^4k^0}\right)^{-2}\left(\dfrac{y^{-2}}{x}\right)$

26. Expand: $\dfrac{x^{-2}}{4m^2}\left(\dfrac{4x^2}{m^{-2}} - \dfrac{8m^{-2}k}{x^{-2}}\right)$

27. Simplify by adding like terms: $3x^2y^2m - \frac{m}{x^{-2}y^{-2}} + \frac{4x^2y^2}{m^{-1}} - \frac{3x^4y^4}{xy}$

28. Evaluate: $a^{-3} - a(x - a)$ if $a = -2$ and $2x + 8 = 16$

29. Solve: $-3(-2 - x) - 3^2 - |-2| = -(-2x - 3)$

30. Simplify: $-2\{-[(3^0 - 5) - (2 - 4)] - |-3| + 2\} + \sqrt[3]{27}$

LESSON 61 More on area and volume

61.A area revisited

We will review the concept of area by considering the areas of triangles and parallelograms. Suppose we need a triangle whose area is 10 cm² and whose base is 4 cm. We know that the altitude of the triangle must be 5 because

$$A = \frac{BH}{2} = \frac{(4)(5)}{2} = 10 \text{ cm}^2$$

We have our choice of any triangle whose base is 4 and whose altitude is 5. We show two of these triangles in this figure.

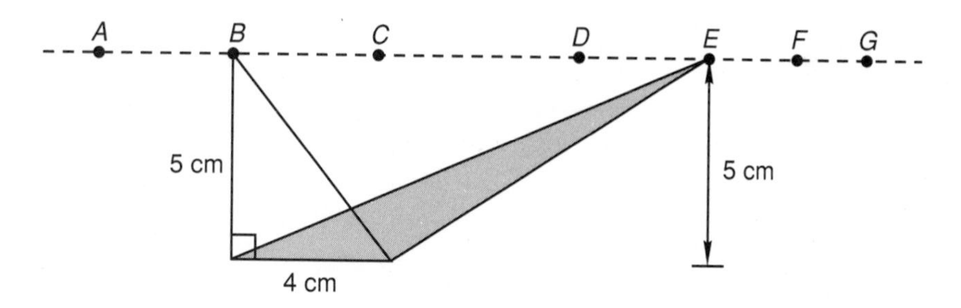

The area of the right triangle on the left is 10 cm². The area of the shaded triangle on the right is also 10 cm². The areas are equal because both triangles have a base of 4 cm and an altitude of 5 cm. If we use the same base, we could move the vertex to any other point on line AE and the area of the triangle would also equal 10 cm².

To find a parallelogram whose base is 6 cm and whose area is 24 cm², we have our choice of an infinite number of parallelograms. The area of the parallelogram is BH, so the height of our parallogram must be 4 cm. We show two possible solutions here.

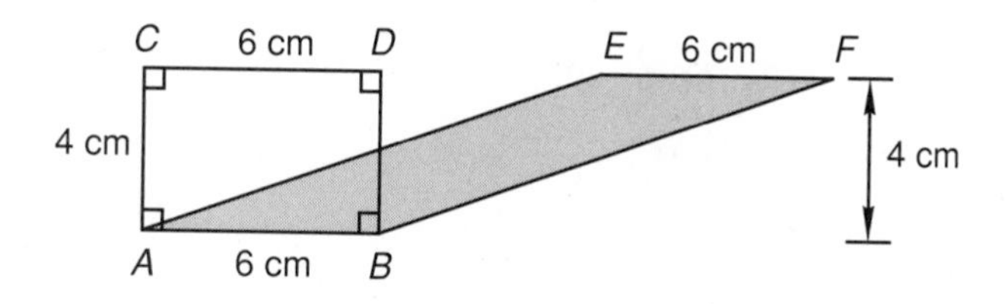

The area of the rectangle (which is also a parallelogram) is 24 cm². The area of the shaded parallelogram on the right is also 24 cm². The areas are equal because both parallelograms have a base of 6 cm and an altitude of 4 cm. If we use $\overline{AB}$ as the base, we can move $\overline{CD}$ left and right any distance we choose along line CF and the parallelogram formed would have an area of 24 cm^2.

61.B cylinders and prisms

We are unable to define some terms in mathematics. These terms are called **primitive terms.** We explain these terms as best we can and then use them to define other terms. The words **point, curve, line,** and **plane** are primitive terms. A curve can be thought of as a continuous string of points. A line is a straight curve. A plane is a flat surface. The fact that a line is also a curve will cause us to realize that a prism is also a cylinder and that a pyramid is also a cone, as we shall see.

An oval is a planar, closed geometric figure that looks like a flattened circle. A circle is a perfectly round oval. A polygon is a planar, closed geometric figure whose sides are line segments.

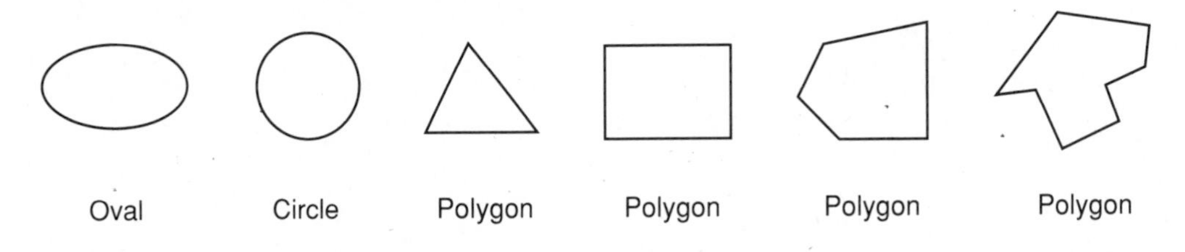

We remember that the top and the bottom of a right solid are called the **bases** of the solid. **The bases are identical planar geometric figures that are parallel to each other.** Here we show five right solids. All of these right solids are also called **right cylinders.**

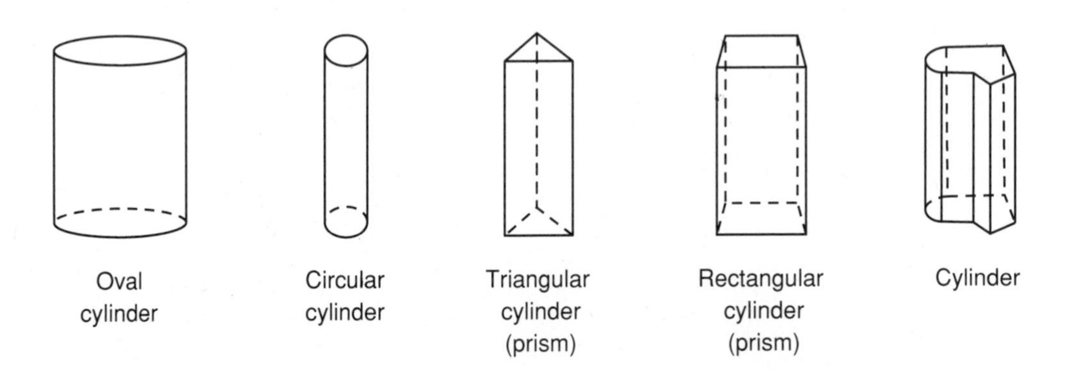

The right cylinder on the left is a right oval cylinder because the top and the bottom are identical ovals that are parallel. The next cylinder is a right circular cylinder. The bases of this cylinder are identical circles on parallel planes. The next cylinder is a triangular cylinder with identical bases that are triangles. Cylinders whose bases are polygons are called **prisms.** A cylinder is formed by moving a **line segment** called an **element** around any closed, flat geometric figure. The element is always parallel to a given line.

Given line

Element

The volume of a cylinder or a prism equals the area of the base times the perpendicular distance between the bases. This is true even if the cylinder is not a right cylinder.

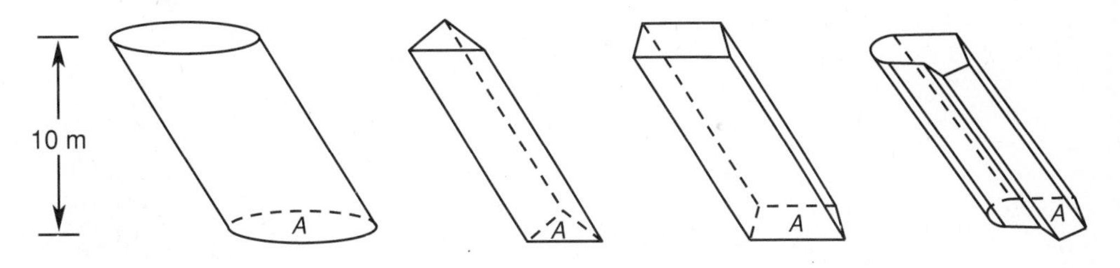

If the perpendicular distances between the bases of each of the cylinders is 10 m and if the area of each of the bases is 10 m^2, the volume of each of the cylinders is 100 m^3.

$$\text{Volume of a cylinder (prism)} = \text{area of base} \times \text{height}$$
$$= 10 \text{ m}^2 \times 10 \text{ m} = 100 \text{ m}^3$$

61.C cones and pyramids

A **cone** is a solid bounded by a closed, flat base and the surface formed by line segments which join the points on the boundary of the base to a fixed point, not on the boundary, that is called the **vertex.** A cone whose base is a polygon is called a **pyramid.** Thus a pyramid is a cone whose base has straight sides.

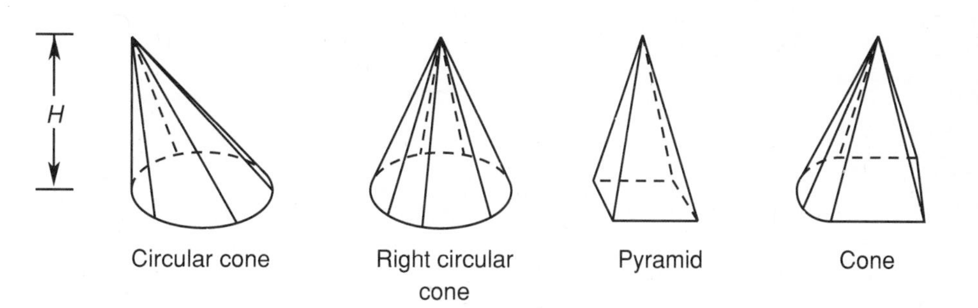

Circular cone | Right circular cone | Pyramid | Cone

The altitude of a cone or pyramid is the perpendicular distance from the base to the vertex.

The volume of a cone (pyramid) is exactly one-third the volume of a cylinder that has the same base and the same height.

A **sphere** is a perfectly round three-dimensional shape. Every point on the surface of a sphere is the same distance from the center. The distance is the radius of the sphere.

Sphere

The volume of a sphere is exactly two-thirds the volume of the right circular cylinder into which the sphere fits. The radius of the cylinder equals the radius of the sphere, and the height of the cylinder equals the diameter of the sphere.

Volume of the sphere equals $\frac{2}{3}$ the volume of the cylinder

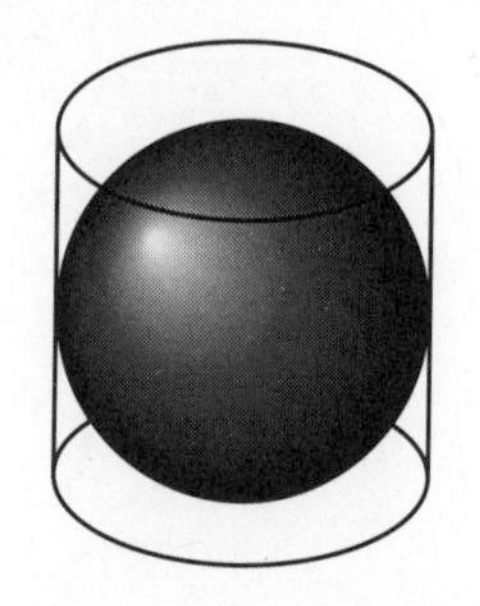

The first proof of this method of finding the volume of a sphere is attributed to the Greek philosopher Archimedes (287–212 B.C.). There is a formula for the volume of a sphere. See if you can use the method of Archimedes to find the formula. Close your eyes and try to remember the diagram above. It will help you recall how to find the volume of a sphere.

To find the surface area of a sphere and the surface area of a right circular cone, we use the following formulas. If the cone is not a right circular cone, the formula for the surface area of the cone will not work.

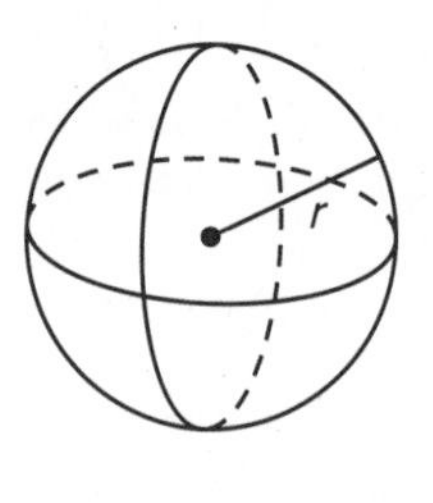

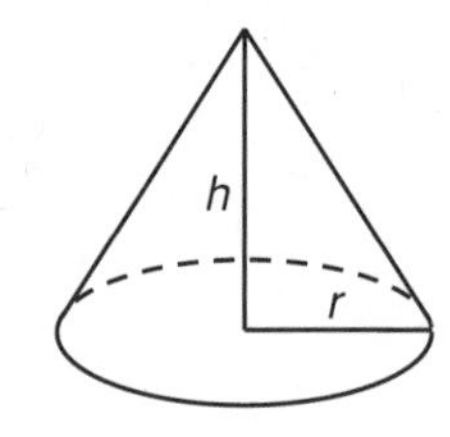

SPHERE

Surface area $= 4\pi r^2$

RIGHT CIRCULAR CONE

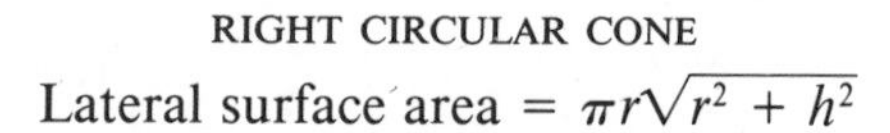

Lateral surface area $= \pi r\sqrt{r^2 + h^2}$

These formulas are not needed often. When they are needed, they can be looked up in a reference book.

example 61.1 Find the volume of this cylinder. The area of the base is 242 m^2.

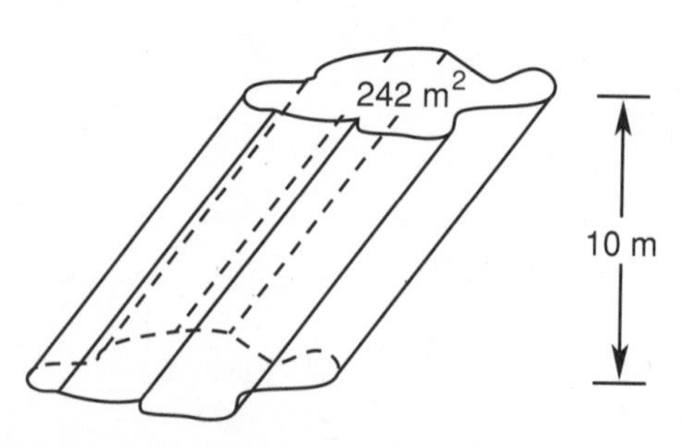

solution The volume of any cylinder equals the area of the base times the altitude.

$$\text{Volume} = 242 \text{ m}^2 \times 10 \text{ m} = \mathbf{2420 \text{ m}^3}$$

example 61.2 Find the volume of this cone. The area of the base is 172 m^2.

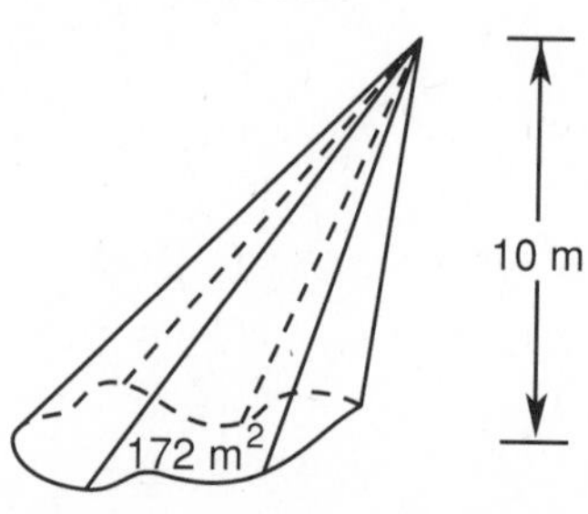

solution The volume of a cone is one-third the volume of the cylinder with the same base and altitude.

$$\text{Volume} = \frac{1}{3}(172\ \text{m}^2)(10) = \mathbf{\frac{1720}{3}\ m^3}$$

example 61.3 Find the volume of a sphere whose radius is 3 centimeters.

solution The volume of a sphere is exactly two-thirds the volume of the right circular cylinder that will contain it.

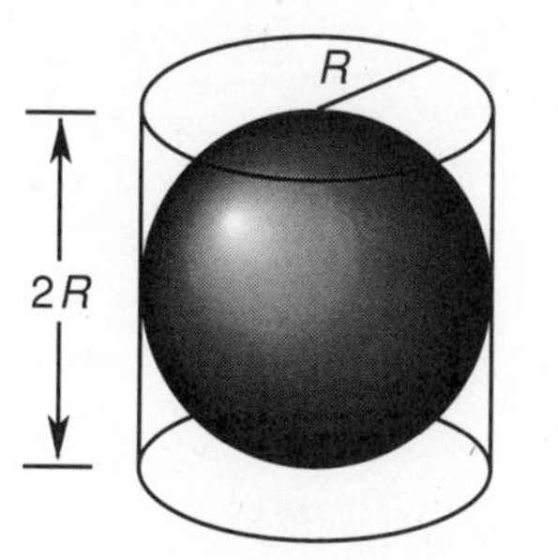

Area of base $= \pi(3)^2 = 9\pi$

Height $= 2R = 6$

$$\text{Volume of cylinder} = 6(9\pi) = 54\pi\ \text{cm}^3$$

The sphere takes up two-thirds of this value.

$$\text{Volume of sphere} = \frac{2}{3}(54\pi\ \text{cm}^3) = \mathbf{36\pi\ cm^3}$$

practice **a.** This figure is the base of a cylinder that is 10 m tall. The cylinder is not a right cylinder. Find the volume of the cylinder.

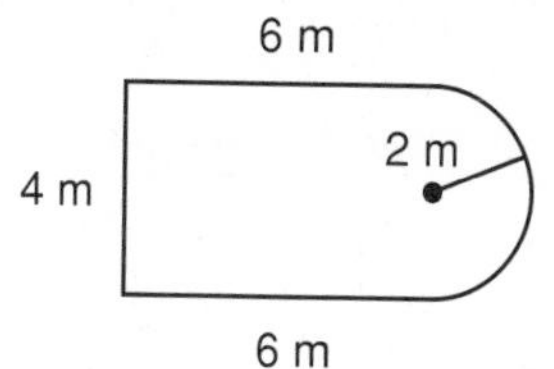

b. Find the volume of a cone whose base is shown in problem **a** and whose height is 6 m.

c. Find the volume of a sphere whose radius is 10 ft.

problem set 61

1. Wilbur was distraught because 230 percent of his number was only 920. What was Wilbur's number?

2. What percent of 240 is 60?

3. Seventy percent of what number is 4900?

4. Jamie increased 240 by 20 percent. What was the number?

5. The average of the first 7 numbers was 21. The average of the next 3 numbers was only 11. What was the overall average of the numbers?

6. Find the volume of a right circular cylinder whose radius is 10 centimeters and whose height is 20 centimeters. Find the volume of a sphere whose radius is 10 centimeters.

Use substitution to solve for x and y.

7. $\begin{cases} 2x + 3y = 7 \\ x = y - 1 \end{cases}$ **8.** $\begin{cases} 3x - y = 3 \\ y = 2x - 1 \end{cases}$ **9.** $\begin{cases} 4x + y = 9 \\ x = 3y - 1 \end{cases}$

Multiply:

10. $(x + 5)^2$

11. $(2x - 3)(3x^2 - 2x + 2)$

12. Given the sets $A = \{0, 5, 10, 12, 13\}$, $B = \{5, 10, 12\}$, and $C = \{10, 12\}$, which of the following is true?

(a) $0 \in C$ (b) $0 \in A$ (c) $6 \notin B$

13. Graph on a rectangular coordinate system: $4y + 8x = 12$

Simplify:

14. $\dfrac{\frac{1}{x}}{a}$

15. $\dfrac{\frac{1}{x+y}}{\frac{a}{b}}$

16. $\dfrac{x+y}{\frac{1}{c}}$

17. $\dfrac{x}{\frac{a}{b}}$

Add:

18. $\dfrac{2}{3x^4} - \dfrac{1}{3x^2y} + \dfrac{a}{2x^3p}$

19. $\dfrac{m}{k(k + c)} + \dfrac{m}{k}$

20. $\dfrac{3x}{2xp^2} - \dfrac{m}{p^3} + \dfrac{4z}{p^4a^2}$

21. Factor: $2p^2bc - 8b^2c^2 + 12p^3b^4c^2$

Simplify:

22. $\dfrac{4x^2 + 4}{4}$

23. $(2m^2y^0)^{-2}(2m^2yx)^4$

24. $\dfrac{(x^3m^0)^{-2}x^2m^5}{m^{-5}m^0x^{-1}}$

25. $\left(\dfrac{m^2x^{-2}}{p^4k^0}\right)^{-2}\left(\dfrac{x^{-2}}{m}\right)$

26. Expand: $\dfrac{p^{-3}}{4m^2}\left(\dfrac{2p^2}{m^{-2}} - \dfrac{8m^{-2}}{a^2b^3p}\right)$

27. Simplify by adding like terms: $\dfrac{4x^3m^4}{y} + 3xx^2y^{-1}m^4 + \dfrac{2x^4m^4}{x^2y}$

28. Evaluate: $a^2 - x(a^{-3} - x)$ if $a = -2$ and $x = 2$

29. Solve: $-2(-3 - x) - 3^2 + 1 - 4 = -(3x + 2)$

30. Simplify: $-x\{-[(2^0 - 3) - (1 - 4)]\} + 3|-2| + \sqrt[3]{-27}$

LESSON 62 Percent word problems

It is absolutely necessary to be able to visualize percent word problems in order to work them effectively. There is no shortcut that can be used as a substitute for understanding the problem. We will use the statement of the problem to draw "before" and "after" diagrams to help us write the percent equation that will give us the missing parts.

example 62.1 Kathy, John, and Susie have only 20 chickens left. If they began with 80 chickens, what percent of the original flock remains?

solution **A diagram should always be drawn as the first step.**

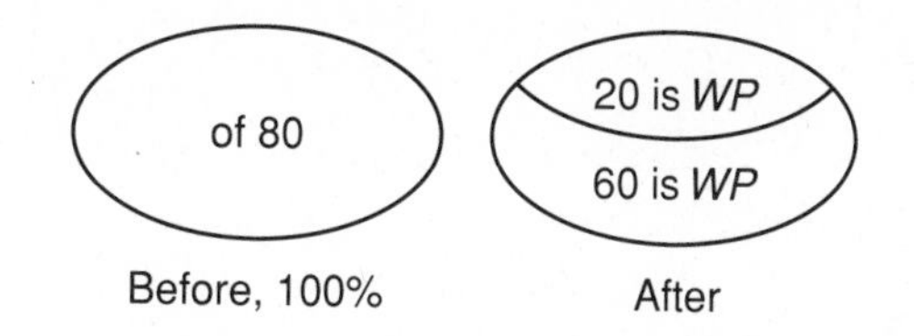

The original flock of 80 is on the left. It is divided into the 20 that remain and the 60 that are missing. We see that we can write two statements that can be used to solve the problem.

(a) 20 is what percent of 80? or (b) 60 is what percent of 80?

We will write an equation to solve question (a).

$$\frac{WP}{100} \times 80 = 20 \longrightarrow \frac{\cancel{100}}{\cancel{80}} \cdot \frac{WP}{\cancel{100}} \times \cancel{80} = 20 \cdot \frac{100}{80} \longrightarrow WP = \frac{2000}{80}$$

$$\longrightarrow \mathbf{WP = 25\%}$$

Thus the other percent is 75 percent. We could also have found 75 percent by solving question (b). Now we can draw the diagram with all numbers in place.

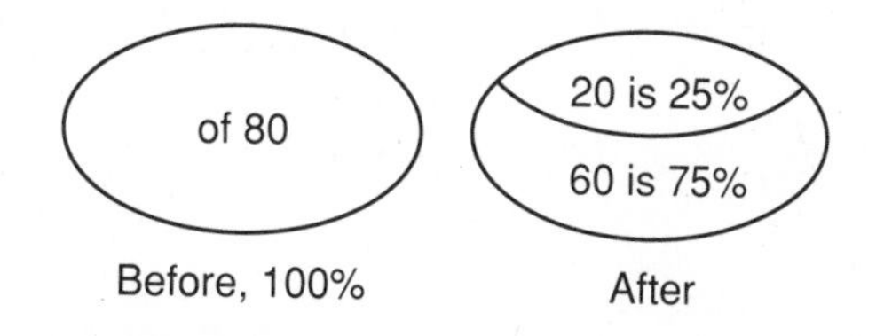

example 62.2 Meme and Jim have 75 thingamabobs. They want to give Hal 20 percent of them. How many thingamabobs do they give Hal?

solution **A diagram should always be drawn as the first step.**

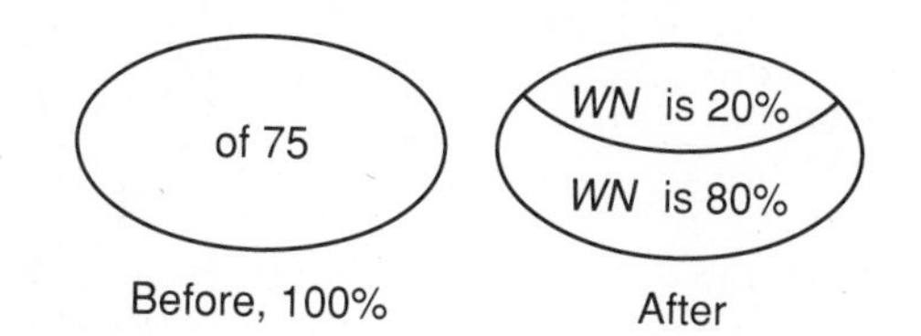

We can use this diagram to help us formulate two questions:

(a) What number is 20% of 75? (b) What number is 80% of 75?

The questions lead to two equations of the form $\frac{P}{100} \times \text{of} = \text{is}$.

$$\text{(a)} \quad \frac{20}{100} \times 75 = WN \qquad \text{(b)} \quad \frac{80}{100} \times 75 = WN$$

We don't have to solve both equations. We will solve equation (a) and subtract this answer from 75 to get the other number.

$$\text{(a)} \quad \frac{20}{100} \times 75 = WN \longrightarrow \frac{1500}{100} = WN \longrightarrow \mathbf{15 = WN}$$

Thus, they give Hal 15 and save 75 − 15 = 60 for themselves. We finish by drawing the diagram with all numbers in place.

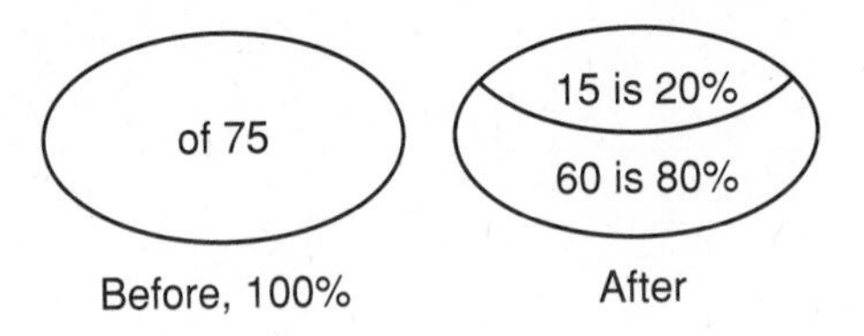

example 62.3 Beau and Christy hide 600 raisins. This is 60 percent more than they hid last month. How many raisins did they hide last month?

solution **A diagram should always be drawn as the first step.**

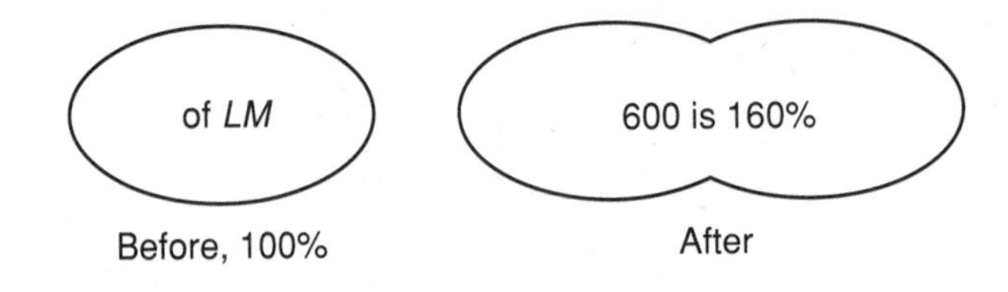

A 60 percent increase gives us 160 percent. The single statement is

600 is 160 percent of last month

$$\frac{160}{100} \times LM = 600 \longrightarrow \frac{\cancel{100}}{\cancel{160}} \cdot \frac{\cancel{160}}{\cancel{100}} \times LM = 600 \cdot \frac{100}{160} \longrightarrow LM = \frac{60{,}000}{160}$$

$$\longrightarrow \mathbf{LM = 375}$$

Thus, last month they hid 375 raisins. We finish by drawing diagrams with all numbers in place.

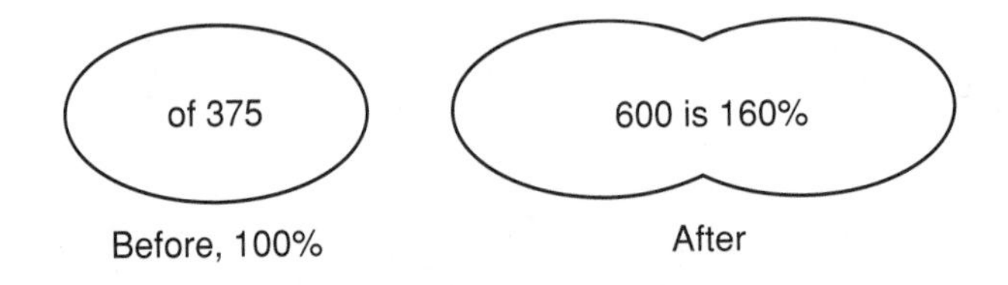

practice Draw the diagrams and then solve the problems:

a. Christopher and Chad collected 40 items. When they added Cathy's items, their total increased by 150%. What was their new total after Cathy's items were added?

b. Jane and Faye have 32 bagatelles left. If they began with 160 bagatelles, what percent of the original number remains?

c. Caesar's legion had an inventory of 2400 pairs of thongs. If they normally wear out 30 percent of their supply by the end of the fiscal year, how many pairs of thongs remain at the end of the fiscal year?

problem set 62

1. Six times a number is increased by 7. Then this sum is multiplied by 4, and the result is 10 larger than 30 times the opposite of the number. What is the number?

2. In the first group of 5, the average was 6.5. In the second group of 15, the average was 4.5. What was the overall average of the two groups?

3. Find the volume of the right prism whose base is shown and whose sides are 10 meters high. Find the volume of a pyramid 11 meters high that has the same base. Dimensions are in meters.

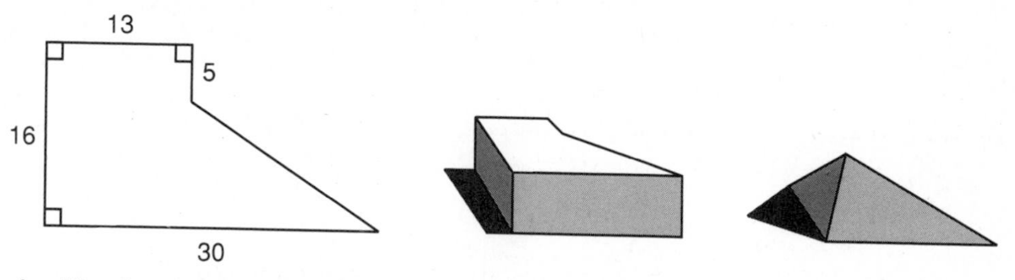

4. Use four unit multipliers to convert 420 kilometers to feet.

5. $5\frac{1}{8}$ of what number is $\frac{3}{16}$?

6. What decimal part of 0.004 is 0.00008?

Solve:

7. $3\frac{2}{5}p + 1\frac{1}{2} = -2\frac{3}{8}$

8. $-(-k) - (-2)(2k - 5) + 7 = -(2k + 4)$

9. Graph $2 < x \leq 4$ on a number line.

10. The numbers 0, 1, 5, 0, 7, 5, 2, 0, and 7 make up set K. Designate set K using set notation.

11. Given $A = \{1, 2, 3\}$ and $B = \{2, 3, 4, 0\}$, which of the following statements are true and which are false?

(a) $7 \in B$ (b) $3 \notin A$ (c) $0 \in B$

Use substitution to solve for x and y.

12. $\begin{cases} x = -19 - 6y \\ 2x + 3y = -11 \end{cases}$

13. $\begin{cases} 2x - 3y = 5 \\ x = -2y - 8 \end{cases}$

14. $\begin{cases} y = 4x + 9 \\ 3x + y = -12 \end{cases}$

Multiply:

15. $(5 + 3x)(8 - 2x)$

16. $(4x + 2)^2$

Graph on a rectangular coordinate system:

17. $y = -3$

18. $3y + x = -9$

Simplify:

19. $\dfrac{x}{\dfrac{1}{a+b}}$

20. $\dfrac{\dfrac{1}{a+b}}{x}$

21. $\dfrac{\dfrac{a}{b}}{\dfrac{1}{x}}$

22. $\dfrac{\dfrac{a+b}{x}}{\dfrac{1}{x}}$

Add:

23. $\dfrac{x}{y} + \dfrac{1}{y+1}$

24. $1 + \dfrac{x}{y}$

25. $y - \dfrac{1}{y}$

26. Factor: $10x^2y^5z - 5x^5y^2z^5 - 10x^4y^4z^4$

Simplify:

27. $\dfrac{(x^2y^0m)(m^{-2}y)}{m^2(my^{-2})}$

28. $\dfrac{(x^0y^2)^{-2}y^5x}{x^2x^{-5}yy^{-3}}$

29. $\left(\dfrac{xy^{-2}}{m^2}\right)^{-3}\dfrac{(y^{-2})^0}{(2x^2)^{-3}}$

30. $-2\{[(3 - 5) - (2^0 - 6) - 2] - [(4 - 3) - 2(-3)]\} + \sqrt[3]{-8}$

LESSON 63 Rearranging before substitution

In every substitution problem encountered thus far, one of the equations has expressed x in terms of y as in the bottom equation in (a), or y in terms of x as in the top equation in (b).

$$\text{(a)} \begin{cases} 2x + 3y = 5 \\ x = 2y + 3 \end{cases} \qquad \text{(b)} \begin{cases} y = 2x + 4 \\ 2x - y = 7 \end{cases}$$

If neither of the equations is in one of these forms, we begin by rearranging one of the equations.

example 63.1 Use substitution to solve the system: $\begin{matrix}\text{(a)} \\ \text{(b)}\end{matrix} \begin{cases} x - 2y = -1 \\ 2x - 3y = 4 \end{cases}$

solution To use substitution to solve this system, it is necessary to rearrange one of the equations. We choose to solve for x in equation (a) because the x term in this equation has a coefficient of 1, and thus we can solve this equation for x in just one step.

$$\begin{array}{rcll} x - 2y &=& -1 & \text{equation (a)} \\ 2y && 2y & \text{add } 2y \text{ to both sides} \\ \hline x &=& 2y - 1 & \end{array}$$

Now we can substitute the expression $2y - 1$ for x in equation (b) and complete the solution.

$$\begin{array}{rll} 2x - 3y = 4 & & \text{equation (b)} \\ 2(2y - 1) - 3y = 4 & & \text{substituted } 2y - 1 \text{ for } x \\ 4y - 2 - 3y = 4 & & \text{multiplied} \\ y - 2 = 4 & & \text{added like terms} \\ \mathbf{y = 6} & & \text{added 2 to both sides} \end{array}$$

We can find the value of x by replacing the variable y with the number 6 in either of the original equations. We will use both of the original equations to demonstrate that either one can be used to find x.

USING EQUATION (a)	USING EQUATION (b)
$x - 2y = -1$	$2x - 3y = 4$
$x - 2(6) = -1$	$2x - 3(6) = 4$
$x - 12 = -1$	$2x - 18 = 4$
$\mathbf{x = 11}$	$\mathbf{x = 11}$

Thus the ordered pair of x and y that satisfies both equations is **(11, 6)**.

example 63.2 Solve: (a) $\begin{cases} 2x - y = 10 \\ 4x - 3y = 16 \end{cases}$ (b)

solution We will first solve equation (a) for y and then substitute the resulting expression for y in equation (b).

$$\begin{array}{rll} 2x - y = & 10 & \\ -2x \quad & -\ 2x & \text{added } -2x \text{ to both sides} \\ \hline -y = & 10 - 2x & \\ y = & -10 + 2x & \text{multiplied both sides by } -1 \end{array}$$

Now we substitute $-10 + 2x$ for y in equation (b).

$$\begin{array}{rl} 4x - 3y = 16 & \text{equation (b)} \\ 4x - 3(-10 + 2x) = 16 & \text{substituted } -10 + 2x \text{ for } y \\ 4x + 30 - 6x = 16 & \text{multiplied} \\ -2x + 30 = 16 & \text{added like terms} \\ -2x = -14 & \text{added } -30 \text{ to both sides} \\ \mathbf{x = 7} & \text{divided by } -2 \end{array}$$

To finish the solution, we can use either of the original equations to solve for y.

USING EQUATION (a)	USING EQUATION (b)
$2x - y = 10$	$4x - 3y = 16$
$2(7) - y = 10$	$4(7) - 3y = 16$
$14 - y = 10$	$28 - 3y = 16$
$-y = -4$	$-3y = -12$
$\mathbf{y = 4}$	$\mathbf{y = 4}$

The solution is the ordered pair **(7, 4).**

example 63.3 Solve: (a) $\begin{cases} 4x - 2y = 38 \\ 2x + y = 25 \end{cases}$ (b)

solution We will first solve equation (b) for y and then substitute the resulting expression for y in equation (a).

$$\begin{array}{rll} 2x + y = & 25 & \text{equation (b)} \\ -2x \quad & -\ 2x & \text{add } -2x \text{ to both sides} \\ \hline y = & 25 - 2x & \end{array}$$

Now we substitute $25 - 2x$ for y in equation (a).

$$\begin{array}{rl} 4x - 2y = 38 & \text{equation (a)} \\ 4x - 2(25 - 2x) = 38 & \text{substituted} \\ 4x - 50 + 4x = 38 & \text{multiplied} \\ 8x = 88 & \text{added like terms} \\ \mathbf{x = 11} & \text{divided both sides by 8} \end{array}$$

Now we can use either of the original equations to solve for y.

USING EQUATION (a)	USING EQUATION (b)
$4x - 2y = 38$	$2x + y = 25$
$4(11) - 2y = 38$	$2(11) + y = 25$
$44 - 2y = 38$	$22 + y = 25$
$-2y = -6$	$\mathbf{y = 3}$
$\mathbf{y = 3}$	

The solution is the ordered pair **(11, 3)**.

practice Use substitution to solve for x and y:

a. $\begin{cases} x - 3y = -7 \\ 2x - 3y = 4 \end{cases}$ **b.** $\begin{cases} 4x - y = 41 \\ 2x + y = 25 \end{cases}$

problem set 63 Use "before" and "after" diagrams as aids in working Problems 1–4. After working the problem, fill in the missing parts in the diagram.

1. When the votes were counted, it was found that 15 percent of the people in Brenham had voted. If 2100 had voted, how many people lived in Brenham?
2. T-Willy got 128 pounds of honey from his hive. If this was a 60 percent increase from last year, how much honey did he get last year?
3. The store owner gave a 35 percent discount, yet Joe and Carol still had to pay \$247 for the camera. What was the original price of the camera?
4. The weight of the elephant was 1040 percent of the weight of the bear. If the elephant weighed 20,800 pounds, what did the bear weigh?
5. The gallimaufry contained things large and small in the ratio of 7 to 2. If the total was 1098 items, how many were large?
6. Given: $A = \{1, 5, 7, 0\}$ and $B = \{0, 1, 5\}$; which of the following are true and which are false?

   (a) $1 \notin A$ (b) $7 \notin B$ (c) $0 \in A$
7. There were 60 pumps in the first group, and they pumped 1000 gallons each. There were 240 pumps in the second group, and they pumped 800 gallons each. What was the average number of gallons pumped by all the pumps?
8. Use six unit multipliers to convert 300,000 square centimeters to square miles.
9. Express in cubic centimeters the volume of the prism whose base is shown and whose sides are 10 meters high. Also, find the volume of a pyramid 10 meters high that has the same base. Dimensions are in meters.

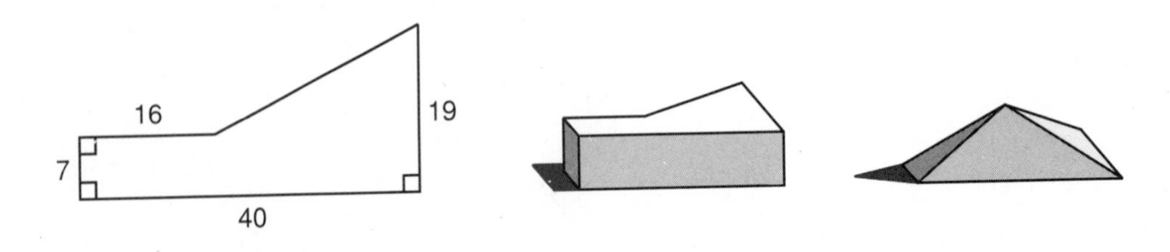

10. Use substitution to solve for x and y: $\begin{cases} 3x - 5y = 36 \\ x + 3y = -16 \end{cases}$

Multiply:

11. $(2x - 4)(x - 4)$

12. $(4 - 2x)(x^2 + 3x + 2)$

Graph on a rectangular coordinate system:

13. $y = -2$

14. $y = -2x$

15. $y + 2x + 2 = 0$

Simplify:

16. $\dfrac{\dfrac{a}{b}}{a + b}$

17. $\dfrac{\dfrac{a}{c}}{d + c}$

18. Find the volume of a sphere whose radius is 8 m.

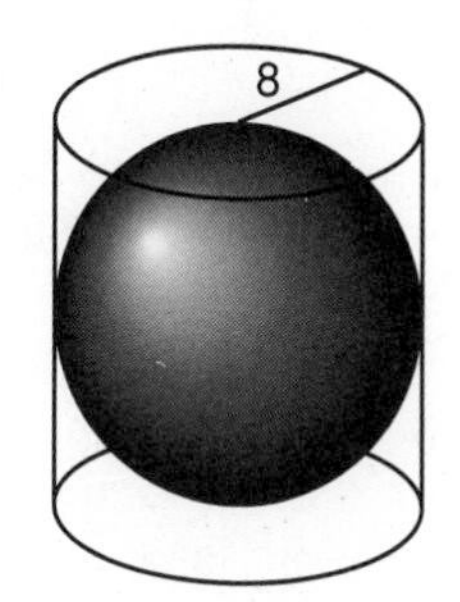

Add:

19. $\dfrac{x}{x + y} + y$

20. $\dfrac{a}{x^2c} + c^3$

21. $\dfrac{m}{c^2} - \dfrac{1}{c} + \dfrac{b}{c + b}$

22. Factor: $3x^3yp^6 - 3x^2y^2p^2 + 9x^2yp^5$

Simplify:

23. $\dfrac{x^2ym + 3x^2ymz}{x^2ym}$

24. $\dfrac{(x^{-2})^0(x^2y^{-2})^{-2}}{p^0x^{-4}(y^2)^{-2}}$

25. $\left(\dfrac{x^2p^2}{4y^{-2}}\right)^{-2}\dfrac{(y^{-2})^2}{2y^0p^5}$

26. $\dfrac{(m^2xy)mxy}{(m^2x^2y)^{-2}(xy)}$

27. Expand: $\dfrac{-x^{-2}}{y^2}\left(-x^2y^2 + \dfrac{3x^4}{xy^2}\right)$

28. Simplify by adding like terms: $xym - \dfrac{3x^2y^2m}{xy} + \dfrac{4x^{-2}y^2}{x^{-3}m^{-1}y} - \dfrac{3xy}{m^{-1}}$

29. Evaluate: $x - xy(y^0 - x^{-3})$ if $x = -2$ and $y = 3$

30. Solve: $-2(-3 - |-2|)x - 2 - 3x = -2 - \dfrac{1}{(-2)^{-3}}$

LESSON 64 Subsets

64.A subsets

If all the members of one set are also members of a second set, the first set is said to be a *subset* of the second set. If we have the two sets

$$B = \{1, 2\} \qquad A = \{1, 2, 3\}$$

Then we can say that set B is a subset of set A because all the members of set B are also members of set A. We use the symbol $\subset$ to mean **is a subset of.** Therefore, we can write

$$B \subset A$$

which is read as "set B is a subset of set A" or as "B is a subset of A." If we also consider set C, where

$$C = \{3, 2, 1\}$$

we would say that $C \subset$ A, read "set C is a subset of set A" because all the members of set C are also members of set A. Sets A and C are said to be **equal sets** because they have the **same members.** Set C is said to be an **improper subset** of set A and set A is said to be an **improper subset** of set C because **equal sets are defined to be improper subsets of each other.** Conversely, set B is a **proper subset** of set A because every member of set B is also a member of set A and the sets are not equal sets.

The **set that has no members** is defined to be a **proper subset** of every set that has members and to be an **improper subset** of itself. This set is called the **empty set** or the **null set** and can be designated by using either of the symbols shown here.

$$\{\ \ \} \text{ is the empty set} \qquad \emptyset \text{ is the null set}$$

Thus we can say that $\{\ \ \} \subset A$ or $\emptyset \subset A$, read "the empty set is a subset of set A" or "the null set is a subset of set A" because **this set is considered to be a subset of every set.** The slash can be used to negate the symbol $\subset$, as

$$A \not\subset B$$

read "set A is not a subset of set B" because all the members of set A are not members of set B.

example 64.1 Given the sets $D = \{0, 1, 2\}$, $E = \{1, 2, 3, \ldots\}$, and $G = \{1, 3, 5\}$, tell which of the following assertions are true and which are false and why.

solution

(a) $E \subset G$ False All members of set E are not members of set G

(b) $G \subset E$ True All members of set G are members of set E

(c) $D \subset E$ False 0 is not a member of set E

64.B subsets of the set of real numbers

The set of **real numbers** is said to be an **infinite set** because there is an infinite number of members of this set. An infinite set has an infinite number of subsets, but we normally restrict our attention to five subsets of the set of real numbers. These are the sets of **natural numbers, whole numbers, integers, rational numbers,** and **irrational numbers.** It would seem reasonable to assume that all authorities use the same definitions for these sets of numbers, but unfortunately this is not the case. In this book we will use the definitions used by almost all authors of recent algebra textbooks. We define the set of natural numbers (counting numbers) as follows:

$$\text{Natural numbers } N = \{1, 2, 3, \ldots\}$$

If we list the number zero in addition to the set of natural numbers, we have designated the set of whole numbers.

$$\text{Whole numbers } W = \{0, 1, 2, 3, \ldots\}$$

Many students want to argue that numbers such as -3 and -15 should also be called whole numbers because they are not fractions and they are not decimal numbers. Maybe they should be called whole numbers, but if we use the definitions given above, we must call these numbers **integers** and we cannot call them whole numbers. We designate the set of integers by listing the set of whole numbers and including the opposite of every natural number.

$$\text{Integers } J = \{\ldots, -3, -2, -1, 0, 1, 2, 3, \ldots\}$$

Every real number is either a rational number or an irrational number. A rational number is a number that **can be** represented as a quotient† (fraction) of integers. Thus each of the following numbers is a rational number.

$$\frac{1}{4} \qquad \frac{-3}{17} \qquad -10 \qquad 0.013$$

The numbers $\frac{1}{4}$ and $\frac{-3}{17}$ are already expressed as fractions of integers. The number -10 can be written as a quotient of integers in many ways such as

$$\frac{-100}{10} \quad \text{and} \quad \frac{3000}{-300} \quad \text{and} \quad \frac{270}{-27}$$

and the number 0.013 can be written as a fraction of integers in many ways, two of which are

$$\frac{13}{1000} \quad \text{and} \quad \frac{39}{3000}$$

We have noted that the positive integral roots of some counting numbers are counting numbers.

$$\sqrt{4} = 2 \qquad \sqrt[3]{27} = 3 \qquad \sqrt[4]{32} = 2 \qquad \sqrt[3]{125} = 5$$

If the square root or cube root or any counting number root of a counting number is not also a counting number, the root is an irrational number. The number π is also an irrational number. The following numbers are all irrational numbers:

$$\pi \qquad \sqrt{3} \qquad \sqrt[3]{3} \qquad \sqrt[15]{3} \qquad \sqrt{21} \qquad \sqrt[9]{21} \qquad \sqrt[4]{20}$$

In the following, we summarize what we have said about real numbers.

Real numbers
(The coordinates of all points on the number line)

Q = **Rational numbers**	P = **Irrational numbers**
(All rational numbers that **can** be expressed as fractions of integers)	(All real numbers that **cannot** be expressed as fractions of integers)

In this book when we ask to which sets a particular number belongs, we restrict

† A fraction of integers whose denominator is not zero.

the possible answers to the sets of natural numbers, whole numbers, integers, rational numbers, irrational numbers, and real numbers. Problems such as the following will give practice in distinguishing between these sets.

example 64.2 $\frac{1}{2} \in$ {What sets of numbers?}

solution This asks that we identify the sets of which the number $\frac{1}{2}$ is a member. If we restrict our reply to the sets discussed above, the answer is the sets of **rationals** and **reals.**

example 64.3 $5 \in$ {What sets of numbers?}

solution The **naturals, wholes, integers, rationals,** and **reals.**

example 64.4 Tell whether the following statements are true or false and tell why.

solution

(a)	{Reals} ⊂ {Integers}	**False**	The reals are not a subset of the integers. The integers are a subset of the reals.
(b)	{Irrationals} ⊂ {Reals}	**True**	All irrational numbers are also real numbers.
(c)	{Irrationals} ⊂ {Rationals}	**False**	The irrationals are not a subset of the rationals. In fact, no irrational number is also a rational number and vice versa.
(d)	{Wholes} ⊂ {Naturals}	**False**	It's the other way around.
(e)	{Integers} ⊂ {Reals}	**True**	All five sets just discussed are subsets of the real numbers.

practice

a. $\frac{11}{6} \in$ {What sets of numbers}?

b. $0.62 \in$ {What sets of numbers}?

c. $4 \in$ {What sets of numbers}?

problem set 64

1. The patricians and plutocrats controlled the society. If they were in the ratio of 2 to 13 and a total of 315 belonged, how many were plutocrats?

Use diagrams as an aid in solving Problems 2 and 3.

2. The girls' weightlifting team outlifted the boys' team by 140 percent. If the boys' team lifted 1400 pounds, how many pounds did the girls' team lift?

3. Harry and Jack raised roses and petunias in their garden. If 15 percent of the flowers were roses and there were 120 roses, how many flowers were there in all?

4. $7\frac{2}{5}$ of what number is $1\frac{7}{10}$?

5. What fraction of $14\frac{1}{4}$ is $\frac{3}{8}$?

Solve:

6. $7\frac{2}{5}x + 5\frac{1}{3} = \frac{1}{15}$

7. $-[-(-2p)] - 3(-3p + 15) = -(-4)(p - 12)$

8. Graph $x \not\geq -2$ on a number line.

9. Evaluate: $a^{-3}(a^{-2} - 2a)$ if $a = -2$

10. $3\sqrt{2} \in$ {What sets of numbers}?

11. Find the volume in cubic feet of a prism whose base is shown and whose height is 2 yards. Also, find the volume of a pyramid 2 yards high that has the same base. Dimensions are in feet.

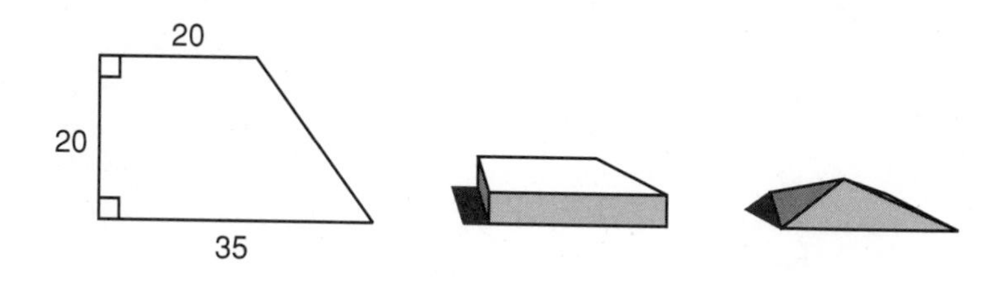

Use substitution to solve for x and y:

12. $\begin{cases} 4x + y = -5 \\ 2x - y = -1 \end{cases}$

13. $\begin{cases} x - 3y = -7 \\ 3x + y = -1 \end{cases}$

14. $\begin{cases} 4x - y = -7 \\ 2x + 2y = 4 \end{cases}$

Multiply:

15. $(4x - 3)(x + 2)$

16. $(4x + 3)^2$

Graph on a rectangular coordinate system:

17. $y = -3x$

18. $y = 3x$

19. $4 + 3x - y = 0$

Simplify:

20. $\dfrac{\frac{a}{b}}{c + x}$

21. $\dfrac{a}{\frac{b}{c + x}}$

22. $\dfrac{1}{\frac{1}{a + b}}$

Add:

23. $\dfrac{4}{xyc} - \dfrac{5m}{xy(c + 1)} - \dfrac{3k}{xy^2}$

24. $k + \dfrac{1}{k}$

25. $my + \dfrac{p}{y}$

26. Factor: $20x^2m^5k^6 - 10xm^4k^4 + 30x^5m^4k^6$

Simplify:

27. $\dfrac{(x^2y^0)^2y^0k^2}{(2x^2k^5)^{-4}y}$

28. $\dfrac{a^0x^2x^0}{m^2y^0m^{-2}}$

29. $\left(\dfrac{p^2x}{y}\right)^2\left(\dfrac{x^{-2}y}{p^2}\right)^3$

30. $-2\{[(-2 - 2) - 3^0(-2 - 1)] - [-2(-3 + 5) - 2]\} - |-2| + \sqrt[3]{-8}$

LESSON 65 Square roots · Evaluation using plus or minus

65.A square roots

We remember that if n is an even counting number, every positive number has one positive nth root and one negative nth root. For example,

$$(4)(4) = 16 \quad \text{and} \quad (-4)(-4) = 16$$

Thus both $+4$ and -4 are square roots of 16.

If n is an odd counting number, every negative number has exactly one nth root. For example,

$$(-3)(-3)(-3) = -27$$

Thus -3 is the only cube root of -27.

To this point, we have considered roots that are integers, such as $\sqrt{16}$ and $\sqrt[3]{-27}$, and roots that are not integers. We remember that if n is any counting number and the nth root is not an integer, the root is an irrational number.

Square roots are the most important roots because they occur often in the solutions of scientific problems. **We remember also that the radical sign indicates the positive square root only.**

$$2^2 = 4 \quad \text{and} \quad (-2)^2 = 4 \quad \text{but} \quad \sqrt{4} = 2 \text{ only}$$

If we wish to indicate the negative square root of 4, we must write

$$-\sqrt{4} = -2$$

DEFINITION OF SQUARE ROOT

If x is greater than zero, then $\sqrt{x}$ is the unique positive real number such that

$$(\sqrt{x})^2 = x$$

This definition says that the square root of a given positive number is that positive number which multiplied by itself equals the given positive number. Thus the square root of 2 times the square root of 2 is 2. Also $\sqrt{2.42}$ times $\sqrt{2.42}$ must equal 2.42. If a is a positive number, the square root of a times the square root of a must equal a. **We use the notation ($a > 0$) to state that a is a positive number because all numbers that are greater than zero are positive numbers.**

(a) $\sqrt{2}\sqrt{2} = \mathbf{2}$ (b) $\sqrt{a}\sqrt{a} = \boldsymbol{a}$ $(a > 0)$

(c) $\sqrt{2.42}\sqrt{2.42} = \mathbf{2.42}$ (c) $\sqrt{x}\sqrt{x} = \boldsymbol{x}$ $(x > 0)$

The square root of 2 is not a whole number, so it is an irrational number. An **irrational number** cannot be written as a decimal number that has a finite number of digits. The **decimal approximation** often used for the square root of 2 is 1.414.

$$\sqrt{2} \approx 1.414 \qquad (\approx \text{ is read "approximately equal to"})$$

A more exact approximation is given by a particular pocket calculator as

$$\sqrt{2} \approx 1.414213562$$

and a table of square roots gives an approximation to 19 decimal places as

$$\sqrt{2} \approx 1.4142135623730950488$$

But we cannot write the square root of 2 exactly by using decimal numerals because an exact representation would require that we use an infinite number of digits.

In Lesson 64 we said that an irrational number is a number that cannot be written as a fraction of integers. **Another way to define an irrational number is to say that an irrational number is a number whose decimal representation is a nonrepeating decimal numeral of infinite length. Thus the $\sqrt{2}$ is an irrational number. The square roots of the following counting numbers are not counting numbers so all of these are irrational numbers.**

$$\sqrt{2} \quad \sqrt{3} \quad \sqrt{5} \quad \sqrt{6} \quad \sqrt{7} \quad \sqrt{10} \quad \sqrt{11} \quad \sqrt{12} \quad \sqrt{13}$$

The easiest way to find an approximation for the square root of a positive real number is to use a table of square roots or a pocket calculator. The use of the square root key on a calculator gives a quick numerical answer but provides little understanding. If we use a cut-and-try approach to find the square roots of a few numbers, we can get a better feel for these irrational numbers.[†] Let's use the multiplication key on a calculator and use the cut-and-try method to find an approximation for the square root of 5.

Now, $2 \times 2 = 4$ and $3 \times 3 = 9$, so $\sqrt{5}$ lies between 2 and 3 and is closer to 2.

Try 2.1	$2.1 \times 2.1 = 4.41$	not large enough
Try 2.2	$2.2 \times 2.2 = 4.84$	still not large enough
Try 2.3	$2.3 \times 2.3 = 5.29$	too large
Try 2.25	$2.25 \times 2.25 = 5.0625$	a little too large
Try 2.24	$2.24 \times 2.24 = 5.0176$	still too large
Try 2.23	$2.23 \times 2.23 = 4.9729$	now too small

Thus we see that $\sqrt{5}$ is a number somewhere between 2.23 and 2.24. We could continue this cut-and-try procedure to find an approximation of $\sqrt{5}$ that is accurate to as many decimal places as we desire.

Although the decimal representation of $\sqrt{5}$ is a nonrepeating decimal numeral of infinite length, $\sqrt{5}$ is a real number and can be associated with a specific point on the number line. Here we graph $\sqrt{5}$ and also graph several other irrational numbers.

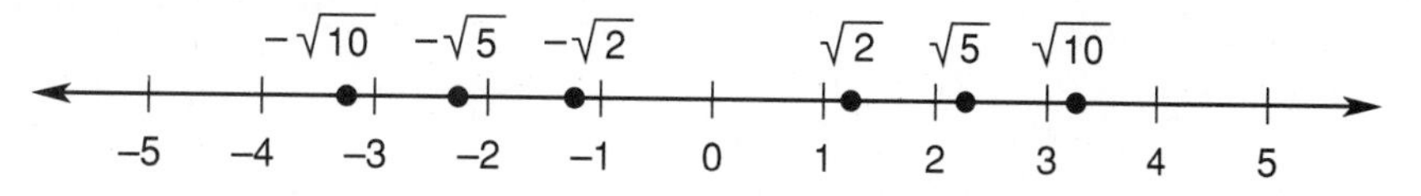

65.B plus or minus

The equation $x^2 = 4$ has two solutions because 2 times 2 equals 4 and (-2) times (-2) also equals 4. The solution to this equation can be written two different ways. The first way is

$$x = 2 \qquad \text{or} \qquad x = -2$$

The second way is

$$x = \pm 2$$

[†] There are several rote methods that can be used to find the square root of a number. Instead of using one of these, we use cut and try because this method emphasizes the meaning of square root.

Equations in which the highest power of the variable is 2 are called **quadratic equations.** Sometimes the solution of a quadratic equation is one number plus or minus another number. The solution can look like this

$$7 \pm 2$$

This notation simplifies into two different numbers: 7 plus 2 equals 9, and 7 minus 2 equals 5.

$$7 \pm 2 \qquad \text{means} \qquad 9 \text{ and } 5$$

We customarily write the two answers separated by a comma.

$$\mathbf{9, 5}$$

example 65.1 Evaluate: $-2^2 + (-3)^2 \pm \sqrt{4}$

solution The first term has a value of -4. The second term has a value of $+9$, so we have

$$-4 + 9 \pm \sqrt{4} \qquad \text{simplified}$$
$$= 5 \pm 2 \qquad \text{simplified}$$
$$= \mathbf{7, 3}$$

Thus the expression we were asked to evaluate has two values, 7 and 3.

practice Use the cut-and-try method to find these roots accurate to one decimal place.

a. $\sqrt{3}$ **b.** $\sqrt{8}$

Evaluate:

c. $8 \pm \sqrt{9}$ **d.** $14 \pm \sqrt{16}$

problem set 65 Use "before" and "after" diagrams to help with problems 1 to 3.

1. The doctor increased the dosage to 128 percent of the original dosage. If the new dosage was 3840 units, what was the original dosage?

2. Odessa could afford the coat because it was sold for 28 percent less than the original price. If the sale price was \$324, what was the original price?

3. When the Huns debouched from the Alpine passes, Attila found that 18 percent of the spearpoints were dull. If 720 spearpoints were dull, how many spears did the Huns bring with them?

4. $5\frac{1}{2} \in$ {What sets of numbers}? **5.** $-2 \in$ {What sets of numbers}?

6. If $x = 36$, evaluate $\sqrt{x} - 6$.

7. Use the cut-and-try method to find $\sqrt{8}$ accurate to one decimal place.

Use substitution to solve for x and y:

8. $\begin{cases} 3x - 2y = -1 \\ y = x - 1 \end{cases}$ **9.** $\begin{cases} 5x - 3y = 1 \\ 7x - y = -5 \end{cases}$ **10.** $\begin{cases} 5x + 2y = -21 \\ -2x + y = 3 \end{cases}$

Multiply:

11. $(2x - 5)(2x + 5)$ **12.** $(3x - 2)^2$

Graph on a rectangular coordinate system:

13. $y = -3$ **14.** $y = -3x$ **15.** $3y + 9x + 9 = 0$

Simplify:

16. $\dfrac{\dfrac{m}{x+y}}{\dfrac{a}{x+y}}$ **17.** $\dfrac{\dfrac{m}{x+y}}{\dfrac{m}{x}}$ **18.** $\dfrac{m}{\dfrac{1}{m}}$

Add:

19. $\dfrac{a}{x^2y} + \dfrac{m+c}{x+y}$ **20.** $1 + \dfrac{y}{x}$ **21.** $y + \dfrac{y}{x}$

22. Factor: $9x^3ym^5 + 6m^2y^4p^4 - 3y^3m^3$

Simplify:

23. $\dfrac{4kp + 4kpx}{4kp}$ **24.** $mx(x^0y)m^2x^2(y^2)$

25. $\dfrac{a^{-2}p^2a(a^0)^2}{(a^{-3})^2(p^{-2})^{-2}}$ **26.** $\dfrac{(mx)(mx^0)}{3^{-2}xyyx^{-3}m}$

27. Expand: $\dfrac{-x^{-3}}{y}\left(\dfrac{x^3}{y^{-1}} - \dfrac{3x^{-3}}{y^2}\right)$

28. Simplify by adding like terms: $-2^0x^2y^2x^{-2} + \dfrac{3y^2}{x^2} - \dfrac{4x^{-2}}{y^{-2}} + 5y^2$

29. Evaluate: $-x^0 - x(x^0 - y)$ if $x = -3$ and $y - 2 = -4$

30. Solve: $-k(-2 - 3) - (-2)(-k - 5) = -2 - (-2k + 4) + \sqrt[3]{-27}$

LESSON 66 Product of square roots rule

Square roots of many numbers, such as $\sqrt{50}$, $\sqrt{200}$, and $\sqrt{147}$, can be written in a simplified form. To write one of these numbers in a simplified form, we use the following rule.

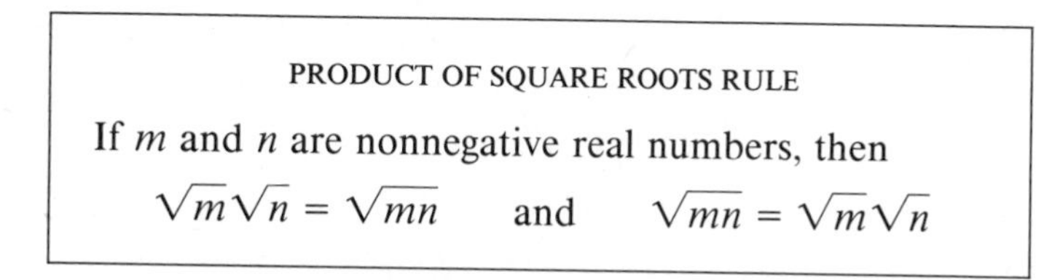

PRODUCT OF SQUARE ROOTS RULE

If m and n are nonnegative real numbers, then

$$\sqrt{m}\sqrt{n} = \sqrt{mn} \quad \text{and} \quad \sqrt{mn} = \sqrt{m}\sqrt{n}$$

This rule can be generalized to the product of any number of factors, and we say that the square root of any product may be written as the product of the square roots of the factors of the product. For example,

$\sqrt{2 \cdot 5 \cdot 5}$ can be written as $\sqrt{2}\sqrt{5}\sqrt{5}$

and $\sqrt{3 \cdot 3 \cdot 3 \cdot 5}$ can be written as $\sqrt{3}\sqrt{3}\sqrt{3}\sqrt{5}$

We will use this rule in the following problems to help us simplify the expressions $\sqrt{50}$, $\sqrt{200}$, $\sqrt{147}$, and $\sqrt{108}$.

example 66.1 Simplify: $\sqrt{50}$

solution We will first write 50 as a product of prime factors.

$$\sqrt{5 \cdot 5 \cdot 2}$$

Now we use the product of square roots rule to write the square root of the product as a product of square roots.

$$\sqrt{5}\sqrt{5}\sqrt{2}$$

Now, by definition $\sqrt{5}\sqrt{5} = 5$, so we have

$$\sqrt{5}\sqrt{5}\sqrt{2} = \mathbf{5\sqrt{2}}$$

example 66.2 Simplify: $\sqrt{200}$

solution First we write 200 as a product of prime factors.

$$\sqrt{200} = \sqrt{2 \cdot 2 \cdot 2 \cdot 5 \cdot 5}$$

Now the square root of the product is written as the product of square roots.

$$\sqrt{2 \cdot 2 \cdot 2 \cdot 5 \cdot 5} = \sqrt{2}\sqrt{2}\sqrt{2}\sqrt{5}\sqrt{5}$$

Since by definition $\sqrt{2}\sqrt{2} = 2$ and $\sqrt{5}\sqrt{5} = 5$, we can simplify as

$$(\sqrt{2}\sqrt{2})\sqrt{2}(\sqrt{5}\sqrt{5}) = (2)\sqrt{2}(5) = \mathbf{10\sqrt{2}}$$

example 66.3 Simplify: $\sqrt{147}$

solution First we write 147 as a product of prime factors.

(a) $\sqrt{147} = \sqrt{3 \cdot 7 \cdot 7}$ write as product of prime factors

(b) $\sqrt{3 \cdot 7 \cdot 7} = \sqrt{3}\sqrt{7}\sqrt{7}$ root of product equals product of roots

(c) $\sqrt{3}(\sqrt{7}\sqrt{7}) = \mathbf{7\sqrt{3}}$ definition of square root

example 66.4 Simplify: $\sqrt{108}$

solution First we write 108 as a product of prime factors.

(a) $\sqrt{108} = \sqrt{2 \cdot 2 \cdot 3 \cdot 3 \cdot 3}$ write as product of prime factors

(b) $\sqrt{2 \cdot 2 \cdot 3 \cdot 3 \cdot 3} = \sqrt{2}\sqrt{2}\sqrt{3}\sqrt{3}\sqrt{3}$ root of product equals product of roots

(c) $(\sqrt{2}\sqrt{2})(\sqrt{3}\sqrt{3})\sqrt{3} = 2 \cdot 3\sqrt{3} = \mathbf{6\sqrt{3}}$ definition of square root

practice Simplify:

a. $\sqrt{75}$ **b.** $\sqrt{200}$ **c.** $\sqrt{189}$

problem set 66

1. If the product of -3 and the opposite of a number is decreased by 7, the result is 1 greater than the number. What is the number?

Draw diagrams for the following percent problems and work the problems.

2. Between Karnak and Edfu, the Pharaoh kept 1020 white goats. If these goats represented 17 percent of the total flock, how many goats did the Pharaoh have?

3. After the temple was destroyed, Amenhotep found 1200 precious stones in the ruins. If 3 percent of these stones were rubies, how many rubies did Amenhotep find?

4. 1.05 of what number is 4.221?

5. What fraction of $3\frac{1}{8}$ is $\frac{7}{4}$?

Solve:

6. $-3\frac{1}{2} + 1\frac{2}{5}p - 4\frac{2}{3} = 0$

7. $0.3k - 0.2 + 0.2k - 0.05 = -2(k - 3)$

8. Graph $4 < x \leq 7$ on a number line.

9. $\frac{\sqrt{3}}{2} \in$ {What sets of numbers}?

10. $4 \in$ {What sets of numbers}?

Use substitution to solve for x and y:

11. $\begin{cases} 5x - 4y = -6 \\ x - 2y = -6 \end{cases}$

12. $\begin{cases} x - 2y = 7 \\ 2x - 3y = -4 \end{cases}$

13. $\begin{cases} 4x + y = 14 \\ 2x - 2y = 22 \end{cases}$

14. If $x + 4 = 20$, evaluate: $\sqrt{x} + \frac{1}{4^{-2}}$

15. Find the volume in cubic feet of the right prism whose base is shown and whose height is 2 yards. Also, find the volume of a pyramid 2 yards high that has the same base. Dimensions are in feet.

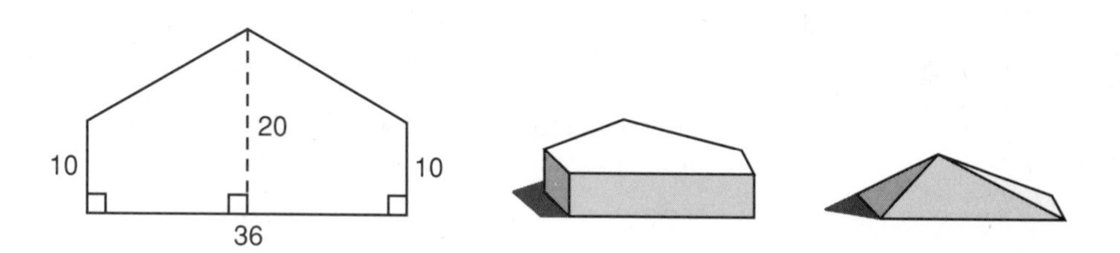

16. Use 12 unit multipliers to convert 8 cubic yards to cubic meters.

17. Seventy sixth-graders had an average weight of 90 pounds. Thirty seventh-graders had an average weight of 100 pounds. What was the average weight of all these students?

18. Find the volume of a circular cylinder whose height is 40 m and whose radius is 20 m. Find the volume of a sphere whose radius is 20 m.

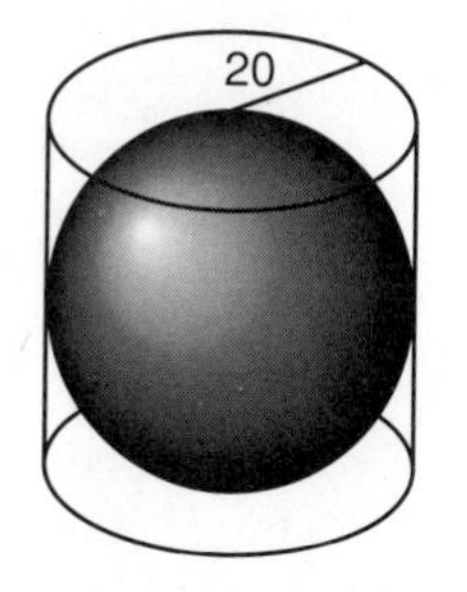

Graph on a rectangular coordinate system:

19. $y = 2$

20. $y = 2x$

21. $y - 2x = 2$

Simplify:

22. $\dfrac{a}{\dfrac{1}{a + b}}$

23. 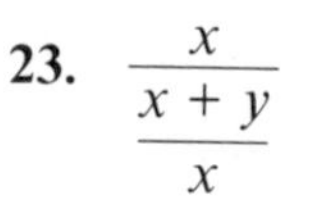

$\dfrac{x}{\dfrac{x + y}{x}}$

Add:

24. $\dfrac{1}{x} + \dfrac{3}{x + y}$ 25. $x + \dfrac{1}{y}$ 26. $1 + \dfrac{1}{y}$

27. Factor: $15m^2x^5k^4 - 5m^6x^6k^6 + 20m^4xk^5$

Simplify:

28. $\dfrac{3x^2m^5(2x^4m^2)}{3x^2m^5m^{-4}}$ 29. $\dfrac{4(p^{-2})^0(p^5)}{4^{-1}p^6x^{-5}}$

30. $\left(\dfrac{3x^{-2}}{y^{-3}}\right)^{-2}\left(\dfrac{x^4}{y^6}\right)^2$ 31. $-2^0 - 2^2(2^0) - (2^0)^{-3} - \sqrt[3]{-27}$

LESSON 67 Domain

In problems in mathematics (and in physics, chemistry, and other mathematically based disciplines) the numbers that may be used as replacements for the variables are often restricted by the nature of the problem or by a restriction stated in the problem. For instance, if a person goes to the store with 25 cents to buy eggs, and eggs cost 10 cents each, the total amount of money that can be spent on eggs can be represented by the equation

$$\text{Total cost} = 10N_E \quad \text{cents}$$

where N_E represents the number of eggs bought. The customer may buy no eggs or 1 egg or 2 eggs. Thus the total cost of the eggs is as shown in (a), (b), or (c).

(a) 0 EGGS	(b) 1 EGG	(c) 2 EGGS
Cost = 10(0)	Cost = 10(1)	Cost = 10(2)
Cost = 0 cents	Cost = 10 cents	Cost = 20 cents

In (a) we use 0 as the replacement for the variable, and in this case the cost is 0 cents. In (b) we use 1 as the replacement for the variable and find the cost to be 10 cents. In (c) we use 2 as the replacement for the variable and find that the cost is 20 cents. We cannot use 3 as a replacement for the variable because the buyer has only 25 cents to spend. We cannot use $2\frac{1}{2}$ as a replacement for the variable because only whole eggs are sold at the market. Neither could we use -4 as a replacement for the variable because buying -4 eggs makes no sense. We are restricted by the statement of the problem to using only the whole numbers

$$\{0, 1, 2\}$$

as replacements for the variables in the equation.

The set of numbers that constitutes the set of permissible replacement values for the variables in a particular equation or inequality is called the *domain* for that equation or inequality.

Since **every equation and every inequality have a domain**[†] and since this is an important concept, it is customary in courses in algebra to include problems in

---

[†] If the domain is not stated, it is implied.

which the domain is specified. These problems should help with the concept of domain. The domains for the problems in this book were chosen by the author, sometimes with a purpose and sometimes just arbitrarily so that the problems would have specified domains. We will use the capital letter D as the symbol for domain in this book and will indicate the domains as sets by enclosing them within braces. For instance,

(a) $D = \{0, 1, 2\}$ (b) $D = \{\text{Reals}\}$ (c) $D = \{\text{Positive integers}\}$

The domains specified here are (a) the numbers 0, 1, and 2, (b) the set of real numbers, and (c) the set of positive integers.

example 67.1 Graph: $x \ngeq 3$; $D = \{\text{Integers}\}$

solution We are asked to indicate the **integers** that are less than 3.

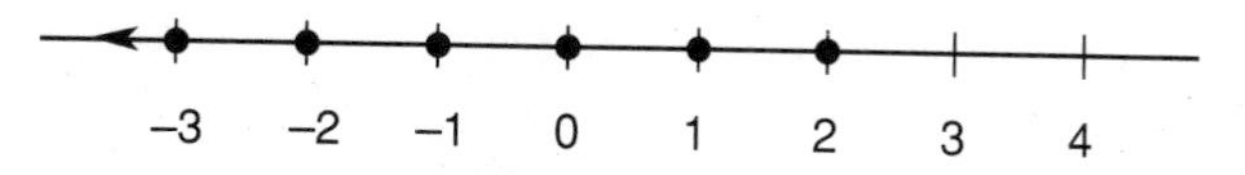

On the number line we have indicated the integers whose values are less than 3. Note that it is not necessary to place an open circle at 3. The arrow on the left indicates an infinite continuation.

example 67.2 Graph: $x \geq -1$; $D = \{\text{Reals}\}$

solution We are asked to indicate all **real numbers** that are greater than or equal to -1.

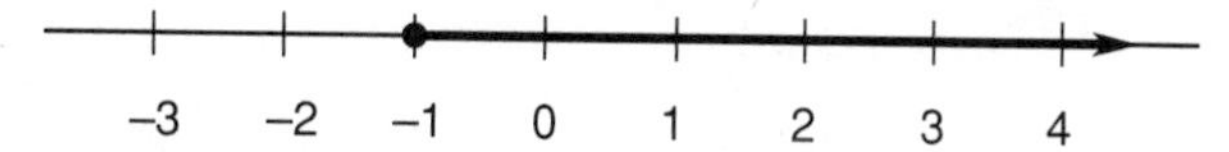

The graph indicates all real numbers that are greater than or equal to -1. The solid circle at -1 indicates that -1 is a member of the solution set.

example 67.3 Graph: $x < -1$; $D = \{\text{Positive integers}\}$

solution The solution is $\emptyset$, the null set, or { }, the empty set, because there are *no* positive integers that are less than -1.

example 67.4 Graph: $x \geq -5$; $D = \{\text{Positive integers}\}$

solution The graph below indicates the numbers that are greater than or equal to -5 and that are also members of the set of positive integers.

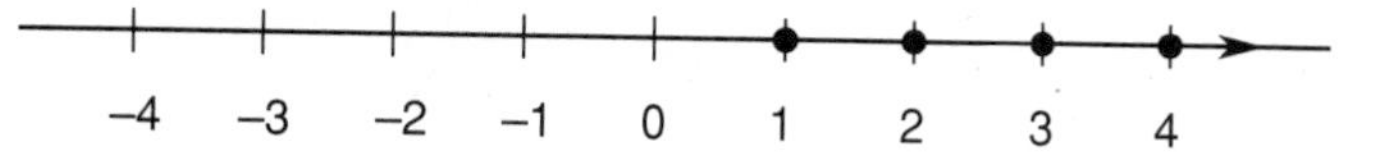

practice Graph on a number line:

a. $x \ngeq -2$; $D = \{\text{Positive integers}\}$

b. $x \nless 4$; $D = \{\text{Reals}\}$

problem set 67

1. The number of bacteria increased by 280 percent overnight. If there were 30,000 bacteria yesterday, how many bacteria were present this morning?

2. When Charles inspected the troops that survived, he found that 3600 were still alive. If 40 percent died in the fight, how many troops did he begin with?

3. Edna and Mabel climbed 40 percent of the mountains in the whole country. If they climbed 184 mountains, how many mountains were in the country?

Simplify:

4. $\sqrt{72}$ 5. $3\sqrt{75}$ 6. $4\sqrt{324}$

7. Multiply: $(4x - 3)(12x + 2)$

Graph on a rectangular coordinate system:

8. $y = -2$ 9. $y = -2x - 5$ 10. $y = -2x + 5$

Graph on a number line:

11. $x \ngtr 4$; $D = \{\text{Integers}\}$

12. Find the surface area of this rectangular solid in square inches. Dimensions are in feet.

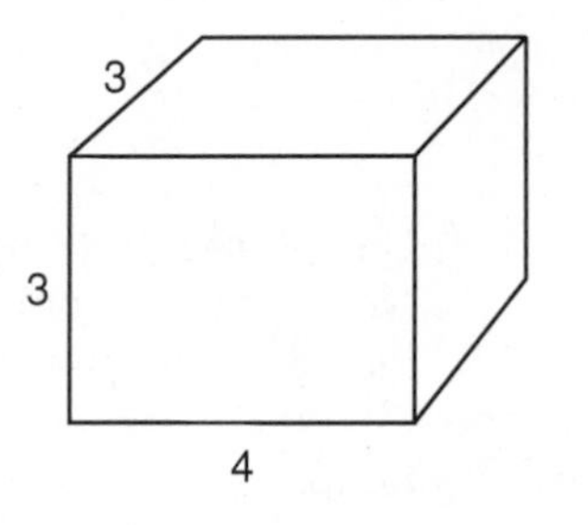

Use substitution to solve for x and y:

13. $\begin{cases} 3x - 2y = 15 \\ 5x + y = 12 \end{cases}$ 14. $\begin{cases} y + 2x = 12 \\ x + 2y = 12 \end{cases}$

15. Use the cut-and-try method to find $\sqrt{27}$ to one decimal place.

Simplify:

16. $\dfrac{\frac{x}{y}}{\frac{1}{y}}$ 17. $\dfrac{x}{\frac{x}{y}}$ 18. $\dfrac{1}{\frac{1}{y}}$

Add:

19. $\dfrac{4}{a^2x} + \dfrac{7}{x(x + a)}$ 20. $2 + \dfrac{3}{y}$ 21. $1 + \dfrac{x}{y}$

22. Factor: $40x^4ym^7z - 20x^5y^5m^2z + 20xy^2m$

Simplify:

23. $\dfrac{4x + 4x^2}{4x}$ 24. $\dfrac{kp^{-2}k(p^0)^2}{kp(k)(p^{-2})^2}$ 25. $\left(\dfrac{3m^2}{y^{-4}}\right)^2\left(\dfrac{m}{y}\right)$

26. $\dfrac{2p^2x^{-4}(x)(x^2)}{y^{-4}(p^2)^{-2}x}$ 27. $-|-3^0| - 3^0(-2)(-3)(-2 - 3)$

28. Simplify by adding like terms: $-\dfrac{3x^2y^{-2}}{x^{-2}y^{-2}} - 2x^4yy^{-1} + 4x^3xyy^{-1} - \dfrac{2x^2}{x^{-2}}$

29. Expand: $-\dfrac{x^{-2}}{y^4}\left(x^2y^4 - \dfrac{3x^{-2}}{y^4}\right)$

30. Evaluate: $x - (x^2)^0(x - y) - |x - y|$ if $x = -2$ and $y = -3$

LESSON 68 *Additive property of inequality*

We restate the additive property of equality here. Note that we write $a + c = b + c$ and also write $c + a = c + b$. We do this to emphasize that the order of the addends does not affect the result.

> ADDITIVE PROPERTY OF EQUALITY
>
> If a, b, and c are any real numbers such that
>
> $$a = b$$
>
> then $a + c = b + c$ and $c + a = c + b$

We have learned that we can use the additive property of equality to help us solve some equations. For example, we can solve $x + 4 = 8$ by adding -4 to both sides of the equation.

$$\begin{array}{rcr} x + 4 & = & 8 \\ -4 & & -4 \\ \hline x & = & 4 \end{array}$$

The **additive property of inequality** is stated in the same way as the additive property of equality except that we use the $>$ symbol rather than the $=$ sign.

> ADDITIVE PROPERTY OF INEQUALITY
>
> If a, b, and c are any real numbers such that
>
> $$a > b$$
>
> then
>
> $a + c > b + c$ and also $c + a > c + b$

This statement can be used to prove that the same quantity can be added to both sides of an inequality without changing the solution set of the inequality.

example 68.1 Graph: $x + 2 < 0$; $D = \{\text{Reals}\}$

solution We isolate x by adding -2 to both sides of the inequality, as permitted by the additive property of inequality.

$$\begin{array}{rcrl} x + 2 & < & 0 & \text{given} \\ -2 & & -2 & \text{add } -2 \text{ to both sides} \\ \hline x & < & -2 & \end{array}$$

Now we graph the inequality $x < -2$.

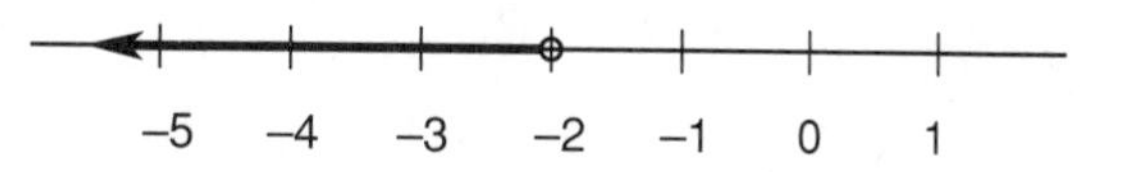

The open circle at -2 indicates that -2 is not a member of the solution set.

example 68.2 Graph: $x - 3 \geq -5$; $D = \{\text{Integers}\}$

solution We isolate x by adding $+3$ to both sides of the inequality, as permitted by the additive property of inequality.

$$\begin{array}{rrl} x - 3 \geq & -5 & \text{given} \\ +3 & +3 & \text{add } +3 \text{ to both sides} \\ \hline x \quad\geq & -2 & \end{array}$$

Now we graph the solution $x \geq -2$ and remember that the domain is the set of integers.

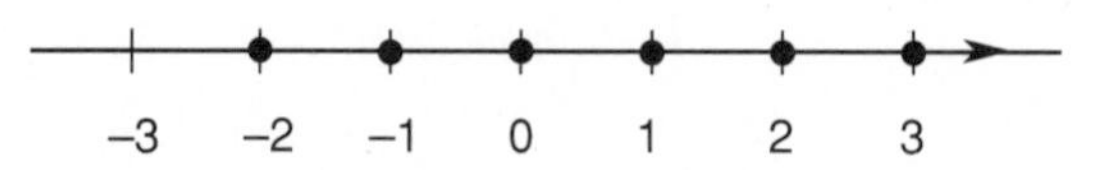

Note that -2 is a solution to the inequality $x \geq -2$.

practice Graph on a number line:

a. $x - 5 \not< 0$; $D = \{\text{Integers}\}$

b. $x + 2 < 5$; $D = \{\text{Integers}\}$

problem set 68

1. The postprandial exercises were situps and pushups in the ratio of 7 to 2. If Hominoid did 9180 exercises, how many were pushups?

2. We estimate that the giant pyramid of Cheops near Cairo contains 2,300,000 blocks of stone. If the builders only used 80 percent of the available blocks, how many blocks were available?

3. When Hannibal increased his army by 17 percent, he found that he had 5850 soldiers. How many soldiers did he have before the increase?

4. $2\frac{1}{8}$ of what number is $\frac{1}{16}$?

5. What fraction of $2\frac{1}{7}$ is $\frac{3}{14}$?

Solve:

6. $3\frac{1}{5}x + \frac{2}{3} = 5\frac{1}{2}$

7. $-|-2| - 2^2 - (-3 - k) = -2(k - |-3|)$

8. Write an inequality that describes this graph. Remember to designate the domain.

9. Find the volume in cubic inches of the right prism whose base is shown and whose height is 1 yard. Also, find the surface area. Then find the volume of a pyramid 1 yard high that has the same base. Dimensions are in inches.

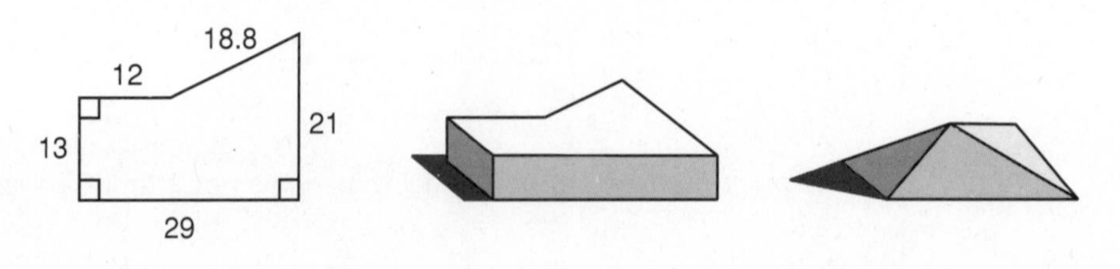

Use substitution to solve for x and y:

10. $\begin{cases} 3x - y = 1 \\ x + 2y = 5 \end{cases}$ **11.** $\begin{cases} 3x + y = -12 \\ 2x - 3y = 3 \end{cases}$ **12.** $\begin{cases} 5x - 4y = 12 \\ 9x + y = 38 \end{cases}$

Simplify:

13. $4\sqrt{50}$ **14.** $6\sqrt{45}$ **15.** $2\sqrt{12}$ **16.** $5\sqrt[5]{-32}$

17. Multiply: $(2x - 5)^2$

Graph on a rectangular coordinate system:

18. $y = -4$ **19.** $y = 3x - 2$ **20.** $2y + 3x = 4$

Simplify:

21. $\dfrac{x}{\dfrac{1}{xy + b}}$ **22.** $\dfrac{\dfrac{x}{a}}{\dfrac{1}{b}}$

23. $-3^0 - 3^2 - 3(2^0 - 1) + \sqrt[3]{27}$

Add:

24. $\dfrac{1}{x^2} + \dfrac{m}{x^3y} + \dfrac{c}{y}$ **25.** $y + \dfrac{1}{x}$ **26.** $1 + \dfrac{1}{xy}$

27. Factor: $30a^2b^3c^4 - 15ab^4c^5 + 45ab^4c^4$

Simplify:

28. $\dfrac{xx(x^{-2})^2y(y)}{x^{-2}x^0xx^{-3}}$ **29.** $\dfrac{4(2x^2y^4p^{-2})}{4x(y^2)^{-2}p^0}$ **30.** $\dfrac{(x^{-2})^{-2}(2y^2)^2(y)}{x^{-2}x^{-5}x^0xx^{-2}}$

LESSON 69 Addition of radical expressions · Weighted average

69.A addition of radical expressions

In Lesson 16, we found that like terms may be added by adding the numerical coefficients of the terms. Radical expressions that have the same index and the same radicand designate the same number and are like terms. Thus the rule for adding like terms can be used to add like radical expressions. We add like radical expressions by adding the numerical coefficients.

example 69.1 Add: $4\sqrt{2} - 5\sqrt{2} + 12\sqrt{2}$

solution We add these like terms by adding their numerical coefficients.

$$4\sqrt{2} - 5\sqrt{2} + 12\sqrt{2} = (4 - 5 + 12)\sqrt{2} = \mathbf{11\sqrt{2}}$$

example 69.2 Add: $4\sqrt{3} + 3\sqrt{5} - 6\sqrt{3}$

solution Only like radical terms may be added.

$$4\sqrt{3} - 6\sqrt{3} + 3\sqrt{5} = (4 - 6)\sqrt{3} + 3\sqrt{5} = \mathbf{-2\sqrt{3} + 3\sqrt{5}}$$

example 69.3 Add: $-3\sqrt{2} + 5\sqrt{3} - 2\sqrt{2} + 8\sqrt{3}$

solution We omit the intermediate step and write the answer directly by simply adding the coefficients of like radical terms.

$$\mathbf{-5\sqrt{2} + 13\sqrt{3}}$$

example 69.4 Add: $4\sqrt{3} - 2\sqrt{2} + 6\sqrt{5}$

solution No two of these radical terms are like radical terms, so no addition is possible.

69.B weighted average

Susan's scores improved on each test. Her scores were 60, 71, and 91. The average of her scores was

$$\text{Average} = \frac{60 + 71 + 91}{3} = \frac{222}{3} = 74$$

The teacher did not think this was a fair grade since every test covered all previous material. The teacher thought that the second test was twice as important as the first test and the third test was four times as important as the first test. So she gave Susan one 60, two 71s, and four 91s.

$$\text{Weighted average} = \frac{60 + 71 + 71 + 91 + 91 + 91 + 91}{7} = \frac{566}{7} \approx 80.9$$

The weighted average of 80.9 was a fairer score than the real average of 74. The teacher gave the first test a **weight** of 1 because she counted it once. She gave the second test a weight of 2 because she counted it twice. She gave the third test a weight of 4 because she counted it four times. If we rearrange our numbers, we can use this problem to help us define weighted average.

$$\text{Weighted average} = \frac{1(60) + 2(71) + 4(91)}{1 + 2 + 4} = \frac{566}{7} \approx 80.9$$

This shows us that the weighted average is the sum of the products of the scores S and their weights W, divided by the sum of the weights.

$$\text{Weighted average} = \frac{(W_1)(S_1) + W_2(S_2) + W_3(S_3) + \cdots + W_n(S_n)}{W_1 + W_2 + W_3 + \cdots W_n}$$

example 69.5 Jim's scores were 60, 70, 80, and 90. What was his weighted average if the tests were weighted 1, 2, 4, and 6 in that order.

solution There were one 60, two 70s, four 80s, and six 90s.

$$\text{Weighted average} = \frac{1(60) + 2(70) + 4(80) + 6(90)}{1 + 2 + 4 + 6} = \frac{1060}{13} \approx \mathbf{81.54}$$

We divided by 13 because there were 13 scores. **We note that 13 is the sum of the weights.**

practice Add:

a. $3\sqrt{3} - 2\sqrt{2} + 5\sqrt{3}$

b. $12\sqrt{7} + 6\sqrt{7} - 20\sqrt{7}$

c. $\sqrt{2} + 3\sqrt{2} - 4\sqrt{2} + 6\sqrt{3}$

d. The test scores were 70, 80, and 100. The weights were 1, 5, and 7. What was the weighted average?

problem set 69

1. The troll became incensed when he saw the billy goats prancing across the bridge. Finally, he tore the bridge down—but not before 18 percent of the goats had crossed. If 45 goats had crossed, how many goats were there?

2. Jaime's scores were 75, 80, 88, and 93. What was his weighted average if the tests were weighted 1, 2, 3, and 4 in that order?

3. A 130 percent increase in the doll population resulted in a total of 1610 dolls. How many dolls were present before the population increased?

Simplified:

4. $5\sqrt{80}$

5. $3\sqrt{120}$

6. Add: $7\sqrt{5} - \sqrt{5} + 5\sqrt{3} - 3\sqrt{3}$

7. Find the volume in cubic inches and the surface area in square inches of the cylinder whose base is shown and whose sides are 2 feet high. Find the volume of a cone 2 feet high that has the same base. Dimensions are in inches.

8. Multiply: $(3p - 4)(2p + 5)$

Graph on a rectangular coordinate system:

9. $x = -\frac{1}{2}$

10. $y = -\frac{1}{2}x$

11. $2y = x - 8$

12. Graph on a number line: $x + 3 > -7$; $D = \{\text{Positive integers}\}$

Use substitution to solve for x and y:

13. $\begin{cases} x + y = 10 \\ -x + y = 0 \end{cases}$

14. $\begin{cases} 3x - 3y = 3 \\ x - 5y = -3 \end{cases}$

15. $\begin{cases} 3x - y = 8 \\ x - 3y = -8 \end{cases}$

Simplify:

16. $\dfrac{\frac{a}{x}}{\frac{1}{a^2}}$

17. $\dfrac{\frac{a}{a + b}}{a}$

18. $\dfrac{\frac{x}{y}}{\frac{1}{y}}$

Add:

19. $\dfrac{a}{x + y} + \dfrac{5}{x^2}$

20. $1 + \dfrac{a}{b}$

21. $x + \dfrac{1}{x}$

22. Factor: $4x^2y^2z - 8x^2y^2z^5$

23. Expand: $-3x^{-2}y^2\left(\dfrac{y^{-2}}{x^{-2}} + 4x^2y\right)$

Simplify:

24. $\dfrac{4kx - 4kx^2}{4kx}$

25. $\dfrac{m^0(p^{-2})^2x^2y^4}{(y^{-2})^2yy^0x^{-2}}$

26. $\left(\dfrac{2x^{-2}y}{p}\right)^2\left(\dfrac{p^2x}{2}\right)^{-2}$

27. $\dfrac{x^2x^{-2}x^0y^2}{y^2(x^{-4})^2}$

28. Simplify by adding like terms: $\dfrac{3a^2x}{m} + \dfrac{5xm^{-1}}{a^{-2}} - \dfrac{4aax^{-1}}{x^{-2}m}$

29. Evaluate: $-x^{-4} - x^2(x - m)$ if $x = -2$ and $m + 3 = 6$

30. Simplify: $-3^0 - 3(-2 - 2^0)(-8^0 - 5) + \sqrt[4]{16}$

LESSON 70 Simplification of radical expressions

In Lesson 66 we learned to simplify expressions such as $\sqrt{50}$ by using the product of square roots rule.

$$\sqrt{50} = \sqrt{5 \cdot 5 \cdot 2} = \sqrt{5}\sqrt{5}\sqrt{2} = \mathbf{5\sqrt{2}}$$

We can use the same procedure to simplify expressions such as

$$\sqrt{8} - \sqrt{50} + \sqrt{98}$$

if we first simplify each expression and then add the like radical terms. We begin by writing each radicand as a product of prime factors.

$$\sqrt{2 \cdot 2 \cdot 2} - \sqrt{5 \cdot 5 \cdot 2} + \sqrt{7 \cdot 7 \cdot 2}$$

Now we write the roots of products as products of roots.

$$\sqrt{2}\sqrt{2}\sqrt{2} - \sqrt{5}\sqrt{5}\sqrt{2} + \sqrt{7}\sqrt{7}\sqrt{2}$$

To finish, we simplify and add like radical terms.

$$2\sqrt{2} - 5\sqrt{2} + 7\sqrt{2} = \mathbf{4\sqrt{2}}$$

example 70.1 Simplify: $\sqrt{18} + \sqrt{8}$

solution

$\sqrt{18} + \sqrt{8}$	given
$\sqrt{2 \cdot 3 \cdot 3} + \sqrt{2 \cdot 2 \cdot 2}$	write each radicand as a product of prime factors
$\sqrt{2}\sqrt{3}\sqrt{3} + \sqrt{2}\sqrt{2}\sqrt{2}$	write roots of products as products of roots
$3\sqrt{2} + 2\sqrt{2} = \mathbf{5\sqrt{2}}$	simplify and add like radical terms

example 70.2 Simplify: $8\sqrt{27} - 3\sqrt{75}$

solution The radicals in this problem have coefficients, so we will use parentheses to help us prevent errors. We can simplify $\sqrt{27}$ and $\sqrt{75}$ as

$$\sqrt{27} = \sqrt{3 \cdot 3 \cdot 3} = \sqrt{3}\sqrt{3}\sqrt{3} = \mathbf{3\sqrt{3}}$$

and

$$\sqrt{75} = \sqrt{5 \cdot 5 \cdot 3} = \sqrt{5}\sqrt{5}\sqrt{3} = \mathbf{5\sqrt{3}}$$

and now we replace $\sqrt{27}$ with $3\sqrt{3}$ and replace $\sqrt{75}$ with $5\sqrt{3}$.

$$8\sqrt{27} - 3\sqrt{75} = 8(3\sqrt{3}) - 3(5\sqrt{3}) = 24\sqrt{3} - 15\sqrt{3} = \mathbf{9\sqrt{3}}$$

example 70.3 Simplify: $\sqrt{27} - 3\sqrt{18} - 6\sqrt{45}$

solution

$\sqrt{27} - 3\sqrt{18} - 6\sqrt{45}$	given
$\sqrt{3 \cdot 3 \cdot 3} - 3\sqrt{3 \cdot 3 \cdot 2} - 6\sqrt{3 \cdot 3 \cdot 5}$	products of prime factors
$\sqrt{3}\sqrt{3}\sqrt{3} - 3\sqrt{3}\sqrt{3}\sqrt{2} - 6\sqrt{3}\sqrt{3}\sqrt{5}$	roots of products as products of roots
$3\sqrt{3} - 3 \cdot 3\sqrt{2} - 6 \cdot 3\sqrt{5}$	definition of square root
$\mathbf{3\sqrt{3} - 9\sqrt{2} - 18\sqrt{5}}$	simplify

No further simplification is possible since no two of the radical terms are like radical terms.

practice Simplify:

a. $\sqrt{8} + \sqrt{98}$

b. $4\sqrt{48} - 5\sqrt{75}$

problem set 70

1. The opposite of a number is tripled and then decreased by 7. The result is 3 greater than twice the number. What is the number?

2. Rubella found 60 escargots in the dell. This was only 80 percent of her largest find. What was the size of her largest find?

3. When the moot assembled, the village leader found that only 37 percent of those who attended had oil for their lamps. If 300 people attended the moot, how many had oil for their lamps?

4. $4\frac{2}{7}$ of what number is $20\frac{1}{2}$?

5. What decimal part of 20.2 is 1.01?

Solve:

6. $2\frac{1}{8}x - \frac{1}{5} = (5^2)(2^{-3})$

7. $0.003k + 0.188 - 0.001k = 0.2k - 0.01$

8. Graph on a number line: $x - 3 \nless -5$; $D = \{\text{Positive integers}\}$

Use substitution to solve for x and y:

9. $\begin{cases} 5x - y = 18 \\ 4x - 3y = 10 \end{cases}$

10. $\begin{cases} x + 2y = 0 \\ 3x + y = -10 \end{cases}$

11. $\begin{cases} 5x + 4y = -28 \\ x - y = -2 \end{cases}$

12. Find the volume in cubic centimeters of the prism whose base is shown and whose sides are 1 meter high. Also find the surface area if the prism is a right prism. Then find the volume of a pyramid 1 meter high that has the same base. Dimensions are in centimeters.

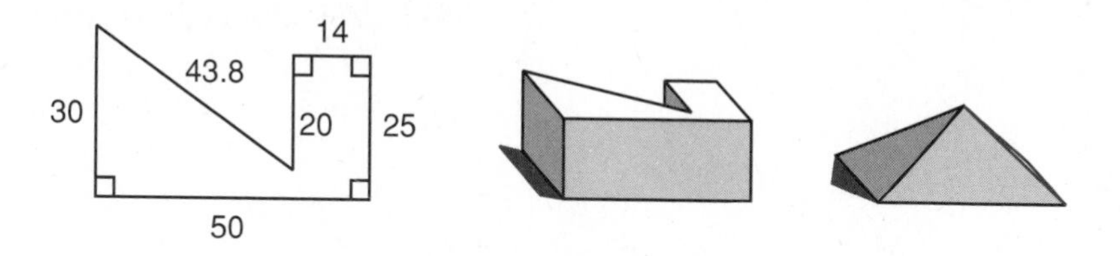

13. If $x + \sqrt{36} = 12$, evaluate $x^2 - 6$.

Simplify:

14. $5\sqrt{20} - 6\sqrt{32}$

15. $2\sqrt{45} - 3\sqrt{20}$

16. Expand: $(4x + 5)^2$

Graph on a rectangular coordinate system:

17. $y = 2x + 2$ **18.** $y = 2x - 2$

19. $3\sqrt{2} \in$ {What sets of numbers}?

Simplify:

20. $\dfrac{a}{\dfrac{1}{x}}$ **21.** $\dfrac{\dfrac{a}{b}}{\dfrac{1}{c}}$ **22.** $\dfrac{\dfrac{a}{b}}{c}$

Add:

23. $\dfrac{a}{x^2y} + \dfrac{b}{x + y}$ **24.** $k + \dfrac{1}{y^2}$ **25.** $m + \dfrac{1}{m^2}$

26. Factor: $12x^2y^3p^4 - 4x^3y^2p^6 + 16x^4y^4p^4$

Simplify:

27. $(3x^2y^5m^2)^2(x^2y)^{-2}$ **28.** $\dfrac{(2xy)^3(xy)^{-2}}{x^2y^0y^{-4}}$

29. $\dfrac{(4x^{-2})^2(x^2y^0)^{-3}}{x^2yy^{-2}y^4}$ **30.** $[(-3 - 4^0) - (-3 - 2)] - \sqrt{25}$

LESSON 71 Elimination

71.A review of equivalent equations

We have said that equivalent equations are equations that have the same solutions. Thus, the solution sets for equivalent equations must be equal sets. The number 2 is a solution to $x + 4 = 6$ and is also a solution to $x^2 - 4 = 0$.

(a) $x + 4 = 6$ (b) $x^2 - 4 = 0$

$(2) + 4 = 6$ $(2)^2 - 4 = 0$

$6 = 6$ True $4 - 4 = 0$ True

But these equations are not equivalent equations because equation (b) has another solution which is not a solution to equation (a). The other solution of equation (b) is -2.

(a) $x + 4 = 6$ (b) $x^2 - 4 = 0$

$(-2) + 4 = 6$ $(-2)^2 - 4 = 0$

$2 = 6$ False $4 - 4 = 0$ True

We remember that if very term of a particular equation is multiplied by the same nonzero quantity, the resulting equation is an equivalent equation to the original equation. On the left below, we write the equation $x + y = 6$. On the right, we write the equation, $2x + 2y = 12$, which is the original equation with every term having been multiplied by 2. The ordered pair (4, 2) is a solution to both equations.

$$x + y = 6 \qquad 2x + 2y = 12$$

$$(4) + (2) = 6 \qquad 2(4) + 2(2) = 12$$

$$6 = 6 \quad \text{True} \qquad 8 + 4 = 12 \quad \text{True}$$

Of course, there is an infinite number of ordered pairs of x and y that will satisfy either of these equations, but it can be shown that any ordered pair that satisfies either one of the equations will satisfy the other equation, and thus we say that these equations are equivalent equations!

71.B elimination

Thus far, we have been using the substitution method to solve systems of linear equations in two unknowns. Now we will see that these equations can also be solved by using another method. **This new method is called the *elimination method* and is sometimes called more meaningfully the *addition method*.** To solve the following system of equations by using elimination,

$$\begin{array}{ll} \text{(a)} & \left\{ \begin{array}{l} x + 2y = 8 \\ 5x - 2y = 4 \end{array} \right. \\ \text{(b)} & \end{array}$$

we first assume that values of x and y exist that will make both of these equations true equations and that x and y in the equations represent these numbers.[†] Thus $x + 2y$ equals the number 8, and $5x - 2y$ equals the number 4. The additive property of equality permits the addition of equal quantities to both sides of an equation. Thus we can add $5x - 2y$ to the left side of equation (a), and add 4 to the right side of equation (a).

$$\begin{array}{lrcr} \text{(a)} & x + 2y & = & 8 \\ \text{(b)} & 5x - 2y & = & 4 \\ \hline & 6x \qquad & = & 12 \end{array}$$

By doing this we have eliminated the variable y. Now we can solve the equation $6x = 12$ for x, find that $\boldsymbol{x = 2}$, and use this value for x in *either* of the original equations to find that $\boldsymbol{y = 3}$.

IN EQUATION (a)	IN EQUATION (b)
$x + 2y = 8$	$5x - 2y = 4$
$(2) + 2y = 8$	$5(2) - 2y = 4$
$2y = 6$	$-2y = -6$
$y = 3$	$y = 3$

example 71.1 Solve by using the elimination method:

$$\begin{array}{ll} \text{(a)} & \left\{ \begin{array}{l} 2x - y = 13 \\ 3x + 4y = 3 \end{array} \right. \\ \text{(b)} & \end{array}$$

solution If we add the equations in their present form,

[†] If values of x and y do not exist that will simultaneously make both equations true statements, the attempted solution will degenerate into an equation that contains only numbers such as $2 = 4$, $4 + 2 = 6$, $0 = 0$, or $0 = 5$. The reasons for results like this will be discussed in Lesson 119.

$$
\begin{array}{lrl}
\text{(a)} & 2x - y &= 13 \\
\text{(b)} & 3x + 4y &= 3 \\ \hline
 & 5x + 3y &= 16
\end{array}
$$

we find that we have accomplished nothing because we have not eliminated one of the variables. But by proper use of the multiplicative property of equality we can change the equations into equivalent equations that, when added, will result in the elimination of one of the variables. We choose to eliminate the variable y, and thus we will multiply every term in equation (a) by 4 and every term in equation (b) by 1.[†] Now the equations are added, and we find that we have eliminated the variable y.

$$
\begin{array}{llllrl}
\text{(a)} & 2x - y = 13 & \longrightarrow (4) \longrightarrow & 8x - 4y = 52 & \text{multiplied by 4} \\
\text{(b)} & 3x + 4y = 3 & \longrightarrow (1) \longrightarrow^{\ddagger} & 3x + 4y = 3 & \text{multiplied by 1} \\
 & & & 11x = 55 & \text{added} \\
 & & & \mathbf{x = 5} & \text{divided by 11}
\end{array}
$$

The number 5 can now be used to replace x in either of the original equations or either of the equivalent equations to find the corresponding value of y. We will demonstrate this by replacing x with 5 in both equation (a) and equation (b).

EQUATION (a)	EQUATION (b)
$2(5) - y = 13$	$3(5) + 4y = 3$
$10 - y = 13$	$15 + 4y = 3$
$-y = 3$	$4y = -12$
$\mathbf{y = -3}$	$\mathbf{y = -3}$

example 71.2 Solve by using the elimination method:

$$
\begin{array}{ll}
\text{(a)} & \left\{ \begin{array}{l} 2x - 3y = 5 \\ 3x + 4y = -18 \end{array} \right. \\
\text{(b)} &
\end{array}
$$

solution There are many ways that the multiplicative property of equality can be used to form equivalent equations that when added will result in one of the variables being eliminated. We will show one way here and then repeat the problem and show another way. Look at the x terms in both equations. If we multiply the x term in the top equation by -3, the product will be $-6x$. If we multiply the x term in the bottom equation by 2, the product will be $+6x$. Of course, we must multiply every term in the equations by -3 and by 2, as required by the multiplicative property of equality. Now if we add the equations we can eliminate x since the sum of $+6x$ and $-6x$ is zero.

$$
\begin{array}{llllrl}
\text{(a)} & 2x - 3y = 5 & \longrightarrow (-3) \longrightarrow & -6x + 9y = -15 & \text{multiplied by } -3 \\
\text{(b)} & 3x + 4y = -18 & \longrightarrow (2) \longrightarrow & 6x + 8y = -36 & \text{multiplied by 2} \\
 & & & 17y = -51 & \\
 & & & \mathbf{y = -3} &
\end{array}
$$

[†] This will leave equation (b) unchanged. We say that we multiply by 1 to establish a general procedure for this type of problem.

[‡] The notations $\rightarrow (4) \rightarrow$ and $\rightarrow (1) \rightarrow$ are just bookkeeping notations and have no mathematical meaning. We find them convenient to help us remember the number that we have used as a multiplier.

Now we will use -3 for y in the original equation (a) to find the corresponding value for x.

$$\text{(a)}\quad 2x - 3y = 5 \longrightarrow 2x - 3(-3) = 5 \longrightarrow 2x + 9 = 5$$
$$\longrightarrow 2x = -4 \longrightarrow \mathbf{x = -2}$$

The solution is the ordered pair **(−2, −3).**

example 71.3 Solve by using the elimination method, but this time eliminate y.

$$\begin{matrix}\text{(a)} \\ \text{(b)}\end{matrix}\quad \begin{cases} 2x - 3y = 5 \\ 3x + 4y = -18 \end{cases}$$

solution Look at the y terms in both equations. One of them already has a minus sign. If we multiply the y term in equation (a) by $+4$, the product will be $-12y$; and if we multiply the y term in equation (b) by $+3$, the product will be $+12y$; and, of course, the sum of $-12y$ and $+12\,y$ is zero.

$$\begin{array}{llllll} \text{(a)} & 2x - 3y = 5 & \longrightarrow (4) \longrightarrow & 8x - 12y = 20 & \text{multiplied by 4} \\ \text{(b)} & 3x + 4y = -18 & \longrightarrow (3) \longrightarrow & \underline{9x + 12y = -54} & \text{multiplied by 3} \\ & & & 17x \qquad\quad = -34 & \end{array}$$

$$\mathbf{x = -2}$$

Now we could use $x = -2$ in either of the original equations or either of the equivalent equations to find that the corresponding value of y is **−3.** Again we find that the solution is the ordered pair **(−2, −3).**

example 71.4 Use elimination to solve the system:

$$\begin{matrix}\text{(a)} \\ \text{(b)}\end{matrix}\quad \begin{cases} 2x + 5y = -7 \\ 3x - 4y = 1 \end{cases}$$

solution Since one of the y terms already has a minus sign, we can find satisfactory equivalent equations by multiplying both equations by positive numbers.

$$\begin{array}{llllll} \text{(a)} & 2x + 5y = -7 & \longrightarrow (4) \longrightarrow & 8x + 20y = -28 & \text{multiplied by 4} \\ \text{(b)} & 3x - 4y = 1 & \longrightarrow (5) \longrightarrow & \underline{15x - 20y = 5} & \text{multiplied by 5} \\ & & & 23x \qquad\quad = -23 & \end{array}$$

$$\mathbf{x = -1}$$

Now we will use -1 for x in equation (a) and find the corresponding value of y.

$$\text{(a)}\quad 2x + 5y = -7 \longrightarrow 2(-1) + 5y = -7 \longrightarrow -2 + 5y = -7$$
$$\longrightarrow 5y = -5 \longrightarrow \mathbf{y = -1}$$

Thus the solution is the ordered pair **(−1, −1).**

practice Use elimination to solve:

a. $\begin{cases} 4y + 3x = -7 \\ 3y + 2x = -6 \end{cases}$

b. $\begin{cases} 5x + 2y = -3 \\ 2x + 3y = -10 \end{cases}$

problem set 71

1. For some strange reason Jim's new diet caused him to gain weight rather than lose weight. If his weight increased 35 percent to 297 pounds, what did he weigh before he began his diet?

2. Only 13 percent of the tribe did not want Sleeping Bear to be chief. If there were 3000 members of the tribe, how many wanted Sleeping Bear?

3. The potato bugs decimated the potato crop, and the harvest was down 22 percent from last year. If 3900 tons were harvested, what was the harvest last year?

Simplify:

4. $6\sqrt{45} + \sqrt{180}$ 5. $2\sqrt{8} - 3\sqrt{32}$ 6. $2\sqrt{12} - 3\sqrt{18}$

Graph on a rectangular coordinate system:

7. $y = -\frac{1}{2}x + 3$ 8. $x = -2\frac{1}{2}$

9. Use substitution to solve for x and y: $\begin{cases} 4x + y = 25 \\ x - 3y = -10 \end{cases}$

10. Graph on a number line: $x - 3 \not> 1$; $D = \{\text{Reals}\}$

11. Find the volume in cubic inches of the cylinder whose base is shown and whose sides are 5 feet high. Find the volume of a cone 5 feet high that has the same base. Dimensions are in inches. The 12-inch diameter is perpendicular to the 8-inch diameter.

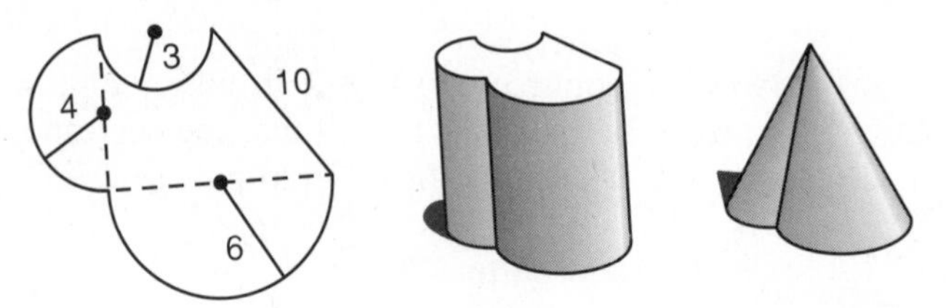

12. Use six unit multipliers to convert 200 cubic meters to cubic inches.

13. Harriet was rated at her job with a pretest and a posttest. The posttest score counted double. If she scored 88 on the pretest and 93 on the posttest, what was her weighted average?

Use elimination to solve for x and y:

14. $\begin{cases} 2x - 4y = -4 \\ 3x + 2y = 18 \end{cases}$ 15. $\begin{cases} 3x - y = 7 \\ 2x + 2y = 10 \end{cases}$

Simplify:

16. $\dfrac{a}{\dfrac{1}{x^2a}}$ 17. $\dfrac{a}{\dfrac{a^2}{a + b}}$

Add:

18. $\dfrac{m}{x^2a} + \dfrac{3}{a(a + x)}$ 19. $4x + \dfrac{1}{y}$ 20. $1 + \dfrac{x}{y}$

Simplify:

21. $\dfrac{12mx + 12mxy}{12mx}$ 22. $\dfrac{x^2(y^{-2})^2xx^4(p^0)^2}{(x^{-3}y^{-2})^2y}$ 23. $(4x^2y^3p^4)^3$

24. Expand: $\left(\dfrac{y^{-5}}{x^2} - \dfrac{3y^5x^{-2}}{p}\right)\dfrac{x^{-2}}{y^5}$

25. Evaluate: $-x^0 - x^2 - a(x - a) - |x^{-3}|$ if $x = -2$ and $a = \sqrt{16}$

26. Simplify: $-2\{[(-2 - 3) - (-2^0 - 2) - 2] - 2\}$

Evaluate:

27. $\dfrac{x^3 - a}{a^3 + 8}$ if $a = -2$ and $x + 1 = 2$

28. $x^2 - a^2$ if $x + 2 = 4$ and $a - 3 = 7$

29. $x^2 + 2xy + y^2$ if $x + 2 = 7$ and $y = -4$

LESSON 72 More about complex fractions

In Lesson 64 we defined a **rational number** to be a number that **can** be expressed as a fraction of integers. Thus the following are all rational numbers.

$$13 \qquad \frac{4}{7} \qquad \frac{-5}{14} \qquad \frac{6}{-13} \qquad 2\frac{1}{3}$$

The number 13 is a rational number because the number 13 can be expressed as a fraction of integers by writing, say, $\frac{52}{4}$. Of course, the mixed number $2\frac{1}{3}$ can be expressed as a fraction of integers as $\frac{7}{3}$.

An algebraic expression containing variables that is written in fractional form is called a **rational expression** because it has the same form as a rational number that is written as a fraction of integers. Thus all the following are **rational expressions.**

$$\frac{x + y}{4} \qquad \frac{a}{-b + c} \qquad 5 \qquad \frac{-7}{x} \qquad \frac{a + b}{14 - x}$$

The number 5 is a rational number and is also considered to be a rational expression because *rational expression* is a general term that describes both rational numbers and rational expressions that include variables and/or numbers. Of course, the denominators of none of these expressions can equal zero. We recognize this fact, and in this section we will omit the restrictive notations that are normally used to emphasize that division by zero is not permissible.

We have used the **denominator-numerator same-quantity theorem** to help us change denominators as required so that rational expressions may be added in three steps as shown here by adding a over b to x over y.

$\dfrac{a}{b} + \dfrac{x}{y} \longrightarrow$ (1) $\dfrac{\quad}{by} + \dfrac{\quad}{by}$ LCM used as new denominator

(2) $\dfrac{ay}{by} + \dfrac{bx}{by}$ new numerators determined

(3) $\dfrac{ay + bx}{by}$ added

We have also used this same theorem to help us simplify fractions of rational expressions (complex fractions).

$$\frac{\frac{x}{y}}{\frac{b}{c}} = \frac{\frac{x}{y} \cdot \frac{c}{b}}{\frac{b}{c} \cdot \frac{c}{b}} = \frac{\frac{xc}{yb}}{\frac{bc}{bc}} = \frac{\frac{xc}{yb}}{1} = \boldsymbol{\frac{xc}{yb}}$$

We will use both of these procedures when we simplify expressions that are fractions of sums of rational expressions. These expressions are also called **complex fractions.**

example 72.1 Simplify: $\dfrac{\dfrac{x}{y}+\dfrac{1}{y}}{\dfrac{x}{y}-\dfrac{1}{y}}$

solution The simplification is performed in two steps. The first step is to add the two expressions in the numerator and add the two expressions in the denominator.

$$\frac{\dfrac{x}{y}+\dfrac{1}{y}}{\dfrac{x}{y}-\dfrac{1}{y}}=\frac{\dfrac{x+1}{y}}{\dfrac{x-1}{y}}$$

Now use the **denominator-numerator same-quantity theorem** by multiplying both the numerator and denominator by $\frac{y}{x-1}$, which is the reciprocal of the denominator $\frac{x-1}{y}$.

$$\frac{\dfrac{x+1}{\cancel{y}}\cdot\dfrac{\cancel{y}}{x-1}}{\dfrac{\cancel{x-1}}{\cancel{y}}\cdot\dfrac{\cancel{y}}{\cancel{x-1}}}=\mathbf{\frac{x+1}{x-1}}$$

example 72.2 Simplify: $\dfrac{1+\dfrac{1}{x}}{7}$

solution First we add in the numerator.

$$\frac{1+\dfrac{1}{x}}{7}=\frac{\dfrac{x+1}{x}}{7}$$

Now we finish by multiplying the denominator and the numerator by the reciprocal of 7, which is $\frac{1}{7}$.

$$\frac{\dfrac{x+1}{x}\cdot\dfrac{1}{7}}{\dfrac{\cancel{7}}{1}\cdot\dfrac{1}{\cancel{7}}}=\mathbf{\frac{x+1}{7x}}$$

example 72.3 Simplify: $\dfrac{\dfrac{1}{x}}{1-\dfrac{1}{x}}$

solution We first add in the denominator.

$$\frac{\dfrac{1}{x}}{1-\dfrac{1}{x}}=\frac{\dfrac{1}{x}}{\dfrac{x-1}{x}}$$

Now we finish by multiplying the denominator and the numerator by $\frac{x}{x-1}$, which is the reciprocal of $\frac{x-1}{x}$.

$$\frac{\frac{1}{\cancel{x}} \cdot \frac{\cancel{x}}{x-1}}{\frac{\cancel{x-1}}{\cancel{x}} \cdot \frac{\cancel{x}}{\cancel{x-1}}} = \mathbf{\frac{1}{x-1}}$$

example 72.4 Simplify: $\dfrac{\frac{a}{b} + 1}{\frac{x}{b} + 4}$

solution First we add in the numerator and add in the denominator.

$$\frac{\frac{a}{b} + 1}{\frac{x}{b} + 4} = \frac{\frac{a+b}{b}}{\frac{x+4b}{b}}$$

Now we finish by multiplying both the denominator and the numerator by the reciprocal of the denominator.

$$\frac{\frac{a+b}{\cancel{b}} \cdot \frac{\cancel{b}}{x+4b}}{\frac{\cancel{x+4b}}{\cancel{b}} \cdot \frac{\cancel{b}}{\cancel{x+4b}}} = \mathbf{\frac{a+b}{x+4b}}$$

practice Simplify:

a. $\dfrac{\frac{1}{w} + \frac{c}{w}}{\frac{1}{c}}$ b. $\dfrac{\frac{1}{m} + 5}{\frac{2}{m} - \frac{x}{m}}$ c. $\dfrac{8 + \frac{1}{y}}{1 + \frac{1}{x}}$

problem set 72

1. If a number is multiplied by 7 and this product is increased by 42, the result is 87 greater than twice the opposite of the number. Find the number.
2. When Batman's entourage joined the motorcade, the total number of vehicles increased 130 percent. If the final count was 345, how many vehicles were present in the beginning?
3. The new drug saved lives but produced side effects in 37 percent of the people who took it. If 1110 people showed side effects, how many took the new drug?
4. $24\frac{1}{2}$ of what number is 120?
5. What fraction of 105 is $5\frac{1}{3}$?

Solve:

6. $20\frac{1}{4}x + 5\frac{1}{2} = 7\frac{1}{16}$
7. $-(-3)^3 - 2^2 = -2(-3k - 4)$
8. Graph on a number line: $x - 3 \not< -2$; $D = \{\text{Positive integers}\}$
9. True or false: $\{\text{Integers}\} \subset \{\text{Reals}\}$

Use substitution to solve for x and y:

10. $\begin{cases} x + 2y = 15 \\ 3x - y = 10 \end{cases}$
11. $\begin{cases} 4x - 3y = 14 \\ x + 3y = -4 \end{cases}$

Graph on a rectangular coordinate system:

12. $x = 3\frac{1}{2}$ **13.** $y = -2x + 4$

Use elimination to solve for x and y:

14. $\begin{cases} 5x - 2y = 10 \\ 7x - 3y = 13 \end{cases}$ **15.** $\begin{cases} 5x + 3y = 1 \\ 7x + 3y = 5 \end{cases}$ **16.** $\begin{cases} 14x - 2y = 12 \\ x + 2y = 3 \end{cases}$

Simplify:

17. $\dfrac{\frac{a}{b}}{\frac{x}{y}}$ **18.** $\dfrac{3 - \frac{a}{b}}{\frac{1}{b} + b}$

19. Find the volume in cubic meters of the right cylinder whose base is shown and whose sides are 4 centimeters high. If the cylinder is a right cylinder, find the surface area. Find the volume of a cone 4 centimeters high that has the same base. Dimensions are in meters.

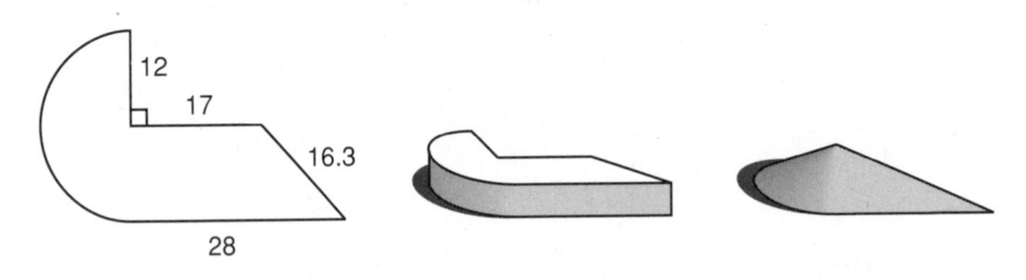

20. Evaluate: $(\sqrt{x})^{-5} - 19$ if $x + 19 = 23$

21. Use 12 unit multipliers to convert 12,000 cubic meters to cubic yards.

22. Roland averaged 4 jousts per journey on his first three journeys. He averaged 8 jousts per journey on his next 27 journeys. What was his overall average of jousts per journey?

23. $4\sqrt{8} - 3\sqrt{12}$ **24.** $2\sqrt{75} - 4\sqrt{243}$

25. Simplify by adding like terms: $-x^2y + 3yx^2 - \dfrac{4y^3x}{y^2x^{-1}} - \dfrac{7x^{-2}}{x^{-4}y^{-1}}$

Simplify:

26. $(4x^{-2}y^2m)^{-2}y$ **27.** $\dfrac{(x^{-2}y^2p)^2(x^2yp)^{-4}}{(xyp^2)^2}$ **28.** $\left(\dfrac{x^{-1}}{y^{-1}}\right)^{-2}\left(\dfrac{y^2}{x^2}\right)^{-4}$

29. $\dfrac{x^{-2}y^{-2}(p^0)^2}{(x^2y^{-2}p^3)^{-2}}$ **30.** $-3^2 - \dfrac{1}{(-3)^{-3}} + (-3)^0$

LESSON 73 Factoring trinomials

To begin a quick review of the nomenclature of polynomials in one variable, we say that **a monomial is a single expression of the form ax^n, where a is any real number and**

n **is any whole number.** Thus the following expressions are monomials.

$$4 \qquad 6x^2 \qquad -2x^{15} \qquad 4.163x^4$$

The number 4 can be classified as a monomial because it can be thought of as $4x^0$, and if x is any nonzero real number, then x^0 equals 1, so $4x^0 = 4 \cdot 1 = 4$.

A binomial is the indicated algebraic sum of two monomials and a trinomial is the indicated algebraic sum of three monomials. We use the word *polynomial* as the general descriptive term to describe monomials, binomials, trinomials, and algebraic expressions that are the indicated sum of four or more monomials.

We are familiar with the vertical format for multiplying binomials as shown here.

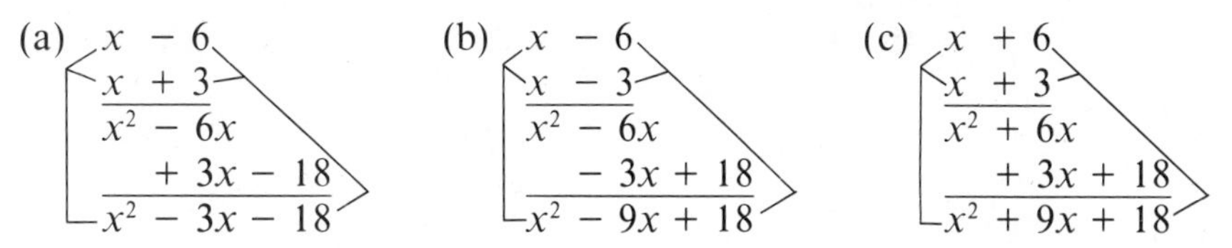

In each of these three examples the product is a trinomial. We call these trinomials **quadratic trinomials in *x*,** or, more simply, **quadratic trinomials.** The word *quadratic* tells us that the highest power of the variable is 2.

To reverse the process and factor the trinomials into a product of binomials, we must observe the pattern that developed when we did the multiplications. Note that:

1. The first term of the trinomial is the product of the first terms of the binomials.
2. The last term of the trinomial is the product of the last terms of the binomials.
3. The *coefficient* of the middle term of the trinomial is the *sum* of the last terms of the binomials.
4. If all signs in the trinomial are positive, all signs in both binomials are positive. If a negative sign appears in the trinomial, at least one of the terms of the binomials is negative.

We use these observations to help us factor trinomials. To factor the trinomial

$$x^2 - 3x - 18$$

we first write down two sets of parentheses to form an indicated product.

$$(\qquad)(\qquad)$$

Since the first term in the trinomial is the product of the first terms of the binomials, we enter x as the first term of each binomial.

$$(x \qquad)(x \qquad)$$

Now, the product of the last terms of the binomials must equal -18, their sum must equal -3, and at least one of them must be negative. There are six pairs of integral† factors of -18:

$$(-18)(1) = -18 \qquad (2)(-9) = -18 \qquad (3)(-6) = -18$$

$$(18)(-1) = -18 \qquad (-2)(9) = -18 \qquad (-3)(6) = -18$$

† Since $(2\sqrt{3})(3\sqrt{3}) = 18$, both $2\sqrt{3}$ and $3\sqrt{3}$ are factors of 18. We will not consider nonintegral factors such as these and will concentrate on factors that are integers. The process of factoring a polynomial into expressions all of whose coefficients are integers is defined as factoring over the set of integers.

Their sums are

$$(-18) + (1) = -17 \qquad (2) + (-9) = -7 \qquad (3) + (-6) = -3$$

$$(18) + (-1) = \ \ 17 \qquad (-2) + (9) = \ \ 7 \qquad (-3) + (6) = \ \ 3$$

Note that while all six pairs have a product of -18, only one pair, 3, -6, sums to -3. Therefore, the last terms of the binomials are 3 and -6, and so $(x + 3)$ and $(x - 6)$ are the factors of $x^2 - 3x - 18$ because

$$(x + 3)(x - 6) = x^2 - 3x - 18$$

Thus, the general approach to factoring a quadratic trinomial that has a leading coefficient of 1 is to determine the pairs of integral factors of the last term of the trinomial whose sum equals the coefficient of the *middle term*. To factor $x^2 - 8x + 16$, we list the factors of $+16$ and see which pair, if any, sum to -8. If no pair of integral factors has a sum of -8, the trinomial cannot be factored over the integers.

PRODUCT	SUM
$(16)(1) = 16$	$(16) + (1) = 17$
$(-16)(-1) = 16$	$(-16) + (-1) = -17$
$(2)(8) = 16$	$(2) + (8) = 10$
$(-2)(-8) = 16$	$(-2) + (-8) = -10$
$(4)(4) = 16$	$(4) + (4) = 8$
$(-4)(-4) = 16$	$(-4) + (-4) = -8$

Thus we find that the factors of $x^2 - 8x + 16$ are $(x - 4)$ and $(x - 4)$ because the product of the first term is x^2, the product of the last terms is $+16$, and the sum of the last terms is -8. This may seem to be a complicated procedure, but there is no shortcut until one becomes sufficiently familiar with the process to perform some of the calculations mentally. We will check our solution by multiplying the factors.

$$\begin{array}{l} x - 4 \\ \underline{x - 4} \\ x^2 - 4x \\ \underline{\quad\ - 4x + 16} \\ x^2 - 8x + 16 \end{array} \qquad \text{Check}$$

example 73.1 Factor: $x^2 - 14x - 15$

solution The last term of the trinomial is -15, so the products of the last terms in the binomial must be -15. Four pairs of integral factors have a product of -15.

$$(3)(-5) = -15 \qquad (-3)(5) = -15 \qquad (-15)(1) = -15 \qquad (15)(-1) = -15$$

but only one pair sums to -14.

$$(3) + (-5) = -2 \qquad (-3) + (5) = 2 \qquad \mathbf{(-15) + (1) = -14} \qquad (15) + (-1) = 14$$

Thus the constant terms of the binomials are -15 and 1 because these are the only two factors whose product is -15 and whose sum is -14. So $x^2 - 14x - 15$ in factored form is $\mathbf{(x - 15)(x + 1)}$. We will check by multiplying the two factors.

$$\begin{array}{l} x - 15 \\ \underline{x + \ 1} \\ x^2 - 15x \\ \underline{\quad\ + \ x - 15} \\ x^2 - 14x - 15 \end{array} \qquad \text{Check}$$

example 73.2 Factor: $x^2 + 3x - 10$

solution The constant term is -10, which has four pairs of integral factors. They are 1 and -10, 10 and -1, -5 and 2, and 5 and -2. The only pair whose sum is $+3$ is the pair 5 and -2, so

$$x^2 + 3x - 10 = \mathbf{(x + 5)(x - 2)}$$

We note that the trinomial in this problem was written as $x^2 + 3x - 10$ with the powers of the variable x in descending order. **If the trinomial is not in this form, the first step in factoring is to write the trinomial in descending powers of the variable.**

example 73.3 Factor: $-5x + x^2 + 6$

solution We begin by writing the trinomial in descending powers of the variable as

$$x^2 - 5x + 6$$

The minus sign in the middle term indicates that at least one of the constant terms is a negative number. The last term, $+6$, is a positive number and is a product of the constant terms, so both of the constants must be negative since their product is positive. Two pairs of negative integers have a product of $+6$,

$$(-3)(-2) = 6 \qquad \text{and} \qquad (-1)(-6) = 6$$

but only the first pair sums to -5.

$$(-3) + (-2) = -5$$

Thus $$x^2 - 5x + 6 = \mathbf{(x - 3)(x - 2)}$$

example 73.4 Factor: $x^2 + 5 + 6x$

solution We begin by writing the trinomial in descending powers of the variable

$$x^2 + 6x + 5$$

There are no minus signs in the trinomial, so all constants in the binomial factors will be positive. The constants therefore are positive integers whose product is $+5$ and whose sum is $+6$. The constants are $+5$ and $+1$ because

$$(5)(1) = 5 \qquad \text{and} \qquad 5 + 1 = 6$$

Thus $$x^2 + 6x + 5 = \mathbf{(x + 5)(x + 1)}$$

practice Factor:

a. $x^2 - x - 42$ **b.** $x^2 + x - 42$ **c.** $x^2 - 6x - 16$

problem set 73

1. Use 12 unit multipliers to convert 600 cubic yards to cubic meters.

2. What is the volume and the surface area of the right solid whose base is shown if the sides are 1 foot high? What is the volume of a cone that has the same base and the same altitude? Dimensions are in feet.

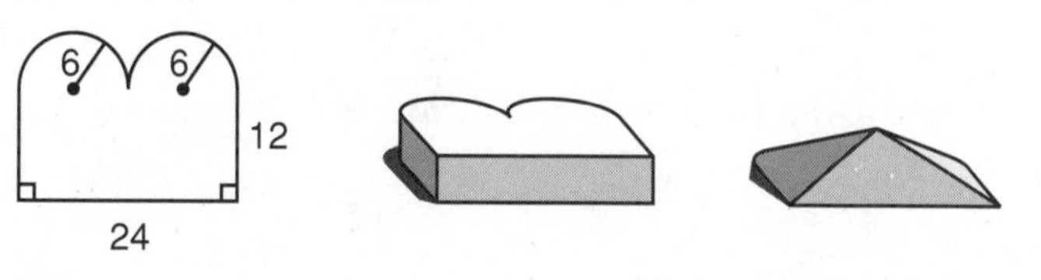

3. If $x - 5 = -8$, evaluate $(x^2 - 5)(x^{-3} - x)$.

Factor. Remember that the product of the constant terms must equal the constant in the trinomial and the sum of the constant terms must equal the coefficient of the x term.

4. $x^2 + 6x - 16$ **5.** $x^2 - 6x + 9$ **6.** $x^2 - 6x - 27$

7. $p^2 - p - 20$ **8.** $x^2 - 2x - 15$ **9.** $p^2 - 4p - 21$

10. $p^2 + p - 20$ **11.** $k^2 - 3k - 40$ **12.** $m^2 + 9m + 20$

First rearrange in descending order of the variable. Then factor.

13. $x^2 + 33 + 14x$ **14.** $-13p + p^2 + 36$ **15.** $-30 + m^2 - m$

16. $11n + n^2 + 18$ **17.** $x^2 + 27 + 12x$ **18.** $x^2 + 90 - 19x$

19. $x^2 + x - 132$ **20.** $a^2 + 90 - 47a$ **21.** $10m + m^2 + 16$

Use substitution to solve for x and y:

22. $\begin{cases} 3x + y = 9 \\ x - 4y = -10 \end{cases}$ **23.** $\begin{cases} 2x + 5y = 7 \\ x + 3y = 4 \end{cases}$

Use elimination to solve for x and y:

24. $\begin{cases} 3x + 4y = -7 \\ 3x - 3y = 21 \end{cases}$ **25.** $\begin{cases} 2x - 2y = -2 \\ 4x - 5y = -9 \end{cases}$

Simplify:

26. $7\sqrt{20} - 5\sqrt{32}$ **27.** $2\sqrt{18} - 5\sqrt{8} + 4\sqrt{50}$

Simplify:

28. $\dfrac{1 + \dfrac{1}{y}}{\dfrac{1}{y}}$ **29.** $\dfrac{\dfrac{a}{b} - 4}{\dfrac{x}{b} - b}$ **30.** $\dfrac{\dfrac{a}{x} - a}{x + \dfrac{y}{x}}$

LESSON 74 *Trinomials with common factors · Subscripted variables*

74.A trinomials with common factors

As the first step in factoring, we always check the terms to see if they have a common factor. If they do, we begin by factoring out this common factor. Then we finish by factoring one or both of the resulting expressions.

example 74.1 Factor: $x^3 + 6x^2 + 5x$

solution If we first factor out the greatest common factor x, we find

$$x^3 + 6x^2 + 5x = x(x^2 + 6x + 5)$$

and now the trinomial can be factored as in the last lesson to get

$$\mathbf{x(x + 5)(x + 1)}$$

example 74.2 Factor: $4bx^3 - 4bx^2 - 80bx$

solution Here we see that the greatest common factor of all three terms is $4bx$, and if we factor our $4bx$, we find

$$4bx^3 - 4bx^2 - 80bx = 4bx(x^2 - x - 20)$$

Now the trinomial can be factored, and the final result is

$$\mathbf{4bx(x - 5)(x + 4)}$$

example 74.3 Factor: $-x^2 + x + 20$

solution **To factor trinomials in which the coefficient of the second-degree term is negative, it is helpful first to factor out a negative quantity.** Here we will factor out (-1).

$$-x^2 + x + 20 = (-1)(x^2 - x - 20)$$

Now we factor the trinomial to get

$$(-1)(x - 5)(x + 4)$$

Finally, the (-1) can be multiplied by either of the binomials to yield two possible final results,

$$\mathbf{(-x + 5)(x + 4)} \qquad \text{or} \qquad \mathbf{(x - 5)(-x - 4)}$$

either of which is correct.

example 74.4 Factor: $-3x^3 - 6x^2 + 72x$

solution First we factor out the greatest common factor $-3x$, and then we factor the trinomial.

$$-3x(x^2 + 2x - 24) = \mathbf{-3x(x + 6)(x - 4)}$$

Thus we again find that the original trinomial has three factors.

74.B subscripted variables

We have used the letter N to represent an unknown number but have always used x and y as variables in systems of two equations such as

$$\text{(a)} \begin{cases} 5x + 10y = 125 \\ x + y = 16 \end{cases} \qquad \text{(b)} \begin{cases} 5x + 25y = 290 \\ x = y + 2 \end{cases}$$

We have solved these systems by using either the substitution method or the elimination method. In Lesson 86, we will look at word problems about coins: nickels, dimes, and quarters. In the equations, we will use N_N for the number of nickels, N_D for the number of dimes, and N_Q for the number of quarters. We will solve the equations by using either the substitution method or the elimination method.

example 74.5 Use elimination to solve: $\begin{cases} 5N_N + 10N_D = 125 \\ N_N + N_D = 16 \end{cases}$

solution We will multiply the bottom equation by -5 and add it to the top equation.

$$\begin{array}{rcl} 5N_N + 10N_D = 125 & \longrightarrow (1) \longrightarrow & 5N_N + 10N_D = 125 \\ N_N + N_D = 16 & \longrightarrow (-5) \longrightarrow & -5N_N - 5N_D = -80 \\ & & \hline 5N_D = 45 \\ & & \mathbf{N_D = 9} \end{array}$$

Since $N_N + N_D = 16$, $\mathbf{N_N = 7}$.

example 74.6 Use substitution to solve: $\begin{cases} 5N_N + 25N_Q = 290 \\ N_Q = N_N + 2 \end{cases}$

solution We will replace N_Q in the top equation with $N_N + 2$ and then solve.

$5N_N + 25(N_N + 2) = 290$	replaced N_Q with $N_N + 2$
$5N_N + 25N_N + 50 = 290$	multiplied
$30N_N + 50 = 290$	simplified
$30N_N = 240$	added -50 to both sides
$\mathbf{N_N = 8}$	divided by 30

Now, since $N_Q = N_N + 2$, $\mathbf{N_Q = 10}$.

practice Factor:

a. $-4x^3 - 28x^2 - 48x$

b. $-x^2 + 24 + 2x$

Use substitution to solve:

c. $\begin{cases} 6N_N + 24N_Q = 360 \\ N_Q = N_N + 5 \end{cases}$

Use elimination to solve:

d. $\begin{cases} 6N_N + 12N_D = 180 \\ N_N + N_D = 12 \end{cases}$

problem set 74

Factor. First rearrange in descending order of the variable if necesary.

1. $x^2 - 3x - 10$ **2.** $x^2 + 12 + 7x$ **3.** $-30 - x + x^2$

4. $x^2 + 10 + 7x$ **5.** $x^2 + 12 + 8x$ **6.** $4 - 4x + x^2$

7. $x^2 + 14 + 9x$ **8.** $x^2 - 14 - 5x$ **9.** $-3x - 18 + x^2$

10. $6x + 8 + x^2$ **11.** $-8 + 2x + x^2$ **12.** $-8 - 2x + x^2$

Each of the following has a common factor. Factor the common factor first.

13. $2x^2 + 10x + 12$ **14.** $5x^2 + 30x + 40$

15. $x^3 - x^2 - 20x$ **16.** $ax^2 + 6ax + 9a$

17. $abx^2 - 6ab + abx$ **18.** $x^3 + 20x + 9x^2$

Factor a negative common factor from each of these first.

19. $-b^3 + 5b^2 + 24b$ **20.** $-3m^2 - 30m - 48$ **21.** $-2p^2 + 110 + 12p$

22. Use 10 unit multipliers to convert 10,000 square kilometers to square miles.

23. Find the perimeter of the figure shown. Dimensions are in feet.

10

10

24. If $x + 24 = 21$, evaluate $x^2 - 2x^{-2} - 4$.

Simplify:

25. $\dfrac{18x}{3^2 - 9}$

26. $\dfrac{\dfrac{1}{x}}{\dfrac{1}{x} - 1}$

27. $5\sqrt{8} - 14\sqrt{50}$

28. $\dfrac{\dfrac{a}{x} + x}{\dfrac{1}{x} - 1}$

29. $\dfrac{\dfrac{1}{a}}{a + \dfrac{1}{a}}$

30. $\dfrac{1 + \dfrac{4}{x}}{7}$

LESSON 75 Factors that are sums

Sometimes a trinomial has a common factor that is a sum, as we see in the following examples.

example 75.1 Factor: $(a + b)x^2 - (a + b)x - 6(a + b)$

solution Each of the terms has the sum $(a + b)$ as a factor. If we factor out $(a + b)$, we get

$$(a + b)(x^2 - x - 6)$$

Now we finish by factoring the trinomial:

$$\mathbf{(a + b)(x - 3)(x + 2)}$$

example 75.2 Factor: $(x + y)x^2 + 9x(x + y) + 20(x + y)$

solution First we factor out $(x + y)$.

$$(x + y)(x^2 + 9x + 20)$$

Now we finish by factoring the trinomial:

$$\mathbf{(x + y)(x + 4)(x + 5)}$$

example 75.3 Factor: $m(x - 1)x^2 + 7mx(x - 1) + 10m(x - 1)$

solution The greatest common factor of each term of the trinomial is $m(x - 1)$. If we begin by factoring this term, we get

$$m(x - 1)(x^2 + 7x + 10)$$

Now we complete the solution by factoring the trinomial $x^2 + 7x + 10$, and the result is

$$\mathbf{m(x - 1)(x + 2)(x + 5)}$$

practice Factor. Begin by factoring the greatest common factor.

a. $(a + b)x^2 + 8x(a + b) + 15(a + b)$

b. $(m - b)x^2c - (m - b)2xc - (m - b)24c$

problem set 75

Factor. Rearrange in descending order of the variable if necessary.

1. $m^2 + 10m + 16$
2. $-48 - 8n + n^2$
3. $y^2 + 56 - 15y$
4. $p^2 - 55 - 6p$
5. $12t + 35 + t^2$
6. $y^2 + 50 + 51y$
7. $77 - 18r + r^2$
8. $m^2 + 21m + 90$
9. $55 + v^2 + 16v$
10. $-63 - 2h + h^2$
11. $-30 - 13x + x^2$
12. $w^2 + 22 - 13w$

Begin by factoring out -1.

13. $-x^2 - 12 + 7x$
14. $-s^2 - 15 - 8s$
15. $-a^2 + 40 - 3a$

Begin by factoring the greatest common factor.

16. $2x^3 + 30x + 16x^2$
17. $4a^2 - 160 + 12a$
18. $abx^2 - 24ab - 5abx$
19. $(x - 1)x^2 + 7x(x - 1) + 10(x - 1)$
20. Write a conjunction that describes this graph. Specify the domain.

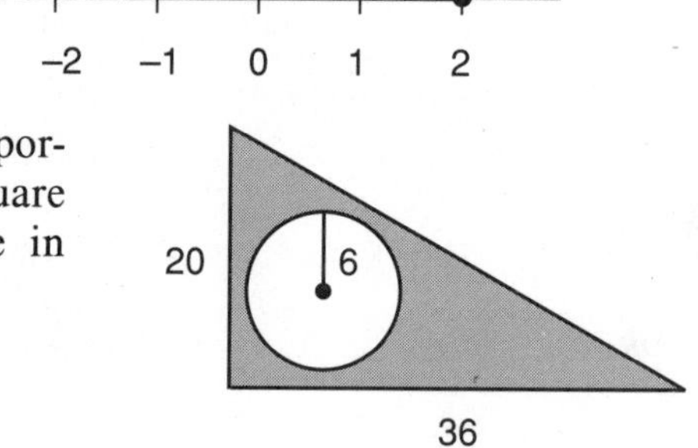

21. Find the area of the shaded portion of the right triangle in square centimeters. Dimensions are in meters.

Use substitution to solve for x and y:

22. $\begin{cases} x + y = 10 \\ x + 2y = 15 \end{cases}$
23. $\begin{cases} 10N_D + 25N_Q = 495 \\ N_Q = N_D + 10 \end{cases}$

Use elimination to solve for x and y:

24. $\begin{cases} 5x - 2y = 9 \\ 3x - y = 6 \end{cases}$
25. $\begin{cases} 2x - 2y = 2 \\ 3x + y = 7 \end{cases}$

Simplify:

26. $3\sqrt{98} - 4\sqrt{50}$
27. $2\sqrt{45} - 2\sqrt{180}$
28. $\dfrac{a + \dfrac{b}{a}}{\dfrac{1}{a} - 4}$
29. $\dfrac{\dfrac{m}{p} + p}{\dfrac{1}{p} - x}$
30. Solve: $-0.003k - 0.03k - 0.3k - 666 = 0$

LESSON 76 *Factoring the difference of two squares*

Since each of the terms in the following binomials is a perfect square,

$$x^2 - y^2 \qquad 4p^2 - 25 \qquad m^2 - 16$$

these binomials are sometimes called the **difference of two squares.** They can be generated by multiplying the sum and difference of two monomials.

$$\begin{array}{lll}
x + y & 2p + 5 & m + 4 \\
\underline{x - y} & \underline{2p - 5} & \underline{m - 4} \\
x^2 + xy & 4p^2 + 10p & m^2 + 4m \\
\underline{\quad - xy - y^2} & \underline{\quad - 10p - 25} & \underline{\quad - 4m - 16} \\
x^2 \qquad - y^2 & 4p^2 \qquad - 25 & m^2 \qquad - 16
\end{array}$$

We note in each case that the middle term is eliminated because the numerical coefficients of the addends that would form the middle term have the same absolute value but are opposite in sign.

If we are asked to factor a binomial that is the difference of two squares, such as

$$9m^2 - 49$$

the problem is a problem in recognition. There is no procedure to follow. We recognize that each term of the binomial is a perfect square and that the binomial can be written as

$$(3m)^2 - (7)^2$$

Now from the pattern developed above, we can write

$$9m^2 - 49 = \mathbf{(3m + 7)(3m - 7)}$$

example 76.1 Factor: $-4 + x^2$

solution **We recognize that both of the terms are perfect squares.** We begin by reversing the order of the terms and writing the squared terms as

$$x^2 - 4 = (x)^2 - (2)^2$$

and now we can write the factored form as

$$\mathbf{(x + 2)(x - 2)}$$

example 76.2 Factor: $49m^2 - a^2$

solution **We recognize that each of the terms is a perfect square** and that the binomial can be written as

$$(7m)^2 - (a)^2$$

The factored form of this binomial is

$$\mathbf{(7m + a)(7m - a)}$$

example 76.3 Factor: $-36a^2 + 25y^2$

solution **We recognize that both of the terms are perfect squares.** We begin by reversing the order of the terms and writing the squared terms as

$$25y^2 - 36a^2 = (5y)^2 - (6a)^2$$

Now we write the factored form as

$$(5y + 6a)(5y - 6a)$$

example 76.4 Factor: $-36x^6y^4 + 49a^2$

solution **Again we recognize that both of the terms are perfect squares.** We write down the answer by inspection as

$$(7a + 6x^3y^2)(7a - 6x^3y^2)$$

practice Factor:

a. $64x^2 - 81y^2$ **b.** $-25 + 100m^2$ **c.** $y^4x^2 - 169z^{10}$

problem set 76

1. Rosemary saw 900 of them in all. If this number was 150 percent greater than the number she expected to see, how many did she expect to see?
2. The ratio of grimaces to smiles was 13 to 5. If 360 students were either grimacing or smiling, how many were smiling?
3. Hannibal noted that the average weight of the first 4 animals was 2000 pounds. The average weight of the next 96 animals was only 100 pounds. What was the average weight of all the animals?
4. Draw a graph which is described by the conjunction $-6 \le x \le 3$.
5. Use 10 unit multipliers to convert 25,000 square miles to square kilometers.
6. Find the surface area of this right circular cylinder in square centimeters. Find the volume in cubic centimeters. Find the volume in cubic centimeters of a sphere whose diameter is 30 m. Dimensions are in meters.

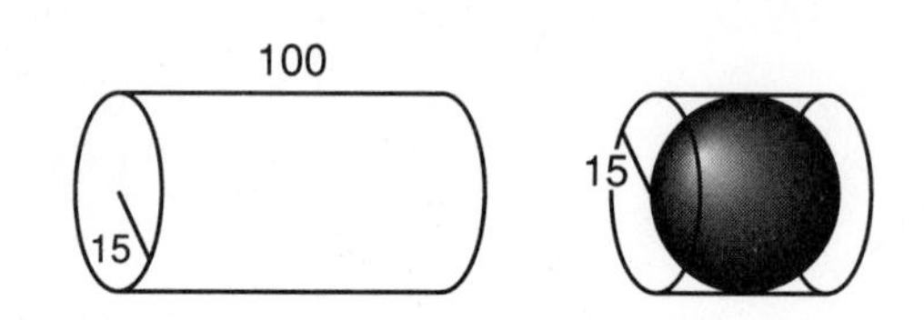

Factor:

7. $4p^2x^2 - k^2$ **8.** $-4m^2 + 25p^2x^2$ **9.** $-9x^2 + 4y^2$

10. $9k^2a^2 - 49$ **11.** $p^2 - 4k^2$ **12.** $36a^2x^2 - k^2$

Factor. Always factor the common factor first.

13. $x^2 - x - 20$ **14.** $4x^2 - 4x - 80$ **15.** $2b^2 - 48 - 10b$

16. $-90 - 39x + 3x^2$ **17.** $(a + b)x^2 + 7(a + b)x + 10(a + b)$

18. $pm^2 + 9pm + 20p$ **19.** $5k^2 + 30 + 25k$ **20.** $-x^2 - 8x - 7$

21. $5m^2 + 5 - 10m$

Use substitution to solve for x and y:

22. $\begin{cases} x + 2y = 12 \\ 3x + y = 16 \end{cases}$ **23.** $\begin{cases} 2x - y = 9 \\ 3x + y = 6 \end{cases}$

Use elimination to solve:

24. $\begin{cases} 5x - 2y = 3 \\ 2x - 3y = -1 \end{cases}$ **25.** $\begin{cases} N_P + N_N = 175 \\ N_P + 5N_N = 475 \end{cases}$

Simplify:

26. $3\sqrt{125} + 2\sqrt{45}$ **27.** $5\sqrt{12} - 2\sqrt{27}$ **28.** $\dfrac{\frac{1}{x} + 1}{\frac{y}{x} + x}$

29. Add: $\dfrac{x}{x(x + y)} + \dfrac{1}{x}$

30. Solve: $-[2(-3 - k)] = -4(-3) - |-3|k$

LESSON 77 *Scientific notation*

In science courses, it is sometimes necessary to use extremely large numbers and/or extremely small numbers. For example, to calculate the number of molecules in 1000 liters of a gas, it would be necessary to multiply 1000 times 1000 times a very large number such as 26,890,000,000,000,000,000, which represents the number of molecules in a cubic centimeter of gas. Besides requiring a lot of paper, multiplying these numbers in their present form is cumbersome and often leads to errors since it is easy to miscount the number of zeros. If we use a type of mathematical shorthand called **scientific notation,** however, computations such as the above can be performed easily and accurately.

To write a number in scientific notation, the numerator and the denominator are multiplied by the required power of 10 that will place the decimal point immediately to the right of the first nonzero digit in the number **(denominator-numerator same-quantity theorem).** For example, if we wish to write the number

$$0.0000416$$

in scientific notation, we would like to place the decimal point between the 4 and the 1.

$$4.16$$

To accomplish this, we multiply the number by 10^5, and we must also divide by 10^5 to keep from changing the value of the expression

$$0.0000416 = \frac{0.0000416}{1}\,\frac{10^5}{10^5} = \frac{4.16}{10^5}$$

Now, if we remember that $\frac{1}{10^5}$ can be written as 10^{-5}, we can write

$$0.0000416 = \frac{4.16}{10^5} = 4.16 \times 10^{-5}$$

We have described the algebraically correct procedure, but since scientific notation is used so often, we prefer to use another thought process. This thought process is much easier to use and not quite so rigorous.

When we look at numbers written in scientific notation, such as

$$4.16 \times 10^{+b} \qquad \text{and}^{\dagger} \qquad 4.16 \times 10^{-b}$$

we think of 10^{+b} as a decimal point indicator that tells us that the true location of the decimal point is really b places to the right of where it is written and 10^{-b} as a decimal point indicator that tells us that the true location of the decimal point is really b places to the left of where it is written. If we use this thought process, the 10^{-7} in the notation

$$4.165 \times 10^{-7}$$

tells us that the **true location** of the decimal point **is really** seven places **to the left** of where it is written, giving

$$0.0000004165$$

as the number being designated. In a like manner, the exponential expression 10^7 in the notation

$$4.165 \times 10^{7}$$

tells us that the **true location** of the decimal point **is really** seven places **to the right** of where it is written, giving

$$41{,}650{,}000$$

which is the number being designated.

It is helpful to use a two-step procedure to write a number in scientific notation. The first step is to place the decimal point immediately to the right of the first nonzero digit in the number. Then we follow this notation with the power of 10 that designates the true location of the decimal point. If we use this procedure to write

$$714{,}600{,}000$$

in scientific notation, we begin by placing the decimal point immediately to the right of the first nonzero digit (which is 7) and dropping the terminal zeros.

$$7.146$$

Now we follow this with $\times\ 10^8$ to indicate that the **true location of the decimal point is really eight places to the right of where we have written it.**

$$\mathbf{7.146 \times 10^{8}}$$

example 77.1 Write 0.000316 in scientific notation.

solution **We always begin by writing the decimal point immediately after the first nonzero digit.**

$$3.16 \times 10^{?}$$

Now we must choose an exponent for 10 that tells us what the true location of the decimal point **really is.** Since it **really is** four places to the left of where we have written it, the proper exponent is -4. Thus

$$0.000316 \qquad \text{equals} \qquad \mathbf{3.16 \times 10^{-4}}$$

example 77.2 Write 0.000316×10^{-7} in scientific notation.

† The replacements for b are restricted to positive integers.

solution We begin by writing 0.000316 in scientific notation.

$$0.000316 = 3.16 \times 10^{-4}$$

Thus we can rewrite the original expression as

$$3.16 \times 10^{-4} \times 10^{-7}$$

and this simplifies to

$$\mathbf{3.16 \times 10^{-11}}$$

example 77.3 Write 0.000316×10^7 in scientific notation.

solution We write 0.000316 as 3.16×10^{-4} and simplify.

$$0.000316 \times 10^7 = 3.16 \times 10^{-4} \times 10^7 = \mathbf{3.16 \times 10^3}$$

example 77.4 Write the following numbers in scientific notation.

(a) 47,800 (b) $47{,}800 \times 10^{-7}$ (c) $47{,}800 \times 10^7$

solution

(a) 47,800 equals $\mathbf{4.78 \times 10^4}$

(b) $47{,}800 \times 10^{-7}$ equals $\mathbf{4.78 \times 10^{-3}}$

(c) $47{,}800 \times 10^7$ equals $\mathbf{4.78 \times 10^{11}}$

A careful study of the four preceding examples is recommended. Often the student searches for a simple rule or shortcut that can be used to solve the problem. No such shortcut exists for writing a number in scientific notation, as the examples will verify.

practice Write in scientific notation:

a. 49,900 **b.** $49{,}900 \times 10^{-11}$ **c.** 0.000499

problem set 77

1. The ratio of withs to withouts was 3 to 11. If 5600 were huddled in the forest, how many were with?

2. The cost of building a house increased 20 percent every year. If it cost $74,000 to build a house one year, what would it cost the next year to build the same house?

3. The fine for sedition was reduced 30 percent. If the new fine was $4900, what was the amount of the fine before the reduction?

4. The ratio of bivalves to other crustaceans was 9 to 1. If there was a total of 130 crustaceans on the table, how many were bivalves?

5. Graph on a number line: $x - 3 \not> 4$; $D = \{\text{Reals}\}$

6. Multiply: $(4 + x)(x^2 + 2x + 3)$

7. Add: $\dfrac{1}{xc^2} + \dfrac{b}{x(c + x)} + \dfrac{5}{x^2c^2}$

8. Simplify: $\dfrac{x + \dfrac{a}{b}}{1 - \dfrac{1}{b}}$

9. Solve: $0.4x - 4 - 0.4 = -0.2(4 - x)$

10. If $\frac{x+9}{2} = 18$, evaluate $\left(\frac{x}{3}\right)^2$.

11. Find the surface area of this right prism in square centimeters. Find the volume in cubic centimeters. What is the volume in cubic centimeters of a pyramid 20 meters high that has the same triangular base? Dimensions are in meters.

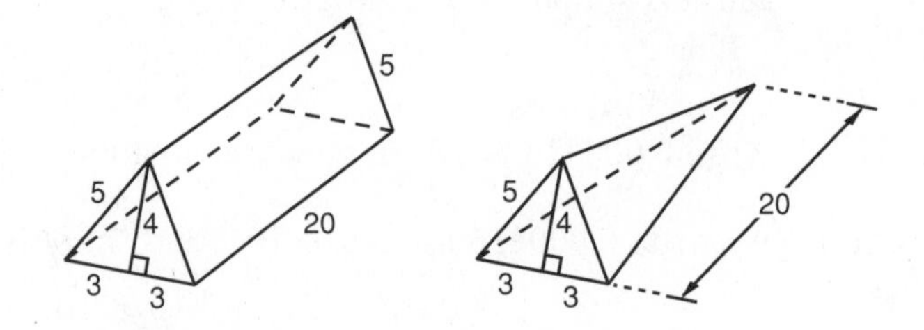

12. The average score of the first 6 games was 7.80 points per game. The average score of the next 4 games was 11.2 points per game. What was the average score for all the games?

13. Write 0.000478 in scientific notation.

14. Graph $y = -2x - 3\frac{1}{2}$ on rectangular coordinate system.

Use substitution to solve:

15. $\begin{cases} 7x + y = -18 \\ 4x - 2y = 0 \end{cases}$

Use elimination to solve:

16. $\begin{cases} N_D + N_Q = 40 \\ 10N_D + 25N_Q = 475 \end{cases}$

17. Expand: $\frac{2x^2}{y^2}\left(\frac{-x^2}{2y^{-2}} + \frac{x^2a^4}{a^{-2}4^{-2}}\right)$

18. Simplify: $\frac{x^{-3}xy^2(y)^{-2}x^{-4}}{(x^0yy^{-2})^2(x^2y^{-3})^{-2}}$

Simplify:

19. $4\sqrt{60} - 7\sqrt{135}$

20. $4\sqrt{80} + 8\sqrt{45}$

Factor the trinomials. If there is a common factor, factor it as the first step.

21. $x^2 + 9x + 20$
22. $x^2 + 15 + 8x$
23. $x^2 + 28 + 11x$
24. $x^3 + 10x^2 + 24x$
25. $ax^2 - 2ax - 15a$
26. $5x^2 - 140 + 15x$

Factor the binomials. If there is a common factor, factor it as the first step.

27. $5x^2 - 5y^2$
28. $45x^2 - 20m^2$
29. $4a^2 - 9b^2$
30. $49a^2p^2 - k^2$

LESSON 78 ***Closure***

The concept of **closure** of a given set under a given operation is interesting, important, and simple. In fact, it is so simple that the simplicity itself often causes students difficulty, for they believe that surely there must be more to it than there is. The word **closure** will appear in almost all higher-level algebra books and will not cause apprehension if we devote sufficient time to understanding what it means now. We will try to be as straightforward as possible with the explanation.

Suppose we want to use a particular operation and work only with a designated set of numbers. If we restrict ourselves to this set of numbers, it seems reasonable to say that we also restrict our answers to members of this set of numbers. If we don't do this, then we are not restricting ourselves to working with the designated set of numbers because we are accepting numbers for answers that are outside of that set.

If we use two of the numbers from this set in an operation, and the answer is always a member of the set, we say that the set is **closed** under the operation. A set cannot be partially closed under an operation. It is either closed or it is not closed. **If the use of any two of the members of the set in an operation results in an answer that is not a member of the set, we say that the set is *not closed* under the operation.** Let's investigate the set of positive integers. We will draw a circle and consider it to be an enclosure.

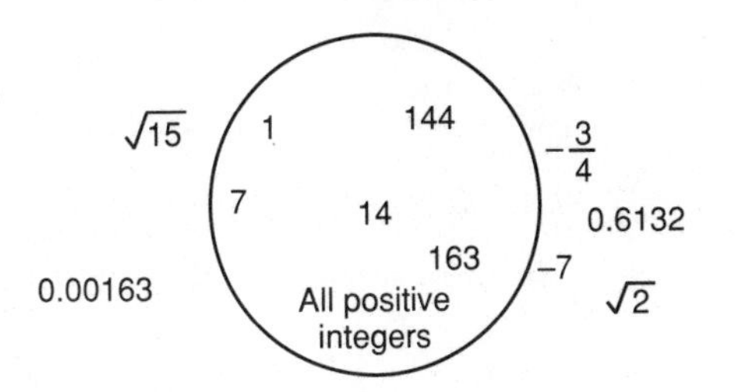

We mentally place all the positive integers inside the enclosure and all the other numbers outside the enclosure.

Now let's see if the set of positive integers is closed under the operation of addition.

$$2 + 5 = 7 \qquad 51 + 93 = 144 \qquad 7 + 13 = 20$$

In fact, the sum of any two positive integers is also a positive integer and lies inside our enclosure. Thus we say that the set of positive integers is closed under the operation of addition.

Let's try subtraction.

$$20 - 6 = 14 \qquad 14 - 2 = 12 \qquad 8 - 12 = -4$$

The first two answers are positive integers and thus lie inside the enclosure, but the third answer does not lie inside the enclosure because -4 is not a member of the set of positive integers. **This one example is all that is needed to state that the set of positive integers is not closed under the operation of subtraction.**

Let's try multiplication.

$$4 \cdot 2 = 8 \qquad 71 \cdot 3 = 213 \qquad 48 \cdot 2 = 96$$

All these products are positive integers. It can be proved (a topic for a more advanced course) that all products of positive integers are also positive integers. Thus we say that the set of positive integers is closed under the operation of multiplication.

What about division? One example of a quotient of two integers that is not an integer is $\frac{1}{2}$. Thus we say that the set of positive integers is not closed under the operation of division.

In the following five examples, we will begin by writing down A, S, M, and D to represent the operations of addition, subtraction, multiplication, and division. When we investigate a particular operation and find that the set is not closed, we will draw a slash line through the letter that represents the particular operation. We will use a check mark to designate closure.

example 78.1 For what operations is the set {0, 1} closed?

solution

ADDITION:	Not closed. 1 + 1 = 2	$\not{A}SMD$†
SUBTRACTION:	Not closed. 0 − (1) = −1	$\not{A}\not{S}MD$
MULTIPLICATION:	Closed. The product of any two numbers of this set is also a member of the set.	$\not{A}\not{S}\overset{\checkmark}{M}D$
DIVISION:	Not closed. $\frac{1}{0}$ is not defined.	$\not{A}\not{S}\overset{\checkmark}{M}\not{D}$

Thus **the set {0, 1} is closed for the operation of multiplication and is not closed for the operations of addition, subtraction, and division.**

example 78.2 For what operations is the set of negative integers closed?

solution

ADDITION:	Closed. The sum of any two negative integers is a negative integer.	$\overset{\checkmark}{A}SMD$
SUBTRACTION:	Not closed. −4 − (−8) = +4	$\overset{\checkmark}{A}\not{S}MD$
MULTIPLICATION:	Not closed. (−4)(−2) = +8	$\overset{\checkmark}{A}\not{S}\not{M}D$
DIVISION:	Not closed. $\frac{-4}{-2} = +2$	$\overset{\checkmark}{A}\not{S}\not{M}\not{D}$

Thus **the set of negative integers is closed for the operation of addition and is not closed for subtraction, multiplication, or division.**

example 78.3 For what operations is the set {−1, 0, 1} closed?

solution

ADDITION:	Not closed. 1 + 1 = 2	$\not{A}SMD$
SUBTRACTION:	Not closed. 1 − (−1) = 2	$\not{A}\not{S}MD$
MULTIPLICATION:	Closed. The product of any combination of the numbers. −1, 0, 1 is one of these numbers.	$\not{A}\not{S}\overset{\checkmark}{M}D$
DIVISION:	Not closed. $\frac{-1}{0}$ is not defined.	$\not{A}\not{S}\overset{\checkmark}{M}\not{D}$

Thus **the set {−1, 0, 1} is closed for the operation of multiplication but is not closed for addition, subtraction, or division.**

example 78.4 For what operations is the set of integers closed?

solution

ADDITION:	Closed. The sum of any two integers is an integer.	$\overset{\checkmark}{A}SMD$
SUBTRACTION:	Closed. The difference of any two integers is an integer.	$\overset{\checkmark}{A}\overset{\checkmark}{S}MD$
MULTIPLICATION:	Closed. The product of any two integers is an integer.	$\overset{\checkmark}{A}\overset{\checkmark}{S}\overset{\checkmark}{M}D$
DIVISION:	Not closed. $\frac{7}{13}$ is not an integer.	$\overset{\checkmark}{A}\overset{\checkmark}{S}\overset{\checkmark}{M}\not{D}$

† Note that if a given number is a member of a set, we may use the number more than once—in fact, as often as we desire.

Thus **the set of integers is closed for the operations of addition, subtraction, and multiplication but is not closed for division.**

example 78.5 For what operations is the set of real numbers closed?

solution

ADDITION:	Closed. The sum of any two real numbers is a real number.	$\overset{\checkmark}{\text{A}}$SMD
SUBTRACTION:	Closed. The difference of any two real numbers is a real number.	$\overset{\checkmark}{\text{A}}\overset{\checkmark}{\text{S}}$MD
MULTIPLICATION:	Closed. The product of any two real numbers is a real number.	$\overset{\checkmark}{\text{A}}\overset{\checkmark}{\text{S}}\overset{\checkmark}{\text{M}}$D
DIVISION:	Not closed. $\frac{23}{0}$ is not defined.	$\overset{\checkmark}{\text{A}}\overset{\checkmark}{\text{S}}\overset{\checkmark}{\text{M}}$Ø

Thus **the set of real numbers is closed for the operations of addition, subtraction, and multiplication but is not closed for division.**

practice For what operations are the following sets closed?

a. {Real numbers} **b.** $\{-3, -2, -1, 0\}$ **c.** {Whole numbers}

problem set 78

1. Three percent of the caterpillars metamorphosed into butterflies. If Ramona could count 120 butterflies, how many caterpillars had there been?
2. Sakahara socked it to them. If 4800 were present and Sakahara socked 34 percent of them, how many did he sock?
3. Muhammad counted the tents and found that 784 were patched. If there were 1400 tents in all, what percent were patched? What percent were not patched?
4. What fraction of 210 is $5\frac{1}{4}$?
5. $-3 \in$ {What sets of numbers}?
6. Graph $x = 2\frac{1}{2}$ on a rectangular coordinate system.
7. Add: $\frac{4}{x^2c} + \frac{5}{xc} + \frac{6}{x(c + x)}$
8. Simplify: $\frac{x^{-2}y^0(x^{-2})^{-2}y^2}{(y^2x^{-4})^2(y^3x^2)}$

Simplify:

9. $\frac{1}{-3^{-3}}$
10. $\frac{5}{-3^{-2}} - \sqrt[5]{-32}$
11. Simplify by adding like terms: $\frac{4ax^2}{x} - \frac{3a^{-4}x}{a^{-5}} - \frac{2x}{a^{-1}} + \frac{6a^2a^{-1}}{x}$

Write these numbers in scientific notation:

12. 0.00123×10^{-5}
13. 0.00123×10^8

For what operations are the following sets closed?

14. {Negative integers}
15. $\{-1, 0, 1, 2\}$
16. Graph $y = -\frac{1}{3}x - 3$ on a rectangular coordinate system.

Use substitution to solve:

17. $\begin{cases} y + 3x = -2 \\ 2x - 4y = 22 \end{cases}$

Use elimination to solve:

18. $\begin{cases} 4x - 5y = -1 \\ 2x + 3y = 5 \end{cases}$

19. Expand: $\frac{4x^2}{y^2}\left(\frac{x^{-2}}{4y^2} - \frac{a^{-2}x^{-1}}{y^{-1}}\right)$

20. Simplify: $15\sqrt{8} - 30\sqrt{18} + 4\sqrt{50}$

Factor these trinomials. If there is a common factor, factor it as the first step.

21. $x^2 + 3x - 10$
22. $4x + x^2 - 21$
23. $18 + 9x + x^2$
24. $5x^2 - 15x - 50$
25. $x^3 - 3x^2 + 2x$
26. $18x - x^3 + 3x^2$

Factor these binomials. If there is a common factor, factor it as the first step.

27. $b^3x^2 - 4b^3$
28. $16x^2 - a^2$
29. $-m^2 + 9p^2$

30. Find the surface area of this right circular cylinder in square centimeters. Find the volume in cubic centimeters. Find the volume in cubic centimeters of a cone 100 cm high that has the same circular base. Find the volume in cubic centimeters of a sphere whose radius is 10 m. Dimensions are in meters.

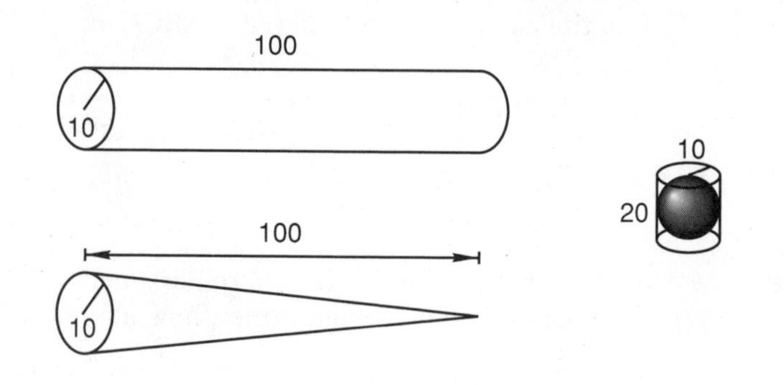

LESSON 79 Consecutive integers

If we use the letter N to designate an unspecified integer and then look at the number line,

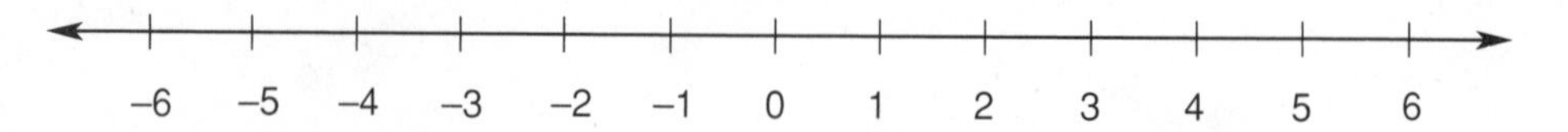

we see that the integer to the right of any given integer is one greater than the given integer. Thus we may use

$$N + 1$$

to designate the next greater integer, $N + 2$ to designate the next greater integer, etc. Integers that are 1 unit apart are called **consecutive integers.**

example 79.1 Find three consecutive integers such that the sum of the first and third is 146.

solution We will designate the consecutive integers as

$$N \qquad N + 1 \qquad \text{and} \qquad N + 2$$

The sum of the first integer and the third integer is 146, so we write

$$N + N + 2 = 146 \qquad \text{equation}$$

and now we solve for N:

$$2N + 2 = 146 \qquad \text{added}$$
$$2N = 144 \qquad \text{simplified}$$
$$N = 72 \qquad \text{divided}$$

Thus $N + 1 = 73$, and $N + 2 = 74$. The desired integers are **72, 73,** and **74.**

Check:
$$72 + 74 = 146$$
$$146 = 146 \qquad \text{Check}$$

example 79.2 Find three consecutive integers such that twice the sum of the first two is 2 less than 3 times the third.

solution We designate the integers as

$$N \qquad N + 1 \qquad \text{and} \qquad N + 2$$

We write the equation as

$$2(N + N + 1) + 2 = 3(N + 2)$$

Note that we added 2 to the sum of the first two integers because this sum was 2 less than. Now we solve.

$$2(2N + 1) + 2 = 3(N + 2) \qquad \text{simplified}$$
$$4N + 2 + 2 = 3N + 6 \qquad \text{multiplied}$$
$$4N + 4 = 3N + 6 \qquad \text{simplified}$$
$$N = 2 \qquad \text{solved}$$

Thus, the integers are **2, 3,** and **4.**

Check: $2(2 + 3) + 2 = 3(4) \longrightarrow 12 = 12$ Check

example 79.3 Find four consecutive integers such that 6 times the sum of the first and fourth is 26 less than 10 times the third.

solution We will use N, $N + 1$, $N + 2$, and $N + 3$ to designate the four integers. When we write the equation, we add 26 because 6 times the sum is 26 less than. We want it to be equal to.

$$6(N + N + 3) + 26 = 10(N + 2) \qquad \text{equation}$$
$$6(2N + 3) + 26 = 10(N + 2) \qquad \text{simplified}$$
$$12N + 18 + 26 = 10N + 20 \qquad \text{multiplied}$$
$$12N + 44 = 10N + 20 \qquad \text{simplified}$$
$$2N = -24 \qquad \text{simplified}$$
$$N = -12 \qquad \text{divided}$$

Thus the integers are **−12, −11, −10,** and **−9.**

Check: $6(-12 - 9) + 26 = 10(-10) \longrightarrow -100 = -100$ Check

a. Find three consecutive integers such that the sum of the first and third is 142.

b. Find four consecutive integers such that 8 times the sum of the first and the third is 40 greater than 10 times the fourth.

problem set 79

1. For what operations is the set of negative integers closed?

2. For what operations is the set of whole numbers closed?

3. Find four consecutive integers such that twice the sum of the first and third is 11 greater than 3 times the second.

4. If $\frac{x + 7}{2} = 5$, evaluate $\left(\frac{27}{x}\right)^2$.

Solve:

5. $31\frac{1}{5}x - 2\frac{3}{5} = 14$

6. $-x + 4 - (-2)(-x - 5) = -(-2x + |4|)$

7. True or false? $\{\text{Rationals}\} \subset \{\text{Integers}\}$

8. Graph $y = -\frac{1}{3}x + 2$ on a rectangular coordinate system.

9. Simplify: $\frac{1}{(-3)^{-2}} + 3^0 + 2^0 - \sqrt[3]{-8}$

Write the following numbers in scientific notation:

10. $430{,}000 \times 10^{-2}$

11. 4300×10^7

12. Seventy-eight percent of the more successful students tended to be serendipitous. If there were 18,400 more successful students in the district, how many tended to be serendipitous?

Add:

13. $\frac{a}{x^2} + \frac{2}{ax^2} + \frac{b}{cx^3}$

14. $\frac{a}{x} + \frac{b}{x + 6}$

15. Simplify: $\frac{aa(a^{-3})^0a^{-2}}{a^4(x^2)^{-2}}$

16. Simplify by adding like terms: $\frac{a^2x^5}{y} - 3aax^6x^{-1}y^{-1} + \frac{4a^2y^{-1}}{x^{-5}} - \frac{3aax^3y^{-2}}{y}$

Use substitution to solve:

17. $\begin{cases} N_N = N_Q + 15 \\ 5N_N + 25N_Q = 525 \end{cases}$

Use elimination to solve:

18. $\begin{cases} 2x + 2y = 14 \\ 3x - 2y = -4 \end{cases}$

19. Expand: $\frac{3x^2y^{-2}}{ax^{-1}}\left(\frac{x^3}{y^2} - \frac{2x^{-2}a}{y^2}\right)$

20. Simplify: $3\sqrt{20} - 2\sqrt{80} + 2\sqrt{125}$

Factor the trinomials:

21. $x^2 - 5x - 14$

22. $-x^3 + 4x^2 + 12x$

23. $ax^2 + 7xa + 10a$

24. $24x + 2x^2 + 70$

25. $24 + 27x + 3x^2$

26. $-px + px^2 - 2p$

Factor the binomials. If there is a common factor, factor it as the first step.

27. $-9x^2 + m^2$ **28.** $4x^2 - 9m^2$

29. $125m^2 - 5x^2$ **30.** $-72k^2 + 2x^2$

LESSON 80 Consecutive odd and consecutive even integers · Fraction and decimal word problems

80.A consecutive odd and consecutive even integers

In Lesson 79, we noted that if we use N to represent some unknown integer, then the next larger integer is $N + 1$, the next is $N + 2$, etc. This is because consecutive integers are 1 unit apart on the number line. **Consecutive even integers are different because consecutive integers are 2 units apart.** If we look at the number line,

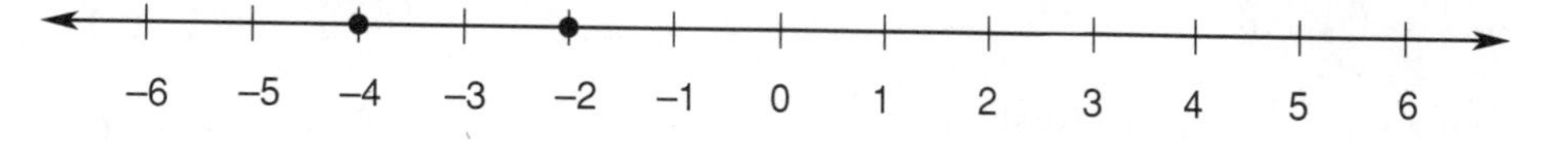

we see that -4 and -2 are consecutive even integers and that they are 2 units apart. **Consecutive odd integers are also 2 units apart.** The numbers -3 and -1 are consecutive odd integers and on the number line, and we see that they are 2 units apart.

−6 −5 −4 −3 −2 −1 0 1 2 3 4 5 6

Thus, if we use N to designate an unspecified odd/even integer, the next greater odd/even integer is $N + 2$, the next is $N + 4$, etc.

example 80.1 Find three consecutive even integers such that the sum of the first and second equals the sum of the third and -10.

solution We will represent the unknown even integers as

$$N \qquad N + 2 \qquad N + 4$$

Thus,

$$N + N + 2 = N + 4 + (-10)$$

Now we solve and get

$$2N + 2 = N - 6$$

$$N = -8$$

so the integers are **−8, −6, and −4.**

Check:

$$(-8) + (-6) = (-4) + (-10)$$

$$-14 = -14 \qquad \text{Check}$$

example 80.2 Find three consecutive odd integers such that the sum of the first and third is 7 greater than the second decreased by 18.

solution We also use $N, N + 2, N + 4$, etc. to represent consecutive odd integers. Thus, we can write the problem as

$$N + N + 4 - 7 = N + 2 - 18$$

and solve:

$$2N - 3 = N - 16$$

$$N = \mathbf{-13}$$

So $N + 2 = \mathbf{-11}$ and $N + 4 = \mathbf{-9}$.

Check:

$$(-13) + (-9) - 7 = (-11) - 18$$

$$-29 = -29 \qquad \text{Check}$$

example 80.3 Find four consecutive odd integers such that the sum of the first and fourth is 25 greater than the opposite of the third.

solution We will use $N, N + 2, N + 4$, and $N + 6$ to represent the unknown integers. Thus, we can write

$$N + N + 6 - 25 = -(N + 4)$$

and solve:

$$2N - 19 = -N - 4$$

$$3N = 15$$

$$N = \mathbf{5}$$

So $N + 2 = \mathbf{7}$, $N + 4 = \mathbf{9}$, and $N + 6 = \mathbf{11}$.

Check:

$$5 + 11 - 25 = -9$$

$$-9 = -9 \qquad \text{Check}$$

80.B fraction and decimal word problems

We have drawn diagrams to help us visualize percent word problems. Since problems involving fractional and decimal parts are essentially the same as percent problems, similar diagrams can be used to help with these. The "before" diagram for these problems will represent 1 instead of 100 percent. The equations we will use are the equations for fractional and decimal parts of a number.

$$F \times \text{of} = \text{is} \qquad \text{or} \qquad D \times \text{of} = \text{is}$$

example 80.4 Lopez used a 5-iron, but the ball covered only $\frac{4}{5}$ of the required distance. If she hit the ball 112 yards, what was the required distance?

solution The before diagram represents 1 instead of 100 percent.

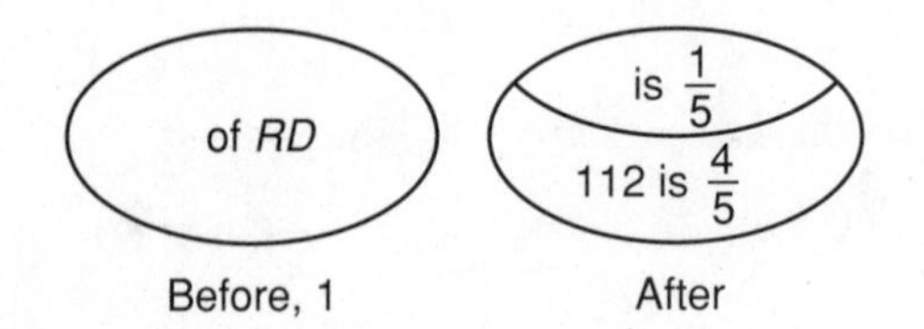

From this, we see that $\frac{4}{5}$ of the required distance is 112.

$$F \times \text{of} = \text{is} \longrightarrow \frac{4}{5}RD = 112 \longrightarrow \boldsymbol{RD = 140 \text{ yards}}$$

example 80.5 McAbee guessed that the total was 30.24, but this was 7.2 times the total. What was the total?

solution The following diagram shows that 30.24 is 7.2 of the total.

of T — 30.24 is 7.2

Before, 1 — After

We use this to solve.

$$D \times \text{of} = \text{is} \longrightarrow 7.2T = 30.24 \longrightarrow T = \frac{30.24}{7.2} \longrightarrow \boldsymbol{T = 4.2}$$

example 80.6 Three-fourths of the tickets had been sold, and there were 420 tickets left. How many tickets were printed?

solution

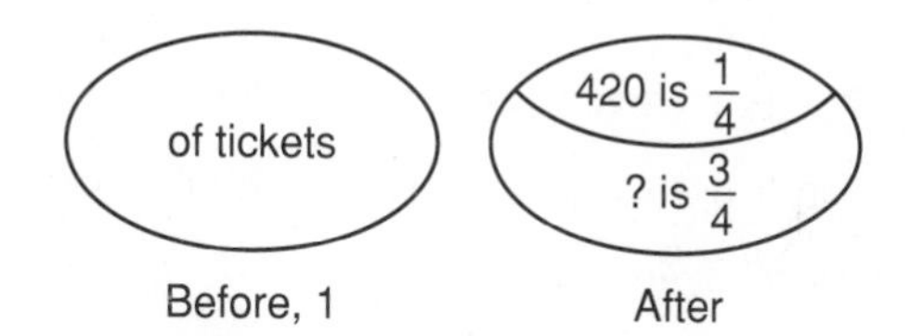

If $\frac{3}{4}$ have been sold, then $\frac{1}{4}$ are left, so we say that $\frac{1}{4}$ of the tickets printed equals 420. Now we write the equation and solve.

$$\frac{1}{4}T = 420 \longrightarrow \frac{\frac{1}{4}T}{\frac{1}{4}} = \frac{420}{\frac{1}{4}} \longrightarrow \boldsymbol{T = 1680}$$

practice

a. Find three consecutive even integers such that the sum of the first and third equals the sum of the second and -14.

b. The golfer used a 6-iron, but the ball traveled only $\frac{5}{6}$ of the required distance. If the golfer hit the ball 180 yards, what was the required distance?

c. Seven-eighths of the tickets had been sold, and there were 560 tickets left. How many tickets were printed?

problem set 80

1. Find four consecutive odd integers such that the sum of the second and third is 19 greater than the fourth.

2. Nolan guessed that the total was 46.08, but this was 6.4 times the total. What was the total?

3. For what operations is the set $\{4, 3, 2\}$ closed?

4. If the sum of three numbers is 495 and the first two numbers are 101.7 and 173.8, what is the average of the three numbers?

5. The figure is the base of a right cylinder that is 6 feet high. How many 1-inch sugar cubes will this cylinder hold? Find the surface area in square inches. Find the volume in cubic inches of a cone 6 feet high that has the same base. Dimensions are in feet. The sides that look parallel are parallel.

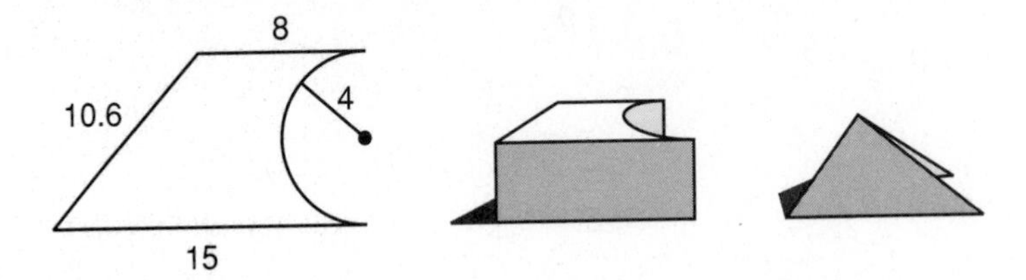

6. Graph on a number line: $4 \leq x < 10$; $D = \{\text{Integers}\}$

7. Add: $\dfrac{1}{xc^2} + \dfrac{a}{xc} + \dfrac{m}{c(x + c)}$

8. Simplify: $\dfrac{x + \dfrac{1}{y}}{\dfrac{x^2}{y} - 5}$

9. Solve: $-2|-2| - 2^2 - 3(-2 - x) = -2(x - 3 - 2)$

Write these numbers in scientific notation:

10. 7000×10^{-15}

11. 0.000007×10^{-15}

12. For what operations is the set of positive real numbers closed?

Graph on a rectangular coordinate system:

13. $y = 2x - 2$

14. $y = -4$

Use substitution to solve:

15. $\begin{cases} x + 3y = 16 \\ 2x - y = 4 \end{cases}$

Use elimination to solve:

16. $\begin{cases} N_N + N_D = 500 \\ 5N_N + 10N_D = 3000 \end{cases}$

17. Expand: $\dfrac{3a^2x}{y^2}\left(\dfrac{y^{-2}a^{-2}}{x^{-1}} - \dfrac{4ax}{y}\right)$

18. Simplify: $\dfrac{(2x^0x^{-3})^{-2}(y^{-5})^{-2}x}{(x^2y)(xy^2)}$

Simplify:

19. $2\sqrt{60} - 2\sqrt{135}$

20. $2\sqrt{75} - 6\sqrt{27}$

Factor the trinomials. If there is a common factor, factor it as the first step.

21. $x^2 - 6x + 9$

22. $2x^2 - 8x + 8$

23. $2x^2 + 8x + 8$

24. $2x^2 + 20x + 50$

25. $3x^2 - 30x + 75$

26. $ax^2 - 12ax + 36a$

Factor the binomials. If there is a common factor, factor it as the first step.

27. $4x^2 - 49$

28. $k^2 - 9x^2y^2$

29. $3p^2 - 12k^2$

30. $-4m^2 + k^2$

LESSON 81 Rational equations

Often we need to solve an equation in which some of the terms of the equation have denominators that are numbers other than the number 1. The equations

$$\frac{y}{2}+\frac{1}{4}=\frac{y}{6} \qquad \text{and} \qquad \frac{3y}{2}+\frac{8-4y}{7}=3$$

are equations of this type. Because the terms of the equations are all rational expressions, we call these equations **rational equations. There are many ways that these equations can be solved, but the most straightforward method of attack is to eliminate the denominators first by a judicious application of the multiplicative property of equality.** Since Lesson 47 we have been adding rational expressions by using the least common multiple of the original denominators as the new denominators. To eliminate the denominators in these equations, we will again use the least common multiple, but we will use it in a different way. As permitted by the multiplicative property of equality, we will **multiply the numerator of every term in the equation by the least common multiple of all the denominators of the terms of the equation.** Since every denominator is guaranteed to be a factor of the least common multiple, we are able to eliminate the denominators in one step. The remainder of the solution is straightforward.

example 81.1 Solve: $\frac{y}{2}+\frac{1}{4}=\frac{y}{6}$

solution **The least common multiple of the denominators is 12. We will multiply the numerator of every term by 12 and cancel the denominators.**

$$\frac{12\cdot y}{2}+\frac{12\cdot 1}{4}=\frac{12\cdot y}{6}$$

Now divide and find

$$6y+3=2y$$

Now solve for y:

$$4y=-3$$

$$\mathbf{y=-\frac{3}{4}}$$

example 81.2 Solve: $\frac{2x}{7}-\frac{3x}{2}=\frac{1}{3}$

solution We begin by multiplying each numerator by 42, which is the **least common multiple of the denominators.**

$$\frac{2x(42)}{7}-\frac{3x(42)}{2}=\frac{1(42)}{3}$$

Now we cancel the denominators and solve.

$$\frac{2x(\overset{6}{\cancel{42}})}{\cancel{7}}-\frac{3x(\overset{21}{\cancel{42}})}{\cancel{2}}=\frac{1(\overset{14}{\cancel{42}})}{\cancel{3}} \longrightarrow 12x-63x=14$$

$$\longrightarrow -51x=14 \longrightarrow \mathbf{x=-\frac{14}{51}}$$

example 81.3 Solve: $\frac{3y}{2} + \frac{8 - 4y}{7} = 3$

solution The beginner often makes mistakes when trying to eliminate denominators when one or more of the terms has a binomial expression as the numerator. A simple ploy that will prevent this common mistake is to enclose the binomial in the numerator in parentheses first.

$$\frac{3y}{2} + \frac{(8 - 4y)}{7} = 3$$

Now multiply every numerator by the **least common multiple of the denominators,** which is 14.

$$\frac{14 \cdot 3y}{2} + \frac{14 \cdot (8 - 4y)}{7} = 14 \cdot 3$$

Now divide, simplify, and complete the solution.

$$21y + 2(8 - 4y) = 42 \longrightarrow 21y + 16 - 8y = 42$$

$$\longrightarrow 13y = 26 \longrightarrow \mathbf{y = 2}$$

example 81.4 Solve: $\frac{x + 1}{4} - \frac{3}{2} = \frac{2x - 9}{10}$

solution First we enclose the binomials in parentheses.

$$\frac{(x + 1)}{4} - \frac{3}{2} = \frac{(2x - 9)}{10}$$

Now we multiply every numerator by the **least common multiple of the denominators,** which is 20.

$$\frac{20(x + 1)}{4} - \frac{20 \cdot 3}{2} = \frac{20(2x - 9)}{10}$$

Now we divide, simplify, and complete the solution.

$$5(x + 1) - 30 = 2(2x - 9) \longrightarrow 5x + 5 - 30 = 4x - 18$$

$$\longrightarrow 5x - 25 = 4x - 18 \longrightarrow \mathbf{x = 7}$$

practice Solve:

a. $\frac{z}{4} - \frac{1}{3} = \frac{z}{2}$

b. $\frac{y + 2}{3} - \frac{5}{2} = \frac{2y - 4}{8}$

problem set 81

1. Find three consecutive integers such that 3 times the sum of the first two is 49 less than the third.

2. Find four consecutive odd integers such that 4 times the sum of the first and third is 4 larger than 4 times the fourth.

3. When the pilgrims counted noses, they got a count that was 128 percent too high. If they counted 9120 noses, what was the correct count?

4. Four thousand two hundred carnations were reserved for use on the float. If this was only $\frac{7}{10}$ of the flowers needed, how many flowers would be on the float?

5. What fraction of 36 is 9?

6. $-2\sqrt{3} \in$ {What sets of numbers}?

7. Add: $\dfrac{a}{x^2y} + \dfrac{b}{x^2y^2} + \dfrac{c}{x^2y^3}$

8. Simplify: $\dfrac{x^0x^2y^{-2}(x^0)^{-2}}{(x^2)^{-3}(y^{-2})^3y^0}$

Simplify:

9. -3^{-3}

10. $3^0 - (2^0 - 3)(-3 - 2) - \sqrt[3]{-8}$

11. Simplify by adding like terms: $\dfrac{m^2xy}{x} - \dfrac{4m^3y}{m} + \dfrac{my}{m^{-1}} - \dfrac{3x^0y}{m^{-2}}$

Write in scientific notation:

12. 0.0003×10^{-15}

13. 4000×10^{-40}

14. Graph $y = -3x + 4$ on a rectangular coordinate system.

Use substitution to solve:

15. $\begin{cases} N_Q = N_D + 300 \\ 10N_D + 25N_Q = 8200 \end{cases}$

Use elimination to solve:

16. $\begin{cases} 4x - 3y = -3 \\ 2x + 4y = -18 \end{cases}$

17. Expand: $\left(\dfrac{ax^2}{y^2} - \dfrac{3x}{xy^2}\right)\dfrac{2x^{-3}}{y^2}$

18. Simplify: $3\sqrt{28} - 5\sqrt{56} + 2\sqrt{63}$

19. The average of five numbers is 790.6. If the first four numbers are 80.2, 91.6, 123, and 204.7, what is the sum of the five numbers?

20. Find the number of 1-inch-square tiles necessary to cover this figure. Dimensions are in feet.

6

10

21. Evaluate: $\left(\sqrt[3]{\dfrac{x}{4}}\right)^{-2}(x^{-1}x)$ if $\dfrac{x - 12}{2} = 10$

22. Solve: $\dfrac{y}{7} + \dfrac{y + 1}{4} = 6$

Factor. Always look for a common factor first.

23. $x^2 - 9x + 20$

24. $2ax^2 - 20ax + 42a$

25. $13mx + 42m + mx^2$

26. $16x^2 - 9a^2$

27. $25m^2 - 4$

28. $-36k^2 + 9m^2y^2$

29. If we exclude division by zero, for what operations is the set of real numbers closed?

LESSON 82 Systems of equations with subscripted variables

In Lesson 96 we will introduce uniform motion word problems. These problems will require the use of variables that represent rate or speed and variables that represent time. Rather than use the usual variables x, y, and z, we will use variables that are easy to associate with the words in the problem. If we need a variable to represent the rate of Mike, we will use R_M, which can be read as "the rate of Mike" or "R sub M." If we need a variable to represent Joanie's rate, we will use R_J, which can be read as "the rate of Joanie" or "R sub J." In the same way, Bud's time and Sadie's time will be represented by the variables T_B and T_S, which can be read as "the time of Bud" and "the time of Sadie" or as "T sub B" and "T sub S." The first problem we will solve involves the rates and times of Anne and Pat and uses the variables R_A, T_A and R_P, T_P.

example 82.1 Solve the following system of equations for R_A.

$$R_AT_A + R_PT_P = 320 \qquad R_P = 50 \qquad T_P = 4 \qquad T_A = 3$$

solution In the equation on the left we will substitute 50, 4, and 3 for R_P, T_P, and T_A, respectively,

$$R_A(3) + 50(4) = 320$$

and solve.

$$3R_A + 200 = 320 \longrightarrow 3R_A = 120 \longrightarrow \mathbf{R_A = 40}$$

The equations for the next example come from a story problem about a turtle and a rabbit. Thus R_R and T_R stand for the rate of the rabbit and the time of the rabbit and R_T and T_T stand for the rate of the turtle and the time of the turtle.

example 82.2 Solve the following system of equations for T_R and T_T.

$$R_TT_T + 120 = R_RT_R \qquad R_T = 2 \qquad R_R = 10 \qquad T_T = T_R$$

solution We begin by substituting 2 and 10 for R_T and R_R in the first equation.

$$\text{(a)} \quad 2T_T + 120 = 10T_R$$

We have used the first three given equations thus far. The remaining given equation is $T_T = T_R$. We can use this equation to change T_T to T_R in equation (a) or to change T_R to T_T. We choose to change T_T to T_R. We do this and then complete the solution.

$$\begin{array}{rcll} 2T_R + 120 &=& 10T_R & \text{substituted } T_R \text{ for } T_T \\ -2T_R \quad\quad && -2T_R & \text{add } -2T_R \text{ to both sides} \\ \hline 120 &=& 8T_R & \\ \dfrac{120}{8} &=& \dfrac{\not{8}T_R}{\not{8}} & \text{divide both sides by 8} \\ \mathbf{15} &=& \mathbf{T_R} & \end{array}$$

And since $T_T = T_R$, $\qquad \mathbf{T_T = 15}$

Now we will solve the equations from a problem in which Little Brother and Sis take a trip. We will use R_L and T_L for the rate and time of Little Brother and R_S and T_S for the rate and time of Sis.

example 82.3 Solve the following system of equations for T_L and T_S.

$$R_LT_L = R_ST_S \qquad R_L = 40 \qquad R_S = 80 \qquad T_S = T_L - 5$$

solution We begin with the first equation by replacing R_L with 40 and R_S with 80.

$$40T_L = 80T_S$$

Now we will replace T_S with $T_L - 5$ and multiply using the distributive property.

$$40T_L = 80(T_L - 5) \longrightarrow 40T_L = 80T_L - 400$$

Now we complete the solution.

$$\begin{array}{rcll} 40T_L & = & 80T_L - 400 & \\ -40T_L & & -40T_L & \text{add } -40T_L \text{ to both sides} \\ \hline 0 & = & 40T_L - 400 & \\ +400 & & +400 & \text{add 400 to both sides} \\ \hline 400 & = & 40T_L & \end{array}$$

$$\frac{400}{40} = \frac{\cancel{40}T_L}{\cancel{40}} \qquad \text{divide both sides by 40}$$

So $\boldsymbol{T_L = 10}$, and since $T_S = T_L - 5$, $T_S = 10 - 5 \longrightarrow \boldsymbol{T_S = 5}$.

The following equations are from a problem about a freight train and an express train. Thus the rate and time of the freight are symbolized by R_F and T_F and those for the express by R_E and T_E.

example 82.4 Solve the following system of equations for R_F and R_E.

$$R_FT_F = R_ET_E \qquad T_F = 16 \qquad T_E = 12 \qquad R_E = R_F + 15$$

solution In the equation on the left we will substitute 16, 12, and $R_F + 15$ for T_F, T_E, and R_E, respectively.

$$R_FT_F = R_ET_E \longrightarrow R_F(16) = (R_F + 15)(12) \longrightarrow 16R_F = 12R_F + 180$$

$$\longrightarrow 4R_F = 180 \longrightarrow \boldsymbol{R_F = 45}$$

And since $R_E = R_F + 15$,

$$R_E = 45 + 15 \longrightarrow \boldsymbol{R_E = 60}$$

practice

a. $R_AT_A + R_PT_P = 460$, $R_P = 50$, $T_P = 4$, $T_A = 2$. Find R_A.

b. $R_TT_T + 200 = R_RT_R$, $R_T = 10$, $R_R = 15$, $T_R = T_T + 10$. Find T_R and T_T.

problem set 82

1. Find four consecutive even integers such that 4 times the sum of the first and fourth is 8 greater than 12 times the third.

2. Galileo tried for a reasonable result but got $4\frac{3}{5}$. If this was $2\frac{3}{10}$ of his goal, what number was he trying for?

3. In the eighth grade Paul wrestled at 125 pounds. If his weight increased 16 percent in 1 year, at what weight did he wrestle in the ninth grade?

4. With the new tractor Carolyn could plow 67 percent of the farm in 2 weeks. If she plowed 268 acres, how large was the farm?

5. Seven-eighths of the workers in London's fish market used scurrilous language. If 400 did not use scurrilous language, how many worked in the fish market?

6. $2 \in$ {What sets of numbers}?

Solve:

7. $3\frac{1}{8}x - 4\frac{2}{5} = 7\frac{1}{2}$

8. $-[-2(x - 4) - |-3|] = -2x - 8$

Write the following numbers in scientific notation:

9. 0.000135×10^{-17}

10. $135{,}000 \times 10^{-17}$

11. Add: $\dfrac{a}{xy} + \dfrac{4}{x(x + y)}$

12. Simplify: $\dfrac{a^0(2x)^{-2}}{a^2(4a^0)^2}$

13. Simplify by adding like terms: $3x^2y - \dfrac{4x^{-2}y^{-2}}{x^{-4}y^{-3}} + \dfrac{5xx}{y^{-1}} - \dfrac{3x^2y^2}{y^{-2}}$

Use substitution to solve:

14. $\begin{cases} x + 3y = 16 \\ 2x - 3y = -4 \end{cases}$

Use elimination to solve:

15. $\begin{cases} N_N + N_D = 22 \\ 5N_N + 10N_D = 135 \end{cases}$

16. Expand: $\dfrac{x^{-2}y^{-2}}{m^2}\left(x^2y^2m^2 - \dfrac{3x^4y^{-4}}{m^{-2}}\right)$

17. Simplify: $5\sqrt{45} - 3\sqrt{180} + 2\sqrt{20}$

Solve:

18. $\dfrac{x}{4} - \dfrac{x + 2}{3} = 12$

19. $\dfrac{2y}{4} - \dfrac{y}{7} = \dfrac{y - 3}{2}$

20. $\dfrac{p}{6} - \dfrac{2p}{5} = \dfrac{4p - 5}{15}$

21. $R_AT_A + R_PT_P = 500$, $R_P = 25$, $T_P = 9$, $T_A = 5$. Find R_A.

22. Evaluate: $\dfrac{\sqrt[3]{x}}{5}$ if $\dfrac{x + 20}{5} = -21$

23. Use nine unit multipliers to convert 23,000 cubic meters to cubic feet.

24. Find the surface area of this pyramid that has four equal triangular faces and a square bottom. Units are in inches.

6

4

Factor. Always begin by factoring the common monomial factor if there is one.

25. $p^2 - 55 - 6p$

26. $-30 - 13x + x^2$

27. $2m^2 - 24m + 70$

28. $-x^3 + 14x^2 - 40x$

29. $4m^2 - 49x^2p^2$

30. For what operations is the set of real numbers closed?

LESSON 83 Operations with scientific notation

In Lesson 77 we introduced the topic of scientific notation and discussed the method of writing a number in scientific notation. Scientific notation is particularly useful in problems that require the multiplication and division of very large or very small numbers. We will discuss multiplication first.

multiplication We begin by multiplying the numbers 4,000,000 and 20,000,000 by using scientific notation. As the first step we write both numbers in scientific notation.

$$4{,}000{,}000 = 4 \times 10^6 \qquad 20{,}000{,}000 = 2 \times 10^7$$

Then we note that the numbers are to be multiplied by writing

$$(4 \times 10^6)(2 \times 10^7)$$

Since the order of multiplication of real numbers does not affect the value of the product, we may rearrange the order of the multiplication and place the powers of 10 last.

$$(4 \cdot 2)(10^6 \cdot 10^7)$$

Now we multiply 4 by 2 and get 8 and multiply the powers of 10 by using the product rule for exponents to get 10^{13}. Thus our answer is the number

$$\mathbf{8 \times 10^{13}}$$

example 83.1 Write the numbers 0.003×10^{-4} and 2×10^{20} in scientific notation and then multiply.

solution First we write the numbers in scientific notation.

$$(0.003 \times 10^{-4})(2 \times 10^{20}) = (3 \times 10^{-7})(2 \times 10^{20})$$

Next we rearrange the order of the factors and then we multiply:

$$(3 \cdot 2)(10^{-7} \cdot 10^{20}) = \mathbf{6 \times 10^{13}}$$

example 83.2 Multiply: $(0.00004 \times 10^{-5})(700{,}000)$

solution We write the numbers in scientific notation, rearrange the order of factors, and then multiply.

$$(4 \times 10^{-10})(7 \times 10^5) = (4 \cdot 7)(10^{-10} \cdot 10^5) = 28 \times 10^{-5}$$
$$= \mathbf{2.8 \times 10^{-4}}$$

division A similar procedure is used to divide numbers written in scientific notation. The powers of 10 are handled separately from the other numbers. To divide 20,000,000 by 4,000,000, we first write both numbers in scientific notation.

$$\frac{20{,}000{,}000}{4{,}000{,}000} = \frac{2 \times 10^7}{4 \times 10^6}$$

We can think of this expression as a product of fractions, which we simplify as follows.

$$\left(\frac{2}{4}\right)\left(\frac{10^7}{10^6}\right) = 0.5 \times 10^1 = \mathbf{5}$$

example 83.3 Divide 0.0016 by 400,000.

solution We write both numbers in scientific notation as the numerator and denominator of a fraction,

$$\frac{1.6 \times 10^{-3}}{4 \times 10^{5}}$$

which we think of as a product of fractions. Then we simplify both fractions.

$$\left(\frac{1.6}{4}\right)\left(\frac{10^{-3}}{10^{5}}\right) = 0.4 \times 10^{-8}$$

Now to finish we write 0.4×10^{-8} in scientific notation as

$$\mathbf{4 \times 10^{-9}}$$

multiplication and division The procedure for simplifying a problem such as

$$\frac{(0.06 \times 10^{5})(300{,}000)}{(1000)(0.00009)}$$

is first to simplify both the numerator and denominator by using scientific notation. Next we rearrange the expression into a product of fractions and then simplify each fraction.

$$\frac{(6 \times 10^{3})(3 \times 10^{5})}{(1 \times 10^{3})(9 \times 10^{-5})} = \frac{6 \cdot 3}{1 \cdot 9} \times \frac{10^{3} \cdot 10^{5}}{10^{3} \cdot 10^{-5}} = \frac{18}{9} \times \frac{10^{8}}{10^{-2}} = \mathbf{2 \times 10^{10}}$$

example 83.4 Simplify: $\dfrac{(0.0007 \times 10^{-23})(4000 \times 10^{6})}{(0.00004)(7{,}000{,}000)}$

solution We will begin by writing every number in scientific notation. Next we rearrange the expression into a product of two fractions and then simplify each fraction.

$$\frac{(7 \times 10^{-27})(4 \times 10^{9})}{(4 \times 10^{-5})(7 \times 10^{6})} = \frac{7 \cdot 4}{4 \cdot 7} \times \frac{10^{-27} \cdot 10^{9}}{10^{-5} \cdot 10^{6}} = \frac{28}{28} \times \frac{10^{-18}}{10^{1}} = \mathbf{1 \times 10^{-19}}$$

example 83.5 Simplify: $\dfrac{(20 \times 10^{-45})(400 \times 10^{20})}{(100{,}000)(0.0008 \times 10^{-15})}$

solution First we write all numbers in scientific notation.

$$\frac{(2 \times 10^{-44})(4 \times 10^{22})}{(1 \times 10^{5})(8 \times 10^{-19})}$$

Now we group the exponentials and the other numbers and simplify.

$$\frac{2 \cdot 4}{1 \cdot 8} \times \frac{10^{-44} \cdot 10^{22}}{10^{5} \cdot 10^{-19}} = \frac{8}{8} \times \frac{10^{-22}}{10^{-14}} = \mathbf{1 \times 10^{-8}}$$

practice Simplify:

a. $\dfrac{(0.07 \times 10^{2})(800{,}000)}{(10{,}000)(0.0000004)}$ **b.** $\dfrac{(0.04 \times 10^{-9})(50 \times 10^{16})}{(0.000004)(50{,}000)}$

problem set 83

1. Cindy had three consecutive even integers. Frank and Mark found that the product of their sum and 3 was 20 greater than 8 times the third integer. What were Cindy's integers?

2. Find four consecutive odd integers such that 6 times the sum of the first and the third is 3 more than 5 times the opposite of the fourth.

3. The fairies outnumbered the hamadryads by 130 percent. If there were 460 fairies in the clearing, how many hamadryads were present?

4. Lancelot paid for halberds only 86 percent of what he paid for cuirasses. If he paid 43 farthings for halberds, how much did the cuirasses cost?

5. Querulous was not satisfied because the sinecure did not pay enough. It paid 4125 pounds, but this was only five-thirteenths of what he expected. How much did Querulous expect?

6. Graph on a number line: $x \not\leq 4$; $D = \{\text{Positive integers}\}$

7. Add: $\dfrac{3}{a} + \dfrac{4}{a^2} + \dfrac{7}{a^2(a + x)}$

8. Solve: $-0.013 - 0.013x + 0.026 = 0.039$

9. Simplify: $\dfrac{4 + \dfrac{1}{y^2}}{\dfrac{x}{y} + \dfrac{m}{y^2}}$

10. Evaluate: $\sqrt[4]{x + 9} + \left(\dfrac{\sqrt{x}}{2}\right)^{-2}$ if $\dfrac{x - 9}{3} = 21$

11. The figure shows the base of a cylinder that is 10 ft high. Dimensions are in feet. Find the volume in cubic inches. If the cylinder is a right cylinder, find the surface area in square inches. Find the volume of a cone 10 ft high that has the same base.

12. Only $\frac{2}{17}$ of the teachers were sciolists. If 3000 were not sciolists, how many teachers were there in all?

13. Graph $y = -4\frac{1}{2}$ on a rectangular coordinate system.

Use substitution to solve:

14. $\begin{cases} N_N = N_D + 12 \\ 5N_N + 10N_D = 510 \end{cases}$

Use elimination to solve:

15. $\begin{cases} 7x - 4y = 29 \\ 3x + 5y = -1 \end{cases}$

16. Expand: $\dfrac{x^{-2}}{y^2 a}\left(\dfrac{y^2 a^{-3}}{x^{-2}} + \dfrac{3x^{-4}}{y^{-2}a^{-4}}\right)$

17. Simplify: $\dfrac{[x^2(y^5)^{-2}]^{-3}}{(x^0 y^2)y^{-2}}$

18. Simplify: $3\sqrt{8} - 5\sqrt{18} + 6\sqrt{72} - 3\sqrt{50}$

Solve:

19. $\dfrac{3x}{2} - \dfrac{5 - x}{3} = 7$

20. $\dfrac{2x - 3}{5} - \dfrac{2x}{10} = \dfrac{1}{2}$

21. $R_F T_F = R_S T_S$, $T_S = 6$, $T_F = 5$, $R_F - 16 = R_S$. Find R_S and R_F.

22. $R_M T_M = R_R T_R$, $R_M = 8$, $R_R = 2$, $T_R = 5 - T_M$. Find T_M and T_R.

23. $R_G T_G + R_B T_B = 100$, $R_G = 4$, $R_B = 10$, $T_B = T_G + 3$. Find T_G and T_B.

Simplify:

24. $\dfrac{(0.08 \times 10^7)(900{,}000)}{(20{,}000)(0.000003)}$

25. $\dfrac{(0.0006 \times 10^{-31})(8000 \times 10^9)}{(0.0000002)(400{,}000)}$

Factor. Always begin by factoring the common factor if there is one.

26. $x^3 + 9x^2 + 8x$

27. $-ax^2 + 48a - 13xa$

28. $bcx^2 - a^2cb$

29. $-m^3 + k^2m$

30. For what operations is the set of negative real numbers closed?

LESSON 84 Graphical solutions

In Lessons 50 and 53, we noted that the degree of a term of a polynomial is the sum of the exponents of the variables in the term. Thus $2xy^5$ is a sixth-degree term, $4x$ is a first-degree term, and xy is a second-degree term.

Also, we remember that the degree of a polynomial is the same as the degree of its highest-degree term and that the degree of a polynomial equation is the same as the degree of the highest-degree term in the equation. Thus

$x^4y + 4y$ is a fifth-degree polynomial

$x^4y + 4y = x$ is a fifth-degree polynomial equation

$x + 2y = 4$ is a first-degree polynomial equation

In Lesson 91 we will learn to solve second-degree polynomial equations by factoring, but until then, we will continue to concentrate on first-degree equations. We have learned that the graph of a first-degree polynomial equation in two unknowns is a straight line and that we call this kind of equation a linear equation. We have learned to find the solution to a system of two linear equations by using the **substitution method** and the **elimination method.** Here we will see that we can find the solution to a system of two linear equations by **graphing** each of the equations and visually estimating the coordinates of the point where the two lines cross. The coordinates of this point will satisfy both equations. The shortcoming of this method is that it is inexact because the coordinates of the crossing point must be estimated.

example 84.1 Solve by graphing: $\begin{cases} y = x + 1 \\ y = -2x + 4 \end{cases}$

solution We choose values for x and use the equations to find the paired values of y.

$y = x + 1$

x	0	2	−3
y	1	3	−2

$y = -2x + 4$

x	0	2	4
y	4	0	−4

It appears that the lines cross at $x = 1$ and $y = 2$, so **(1, 2)** is our solution.

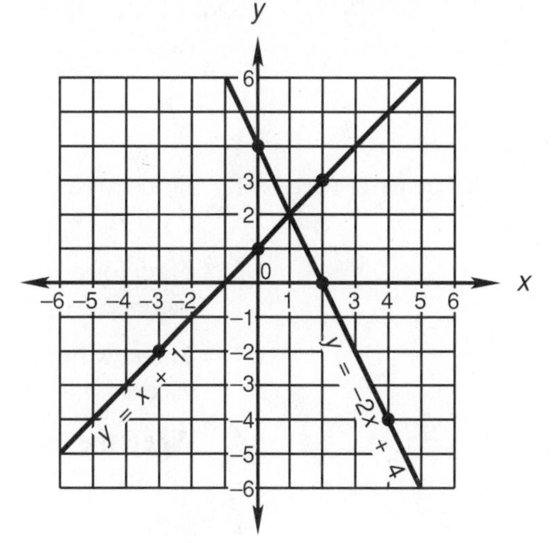

example 84.2 Solve by graphing: $\begin{cases} y = 2 \\ y = x \end{cases}$

solution We graph the lines as the first step.

$y = x$

x	0	4
y	0	4

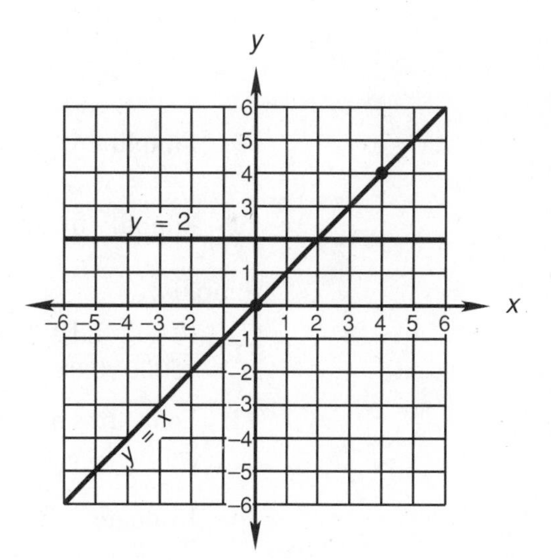

It appears that the lines cross at $x = 2$, $y = 2$, so our graphical solution is **(2, 2)**.

example 84.3 Solve by graphing: $\begin{cases} y = x + 2 \\ y = x - 1 \end{cases}$

solution It appears from the graph that the lines are parallel and thus never intersect. If this is so, no point lies on both of the lines and no ordered pair will satisfy both equations. An attempt at solving this system by substitution or elimination will degenerate into a false numerical statement. We will look into this in detail in Lesson 119.

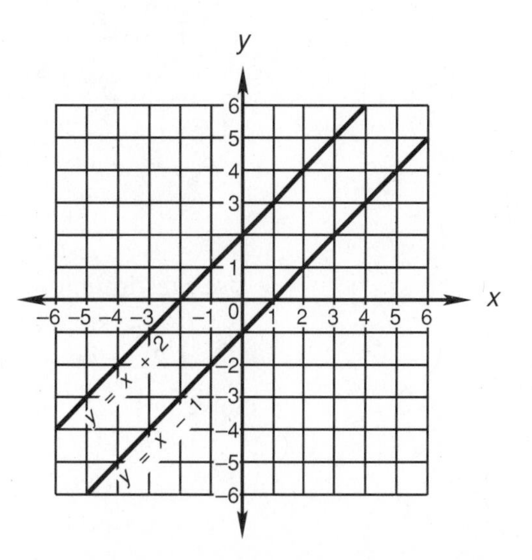

practice Solve by graphing:

a. $\begin{cases} y = x - 1 \\ y = -2x + 5 \end{cases}$ b. $\begin{cases} y = x \\ y = 4 \end{cases}$

problem set 84

1. Find three consecutive even integers such that 4 times the sum of the first and third is 16 greater than 7 times the second.

2. Find four consecutive integers such that if the sum of the first and third is increased by 10, the result is 6 greater than 4 times the fourth.

3. The new hog food supplement increased the weight gain by $\frac{2}{5}$. If the weight gain used to be 300 pounds, what was the new weight gain?

4. When the leprechauns ran into the forest, the number of little people present was decreased by 35 percent. If 105 ran into the forest, how many were left?

5. What fraction of $7\frac{2}{5}$ is $49\frac{1}{3}$?

6. $-\frac{\sqrt{3}}{2} \in$ {What sets of numbers}?

7. Simplify: $\dfrac{\frac{x}{y} - 1}{\frac{x}{y} + m}$

8. Simplify: $\dfrac{x^2(2y^{-2})^{-3}}{(4x^2)^{-2}}$

9. Simplify: $\dfrac{1}{-3^{-2}} + (-3^0)(-3 - 5)$

10. Simplify by adding like terms: $\dfrac{ax^{-4}}{(x^{-2})^2} + \dfrac{3a^{-2}a^3}{a^0} - \dfrac{6a^5}{(a^{-2})^{-2}} + 3a$

Simplify:

11. $\dfrac{(0.003 \times 10^7)(700{,}000)}{(5000)(0.0021 \times 10^{-6})}$

12. $\dfrac{(0.0007 \times 10^{-10})(4000 \times 10^5)}{(0.0004)(7000)}$

13. Four-fifths of the delegates crowded into the convention hall. If 140 could not get in, how many attended the convention?

14. Evaluate: $\dfrac{\sqrt[3]{x + 2} + 11}{3}$ if $\dfrac{x - 22}{5} = (2)^3$

15. Find the volume of this rectangular prism in cubic centimeters and the surface area in square centimeters. What is the volume of a rectangular pyramid 2 meters high that has the same base, which measures 3 meters by 4 meters? Dimensions are in meters.

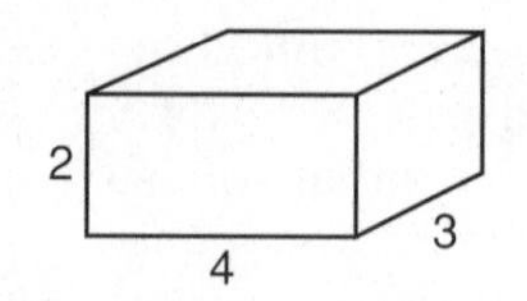

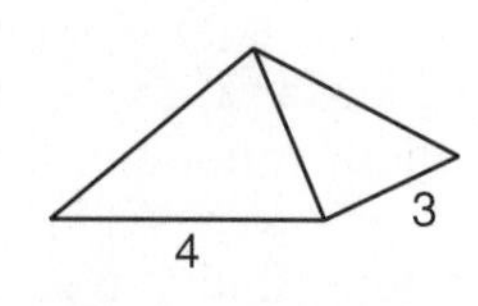

Use substitution to solve:

16. $\begin{cases} x + 5y = 17 \\ 2x - 4y = -8 \end{cases}$

Use elimination to solve:

17. $\begin{cases} N_N + N_D = 30 \\ 5N_N + 10N_D = 250 \end{cases}$

Solve:

18. $\dfrac{x}{3} + \dfrac{5x + 3}{2} = 5$

19. $\dfrac{y + 3}{2} - \dfrac{4y}{3} = \dfrac{1}{6}$

20. Expand: $\dfrac{x^{-2}}{a^2y^{-2}}\left(\dfrac{x^4a^5}{y^4} - \dfrac{3x^{-4}}{a^{-4}y^2}\right)$

21. Simplify: $4\sqrt{28} - 3\sqrt{63} + \sqrt{175}$

22. $R_GT_G + R_BT_B = 120$, $R_G = 4$, $R_B = 10$, $T_G = T_B + 2$. Find T_G and T_B.

23. $R_KT_K = R_NT_N$, $R_K = 6$, $R_N = 3$, $T_K = T_N - 8$. Find T_K and T_N.

Solve by graphing:

24. $\begin{cases} y = x - 6 \\ y = -x \end{cases}$

25. $\begin{cases} y = x + 1 \\ y = -x - 1 \end{cases}$

Factor. Always look for a common factor.

26. $ax^2 + 6a - 7ax$

27. $-mx^2 - 8m - 6mx$

28. $mx^2 - 9ma^2$

29. $-k^2 + 4m^2x^2$

30. For what operations is the set of negative even integers closed?

LESSON 85 Writing the equation of a line

85.A writing the equation of a line

We remember that the graph of a vertical line is everywhere equidistant from the y axis. In the figure on the left below, every point on line A is 3 units to the left of the y axis, and the equation of this vertical line is

$$x = -3$$

Every point on line B is 5 units to the right of the y axis, and the equation of this line is

$$x = 5$$

The equation of every vertical line has this form, and if we use k to represent the number, we can say that the general form of a vertical line is

$$x = \pm k$$

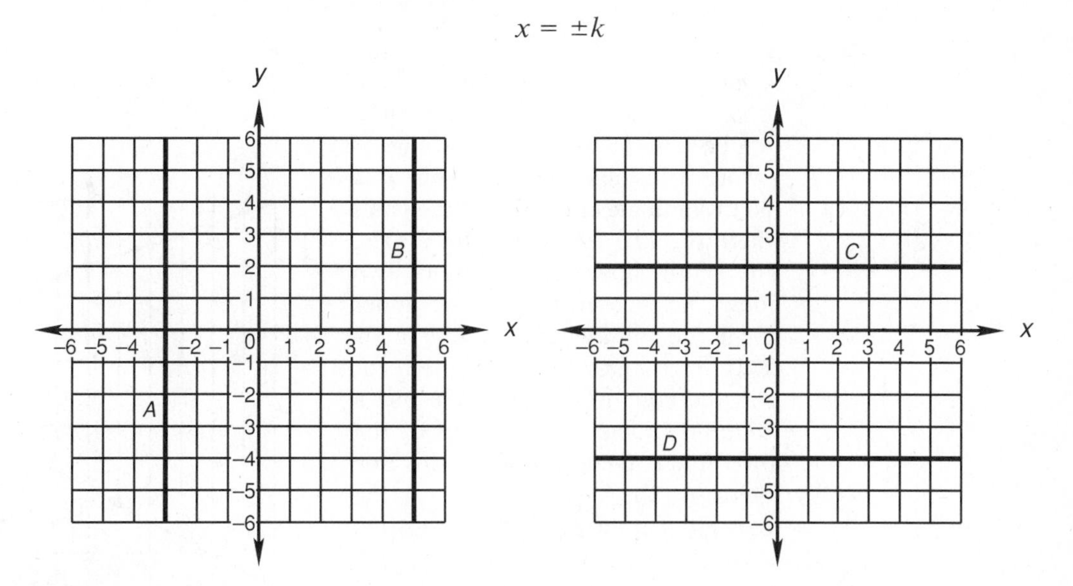

The graph of a horizontal line is everywhere equidistant from the x axis. Every point on line C is 2 units above the x axis, and the equation of this line is

$$y = +2$$

Every point on line D is 4 units below the x axis, and thus the equation of this line is

$$y = -4$$

If we use k to represent the number, we can say that the general form of a horizontal line is

$$y = \pm k$$

Thus, we see that the equations of vertical and horizontal lines can be determined by inspection. These equations contain an x and one number or a y and one number.

$$x = +5 \qquad x = -3 \qquad y = +2 \qquad y = -4$$

The equation of a line that is neither vertical nor horizontal cannot be determined by inspection. However, the equations of these lines can be written in what we call the **slope-intercept form.** The following equations are equations of three different lines written in slope-intercept form.

(a) $y = -6x + 2$ (b) $y = \frac{2}{3}x - 5$ (c) $y = 0.007x + 3$

We note that each equation contains an equals sign, a y, an x, and two numbers. **The only difference in the equations is that the numbers are different.**

We use the letters m and b when we write this equation without specifying the two numbers.

$$y = mx + b$$

Since the equation of any line that is not a vertical line or a horizontal line can be written in this form, the problem of finding the equation of a given line is reduced to the problem of finding the two numbers which will be the values of m and b in the equation.

85.B intercept

In the slope-intercept form of the equation, $y = mx + b$, we call the constant b the **intercept** of the equation because b represents the value of y when x has a value of 0. **Thus b is the y coordinate of the line at the point where the line intercepts the y axis.** The figure shows the graphs of two lines. The y coordinate of the upper line at the point where this line intercepts the y axis is $+4$, so the intercept b in the equation of this line has a value of 4. The y coordinate of the lower line at the point where this line intercepts the y axis is -3, so the intercept b in the equation of this line has a value of -3.

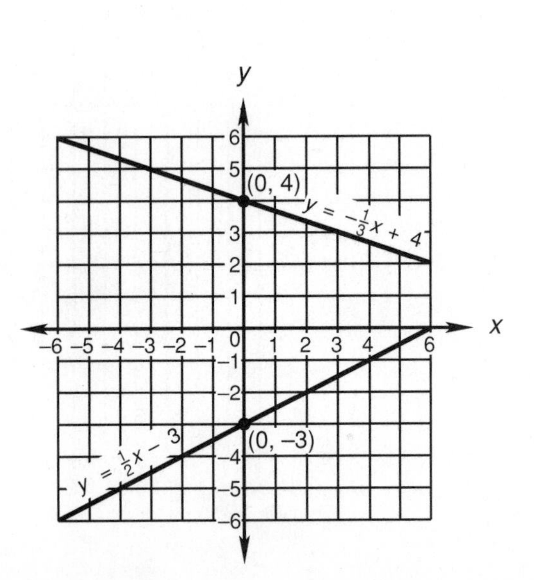

85.C slope

In the slope-intercept form, $y = mx + b$, we call the constant m the slope of the line. Thus, in the equation $y = -2x + 6$, we say that the slope of this line is -2 because the coefficient of x is -2. **We note that the slope has both a sign and a magnitude** (absolute value). A line represented by a line segment that points toward the upper right part of the coordinate plane has a positive slope. A line represented by a line segment that points toward the lower right part of the coordinate plane has a negative slope. As a mnemonic to help us remember this, we will use the little man and his car. **He always comes from the left side, as shown here.**

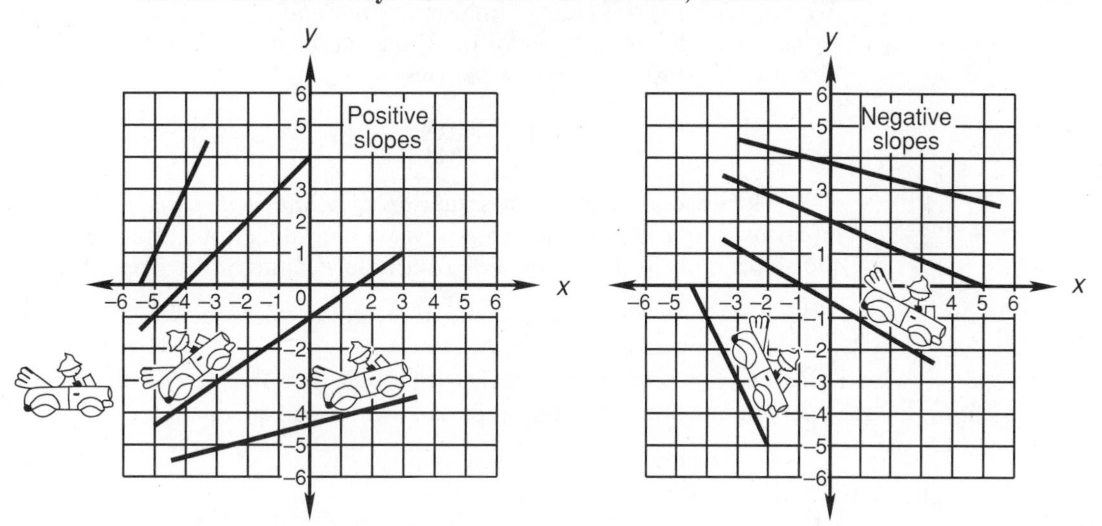

He sees the first set of lines as uphill with positive slopes and the second set of lines as downhill with negative slopes.

The magnitude, or absolute value, of the slope is defined to be the ratio of the absolute value of the change in the y coordinate to the absolute value of the change in the x coordinate as we move from one point on the line to another point on the line.

$$|m| = \frac{|\text{change in } y|}{|\text{change in } x|}$$

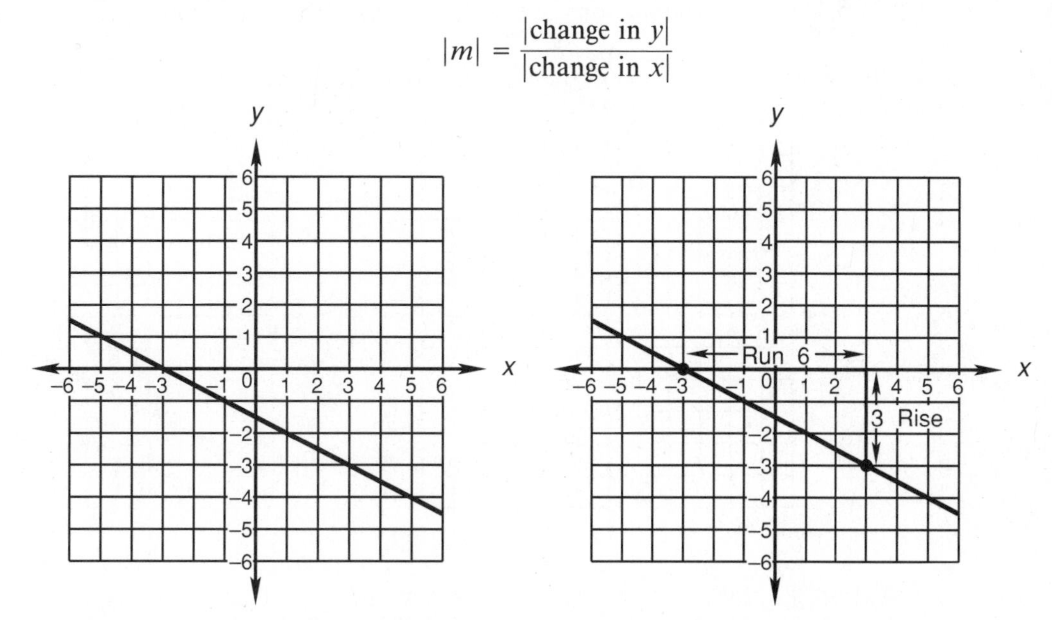

The figure on the left shows the graph of a line that has a negative slope. To find the magnitude of the slope of this line, we arbitrarily choose two points on the line and connect the two points with lines drawn parallel to the coordinate axes. This has been done in the figure on the right.

The length of the horizontal line is 6 and is the difference of the x coordinate of the two points. The length of the vertical line is 3 and is the difference of the y coordinates of the two points. Since the magnitude of the slope is the ratio of the absolute value of the change in the y coordinate to the absolute value of the change in the x coordinate, we see that the magnitude, or absolute value, of the slope of this line is $\frac{1}{2}$.

$$|m| = \frac{|\text{change in } y|}{|\text{change in } x|} \longrightarrow |m| = \frac{3}{6} \longrightarrow |m| = \frac{1}{2}$$

We have labeled the change in x as the **run** and the change in y as the **rise.** Using these words, the magnitude of the slope can be defined as the **absolute value of the rise over the absolute value of the run.**

$$|\text{Slope}| = \frac{|\text{rise}|}{|\text{run}|} \qquad \text{or} \qquad |m| = \frac{|\text{rise}|}{|\text{run}|}$$

The general form of the equation of a line is $y = mx + b$, and to write the equation of this line, we need to know (1) the value of the intercept b, (2) the sign of the slope, and (3) the magnitude or absolute value of the slope. We see that

1. The y value of the coordinate of the point where the line intercepts the y axis is approximately -1.4, so $b = -1.4$.
2. The line points to the lower right and thus the *sign* of the slope is negative.
3. The magnitude of the slope is $\frac{3}{6}$, which is equivalent to $\frac{1}{2}$.

So the equation of this line is

$$y = -\frac{1}{2}x - 1.4$$

Not everyone will read the intercept on the graph as -1.4. Some will say -1.5 or some other number close to -1.4. Also, not everyone will compute the magnitude of the slope to be exactly $\frac{1}{2}$. However, the value found for these numbers should be close to -1.4 and $-\frac{1}{2}$.

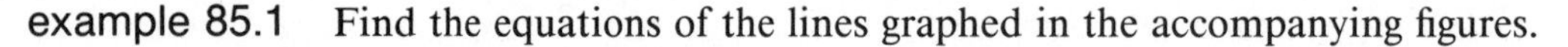

example 85.1 Find the equations of the lines graphed in the accompanying figures.

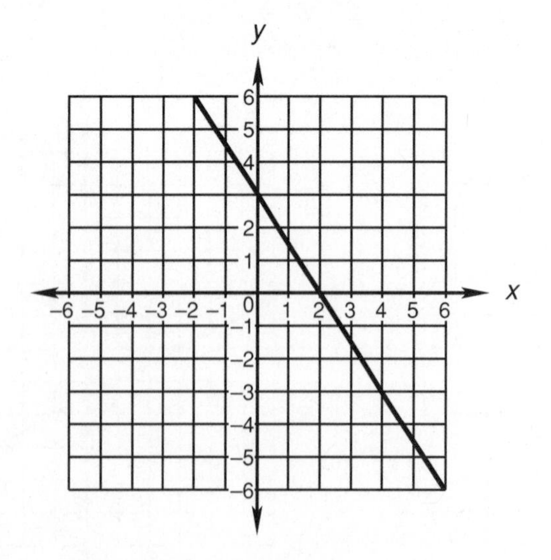

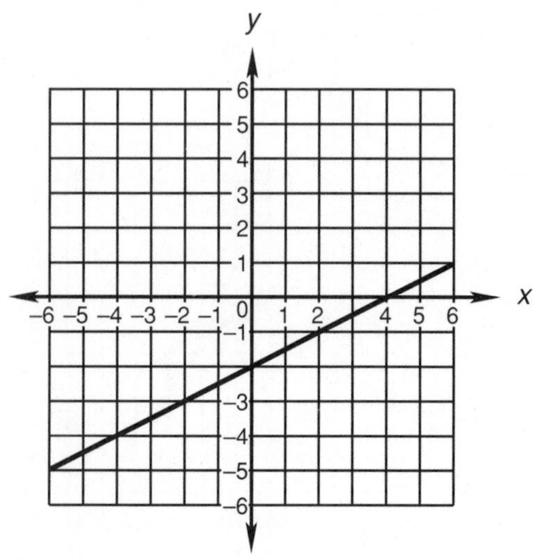

solution The desired equation is $y = mx + b$, and we need to find m and b.

By inspection, $b = +3$.
By inspection, the sign of m is $-$.

The desired equation is $y = mx + b$, and we need to find m and b.

By inspection, $b = -2$.
By inspection, the sign of m is $+$.

Now we need to find the magnitudes or absolute values of the slopes. We will arbitrarily choose two points on each of the lines, draw the triangles, and compute the slopes.

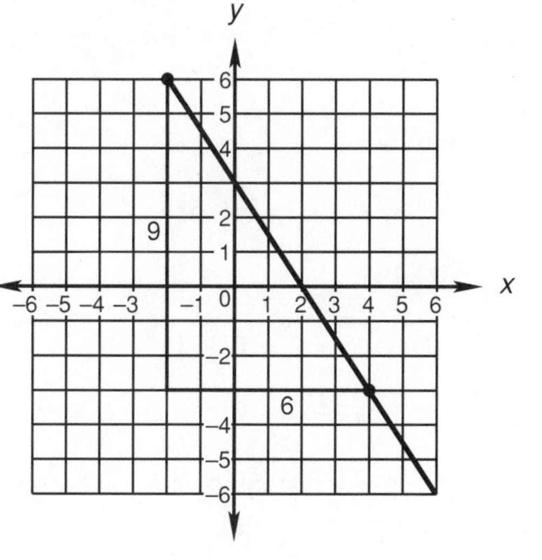

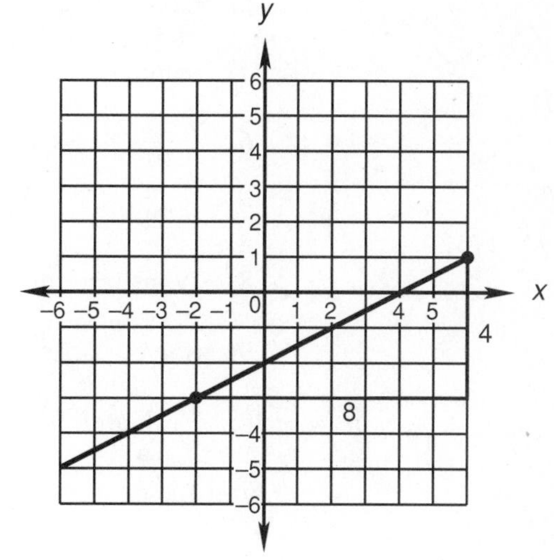

$$|m| = \frac{9}{6} = \frac{3}{2}$$

So $b = +3$ and $m = -\frac{3}{2}$.
Using these values in $y = mx + b$ yields

$$y = -\frac{3}{2}x + 3$$

$$|m| = \frac{4}{8} = \frac{1}{2}$$

So $b = -2$ and $m = +\frac{1}{2}$.
Using these values in $y = mx + b$ yields

$$y = \frac{1}{2}x - 2$$

practice Find the equations of these lines.

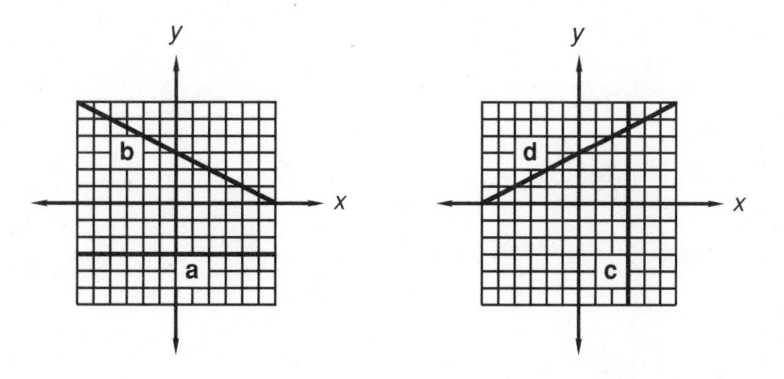

problem set 85

1. Leona found three consecutive integers such that the product of 5 and the sum of the first two was 7 greater than the opposite of the third. What were her integers?

2. Robin and Andrea found four consecutive even integers such that the product of 3 and the sum of the first and second was 18 greater than 5 times the third. What were the integers?

3. The salesperson reduced the price 14 percent to be able to sell the car for \$3440. What was the original price of the car? How much was the price reduced?

4. When the next rise occurred in the flood tide, the amount of inundated land increased 270 percent. If the former inundation was 1400 acres, how many acres were under water after the rise?

5. Ezekiel had 7452 jeroboams on display. If this was 1.62 times the total number of jeroboams in the next country, how many jeroboams were in the next country?

6. $\frac{7\sqrt{2}}{3} \in$ {What sets of numbers}?

7. Simplify: $-2(4 - 1)(-1 - 2^0) + |-3 + 5|$

8. What fraction of $2\frac{1}{4}$ is $7\frac{3}{8}$?

9. Solve: $2.2x - 0.1x + 0.02x = -2 - 0.332$

10. Evaluate: $\frac{\sqrt[5]{x + 7}}{3}$ if $\frac{x - 50}{10} = -30$

Find the equations of these lines. Each small square measures 1 unit.

11.

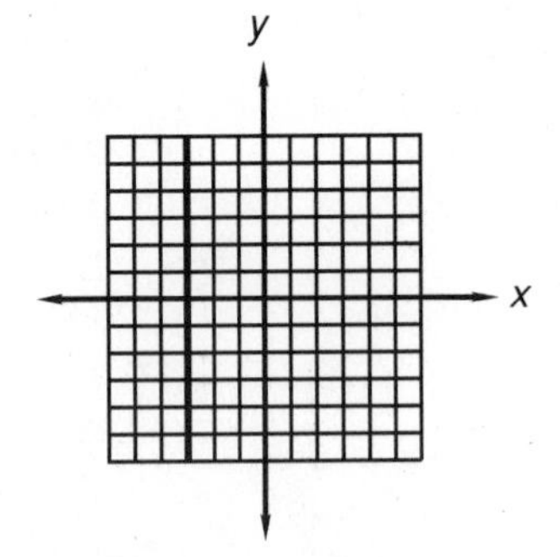

12.

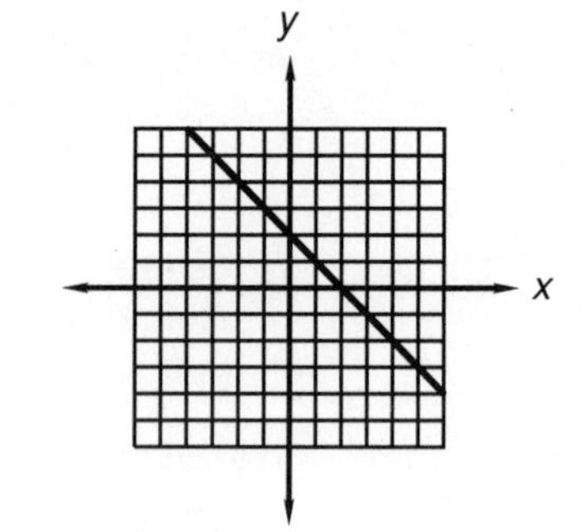

13. Find the area of this figure in square centimeters. Dimensions are in meters.

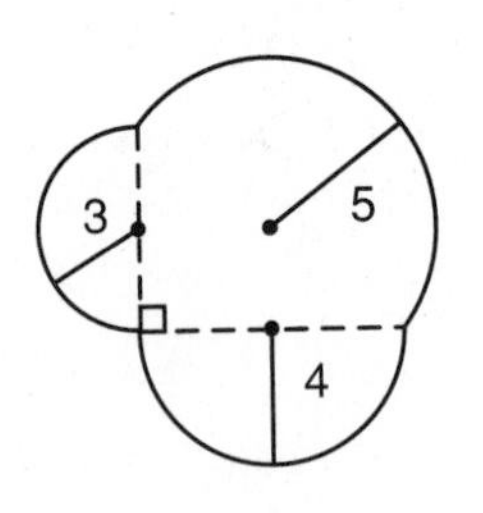

Simplify:

14. $\frac{(0.0056 \times 10^{-5})(100{,}000 \times 10^{-14})}{8000 \times 10^{15}}$

15. 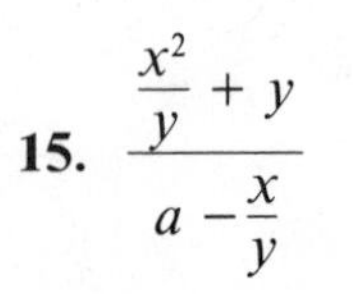

$$\frac{\frac{x^2}{y} + y}{a - \frac{x}{y}}$$

16. Add: $\frac{4}{x + y} + \frac{6}{x} - \frac{4}{ax}$

17. Expand: $\frac{x^{-2}a}{y^2}\left(\frac{a^4y^{-2}}{x} - \frac{3x^2a}{y^2}\right)$

18. Simplify by adding like terms: $a^2xy - \frac{3a^2x}{y^{-1}} + \frac{4x}{y^{-1}a^2} + \frac{5x^{-1}y}{a^2}$

19. Solve by graphing: $\begin{cases} y = x + 4 \\ y = -x + 2 \end{cases}$

Use substitution to solve:

20. $\begin{cases} 5N_N + 10N_D = 450 \\ N_D = N_N + 30 \end{cases}$

Use elimination to solve:

21. $\begin{cases} 5x - 2y = 7 \\ 4x + y = 3 \end{cases}$

Solve:

22. $\frac{p}{4} - \frac{p+2}{6} = -4$

23. $\frac{k+2}{5} - \frac{k}{10} = \frac{3}{20}$

24. Simplify: $3\sqrt{72} - 14\sqrt{18} + 6\sqrt{8}$

25. $R_T T_T = R_B T_B$, $R_T = 80$, $R_B = 20$, $T_B = T_T + 18$. Find T_B and T_T.

26. $R_L T_L = R_S T_S$, $R_L = 120$, $R_S = 280$, $T_S = 20 - T_L$. Find T_S and T_L.

Factor. Always factor the common factor first.

27. $4x^2 + 40x + 100$

28. $-x^3 - 30x - 11x^2$

29. $ax^2 - 35a + 2ax$

30. For what operations is the set of positive odd integers closed?

LESSON 86 *Coin problems*

When we solve a problem, we read the problem and search for statements about quantities that are equal. Each time we find a statement of equality, we write it as an algebraic equation. When we have the same number of independent equations as we have variables, we solve the equations by using substitution or elimination or by graphing.

The word problems that we have been working thus far have contained only one statement about quantities that are equal. We have worked these problems by using one equation and one variable. Now we will begin solving word problems that contain two statements about quantities that are equal. We will turn each of the statements into an equation that has two variables. Then we will solve the equations by using either substitution or elimination. The first problems of this kind are called **coin problems.**

In coin problems one statement will be about the number of coins. This statement will be like one of the following:

(a) The number of nickels plus the number of dimes equals 40.

$$N_N + N_D = 40$$

(b) There were 6 more nickels than dimes.

$$N_N = N_D + 6$$

The other statement will be about the value of the coins. Two examples are:

(c) The value of the nickels plus the value of the dimes equals \$4.65.

$$5N_N + 10N_D = 465$$

(d) The value of the dimes and quarters equaled \$25.10.

$$10N_D + 25N_Q = 2510$$

To avoid decimal numbers, we will often write all values in cents as in (c) and (d).

example 86.1 Jack and Betty have 28 coins that are nickels and dimes. If the value of the coins is \$1.95, how many coins of each type do they have?

solution This is a typical problem about coins. It says that the number of nickels plus the number of dimes equals 28 and that the value of the nickels plus the value of the dimes equals 195 cents.

$$\text{(a)}\quad N_N + N_D = 28$$

$$\text{(b)}\quad 5N_N + 10N_D = 195$$

The values of N_N and N_D that will simultaneously satisfy both equations may be found using either the substitution method or the elimination method.

SUBSTITUTION

$$5(28 - N_D) + 10N_D = 195$$
$$140 - 5N_D + 10N_D = 195$$
$$140 + 5N_D = 195$$
$$5N_D = 55$$
$$\mathbf{N_D = 11}$$

and since $N_N + N_D = 28$

$$\mathbf{N_N = 17}$$

by inspection.

ELIMINATION

$$\begin{aligned} -5N_N - 5N_D &= -140 \\ 5N_N + 10N_D &= 195 \\ \hline 5N_D &= 55 \end{aligned}$$

$$\mathbf{N_D = 11}$$

and since $N_N + N_D = 28$

$$\mathbf{N_N = 17}$$

by inspection.

example 86.2 Toy has \$4.45 in quarters and dimes. She has 8 more quarters than dimes. How many coins of each type does she have?

solution This problem has two statements about things that are equal. The first is that the value of the quarters plus the value of the dimes equals 445 pennies. We write this as

$$\text{(a)}\quad 25N_Q + 10N_D = 445$$

Since she has 8 more quarters, we add 8 to the number of dimes so the two sides of the equation will be equal.

$$\text{(b)}\quad N_D + 8 = N_Q$$

To solve, we will substitute equation (b) into equation (a).

$$25(N_D + 8) + 10N_D = 445 \quad \text{substituted}$$
$$25N_D + 200 + 10N_D = 445 \quad \text{multiplied}$$
$$35N_D + 200 = 445 \quad \text{simplified}$$
$$35N_D = 245 \quad \text{added } -200 \text{ to both sides}$$
$$\mathbf{N_D = 7}$$

Thus, $\mathbf{N_Q} = (7) + 8 = \mathbf{15}$

example 86.3 Orlando had a hoard of 22 nickels and dimes whose value was \$1.35. How many coins of each type did he have?

solution The two statements of equality are:

(a) The number of nickels plus the number of dimes equaled 22.

$$\text{(a)} \quad N_N + N_D = 22$$

(b) The value of the nickels plus the value of the dimes equaled 135 pennies.

$$\text{(b)} \quad 5N_N + 10N_D = 135$$

This time we will use elimination. We multiply each term in (a) by -5 to get (a′), which we add to (b).

$$\begin{aligned} \text{(a)} \quad -5N_N - 5N_D &= -110 \\ \text{(b)} \quad 5N_N + 10N_D &= 135 \\ \hline 5N_D &= 25 \end{aligned}$$

$$\mathbf{N_D = 5}$$

so $$\mathbf{N_N = 17}$$

practice Solve:

a. Ahmad and Regina have 36 coins that are nickels and dimes. If the value of the coins is \$2.90, how many coins of each type do they have?

b. Emil has \$6.45 in quarters and dimes. He has 9 more quarters than dimes. How many coins of each type does he have?

problem set 86

1. Heidi and Micah have 51 dimes and nickels. If the value of the coins is \$4.10, how many more dimes than nickels are there?

2. Find the volume in cubic inches of a cylinder 8 feet high if the base is as shown. If the cylinder is a right cylinder, find the surface area. Find the volume of a cone 8 feet high that has this same base. Dimensions are in feet.

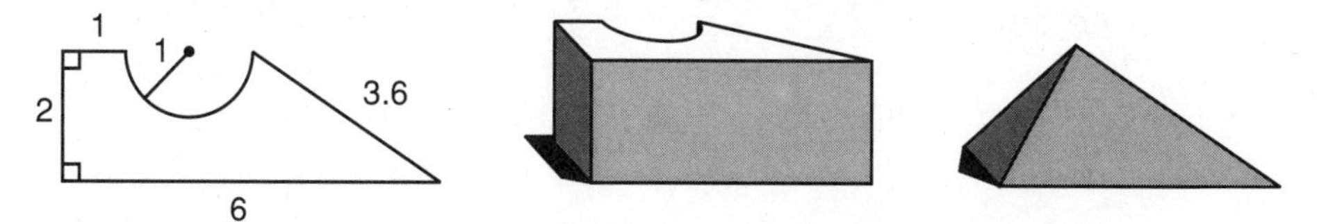

3. For 10 days the business averaged \$650.50 in transactions per day. For the following 20 days the average was \$874.75. What was the overall average for all 30 days?

4. A paroxysm of laughter escaped a few. If the ratio of the laughers to the stolid was 2 to 17 and 7600 were in the throng, how many did not laugh?

Simplify:

5. $\dfrac{(0.0016 \times 10^{-7})(3000 \times 10^{5})}{1{,}200{,}000}$

6. $\dfrac{(0.003 \times 10^{-5})(700 \times 10^{14})}{21{,}000{,}000}$

7. Find the equations of lines A and B.

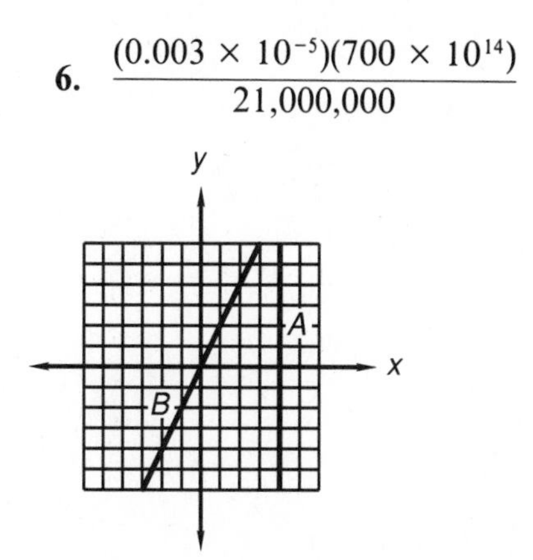

8. For what operations is the set of counting numbers closed?

Solve by graphing:

9. $\begin{cases} y = x + 2 \\ y = -x \end{cases}$

Use elimination to solve:

10. $\begin{cases} 3x + 2y = 11 \\ 2x - 3y = 16 \end{cases}$

11. Simplify: $4\sqrt{8} - 5\sqrt{32} + 6\sqrt{18}$

12. $3\frac{1}{4}$ of what number is $15\frac{1}{2}$?

13. Graph on a number line: $-3 \leq x < 2$; $D = \{\text{Positive integers}\}$

14. $R_M T_M + R_S T_S = 170$, $R_M = 20$, $R_S = 30$, $T_M = T_S + 1$. Find T_M and T_S.

15. $R_P T_P = R_M T_M$, $R_P = 45$, $R_M = 15$, $T_P = T_M + 8$. Find T_P and T_M.

16. $R_G T_G + 10 = R_P T_P$, $T_G = 4$, $T_P = 2$, $R_P = R_G + 45$. Find R_P and R_G.

Solve:

17. $\frac{x}{3} - 2 = \frac{4 - x}{5}$

18. $\frac{x}{4} - \frac{1}{2} = \frac{2 - x}{8}$

19. $-\frac{x + 2}{3} - \frac{2x + 8}{7} = 4$

20. $\frac{x}{4} - \frac{2x + 5}{2} = 7$

Add:

21. $\frac{1}{a^2} + \frac{2b}{a^3} - \frac{3b}{4a^3}$

22. $\frac{4}{a + b} + \frac{6}{a^2}$

23. Evaluate: $x(x^{-5} - y) - x^2$ if $x = -2$ and $y + 3 = \sqrt{81} - 3$

Simplify:

24. $\dfrac{x + \frac{x}{y}}{\frac{ax}{y} + 1}$

25. $\dfrac{\frac{ab}{c} - \frac{1}{c^2}}{4 - \frac{a}{c^2}}$

26. -4^{-2}

Factor:

27. $28x + 11x^2 + x^3$

28. $-xy^2 + 4a^2x$

29. Expand: $\frac{x^{-2}}{y^2}\left(x^2y^2 - \frac{3a^0x^2}{y^2}\right)$

Simplify:

30. $\dfrac{(x^2)^{-3}y^2p^0x^4}{[(x^2)^{-3}y]^{-2}x^{-4}}$

31. $\dfrac{x^2yyy^3}{(x^2y)^0}$

LESSON 87 Multiplication of radicals

We have used the product of square roots rule to help us simplify radical expressions such as the square root of 50.

$$\sqrt{50} = \sqrt{5 \cdot 5 \cdot 2} = \sqrt{5}\sqrt{5}\sqrt{2} = \mathbf{5\sqrt{2}}$$

The rule is restated here.

THE PRODUCT OF SQUARE ROOTS RULE

If m and n are nonnegative real numbers, then

$$\sqrt{n}\sqrt{m} = \sqrt{mn} \quad \text{and} \quad \sqrt{mn} = \sqrt{m}\sqrt{n}$$

We can use this rule to multiply radical expressions whose radicands are different.

$$\sqrt{2}\sqrt{3} = \mathbf{\sqrt{6}}$$

example 87.1 Simplify: $4\sqrt{3} \cdot 3\sqrt{2}$

solution Since the order of the factors does not affect the product in the multiplication of real numbers, we will rearrange the factors as

$$4 \cdot 3\sqrt{3}\sqrt{2}$$

and 4 is multiplied by 3 and $\sqrt{3}$ is multiplied by $\sqrt{2}$ to yield

$$\mathbf{12\sqrt{6}}$$

example 87.2 Simplify: $4\sqrt{3} \cdot 6\sqrt{6}$

solution Again we will rearrange the order of the indicated multiplications to get

$$4 \cdot 6\sqrt{3}\sqrt{6}$$

and multiply 4 by 6 and $\sqrt{3}$ by $\sqrt{6}$ to get

$$24\sqrt{18}$$

and $\sqrt{18}$ can be written as $3\sqrt{2}$, we can write

$$24\sqrt{18} = 24(3\sqrt{2}) = \mathbf{72\sqrt{2}}$$

example 87.3 Simplify: $4\sqrt{3}(5\sqrt{2} + 6\sqrt{3})$

solution The notation indicates that $4\sqrt{3}$ is to be multiplied by both terms within the parentheses:

$$4\sqrt{3}(5\sqrt{2} + 6\sqrt{3}) = 4\sqrt{3} \cdot 5\sqrt{2} + 4\sqrt{3} \cdot 6\sqrt{3}$$

Now we rearrange the order of the factors in each term to get

$$4 \cdot 5\sqrt{3}\sqrt{2} + 4 \cdot 6\sqrt{3}\sqrt{3}$$

and lastly we perform the indicated multiplications.

$$\mathbf{20\sqrt{6} + 72}$$

example 87.4 Simplify: $4\sqrt{2}(3\sqrt{2} + 5)$

solution First we multiply as indicated

$$4\sqrt{2} \cdot 3\sqrt{2} + 4\sqrt{2} \cdot 5$$

and now simplify.

$$\mathbf{24 + 20\sqrt{2}}$$

practice Multiply and simplify:

a. $5\sqrt{12} \cdot 4\sqrt{3}$

b. $3\sqrt{2}(4\sqrt{3} + 5\sqrt{6})$

problem set 87

1. When the piggy bank was opened, it yielded \$4.75 in nickels and pennies. If there were 175 coins in all, how many were nickels and how many were pennies?

2. The nickels and dimes all fell on the floor. There were 12 more nickels than dimes, and the total value of the coins was \$5.10. How many nickels and how many dimes were on the floor?

3. The room was a mess. In 1 hour Gretchen picked up 80 percent of the toys on the floor. If she picked up 128 toys, how many toys were on the floor to begin with?

4. Ramses cogitated. He thought of three consecutive even integers and found that 3 times the sum of the first two was 58 less than 14 times the opposite of the third. What were his integers?

5. Simplify: $\dfrac{(0.00032 \times 10^{-5})(4000 \times 10^{7})}{(160{,}000)(0.00002)}$

6. Find the equation of line 6.

7. Find the equation of line 7.

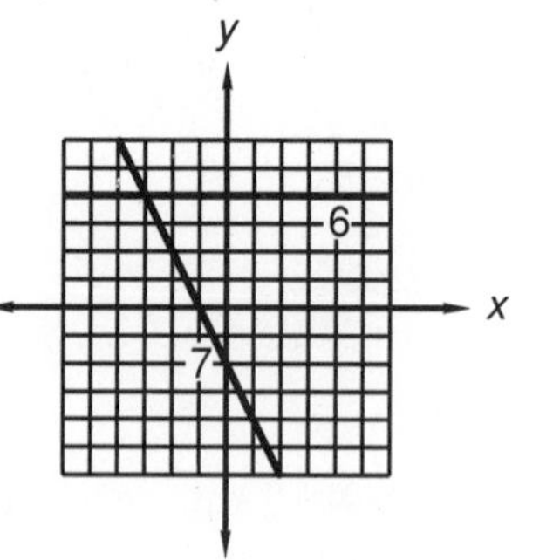

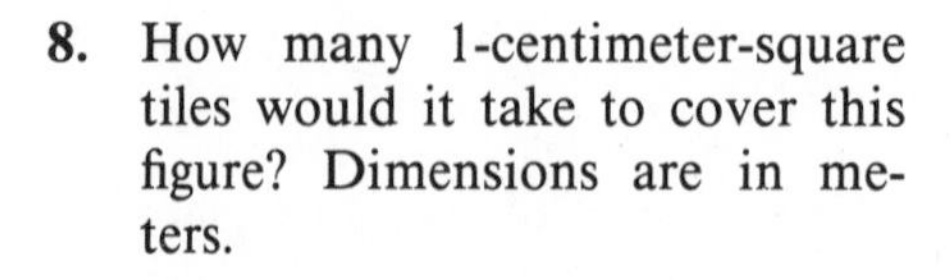
8. How many 1-centimeter-square tiles would it take to cover this figure? Dimensions are in meters.

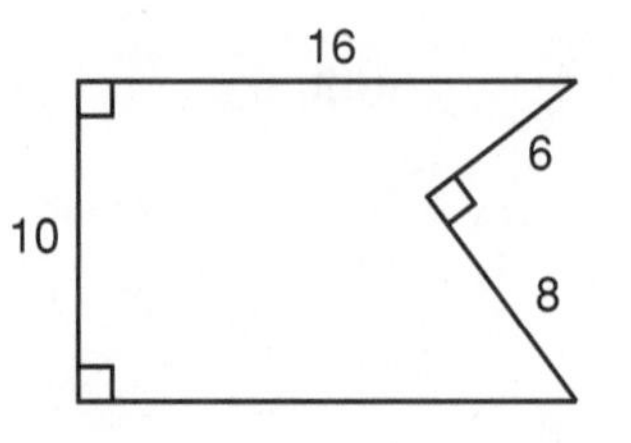

9. Describe this graph by writing a negated inequality and stating the domain.

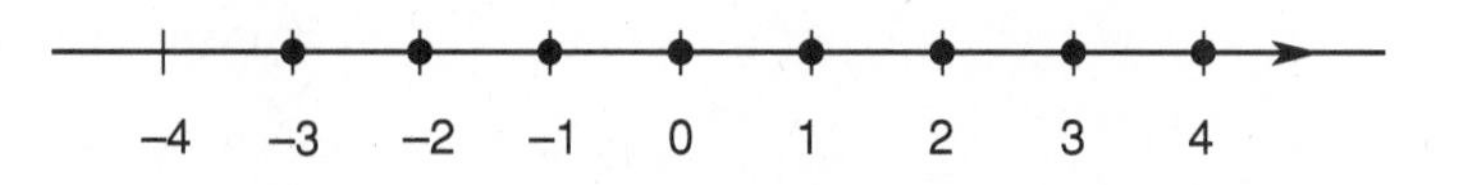

Simplify:

10. $5\sqrt{2}(2\sqrt{3} + \sqrt{6})$

11. $5\sqrt{8} - 4\sqrt{32}$

Solve by graphing:

12. $\begin{cases} y = -2x - 2 \\ y = -4 \end{cases}$

Solve by elimination:

13. $\begin{cases} 4x + 3y = -14 \\ 3x + 2y = -10 \end{cases}$

14. What fraction of $3\frac{1}{8}$ is $1\frac{1}{8}$?

15. $-4 \in$ {What sets of numbers}?

Solve:

16. $\frac{3 + x}{4} - \frac{x}{3} = 5$

17. $\frac{1}{2} - \frac{2x}{5} = 7$

18. $R_M T_M = R_K T_K$, $R_M = 30$, $R_K = 10$, $T_M = 16 - T_K$. Find T_M and T_K.

Add:

19. $\frac{1}{x^2} - \frac{3a}{x - a} + \frac{2}{x}$

20. $-\frac{a}{x} + \frac{b}{x^2c} - \frac{d}{x^3c^2}$

Simplify:

21. $\dfrac{a + \frac{1}{a}}{\frac{3}{a^2} - b}$

22. $\dfrac{\frac{abx}{c} - 4}{\frac{1}{c} + a}$

23. Evaluate: $-x^2(x^{-2} - y) - |x - y^{-4}|$ if $x = -3$ and $y = -2$

Simplify:

24. $-2^0[(3^0 - 5) - 2^2(2 - 3) + 5]$

25. $-3[(-2 - 2) - 2^0 - (4 - 3)(-2)]$

26. Solve: $-0.2 - 0.02 - 0.02x = 0.4(1 - x) - 0.012$

27. Expand: $\frac{3x^{-2}}{a^2y^5}\left(\frac{x^2a^2}{y^5} - \frac{x^{-3}y}{a}\right)$

Simplify:

28. $\dfrac{(a^{-3})^0(a^2)^0(a^{-2})^{-2}}{a^4(x^{-5})^{-2}xx^2}$

29. $\dfrac{(x^2y)xy^{-2}}{x^2yy}$

30. For what operations is the set of rational numbers closed?

LESSON 88 *Division of polynomials*

In the problem sets since Lesson 51 we have been finding the product of two polynomials. Now we investigate the inverse process, the division of polynomials.

The simplest type of polynomial division is the division of a polynomial by a monomial. The desired result can be obtained by dividing each term of the polynomial by the monomial as shown in the following examples.

example 88.1 Divide $3x^3 + 7x^2 - x$ by x.

solution We will divide each of the three terms by x.

$$\frac{3x^3}{x} + \frac{7x^2}{x} - \frac{x}{x} = \mathbf{3x^2 + 7x - 1}$$

example 88.2 Divide $12x^{12} - 8x^8 + 4x^6$ by $4x^4$.

solution We will divide every term by $4x^4$.

$$\frac{12x^{12}}{4x^4} - \frac{8x^8}{4x^4} + \frac{4x^6}{4x^4} = \mathbf{3x^8 - 2x^4 + x^2}$$

Before considering the method of dividing a polynomial by a binomial, we will complete a long division problem involving the natural numbers, an algorithm with which we are familiar. The method of dividing a polynomial by a binomial is very similar.

$$\begin{array}{r|l} & \;4 \\ \hline 12 & 49 \\ & \underline{48} \\ & \;\;1 \end{array} \qquad \text{so} \qquad 49 \div 12 = 4\frac{1}{12}$$

We multiplied 4 by 12 and recorded the product of 48 below the 49. Next we **mentally changed the sign** of the $+48$ to -48 and then added algebraically to find the remainder of 1.

example 88.3 Divide $-2x^2 + 3x^3 + 5x + 50$ by $-3 + x$.

solution The first step is to write both polynomials in descending powers of the variable and use the same format for division as we used above.

$$x - 3 \overline{)\, 3x^3 - 2x^2 + 5x + 50}$$

To determine the first term of the quotient, divide the first term of the divisor into the first term of the dividend; in this case, divide x into $3x^3$ and get $3x^2$. Record as indicated.

$$\begin{array}{r|l} & 3x^2 \\ \hline x - 3 & 3x^3 - 2x^2 + 5x + 50 \end{array}$$

Now multiply the term $3x^2$ by $x - 3$ and record as shown below.

$$\begin{array}{r|l} & 3x^2 \\ \hline x - 3 & 3x^3 - 2x^2 + 5x + 50 \\ & \underline{3x^3 - 9x^2} \end{array}$$

Now **mentally change the sign** of both $3x^3$ and $-9x^2$ and add algebraically.

$$\begin{array}{r|l} & 3x^2 \\ \hline x - 3 & 3x^3 - 2x^2 + 5x + 50 \\ & \underline{3x^3 - 9x^2} \\ & \qquad + 7x^2 \end{array}$$

Now bring down the $+ 5x$.

$$\begin{array}{r|l} & 3x^2 \\ \hline x - 3 & 3x^3 - 2x^2 + 5x + 50 \\ & \underline{3x^3 - 9x^2} \\ & \qquad + 7x^2 + 5x \end{array}$$

Now divide the x of $x - 3$ into $7x^2$, get $7x$, and record as shown.

$$
\begin{array}{r|l}
 & 3x^2 + 7x \\ \hline
x - 3 & 3x^3 - 2x^2 + 5x + 50 \\
 & \underline{3x^3 - 9x^2} \\
 & \qquad 7x^2 + 5x
\end{array}
$$

Multiply $7x$ by $x - 3$ and record.

$$
\begin{array}{r|l}
 & 3x^2 + 7x \\ \hline
x - 3 & 3x^3 - 2x^2 + 5x + 50 \\
 & \underline{3x^3 - 9x^2} \\
 & \qquad 7x^2 + 5x \\
 & \qquad \underline{7x^2 - 21x}
\end{array}
$$

Now **mentally change the sign** of both $7x^2$ and $-21x$ and add algebraically. Repeat the procedure until the remainder of 128 is obtained.

$$
\begin{array}{r|l}
 & 3x^2 + 7x + 26 \\ \hline
x - 3 & 3x^3 - 2x^2 + 5x + 50 \\
 & \underline{3x^3 - 9x^2} \\
 & \qquad 7x^2 + 5x \\
 & \qquad \underline{7x^2 - 21x} \\
 & \qquad\qquad 26x + 50 \\
 & \qquad\qquad \underline{26x - 78} \\
 & \qquad\qquad\qquad 128
\end{array}
$$

Thus we find that

$$\frac{3x^3 - 2x^2 + 5x + 50}{x - 3} = \mathbf{3x^2 + 7x + 26 + \frac{128}{x - 3}}$$

example 88.4 Divide $2x^3 - x - x^2 + 4$ by $-2 + x$.

solution As the first step, we rearrange both expressions in descending powers of the variable and write

$$
\begin{array}{r|l}
\hline
x - 2 & 2x^3 - x^2 - x + 4
\end{array}
$$

Now we divide using the same procedure we used in the last example.

$$
\begin{array}{r|l}
 & 2x^2 + 3x + 5 \\ \hline
x - 2 & 2x^3 - x^2 - x + 4 \\
 & \underline{2x^3 - 4x^2} \\
 & \qquad 3x^2 - x \\
 & \qquad \underline{3x^2 - 6x} \\
 & \qquad\qquad 5x + 4 \\
 & \qquad\qquad \underline{5x - 10} \\
 & \qquad\qquad\qquad 14
\end{array}
$$

Thus,

$$(2x^3 - x - x^2 + 4) \div (-2 + x) = \mathbf{2x^2 + 3x + 5 + \frac{14}{x - 2}}$$

practice Divide:

a. $(5x^3 - 9x^2 + x) \div x$

b. $(-3x^2 + 6x^3 + x - 40) \div (-2 + x)$

problem set 88

1. There were 40 dimes and quarters in the drawer. Peggy counted them and found that their total value was \$4.75. How many coins of each type were there?

2. Juan had \$5.25 in nickels and quarters. If he had 15 more nickels than quarters, how many coins of each type did he have?

3. Find four consecutive odd integers such that the product of -3 and the sum of the first and fourth is 30 less than 10 times the opposite of the third.

4. When the tractorcade approached the town square, the driver of the lead vehicle was told that only 64 percent of the tractors had enough fuel. If 224 tractors had enough fuel, how many tractors were in the tractorcade?

5. Simplify: $\dfrac{(0.0003 \times 10^{-8})(8000 \times 10^{6})}{0.004 \times 10^{5}}$

6. Find the equation of line 6.

7. Find the equation of line 7.

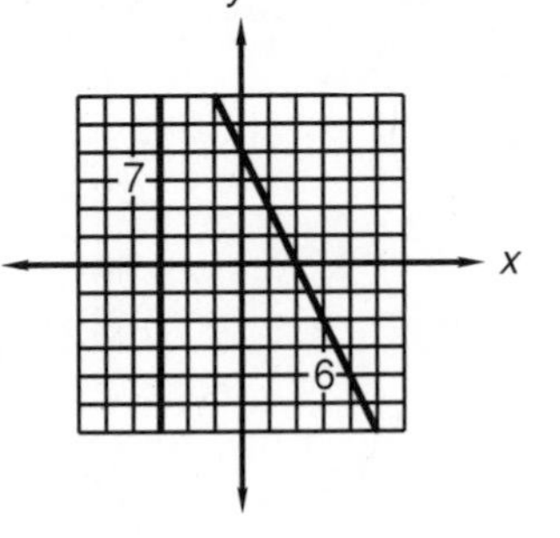

Simplify:

8. $4\sqrt{3} \cdot 6\sqrt{6}$

9. $3\sqrt{2}(7\sqrt{2} - \sqrt{6})$

10. $5\sqrt{45} - \sqrt{180}$

11. Evaluate: $\dfrac{\sqrt[3]{x - a}}{2}$ if $x = -100$ and $a = -x - 173$

12. The figure is the base of a cylinder 5 feet high. How many 1-inch sugar cubes will the container hold? If the cylinder is a right cylinder, what is the surface area? Find the volume of a cone 5 feet high that has the same base. Dimensions are in feet. Express all answers as inches in expanded form.

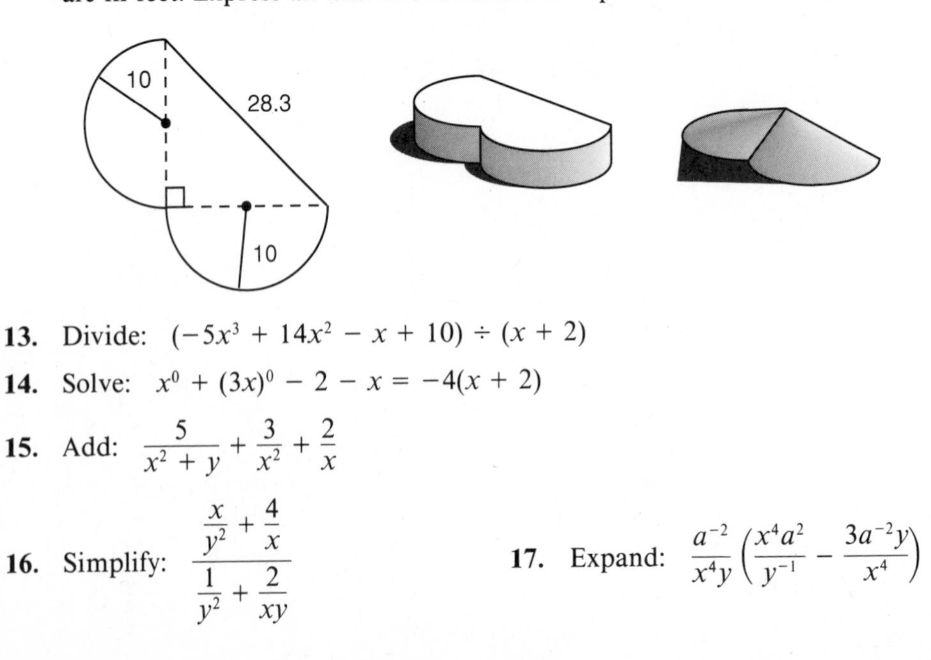

13. Divide: $(-5x^3 + 14x^2 - x + 10) \div (x + 2)$

14. Solve: $x^0 + (3x)^0 - 2 - x = -4(x + 2)$

15. Add: $\dfrac{5}{x^2 + y} + \dfrac{3}{x^2} + \dfrac{2}{x}$

16. Simplify: $\dfrac{\dfrac{x}{y^2} + \dfrac{4}{x}}{\dfrac{1}{y^2} + \dfrac{2}{xy}}$

17. Expand: $\dfrac{a^{-2}}{x^4y}\left(\dfrac{x^4a^2}{y^{-1}} - \dfrac{3a^{-2}y}{x^4}\right)$

18. Simplify by adding like terms: $\frac{4x^2y}{xy} - \frac{3x^{-3}y^0}{x^{-4}} + \frac{2}{x^{-1}} - \frac{4x}{y}$

Solve by graphing:

19. $\begin{cases} y = 3x - 2 \\ y = -x + 2 \end{cases}$

Solve by substitution:

20. $\begin{cases} 4x - 5y = -26 \\ x - y = -6 \end{cases}$

Solve by elimination:

21. $\begin{cases} 4x - 5y = 45 \\ 2x - 3y = 25 \end{cases}$

Solve:

22. $\frac{x}{2} - \frac{3 + x}{4} = \frac{1}{6}$

23. $\frac{3x}{2} - \frac{4x + 1}{5} = 4$

24. $R_MT_M + 6 = R_DT_D$, $R_M = 3$, $R_D = 12$, $T_M = 4 + T_D$. Find T_M and T_D.

25. Simplify: $\frac{xx^{-3}x^4(y^2)}{x^2(2x)^{-3}}$

Factor. Always factor the common factor first.

26. $ax^2 + 4ax + 4a$

27. $-10 - 3x + x^2$

28. $-4ax^2 + 9a$

29. $20x + 12x^2 + x^3$

30. For what operations is the set of negative rational numbers closed?

LESSON 89 More on systems of equations

When we solve systems of equations using the substitution method, sometimes it is necessary to rearrange one of the equations before the necessary substitutions can be accomplished. In the examples shown here, we will have to rearrange the equation that relates the time variables before we can substitute.

example 89.1 $R_ET_E = R_WT_W$, $R_E = 200$, $R_W = 250$, $T_E + T_W = 9$. Find T_E and T_W.

solution We begin by replacing R_E with 200 and R_W with 250 to find

$$200T_E = 250T_W$$

Now we must use the equation $T_E + T_W = 9$ to substitute for T_E or for T_W. The equation cannot be used in its present form and must be solved for T_E by adding $-T_W$ to both sides or be solved for T_W by adding $-T_E$ to both sides. We will work the problem both ways to show that the final results will be the same.

$$\begin{array}{rcl} T_E + T_W &=& 9 \\ -T_W && -T_W \\ \hline T_E &=& 9 - T_W \end{array}$$

Now we will substitute $9 - T_W$ for T_E and solve the resulting equation for T_W.

$$200(9 - T_W) = 250T_W$$

$$\begin{array}{rcl} 1800 - 200T_W &=& 250T_W \\ +200T_W && +200T_W \\ \hline 1800 &=& 450T_W \end{array}$$

$$\begin{array}{rcl} T_E + T_W &=& 9 \\ -T_E && -T_E \\ \hline T_W &=& 9 - T_E \end{array}$$

Now we will substitute $9 - T_E$ for T_W and solve the resulting equation for T_E.

$$200T_E = 250(9 - T_E)$$

$$\begin{array}{rcl} 200T_E &=& 2250 - 250T_E \\ 250T_E && +250T_E \\ \hline 450T_E &=& 2250 \end{array}$$

so $T_W = \frac{1800}{450}$ or $\boldsymbol{T_W = 4.}$

Since $T_E + T_W = 9$, we can solve for T_E by replacing T_W with 4.

$$\begin{array}{rcl} T_E + 4 &=& 9 \\ -4 && -4 \\ \hline \boldsymbol{T_E} &=& \boldsymbol{5} \end{array}$$

so $T_E = \frac{2250}{450}$ or $\boldsymbol{T_E = 5.}$

Since $T_E + T_W = 9$, we can solve for T_W by replacing T_E with 5.

$$\begin{array}{rcl} 5 + T_W &=& 9 \\ -5 && -5 \\ \hline \boldsymbol{T_W} &=& \boldsymbol{4} \end{array}$$

Thus we see that while one procedure leads to our solving for T_E first and the other leads to our solving for T_W first, both procedures finally yield the same answers.

example 89.2 $R_1T_1 + R_2T_2 = 360$, $R_1 = 30$, $R_2 = 40$, $T_1 + T_2 = 10$. Find T_1 and T_2.

solution The time equation cannot be used in its present form. We decide to solve the time equation for T_1.

$$\begin{array}{rcl} T_1 + T_2 &=& 10 \\ -T_2 && -T_2 \\ \hline T_1 &=& 10 - T_2 \end{array}$$

Now we replace R_1 with 30, R_2 with 40, and T_1 with $10 - T_2$ and then solve.

$$\begin{aligned} 30(10 - T_2) + 40T_2 &= 360 && \text{substituted} \\ 300 - 30T_2 + 40T_2 &= 360 && \text{multiplied} \\ 300 + 10T_2 &= 360 && \text{simplified} \\ 10T_2 &= 60 && \text{added } -300 \text{ to both sides} \\ \boldsymbol{T_2} &\boldsymbol{= 6} \end{aligned}$$

Thus, $\boldsymbol{T_1 = 4}$

practice Find the values of all the unknowns:

a. $R_1T_1 = R_2T_2$, $T_1 + T_2 = 60$, $R_1 = 3$, $R_2 = 6$

b. $R_0T_0 = R_ST_S + 16$, $R_0 + R_S = -1$, $T_0 = 8$, $T_S = 4$

problem set 89

1. The big sack contained \$30 in nickels and dimes. If there were 500 coins in the sack, how many were nickels and how many were dimes?
2. When the box broke open, \$82 in quarters and dimes fell out. If there were 300 more quarters than dimes, how many dimes and how many quarters were there?
3. Find four consecutive integers such that 3 times the sum of the first and third is 84 greater than the opposite of the second.
4. Ninety percent of the people who voted voted for Sammy because he was 7 feet tall. If 1930 people voted, how many voted for Sammy?
5. Simplify: $\dfrac{(0.0072 \times 10^{-4})(100{,}000)}{6000 \times 10^{-24}}$

6. Find the equation of line 6.

7. Find the equation of line 7.

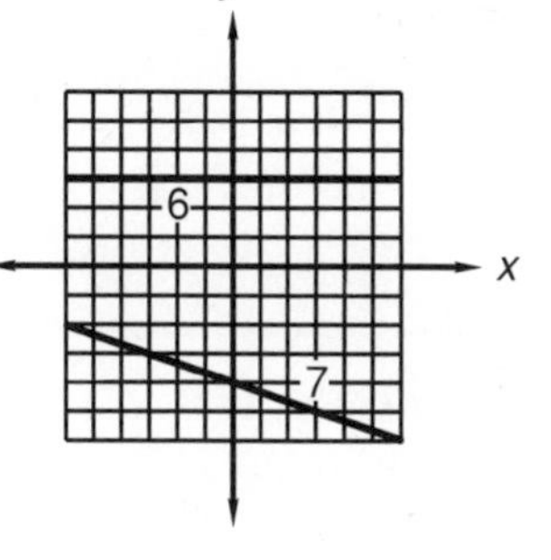

Simplify:

8. $3\sqrt{2}(4\sqrt{2} + 6\sqrt{6})$

9. $5\sqrt{3}(2\sqrt{3} - 6\sqrt{12})$

Divide:

10. $(x^3 - 2x^2 + 4x) \div (x - 2)$

11. $(2x^3 - 3x^2 + 2x - 4) \div (x + 3)$

12. 0.05 of what number is 0.0009?

Add:

13. $\dfrac{x}{24a} + \dfrac{y}{70a^2}$

14. $\dfrac{k}{42} - \dfrac{3x}{18}$

15. Simplify: $\dfrac{\dfrac{x}{y} - 1}{\dfrac{a}{y} + b}$

16. Simplify by adding like terms: $\dfrac{x^3y}{xy^{-1}} + 3xxy^2 - \dfrac{2x^4x}{x^2xy^{-2}} - \dfrac{5x^2}{xy}$

17. Solve: $(3x)^0(-2 - 3x) - x = -3(-2 - 3)$

Solve by graphing:

18. $\begin{cases} y = -2 \\ y = 2x - 2 \end{cases}$

Solve by substitution:

19. $\begin{cases} x + y = 2 \\ 2x - 3y = -1 \end{cases}$

Solve by elimination:

20. $\begin{cases} 3x - y = 8 \\ x + 2y = 12 \end{cases}$

21. Expand: $\left(x^2y - \dfrac{3x^{-4}a}{y}\right)\dfrac{x^{-2}}{y}$

22. Solve: $\dfrac{4 + x}{2} - \dfrac{1}{3} = \dfrac{1}{6}$

23. Evaluate: $-b \pm \sqrt{b^2}$ if $b - 40 = \sqrt{81}$

24. $R_TT_T + R_JT_J = 180$, $R_J + R_T = 20$, $T_T = 8$, $T_J = 10$. Find R_J and R_T.

25. Simplify: $\dfrac{(x^2)^{-2}(y^0)^2yy^3}{(y^{-2})^3yy^4y^{-1}x}$

Factor. Always factor the greatest common factor first.

26. $max^2 + 9xma + 14ma$

27. $-x^3 - 35x - 12x^2$

28. For what operations is the set of irrational numbers closed?

29. Find the surface area of this right circular cylinder in square inches. Find the volume of a sphere whose radius is 8 inches. Dimensions are in inches.

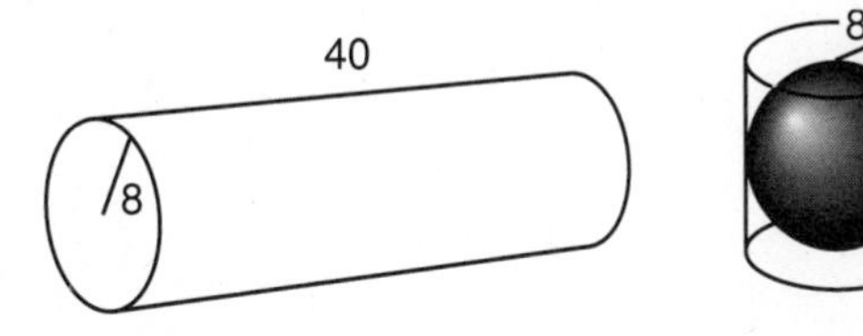

30. Expand: $(a + \sqrt{2}b)^2$

LESSON 90 More on division of polynomials

We remember that the first step in dividing polynomials is to rearrange the terms so that both expressions are written in descending order of the variable. Division can be accomplished without this step, but it is much more difficult. In this lesson, we discuss another helpful procedure that can be used in polynomial division.

example 90.1 Divide $-2x + 5 + 3x^3$ by $-3 + x$.

solution We begin by writing both polynomials in descending order of the variable and using the format for long division.

$$x - 3\overline{)3x^3 - 2x + 5}$$

We note that the dividend has an x^3 term and an x term but no x^2 term. A good ploy to avoid confusion is to insert an x^2 term with zero as its coefficient, as shown below. Of course, zero multiplied by x^2 equals zero so the polynomial is really unchanged. Now we perform the division using the same procedure we learned in Lesson 88.

$$\begin{array}{r} 3x^2 + 9x + 25 \\ x - 3\overline{)3x^3 + 0x^2 - 2x + 5} \\ \underline{3x^3 - 9x^2} \qquad\qquad \\ 9x^2 - 2x \qquad \\ \underline{9x^2 - 27x} \qquad \\ 25x + 5 \\ \underline{25x - 75} \\ 80 \end{array}$$

Thus, $$\frac{3x^3 - 2x + 5}{x - 3} = \mathbf{3x^2 + 9x + 25 + \frac{80}{x - 3}}$$

example 90.2 Divide $-4 + x^3$ by $-3 + x$.

solution Again we begin by rearranging each polynomial and using the long division format.

$$x - 3\overline{)x^3 - 4}$$

This time, we see that the dividend does not have an x^2 term or an x term. We will insert these terms and give each of them a coefficient of zero. Then we will divide.

$$\begin{array}{r} x^2 + 3x + 9 \\ x - 3\overline{)x^3 + 0x^2 + 0x - 4} \\ \underline{x^3 - 3x^2} \qquad\qquad \\ 3x^2 + 0x \qquad \\ \underline{3x^2 - 9x} \qquad \\ 9x - 4 \\ \underline{9x - 27} \\ 23 \end{array}$$

Thus, $$(-4 + x^3) \div (-3 + x) = \mathbf{x^2 + 3x + 9 + \frac{23}{x - 3}}$$

practice Divide:

a. $\dfrac{x^3 - 5}{x - 2}$

b. $(3x^3 - 5x + 4) \div (x - 4)$

problem set 90

1. There were 143 more Susan B. Anthony dollars than there were quarters. If the total value of the coins was \$153, how many dollars were there?

2. The collection had 60 coins that were nickels and dimes. If the total value of the coins was \$5, how many nickels were in the collection?

3. Forty percent of the crop was destroyed by the thunderstorm. If the farm consisted of 570 acres, how many acres were affected by the thunderstorm?

4. Find three consecutive integers such that -4 times the sum of the first and third is 13 less than 7 times the opposite of the second.

5. Simplify: $\dfrac{(0.00035 \times 10^{-8})(2000 \times 10^{-3})}{(0.0007 \times 10^{6})(2{,}000{,}000)}$

6. Find the equation of line 6.

7. Find the equation of line 7.

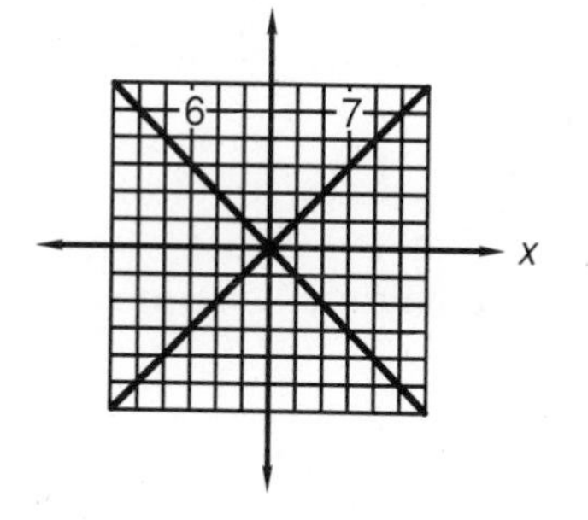

Simplify:

8. $5\sqrt{2}(3\sqrt{2} - 2\sqrt{12})$

9. $3\sqrt{5}(2\sqrt{5} - 6\sqrt{15})$

10. Evaluate: $-b \pm \sqrt{b^2 - 4a}$ if $a = 3$ and $b = 4$

11. Divide: $(3x^3 - x - 7) \div (x - 5)$

Solve by graphing:

12. $\begin{cases} x = 2 \\ y = 2x - 4 \end{cases}$

Solve by substitution:

13. $\begin{cases} 2x + y = 1 \\ 5x - 2y = 7 \end{cases}$

Solve by elimination:

14. $\begin{cases} 7x + 2y = 3 \\ 3x - 3y = 9 \end{cases}$

Simplify:

15. $3\sqrt{2}(4\sqrt{2} - \sqrt{12})$

16. $2\sqrt{2}(5\sqrt{2} - 3\sqrt{20})$

17. $2\sqrt{3}(\sqrt{3} - 2\sqrt{15})$

18. Solve: $2\frac{1}{4}p - 3\frac{1}{8} = \frac{1}{5}$

19. Graph: $x \not< 3$; $D = \{\text{Positive integers}\}$

Solve:

20. $\dfrac{x}{4} - \dfrac{x - 2}{3} = \dfrac{1}{2^{-3}}$

21. $\dfrac{x}{3} - \dfrac{x + 2}{4} = 5$

Add:

22. $\dfrac{3}{a + b} - \dfrac{4}{b} + \dfrac{6}{b^2}$

23. $\dfrac{3}{a^2x} + \dfrac{2b}{a(x + a)}$

24. Simplify: $\dfrac{x - \dfrac{4}{y}}{\dfrac{a}{y} + 3}$

25. Evaluate: $x^2 - x^0 - (x^0)^2 + ax(x - a)$ if $x = -2$ and $a + 2 = 6$

26. Solve: $(-2 - 3)x^0 - 2(-2)x = 3(x^0 - 2)$

27. Factor: $15ax + 56a + ax^2$

28. $R_PT_P = R_DT_D - 90$, $R_P = 30$, $R_D = 30$, $T_P + T_D = 9$. Find T_P and T_D.

29. Expand: $\dfrac{a^2y^{-2}}{x^4}\left(\dfrac{a^{-4}y^{-2}}{x} - \dfrac{3a^{-2}x}{y}\right)$

30. Simplify: $\dfrac{xxx^{-2}(y^0)(3y^{-2})^{-1}}{(2x^{-4})^2y}$

LESSON 91 Solution of quadratic equations by factoring

91.A quadratic equations

Quadratic equations are second-degree polynomial equations. *Second-degree* means that the greatest exponent of x in any term is 2. Both of these equations are quadratic equations in x because the greatest exponent of x is 2.

$$4 - 3x = 2x^2 \qquad 3x^2 - 2x + 4 = 0$$

The equation on the right is in **standard form** because all nonzero terms are on the left of the equals sign and the terms are written in descending order of the variable. The coefficient of x^2 cannot be zero, but either of the other two numbers can be zero. Thus each of the following equations is also a quadratic equation in x.

$$4x^2 = 0 \qquad 4x^2 + 2x = 0 \qquad 4x^2 - 3 = 0$$

To designate a general quadratic equation, we use the letter a to represent the coefficient of x^2, the letter b to represent the coefficient of x, and the letter c to represent the constant term. Using these letters to represent the constants in the equation, we can write a general quadratic equation in standard form as

$$ax^2 + bx + c = 0$$

If we let $a = 1$, $b = -3$, and $c = -10$, we have the equation

$$x^2 - 3x - 10 = 0$$

If we substitute either 5 or -2 for the variable x in the quadratic equation $x^2 - 3x - 10 = 0$, the equation will be transformed into a true equation as shown here.

IF $x = 5$	IF $x = -2$
$(5)^2 - 3(5) - 10 = 0$	$(-2)^2 - 3(-2) - 10 = 0$
$25 - 15 - 10 = 0$	$4 + 6 - 10 = 0$
$0 = 0$	$0 = 0$

The numbers 5 and -2 are the only numbers that will satisfy the equation above. **Every quadratic equation has at most† two numbers that will make the equation a true statement.** For that matter, every third-degree polynomial equation in one variable has at most three numbers that will satisfy the equation; every fourth-degree polynomial equation in one variable has at most four numbers that will satisfy the equation. To generalize, we can say that every nth-degree polynomial equation in one variable (n is a natural number) has at most n roots.

91.B solution of quadratic equations by factoring

Some quadratic equations can be solved by using the zero factor theorem.

> ZERO FACTOR THEOREM
>
> If p and q are any real numbers and if $p \cdot q = 0$, then either $p = 0$ or $q = 0$, or both p and q equal 0.

This says that **if the product of two real numbers is zero, one or both of the numbers are zero.** Thus if we indicate the product of 4 and an unspecified number by writing

$$4(\quad) = 0$$

the only number that we can place in the parentheses that will make the equation a true equation is the number zero.

In the same way, if we indicate the product of two unspecified numbers by writing

$$(\quad)(\quad) = 0$$

the quantity in the first parentheses or the quantity in the second parentheses or both quantities must equal zero or the product will not equal zero.

Now let's look at the equation

$$(x - 3)(x + 5) = 0$$

Here we have two quantities multiplied and the product is equal to zero. From the **zero factor theorem,** we know that at least one of the quantities must equal zero if the product is to equal zero, so either

$$x - 3 = 0 \qquad \text{or} \qquad x + 5 = 0$$

But if $x - 3 = 0$, $\boldsymbol{x = 3}$ and if $x + 5 = 0$, $\boldsymbol{x = -5}$

Thus the two values of x that satisfy the condition stated are 3 and -5.

We can use the zero factor theorem to help us solve quadratic equations that can be factored. We do this by first writing the equation in standard form and factoring the polynomial. Then we set each of the factors equal to zero and solve for the values of the variable.

example 91.1 Use the factor method to find the roots of $x^2 - 18 = 3x$.

† We say *at most* because some quadratic equations have only one root. For instance, the only root of the equation $x^2 - 4x + 4 = 0$ is the number $+2$.

solution First we write the equation in standard form and then factor.

$$x^2 - 3x - 18 = 0 \longrightarrow (x + 3)(x - 6) = 0$$

Since the product $(x + 3)(x - 6)$ equals zero, by the zero factor theorem, we know that one or both of these factors must equal zero.

If $x + 3 = 0$, $\boldsymbol{x = -3}$ If $x - 6 = 0$, $\boldsymbol{x = +6}$

To check, we will use -3 and $+6$ as values for x in the original equation.

IF $x = -3$	IF $x = 6$
$(-3)^2 - 18 = 3(-3)$	$(6)^2 - 18 = 3(6)$
$9 - 18 = -9$	$36 - 18 = 18$
$-9 = -9$ Check	$18 = 18$ Check

example 91.2 Find the roots of $-25 = -4x^2$.

solution First we write the equation in standard form, $4x^2 - 25 = 0$, and then we factor to get $(2x - 5)(2x + 5) = 0$. For this to be true, either $2x - 5$ equals zero or $2x + 5$ equals zero.

IF $2x - 5 = 0$	IF $2x + 5 = 0$
$2x = 5$	$2x = -5$
$\boldsymbol{x = \frac{5}{2}}$	$\boldsymbol{x = -\frac{5}{2}}$

To check, we will use $\frac{5}{2}$ and $-\frac{5}{2}$ as values for x in the original equation.

IF $x = \frac{5}{2}$	IF $x = -\frac{5}{2}$
$-25 = -4\left(\frac{5}{2}\right)^2$	$-25 = -4\left(-\frac{5}{2}\right)^2$
$-25 = -4\left(\frac{25}{4}\right)$	$-25 = -4\left(\frac{25}{4}\right)$
$-25 = -25$ Check	$-25 = -25$ Check

example 91.3 Find the values of x that satisfy $x - 56 = -x^2$.

solution First we rewrite the equation in standard form.

$$x^2 + x - 56 = 0$$

Now we factor.

$$(x + 8)(x - 7) = 0$$

For this to be true, either $x + 8$ equals zero or $x - 7$ equals zero.

If $x + 8 = 0$, $\boldsymbol{x = -8}$ If $x - 7 = 0$, $\boldsymbol{x = 7}$

To check, we will use -8 and $+7$ as replacements for x in the original equation.

IF $x = -8$	IF $x = 7$
$(-8) - 56 = -(-8)^2$	$(7) - 56 = -(7)^2$
$-8 - 56 = -64$	$7 - 56 = -49$
$-64 = -64$ Check	$-49 = -49$ Check

example 91.4 Solve $x^2 - x = 42$.

solution First we write the equation in standard form

$$x^2 - x - 42 = 0$$

and now we factor.

$$(x - 7)(x + 6) = 0$$

The zero factor theorem says that if a product of two factors equals zero, either one factor or the other factor must equal zero.

If $x - 7 = 0$, $\boldsymbol{x = 7}$ If $x + 6 = 0$, $\boldsymbol{x = -6}$

Finally, we check our answers.

$$(7)^2 - (7) = 42 \qquad (-6)^2 - (-6) = 42$$

$$49 - 7 - 42 = 0 \qquad 36 + 6 - 42 = 0$$

$$0 = 0 \quad \text{Check} \qquad 0 = 0 \quad \text{Check}$$

practice Solve by factoring:

a. $x^2 = -5x + 24$ **b.** $-48 + x^2 = -8x$

problem set 91

1. The bowl contained 150 coins. If they were all pennies and nickels and their total value was \$2.70, how many coins of each type were there?
2. The second bowl also contained \$2.70 in pennies and nickels. If there were 54 more pennies than nickels in this bowl, how many were pennies and how many were nickels?
3. When the home team scored, 78 percent of the crowd stood and cheered, and the rest were dejected. If 8800 were dejected, how many stood and cheered?
4. Find four consecutive odd integers such that 4 times the sum of the first and fourth is 3 greater than 7 times the third.
5. Simplify: $\dfrac{(0.016 \times 10^{-5})(300 \times 10^{6})}{(20{,}000 \times 10^{4})(400 \times 10^{-8})}$
6. Find the equation of line 6.
7. Find the equation of line 7.

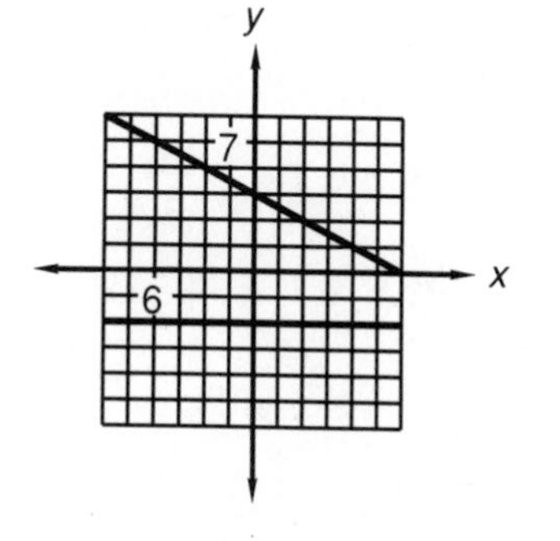

Simplify:

8. $5\sqrt{5}(2\sqrt{10} - 3\sqrt{5})$ **9.** $4\sqrt{7}(2\sqrt{7} - 3\sqrt{14})$

Divide:

10. $(2x^3 - 5x + 4) \div (x + 2)$ **11.** $(3x^3 - 4) \div (x - 5)$

12. This is the base of a prism 10 feet tall. How many 1-inch cubes will this prism hold? What is the volume in cubic inches of a pyramid with the same altitude and base? Dimensions are in inches.

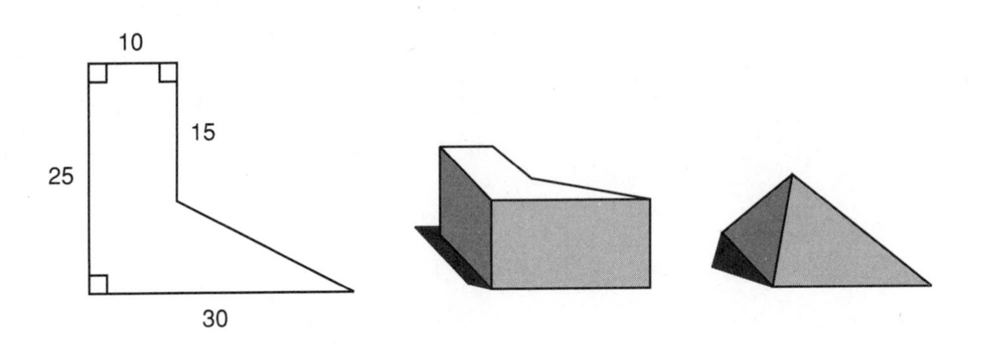

13. For what operations is the set of positive rational numbers closed?

14. Evaluate: $\dfrac{-b \pm \sqrt{b^2}}{2a}$ if $b - 5 = 17$ and $a = \dfrac{b}{2}$

15. The sum of four numbers is 396.80. The first two numbers are 96.8 and 100.1. Find the average of the four numbers.

16. Expand: $(x - \sqrt{3}\,y)^2$

Solve by factoring:

17. $28 = x^2 - 3x$

18. $x^2 = 25$

19. $x^2 - 6 = x$

20. $-x^2 - 8x = 16$

Solve by graphing:

21. $\begin{cases} y = 2x - 3 \\ x = -4 \end{cases}$

Use elimination to solve:

22. $\begin{cases} 3x + 5y = 16 \\ 4x - 3y = 2 \end{cases}$

23. Graph on a number line: $x + 2 \not\leq 4$; $D = \{\text{Positive integers}\}$

24. Solve: $4\frac{3}{5}x - 2^{-2} = \dfrac{1}{10}$

25. Simplify: $-(-4 - 2^0) - |-2| + \dfrac{1}{-2^{-2}}$

26. $R_P T_P + R_K T_K = 170$, $T_P = 2$, $T_K = 3$, $R_P = R_K + 10$. Find R_K and R_P.

27. Solve: $\dfrac{2x}{3} - \dfrac{2x - 4}{5} = 7$

28. Add: $\dfrac{1}{a} + \dfrac{2}{a^2} + \dfrac{3}{a + x}$

29. Solve: $(-2x^0 - 3)2 - 3x = -2(x^0 - 2)$

30. Simplify: $\dfrac{axx^{-12}y^{-2}(a^4)^{-2}}{a^{-4}(a^2)a(a^{-4}x^2)}$

LESSON 92 Value problems

We remember that when we read word problems, we look for word statements about quantities that are equal. Then we transform each of these word statements into an algebraic equation which makes the same statement of equality. We use as many variables as are necessary. When we have written as many independent equations as we have variables, we solve the equations by using the substitution method or the elimination method. When we write the equations, instead of x and y, we use meaningful variables so that we can remember what these variables represent. We have used two equations in two variables to solve coin problems. These problems are of a genre called **value problems.** We will look at other types of value problems in this lesson. They are very similar to coin problems.

example 92.1 Airline fares for a flight from Tifton to Adel are \$30 for first class and \$25 for tourist class. If a flight had 52 passengers who paid a total of \$1360, how many first-class passengers were on the trip?

solution There are two statements of equality here. The number of first-class passengers plus the number of tourist-class passengers equals 52.

$$\text{(a)} \quad N_F + N_T = 52$$

The cost of the first-class tickets plus the cost of the tourist tickets equals 1360.

$$\text{(b)} \quad 30N_F + 25N_T = 1360$$

We will solve these equations by using elimination. We will multiply (a) by -30 to get (a′), which we then add to (b):

$$\begin{array}{lrcl} \text{(a}'\text{)} & -30N_F - 30N_T & = & -1560 \\ \text{(b)} & 30N_F + 25N_T & = & 1360 \\ \hline & -5N_T & = & -200 \end{array} \longrightarrow \frac{-5N_T}{-5} = \frac{-200}{-5} \longrightarrow N_T = 40$$

So $\boldsymbol{N_T = 40}$, and $\boldsymbol{N_F = 12}$ because there were 52 in all.

example 92.2 Wataksha's dress shop sold less expensive dresses for \$20 each and more expensive ones for \$45 each. They took in \$1375 and sold 20 more of the less expensive dresses than the more expensive dresses. How many of each kind did they sell?

solution Again we have statements of equality. The first is that the value of the less expensive dresses plus the value of the expensive dresses equals \$1375.

$$\text{(a)} \quad 20N_C + 45N_E = 1375$$

The other is that they sold 20 more of the less expensive dresses. Thus, we add 20 to the number of expensive dresses to get a statement of equality.

$$\text{(b)} \quad N_C = N_E + 20$$

We will use substitution to solve.

$$\begin{aligned} 20(N_E + 20) + 45N_E &= 1375 && \text{substituted} \\ 20N_E + 400 + 45N_E &= 1375 && \text{multiplied} \\ 65N_E + 400 &= 1375 && \text{simplified} \\ 65N_E &= 975 && \text{added } -400 \text{ to both sides} \end{aligned}$$

$$N_E = 15 \qquad \text{divided}$$
$$N_C = 35 \qquad 20 \text{ more than } N_E$$

practice Tickets for good seats cost \$8, and tickets for the other seats cost \$3. If 18 tickets were sold for a total of \$119, how many tickets for good seats were sold?

problem set 92

1. Find the surface area of this right circular cylinder in square centimeters. What is the volume? What is the volume of a circular cone 20 meters high that has the same base? Find the volume of a sphere whose radius is 24 meters. Dimensions are in meters.

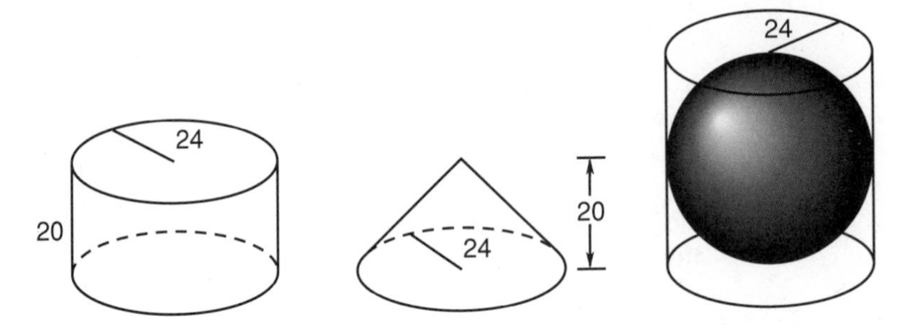

2. Oatmeal cookies cost 45 cents each and fig bars cost 30 cents each. If the group spent \$7.35, how many oatmeal cookies were purchased if they numbered 7 less than the number of fig bars purchased?

3. When the mob stormed the Bastille, only 23 percent had a weapon of any kind. If 1610 had a weapon, how many were in the mob?

4. Find three consecutive odd integers such that 4 times the sum of the first two is 62 less than the product of -30 and the third.

5. Simplify: $\dfrac{(0.0006 \times 10^{-23})(300 \times 10^{14})}{90{,}000 \times 10^{25}}$

6. Find the equations of lines (a) and (b).

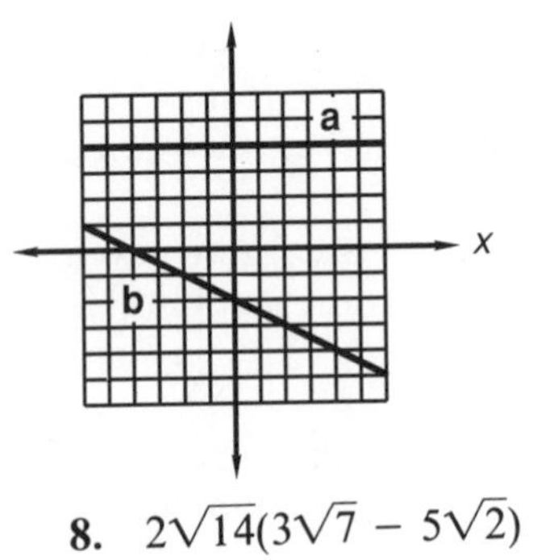

Simplify:

7. $3\sqrt{5}(5\sqrt{10} - 2\sqrt{5})$

8. $2\sqrt{14}(3\sqrt{7} - 5\sqrt{2})$

Divide:

9. $(x^2 - x - 6) \div (x + 2)$

10. $(3x^3 - 1) \div (x + 4)$

Solve by factoring:

11. $x^2 - 12x + 35 = 0$

12. $-35 = x^2 + 12x$

13. $x^2 = 12x - 32$

14. $17x = -x^2 - 60$

15. $4x^2 - 9 = 0$

16. $-49 = -9p^2$

17. $x^2 + 25 = -10x$

18. $x^2 - 11x + 24 = 0$

19. $-9x^2 + 4 = 0$

Solve by graphing:

20. $\begin{cases} y = 2x - 2 \\ y = -x + 4 \end{cases}$

Use elimination to solve:

21. $\begin{cases} 3x + 4y = 28 \\ 2x - 3y = -4 \end{cases}$

22. $2\frac{1}{4}$ is what part of $1\frac{2}{5}$?

23. Some of the sounds were susurrant, but $\frac{3}{17}$ of the sounds were plangent. If 2800 sounds were susurrant, how many sounds were there in all?

24. $R_M T_M + 10 = R_T T_T$, $R_M = 20$, $R_T = 55$, $T_M + T_T = 7$. Find T_M and T_T.

25. Solve: $-2(-x^0 - 3x^0) = -2(x + 5)$

26. Simplify: $-3^0 - |-3^0| - 3^2 + (-3)^2$

27. Evaluate: $|-x^2| + (-x)(-y)$ if $x = -3$ and $y = 4$

28. Simplify: (a) $\dfrac{1}{-3^{-2}}$ (b) $\dfrac{1}{(-3)^{-2}}$ (c) $-(-3)^{-2}$

Simplify:

29. $\dfrac{\dfrac{xy}{a} + \dfrac{1}{y}}{\dfrac{x}{y} - \dfrac{1}{a}}$

30. $\dfrac{a^4b^4(2ab)^2}{(3b^{-2})^{-2}}$

LESSON 93 Slope-intercept method of graphing

Thus far, we have graphed lines by finding ordered pairs of x and y that lie on the line. To graph $y = -\frac{3}{5}x + 2$, we choose values for x and write them in a table.

x	0	5	−5
y			

Then we use each of these numbers one at a time in the equation and find the corresponding values of y.

x	0	5	−5
y	2	−1	5

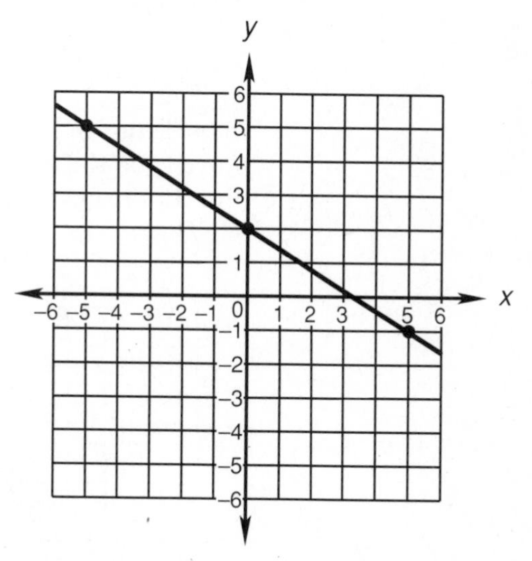

We finish by graphing the ordered pairs on the coordinate system at the right and drawing the line.

This method is dependable, but it is time-consuming. We can use the slope and the intercept of the line to get an accurate graph in less time. We will demonstrate this method by graphing the same line again.

example 93.1 Graph $y = -\frac{3}{5}x + 2$.

solution We begin by writing the slope in the form of a fraction that has a positive denominator. If we do this, the denominator will be +5 and the numerator will be −3.

$$y = \frac{-3}{+5}x + 2$$

Now we will graph the line in three steps as shown below. As the first step, we graph the intercept (0, 2) in the left-hand figure.

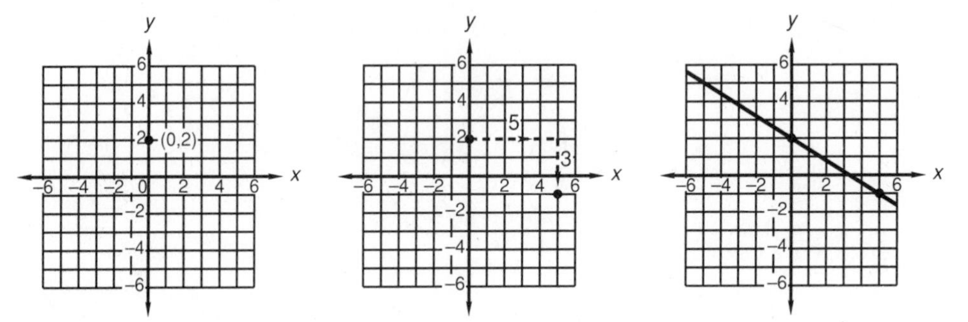

Now in the center figure, from the intercept we move to the right (the positive x direction) a distance of 5 (the denominator of the slope). Then we move up or down the distance indicated by the numerator of the slope. We move down 3 since our numerator is −3. We graph this new point, and in the figure on the right we draw the line through the two points.

example 93.2 Use the slope-intercept method to graph the equation $x - 2y = 4$.

solution As the first step we write the equation in slope-intercept form by solving for y.

$$x - 2y = 4 \longrightarrow -2y = -x + 4 \longrightarrow 2y = x - 4 \longrightarrow y = \frac{1}{2}x - 2$$

Now we write the slope $\frac{1}{2}$ as a fraction with a positive denominator.

$$y = \frac{+1}{+2}x - 2$$

In the figure on the left we graph the intercept (0, −2). In the figure in the middle we move from the intercept an x distance of +2 (to the right) and a y distance of +1 (up). In the figure on the right we draw the line through the two points.

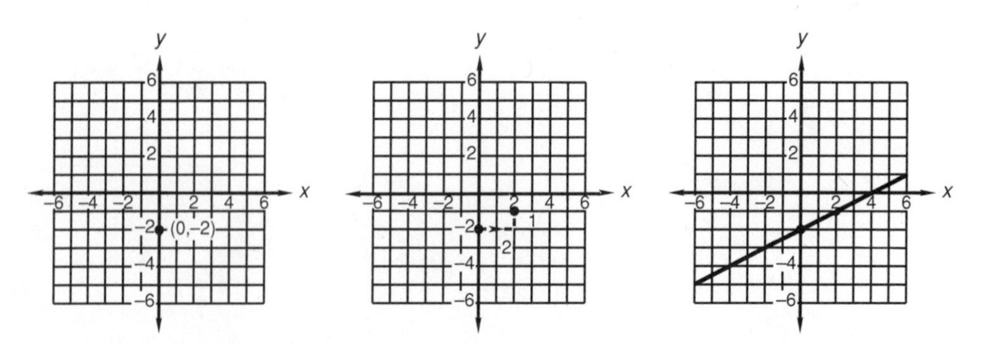

When the points are close together as in this case, it is difficult to draw the line accurately. To get another point, we multiply the denominator and the numerator of the slope by a convenient integer and use the new form of the slope to get the second point. For the line under discussion, we will multiply the slope by $\frac{2}{2}$ and get

$$\frac{+1}{+2} \cdot \frac{(2)}{(2)} \longrightarrow \frac{+2}{+4}$$

In the figures below we use the same intercept but move an x distance of $+4$ and a y distance of $+2$ to find the new point.

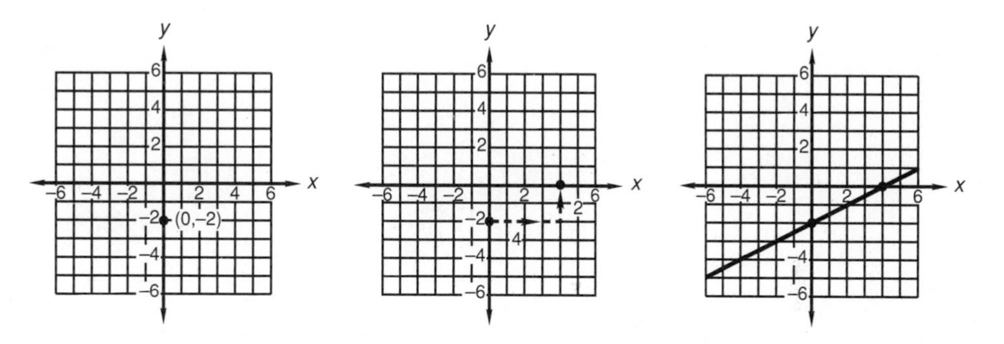

practice Use the slope-intercept method to graph the equation:

a. $y = -\frac{2}{3}x + 2$

b. $5 + 3y = x$

problem set 93

1. The total value of the pennies and nickels was $14.50. Emet and Callaway counted the coins and found that the total was 450 coins. How many coins of each type did they have?

2. Ice cream bars cost 30 cents and whifferdils cost 50 cents. Gary and Cavender treated all the kids, and they spent $13.50. How many kids had whifferdils if these numbered 5 less than those who had ice cream bars?

3. Most of the students thought that they were chic, but only 40 percent could pronounce the word correctly. If 1440 could pronounce *chic* correctly, how many mispronounced this word?

4. Find four consecutive odd integers such that the opposite of the sum of the first two is 4 greater than the product of the fourth and -4.

5. What is the volume of the cylinder whose base is the figure shown and whose sides are 5 inches high? If the cylinder is a right cylinder, what is its surface area? What is the volume of a cone 5 inches high that has the same base? Dimensions are in inches.

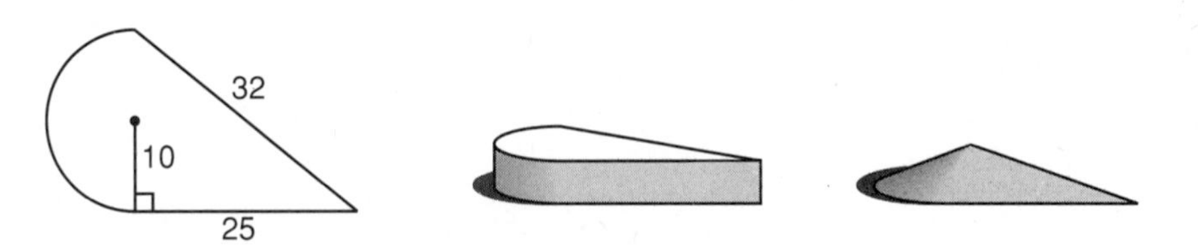

6. For what operations is the set of negative irrational numbers closed?

Use the slope-intercept method to graph the following lines:

7. $y = -\frac{3}{2}x + 3$

8. $y = -\frac{1}{2}x + 2$

9. Simplify: $\dfrac{(4000 \times 10^{-23})(0.00035 \times 10^{15})}{5000 \times 10^{5}}$

10. Find the equations of lines (a) and (b).

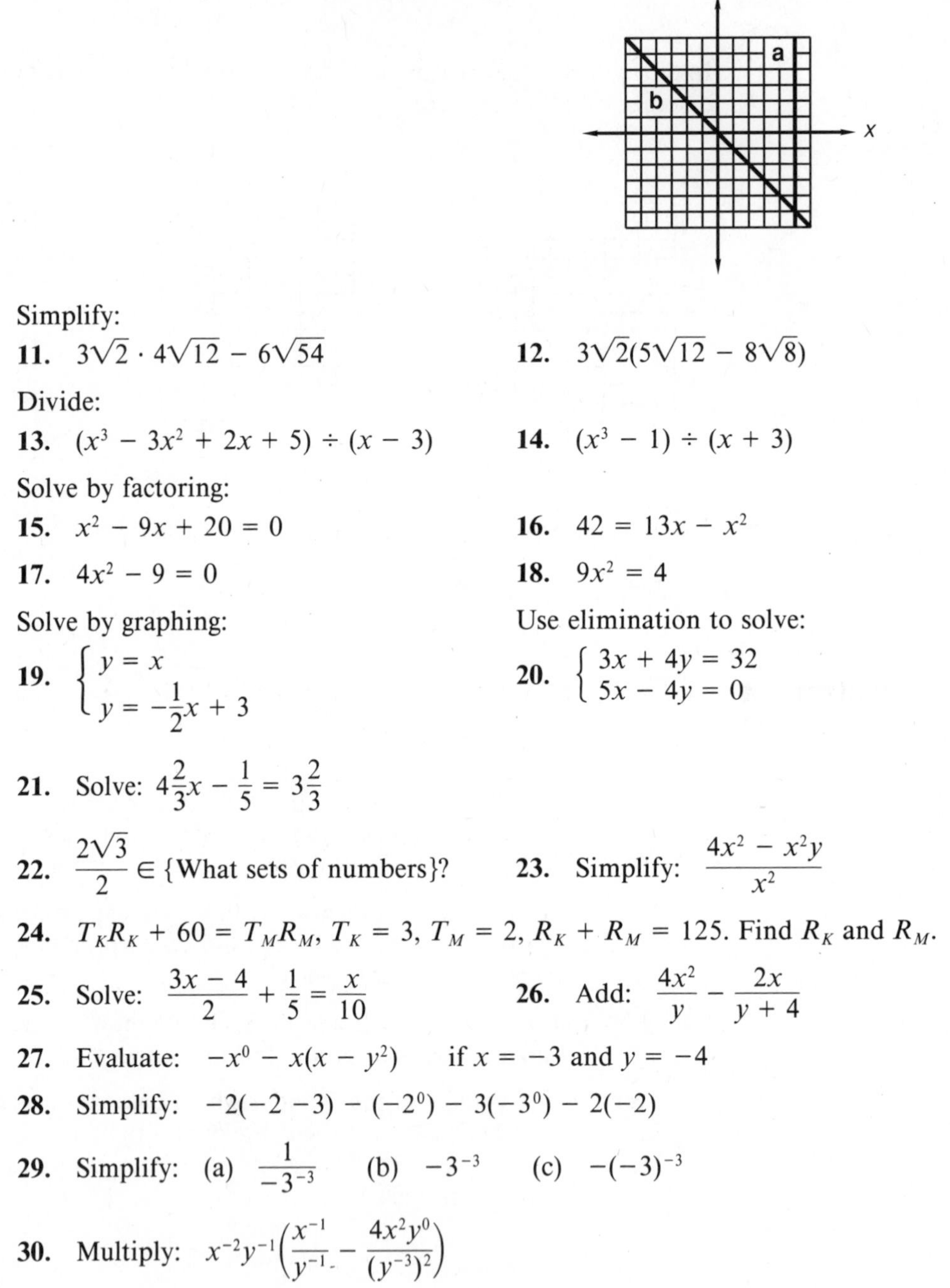

Simplify:

11. $3\sqrt{2} \cdot 4\sqrt{12} - 6\sqrt{54}$

12. $3\sqrt{2}(5\sqrt{12} - 8\sqrt{8})$

Divide:

13. $(x^3 - 3x^2 + 2x + 5) \div (x - 3)$

14. $(x^3 - 1) \div (x + 3)$

Solve by factoring:

15. $x^2 - 9x + 20 = 0$

16. $42 = 13x - x^2$

17. $4x^2 - 9 = 0$

18. $9x^2 = 4$

Solve by graphing:

19. $\begin{cases} y = x \\ y = -\frac{1}{2}x + 3 \end{cases}$

Use elimination to solve:

20. $\begin{cases} 3x + 4y = 32 \\ 5x - 4y = 0 \end{cases}$

21. Solve: $4\frac{2}{3}x - \frac{1}{5} = 3\frac{2}{3}$

22. $\dfrac{2\sqrt{3}}{2} \in$ {What sets of numbers}?

23. Simplify: $\dfrac{4x^2 - x^2y}{x^2}$

24. $T_K R_K + 60 = T_M R_M$, $T_K = 3$, $T_M = 2$, $R_K + R_M = 125$. Find R_K and R_M.

25. Solve: $\dfrac{3x - 4}{2} + \dfrac{1}{5} = \dfrac{x}{10}$

26. Add: $\dfrac{4x^2}{y} - \dfrac{2x}{y + 4}$

27. Evaluate: $-x^0 - x(x - y^2)$ if $x = -3$ and $y = -4$

28. Simplify: $-2(-2 - 3) - (-2^0) - 3(-3^0) - 2(-2)$

29. Simplify: (a) $\dfrac{1}{-3^{-3}}$ (b) -3^{-3} (c) $-(-3)^{-3}$

30. Multiply: $x^{-2}y^{-1}\left(\dfrac{x^{-1}}{y^{-1}} - \dfrac{4x^2y^0}{(y^{-3})^2}\right)$

LESSON 94 Word problems with two statements of equality

The coin problems and the general value problems we have studied thus far have contained two statements about quantities that are equal. We have solved these problems by using two equations in two unknowns (variables). Many other problems contain two statements of equality and are solved the same way. We will look at some of these in this lesson. We will write the equations and give the answers. Use substitution or elimination to see if you get the same answers.

example 94.1 Together Charles and Nelle picked 92 quarts of berries. If Charles picked 6 more quarts than Nelle picked, how many quarts did each of them pick?

solution (a) The number Charles picked plus the number Nelle picked equaled 92.

$$N_C + N_N = 92$$

(b) The number Nelle picked plus 6 equaled the number Charles picked. **To avoid adding 6 to the wrong side, we begin with an equation that we know is incorrect.**

$$N_N = N_C \qquad \text{incorrect}$$

Charles picked 6 more than Nelle picked, so we add 6 to the number that Nelle picked to get the correct statement of equality.

$$N_N + 6 = N_C \qquad \text{correct}$$

ANSWER $\mathbf{N_C = 49}$ $\mathbf{N_N = 43}$

example 94.2 The number of boys in Sarah's class exceeded the number of girls by 7. If there were a total of 29 pupils in the class, how many were boys and how many were girls?

solution (a) The number of boys exceeded the number of girls by 7. Thus we add 7 to the number of girls.

$$N_B = N_G \qquad \text{incorrect}$$

$$N_B = N_G + 7 \qquad \text{correct}$$

(b) The number of boys plus the number of girls equaled 29.

$$N_B + N_G = 29$$

ANSWER $\mathbf{N_B = 18}$ $\mathbf{N_G = 11}$

example 94.3 Phillip cut a 38-meter rope into two pieces. The long piece was 9 meters longer than the short piece. What were the two lengths?

solution (a) The length of the long piece plus the length of the short piece equaled 38 meters.

$$L + S = 38$$

(b) The long piece was 9 meters longer than the short piece, so we add 9 to S.

$$L = S \qquad \text{incorrect}$$

$$S + 9 = L \qquad \text{correct}$$

ANSWER $\mathbf{L = 23.5}$ $\mathbf{S = 14.5}$

example 94.4 The sum of two numbers is 72. The difference of the numbers is 26. What are the numbers?

solution (a) The large number plus the small number equals 72.

$$L + S = 72$$

(b) The large number minus the small number equals 26.

$$L - S = 26$$

ANSWER $\mathbf{L = 49 \qquad S = 23}$

example 94.5 The greater of two numbers is 16 greater than the smaller. When added together, their sum is 4 less than 3 times the smaller. What are the numbers?

solution (a) The greater number is 16 greater than the smaller.

$$G = S \qquad \text{incorrect}$$

$$G - 16 = S \qquad \text{correct}$$

(b) The sum is 4 less than 3 times the smaller.

$$G + S + 4 = 3S$$

ANSWER $\mathbf{G = 36 \qquad S = 20}$

example 94.6 The ratio of two numbers is 5 to 4 and the sum of the numbers is 63. What are the numbers?

solution (a) The ratio of the numbers is 5 to 4.

$$\frac{N_1}{N_2} = \frac{5}{4}$$

(b) The sum of the numbers is 63.

$$N_1 + N_2 = 63$$

ANSWER $\mathbf{N_1 = 35 \qquad N_2 = 28}$

practice

a. The sum of two numbers is 98. The difference is 40. What are the numbers?

b. The number of girls in Marvin's class exceeded the number of boys by 11. If there were 37 pupils in the class, how many were girls and how many were boys?

problem set 94

1. Student grades were based on a weighted average. The final test was weighted at 4 times the weight of a weekly test. If Marion had an average of 92.4 on the 10 regular weekly tests and scored 84 on the final, what was her overall average?

2. Use 12 unit multipliers to convert 1000 cubic yards to cubic meters.

3. Expand: $(3x + 2y)^2$

4. Gertrude cut a 76-meter rope into two pieces. The long piece was 12 meters longer than the short piece. How long was each piece?

5. Shields and Jim sold tickets to the basketball game. Good seats were $5 each, and poor seats cost only \$2 each. If 210 people attended and paid \$660, how many people bought good seats?

Find the solution to the following systems of equations by graphing. Use the slope-intercept method from Lesson 93.

6. $\begin{cases} 3x + y = 6 \\ x = -y \end{cases}$

7. $\begin{cases} y = \frac{2}{3}x - 3 \\ y = -x + 2 \end{cases}$

8. Simplify: $\dfrac{(3000 \times 10^{-5})(0.004 \times 10^{10})}{(200 \times 10^{14})(0.000002)}$

9. Find the equations of lines (a) and (b).

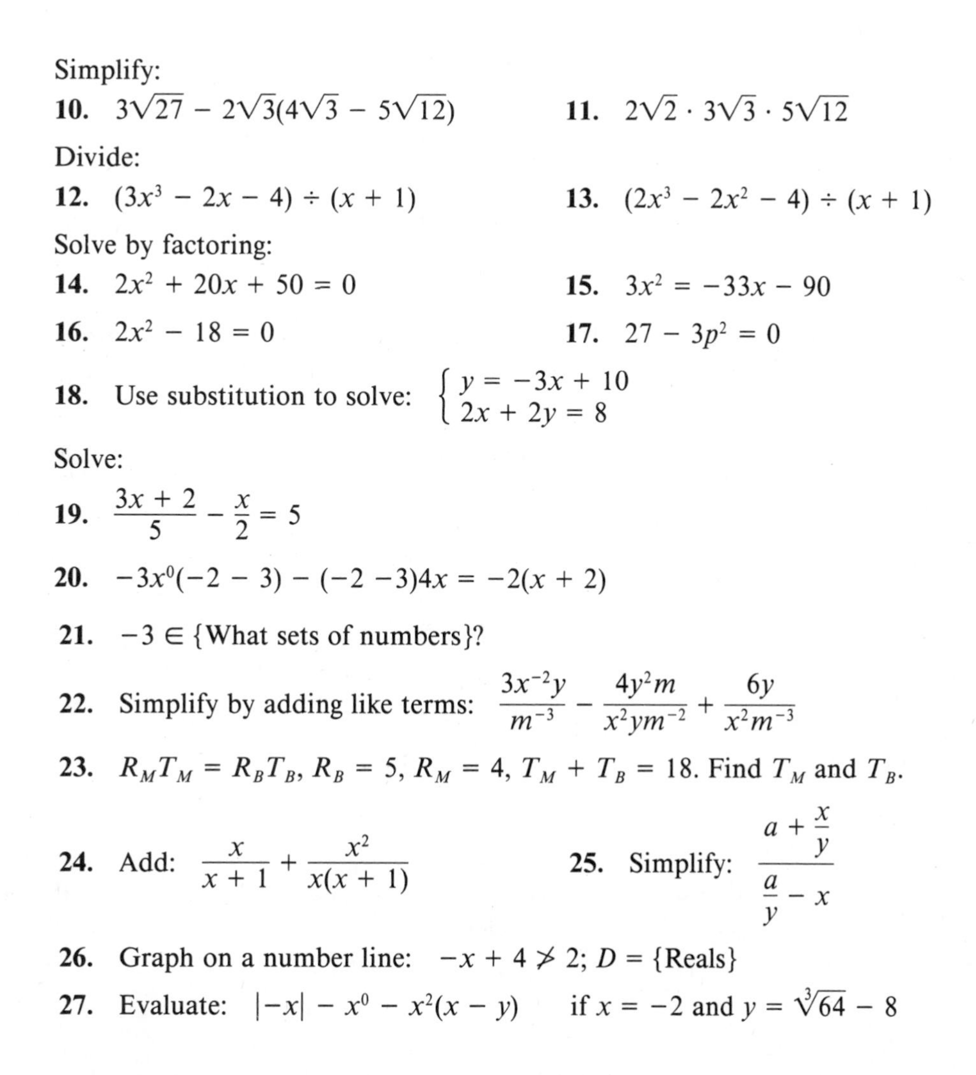

Simplify:

10. $3\sqrt{27} - 2\sqrt{3}(4\sqrt{3} - 5\sqrt{12})$

11. $2\sqrt{2} \cdot 3\sqrt{3} \cdot 5\sqrt{12}$

Divide:

12. $(3x^3 - 2x - 4) \div (x + 1)$

13. $(2x^3 - 2x^2 - 4) \div (x + 1)$

Solve by factoring:

14. $2x^2 + 20x + 50 = 0$

15. $3x^2 = -33x - 90$

16. $2x^2 - 18 = 0$

17. $27 - 3p^2 = 0$

18. Use substitution to solve: $\begin{cases} y = -3x + 10 \\ 2x + 2y = 8 \end{cases}$

Solve:

19. $\dfrac{3x + 2}{5} - \dfrac{x}{2} = 5$

20. $-3x^0(-2 - 3) - (-2 - 3)4x = -2(x + 2)$

21. $-3 \in$ {What sets of numbers}?

22. Simplify by adding like terms: $\dfrac{3x^{-2}y}{m^{-3}} - \dfrac{4y^2m}{x^2ym^{-2}} + \dfrac{6y}{x^2m^{-3}}$

23. $R_M T_M = R_B T_B$, $R_B = 5$, $R_M = 4$, $T_M + T_B = 18$. Find T_M and T_B.

24. Add: $\dfrac{x}{x + 1} + \dfrac{x^2}{x(x + 1)}$

25. Simplify: $\dfrac{a + \frac{x}{y}}{\frac{a}{y} - x}$

26. Graph on a number line: $-x + 4 \not> 2$; $D = \{\text{Reals}\}$

27. Evaluate: $|-x| - x^0 - x^2(x - y)$ if $x = -2$ and $y = \sqrt[3]{64} - 8$

Simplify:

28. $-|-2| - |-2^0| - (-2 - 4)$

29. (a) -3^{-2} (b) $(-3)^{-2}$ (c) $-(-3)^{-2}$

30. $\dfrac{x^{-2}(y^{-2})^2(y^0)^2}{xy^{-2}(x^{-2}y)^{-2}}$

LESSON 95 Multiplicative property of inequality

With one glaring exception, the rules for solving inequalities are the same as the rules for solving equations. The following two rules apply to both equalities and inequalities.

> ADDITION RULE
>
> The same quantity can be added to both sides of an equation or inequality without changing the solution set of the equation or inequality.

> MULTIPLICATION RULE (POSITIVE)
>
> Every term on both sides of an equation or inequality can be multiplied by the same positive number without changing the solution set of the equation or inequality.

The glaring exception occurs when we multiply by a negative number! The truth of a statement of equality is not altered by multiplying by a negative number.

$$5 = 2 + 3 \qquad \text{True}$$

Now multiply every term by -2 and get

$$5(-2) = 2(-2) + 3(-2) \longrightarrow -10 = -4 - 6 \qquad \text{Still true!}$$

But the truth of a statement of inequality is altered!

$$8 > 5 \qquad \text{True}$$

Now we multiply every term by -2 and get

$$-16 > -10 \qquad \text{Now false!}$$

Thus, when every term on both sides of an inequality is multiplied by a negative number, the inequality symbol must be reversed so that the solution set of the inequality will not be changed. To show this, we will repeat the problem.

$$8 > 5 \qquad \text{True}$$

Now we multiply every term by -2 and reverse the inequality symbol.

$$-16 < -10 \qquad \text{Still true!}$$

example 95.1 Graph the solution: $-x \geq 2$; $D = \{\text{Reals}\}$

solution We solve the given inequality for $+x$ by multiplying both sides by (-1) and **reversing the inequality symbol.**

$-x \geq 2$	original inequality
$(-1)(-x) \leq (-1)(2)$	multiplied by -1 and reversed symbol
$x \leq -2$	simplified

Thus we want to graph the solution of $x \leq -2$.

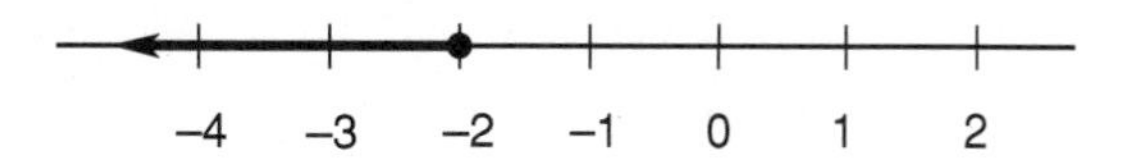

The graph indicates that the number -2 and all real numbers less than -2 satisfy the stated inequality.

example 95.2 Graph the solution: $4 - x \leq 6$; $D = \{\text{Integers}\}$

solution We first isolate $-x$ by adding -4 to both sides. **Note that we do not reverse the inequality symbol when we *add* a negative quantity to both sides of an inequality.**

$$\begin{array}{r} 4 - x \leq 6 \\ \underline{-4 \qquad -4} \\ -x \leq 2 \end{array}$$

Now we **multiply** both sides by -1 and **reverse the inequality symbol** to get

$$x \geq -2$$

And if we graph $x \geq -2$ over the integers, we get

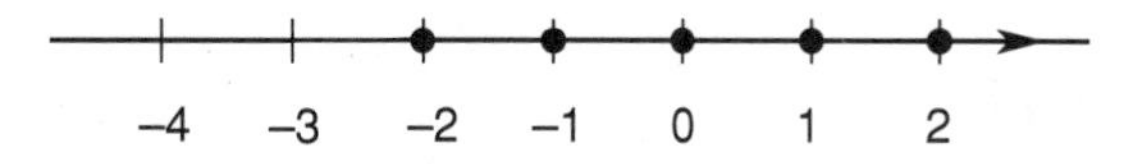

example 95.3 Graph the solution: $-3x + 4 \leq 13$; $D = \{\text{Reals}\}$

solution We add -4 to both sides to get $-3x \leq 9$ and then divide both sides by -3 (or *multiply* both sides by $-\frac{1}{3}$) **and reverse the inequality symbol** to get

$$x \geq -3$$

and we graph the solution to the inequality to indicate all real numbers that are greater than or equal to -3.

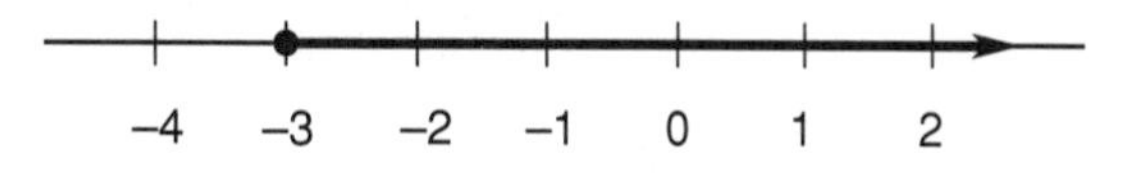

practice Graph on a number line:

a. $-x \geq 5$; $D = \{\text{Reals}\}$

b. $2 - x \not\leq 1$; $D = \{\text{Integers}\}$

c. $-3x + 5 > 1$; $D = \{\text{Integers}\}$

problem set 95

1. Wetumka had \$6.50 in dimes and quarters. If he had 5 more quarters than dimes, how many coins of each type did he have?

2. Seed corn was \$5 a bag, whereas dog food cost only \$3 a bag. Wewoka bought 50 bags and spent \$190. How many bags of dog food did she buy?

3. There were 90 more orchids on the first float than there were on the second float. If there were 630 orchids altogether, how many were on each float?

4. When the first frost came, the number of people wearing shoes jumped 180 percent. If 5600 people now wore shoes, how many people wore shoes before the frost?

5. Graph the solution to this inequality on a number line:

$$-x \geq 3; D = \{\text{Reals}\}$$

6. For what operations is the set of positive integers closed?

7. Evaluate: $-b \pm \sqrt{b^2 - 4ac}$ if $a = -3$, $c = 4$, and $b + 5 = \sqrt{16}$

Solve by graphing:

8. $\begin{cases} y = -x + 1 \\ y = 2x + 7 \end{cases}$

Use elimination to solve:

9. $\begin{cases} 4x - 3y = 14 \\ 5x - 4y = 18 \end{cases}$

10. Simplify: $\dfrac{(0.00004 \times 10^{15})(700 \times 10^{-5})}{14{,}000 \times 10^{-21}}$

11. Find the equations of lines (a) and (b).

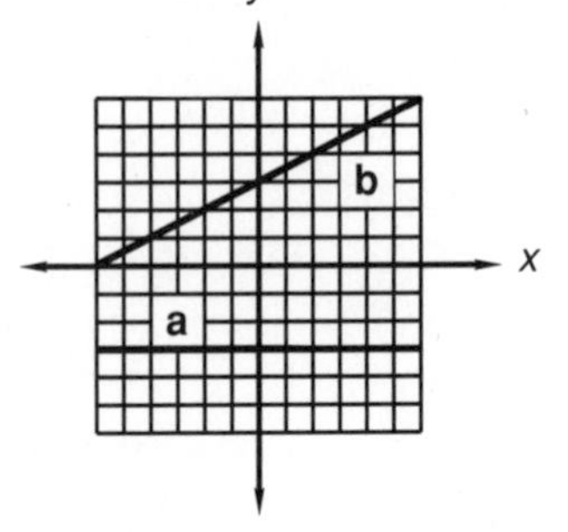

Simplify:

12. $5\sqrt{75} - 2\sqrt{108} + 5\sqrt{12}$

13. $2\sqrt{6}(3\sqrt{6} - 2\sqrt{12})$

14. Divide: $(x^4 - 2x^2 - 4) \div (x + 2)$

Solve by factoring:

15. $21 = 10x - x^2$

16. $-49 = -4x^2$

17. $32 = -x^2 - 12x$

Solve:

18. $\dfrac{4x}{3} - \dfrac{x + 1}{5} = 10$

19. $3\frac{1}{4}x - \frac{2}{3} = 7\frac{1}{8}$

20. Graph on a number line: $4 \leq x < 7$; $D = \{\text{Integers}\}$

21. $-3\frac{1}{3} \in \{\text{What sets of numbers}\}$?

22. Add: $\dfrac{x}{x + 1} + \dfrac{4}{x}$

23. Simplify by adding like terms: $5x^2ym^2 - \dfrac{3xxy^{-2}}{y^{-3}m^{-1}} + \dfrac{2xmy}{y^3y}$

24. $R_HT_H - 125 = R_OT_O$, $T_H = 2$, $T_O = 3$, $R_H + R_O = 85$. Find R_H and R_O.

25. Solve: $3x^0 - 2x^0 - 3(x^0 - 2x) = -2x(4 - 3)$

Simplify:

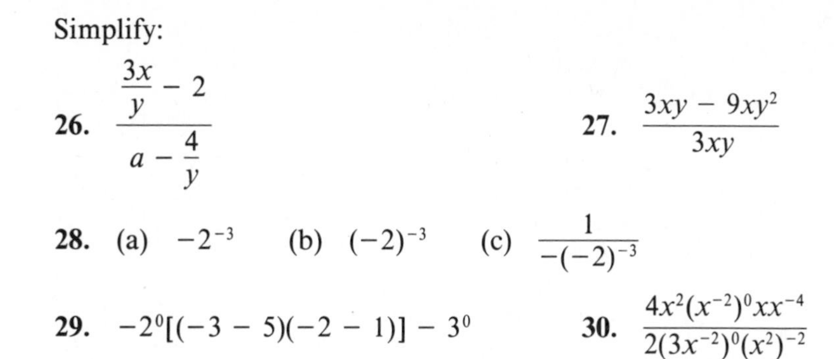

26. $\dfrac{\dfrac{3x}{y} - 2}{a - \dfrac{4}{y}}$ **27.** $\dfrac{3xy - 9xy^2}{3xy}$

28. (a) -2^{-3} (b) $(-2)^{-3}$ (c) $\dfrac{1}{-(-2)^{-3}}$

29. $-2^0[(-3 - 5)(-2 - 1)] - 3^0$ **30.** $\dfrac{4x^2(x^{-2})^0xx^{-4}}{2(3x^{-2})^0(x^2)^{-2}}$

LESSON 96 *Uniform motion problems about equal distances*

Since Lesson 82 we have been using substitution to solve systems of four simultaneous equations involving four unknowns such as

$$R_FT_F = R_ET_E \qquad T_F = 16 \qquad T_E = 12 \qquad R_E = R_F + 15$$

These equations are typical of the equations that we will learn to write in this lesson to help us solve uniform motion word problems. Uniform motion problems are so named because the statement of the problem tells about objects or things that move at a uniform rate or at an average rate.

The statements of equality made in uniform motion problems are statements of equality that concern **distance,** statements of equality that concern **rate,** and statements of equality that concern **time,** and use the relationship that

$$\text{Distance} = \text{rate} \times \text{time} \qquad \text{or} \qquad D = RT$$

The statements about rate and time are not difficult to locate in the wording of the problem, but the beginner often has trouble identifying the statement that defines the relationship that concerns distance. **Since the distance equation is the troublesome one, we will consider this equation to be the key equation for this type of problem, and we will always write this equation first.** Then we will write the equations that concern time and the equations that concern rate. **When we have as many independent equations as we have variables, we will use the substitution method and/or the elimination method to solve for the variables.**

The statements of the distances discussed in the problems can be represented graphically by drawing diagrams in which arrows represent the distances. The problems in this book will usually describe two distances. In this lesson, we will investigate problems that describe two equal distances.

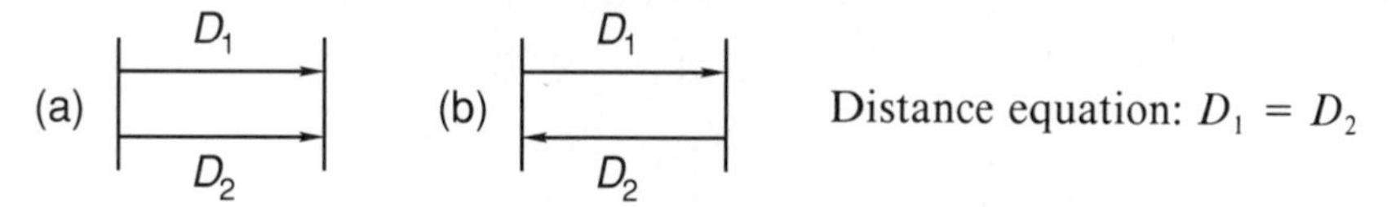

Distance equation: $D_1 = D_2$

One of these diagrams will result when the problem states that two distances are equal. In (a) the objects traveled in the same direction, and in (b) they traveled in opposite directions. Both diagrams give us the same equation.

example 96.1 On Tuesday the express train made the trip in 12 hours. On Wednesday the freight train made the same trip in 16 hours. Find the rate of each train if the rate of the freight train was 15 kilometers per hour less than the rate of the express train.

solution We read the problem and disregard statements about time and rate and look for the statement about distance. This information allows us to draw the diagram and from the diagram get the equation

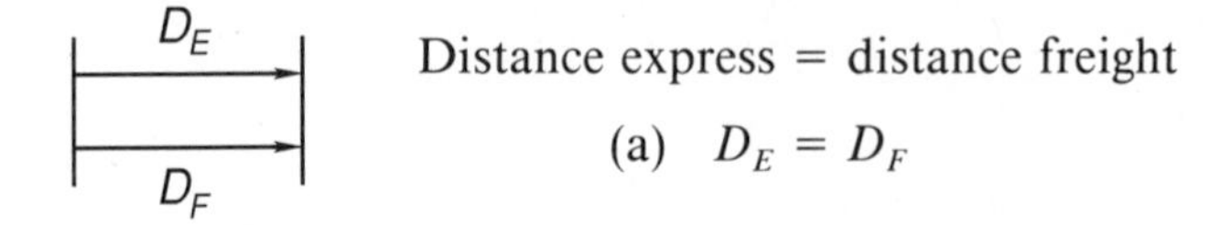

Distance express = distance freight

(a) $D_E = D_F$

This is the distance equation, which is the key equation for this uniform motion problem. Now, the distance the express traveled equals the rate of the express times the time the express traveled, or

$$D_E = R_E T_E$$

and the distance the freight traveled equals the rate of the freight times the time the freight traveled, or

$$D_F = R_F T_F$$

Now, if we substitute $R_E T_E$ for D_E and $R_F T_F$ for D_F in equation (a), we get

$$\boldsymbol{R_E T_E = R_F T_F}$$

which is the distance equation for this problem in final form.

The statement of the problem gives the time of the express as 12 hours and the time of the freight as 16 hours so

$$\boldsymbol{T_E = 12 \qquad T_F = 16}$$

Now we have three equations but have four unknowns, R_E, T_E, R_F, and T_F. **Thus, we need one more equation so that the number of equations will equal the number of unknowns.** We get the final equation from the statement in the problem concerning rates, which says that the rate of the express is 15 kilometers per hour greater than the rate of the freight. **Writing this equation is tricky.** In an effort to avoid the common error of adding 15 to the wrong side, we first write

$$R_F = R_E \qquad \text{incorrect}$$

which we know is incorrect because the rate of the express is greater than the rate of the freight. We add 15 to R_F so that the equation will have equal quantities on both sides.

$$\boldsymbol{R_E = R_F + 15}$$

We have found four equations in four unknowns,

$$R_E T_E = R_F T_F \qquad T_E = 12 \qquad T_F = 16 \qquad R_E = R_F + 15$$

and we use the substitution method to solve for R_F and R_E.

$R_E T_E = R_F T_F$	equation
$(R_F + 15)12 = R_F(16)$	substituted
$12R_F + 180 = 16R_F$	multiplied
$-4R_F = -180$	simplified
$4R_F = 180$	multiplied by -1
$\boldsymbol{R_F = 45}$ **kilometers per hour**	divided

Since $R_E = R_F + 15$,

$$\mathbf{R_E = 60 \text{ kilometers per hour}}$$

example 96.2 The members of the girls club hiked to Lake Tenkiller at 2 miles per hour. Mr. Ali gave them a ride back home at 12 miles per hour. Find their hiking time if it was 5 hours longer than their riding time. How far was it to Lake Tenkiller?

solution We begin by drawing a diagram of the distances traveled and writing the distance equation.

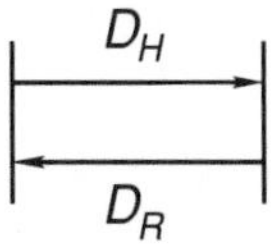

D_H / D_R

$$D_H = D_R \qquad \text{so} \qquad R_H T_H = R_R T_R$$

Next we reread the problem and write the other three equations.

$$R_H = 2 \qquad R_R = 12 \qquad T_H = T_R + 5$$

Now we substitute these equations into the distance equation and solve.

$$2(T_R + 5) = 12T_R \qquad \text{substituted}$$

$$2T_R + 10 = 12T_R \qquad \text{multiplied}$$

$$10 = 10T_R \qquad \text{added } -2T_R \text{ to both sides}$$

$$\mathbf{T_R = 1} \qquad \text{and so} \qquad \mathbf{T_H = 6}$$

Thus the distance is either 2 times 6 or 12 times 1, both of which equal **12 miles.**

example 96.3 Durant drove to the oasis in 2 hours and Madill walked to the oasis in 10 hours. How far is it to the oasis if Durant drove 16 miles per hour faster than Madill walked?

solution We begin by drawing a distance diagram from which we can get the distance equation.

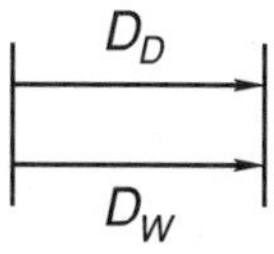

D_D / D_W

$$D_D = D_W \qquad \text{so} \qquad R_D T_D = R_W T_W$$

Now we reread the problem to get the time and rate equations.

$$T_D = 2 \qquad T_W = 10 \qquad R_D = R_W + 16$$

Now we solve.

$$(R_W + 16)2 = R_W(10) \qquad \text{substituted}$$

$$2R_W + 32 = 10R_W \qquad \text{multiplied}$$

$$32 = 8R_W \qquad \text{added } -2R_W \text{ to both sides}$$

$$\mathbf{4 = R_W} \qquad \text{divided}$$

Thus $R_D = 20$, and since $T_D = 2$, the distance equals 2 times 20, or **40 miles.**

practice **a.** Candide and Pangloss walked to the site of the disaster in 9 hours. The next morning they jogged back home in 3 hours. How far did they jog if they jogged 4 miles per hour faster than they walked?

b. On Monday, Voltaire drove to town at 60 miles per hour. On Tuesday, he drove to town at 40 miles per hour. If the total traveling time for both trips was 15 hours, how far was it to town?

problem set 96

1. The sum of four numbers is 12,000.16. The first three numbers are 4200, 1700, and 3400. Find the average of the four numbers.
2. Use 12 unit multipliers to convert 10,000 cubic yards to cubic meters.
3. In the morning the passenger train made the trip in 3 hours. In the afternoon the freight train made the same trip in 7 hours. Find the rate of each if the rate of the freight train was 40 miles per hour less than the rate of the passenger train.
4. Vanessa rode her bike to the conclave at 10 kilometers per hour and then walked back to school at 4 kilometers per hour. If the round trip took her 14 hours, how far was it to the site of the conclave?
5. Fustian phrases obscured 0.62 of the points the speaker tried to make. If he tried to make 50 points, how many was the audience able to comprehend?
6. Solve: $p - (-p) - 5(p - 3) - (2p - 5) = 3(p + 2p)$
7. Graph on a number line: $-4x + 4 \geq 8$; $D = \{\text{Reals}\}$

Solve by graphing:

8. $\begin{cases} y = -2x + 4 \\ y = -2 \end{cases}$

Use elimination to solve:

9. $\begin{cases} 3x + y = 20 \\ 2x - 3y = -5 \end{cases}$

10. Simplify: $\dfrac{0.000030 \times 10^{-18}}{(5000 \times 10^{-14})(300 \times 10^{5})}$

11. Find the equations of line (a) and (b).

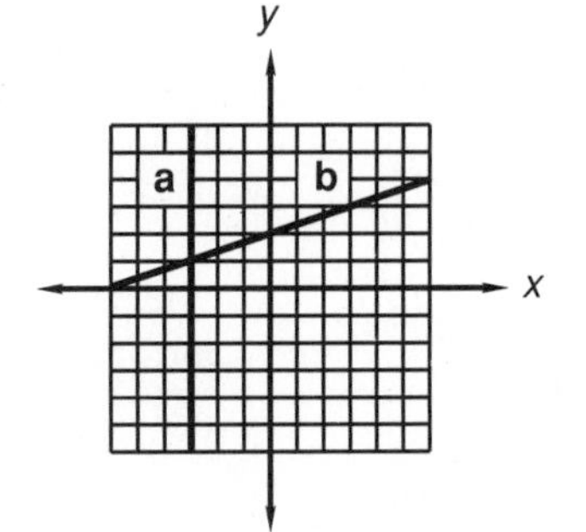

Simplify:

12. $3\sqrt{45} - 2\sqrt{180} + 2\sqrt{80}$
13. $3\sqrt{2}(4\sqrt{20} - 3\sqrt{2})$
14. Divide: $(x^3 - 4) \div (x - 5)$

Solve by factoring:

15. $x^2 = -6x - 8$
16. $9 = 4x^2$
17. $x^2 = -12x - 32$

Solve:

18. $\dfrac{x}{5} - \dfrac{4 + x}{7} = 5$
19. $2\frac{1}{8}x - 3\frac{1}{4} = 2\frac{1}{16}$
20. $7\frac{2}{3}$ is what fraction of $\frac{5}{6}$?
21. True or false? $\{\text{Reals}\} \subset \{\text{Rationals}\}$
22. $R_W T_W + 200 = R_R T_R$, $R_W = 40$, $R_R = 60$, $T_W + T_R = 10$. Find T_W and T_R.

23. Add: $\dfrac{2x}{5} + \dfrac{3x + 5}{6x}$

Simplify:

24. $\dfrac{\dfrac{3a}{b} - 2b}{b - \dfrac{4}{b}}$ **25.** $\dfrac{4 + 4k}{4}$

26. Evaluate: $|-x^2| - |x| + x(x - y^0)$ if $x = -\sqrt{9}$ and $5y = 20$

Simplify:

27. $-3^0 - [(-3 + 5) - (-2 - 5)]$

28. (a) -2^{-2} (b) $\dfrac{1}{-2^{-2}}$ (c) $-(-2)^{-2}$ (d) $\sqrt[7]{-128}$

29. Multiply: $\left(\dfrac{x^4y}{a^2} - \dfrac{x^{-3}y^2}{ya^{-4}}\right)\dfrac{x^{-4}y}{a^{-2}}$

30. Find the volume of a prism in cubic centimeters whose base is the figure shown and whose sides are 1 meter high. If the prism is a right prism, what is the surface area? What is the volume of a pyramid 1 meter high that has the base shown? Dimensions are in centimeters.

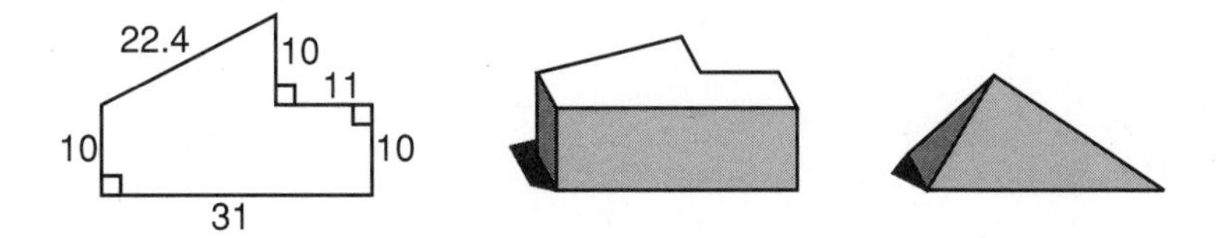

LESSON 97 Properties of the set of real numbers

97.A big words with small meanings

Every discipline has big words with small meanings. These words are not intimidating to people who have used them for a long time. When talking to another doctor, a doctor will use words such as mandible, clavicle, and scapula. The doctor will use the words jawbone, collar bone, and shoulder blade to convey the same meanings when talking to a patient. In mathematics, we use the words **identity, inverse, associative, commutative,** and **distributive** to describe five properties of the set of real numbers. These words have very simple meanings.

97.B identity and inverse

Think of a number. Then add 0 to the number. The result is **identically** the number you thought of. Then multiply your number by 1. The result is **identically** the number you thought of. We give big names with small meanings to the number 0 and the number 1. We say that the number 0 is the **identity for addition** and the number 1 is the **identity for multiplication.** That's all there is to it.

The identity for addition (the additive identity) is the number 0 because

$$4 + 0 = 4 \qquad \text{and} \qquad -13 + 0 = -13$$

The identity for multiplication (the multiplicative identity) is the number 1 because

$$4 \cdot 1 = 4 \qquad \text{and} \qquad (-13) \cdot (1) = -13$$

Pick another number. What number do you add to your number so that the sum is the number 0? The opposite of the number, of course. If the number you picked was +4, you would add −4 to get a sum of 0. If the number you picked was −13, you would add 13 to get a sum of 0 because +13 is the opposite of −13.

We say that −4 is the **additive inverse** of +4 and that +13 is the additive inverse of −13. **Every number has its own special additive inverse.** If you add a number and its additive inverse, the result is zero, which is the additive identity.

The additive inverse of +37 is −37.

The additive inverse of −163 is +163.

The additive inverse of −1485.6 is +1485.6.

The additive inverse of $-p$ is $+p$.

If your number was +4, what number would you multiply by to get a product of +1? You would multiply by 1 over +4, of course. If your number was −13, you would multiply by 1 over −13.

$$4 \cdot \frac{1}{4} = 1 \qquad (-13) \times \left(\frac{1}{-13}\right) = 1$$

We say that the **multiplicative inverse** of −4 is 1 over minus 4, and the multiplicative inverse of −13 is 1 over −13. We see that the multiplicative inverse of a number is just the number "upside down." **Thus, the multiplicative inverse of a number is the same thing as the reciprocal of the number. Every number has its own special multiplicative inverse.** The product of a number and its multiplicative inverse **is always the multiplicative identity.**

The multiplicative inverse of −5 is $-\frac{1}{5}$.

The multiplicative inverse of $-\frac{1}{5}$ is −5.

The multiplicative inverse of 42 is $\frac{1}{42}$.

The multiplicative inverse of $\frac{1}{42}$ is 42.

The multiplicative inverse of $-p$ is $-\frac{1}{p}$.

97.C commutative property

The sum of 2 and 3 equals the sum of 3 and 2. The sum of a and b equals the sum of b and a.

$$2 + 3 = 3 + 2 \qquad a + b = b + a$$

The product of 2 and 3 equals the product of 3 and 2. The product of a and b equals the product of b and a.

$$2 \cdot 3 = 3 \cdot 2 \qquad a \cdot b = b \cdot a$$

If we exchange the order of the numbers when we add two numbers, the sum is the same. If we exchange the order of the numbers when we multiply two numbers, the product is the same. We would call these properties (peculiarities) of the set of real numbers the **exchange property for addition** and the **exchange property for multiplication** except for a historical occurrence.

The first modern universities (so called because "universal" truths were studied) were the universities of Paris and Bologna. These schools began as a collection of scholars and teachers in the 1100s. The teachers and the students came from many countries, and they all spoke Latin because all books were written in Latin. All classes were taught in Latin. The Latin word meaning "to exchange" was *commutare.* Thus, at these universities, they used a form of this word to describe the exchange properties. Today we call these properties the **commutative property of addition** and the **commutative property of multiplication.** What do these big words tell us? They just tell us that

$$4 + 2 = 2 + 4 \qquad \text{and that} \qquad 4 \cdot 2 = 2 \cdot 4$$

If we use letters instead of numbers, we can write

$$a + b = b + a \qquad \text{and} \qquad ab = ba$$

97.D associative[†] property

We remember that addition is a binary operation because only two numbers can be added in one step. If we want to find the sum of three numbers, we must add twice.

$$\begin{aligned} &(4 + 2) + 3 \qquad && 4 + (2 + 3) \\ &= 6 + 3 && = 4 + 5 \\ &= 9 && = 9 \end{aligned}$$

On the left, we "associated" 4 and 2 and found that their sum was 6. Then we added 3 to 6 to get 9. On the right, we "associated" 2 and 3 and found their sum was 5. Then we added 4 to 5 to get 9. Because we can "associate" the first two numbers or the last two numbers of a three-number addition problem, we say that **real numbers are associative in the operation of addition,** and we call this fact the **associative property of addition.** Don't complain. Remember that doctors talk about mandibles, clavicles, and scapulae. Every discipline has its own jargon.

Now guess what? We can do the same thing for multiplication. Multiplication is also a binary operation because only two numbers can be multiplied in one step. If we want to find the product of three numbers, we must multiply twice.

$$\begin{aligned} &(4 \cdot 2)\, 3 \qquad && 4\,(2 \cdot 3) \\ &= 8 \cdot 3 && = 4 \cdot 6 \\ &= 24 && = 24 \end{aligned}$$

On the left we associated the first two numbers for the first multiplication, and on the right we associated the last two numbers. Because we will get the same answer both ways with any three real numbers, we say that the real numbers are **associative under the operation of multiplication,** and we call this fact the **associative property of multiplication.**

97.E distributive property

The product of a number and a sum can be found two different ways. If we consider

$$4\,(3 + 2)$$

[†]In the examples in this book we will use a modified form of the associative property and group any two consecutive addends or factors within a single pair of parentheses.

we find that we get the same answer if we begin by adding 3 + 2 and then multiplying by 4 or if we multiply 4 by 3 and multiply 4 by 2 and then add.

ADDING FIRST	MULTIPLYING FIRST
$4(3 + 2)$	$4(3 + 2)$
$= 4(5)$	$= (4)(3) + (4)(2)$
$= 20$	$= 12 + 8$
	$= 20$

When we multiply first, mathematicians say that we have **distributed** the multiplication over addition, and they call this property the **distributive property.** When we are faced with an expression that has variables inside the parentheses, such as

$$4(b + c)$$

we cannot add first because we don't know what numbers b and c represent. But we can multiply and get

$$4(b + c) = 4b + 4c$$

Before 1965, teachers would tell students to "multiply out." Now many teachers say "use the distributive property to expand." This phrase just means to "multiply out."

97.F field properties

The set of real numbers is closed under the operations of addition and multiplication. The set of real numbers has nine other properties. We say that any set of numbers that is closed under the operations of multiplication and addition and has these nine properties constitutes a **field.** There is no special significance attached to the word field. We have to use some name, and field is possibly as good a name as any other. It is the word that we use to describe a set that, when used in the operations of addition and multiplication, has these nine properties. The chart below gives the names of the properties and both a numerical and an abstract example of each property.

PROPERTIES OF A FIELD

Addition	*Name*	*Multiplication*
$4 + 0 = 4$ $a + 0 = a$	IDENTITY	$4 \cdot 1 = 4$ $a \cdot 1 = a$
$4 + (-4) = 0$ $a + (-a) = 0$	INVERSE	$4 \cdot \frac{1}{4} = 1$ $a \cdot \frac{1}{a} = 1$
$4 + (3 + 2) = (4 + 3) + 2$ $a + (b + c) = (a + b) + c$	ASSOCIATIVE PROPERTY	$(4 \cdot 3) \cdot 2 = 4 \cdot (3 \cdot 2)$ $(a \cdot b) \cdot c = a \cdot (b \cdot c)$
$4 + 2 = 2 + 4$ $a + b = b + a$	COMMUTATIVE PROPERTY	$4 \cdot 2 = 2 \cdot 4$ $ab = ba$
	DISTRIBUTIVE PROPERTY $4(3 + 2) = 4 \cdot 3 + 4 \cdot 2$ $a(b + c) = ab + ac$	

We note that there are four properties listed under addition and that the same four properties are listed under multiplication. These are the **identity,** the **inverse,** the **associative property,** and the **commutative property.** At the bottom of the chart we see the **distributive property** in the middle of the page since the **distributive property includes both addition and multiplication.**

Before the "new math" of the late 1960s, the study of the properties of the set of real numbers was reserved for college-level abstract algebra and advanced math courses. In such courses students study the properties of the set of real numbers and identify other systems that have similar properties.

In algebra I students are often confused by the study of the properties of real numbers because they cannot see why these properties are given so much attention. Why do we need to have an identity for addition? Why is it helpful for every number to have an additive inverse? These are questions that will be answered in college mathematics courses. At this point we will content ourselves with the knowledge that the concepts described by the big words are easily understood, and we will learn not to be intimidated by them. Future homework problem sets will contain problems that will permit practice in associating the big words and their small meanings.

practice

a. What is the additive inverse of -6?

b. What is the multiplicative identity for -6?

c. Use the numbers 4 and 3 to illustrate the commutative property for multiplication.

d. Use the numbers 3, 4, and 5 to illustrate the associative property.

e. Use the numbers 4, 5, and 6 to illustrate the distributive property.

problem set 97

1. Gaskin and Sloan raced to the cotton patch. Gaskin's speed was 12 kilometers per hour, while Sloan's was only 8 kilometers per hour. What was the time of each if Gaskin's time was 5 hours less than Sloan's time?

2. At noon Joyce F. headed for the lake at 30 miles per hour. Her car broke down, and she made the long walk back home at 4 miles per hour. How long did she walk if she was gone for 17 hours? How far did she walk?

3. Tickets to the carnival were \$3 for adults and \$2 for kids. Tommy was a big spender, as he took 77 people to the carnival and spent \$209 for their tickets. How many kids did he pay for?

4. Eighty percent of the children preferred the wild goose ride to the ferris wheel. If 300 children preferred the ferris wheel, how many preferred the wild goose ride?

Use the letters a, b, and c as required to state:

5. The distributive property

6. The associative property of addition

7. What is another name for the multiplicative inverse of a number?

8. Factor: $-m^3 - 11m^2 - 24m$

9. Solve: $-4(x - 2)(-2) + 3(x - 4) = 2x(4 - 2^0)$

10. Graph: $-x + 4 \leq 2$; $D = \{\text{Integers}\}$

Solve by graphing:

11. $\begin{cases} y = x + 2 \\ y = -x \end{cases}$

Use elimination to solve:

12. $\begin{cases} 5x - 2y = 18 \\ 3x + y = 24 \end{cases}$

13. Simplify: $\dfrac{(0.00042 \times 10^{-15})(300{,}000)}{(180{,}000 \times 10^{-14})(7000 \times 10^{-23})}$

14. Find the equations of line (a) and line (b).

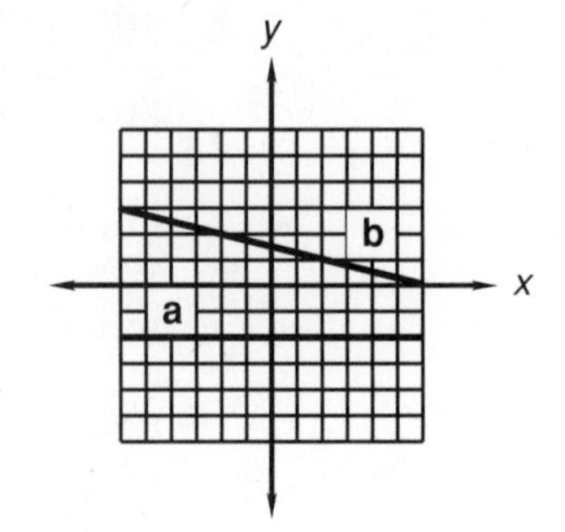

Simplify:

15. $3\sqrt{2} \cdot 4\sqrt{3} \cdot 4\sqrt{6} - 3\sqrt{2}$

16. $3\sqrt{2}(5\sqrt{12} - 6\sqrt{36})$

17. Divide: $(x^4 - x - 4) \div (x - 1)$

Solve by factoring:

18. $24 = -x^2 - 10x$

19. $-4 + 9x^2 = 0$

Solve:

20. $\dfrac{3x}{2} - \dfrac{x - 5}{6} = 3$

21. $3\frac{2}{5}x - \dfrac{1}{10} = \dfrac{1}{20}$

22. Add: $\dfrac{p}{p + 4} - \dfrac{p - 2}{p}$

23. Simplify: $\dfrac{xy - \dfrac{1}{y}}{\dfrac{x}{y} - 4}$

24. $15\frac{2}{5}$ is what fraction of $20\frac{1}{3}$?

25. $\dfrac{5\sqrt{2}}{7} \in$ {What sets of numbers}?

26. Evaluate: $-|x^0| - (x - y)(y - x)$ if $x = -4$ and $y = -5$

Simplify:

27. $-2^0 - 2[(-3 - 2) - (-2)] - [(-3 + 2) - (-2)]$

28. (a) $\dfrac{1}{-3^{-2}}$ (b) $\dfrac{1}{-(-3)^{-2}}$ (c) $-(-2)^{-2}$

29. Multiply: $x^2y^2\left(\dfrac{x^{-2}}{y^{-2}} + 4x^4y^{-2}\right)$

30. (a) Use the letters a, b, and c to state the distributive property.
(b) What is the additive inverse of -2?
(c) What is the multiplicative inverse of $-b$?

LESSON 98 Rational expressions

98.A products of rational expressions

Fractions are multiplied by multiplying the numerators to form the numerator of the product and by multiplying the denominators to form the denominator of the product. Thus to multiply

$$\frac{4x}{5} \cdot \frac{3xa}{y}$$

we multiply the numerators to get $12x^2a$ and multiply the denominators to get $5y$.

$$\frac{4x}{5} \cdot \frac{3xa}{y} = \frac{12x^2a}{5y}$$

Sometimes we encounter indicated products of rational expressions whose simplification is facilitated if the terms are first factored and such canceling as is possible performed before any multiplication is done.

example 98.1 Simplify: $\dfrac{x^2 - 25}{x^2 - 7x} \cdot \dfrac{x^2 + 3x}{x^2 - 2x - 15}$

solution If we multiply the expressions in their present form, we get a very complicated expression for the product.

$$\frac{(x^2 - 25)(x^2 + 3x)}{(x^2 - 7x)(x^2 - 2x - 15)} = \frac{x^4 - 25x^2 + 3x^3 - 75x}{x^4 - 9x^3 - x^2 + 105x}$$

This expression has x raised to the fourth power in both the numerator and the denominator and is very difficult to simplify. If we factor and cancel before we multiply, however, the simplified form can be obtained quickly and easily.

$$\frac{\cancel{(x - 5)}(x + 5)}{\cancel{x}(x - 7)} \cdot \frac{\cancel{x}\cancel{(x + 3)}}{\cancel{(x + 3)}\cancel{(x - 5)}} = \mathbf{\frac{x + 5}{x - 7}}$$

example 98.2 Simplify: $\dfrac{x^2 + x - 6}{x^2 - 4x - 21} \cdot \dfrac{x^2 - 8x + 7}{x^2 - x - 2}$

solution Problems like this one are encountered only in algebra books. **These problems are carefully contrived to give the student practice in factoring and canceling.** Thus we factor and cancel like factors that appear in both the numerator and denominator.

$$\frac{\cancel{(x + 3)}\cancel{(x - 2)}}{\cancel{(x - 7)}\cancel{(x + 3)}} \cdot \frac{\cancel{(x - 7)}(x - 1)}{\cancel{(x - 2)}(x + 1)} = \mathbf{\frac{x - 1}{x + 1}}$$

98.B quotients of rational expressions

In Lesson 59 we learned to simplify expressions such as

$$\frac{\frac{a}{b}}{\frac{c}{d}}$$

by using the **denominator-numerator same-quantity rule** to justify multiplying both the denominator and the numerator by $\frac{d}{c}$, which is the reciprocal of the denominator.

$$\frac{\frac{a}{b}}{\frac{c}{d}} = \frac{\frac{a}{b} \cdot \frac{d}{c}}{\frac{c}{d} \cdot \frac{d}{c}} = \frac{\frac{ad}{bc}}{1} = \mathbf{\frac{ad}{bc}}$$

If the same division problem had been stated by writing

$$\frac{a}{b} \div \frac{c}{d}$$

we see that the same result can be obtained by inverting the divisor and multiplying.

$$\frac{a}{b} \div \frac{c}{d} = \frac{a}{b} \cdot \frac{d}{c} = \mathbf{\frac{ad}{bc}}$$

We can use this procedure to simplify quotients of more complicated rational expressions.

example 98.3 Simplify: $\dfrac{x^2 - 2x}{x^2 + 2x - 8} \div \dfrac{x^2 + 5x}{x^2 + 7x + 12}$

solution As the first step we invert the divisor and change the division symbol to a multiplication dot. Then we factor and cancel as in the two previous examples.

$$\frac{x^2 - 2x}{x^2 + 2x - 8} \cdot \frac{x^2 + 7x + 12}{x^2 + 5x} = \frac{\cancel{x}(\cancel{x - 2})}{(\cancel{x + 4})(\cancel{x - 2})} \cdot \frac{(\cancel{x + 4})(x + 3)}{\cancel{x}(x + 5)} = \mathbf{\frac{x + 3}{x + 5}}$$

example 98.4 Simplify: $\dfrac{x^2 - x - 6}{x^2 - 3x - 10} \div \dfrac{x^2 + 5x + 4}{x^2 - x - 20}$

solution Again as the first step we invert the divisor and indicate multiplication rather than division. Then we factor and cancel.

$$\frac{x^2 - x - 6}{x^2 - 3x - 10} \cdot \frac{x^2 - x - 20}{x^2 + 5x + 4} = \frac{(x - 3)(\cancel{x + 2})}{(\cancel{x - 5})(\cancel{x + 2})} \cdot \frac{(\cancel{x - 5})(\cancel{x + 4})}{(\cancel{x + 4})(x + 1)} = \mathbf{\frac{x - 3}{x + 1}}$$

practice Simplify:

a. $\dfrac{x^2 - x - 6}{x^2 - 6x - 16} \div \dfrac{x^2 - 3x}{x^2 - 3x - 40}$

b. $\dfrac{x^2 + 12x + 36}{x^2 + 13x + 42} \cdot \dfrac{x^2 + 4x - 21}{x^2 + 2x - 24}$

problem set 98

1. Norma and David crawled to the barn and then hopped back to the house. They crawled at 300 centimeters per minute and hopped at 400 centimeters per minute. If the round trip took 7 minutes, how long did they crawl? How far was it to the barn?

2. Annette drove to Shawnee in 4 hours and drove back in 3 hours. What were her speeds if her speed coming back was 11 miles per hour greater than her speed going?

3. When the time came to stand up and be counted, only 92 people stood up. If 460 people were present, what percent stood up and were counted?

4. Hobert and Higgs counted the boys and girls at the assembly. There were 179 students and 13 more boys than girls. How many boys and how many girls were present?

5. Use the letters a, b, and c as necessary to state (a) the associative property of multiplication, and (b) the commutative property of addition.

6. If the reciprocal of a number is $-\frac{1}{9}$, what is the additive inverse of the number?

7. $a + (b + c) = (a + b) + c$ is a statement of which property of the set of real numbers?

8. Five-thirteenths of the citizens believed that the cause of their difficulty was procrastination. If 400 did not agree with this analysis, how many citizens lived in the community?

9. Solve: $-2(3x - 4^0) + 3x - 2^0 = -(x - 3^2)$

10. Graph: $-x - 3 \not> 2$; $D = \{\text{Reals}\}$

Solve by graphing:

11. $\begin{cases} y = 2x - 4 \\ y = -x + 2 \end{cases}$

Use elimination to solve:

12. $\begin{cases} 3x + 5y = -14 \\ -2x + y = 5 \end{cases}$

13. Simplify: $\dfrac{(0.000004)(0.003 \times 10^{21})}{(20{,}000 \times 10^8)(0.002 \times 10^{15})}$

14. Find the equations of lines (a) and (b).

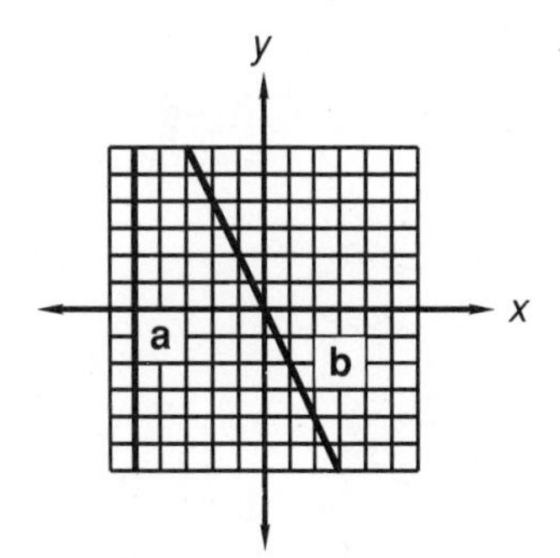

Simplify:

15. $3\sqrt{6} \cdot 2\sqrt{5} - \sqrt{120}$

16. $4\sqrt{12}(3\sqrt{2} - 4\sqrt{3})$

17. Divide: $(3x^3 - 4) \div (x + 3)$

18. Solve by factoring: $40 = -x^2 - 14x$

19. Graph on a number line: $2 > x > -3$; $D = \{\text{Reals}\}$

Solve:

20. $\dfrac{x - 5}{7} + \dfrac{x}{4} = \dfrac{1}{2}$

21. $3\frac{1}{8}x - 2\frac{1}{2} = \frac{1}{8}$

22. Expand: $(3x + 3y)^2$

23. True or false? $\{\text{Wholes}\} \subset \{\text{Naturals}\}$

24. Add: $\dfrac{x}{ya^2} + \dfrac{xa}{a^2y^2}$

25. Simplify: $\dfrac{k + \dfrac{k}{y}}{y + \dfrac{a}{y}}$

26. Evaluate: $-y^0(-y^2 - 4y) - ay$ if $y = -2$ and $a = -5$

27. Simplify: $-2^0[(-3 - 4^0) - (-2 - 3^0) - 2^2]$

Simplify:

28. (a) $\frac{1}{-3^{-3}}$ (b) $\frac{1}{-(-3)^{-3}}$ (c) $-(-3)^{-3}$ (d) $\sqrt[3]{-64}$

29. $\frac{(-4x^{-2})^2}{(-2y^{-2})^2x}$

30. $\frac{x^3 - 4x}{x^2 + 7x + 10} \div \frac{x^2 - 2x}{x^2 - 25}$

LESSON 99 Uniform motion problems of the form $D_1 + D_2 = N$

In the uniform motion problems we have worked up to now, two people or things have traveled equal distances. The distance diagrams have looked like one of the following.

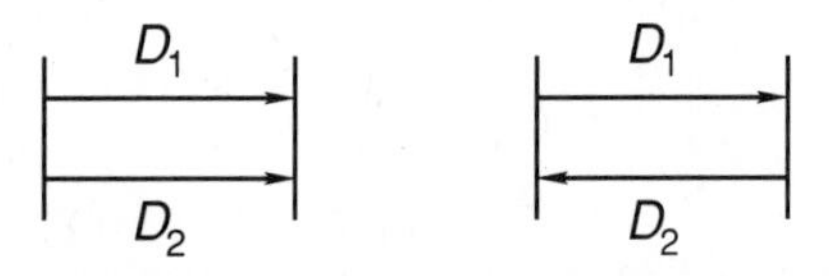

These diagrams have indicated that our distance equation should be of the form $R_1T_1 = R_2T_2$. Now we will consider problems that state that the sum of two distances equals a given number. The diagrams and equations of these problems will have the following forms.

D_1 D_2 352

D_1 D_2 352

$$D_1 + D_2 = 352 \quad \text{so} \quad R_1T_1 + R_2T_2 = 352$$

example 99.1 A southbound bus left Fort Walton Beach at 9 a.m. Two hours later a northbound bus left the same station. If the buses traveled at the same rate and were 352 kilometers apart at 2 p.m., find the rate of the buses.

solution The statement of the problem leads to this distance diagram.

D_N D_S 352

The distance equation is $D_N + D_S = 352$ or $R_NT_N + R_ST_S = 352$. The southbound bus traveled for 5 hours and the northbound bus traveled for 3 hours.

$$T_N = 3 \qquad T_S = 5$$

The rates were the same so $R_N = R_S$.

Now we have four equations in four unknowns.

$$R_NT_N + R_ST_S = 352 \qquad T_N = 3 \qquad T_S = 5 \qquad R_N = R_S$$

We use substitution to solve.

$$R_N(3) + R_N(5) = 352 \longrightarrow 8R_N = 352 \longrightarrow \mathbf{R_N = 44 \text{ kilometers per hour}}$$

and therefore $\mathbf{R_S = 44 \text{ kilometers per hour}}$

example 99.2 A train starts from Toledo at 11 a.m. and heads for Makinaw, 332 kilometers away. At the same time, a train leaves Makinaw and heads for Toledo at 65 kilometers per hour. If the trains meet at 1 p.m., what is the rate of the first train?

solution First we draw the diagram and write the distance equation.

D_1 D_2 332

$$D_1 + D_2 = 332 \quad \text{so} \quad R_1T_1 + R_2T_2 = 332$$

Then we reread the problem and write the other three equations.

$$T_1 = 2 \qquad T_2 = 2 \qquad R_2 = 65$$

Now we substitute and solve:

$R_1(2) + 65(2) = 332$	substituted
$2R_1 + 130 = 332$	multiplied
$2R_1 = 202$	added -130 to both sides
$\mathbf{R_1 = 101}$ **kilometers per hour**	divided by 2

example 99.3 The ships were 400 miles apart at midnight and were headed toward each other. If they collided head-on at 8 a.m., find the speed of both ships if one was 20 miles per hour faster than the other.

solution First we draw the diagram and write the distance equation.

D_F D_S 400

$$D_F + D_S = 400 \quad \text{so} \quad R_FT_F + R_ST_S = 400$$

The other three equations are

$$T_F = 8 \qquad T_S = 8 \qquad R_F = R_S + 20$$

Now we substitute and solve.

$(R_S + 20)(8) + R_S(8) = 400$	substituted
$8R_S + 160 + 8R_S = 400$	multiplied
$16R_S + 160 = 400$	simplified
$16R_S = 240$	added -160 to both sides
$\mathbf{R_S = 15}$ **miles per hour**	divided by 16

Thus $\mathbf{R_F = 35}$ **miles per hour**

practice

a. An eastbound bus left Ukiah at noon. Three hours later a westbound bus left the same station. If the buses traveled at the same rate and were 500 kilometers apart at 6 p.m., find the rate of each bus.

b. A train starts from Peoria at 2 p.m. and heads for Reedley, 666 kilometers away. At the same time, a train leaves Reedley and heads for Peoria at 85 kilometers per hour. If the trains meet at 5 p.m., what is the rate of the first train?

problem set 99

1. The ships were 700 miles apart at midnight and were headed toward each other. If they collided at 10 a.m., find the speed of both ships if one was traveling 30 miles per hour faster than the other.

2. (a) What is the additive identity?
   (b) What is the multiplicative identity?
   (c) If the reciprocal of a number is -9, what is the multiplicative inverse of the same number?

3. Arcelia's final exam was given 3 times the weight of a weekly test. The average of her 7 weekly tests was 91. The overall weighted average was 93. What grade did she make on the final? (This is a two-step problem.)

4. Pitts bought pots for \$5 each, and Joe bought buckets for \$7 each. If they spent \$1140 for 192 utensils, how many of each type did they buy?

5. When the stranger came into the forest, 37 percent of the little people ran to hide. If 2520 refused to hide, how many little people lived in the forest?

Simplify:

6. $\dfrac{4x + 12}{x^2 + 11x + 30} \div \dfrac{x^3 - 4x^2 - 21x}{4x^2 + 20x}$

7. $\dfrac{x^2 + 11x + 24}{x^2 + 3x} \div \dfrac{x^2 + 13x + 40}{4x^2 + 20x}$

8. Nine-sixteenths of the girls believed that saltation was salubrious. If the other 700 girls were undecided, how many girls had made up their minds concerning this topic?

9. Solve: $-p^0(p - 4) - (-p^0)p + 3^0(p - 2) = -p - 6^0$

10. Graph on a number line: $-4 \le x < 1$; $D = \{\text{Integers}\}$

Solve by graphing:

11. $\begin{cases} y = x \\ x = -3 \end{cases}$

Use elimination to solve:

12. $\begin{cases} y = 2x + 4 \\ 2y - x = -1 \end{cases}$

13. Simplify: $\dfrac{(0.00035 \times 10^{15})(200{,}000)}{(1000 \times 10^{-45})(0.00007)}$

14. Find the equations of lines (a) and (b).

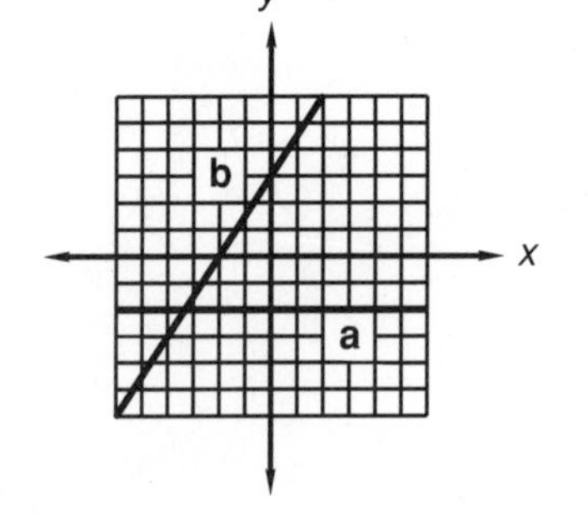

Simplify:

15. $4\sqrt{3} \cdot 5\sqrt{6} + \sqrt{5} \cdot 2$

16. $4\sqrt{12}(3\sqrt{2} - 3\sqrt{12})$

17. Divide: $(2x^3 + 5x^2 - 1) \div (2x + 1)$

Solve by factoring:

18. $x^2 = 7x + 30$

19. $100 = 9p^2$

Solve:

20. $\dfrac{x - 7}{4} - \dfrac{x}{2} = \dfrac{1}{8}$

21. $5\frac{1}{6}x + 2\frac{1}{4} = \frac{3}{8}$

22. Add: $\dfrac{a}{x^2} + \dfrac{a^2}{x^3y} + \dfrac{a^3}{x(x + y)}$

23. Simplify: $\dfrac{1 + \dfrac{1}{y}}{y - \dfrac{1}{y}}$

24. $3\frac{1}{4}$ is what fraction of $\frac{7}{8}$?

25. $0.037 \in$ {What sets of numbers}?

26. Evaluate: $ay^2 - y(-y + a^0)$ if $y = -3$ and $a = -4$

Simplify:

27. $-3[(-2^0 - 5^0) - 2 - (4 - 6)(-2)]$

28. (a) $\dfrac{4x - 4}{4}$ (b) $-(-2)^{-3}$

29. Multiply: $4x^2y^{-1}\left(\dfrac{p^0y}{x^2} - 3x^{-2}y^4\right)$

30. Find the volume in cubic inches of a cylinder whose base is shown and whose sides are 2 feet high. If the cylinder is a right cylinder, find the surface area. What is the volume of a cone with the same height and base? Dimensions are in inches.

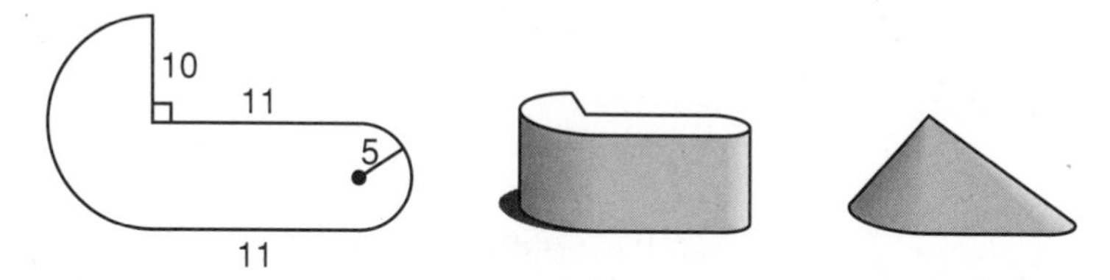

LESSON 100 Difference of two squares theorem

In Lesson 65 we introduced the topic of square roots and noted that every positive real number has both a positive square root and a negative square root. Because 2 times 2 equals 4 and -2 times -2 also equals 4,

$$(2)(2) = 4 \qquad \text{and} \qquad (-2)(-2) = 4$$

we say that the two square roots of 4 are $+2$ and -2. **But when we use the radical to indicate the square root of a positive number, as**

$$\sqrt{4}$$

we are designating the principal or positive square root, which in this case is $+2$. If we wish to indicate the negative square root, we must use a minus sign in front of the expression, as

$$-\sqrt{4} = -2$$

If we are asked to find the numbers that satisfy the equation

$$x^2 = 4$$

we know that the numbers are +2 and −2 because

$$(+2)^2 = 4 \qquad \text{and also} \qquad (-2)^2 = 4$$

The general form of the equation $x^2 = 4$ is

$$p^2 = q^2$$

If we add $-q^2$ to both sides, we can factor

$$p^2 = q^2 \longrightarrow p^2 - q^2 = 0 \longrightarrow (p + q)(p - q) = 0$$

and use the zero factor theorem to solve

$$\text{if} \quad \begin{array}{rcr} p + q &=& 0 \\ -\,q && -q \\ \hline p &=& -q \end{array} \qquad \text{if} \quad \begin{array}{rcr} p - q &=& 0 \\ +\,q && +q \\ \hline p &=& q \end{array}$$

and we find that if $p + q = 0$, then $p = -q$, and if $p - q = 0$, then $p = q$.

The theorem that describes the solution to an equation of the form $p^2 = q^2$ is called the **difference of two squares theorem.**

DIFFERENCE OF TWO SQUARES THEOREM
if p and q are real numbers and if $p^2 = q^2$, then
$p = q \qquad \text{or} \qquad p = -q$

A quadratic equation in x has an x^2 term as the highest power of the variable. A quadratic equation in p has a p^2 term as the highest power of the variable. **It is very important to realize that in a beginning algebra book the ± sign arises only in the solution of a quadratic equation.**

$$\text{If} \quad m^2 = 3 \quad \text{then} \quad m = \pm\sqrt{3}$$

There are two solutions to this equation because $(\sqrt{3})^2$ equals 3 and $(-\sqrt{3})^2$ equals 3. A common mistake of beginners is to use the ± sign whenever the square root symbol is encountered.

$$\sqrt{9} = \pm 3 \qquad \textbf{NO! NO! NO!}$$

The square root sign is used to designate *only* the positive square root of a number.

$$\sqrt{9} = 3 \qquad \text{correct}$$

In this case there is no equation to be solved. The equals sign is written by us to show that $\sqrt{9}$ has the value of +3. If we wish to designate the negative square root of 9, we write

$$-\sqrt{9} = -3$$

example 100.1 Solve: $p^2 = 16$

solution We know that the general equation $p^2 = q^2$ has two solutions, which are $p = q$ and $p = -q$. In the same way the given equation has two solutions, which are

$$\mathbf{p = +4} \qquad \text{and} \qquad \mathbf{p = -4}$$

We usually combine these notations and write

$$\mathbf{p = \pm 4}$$

example 100.2 Solve: $p^2 = 41$

solution To use the form $p^2 = q^2$, it is helpful to write the given equation as

$$p^2 = (\sqrt{41})^2$$

If we do this, we can write the answer as

$$\mathbf{p = \pm\sqrt{41}}$$

example 100.3 Solve: $k^2 = 13$

solution We omit the intermediate step and write the solution by inspection.

$$\mathbf{k = \pm\sqrt{13}}$$

practice Use the difference of two squares theorem to write the answers to the following equations:

a. $p^2 = 169$ **b.** $q^2 = 23$ **c.** $w^2 = 14$

problem set 100

1. At 3 p.m., Brunhilde headed north at 30 kilometers per hour. Two hours later Ludwig headed south at 40 kilometers per hour. At what time will they be 340 kilometers apart?

2. Alphasia headed for the rodeo at 9 a.m. at 30 miles per hour. At 11 a.m. Bubba headed after her at 60 miles per hour. What time was it when Bubba caught Alphasia?

3. Billye H. ran to town at 8 kilometers per hour and then walked back home at 3 kilometers per hour. How far was it to town if the round trip took 11 hours?

4. Marfugge had \$72.50 in quarters and half-dollars. If he had 190 coins in all, how many of each type did he have?

5. Fifteen percent of the seniors voted for Gina. If 289 seniors voted for the other candidates, how many seniors voted in the election?

6. Simplify: $\dfrac{x^3 + 2x^2 - 15x}{x^2 + 5x} \div \dfrac{x^3 - 6x^2 + 9x}{x^2 - 3x}$

Use the difference of two squares theorem to write the answers to the following equations:

7. (a) $p^2 = 49$ (b) $p^2 = 39$ (c) $k^2 = 11$

8. Revulsion overcame $\frac{7}{10}$ of the audience when it was announced that the cute little animal was saprophagous. If 1200 were not affected, how many felt revulsion?

9. Solve: $m - m^0(m - 4) - (-2)m + (-2)(m - 4^0) = m - 6$

10. Graph on a number line: $-4 - x \not> -2$; $D = \{\text{Negative integers}\}$

Solve by graphing:

11. $\begin{cases} y = 2x \\ x = -1 \end{cases}$

Use elimination to solve:

12. $\begin{cases} 3x + 5y = -13 \\ 2x - 3y = 23 \end{cases}$

13. Simplify: $\dfrac{(42{,}000{,}000)(0.0001 \times 10^{-5})}{(7000 \times 10^{14})(200{,}000 \times 10^{-8})}$

14. Find the equations of lines (a) and (b).

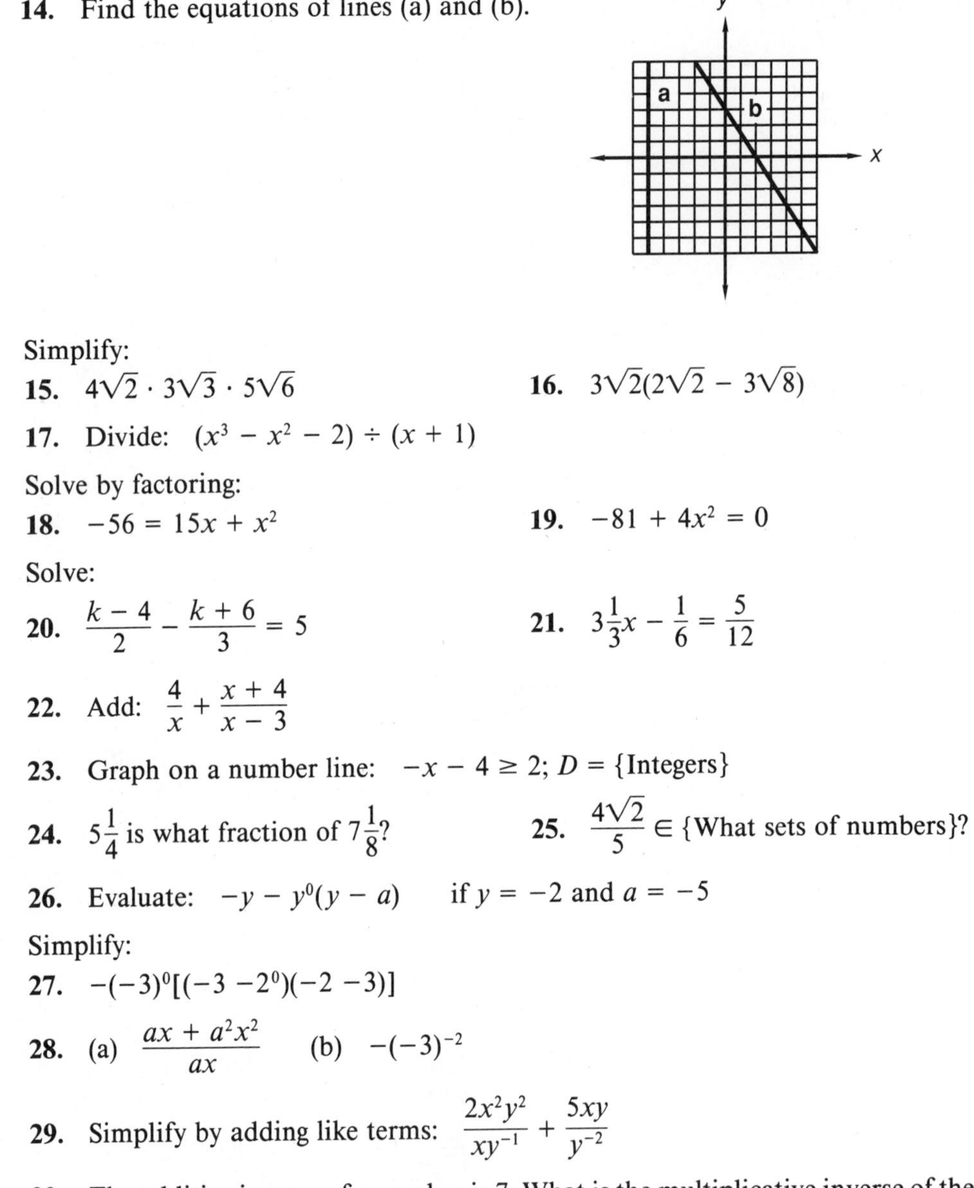

Simplify:

15. $4\sqrt{2} \cdot 3\sqrt{3} \cdot 5\sqrt{6}$

16. $3\sqrt{2}(2\sqrt{2} - 3\sqrt{8})$

17. Divide: $(x^3 - x^2 - 2) \div (x + 1)$

Solve by factoring:

18. $-56 = 15x + x^2$

19. $-81 + 4x^2 = 0$

Solve:

20. $\dfrac{k - 4}{2} - \dfrac{k + 6}{3} = 5$

21. $3\frac{1}{3}x - \frac{1}{6} = \frac{5}{12}$

22. Add: $\dfrac{4}{x} + \dfrac{x + 4}{x - 3}$

23. Graph on a number line: $-x - 4 \geq 2$; $D = \{\text{Integers}\}$

24. $5\frac{1}{4}$ is what fraction of $7\frac{1}{8}$?

25. $\dfrac{4\sqrt{2}}{5} \in$ {What sets of numbers}?

26. Evaluate: $-y - y^0(y - a)$ if $y = -2$ and $a = -5$

Simplify:

27. $-(-3)^0[(-3 - 2^0)(-2 - 3)]$

28. (a) $\dfrac{ax + a^2x^2}{ax}$ (b) $-(-3)^{-2}$

29. Simplify by adding like terms: $\dfrac{2x^2y^2}{xy^{-1}} + \dfrac{5xy}{y^{-2}}$

30. The additive inverse of a number is 7. What is the multiplicative inverse of the number?

LESSON 101 Pythagorean theorem

101.A angles and triangles

It is interesting to note that mathematicians often disagree on definitions and on terminology. Most authors use similar definitions, but not all do. The definition of an angle is a good example. Most agree that two intersecting lines form four angles.

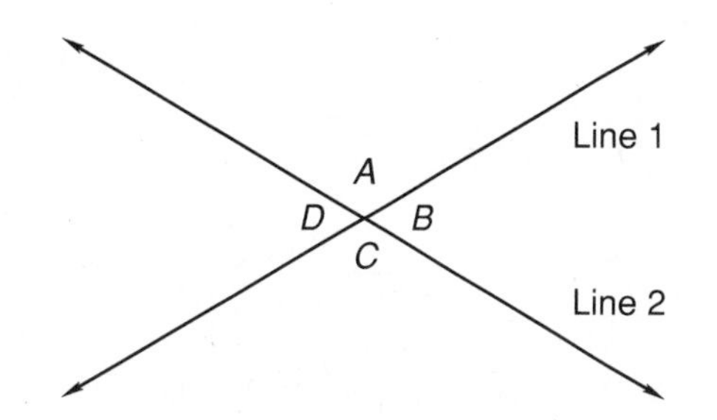

Here are shown two intersecting lines and the four angles formed. If we look only at angle B, we see that it is formed by part of lines 1 and 2. We call these parts **half lines** or **rays.**

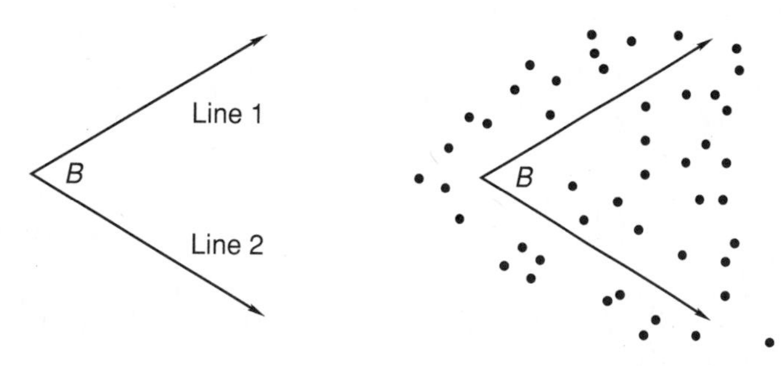

European authors generally define an angle to be the opening between the rays. Thus, to them the angle is the set of points bounded by the rays. American authors tend to define the angle to be the rays themselves. To them the angle is the set of points that make up the rays. Others say that the rays are the sides of the angle but don't say what the angle is. Some don't speak of the opening at all, but define an angle to be a rotation of a ray about its endpoint. A precise definition is not required in this book so we will just say

An angle is formed by two half lines or rays that are in the same plane and that have a common endpoint.

To begin a quick review of angle measures, we remember that if two straight lines intersect and are perpendicular to one another, we define the measure of each of the four angles created to be 90 degrees. We also remember that instead of writing the word *degrees,* it is customary to place a small elevated circle after the number that designates the number of degrees. Thus 90 degrees can be written as 90°, and 47 degrees and 135 degrees can be written as 47° and 135°. We see here two intersecting perpendicular lines with the resulting 90° angles.

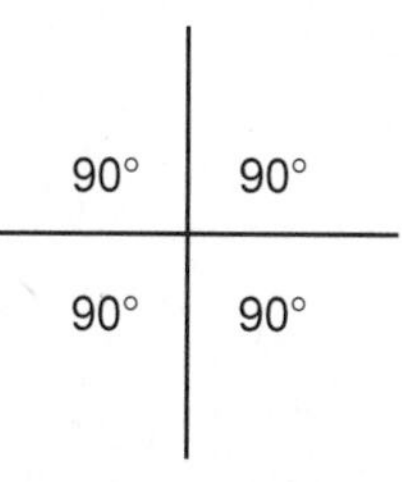

With this definition of a 90° angle and two axioms, it can be proved, by using geometry, that **the sum of the interior angles of any triangle is 180°.** We show three

triangles and note that the sum of the three interior angles in each triangle is 180°.

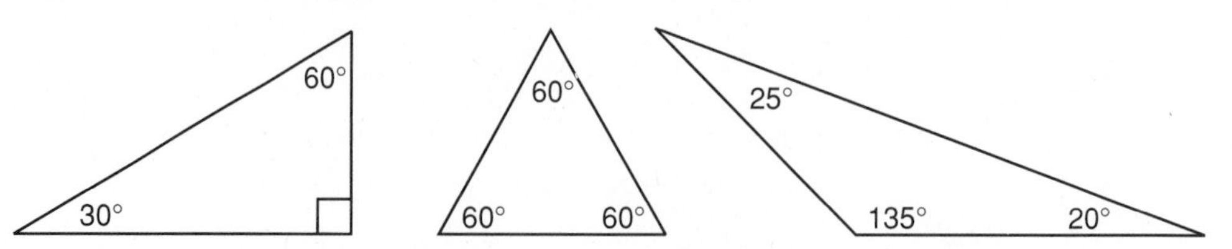

The triangle on the left has one angle that has a measure of 90°. Any triangle that contains a right angle is called a **right triangle,** and the side of the triangle that is opposite the right angle is always the longest side. We call this side of a right triangle the **hypotenuse.** The other two sides are called **legs** or, simply, **sides.** Right triangles have a special property that makes them very useful in mathematics, engineering, and physics.

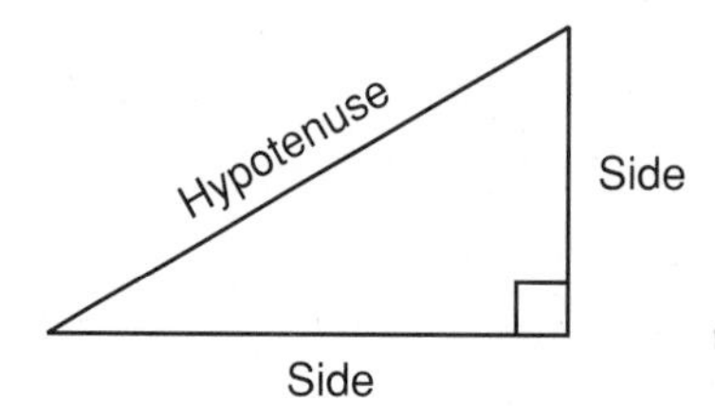

101.B pythagorean theorem

It can be shown that **the square drawn on the hypotenuse of a right triangle has the same area as the sum of the areas of the squares drawn on the other two sides.** While this theorem was known to the Egyptians as early as the Middle Kingdom (ca. 2000 B.C.), the geometric proof of the theorem is normally attributed to a Greek philosopher and mathematician named **Pythagoras.** Pythagoras was born on the Aegean island of Samos and was later associated with a school or brotherhood in the town of Crotona on the Italian peninsula in the sixth century B.C. We call the theorem for which he supposedly developed the proof the **Pythagorean theorem.**

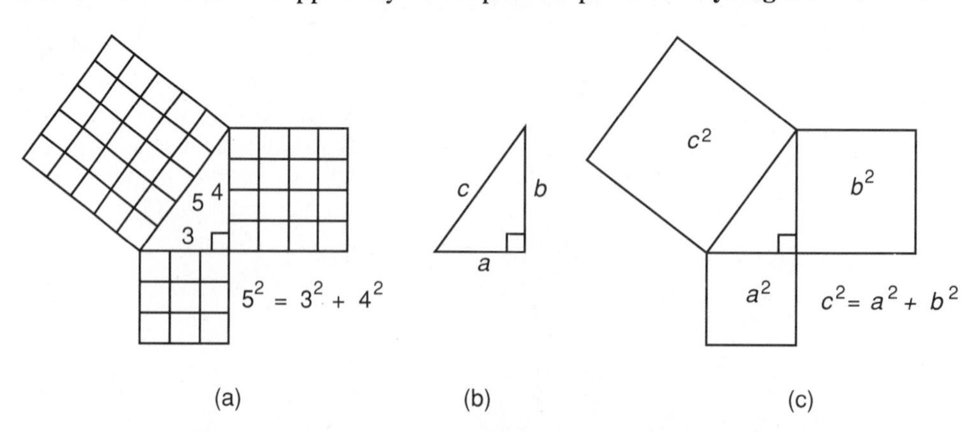

(a) (b) (c)

In (a) we show a right triangle whose sides have lengths of 3 and 4 units, respectively, and whose hypotenuse has a length of 5 units. A square has been drawn on each of the three sides, and since the area of a square whose sides have a length of L is L^2, we see that the areas of the two squares on the sides are 4^2 and 3^2 and that the sum of these two areas equals 25 square units. This is the same as the area of the square drawn on the hypotenuse, which equals 5^2 or 25 square units. Figure (b) shows another triangle whose sides have lengths of a and b and whose hypotenuse has a length of c. The area of a square drawn on the hypotenuse would be c^2, and the areas of the squares drawn on sides a and b would be a^2 and b^2, respectively, as shown in (c). We normally label the hypotenuse as c and the other two sides as a and b. Thus the general algebraic expression of the **Pythagorean theorem** is

$$a^2 + b^2 = c^2$$

where **c is the length of the hypotenuse and a and b represent the lengths of the other two sides.**

This theorem can be used to find the length of a side of a right triangle if the lengths of the other two sides are known.

example 101.1 Given the triangle with the lengths of the sides as shown, use the Pythagorean theorem to find the length of side a.

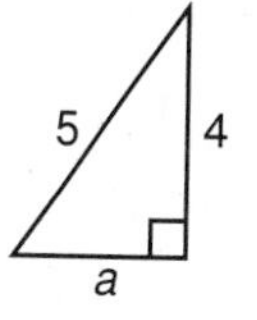

solution The square of the hypotenuse equals the sum of the squares of the other two sides. Thus,

$$5^2 = 4^2 + a^2 \longrightarrow 25 = 16 + a^2 \longrightarrow 9 = a^2$$

We use the difference of two squares theorem to finish the solution.

$$a^2 = 9 \quad \text{which leads to} \quad a = +3 \quad \text{or} \quad a = -3$$

While -3 is a solution to the equation $a^2 = 9$, it is not a solution to the problem at hand because physical lengths are designated by positive numbers. Thus we reject this solution and say that

$$\boldsymbol{a = +3}$$

example 101.2 Find the side p in the triangle shown.

solution We apply the Pythagorean theorem to this triangle to write

$$p^2 = 5^2 + 4^2$$

Now we simplify and use the difference of two squares theorem.

$$p^2 = 41 \longrightarrow p^2 = (\sqrt{41})^2 \longrightarrow p = \sqrt{41} \quad \text{or} \quad p = -\sqrt{41}$$

But sides of triangles do not have negative lengths, so we discard the negative result and say

$$\boldsymbol{p = \sqrt{41}}$$

example 101.3 Find side k.

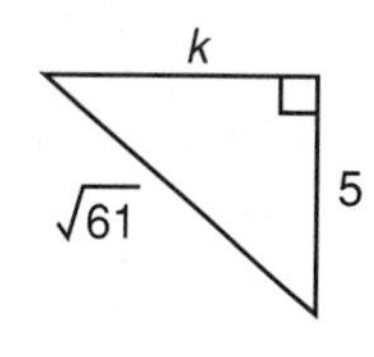

solution We use the Pythagorean theorem to write

$$(\sqrt{61})^2 = k^2 + 5^2$$

and now we simplify.

$$61 = k^2 + 25$$

Now we finish the solution by using the difference of two squares theorem.

$$36 = k^2 \longrightarrow (6)^2 = k^2$$

so $\qquad 6 = k \quad$ or $\quad -6 = k$

Since -6 has no meaning as the length of a side of a triangle, we say

$$\mathbf{k = 6}$$

example 101.4 Find side m.[†]

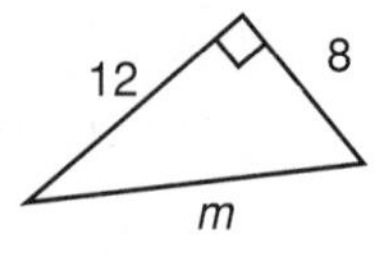

solution By using the Pythagorean theorem we can write

$$m^2 = 12^2 + 8^2$$

and now we simplify and solve.

$$m^2 = 208 \longrightarrow m^2 = (\sqrt{208})^2$$

$$\longrightarrow m = \sqrt{208} \longrightarrow \mathbf{m = 4\sqrt{13}}$$

practice Use the Pythagorean theorem to find the length of the indicated side.

a. Find side p.

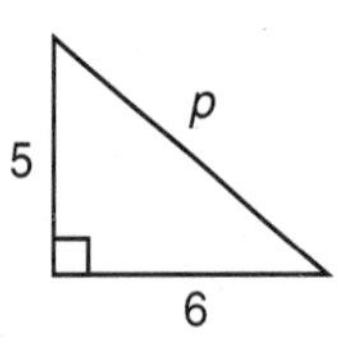

b. Find side f.

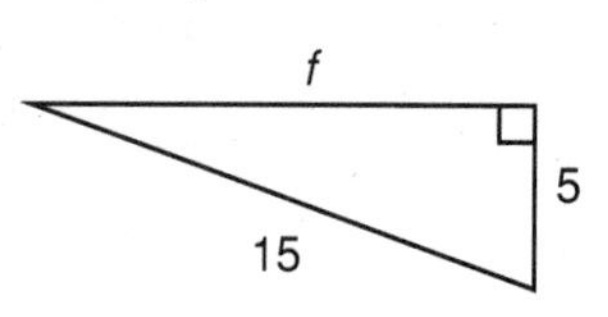

problem set 101

1. If the sum of -4 and the opposite of a number is multiplied by -3, the result is 6 less than the product of the number and 2. What is the number?

2. Find four consecutive integers such that 4 times the sum of the first and fourth is 1 less than 9 times the third.

3. Calvin could see 32. This was only 20 percent of the number that Tooley could see. How many could Tooley see?

4. Wendy drove at 60 miles per hour. Thus, she made the trip in 1 hour less than it took Deborah because Deborah only drove at 50 miles per hour. How long did each of them drive and how long was the trip?

5. Spann found a sack that contained \$9000 in \$5 bills and \$10 bills. Margaret helped count the money and found that there were 1250 bills in all. How many were \$5 bills and how many were \$10 bills?

6. Simplify: $\dfrac{x^2 + x - 20}{x^2 + 6x - 16} \div \dfrac{x^2 - 2x - 8}{x^2 + 10x + 16}$

7. Louwon was elated! Her weighted rating was 120. If her first rating was 96 and her second rating was weighted at double value, what was her second rating?

[†] The Greeks must have drawn some of their right triangles as this one is drawn because the word *hypotenuse* comes from the Greek words *hupo* meaning "under" and *teinein* meaning "to stretch," so hypotenuse means "stretched under."

8. Use 15 unit multipliers to convert 1 cubic mile to cubic kilometers.

9. Use the Pythagorean theorem to find k.

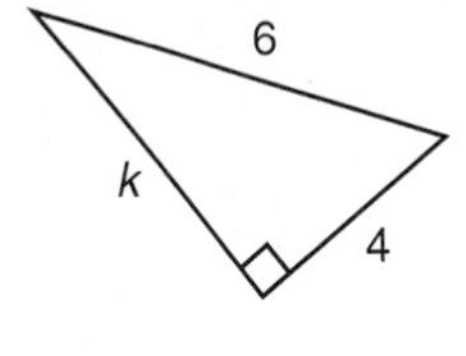

10. Solve: $-(-3)k^0 - 3^0k + (-2)(2 - k) - (-3)(k + 2) = 0$

11. Solve by graphing: $\begin{cases} y = x + 2 \\ y = -x + 4 \end{cases}$

12. Simplify: $\dfrac{(36{,}000 \times 10^{-5})(400{,}000)}{(0.0006 \times 10^{-4})(600 \times 10^{5})}$

13. Find the equations of lines (a) and (b).

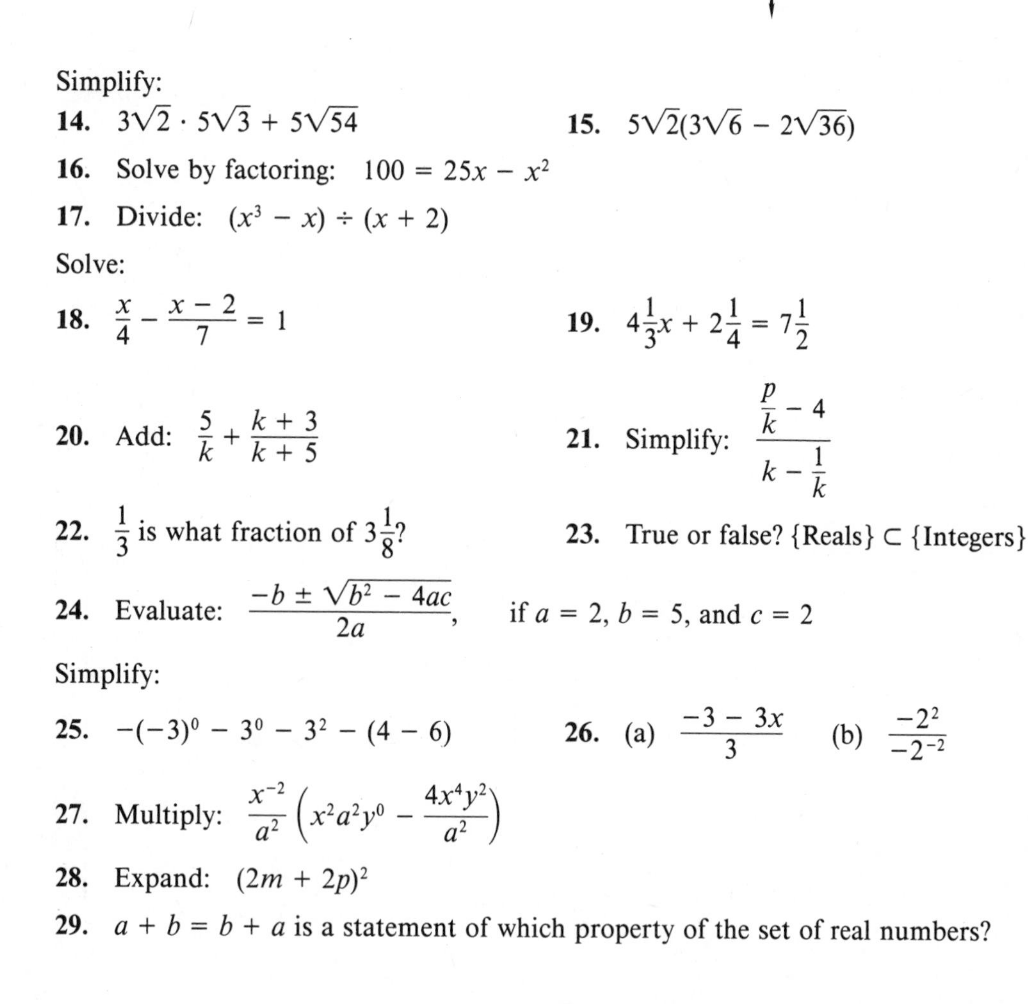

Simplify:

14. $3\sqrt{2} \cdot 5\sqrt{3} + 5\sqrt{54}$

15. $5\sqrt{2}(3\sqrt{6} - 2\sqrt{36})$

16. Solve by factoring: $100 = 25x - x^2$

17. Divide: $(x^3 - x) \div (x + 2)$

Solve:

18. $\dfrac{x}{4} - \dfrac{x - 2}{7} = 1$

19. $4\frac{1}{3}x + 2\frac{1}{4} = 7\frac{1}{2}$

20. Add: $\dfrac{5}{k} + \dfrac{k + 3}{k + 5}$

21. Simplify: $\dfrac{\dfrac{p}{k} - 4}{k - \dfrac{1}{k}}$

22. $\dfrac{1}{3}$ is what fraction of $3\frac{1}{8}$?

23. True or false? {Reals} $\subset$ {Integers}

24. Evaluate: $\dfrac{-b \pm \sqrt{b^2 - 4ac}}{2a}$, if $a = 2$, $b = 5$, and $c = 2$

Simplify:

25. $-(-3)^0 - 3^0 - 3^2 - (4 - 6)$

26. (a) $\dfrac{-3 - 3x}{3}$ (b) $\dfrac{-2^2}{-2^{-2}}$

27. Multiply: $\dfrac{x^{-2}}{a^2}\left(x^2a^2y^0 - \dfrac{4x^4y^2}{a^2}\right)$

28. Expand: $(2m + 2p)^2$

29. $a + b = b + a$ is a statement of which property of the set of real numbers?

30. What is the volume of the cylinder whose base is shown and whose sides are 50 centimeters high? If the cylinder is a right cylinder, what is the surface area? What is the volume of a cone that has the same base and altitude? Dimensions are in meters.

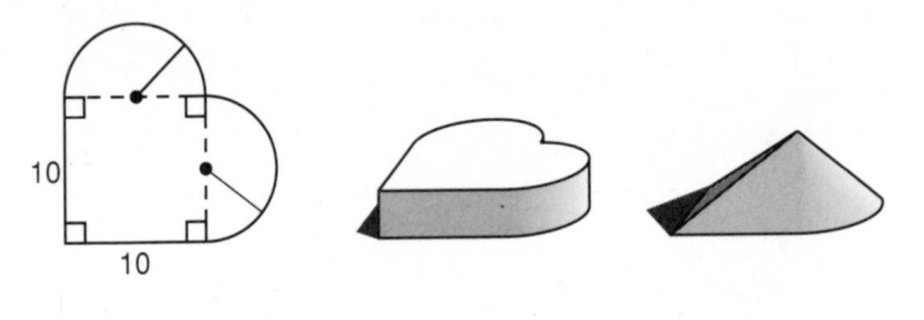

LESSON 102 Distance between two points

In Lesson 101, we discussed the use of the Pythagorean theorem in algebraic form to find the missing side of a triangle. To find side c of the triangle shown here, we write

$$c^2 = 4^2 + 7^2$$

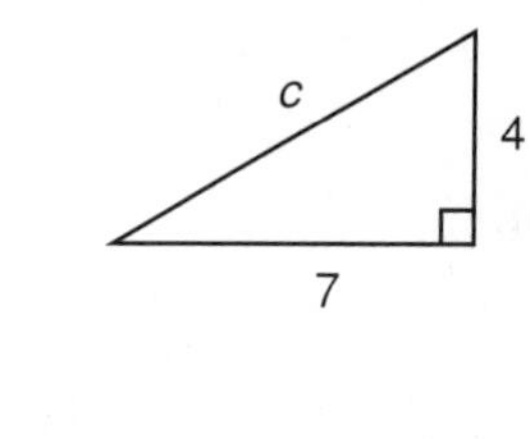

and solve to find that $c = \sqrt{65}$:

$$c^2 = 16 + 49$$

$$c^2 = 65$$

$$\mathbf{c = \sqrt{65}}$$

If we are given the coordinates of two points, we can find the distance between the points by graphing the points, drawing the triangle, and then solving the triangle to find the hypotenuse, which will be the missing side.

example 102.1 Find the distance between the points whose coordinates are (4, 2) and (−3, −2).

solution The first step is to graph the points as done in the figure at the left.

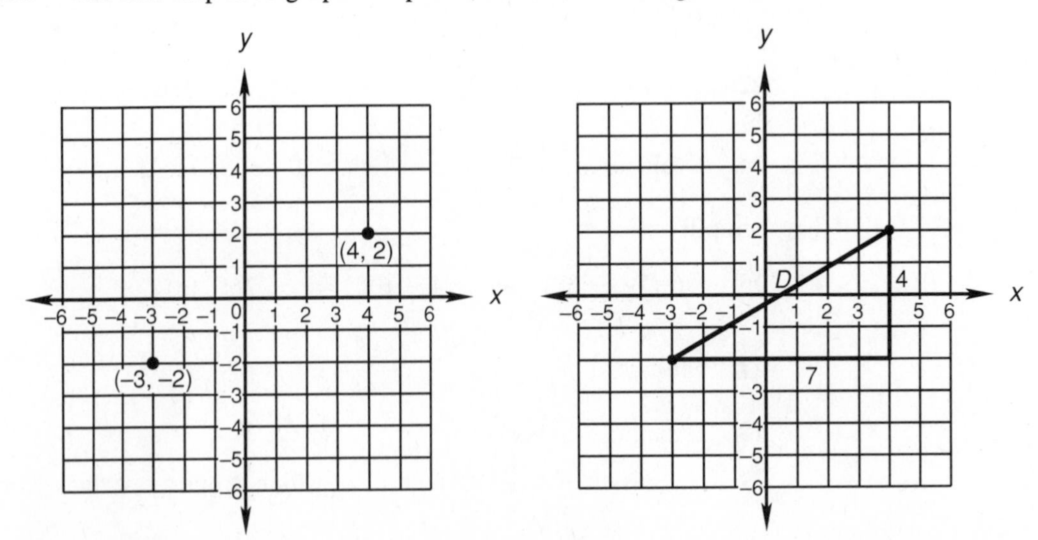

Then we connect the points with a straight line as shown in the figure on the right. We complete the triangle by using a vertical line for one side and a horizontal line for the other side. Next we use the Pythagorean theorem to complete the solution.

$$D^2 = 7^2 + 4^2 \longrightarrow D^2 = 65 \longrightarrow \boldsymbol{D = \sqrt{65}}$$

example 102.2 Find the distance between the points (3, −4) and (−5, 2).

solution We graph the points and draw the required triangle as shown in the figure.

The distance between the points is found by using the Pythagorean theorem.

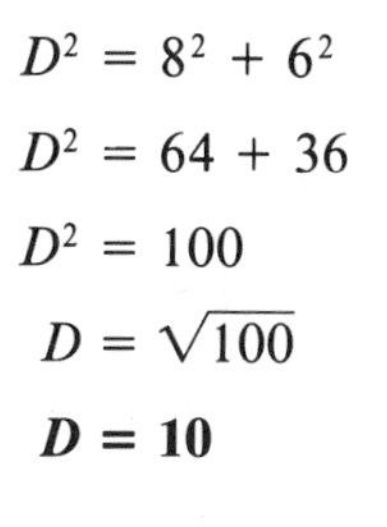

$$D^2 = 8^2 + 6^2$$
$$D^2 = 64 + 36$$
$$D^2 = 100$$
$$D = \sqrt{100}$$
$$\boldsymbol{D = 10}$$

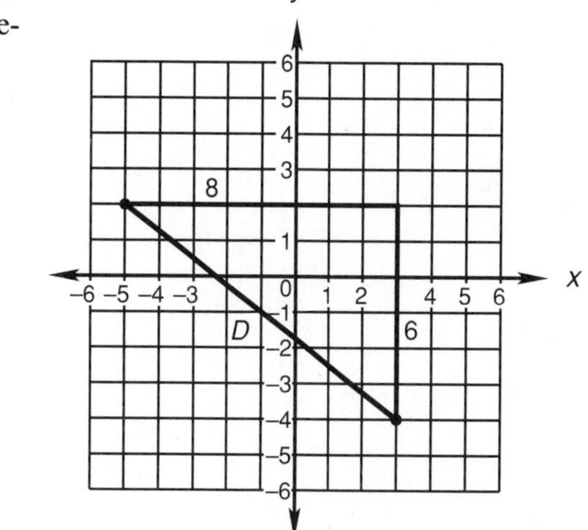

practice Find the distance between (4, −2) and (−4, −4).

problem set 102

1. David added 7 to twice the opposite of a number and then multiplied this sum by 3. Wade got the same result by adding 42 to 3 times the opposite of the number. What was the number?
2. Find four consecutive even integers such that if the sum of the first and the third is multiplied by 3, the result is 10 greater than 5 times the fourth.
3. The free gifts increased the crowd by 250 percent. If 180 people were there at first, how many were there after the gifts were announced?
4. At noon, Sarah headed from Elk City to Idabel at 60 miles per hour. Two hours later, Joan headed from Idabel to Elk City at 46 miles per hour. If it is 332 miles from Elk City to Idabel, what time did they meet?
5. The length of time that the girls ran was 20 percent greater than the length of time the boys ran. If the girls ran for 48 hours, how long did the boys run?
6. Use the Pythagorean theorem to find p.

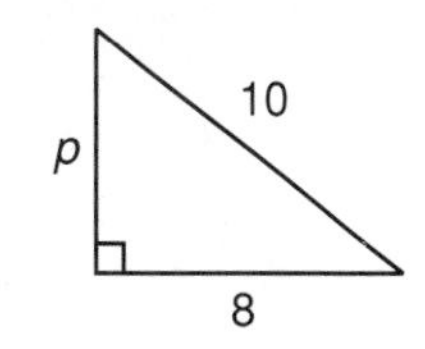

7. Find the distance between (−4, 5) and (2, 3).
8. Find x if $x = \sqrt{b^2 - 4ac}$ and if $b = 11$, $a = 5$, and $\dfrac{c + 10}{2} = 6$.

Simplify:

9. $\dfrac{x^2 + 2x}{4x + 12} \div \dfrac{x^2 - 2x - 8}{x^2 - x - 12}$

10. $\dfrac{(0.00042 \times 10^{-8})(15{,}000)}{(5000 \times 10^{7})(0.0021 \times 10^{14})}$

11. Solve: $-2k^0 - 4k + 6(-k - 2^0) - (-5k) = -(2 - 5)k - 4k$

12. Solve by graphing: $\begin{cases} y = -x \\ y = -4 \end{cases}$

13. Find the equations of lines (a) and (b):

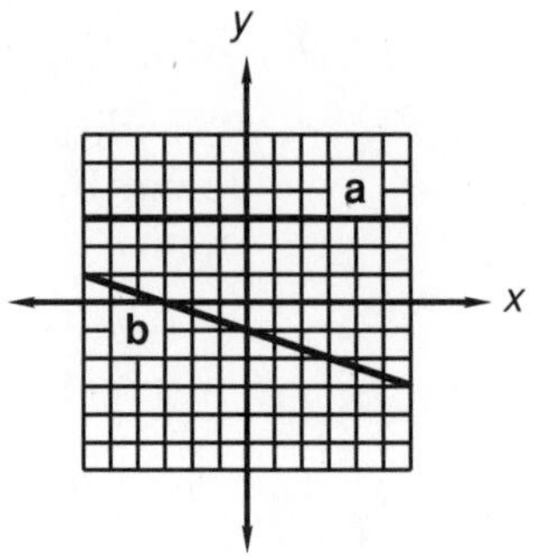

Simplify:

14. $4\sqrt{50} - 3\sqrt{8} + 2\sqrt{3}\sqrt{6}$

15. $3\sqrt{2}(6\sqrt{6} - 4\sqrt{12})$

16. Solve by factoring: $-14 = -x^2 - 5x$

17. Divide: $(x^3 + 6x^2 + 6x + 5) \div (x + 5)$

Solve:

18. $\dfrac{y}{3} - \dfrac{y - 2}{5} = 3$

19. $4\frac{7}{8}p + \dfrac{2}{5} = \dfrac{3}{10}$

20. Add: $\dfrac{6}{m} + \dfrac{4m}{m + 5}$

21. What fraction of $7\frac{1}{4}$ is $\frac{5}{8}$?

22. Simplify: $\dfrac{3x - \dfrac{1}{y}}{\dfrac{2x}{y} - 4}$

23. $-0.061 \in$ {What sets of numbers}?

24. Evaluate: $-x^0a(a - x^0) - a^2$ if $x = -2$ and $a = -4$

Simplify:

25. $-2 - 2^0(-3 - 2) - (-4 + 6)(-5^0 + 2) - 2^2 - \sqrt[5]{-243}$

26. (a) $\dfrac{5x^2 - 5x}{5x}$ (b) $\dfrac{-3^0}{-3^{-2}}$

27. What is another name for the multiplicative inverse of a number?

28. Simplify by adding like terms: $\dfrac{2xxxx}{x^{11}} - 3x^{-7} + \dfrac{4a^0}{x^7}$

29. Graph on a number line: $-x + 2 \not< 3$; $D = \{\text{Integers}\}$

30. The notation $4(6 - 2) = 4 \cdot 6 - 4 \cdot 2$ illustrates which property of the set of real numbers?

LESSON 103 Algebraic proofs

103.A definitions, axioms, and proofs

The development of algebra rests on a foundation of definitions and axioms. We must first define what we mean when we write certain symbols. In Lesson 17 we defined the notation

$$x^4$$

to mean $x \cdot x \cdot x \cdot x$. It was not necessary to choose the notation x^4. We could just as easily have chosen the notation

$$^4x$$

but we didn't. Later on we defined the notation

$$xc$$

to indicate the product of x and c. We could have defined this notation to indicate the quotient of x divided by c, but we didn't. Thus we began our development by carefully defining all notations and symbols.

Unfortunately, we cannot build a mathematical system using definitions alone. To the definitions we must add statements of observations that everyone accepts as true without proof. These statements are called **axioms** or **postulates.** The statements that concern real numbers make up part of what we call the properties of real numbers as we discussed in Lesson 97. For instance, we accept without proof that the order of addition of two real numbers does not affect the resulting sum. We state this abstractly by writing

$$a + b = b + a$$

In Lesson 97, we said that would call this property the commutative property of addition for real numbers. Also, in the same lesson, we stated the associative property of real numbers under addition by writing

$$(a + b) + c = a + (b + c)$$

In mathematics we take pride in holding our list of definitions and axioms to an absolute minimum. Then we use the definitions and axioms to prove all other statements and assertions.

103.B playing the game

The favorite game of mathematicians is a game called **proof.** Every game has rules that must be observed. The rules for proof are:

1. Begin with as few definitions and postulates as possible.
2. Use the definitions and postulates in a step-by-step process to show that another statement is true.
3. Justify each step.

We will begin our study of proofs with some very simple proofs that will help us learn to play the game. The definition of commutative property was carefully restricted to the addition of **exactly two numbers.**

$$4 + 3 = 3 + 4$$

The definition of the associative property was carefully restricted to the addition of

exactly three numbers written in the same order.

$$(4 + 3) + 2 = 4 + (3 + 2)$$

We can use these precise statements to **prove** that we can change the order of addends in a particular addition problem that has more than two addends.

example 103.1 Prove that $4 + 3 + 2 = 3 + 4 + 2$.

solution Why prove the obvious? Because we are learning how to play the game of proof. We will work with the left side of the statement and change its form to that of the right side. First we put parentheses around the 3 and 4 to "associate" them.

$(4 + 3) + 2$ associative property

We justified what we did by noting that we used the associative property. The commutative property tells us we can change the order of two addends, so we exchange the 3 and the 4 inside the parentheses.

$(3 + 4) + 2$ commutative property

We justified this step by noting that we used the commutative property. As the last step, we will remove the parentheses.

$3 + 4 + 2$ removed parentheses

We have shown that, if we accept without proof the commutative property and the associative property, we can prove that another rearrangement is also permissible. The steps in the proofs in this lesson consist of writing parentheses around two numbers, exchanging the numbers, and then removing the parentheses. That's all there is to it.

example 103.2 If x, y, and m represent unspecified real numbers, use the commutative and associative properties to show that

$$x + y + m = m + x + y$$

solution The left-hand side is just like the right-hand side except for the position of the m. We decide to work with the left-hand side and move the m two places to the left. First we exchange the positions of m and y. Then we change the positions of m and x.

$x + y + m$ given

$x + (y + m)$ associative property

$x + (m + y)$ commutative property

Now we must exchange the positions of m and x. First we move the parentheses. Then we can exchange the positions of m and x.

$(x + m) + y$ associative property

$(m + x) + y$ commutative property

$m + x + y$ removed parentheses

example 103.3 If a, b, c, and d represent unspecified real numbers, show that

$$a + b + c + d = b + a + d + c$$

solution The expression on the right is the same as that on the left except that the order of the first two symbols and the last two symbols is reversed. We need only four steps.

$a + b + c + d$	given
$(a + b) + (c + d)$	associative property
$(b + a) + (d + c)$	commutative property (twice)
$b + a + d + c$	removed parentheses

example 103.4 If a, b, c, and d represent unspecified real numbers, show that

$$abcd = bdac$$

solution This one looks more difficult. Let's go about it step by step and feel our way along. We will transform the expression on the left into the same form as the expression that we have on the right. On the right the first factor is b, so let's move b to the front.

$abcd$	given
$(ab)cd$	associative property
$(ba)cd$	commutative property
$bacd$	removed parentheses

Now we have moved b to the front. On the right above we note that c is the last factor, so we will move c to the end.

$bacd$	from above
$ba(cd)$	associative property
$ba(dc)$	commutative property
$badc$	removed parentheses

We wanted to end up with $bdac$. Thus far, we have moved b to the front and c to the end, and all that is amiss is the order of the two middle factors.

$badc$	from above
$b(ad)c$	associative property
$b(da)c$	commutative property
$bdac$	removed parentheses

As you see, these examples are almost the same as those involving addition. In these we use the properties of associativity and commutativity for multiplication instead of the same properties for addition.

example 103.5 If P, Q, and L are unspecified real numbers, show that $PQL = LPQ$.

solution We want to prove that $PQL = LPQ$. We will begin with PQL and change this to LPQ

PQL	given
$P(QL)$	associative property
$P(LQ)$	commutative property
$(PL)Q$	associative property
$(LP)Q$	commutative property
LPQ	removed parentheses

practice If the letters represent unspecified real numbers, use the associative and commutative properties to show that:

a. $xyzo = yoxz$

b. $s + r + p = r + p + s$

problem set 103

1. Steely carried the rock part of the way and Paul carried it the rest of the way. Steely traveled at 20 miles per hour and Paul traveled at 35 miles per hour. If they carried the rock 490 miles in 17 hours, how long did Paul carry the rock?

2. Jill made the trip in 8 hours, while it took Rebecca 10 hours to make the same trip. Find the rate of each if Jill was 10 kilometers per hour faster than Rebecca.

3. Find four consecutive even integers such that 6 times the sum of the first and the fourth is 108 greater than the product of 8 and the opposite of the third.

4. The 3000 hawks and eagles filled the sky. The number of hawks was 1800 greater than 3 times the number of eagles. How many of each kind were there?

5. Twenty-three percent of the newcomers thought there was no difference between a thaumaturge and a prestidigitator. If 3465 people believed that there was a difference, how many newcomers were there in all?

6. If the letters represent unspecified real numbers, use the associative and commutative properties to show that: $s + x + n = x + n + s$

7. The sum of the bank balances in the accounts of 10 prosperous merchants was \$1,500,000. If the average of the first 9 bank balances was \$90,000, what was the amount in the tenth bank account?

8. Use 15 unit multipliers to convert 10 cubic kilometers to cubic miles.

9. Find d.

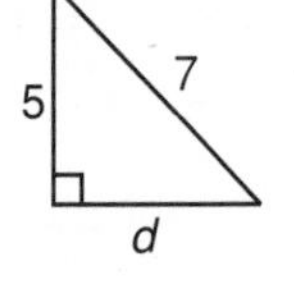

10. Find the distance between $(-3, -4)$ and $(-1, 2)$.

11. Simplify: $\dfrac{x^2 + 5x + 6}{-x^2 - 3x} \div \dfrac{x^2 + 7x + 10}{x^3 + 8x^2 + 15x}$

12. Solve: $-3(-2 - p) - p^0(-4) - 2^0(-5p - 6) = -3 - (-2p)$

13. Solve by graphing: $\begin{cases} y = -2x - 3 \\ x = 2 \end{cases}$

14. Simplify: $\dfrac{(0.000075)(200 \times 10^{-15})}{(0.025 \times 10^{45})(300 \times 10^{-23})}$

15. Find the equations of lines (a) and (b).

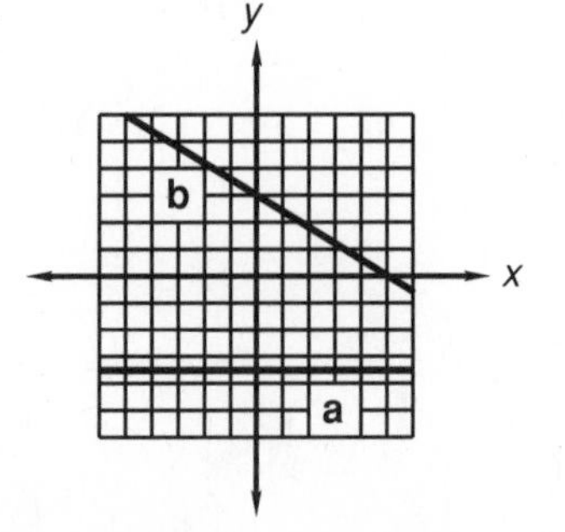

Simplify:

16. $5\sqrt{75} \cdot 2\sqrt{3} + 7\sqrt{3} \cdot 2\sqrt{6}$

17. $4\sqrt{6}(3\sqrt{6} - 2\sqrt{2})$

18. Solve by factoring: $50 = x^2 + 5x$

19. Divide: $(2x^3 + x^2 - 3x) \div (2x + 3)$

Solve:

20. $\frac{x}{4} - \frac{x - 5}{8} = 2$

21. $7\frac{1}{2}x - 4\frac{1}{3} = 14\frac{1}{8}$

22. Add: $\frac{4}{x} + \frac{6x + 2}{x^2} + \frac{3}{x(x + 1)}$

23. Simplify: $\dfrac{x + \frac{x}{y}}{y - \frac{1}{y}}$

24. $7\frac{3}{5}$ is what fraction of $14\frac{1}{8}$?

25. $(3\sqrt{2} - 5) \in$ {What sets of numbers}?

26. Evaluate: $-p^0 - p^2(p - a^0) - ap + |-ap|$ if $a = -3$ and $p = -4$

27. Simplify: (a) $\frac{4xym + 4x^2ym^2}{4xym}$ (b) $\frac{-5^0}{-2^{-4}}$

28. Multiply: $4x^2y\left(\frac{x^{-2}}{4y} - \frac{5x^3y^{-3}}{a^{-4}}\right)$

29. Simplify: $-3^0[(-3^2 + 4)(-2^2 - 2) - (-2) + 4] - \sqrt[3]{-8}$

30. Find x if $x^2 - 13 = \sqrt{9}$.

LESSON 104 Uniform motion—unequal distances

Some uniform motion problems tell us that one person or object traveled a distance that is greater by a specified amount than the distance traveled by another person or object. The distance diagram for these problems usually takes one of the following forms:

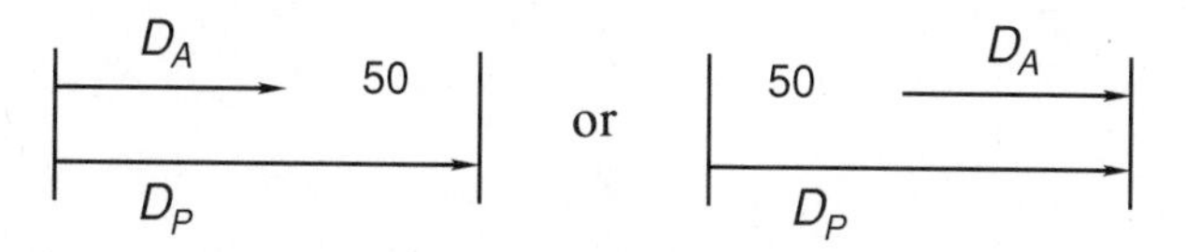

In the picture on the left, both started from the same place and P went 50 farther than A. In the picture on the right, A started out 50 in front of P, and they both ended at the same place. In either case the distance that A traveled plus 50 equals the distance that P traveled. The distance equation for both diagrams is the same.

$$D_A + 50 = D_P \quad \text{so} \quad R_AT_A + 50 = R_PT_P$$

example 104.1 At 8 p.m. Achilles left camp and headed south at 20 kilometers per hour. At 10 p.m.

Patroclos headed south from the same camp. If Patroclos was 50 kilometers ahead by 3 a.m., what was his speed?

solution Since they had the same starting point, the two arrows begin at the same point. Patroclos went farther so his arrow is longer.

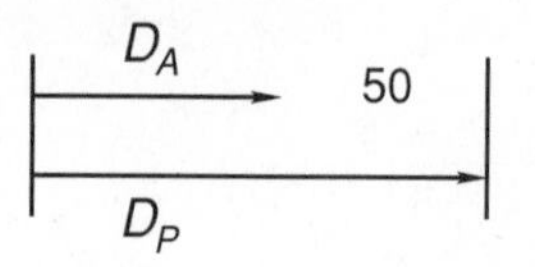

Patroclos went 50 kilometers farther, so we write the distance equation as

$$D_A + 50 = D_P$$

and we substitute R_AT_A for D_A and R_PT_P for D_P to get

$$R_AT_A + 50 = R_PT_P$$

We reread the problem to get the rate and time equations.

$$R_A = 20 \qquad T_A = 7 \qquad T_P = 5$$

Now we solve.

$$\begin{aligned} (20)(7) + 50 &= R_P(5) \\ 140 + 50 &= 5R_P \\ 190 &= 5R_P \\ \textbf{38 kilometers per hour} &= \boldsymbol{R_P} \end{aligned}$$

example 104.2 Rachel has a 15-kilometer head start on Charlene. How long will it take Charlene to catch Rachel if Rachel travels at 70 kilometers per hour and Charlene travels at 100 kilometers per hour?

solution Rachel began 15 kilometers ahead and they ended up in the same place, so the distance diagram is

15 D_R

D_C

We get the distance equation from the diagram as

$$15 + D_R = D_C$$

and we replace D_R with R_RT_R and D_C with R_CT_C to get

$$15 + R_RT_R = R_CT_C$$

Then we reread the problem to get the other three equations:

$$R_R = 70 \qquad R_C = 100 \qquad T_R = T_C$$

Now we solve.

$$\begin{aligned} 15 + 70T_R &= 100T_C \\ 15 &= 30T_C \\ \mathbf{\frac{1}{2}} &= \boldsymbol{T_C} \end{aligned}$$

So Charlene will catch Rachel in $\frac{1}{2}$ hour.

example 104.3 Harry and Jennet jog around a circular track that is 210 meters long. Jennet's rate is 230 meters per minute, while Harry's rate is only 200 meters per minute. In how many minutes will Jennet be a full lap ahead?

solution This problem is simpler if we straighten it out and get the following distance diagram.

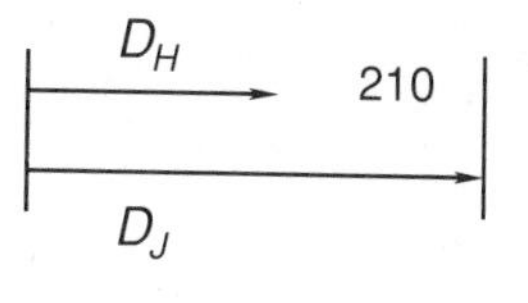

We get the distance equation from this diagram as

$$D_H + 210 = D_J \qquad \text{so} \qquad R_H T_H + 210 = R_J T_J$$

The time equation is $T_H = T_J$, and the rate equations are $R_J = 230$, $R_H = 200$. Thus our four equations are

$$R_H T_H + 210 = R_J T_J \qquad T_H = T_J \qquad R_J = 230 \qquad R_H = 200$$

We use substitution to solve.

$$200T_H + 210 = 230T_H \qquad \text{substituted}$$
$$210 = 30T_H \qquad \text{simplified}$$
$$\mathbf{7\ minutes} = \boldsymbol{T_H} \qquad \text{divided}$$

Thus $\boldsymbol{T_J} = \mathbf{7\ minutes}$ because $T_J = T_H$.

practice

a. At 5 a.m. Napoleon headed south from Waterloo at 4 kilometers per hour. At 7 a.m. Wellington headed south from Waterloo. If Wellington passed Napoleon and was 20 kilometers ahead of Napoleon at 2 p.m., how fast was Wellington traveling?

b. Helen has a 4-kilometer head start on Paris. How long will it take Paris to catch Helen if Helen travels at 6 kilometers per hour and Paris travels at 8 kilometers per hour?

problem set 104

1. Ferris and Julia jog around a circular track that is 500 meters long. Julia's rate is 250 meters per minute, while Ferris's rate is only 230 meters per minute. In how many minutes will Julia be a full lap ahead?

2. Use 12 unit multipliers to convert 25,000 cubic meters to cubic miles.

3. Find x if $x^2 - 20 = 380$.

4. When the car overturned, the jar broke and spilled 450 nickels and quarters all over the freeway. If their value was \$62.50, how many coins of each type were there?

5. When the nurse gave the shots, she noticed that 34 percent of the people winced and the rest were stolid. If 3300 people were stolid, how many shots did she give?

6. Graph on a number line: $-x - 3 < 2$; $D = \{\text{Reals}\}$

Use the Pythagorean theorem to find the missing sides in the triangles shown.

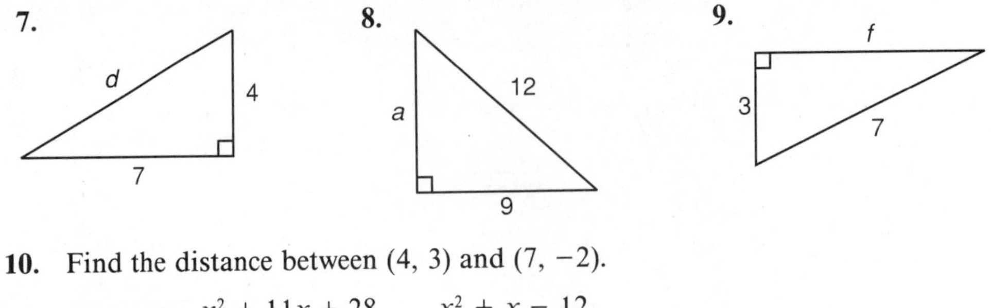

10. Find the distance between (4, 3) and (7, −2).

11. Simplify: $\dfrac{x^2 + 11x + 28}{-x^2 + 5x} \div \dfrac{x^2 + x - 12}{x^3 - 3x^2 - 10x}$

12. Solve: $-x - (-3)(x - 5) - 2^0(2x + 3) = 5x - 7 - 7^0$

13. Solve by graphing: $\begin{cases} y = x - 4 \\ y = -x + 2 \end{cases}$

14. Simplify: $\dfrac{(22{,}000 \times 10^{-7})(500)}{(0.0011)(0.002 \times 10^{14})}$

15. Find the equations of lines (a) and (b).

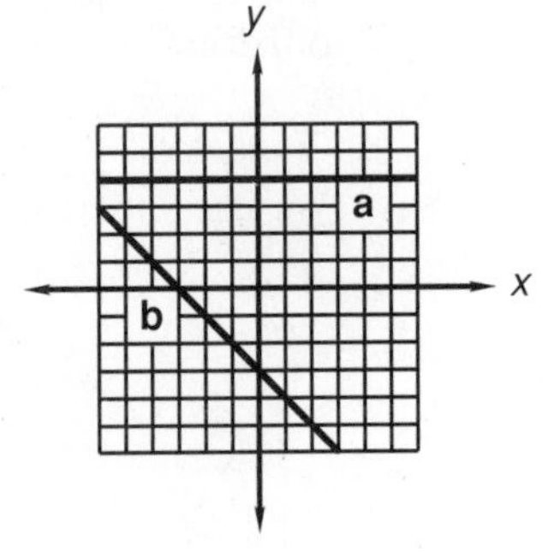

Simplify:

16. $3\sqrt{2} \cdot 4\sqrt{3} \cdot 5\sqrt{12} + 2\sqrt{8}$

17. $3\sqrt{2}(5\sqrt{2} - 4\sqrt{42})$

18. Solve by factoring: $81 = 4x^2$

19. Divide: $(x^3 - 4) \div (x - 4)$

Solve:

20. $\dfrac{p}{6} - \dfrac{p + 2}{4} = \dfrac{1}{3}$

21. $7\frac{1}{9}p + 3\frac{1}{3} = 2\frac{1}{6}$

22. Add: $\dfrac{x}{x + 4} + \dfrac{3}{x} - \dfrac{x + 2}{x^2}$

23. Simplify: $\dfrac{\dfrac{1}{a} + 4}{a^2 + \dfrac{4}{a}}$

24. $7\frac{3}{8}$ is what fraction of 21?

25. True or false? {Integers} ∈ {Reals}?

26. Evaluate: $-p^2 - p^0 + p(-p + a)$ if $p = -3$ and $a = 4$

Simplify:

27. (a) $\dfrac{-2p^2a^2 - p^2a}{-p^2a}$ (b) $\dfrac{-3^2}{-(-3)^{-2}}$

28. $-2^0 - 2[(-3 - 3^0)(-2 + 6)]$

29. Simplify by adding like terms: $\dfrac{1}{(2x)^{-2}y^{-6}} - \dfrac{3x^4}{x^2y^{-6}} - 2x^2y^6$

30. If the letters represent unspecified real numbers, use the associative and commutative properties to prove that $a + c + x = x + c + a$.

LESSON 105 Square roots of large numbers

We have learned to simplify the square root of a positive integer by first writing the integer as a product of prime numbers and then using the product of square roots rule, as we show here.

$$\sqrt{50} = \sqrt{5 \cdot 5 \cdot 2} = \sqrt{5}\sqrt{5}\sqrt{2} = \mathbf{5\sqrt{2}}$$

A similar but slightly different thought process makes the simplification of some of these expressions somewhat easier. Instead of expressing the integer as a product of prime numbers, we express it as a product of a number and the squares of a prime number. For example, in the problem above, as the first step we would write 50 as the product of 25 and 2

$$\sqrt{50} = \sqrt{25 \cdot 2}$$

and then use the product of square roots theorem to complete the simplification.

$$\sqrt{25 \cdot 2} = \sqrt{25}\sqrt{2} = \mathbf{5\sqrt{2}}$$

This thought process is especially helpful when the radicand has a factor that is the square of 10 or 100 or 1000 or 10,000 or any other power of 10.

$$(10)^2 = 100 \qquad (100)^2 = 10{,}000 \qquad (1000)^2 = 1{,}000{,}000$$

$$\text{and} \qquad (10{,}000)^2 = 100{,}000{,}000$$

We note that all of these products have an even number of zeros: 100 has two; 10,000 has four; 1,000,000 has six; and 100,000,000 has eight. **Thus, when we simplify, we always write the number so that one of its factors is an even power of 10.** An even power of 10 is the number 1 followed by an even number of zeros.

example 105.1 Simplify: $\sqrt{50{,}000}$

solution We see four zeros, so we write

$$\sqrt{10{,}000 \cdot 5} = \sqrt{10{,}000}\sqrt{5} = \mathbf{100\sqrt{5}}$$

example 105.2 Simplify: $\sqrt{500{,}000}$

solution It would not help to write $\sqrt{5 \cdot 100{,}000}$ because 100,000 has an odd number of zeros, so we write

$$\sqrt{50 \cdot 10{,}000} = \sqrt{50}\sqrt{10{,}000} = 100\sqrt{50}$$

Now we simplify $\sqrt{50}$ as $\sqrt{5 \cdot 5 \cdot 2} = \sqrt{5}\sqrt{5}\sqrt{2} = 5\sqrt{2}$. Thus

$$100\sqrt{50} = 100(5\sqrt{2}) = \mathbf{500\sqrt{2}}$$

example 105.3 Simplify: $\sqrt{40{,}000{,}000}$

solution Looking for an even number of zeros, we write

$$\sqrt{1{,}000{,}000 \cdot 40} = \sqrt{1{,}000{,}000}\sqrt{40} = 1000\sqrt{40}$$

Now we simplify $\sqrt{40}$ as $\sqrt{2 \cdot 2 \cdot 10} = \sqrt{2}\sqrt{2}\sqrt{10} = 2\sqrt{10}$. Thus $1000\sqrt{40}$ can be written as

$$1000(2\sqrt{10}) = \mathbf{2000\sqrt{10}}$$

example 105.4 Simplify: $\sqrt{700{,}000{,}000}$

solution Using eight zeros in one factor, we can write

$$\sqrt{7 \cdot 100{,}000{,}000} = \sqrt{7}\sqrt{100{,}000{,}000} = \mathbf{10{,}000\sqrt{7}}$$

practice Simplify:

a. $\sqrt{150{,}000}$ **b.** $\sqrt{500{,}000}$

problem set 105

1. Eleanor started out at 60 miles per hour at 9 a.m., 2 hours before Alexi started out to catch her. If she was still 60 miles ahead at 3 p.m., how fast was Alexi driving?

2. Thirty percent of the sailors wanted to turn back, but the rest of the sailors agreed with Aeneas. If 28 sailors agreed with Aeneas, how many sailors were on the ship?

3. It took Perseus 60 hours to get there with white sails and 100 hours to come back with black sails. How far was it if his speed with white sails was 2 miles per hour greater than his speed with black sails?

4. The product of 5 and the sum of a number and -8 is 9 greater than the product of 2 and the opposite of the number. Find the number.

5. Red shoes were \$7 a pair and white shoes were \$3 a pair. Lloyd G. bought 30 pairs for \$130. How many pairs of each color did he buy?

6. Bobby and Joan found four consecutive integers such that 5 times the sum of the second and third was 6 less than 7 times the first. What were their integers?

7. Find g.

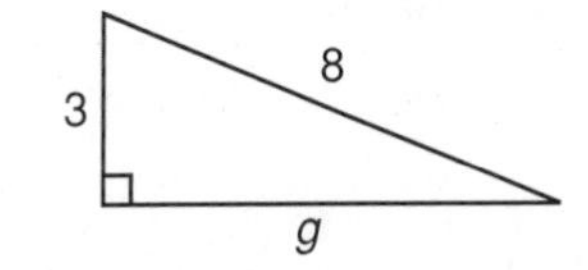

8. Find the distance between $(-4, -2)$ and $(4, -6)$.

9. Simplify: $\dfrac{4x^2 + 8x}{x^2 + 8x + 12} \div \dfrac{4x^2 - 16x}{x^2 + 3x - 18}$

10. Solve: $(-3)x^0 - (-2x) - 4(x - 4) - (2 - x) = 3^0(x - 4)$

11. Solve by graphing: $\begin{cases} y = -2x \\ y = -4 \end{cases}$

12. Simplify: $\dfrac{(1200 \times 10^{-42})(300 \times 10^{14})}{(0.004 \times 10^{5})(3000 \times 10^{-20})}$

13. Find the equations of lines (a) and (b).

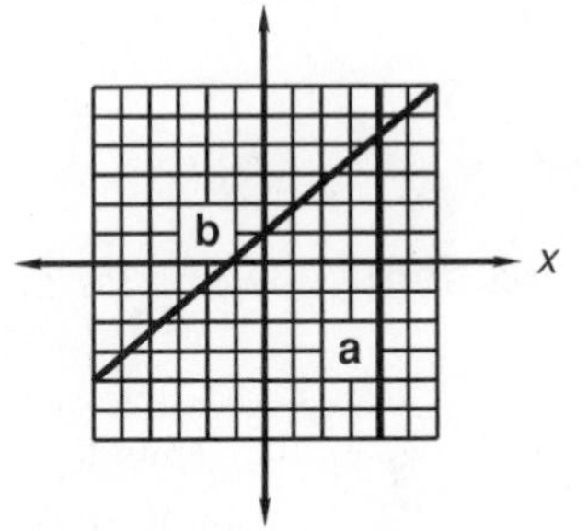

14. Find x if $x = \sqrt{b^2 - 4ac}$ and $a = 2$, $b = 12$, and $4(c + 2) = 48$.

15. For what operations is the set $\{-1, 0, 1\}$ closed?

16. Simplify: $4\sqrt{50{,}000} + 3\sqrt{5{,}000{,}000}$

17. Graph on a number line: $-4 < x \le 2$; $D = \{\text{Integers}\}$

18. Solve by factoring: $64 = 16x - x^2$

19. Divide: $(x^4 + x^3 + 2x + 2) \div (x + 1)$

Solve:

20. $\dfrac{k}{7} - \dfrac{k - 4}{3} = 2$

21. $3\frac{1}{4}k - 2\frac{1}{2} = \frac{3}{4}$

22. Add: $\dfrac{4}{x^2} + \dfrac{x + 4}{x} - \dfrac{3x + 2}{x + 1}$

23. Simplify: $\dfrac{\frac{1}{x} - 4}{y - \frac{1}{x}}$

24. $2\frac{1}{5}$ is what fraction of $28\frac{1}{10}$?

25. $0.0003\sqrt{2} \in$ {What sets of numbers}?

26. Evaluate: $p^2 - p^0(-p^2 - p - x)$ if $p = -4$ and $x = -2$

Simplify:

27. $\dfrac{pa - pa^2}{pa}$

28. $\dfrac{-2^{-2}}{(-2)^2}$

29. $-3^2 - 2^0 - 2[(-2 - 3) - (-4 - 6)]$

30. Multiply: $\dfrac{x^2a}{a^2}\left(\dfrac{a}{x^2} - \dfrac{x^2a}{a^2x}\right)$

LESSON 106 Rounding numbers

106.A place value

In this book we have rounded most answers to two decimal places and have not practiced the skill of rounding numbers. In this lesson we will review the process of rounding to any designated number of digits. Future problem sets will provide problems that allow us to practice rounding.

We use the 10 digits

$$0, 1, 2, 3, 4, 5, 6, 7, 8, 9$$

to write the decimal numerals that we use to represent numbers. The value of a digit in a numeral depends on the position of the digit with respect to the decimal point. For instance, in the numeral

$$40{,}632{,}903.195034$$

the first 9 has a value of 900 because it is in the hundred's place, three places to the left of the decimal point. The second 9 has a value of only $\frac{9}{100}$ because it is written in the hundredths' place, which is two places to the right of the decimal point.

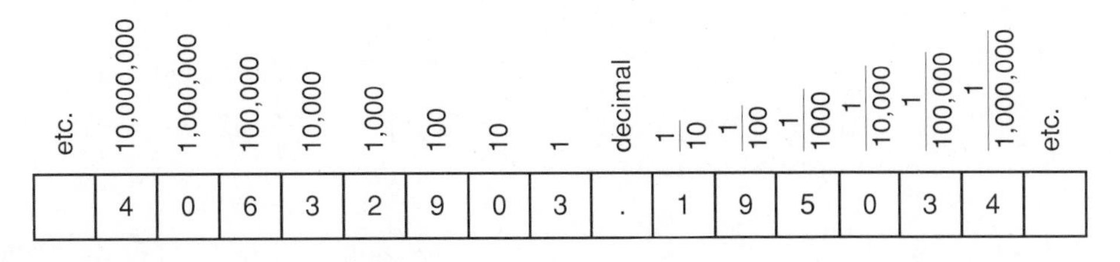

It is important to note that the first place to the right of the decimal point is the tenths' place, whereas the tens' place is not one place but two places to the left of the decimal point. The first place to the left is the units' (ones') place.

106.B rounding numbers

Often we use numbers that are approximations of other numbers. For instance, the circumference of the earth at the equator is 24,874 miles. This measurement is to the nearest mile and is a difficult number to remember. So we say that the circumference is 25,000 miles and say that we have rounded 24,874 to the nearest thousand. This is because 24,874 is closer to 25,000 than it is to 24,000. Thus, when we round, we change the digits on the end of a number to zeros.

Rounding requires three steps, and mistakes can be avoided if a circle and an arrow are used as aids. To demonstrate, we will round 24,874 to the nearest thousand.

1. Circle the digit in the place to which we are rounding and mark the digit to its right with an arrow.

$$2\textcircled{4}\overset{\downarrow}{8}\,7\,4$$

2. Change the arrow-marked digit and all digits to its right to zero.

$$2\textcircled{4}\overset{\downarrow}{0}\,0\,0$$

3. Leave the circled digit unchanged or increase it 1 unit as determined by the following rules:
   (a) If the arrow-marked digit is less than 5, do not change the circled digit.
   (b) If the arrow-marked digit is greater than 5 or is a 5 followed somewhere by a nonzero digit, increase the circled digit 1 unit. This rule applies to the problem we are working, so we finish by writing.

$$\mathbf{2\textcircled{5}\overset{\downarrow}{0}\,0\,0}$$

   (c) If the arrow-marked digit is a terminal 5 or a 5 followed only by zeros, the number is halfway; and the circled digit can be left unchanged or can be increased by 1 as you wish. To be consistent, many people do not change the circled digit if it is even and increase it 1 if it is odd. The procedure to be used in this case is really not important, and we will try to avoid this case in the problem set. Concentrate on remembering rules (a) and (b).

example 106.1 Round 47,258,312.065 to the nearest ten thousand.

solution We circle the ten-thousands' digit and mark the digit to its right with an arrow.

$$47,2\textcircled{5}\overset{\downarrow}{8},312.065$$

Next we change the arrow-marked digit and all digits to its right to zero.

$$47,2\textcircled{5}\overset{\downarrow}{0},000.000$$

Since the arrow-marked digit is greater than 5, we increase the circled digit 1 unit, and our answer is

$$47,2\overset{\downarrow}{\textcircled{6}}0,000.000 \qquad \text{which is} \qquad \mathbf{47{,}260{,}000}$$

example 106.2 Round 104.06245327 to the nearest thousandth.

solution We circle the thousandths' digit and mark the digit to its right with an arrow.

$$104.06\textcircled{2}\overset{\downarrow}{4}5327$$

Then we change the arrow-marked digit and all digits to its right to zero.

$$104.06\textcircled{2}\overset{\downarrow}{0}0000$$

Since the arrow-marked digit is less than 5, we do not change the circled digit. Thus our answer is as follows because the terminal zeros have no value.

104.062

example 106.3 Round 0.00041378546 to the nearest one-hundred-millionth.

solution We circle the one-hundred-millionths' place and mark the digit to its right with an arrow.

$$0.0004137\textcircled{8}\overset{\downarrow}{5}46$$

Now we change to zero the arrow-marked digit and all digits to the right of the arrow-marked digit.

$$0.0004137\textcircled{8}\overset{\downarrow}{0}00$$

The first digit changed to zero was 5, and it was followed by the nonzero digits 4 and 6. Thus we increase the circled digit from 8 to 9 and get

0.00041379

example 106.4 Round 2.0031664567 to five decimal places.

solution The fifth decimal place is the hundred-thousandths' place, which we circle. Then we mark the next digit with an arrow.

$$2.0031\textcircled{6}\overset{\downarrow}{6}4567$$

Since the arrow-marked digit is greater than 5, when we change it to zero we increase the circled digit 1 unit from 6 to 7, and our answer is

2.00317

example 106.5 Round $314.0\overline{364}$ to (a) five decimal places, (b) nine decimal places, (c) the nearest one-hundredth, (d) the nearest ten.

solution The line over the 364 tells us that these digits repeat, so the number is

$$314.0364364364364 \cdots$$

We round this number as specified:

(a) Five decimal places **314.03644**

(b) Nine decimal places **314.036436436**

(c) Nearest hundredth **314.04**

(d) Nearest ten **310**

practice Round:

a. 59,742,004.012 to the nearest ten thousand.

b. 513.129347 to the nearest ten-thousandth.

c. $63.0\overline{149}$ to six decimal places.

problem set 106

1. Boesch had a 40-meter head start. How long did it take Louis to catch up if Louis traveled at 10 meters per second while Boesch traveled at only 6 meters per second?

2. Night came and the mangroves and palmettos closed in on their victim. If 27 percent were mangroves and 511 were palmettos, how many total bushes were on the attack?

3. Robert ran to the redoubt while Wilbur walked to the parapet. Both distances were the same, and Robert's speed was 6 miles per hour while Wilbur's was 8 miles per hour. What was the time of each if Robert's time was 2 hours longer than Wilbur's?

4. Penelope and Miranda found four consecutive odd integers such that 5 times the sum of the first two was 5 less than 19 times the fourth. What were the integers?

5. The 60-foot rope was cut into two pieces. One of the pieces was 10 feet longer than 4 times the other piece. How long were the two pieces?

6. Round:
(a) 104.06253527 to the nearest ten-thousandth.
(b) $413.0\overline{527}$ to the nearest hundred.

7. Find s.

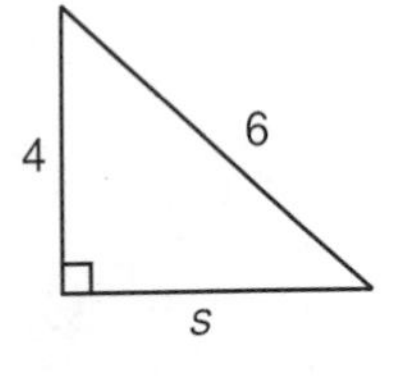

8. Find the distance between $(-8, -4)$ and $(-6, -2)$.

9. Simplify: $\dfrac{x^2 + 8x + 15}{x^2 + 3x} \div \dfrac{x^2 + 3x - 10}{x^3 - 6x^2 + 8x}$

10. Solve: $-2x - 3(x - 2^0) + 2x(-3 - 4^0) = x^0 - 4x - 2$

11. Solve by graphing: $\begin{cases} y = 2x \\ y = -x + 6 \end{cases}$

12. Simplify: $\dfrac{(400 \times 10^5)(0.0008 \times 10^{14})}{(20{,}000 \times 10^{-30})(0.00002)}$

13. Find the equations of lines (a) and (b).

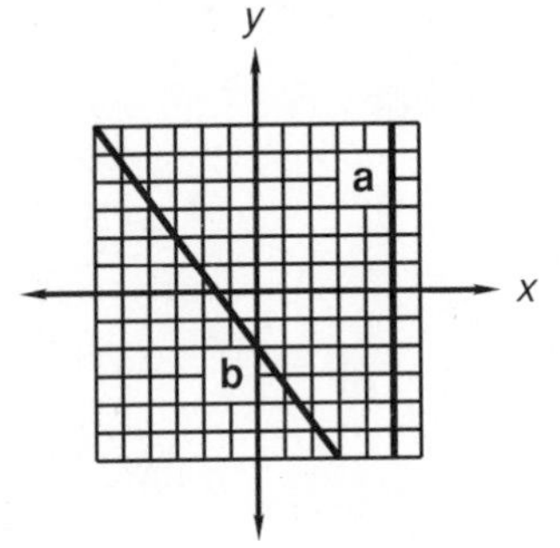

Simplify:

14. $3\sqrt{30{,}000} - 9\sqrt{300} + 3\sqrt{2} \cdot 5\sqrt{6}$ **15.** $3\sqrt{2}(4\sqrt{8} - 3\sqrt{12})$

16. (a) If the reciprocal of a number is $-\frac{1}{101}$, what is the additive inverse of the number?

(b) $(a + x) + y = a + (x + y)$ is a statement of what property of the set of real numbers?

17. Graph on a number line: $-2 \geq -2x + 2$; $D = \{\text{Integers}\}$

18. Solve by factoring: $35 = -12x - x^2$

19. Divide: $(x^3 + 12x + 5) \div (x + 2)$

Solve:

20. $\frac{x}{4} - \frac{x + 2}{6} = 4$ **21.** $5\frac{1}{2}m + \frac{3}{8} = \frac{1}{16}$

22. Add: $\frac{4}{x^2} - \frac{x + 3}{4x} - \frac{2x}{x + 1}$ **23.** Simplify: $\dfrac{\frac{ay}{x} - 4}{\frac{1}{x} + 5}$

24. $3\frac{1}{11}$ is what fraction of 22?

25. $\frac{3 + 4\sqrt{2}}{5} \in \{\text{What sets of numbers}\}$?

26. Evaluate: $-p^2 - p^0(-p + x)$ if $p = -2$ and $x = -4$

Simplify:

27. $\frac{x^2 - ax^2}{x^2}$ **28.** $\frac{(4x)^2(y^{-2})^2xx^2y}{(2x)^2(y^2)^2x^0yx^2}$ **29.** $\frac{-3^{-2}}{3}$

30. $-2^2 - 2[(-3 - 2)(-5 - 4)][-3^0(-2 - 5)]$

LESSON 107 Square root tables

In Lesson 65 we learned how to approximate the square root of a number by a process of cut and try. Of course, the easiest way to find the square root of a particular positive real number is to insert the number into our handy pocket calculator and then hit the square root key.

In this lesson we will learn to use a table of square roots because the use of such a table will allow us to practice scientific notation and will give us a better feel for what we are doing when we take the square root of a number. **The procedure used will also permit us to estimate the square roots of very large and very small numbers.** The problems will also provide excellent practice in using tables. A table of square roots is in Appendix B of this book. Tables of square roots customarily give the square roots of numbers between 1 and 10. Thus our procedure will be:

1. Write the number in scientific notation and round as necessary.
2. If the power of 10 is not an even power of 10, express it as 10 times an even power of 10.

example 107.1 Estimate $\sqrt{416.23}$. Then use the table of square roots in Appendix B to get a more accurate approximation.

solution The square root table in this book has only three digits, so we round to three digits and write the number in scientific notation.

$$\sqrt{4.16 \times 10^2}$$

Now use the product of square roots rule.

$$\sqrt{4.16}\sqrt{10^2}$$

The square root of 4.16 is a little greater than 2 and the square root of 10^2 is 10, so we estimate the answer as

$$(2.1) \times 10 \approx \mathbf{21}$$

If we look up $\sqrt{4.16}$ in the table, we get a more accurate answer.

n	n^2	$\sqrt{n}$
4.10	16.8100	2.02485
4.11	16.8921	2.02731
4.12	16.9744	2.02978
4.13	17.0569	2.03224
4.14	17.1396	2.03470
4.15	17.2225	2.03715
4.16	17.3056	2.03961
4.17	17.3889	2.04206
4.18	17.4724	2.04450
4.19	17.5561	2.04695

$$\sqrt{4.16}\sqrt{10^2} = \mathbf{2.03961 \times 10}$$

example 107.2 Estimate $\sqrt{4160}$. Then improve your estimate by using the table of square roots.

solution As the first step, we round and write the number in scientific notation.

$$\sqrt{4160} = \sqrt{4.16 \times 10^3}$$

Whenever the power of 10 is odd, we use a trick. We write it as 10 times 10 to an even power. We can write 10^3 as $10 \cdot 10^2$.

$$\sqrt{4.16 \times 10^3} = \sqrt{4.16 \cdot 10 \cdot 10^2}$$

Now we use the product of square roots rule and get

$$\sqrt{4.16 \cdot 10 \cdot 10^2} = \sqrt{4.16}\sqrt{10}\sqrt{10^2}$$

The square root of 10 is a little greater than 3, so our estimated answer is

$$(2.1)(3.1)(10) \approx \mathbf{60}$$

We use the table of squares roots to get a more accurate approximation.

n	n^2	$\sqrt{n}$
4.10	16.8100	2.02485
4.11	16.8921	2.02731
4.12	16.9744	2.02978
4.13	17.0569	2.03224
4.14	17.1396	2.03470
4.15	17.2225	2.03715
4.16	17.3056	2.03961
4.17	17.3889	2.04206
4.18	17.4724	2.04450
4.19	17.5561	2.04695

n	n^2	$\sqrt{n}$
9.91	98.2081	3.14802
9.92	98.4064	3.14960
9.93	98.6049	3.15119
9.94	98.8036	3.15278
9.95	99.0025	3.15436
9.96	99.2016	3.15595
9.97	99.4009	3.15753
9.98	99.6004	3.15911
9.99	99.8001	3.16070
10.00	100.0000	3.16228

$$\sqrt{4.16 \cdot 10 \cdot 10^2}$$
$$= \sqrt{4.16}\sqrt{10}\sqrt{10^2}$$
$$= \mathbf{(2.03961)(3.16228) \times 10}$$

We will leave the answer as an indicated multiplication. We are interested in learning to estimate and learning how to read tables. If we had been interested in a numerical answer, we would have used a calculator and hit the square root key.

example 107.3 Estimate $\sqrt{0.0005273}$. Then use the table of square roots.

solution First we write the number in scientific notation.

$$\sqrt{0.0005273} = \sqrt{5.27 \times 10^{-4}}$$

The product of square roots rule gives us

$$\sqrt{5.27 \times 10^{-4}} = \sqrt{5.27} \times \sqrt{10^{-4}}$$

The square root of 5 is about 2.2 and the square root of 10^{-4} is 10^{-2}, so we estimate

$$\mathbf{2.2 \times 10^{-2}}$$

If we use the square root table, we get

n	n^2	$\sqrt{n}$
5.20	27.0400	2.28035
5.21	27.1441	2.28254
5.22	27.2484	2.28473
5.23	27.3529	2.28692
5.24	27.4576	2.28910
5.25	27.5625	2.29129
5.26	27.6676	2.29347
5.27	27.7729	2.29565
5.28	27.8784	2.29783
5.29	27.9841	2.30000

$$\sqrt{5.27 \times 10^{-4}}$$
$$= \sqrt{5.27}\sqrt{10^{-4}}$$
$$= \mathbf{2.29565 \times 10^{-2}}$$

example 107.4 Estimate $\sqrt{0.00005273}$. Then use the table of square roots.

solution We always begin by writing the number in scientific notation.

$$\sqrt{0.00005273} = \sqrt{5.27 \times 10^{-5}}$$

We must be very careful this time because 10^{-5} does not equal 10×10^{-4}. It does equal 10×10^{-6}, however.

$$\sqrt{5.27 \times 10^{-5}} = \sqrt{5.27 \times 10 \times 10^{-6}}$$

Now we use the product of square roots rule to get

$$\sqrt{5.27}\sqrt{10}\sqrt{10^{-6}}$$

and our estimate is

$$\mathbf{(2.1)\ (3.1) \times 10^{-3}}$$

If we use the square root table, we get

n	n^2	$\sqrt{n}$
5.20	27.0040	2.28035
5.21	27.1441	2.28254
5.22	27.2484	2.28473
5.23	27.3529	2.28692
5.24	27.4576	2.28910
5.25	27.5625	2.29129
5.26	27.6676	2.29347
5.27	27.7729	2.29565
5.28	27.8784	2.29783
5.29	27.9841	2.30000

n	n^2	$\sqrt{n}$
9.91	98.2081	3.14802
9.92	98.4064	3.14960
9.93	98.6049	3.15119
9.94	98.8036	3.15278
9.95	99.0025	3.15436
9.96	99.2016	3.15595
9.97	99.4009	3.15753
9.98	99.6004	3.15911
9.99	99.8001	3.16070
10.00	100.0000	3.16228

$$\sqrt{5.27 \times 10^{-5}}$$
$$= \sqrt{5.27 \times 10 \times 10^{-6}}$$
$$= \sqrt{5.27}\sqrt{10}\sqrt{10^{-6}}$$
$$= \mathbf{(2.29565)(3.16228) \times 10^{-3}}$$

example 107.5 Estimate $\sqrt{46{,}200 \times 10^{-17}}$. Then use the table of square roots.

solution First we write the number in scientific notation.

$$\sqrt{46{,}200 \times 10^{-17}} = \sqrt{4.62 \times 10^{-13}}$$

Again we get a negative odd power of 10. We rewrite as usual, but with care because 10^{-13} equals 10×10^{-14}.

$$\sqrt{4.62 \times 10^{-13}} = \sqrt{4.62 \times 10 \times 10^{-14}} = \sqrt{4.62}\sqrt{10}\sqrt{10^{-14}}$$

The square root of 4.62 is about 2.2, so our estimation is

$$\mathbf{(2.2)\ (3.1) \times 10^{-7}}$$

If we use the square root table, we get

n	n^2	$\sqrt{n}$
4.60	21.1600	2.14476
4.61	21.2521	2.14709
4.62	21.3444	2.14942
4.63	21.4369	2.15174
4.64	21.5296	2.15407
4.65	21.6225	2.15639
4.66	21.7156	2.15870
4.67	21.8089	2.16102
4.68	21.9024	2.16333
4.69	21.9961	2.16564

n	n^2	$\sqrt{n}$
9.91	98.2081	3.14802
9.92	98.4064	3.14960
9.93	98.6049	3.15119
9.94	98.8036	3.15278
9.95	99.0025	3.15436
9.96	99.2016	3.15595
9.97	99.4009	3.15753
9.98	99.6004	3.15911
9.99	99.8001	3.16070
10.00	100.0000	3.16228

$$\sqrt{4.62 \times 10^{-15}}$$
$$= \sqrt{4.62 \times 10^{-13}}$$
$$= \sqrt{4.62 \times 10 \times 10^{-14}}$$
$$= \sqrt{4.62}\sqrt{10}\sqrt{10^{-14}}$$
$$= \mathbf{(2.14942)(3.16228) \times 10^{-7}}$$

practice Estimate these square roots. Then use the table of square roots in Appendix B to get a more accurate estimate.

a. $\sqrt{536{,}800}$ b. $\sqrt{0.00006841}$

problem set 107

1. Louis ran to Versailles and then walked back to town. His running rate was 6 kilometers per hour and his walking rate was 3 kilometers per hour. How far was it to Versailles if the round trip took 6 hours?

2. Prince Valiant bought some replacement armor for 540 florins. He bought helmets for 4 florins each and cuirasses for 6 florins each. If he bought 100 pieces of armor, how many helmets did he buy?

3. Roger Goose had a 500-yard head start on Willa. If Willa's speed was 40 yards per second and Roger's speed was only 20 yards per second, how long did it take Willa to catch up?

4. Demosthenes had white pebbles and black pebbles. If 27 percent of his pebbles were white and he had 438 black pebbles, how many pebbles did he have in all?

5. Find four consecutive even integers such that -6 times the sum of the second and fourth is 8 less than 11 times the opposite of the third.

6. If the letters represent unspecified real numbers, use the associative and commutative properties to prove that $abxy = yabx$.

7. Find x if $x = \dfrac{\sqrt{b^2 - 4ac}}{2a}$ and if $a = 5$, $b = 7$, and $4c = 9$.

8. Estimate. Then use the square root table in Appendix B to approximate: $\sqrt{0.003266 \times 10^{-18}}$

9. Find k.

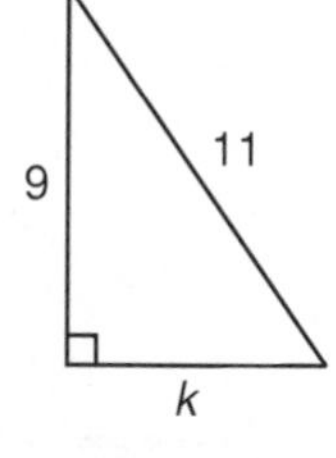

10. Find the distance between $(4, -2)$ and $(-2, 3)$.

11. Simplify: $\dfrac{x^3 + 12x^2 + 35x}{4x^2 + 8x} \div \dfrac{x^2 + 15x + 50}{4x + 8}$

12. Solve: $3x - 2^0(x - 4) + 3(2x + 5) = 7 + x^0$

13. Solve by graphing: $\begin{cases} y = 2x + 4 \\ y = -2x \end{cases}$

14. Simplify: $\dfrac{(30{,}000)(0.000005)}{(1500 \times 10^{-5})(10{,}000)}$

15. Find the equations of lines (a) and (b).

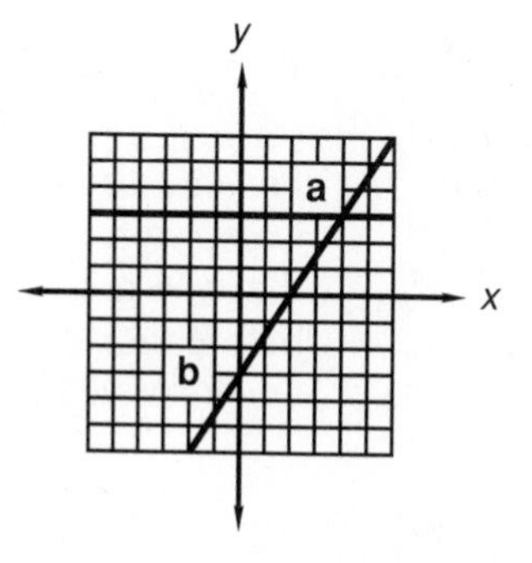

Simplify:

16. $4\sqrt{20{,}000}$

17. $4\sqrt{3} \cdot 5\sqrt{2} - 2\sqrt{2}(3\sqrt{2} - 2\sqrt{12})$

18. Graph: $2 \le x < 5$; $D = \{\text{Positive integers}\}$

19. Solve by factoring: $70 = x^2 + 3x$

Solve:

20. $\frac{2x}{3} - \frac{x + 4}{7} = 2$

21. $3\frac{1}{4}p + 3\frac{1}{4} = 7\frac{1}{8}$

22. Add: $\frac{4}{x^2} - \frac{3x + 2}{x + 1} + \frac{5}{x}$

23. Simplify: $\dfrac{\frac{mx}{4} + 1}{\frac{1}{4} + a}$

24. True or false? $\{\text{Integers}\} \subset \{\text{Naturals}\}$

25. Evaluate: $-p - p(-p - ap)$ if $a = -4$ and $p = -5$

Simplify:

26. $\frac{4 - 4x}{4}$

27. $\frac{-2^{-2}}{-2^2}$

28. $-2[|4 - 2| - (5 - 3)(-2 - 4^0)] - \sqrt[3]{-125}$

29. Multiply: $\frac{a^{-3}y^{-2}}{x}\left(\frac{xa^3}{y^2} - \frac{2ay}{x}\right)$

30. Round $40.\overline{37}$ to the nearest ten-millionth.

LESSON 108 Factorable denominators

Algebra books tend to emphasize the factoring of trinomials and binomials because the ability to factor is important and also because doing these exercises provides experience in manipulating expressions that contain variables. Thus all algebra books contain problems such as this one:

$$\text{Simplify:} \quad \frac{x^2 + x - 20}{x^2 - x - 12} \div \frac{x^2 + 7x + 10}{x^2 + 9x + 14}$$

For the same reasons, algebra books present problems in addition of rational expressions whose denominators are factorable. These problems are designed so that the addition is facilitated if one or more of the denominators are factored before the addition is attempted. The key to these problems is recognizing that they are contrived problems designed to give practice in factoring.

example 108.1 Add: $\frac{6x}{x^2 - x - 12} - \frac{p}{x - 4}$

solution We recognize this problem as a problem designed to give practice in factoring. We begin by factoring the first denominator.

$$\frac{6x}{(x-4)(x+3)} - \frac{p}{x-4}$$

And now we can see that the least common multiple of the denominators is

$$(x-4)(x+3)$$

We use this as our new denominator and add.

$$\frac{}{(x-4)(x+3)} - \frac{}{(x-4)(x+3)} \longrightarrow \frac{6x}{(x-4)(x+3)} - \frac{p(x+3)}{(x-4)(x+3)}$$

$$= \frac{\mathbf{6x - px - 3p}}{\mathbf{x^2 - x - 12}}$$

example 108.2 Add: $\dfrac{7}{x^2 - 5x - 6} - \dfrac{5}{x^2 - 6x}$

solution As the first step we factor both denominators

$$\frac{7}{(x-6)(x+1)} - \frac{5}{x(x-6)}$$

and we see that the least common multiple of the denominators is $x(x-6)(x+1)$. We use this as our new denominator.

$$\frac{}{x(x-6)(x+1)} - \frac{}{x(x-6)(x+1)} \longrightarrow \frac{7x}{x(x-6)(x+1)} - \frac{5(x+1)}{x(x-6)(x+1)}$$

$$= \frac{\mathbf{2x - 5}}{\mathbf{x(x-6)(x+1)}}$$

example 108.3 Add: $\dfrac{4x+2}{x^2 + x - 6} - \dfrac{4}{x^2 + 3x}$

solution We begin by factoring both denominators.

$$\frac{4x+2}{(x+3)(x-2)} - \frac{4}{x(x+3)}$$

We see that the LCM of the denominators is $x(x+3)(x-2)$, so we get

$$\frac{x(4x+2)}{x(x+3)(x-2)} - \frac{4(x-2)}{x(x+3)(x-2)}$$

We now simplify the numerator, remembering that $-4(x-2) = -4x + 8$.

$$\frac{4x^2 + 2x - 4x + 8}{x(x+3)(x-2)} = \frac{\mathbf{4x^2 - 2x + 8}}{\mathbf{x(x+3)(x-2)}}$$

practice Add:

a. $\dfrac{9x}{x^2 - 4x - 21} - \dfrac{7}{x-7}$ **b.** $\dfrac{4}{x^2 - 10x + 9} - \dfrac{5}{x^2 - x}$

problem set 108

1. At the three-quarter pole My Bequest was only 40 feet behind Flying Lady. How long did it take My Bequest at 54 feet per second to catch Flying Lady at 46 feet per second?

2. Find three consecutive even integers such that 4 times the first equals 16 times the sum of the third and the number 2.

3. Ed and Alice walked to the dock at 5 miles per hour, jumped into the boat, and motored to Destin at 15 miles per hour. If the total distance was 20 miles and the trip took 2 hours in all, how far did they go by boat?

4. On the first day of the sale, the girls sold 20 percent of their cookies. If they still had 384 cookies left, how many cookies did they bring to the sale?

5. Charles opened the old trunk and found \$6750 in \$1 bills and \$10 bills. If there were 150 more 1s than 10s, how many of each kind were there?

6. Estimate. Then use the square root table in Appendix B to approximate: $\sqrt{108{,}052 \times 10^{-10}}$

7. Find a.

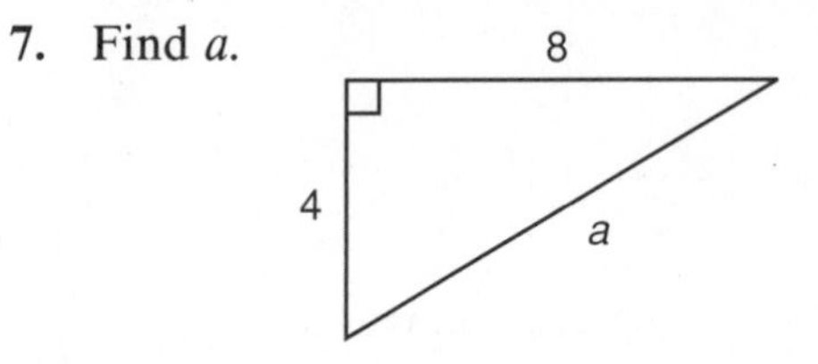

8. Find the distance between $(4, -3)$ and $(-4, 2)$.

9. Simplify: $\dfrac{x^3 + 11x^2 + 24x}{x^2 + 10x + 21} \div \dfrac{4x^2 + 32x}{4x + 40}$

10. Solve: $-2x(4 - 3^0) - (2x - 5) + 3x - 2 = -2^0x$

11. Solve by graphing: $\begin{cases} y = -2x \\ y = -2 \end{cases}$

12. Simplify: $\dfrac{(4000 \times 10^{-40})(0.0003 \times 10^{-21})}{(20{,}000)(3000 \times 10^{-14})}$

13. Find the equations of lines (a) and (b).

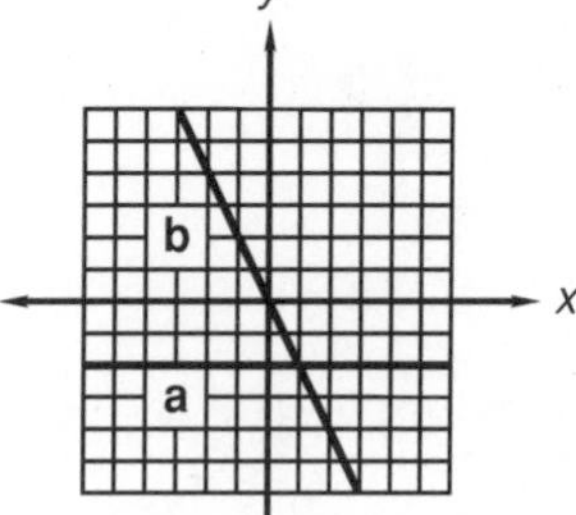

Simplify:

14. $2\sqrt{60{,}000}$

15. $4\sqrt{5} \cdot 2\sqrt{3} + 5\sqrt{3}(\sqrt{3} + 2\sqrt{5})$

16. Graph: $-4 - x \not< 2$; $D = \{\text{Reals}\}$

17. Solve $-81 + 4x^2 = 0$ by factoring.

Solve:

18. $\dfrac{5x}{3} - \dfrac{x - 5}{2} = 14$

19. $4\frac{1}{8}p - \frac{3}{4} = 2\frac{1}{4}$

20. Find the volume in cubic meters of a cylinder whose base is shown and whose height is 3 centimeters. If the cylinder is a right cylinder, find the surface area. What is the volume of a cone that has the same base and altitute? Dimensions are in centimeters.

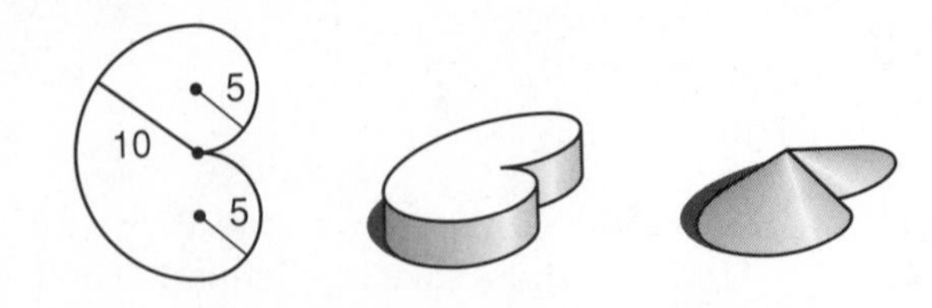

21. The set of negative irrational numbers is closed for what operations?

22. Add: $\dfrac{p}{x^2 - 9} + \dfrac{2x}{x^2 - 3x}$

23. $4^2 \in$ {What sets of numbers}?

Simplify:

24. $\dfrac{-4x^2 - 8x^2a}{-4x^2}$

25. $\dfrac{-3^{-2}}{(-3)^2}$

26. $\dfrac{\dfrac{5p}{x} - 4}{\dfrac{3}{x} - x}$

27. Evaluate: $-x - xk(x - k)$ if $x = -4$ and $k = 5$

Simplify:

28. $\dfrac{x^2a(x^2a)(x^{-2})^2x^0xa^2}{(a^{-3})^2ax^{-2}x^4x}$

29. $2^2[-(-2^0 - 3)(-2 + 5) + 3]$

30. Round $7.\overline{185}$ to the nearest hundred-millionth.

LESSON 109 Absolute value inequalities

We review the concept of absolute value by saying that every real number except zero can be thought of as having two qualities or parts. One of the parts is designated by the plus or minus sign, and the other part is the numerical part. We can think of the numerical part as designating the quality of "bigness" of the number, and we call this quality the **absolute value** of the number. Thus we say that the two numbers

$$3 \quad \text{and} \quad -3$$

both have an absolute value of 3 although one of them is a positive number and one of them is a negative number.

We designate the absolute value of a number by enclosing the number within vertical lines. Thus we designate the absolute value of 3 by writing $|3|$, and we designate the absolute value of -3 by writing $|-3|$. Of course, the absolute value of both of these numbers is 3.

$$|3| = 3 \qquad |-3| = 3$$

It is difficult to describe the absolute value of a number by using words. Most authors do not like to speak of the "bigness" of a number as we have done, for they feel that bigness can be confused with the concept of *greater than* that is used to compare numbers. For this reason, many prefer the formal definition of absolute value used in more advanced courses. This definition uses symbols and does not use words.

The definition is in three parts and we have avoided using the definition thus far because the third part is confusing to some people.

(a) If $x > 0$, $|x| = x$

(b) If $x = 0$, $|x| = 0$

(c) If $x < 0$, $|x| = -x$

Part (a) speaks of the absolute value of positive numbers, which are numbers that are greater than zero. Part (b) describes the absolute value of zero. Part (c) can be confusing because of the minus sign. Part (c) describes the absolute value of negative numbers, which are numbers that are less than zero. We know that the absolute value of a negative number is the opposite of the number, as

$$|-3| = -(-3) = 3$$

Thus when we write

$$\text{If } x < 0, \qquad |x| = -x$$

we are not saying that the absolute value is a negative number but that the absolute value of a negative number is the opposite of the negative number, which is a positive number.

> The absolute value of a nonzero real number is a positive number.[†] The absolute value of zero is zero.

$$|3| = +3 \qquad |-3| = +3 \qquad |0| = 0$$

In an attempt to describe the absolute value of a number by using words, many authors define the absolute value of a number to be the number that describes the distance on the number line from the origin to the graph of the number being considered. If we use this definition, we will find that +3 and −3 have the same absolute value, for they are both 3 units from the origin, and thus both numbers have an absolute value of 3.

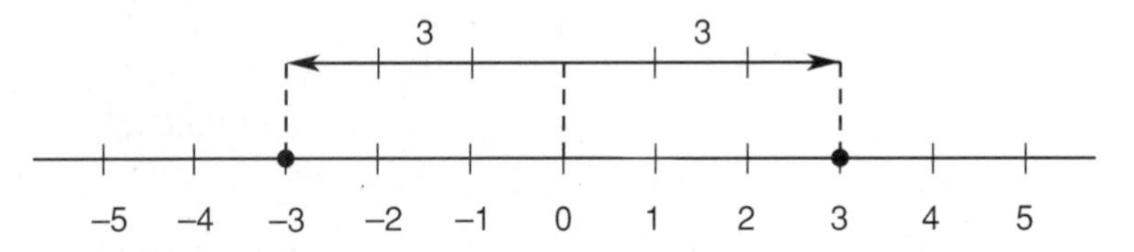

Other authors note that every nonzero real number has an opposite. They say that the absolute value of either member of a pair of opposites is the positive member of the pair. Thus, the absolute value of either

$$3 \quad \text{or} \quad -3$$

is 3, the positive member of the pair. If we remember this, we can see that there are two answers to the following question.

$$|\text{What numbers?}| = 4$$

Here we ask what numbers have an absolute value of 4. Of course, the answers are +4 and −4 because both of these numbers have an absolute value of 4. It is customary to use a single-letter variable as the unknown, so we will restate the question by writing

$$|x| = 4$$

and as we have said, the two values of x that satisfy this condition are +4 and −4.

[†] Some might like to say that the absolute value of a nonzero number is the positive value of the number, but this would be incorrect. A real number that is not zero is either a positive number or a negative number. Numbers do not have two values.

Often it is desirable to display the solution of an absolute value equation or inequality in graphical form. The graph of the solution to the equation $|x| = 4$ is the graph of the numbers $+4$ and -4, as shown here.

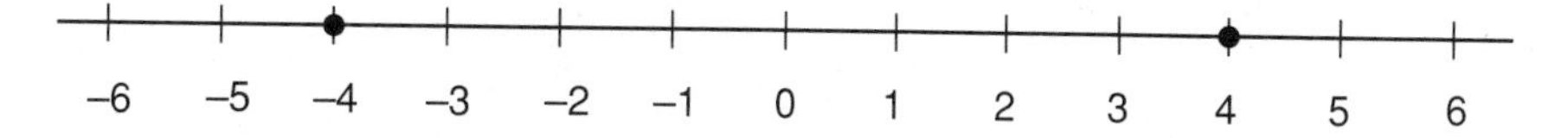

example 109.1 Graph: $|x| > 2$; $D = \{\text{Reals}\}$

solution We are asked to indicate on the number line every real number whose absolute value is greater than 2.

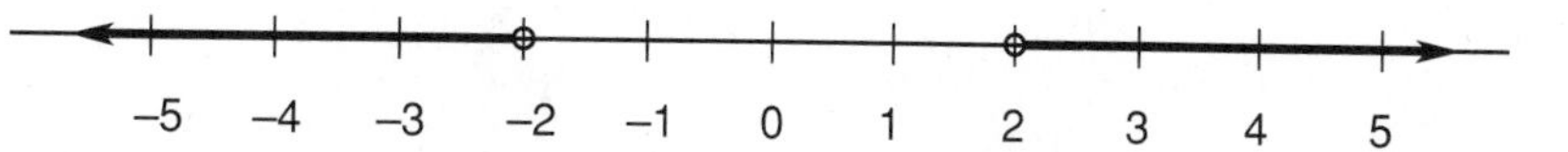

We note that the solution set to this inequality contains many negative numbers. This is correct, for if we think carefully, we will realize that while these negative numbers are less than 2, their absolute values are greater than 2.

example 109.2 Graph: $|x| \leq 3$; $D = \{\text{Integers}\}$

solution We are asked to indicate on the number line the location of all integers whose absolute value is equal to or less than 3.

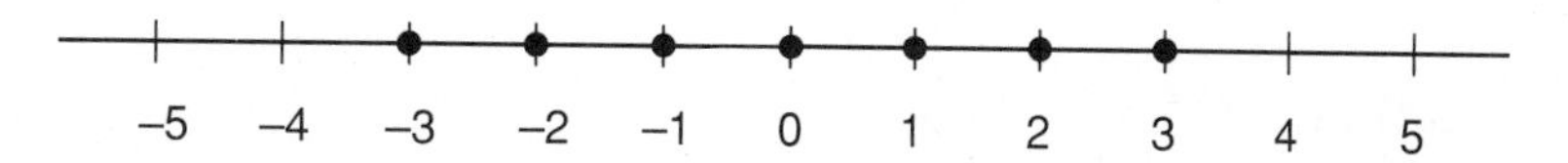

We note that all integers that are greater than or equal to -3 and that are also less than or equal to 3 have an absolute value that is equal to or less than 3.

example 109.3 Graph: $|x| < -4$; $D = \{\text{Reals}\}$

solution

-5 -4 -3 -2 -1 0 1 2 3 4 5

We have not graphed the solution for the given condition because there are no real numbers that satisfy the given condition. The statement $|x| < -4$ asks for the real number replacements for x whose absolute values are less than -4. There are no real numbers that satisfy this condition since the absolute value of any nonzero real number is greater than zero. If we use the formal language of sets, instead of saying that we have not graphed the solution, we say that the solution set is the empty set { }, or the null set $\emptyset$, and we say that the bare number line shown is the graph of this set because this set has no members.

example 109.4 Graph: $|x| > -4$; $D = \{\text{Reals}\}$

solution

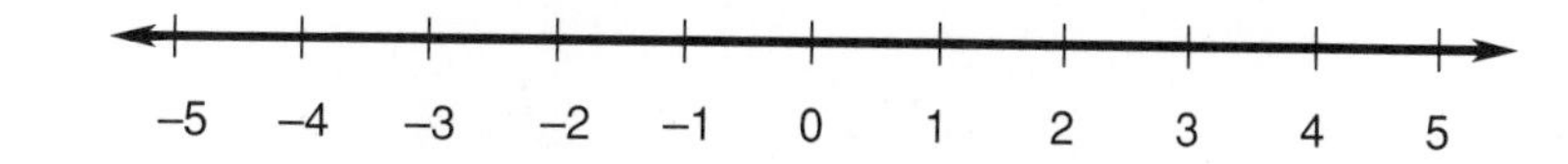

This one is also tricky. The absolute value of every real number is zero or a number greater than zero. Certainly, then, if the absolute value of every real number is equal to or greater than zero, the absolute value of every real number is also greater than -4. Thus the solution to the stated condition is the set of real numbers, which is graphed by indicating the entire number line.

example 109.5 Graph: $-|x| \geq -2$; $D = \{\text{Reals}\}$

solution We begin by multiplying both sides of the inequality by -1 and also **reversing the inequality symbol,** and we find

$$|x| \leq 2$$

which is graphed below.

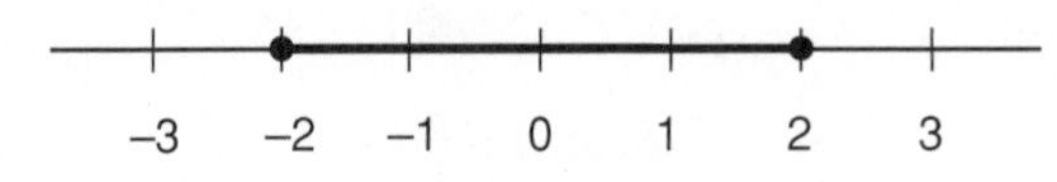

example 109.6 Graph: $-|x| - 2 > -5$; $D = \{\text{Integers}\}$

solution First we must solve the inequality for $+|x|$ by isolating $|x|$ on one side of the inequality.

$-\|x\| - 2 > -5$	given
$-\|x\| > -3$	added + 2 to both sides
$\|x\| < 3$	multiplied both sides by -1 and **reversed the inequality symbol**

–5 –4 –3 –2 –1 0 1 2 3 4 5

practice Graph:

a. $-|x| - 9 > -11$; $D = \{\text{Integers}\}$ **b.** $|x| > 0$; $D = \{\text{Reals}\}$

problem set 109

1. At noon the armadillo left the wild kingdom and headed north at 3 kilometers per hour. Two hours later the raccoon left the same kingdom and headed south at 5 kilometers per hour. At what time will the animals be 38 kilometers apart?

2. Stephanie threw her pennies into the sandbox and then Shannon added her nickels. If they threw in 10 more pennies than nickels and threw in \$29.50 in all, how many nickels were in the sandbox?

3. Twelve percent of the inhabitants had white hair. If 4224 inhabitants had colored hair, how many inhabitants were there in all?

4. Find four consecutive integers such that 5 times the opposite of the first is 5 greater than the product of -3 and the sum of the third and fourth.

5. When the debris was cleared away, there were 52 bricks left. Some were red and the rest were white. The red bricks numbered 16 more than twice the number of white bricks. How many bricks of each color were there?

6. Estimate. Then use the square root table in Appendix B to approximate: $\sqrt{8{,}372{,}150 \times 10^{-15}}$

7. Find p.

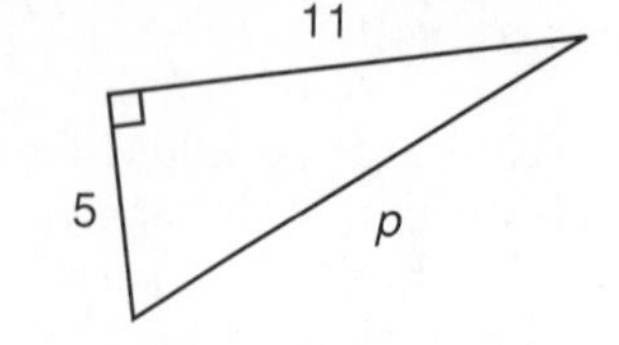

8. Find the distance between $(5, -3)$ and $(7, -2)$.

9. Simplify: $\dfrac{x^3 + x^2 - 12x}{x^2 + 4x} \div \dfrac{x^2 - 11x + 24}{x^2 + 2x - 80}$

10. Solve: $(-x) - 3^0(x - 7) = -(-x - 5)$

11. Solve by graphing: $\begin{cases} x = 2 \\ y = -\dfrac{1}{2}x + 6 \end{cases}$

12. Simplify: $\dfrac{(0.0004 \times 10^{15})(0.06 \times 10^{41})}{(30{,}000{,}000)(400 \times 10^{-21})}$

13. Find the equation of lines (a) and (b).

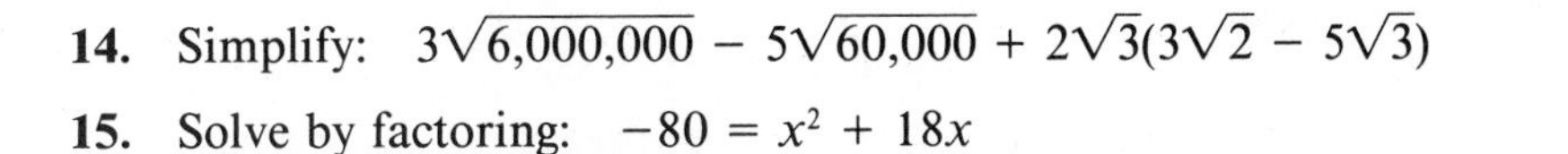

14. Simplify: $3\sqrt{6{,}000{,}000} - 5\sqrt{60{,}000} + 2\sqrt{3}(3\sqrt{2} - 5\sqrt{3})$

15. Solve by factoring: $-80 = x^2 + 18x$

16. Find the volume of a cylinder whose base is shown and whose sides are 3 meters high. If the cylinder is a right cylinder, find the surface area. What is the volume of a cone that has the same base and altitude? Dimensions are in centimeters.

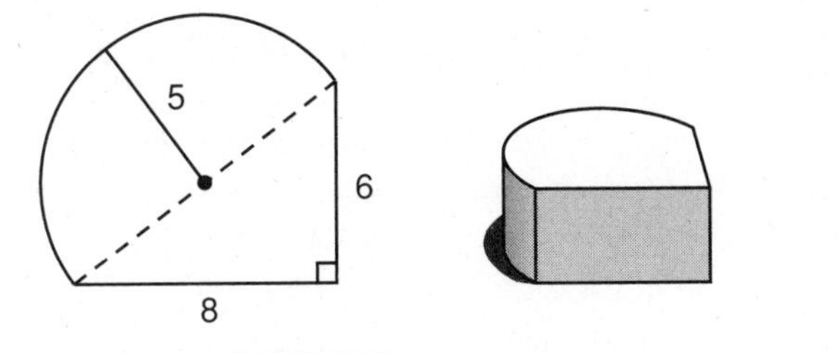

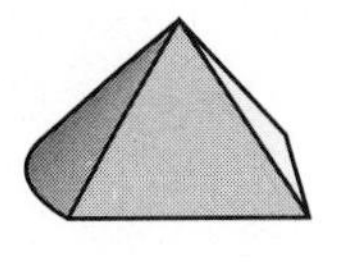

17. Find x if $x = \dfrac{\sqrt{b^2 - 4ac}}{2a}$ and if $a = 10$, $b = 13$, and $8c = 5$.

18. Graph: $-|x| + 3 > 0$; $D = \{\text{Integers}\}$

Add:

19. $\dfrac{4}{x - 4} + \dfrac{5}{x^2 - 16}$

20. $\dfrac{4}{x^2 + 2x + 1} + \dfrac{3x}{x + 1}$

Solve:

21. $\dfrac{5x}{2} - \dfrac{x - 3}{5} = 7$

22. $4\frac{1}{5}k + 2\frac{1}{4} = 7\frac{1}{8}$

23. Evaluate: $-x^0 - x^2 - x(-x + y)$ if $x = -4$ and $y = -3$

Simplify:

24. $\dfrac{-3x - 9x^2}{-3x}$

25. $\dfrac{-3^0 - 4^0}{-2^{-2}}$

26. $\dfrac{4 + \dfrac{y}{x^2}}{y^2 + \dfrac{y}{x^2}}$

27. $0.0013 \in$ {What sets of numbers}?

28. Multiply: $\dfrac{3x^2y^{-2}}{a^{-2}}\left(\dfrac{x^2y}{a} - \dfrac{4x^{-5}y^{-3}}{a^5}\right)$

29. Simplify: $-3^0[(-2 - 2^0) - (-5 + 6^0)] - \sqrt[3]{-64}$

30. What is the product of -42 and the additive identity?

LESSON 110 Rational equations

In Lesson 81 we discussed the solution of rational equations in which all denominators are integers. In these problems we found that the recommended first step is to multiply the numerator of every term by the least common multiple of the denominators. Since every denominator is a factor of the least common multiple, this procedure permits us to cancel every denominator, as we see in the following example.

example 110.1 Solve: $\dfrac{y}{2} + \dfrac{1}{4} = \dfrac{y}{6}$

solution As the first step we will multiply every numerator by 12, the least common multiple of the denominators.

$$12\left(\frac{y}{2}\right) + 12\left(\frac{1}{4}\right) = 12\left(\frac{y}{6}\right) \longrightarrow 6y + 3 = 2y \longrightarrow 4y = -3 \longrightarrow \mathbf{y = -\frac{3}{4}}$$

The denominators of the equation in Example 110.1 are all real numbers. In this lesson we will discuss equations whose denominators contain variables. If there are variables in the denominator, the replacement values of the variables are restricted. For instance, if the given equation is

$$\frac{t-2}{t} = \frac{14}{3t} - \frac{1}{3}$$

The number zero would not be a permissible value for t, for if we substitute the number zero for t, we find

$$\frac{0-2}{0} = \frac{14}{3(0)} - \frac{1}{3} \longrightarrow \mathbf{\frac{-2}{0} = \frac{14}{0} - \frac{1}{3}} \qquad \text{incorrect}$$

which is meaningless, for division by zero is not defined.

If our equation is

$$\frac{n}{n+2} = \frac{3}{5}$$

we cannot accept -2 as a value for n, for if we try to substitute -2 for n, we obtain

$$\frac{(-2)}{(-2)+2} = \frac{3}{5} \longrightarrow \mathbf{\frac{-2}{0} = \frac{3}{5}} \qquad \text{incorrect}$$

an expression in which zero is the denominator of a fraction, and division of a nonzero real number by zero is not defined.

Thus, as our first step in the solution of rational equations whose terms have

variables in one or more denominators, we will list the unacceptable values of the variable, which, of course, are those values of the variable that would cause any denominator to equal zero.

example 110.2 Solve: $\frac{t-2}{t} = \frac{14}{3t} - \frac{1}{3}$

solution ($t \neq 0$). As the next step we multiply every numerator by $3t$, the least common multiple of the denominators. This will allow us to cancel the denominators, and then we will solve the resulting equation.

$$3t\left(\frac{t-2}{t}\right) = 3t\left(\frac{14}{3t}\right) - 3t\left(\frac{1}{3}\right) \longrightarrow 3t - 6 = 14 - t$$

$$\longrightarrow 4t = 20 \longrightarrow \boldsymbol{t = 5}$$

example 110.3 Solve: $\frac{n}{n+2} - \frac{3}{5} = 0$

solution ($n \neq -2$). Now we multiply every term by $(5)(n+2)$, the least common multiple of the denominators, cancel the denominators, and solve.

$$(5)(n+2)\frac{(n)}{(n+2)} - \frac{3}{5}(5)(n+2) = 0 \longrightarrow 5n - 3(n+2) = 0$$

$$\longrightarrow 5n - 3n - 6 = 0 \longrightarrow 2n = 6 \longrightarrow \boldsymbol{n = 3}$$

example 110.4 Solve: $\frac{2}{3n} - \frac{2}{n+4} = 0$

solution ($n \neq 0, -4$). Now we multiply every numerator by $(3n)(n+4)$, cancel the denominators, and solve.

$$(3n)(n+4)\frac{(2)}{(3n)} - \frac{(2)}{(n+4)}(3n)(n+4) = 0 \longrightarrow (n+4)2 - 6n = 0$$

$$\longrightarrow 2n + 8 - 6n = 0 \longrightarrow 8 = 4n \longrightarrow \boldsymbol{n = 2}$$

example 110.5 Solve: $\frac{4}{x} - \frac{7}{x-2} = 0$

solution ($x \neq 0, 2$). Now multiply each term by $x(x-2)$, cancel the denominators, and solve.

$$x(x-2)\frac{4}{x} - \frac{x(x-2)}{1}\frac{7}{x-2} = 0 \longrightarrow 4x - 8 - 7x = 0$$

$$\longrightarrow -3x = 8 \longrightarrow 3x = -8 \longrightarrow \boldsymbol{x = -\frac{8}{3}}$$

example 110.6 Solve: $\frac{4}{p} - \frac{3}{p-4} = 0$

solution ($p \neq 0, 4$). We multiply each term by $p(p-4)$, which is the least common multiple of the denominators, cancel the denominators, and solve.

$$p(p-4)\frac{4}{p} - p(p-4)\frac{3}{p-4} = 0 \longrightarrow 4p - 16 - 3p = 0$$

$$\longrightarrow p - 16 = 0 \longrightarrow \boldsymbol{p = 16}$$

practice Solve:

a. $\frac{5}{t} - \frac{2}{t - 3} = 0$

b. $\frac{8}{y} + \frac{5}{y - 3} = 0$

problem set 110

1. When the battle began, there were 30 percent more brigantines than men-of-war. If there were 260 brigantines, how many brigantines and men-of-war took part in the battle?

2. Frederick headed for Lutzen at 3 kilometers per hour. Later he increased his speed to 4 kilometers per hour. If it was 52 kilometers to Lutzen and the total time of travel was 15 hours, how long did he travel at 3 kilometers per hour?

3. Josephine walked to Brundig and then trotted back home. She walked at 2 miles per hour and trotted at 4 miles per hour. How far was it to Brundig if her walking time was 2 hours longer than her trotting time?

4. If the sum of a number and 10 is multiplied by 5, the result is 2 greater than 7 times the opposite of the number. What is the number?

5. Cookies sold for 10 cents and doughnuts for 20 cents. Pericles bought 25 items for \$3.50. How many doughnuts and how many cookies did he buy?

6. Parking fees were based on a weighted value. The first hour was weighted at 5 times the cost of each of the other hours. What was the charge for 6 hours of parking if the second hour cost \$1?

7. If the multiplicative inverse of a number is -7, what is the multiplicative identity of the number?

Solve:

8. $\frac{p - 4}{p} = \frac{16}{5p} - \frac{1}{5}$

9. $\frac{3}{4n} = \frac{3}{n + 3}$

Graph:

10. $-|x| + 4 > -2$; $D = \{\text{Integers}\}$

11. $-|x| - 4 > -2$; $D = \{\text{Reals}\}$

12. Find p.

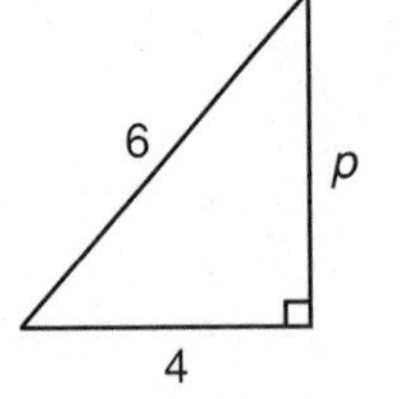

13. Simplify: $\frac{x^2 + 10x + 25}{x^2 + 5x} \div \frac{x^2 + 8x + 15}{x^3 + x^2 - 6x}$

14. Solve by graphing: $\begin{cases} y = x - 2 \\ y = -\frac{1}{2}x + 1 \end{cases}$

15. Find the equations of lines (a) and (b).

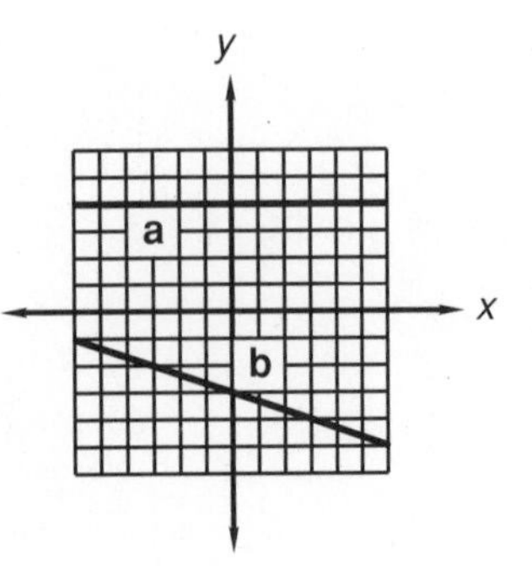

16. Solve: $4\frac{1}{2}x + \frac{1}{2} = \frac{3}{4}$

Simplify:

17. (a) $\frac{3x - 3x^2}{3x}$ (b) $\frac{-4^{-2}}{-(-2)^{-2}}$

18. $\dfrac{a^2 + \dfrac{1}{a}}{ax + \dfrac{b}{a}}$

19. $(-2)[(-2 + 5) + (-2 - 3^0)]$

Add:

20. $\dfrac{4}{a - 2} + \dfrac{6a}{a^2 - 4}$

21. $\dfrac{5}{x + 4} - \dfrac{3}{x^2 + 2x - 8}$

Estimate. Then use the square root table to approximate:

22. $\sqrt{417{,}530 \times 10^{20}}$

23. $\sqrt{417{,}530 \times 10^{-60}}$

24. Simplify: $3\sqrt{2} \cdot 4\sqrt{3} - 4\sqrt{60{,}000} + 2\sqrt{3}(3\sqrt{2} - \sqrt{3})$

25. Divide: $(x^3 - 4) \div (x + 7)$

26. Solve by factoring: $4x^2 - 81 = 0$

27. What fraction of $2\frac{1}{4}$ is $\frac{7}{8}$?

28. $\dfrac{4\sqrt{2}}{5} \in$ {What sets of numbers}?

29. Evaluate: $-x - x^2 + (-x)^3(x - y)$ if $x = -3$ and $y = -5$

30. Simplify by adding like terms: $\dfrac{x}{y} + \dfrac{3x^2y}{y^2x} - \dfrac{4x^0xx^2y^{-2}}{(x)^2yy^{-2}}$

LESSON 111 *Abstract rational equations*

In the preceding lesson we discussed the fact that when we have equations with variables in the denominator, such as

$$\frac{4}{x} + \frac{3}{x - 2} = \frac{7}{2x}$$

the values we may use to replace x are restricted because a denominator can never equal zero. Thus, in the equation above, we cannot use 0 or +2 for x, for either of these will cause at least one denominator to equal zero. We often note the impermissible values of the variable for a problem by listing them using a notation such as $(x \neq 0)$, $(x \neq 2)$, $(x, m \neq 0)$, etc., as we do in the following examples. We will omit these notations in the problem sets.

example 111.1 Find m: $\dfrac{1}{x} + \dfrac{b}{m} = c \qquad (x, m \neq 0)$

solution xm is the least common multiple of the denominators. Thus we begin by multiplying[†] every numerator by xm, the least common multiple of the denominators, and then we cancel the denominators.

$$(\cancel{x}m)\frac{1}{\cancel{x}} + (x\cancel{m})\frac{b}{\cancel{m}} = cxm$$

This leaves us with the following equation.

[†] Permitted by the multiplicative property of equality.

$$m + xb = cxm$$

Now we use the additive property of equality to place all terms with m's on one side and all other terms on the other side.

$$\begin{array}{r} m + xb = cxm \\ \underline{-m \qquad\qquad -m} \\ xb = cxm - m \end{array}$$

Then we factor out the m on the right-hand side and finish by dividing both sides by $(cx - 1)$.

$$xb = m(cx - 1) \longrightarrow \frac{xb}{cx - 1} = \frac{m(cx - 1)}{(cx - 1)} \longrightarrow \mathbf{\frac{xb}{cx - 1} = m} \qquad (cx - 1 \neq 0)$$

example 111.2 Find b: $\frac{a}{b} + \frac{c}{d} = x \qquad (b, d \neq 0)$

solution As the first step we multiply each numerator by bd, the least common multiple of the denominators,

$$(bd)\frac{a}{b} + (bd)\frac{c}{d} = bdx$$

and cancel the denominators.

$$da + bc = bdx$$

Now we use the additive property of equality as necessary to position all terms that contain b on one side of the equation (either side) and all terms that do not contain b on the other side. We decide to position all terms that contain b on the right-hand side of the equation by adding $-bc$ to both sides.

$$\begin{array}{r} da + bc = bdx \\ \underline{-bc \qquad -bc} \\ da \qquad = bdx - bc \end{array}$$

Now factor out the b on the right-hand side,

$$da = b(dx - c)$$

and as a last step divide both sides of the equation by $dx - c$.

$$\frac{da}{(dx - c)} = b\frac{(dx - c)}{(dx - c)} \longrightarrow \mathbf{\frac{da}{dx - c} = b} \qquad (dx - c \neq 0)$$

example 111.3 Find x: $\frac{a}{b} - c = \frac{d}{x} \qquad (b, x \neq 0)$

solution First we eliminate the denominators.

$$(bx)\frac{a}{b} - (bx)c = (bx)\frac{d}{x} \longrightarrow xa - bxc = bd$$

Now, since all x terms are already on one side and all other terms are on the other side, we factor out the x and divide both sides by the coefficient of x.

$$x(a - bc) = bd \longrightarrow \frac{x(a - bc)}{(a - bc)} = \frac{bd}{(a - bc)} \longrightarrow \mathbf{x = \frac{bd}{a - bc}} \qquad (a - bc \neq 0)$$

example 111.4 Find x: $\frac{a}{x} - y + \frac{m}{n} = k \qquad (x, n \neq 0)$

solution First we eliminate the denominators.

$$(\cancel{x}n)\frac{a}{\cancel{x}} - xny + (x\cancel{n})\frac{m}{\cancel{n}} = xnk \longrightarrow na - xny + xm = xnk$$

Now we move all terms that contain an x to the right-hand side, factor out the x, and divide by the coefficient of x.

$$\begin{array}{rcl} na - xny + xm &=& xnk \\ + xny - xm && + xny - xm \\ \hline na &=& xnk + xny - xm \end{array} \longrightarrow na = x(nk + ny - m)$$

$$\longrightarrow \frac{na}{(nk + ny - m)} = \frac{x\cancel{(nk + ny - m)}}{\cancel{(nk + ny - m)}}$$

$$\longrightarrow \mathbf{\frac{na}{nk + ny - m} = x} \qquad (nk + ny - m \neq 0)$$

practice a. Find b: $\frac{3z}{m} + \frac{n}{b} = f$ b. Find m: $\frac{3a}{y} - s + \frac{k}{m} = x$

problem set 111

1. Gold was worth 422 marks a gram and copper was worth 4 marks a gram. If Brother Gregory had an 8-gram mixture of gold and copper that was worth 2122 marks, how many grams of gold did he have?
2. Find three consecutive odd integers such that 4 times the first is 14 less than twice the sum of 2 and the third.
3. Thirty percent of the people did not like the king. If 81,150 people did not like the king, how many people lived in the kingdom?
4. Milton ran north at 8 kilometers per hour. Four hours later Harriet set out in pursuit at 16 kilometers per hour. How long did it take Harriet to catch Milton?
5. The train made the trip in 4 hours. The girls walked it in 48 hours. How far was it if the speed of the train was 55 miles per hour faster than the girls walked?

Solve:

6. $\frac{1 + m}{m} - \frac{3}{m} = 0$
7. $\frac{3}{4x} = \frac{2}{x + 5}$
8. $\frac{x}{5} - \frac{3 + x}{7} = 0$
9. $\frac{2}{x} - \frac{3}{x - 1} = 0$
10. Find n: $\frac{a}{n} - m + \frac{5k}{x} = y$
11. Find d: $\frac{2c}{a} - x = \frac{b}{d}$
12. If the letters represent unspecified real numbers, use the associative and commutative properties to prove that $abxy = xbya$.
13. Find x if $x = \frac{-b \pm \sqrt{b^2 - 4ac}}{2a}$ and if $a = 2$, $b - 1 = 2$, and $c + 1 = -1$.

Add:

14. $\frac{4}{x^2 - 4} + \frac{3x}{x - 2}$
15. $\frac{7}{x + 5} - \frac{2x}{x^2 - 25}$
16. Estimate. Then use the square root table to approximate: $\sqrt{714{,}200 \times 10^{-15}}$

17. Find the distance between $(-4, 2)$ and $(-10, 6)$.

18. Solve: $2x^0(x - 2) - 3x - 4 - [-(-2)] - 7^0 = -2x - 4$

19. Simplify: $\dfrac{(21{,}000 \times 10^{50})(0.0006 \times 10^{15})}{(0.007 \times 10^{20})(9000 \times 10^{-40})}$

20. Solve by factoring: $63 = -x^2 - 16x$

21. Simplify: $4\sqrt{20{,}000} - 15\sqrt{8} + 3\sqrt{2}(4\sqrt{2} - 5)$

Add:

22. $\dfrac{3}{x - 5} + \dfrac{2}{x} + \dfrac{7}{x^2 - 25}$

23. $\dfrac{-x}{x + 5} - \dfrac{3x}{x^2 + 3x - 10}$

24. Evaluate: $-x^2 - x(xy - xy^2)$ if $x = -2$ and $y = -3$

25. Simplify: $\dfrac{(x^2)^{-2}yyx^{-2}}{(x^2y^{-2})^{-3}}$

Graph:

26. $-|x| - 2 < -4$; $D = \{\text{Reals}\}$

27. $-x + 2 \le 7$; $D = \{\text{Integers}\}$

28. Simplify: (a) $\dfrac{4x^2ay - 4xay}{4xay}$ (b) $\dfrac{-2^{-2}}{-(-2^0)^{-3}}$

29. Multiply: $\dfrac{4x^{-2}}{a^2}\left(\dfrac{x^2}{4a^{-2}} - \dfrac{2x^{-2}}{a^4}\right)$

30. A right circular cylinder is 10 meters long. A right prism is cut out of the cylinder as shown. Find the volume of the remaining solid. Dimensions are in meters.

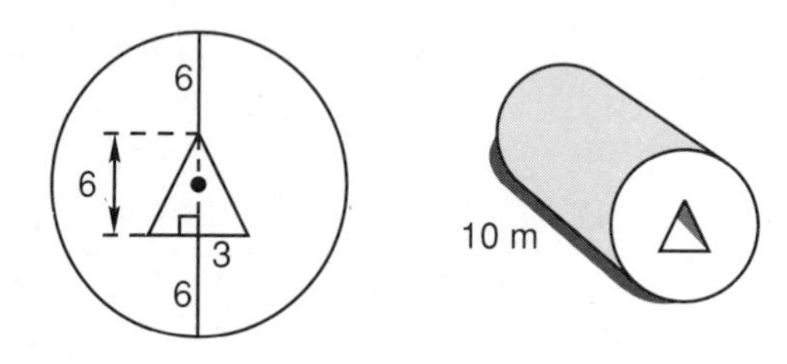

LESSON 112 Equation of a line through two points

112.A linear equations

The graph of a first-degree equation in two unknowns is a straight line. This is the reason we call these equations linear equations. We also use the name *linear equation* to describe first-degree equations in more than two unknowns. The graph of a first-degree equation in three unknowns, such as

$$4x + 2y - z = 5$$

is a plane, and an equation in more than three unknowns cannot be graphed because graphs are restricted to the three dimensions available for graphing.

The standard form of the equation of a straight line is

$$ax + by + c = 0$$

where a, b, and c are constants. The following are equations of straight lines in standard form:

$$4x + y + 1 = 0 \qquad -2x - y - 11 = 0$$

If the equation of a line is written so that y is expressed as a function of x, such as

$$y = mx + b$$

we say that we have written the equation in **slope-intercept form.** In this equation m represents the slope of the line and b represents the y intercept of the line, which is the y coordinate of the point where the line in question crosses the y axis.

112.B equation of a line through two points

Thus far, we have learned how to draw the graph of a given linear equation and have learned how to find a good approximation of the equation of a given line. Both of these exercises have helped us to understand the relationship between the equation of a line and the graph of a line.

Algebra books usually contain three other types of straight line problems that are helpful in exploring this relationship. In the first type we are given the coordinates of two points and asked to find the equation of the line that passes through the two points.

example 112.1 Find the equation of the line that goes through the points (4, 2) and (−5, −3).

solution The slope-intercept form of the desired equation is $y = mx + b$, and **we need to find the values of m and b.** First we graph the two points and draw the line in the figure on the left. If we draw the slope triangle so that the sides of the triangle terminate on these points as we have done in the figure on the right,

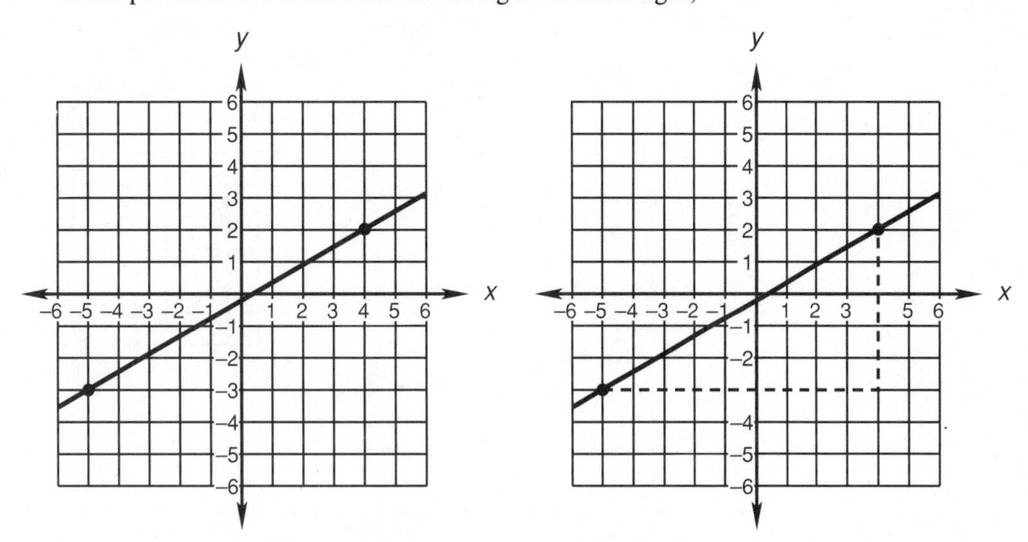

we can determine the **exact value of the slope** to be $+\frac{5}{9}$. We say the exact value because we were given the exact coordinates of two points on the line. We did not estimate their coordinates by looking at a graph.

We can see from the graph that the value of the intercept is approximately −0.3. **This estimated value of the intercept is not acceptable when the exact**

coordinates of two points on the graph are known, for we can find the exact value for the intercept. We know the exact value of the slope, so we can write the desired equation as

$$y = \frac{5}{9}x + b$$

We know the exact values of the coordinates of two points that lie on the line. We can use the coordinates of either of these points for x and y in the equation above and find the exact value of b algebraically.

USING (4, 2)	USING (−5, −3)
$y = \frac{5}{9}x + b$	$y = \frac{5}{9}x + b$
$(2) = \frac{5}{9}(4) + b$	$(-3) = \frac{5}{9}(-5) + b$
$\frac{18}{9} = \frac{20}{9} + b$	$-\frac{27}{9} = -\frac{25}{9} + b$
$-\frac{2}{9} = b$	$-\frac{2}{9} = b$

Now that we have the exact values of m and b, we can write the exact equation of the line that goes through the two points as

$$\mathbf{y = \frac{5}{9}x - \frac{2}{9}}$$

example 112.2 Find the equation of the line that goes through the points (4, −2) and (−3, 4).

solution The general form of the desired equation is $y = mx + b$. We need to find the values of m and b. We graph the points, draw the line, and draw the triangle to find that the slope m is exactly $-\frac{6}{7}$.

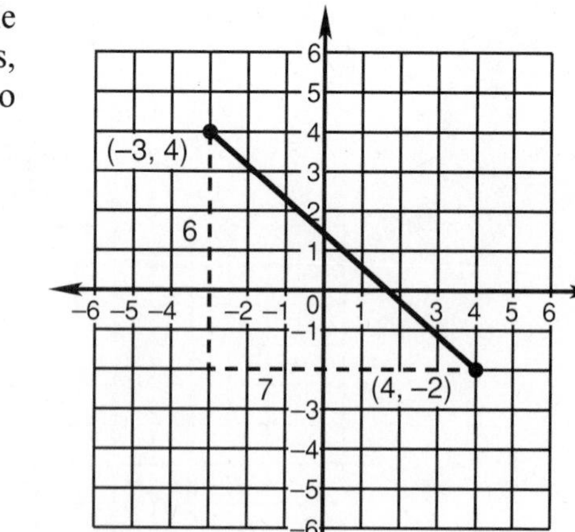

Now we have

$$y = -\frac{6}{7}x + b$$

and we can find the exact value of b by substituting either (4, −2) or (−3, 4) for x and y and solving algebraically for b. We will use the point (4, −2).

$$-2 = -\frac{6}{7}(4) + b$$

$$-\frac{14}{7} = -\frac{24}{7} + b$$

$$\frac{10}{7} = b$$

So the desired equation is

$$y = -\frac{6}{7}x + \frac{10}{7}$$

and the values that we have found for the slope and the intercept are exact.

We see from these two examples that when we are given the coordinates of two points that lie on the line, the exact equation of the line can be determined. Estimated values of the slope and intercept are not acceptable for this type of problem.

example 112.3 Find the equation of the line through the points (4, 3) and (4, −3).

solution When we graph the points and draw the line, we find that the line is a vertical line. Vertical and horizontal lines can be thought of as special cases. By inspection, the equation of this line is $\boldsymbol{x = 4}$.

The change in y is 6 and the change in x is zero. If we try to find the slope of this line by using the rise over the run, we get $\frac{6}{0}$, which has no value. Thus we say that the slope of this line is **undefined.**

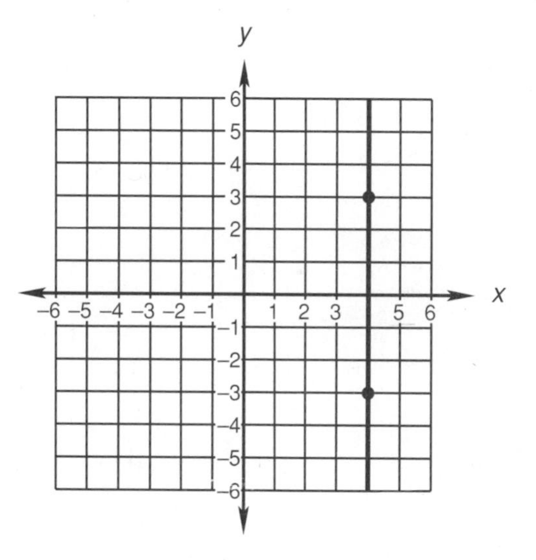

Whenever someone says that the slope of a line is undefined, they are discussing a vertical line. Every line except a vertical line has a defined slope.

practice Find the equations of the lines that go through the following pairs of points:

a. (3, −5) and (2, −1) **b.** (−3, −3) and (6, 5)

problem set 112

1. The passenger train headed north at 70 miles per hour at 6 a.m. At 8 a.m. the freight train headed south from the same station at 30 miles per hour. At what time will the trains be 440 miles apart?
2. The sciolist headed for town at 30 miles per hour. Two hours later, the charlatan began his pursuit at 50 miles per hour. How long did it take the charlatan to catch the sciolist?
3. There were quarters and dimes in profusion. Their value was \$9.55, and there were 64 coins in all. How many were quarters and how many were dimes?

4. Find three consecutive even integers such that if the sum of 5 and the second is multiplied by -7, the result is 11 greater than 5 times the opposite of the third.

5. Seventeen percent of the mob had a propensity for jogging and the rest just wanted to walk. If 3825 wanted to jog, how many were in the mob?

6. Find the equation of the line that goes through the points (2,5) and $(-4, -3)$.

7. If the additive inverse of a number is $\frac{2}{3}$, what is the reciprocal of the same number?

8. Find the volume of the walls of a right cylinder whose base is the shaded area of the figure shown and whose sides are 3 meters high. Dimensions are in meters.

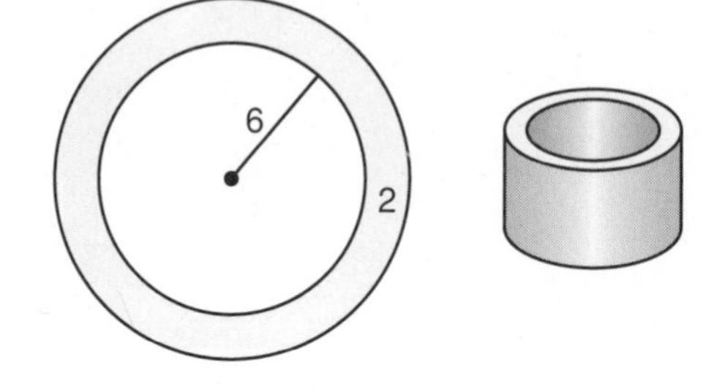

Solve:

9. $\frac{2 + x}{4} + \frac{x}{2} = 5$

10. $\frac{9}{4x} = \frac{5}{x + 11}$

11. $\frac{12}{x} + \frac{1}{4x} = 7$

12. Find y: $\frac{x}{y} + \frac{1}{m} = p$

13. Find c: $\frac{k}{m} + \frac{1}{c} = x$

14. Find b: $\frac{1}{b} + \frac{k}{x} = y$

15. Find m: $\frac{1}{m} + \frac{b}{c} = \frac{x}{y}$

Add:

16. $\frac{4}{x^2 - 25} - \frac{x}{x - 5}$

17. $\frac{3x}{x^2 - x - 6} - \frac{3}{x - 3}$

18. Estimate. Then use the square root table to approximate: $\sqrt{0.000325 \times 10^{-41}}$

19. Find the distance between $(-5, 3)$ and $(4, -2)$.

20. Solve: $-3^0(x - 4) - 2x - (-2x) - [-(-3)] + 5^0 = 2(-x + 2)$

Simplify:

21. $\frac{(35{,}000 \times 10^{-40})(300 \times 10^{15})}{(0.007 \times 10^{15})(15{,}000{,}000)}$

22. $\frac{(y^{-2})^0 y^0 y^2 yyx(xy)^2}{y^2 y^{-2}(y^2)^{-2} axy}$

23. $3\sqrt{30{,}000} - 5\sqrt{27} + 5\sqrt{3}(2\sqrt{3} - 2)$

24. (a) $\frac{6x + 6}{6}$ (b) $\frac{-3^{-2}}{(-2)^2}$

25. Solve by factoring: $-56 = 15x + x^2$

26. Evaluate: $-x^2 - x(xy - y)$ if $x = -3$ and $y = 4$

27. Graph on a number line: $-|x| - 4 > 2$; $D = \{\text{Reals}\}$

28. Simplify by adding like terms: $\frac{x^2y^3a^2}{a^{-2}} + \frac{a^5ay^4}{a^2yx^{-2}} - \frac{4a^2y^5}{y^{-2}x^{-2}}$

29. Multiply: $\frac{4x^2y^{-2}}{a^2}\left(\frac{x^{-2}y^{-2}}{a^{-2}} + \frac{3xy^{-2}}{a^2}\right)$

30. Round $45{,}732.\overline{654}$ to the nearest ten thousand.

LESSON 113 Functions

113.A dependent and independent variables (again)

If we consider the equation

$$y = x + 6$$

we see that this equation matches a value of y with any value of x. We will demonstrate this by replacing x with 4 and then with 23.

If $x = 4$	If $x = 23$
$y = (4) + 6 \longrightarrow \mathbf{y = 10}$	$y = (23) + 6 \longrightarrow \mathbf{y = 29}$

On the left we let $x = 4$ and find that $y = 10$, and on the right we let $x = 23$ and find that $y = 29$. **We see that the value of y depends on the value that we assign to x, so we say that y is the dependent variable and that x is the independent variable. Of course, we could assign a value to y and use the equation to find the value of x, and this would reverse the names. It is customary to use the letter x to designate the independent variable and to use the letter y to designate the dependent variable. We follow this convention in this book.**

113.B functions

We use the word **function** to describe a relationship between two sets in which

1. The first set is the domain and the domain is defined.
2. For each member of the domain there is exactly one answer in the second set.

If we consider the equation

$$y = 4x + 6 \qquad D = \{\text{Reals}\}$$

we see that this equation will give us **one answer** for y for any real number value of x. For example,

If we let $x = 2$	If we let $x = 10$
$y = 4(2) + 6$	$y = 4(10) + 6$
$\mathbf{y = 14}$	$\mathbf{y = 46}$

Thus the equation and the domain define a function because

1. **The domain is specified.**
2. **For any value of the domain the equation will give us one and only one value (answer) for y.**

Now let us consider the ordered pairs

$$(4, 7) \qquad (5, -2) \qquad (7, -4) \qquad (8, 3)$$

and ask if these ordered pairs designate a function. The answer is yes because

1. The domain is specified. The only x values we can use are the x values in the ordered pairs, which are the numbers $\{4, 5, 7, 8\}$.
2. For each value of the domain there is exactly one answer for y. If we let x be 4, then y is 7; if x is 5, then y is -2; if x is 7, then y is -4; if x is 8, then y is 3.

From this example we see that an equation is not necessary to designate† a function. We have fulfilled both of the requirements of a function by listing the ordered pairs.

If we use proper nomenclature to describe a function, we must use the words **domain, image,** and **range.** The domain, as we know, is the set of all permissible replacement values of x. We will use the word **image** instead of using the word **answer** to describe the value of y that is matched with each value of x by the function. We will use the word **range** to describe the set of all images.

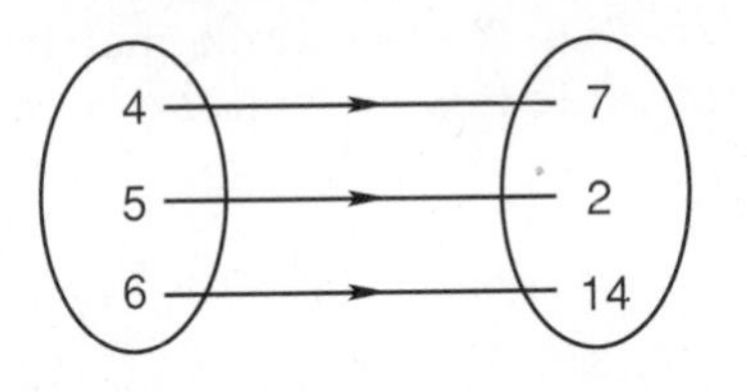

Domain. The set of all permissible values of x.

Range. The set of all images (values of y or "answers").

> DEFINITION:
> FUNCTION, DOMAIN, IMAGE, RANGE
>
> A **function** is a correspondence or mapping between two sets that associates with each element of the first set a unique element of the second set. The first set is called the **domain** of the function. For each element x of the domain, the corresponding element y of the second set is called the **image** of x under the function. The set of all images of the elements of the domain is called the **range** of the function.

There are two accepted definitions of a function. The definition above says that a function is the correspondence or mapping, whereas the other definition says that a function is the set of ordered pairs.

> A **function** is a set of ordered pairs such that no two ordered pairs have the same first element and different second elements.

Perhaps it will help if we show a few functions and a few nonfunctions. Remember, to be a function,

1. **The domain must be specified.**
2. **A way must be provided to find the image for each member of the domain, and there must be exactly one image for each member of the domain.**

---

† The equation itself cannot be called the function because the function is the mapping or pairing between the members of the domain and the answers for each member.

example 113.1 Does the diagram designate a function?

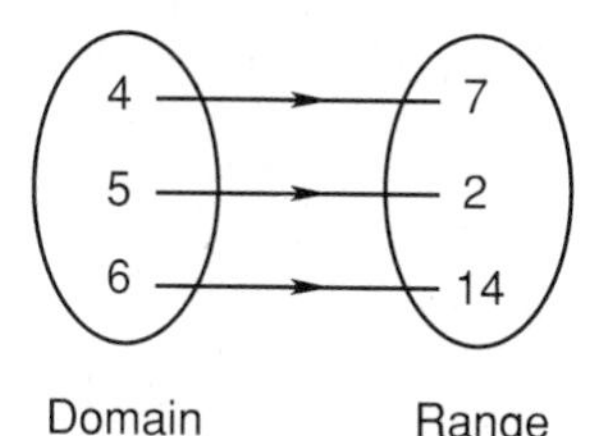

solution **Yes.** The elements of the domain are specified. There is one image and only one image for each member of the domain. We do not have an equation, but we can find the images by looking at the diagram.

example 113.2 Does the diagram designate a function?

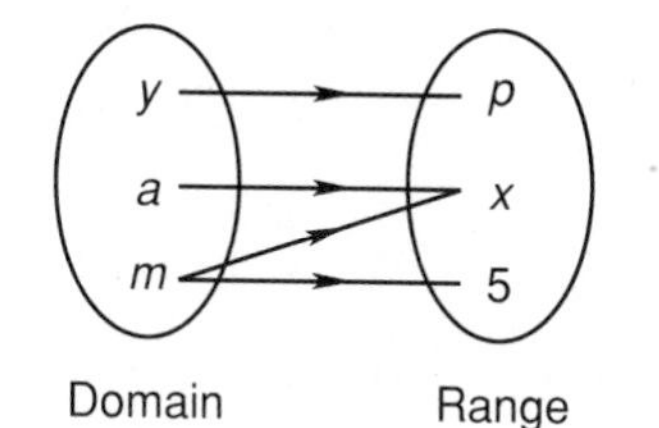

solution **No.** The members of the domain are specified, but m has two images and it is allowed only one.

example 113.3 Does the diagram designate a function?

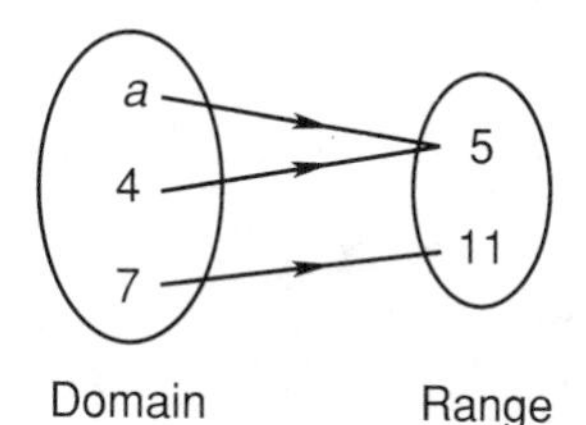

solution **Yes.** The domain is specified, and each member of the domain has exactly one image. True, 5 is the image of both a and 4, but this is permissible.

example 113.4 Does the diagram designate a function?

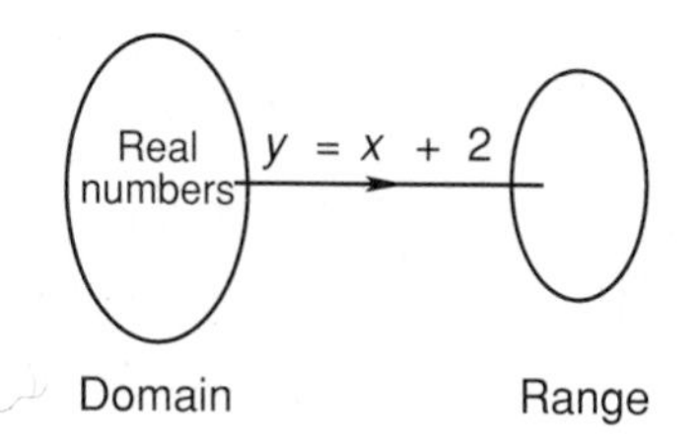

solution **Yes.** The diagram says that we can use any real number for x, so the domain is specified. The members of the range are not specified, but the equation will allow us to find the value of y that is paired with any real number we use for x.

example 113.5 Does either of the sets of ordered pairs shown designate a function?

(a) $\{(4, 6), (7, 2), (4, 5)\}$ (b) $\{(4, 8), (15, 6), (11, 7)\}$

solution The set **(a) does not designate a function,** for 4 has two images. The set **(b) does designate a function.** The domain is specified, and there is exactly one image for every member of the domain.

example 113.6 Does the diagram designate a function?

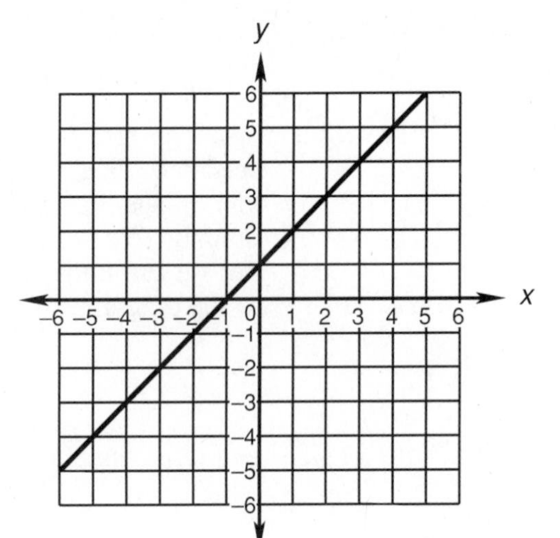

solution **Yes,** it certainly does designate a function. For every value of the x coordinate on the line, there is exactly one corresponding y coordinate. For instance, when x equals 2, it appears that the y coordinate of the ordered pair that lies on the line is about 3. We can't tell from the graph the exact value of the y coordinate, but we do know that an exact value does exist.

example 113.7 Does the diagram designate a function? (The domain is the set of real number such that $x \geq 1$.)

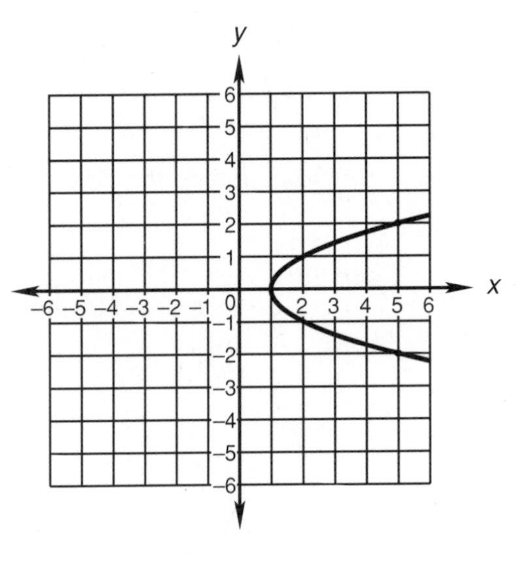

solution **No.** For every member of the domain that is greater than 1, there are two images. For instance, when $x = 5$, the graph shows that y can be either $+2$ or -2.

113.C relations

All functions can also be called *relations* because a relation is a pairing that matches each element of the domain with *one or more images* in the range.

example 113.8 Does the diagram designate a function?

solution **No.** The number 6 has two images. The diagram designates a relation.

example 113.9 Does the diagram designate a function or a relation?

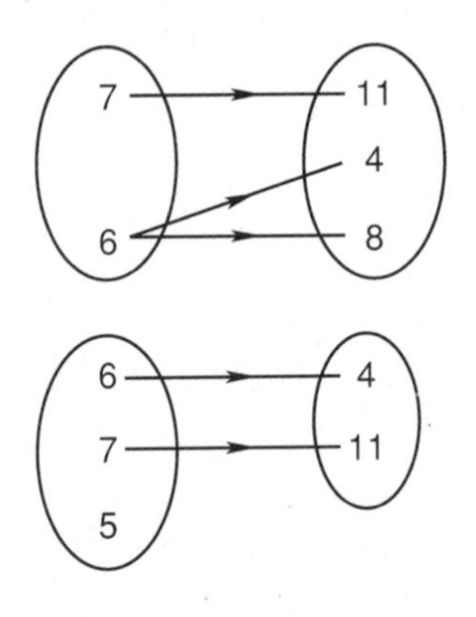

solution **Neither.** For a correspondence to be called a relation, there must be one or more images for each element of the domain. In order to be called a function, there must be exactly one image for each element of the domain. In the diagram the number 5 does not have even one image, and thus the diagram does not designate either a function or a relation.

practice Which of the following diagrams depict functions?

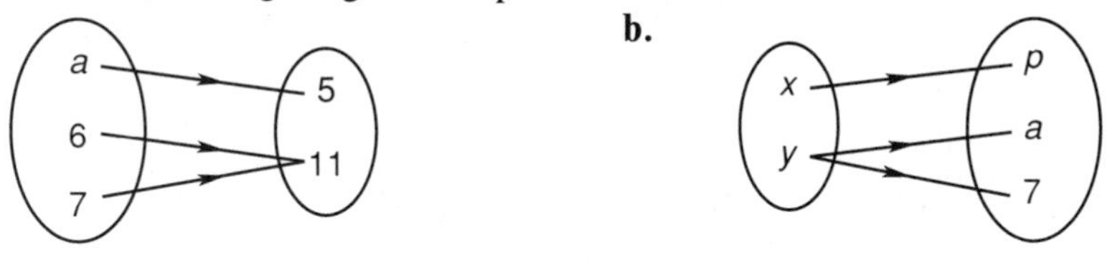

c. What is the domain of the diagram shown in Problem b?

d. What is the range of the diagram shown in Problem a?

problem set 113

1. Miltiades and his army marched from Athens to Marathon, the site of the battle, at 2 miles per hour. After the battle Pheidippides ran back to Athens with the news at 13 miles per hour. If the total traveling time was 15 hours, how far was it from Marathon to Athens?

2. Leonidas walked to Thermopylae at 4 miles per hour and was carried back to Sparta at 2 miles per hour. If the total travel time was 120 hours, what was the distance from Sparta to Thermopylae?

3. The bag broke open and a veritable fortune cascaded forth. There were 5100 gold coins worth \$260,000. If some were \$50 gold coins and the others were \$100 gold coins, how many of each kind were there?

4. The farrago contained 4000 red jelly beans. If there were 20,000 pieces of candy in the farrago, what percent were red jelly beans?

5. Once there were four consecutive integers. The product of the sum of the first and the third and -7 was 4 greater than 12 times the opposite of the fourth. What were the integers?

Find the equations of the lines through the following pairs of points:

6. $(-3, -1)$ and $(4, -6)$ 7. $(2, -5)$ and $(3, -7)$ 8. $(5, -7)$ and $(7, -3)$

9. Which of the following diagrams and sets of ordered pairs depict functions?

(a)

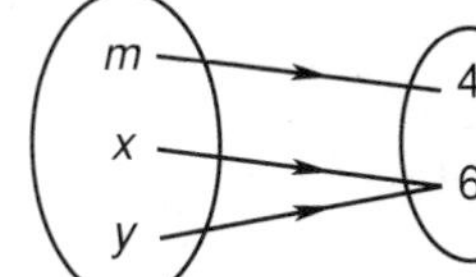

(b), (c)

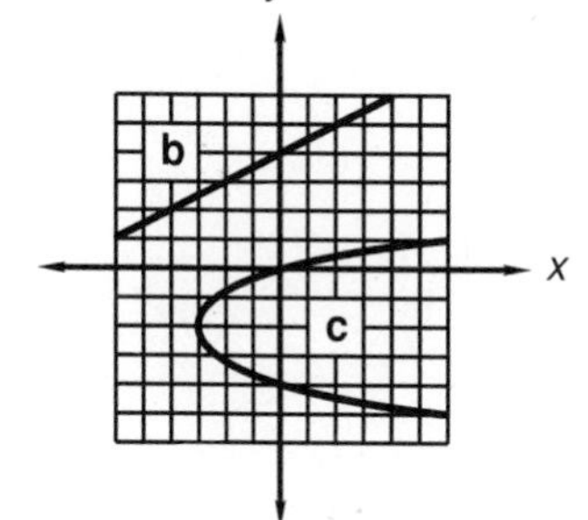

(d) $(4, 6), (5, 7), (3, -2), (6, -3), (2, 7)$
(e) $(4, 6), (5, 7), (3, -2), (6, -3), (5, 3)$
(f) $(-2, 6), (5, 6), (-3, -2), (-8, -6), (4, -6)$

10. The set of positive odd integers is closed for what operations?

11. Find the equations of lines (a) and (b).

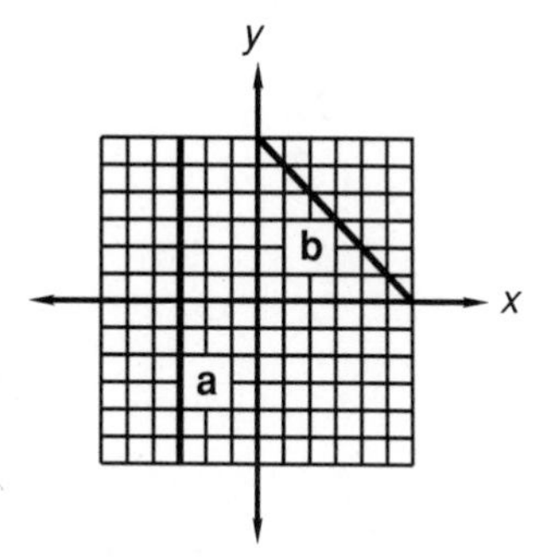

Solve:

12. $\frac{4}{x} - \frac{2}{x - 4} = 0$

13. $\frac{x}{4} - \frac{x + 6}{5} = 1$

14. Find b: $\frac{a}{b} + \frac{1}{c} = d$

15. Find c: $\frac{a}{x} - \frac{1}{c} = \frac{b}{d}$

16. Find c: $\frac{p}{x} + \frac{1}{c} = k$

Add:

17. $\frac{4}{x^2 - 9} - \frac{3}{x + 3}$

18. $\frac{5}{x + 2} - \frac{3x}{x^2 + 5x + 6}$

19. Estimate. Then use the square root table to approximate: $\sqrt{0.0052 \times 10^{-7}}$

20. Graph: $|x| - 3 > 4$; $D = \{\text{Negative reals}\}$

21. Solve: $-(-2x + 4) - 3^0(3 - 3x) - (-2) = 4(3 - x^0)$

Simplify:

22. $\frac{(42{,}000 \times 10^{46})(5000 \times 10^{-20})}{(0.00007 \times 10^{21})(0.0006 \times 10^{-14})}$

23. $\frac{x^2 + 6x + 9}{x^2 + 3x} \div \frac{x^3 + 5x^2 + 6x}{x^2 + 2x}$

24. Find k.

2

4

k

Simplify:

25. $\sqrt{50{,}000} - 25\sqrt{125} + 5\sqrt{5}(\sqrt{5} - 5)$

26. $\frac{xa + \frac{1}{a}}{\frac{x}{a} + a}$

27. $-2^0[(-2 + 3)(-2 - 4) - (-2 - 5)]$

28. $4\sqrt{3} \in$ {What sets of numbers}?

29. Solve: $4\frac{1}{5}m + \frac{3}{4} = \frac{7}{8}$

30. Solve by graphing: $\begin{cases} y = x - 2 \\ y = -x + 2 \end{cases}$

LESSON 114 Functional notation

114.A functions

In the preceding lesson we discussed functions at some length but did not discuss why functions are important.

Algebra books published in the late nineteenth and early twentieth centuries did not contain the word *function*. These books spoke of equations: linear equations, quadratic equations, and other types of equations. The calculus books of the 1930s and 1940s used the word function to describe some equations. The definition used, however, permitted more than one image for each member of the domain. If two images existed for some members of the domain, the function was called a **double-valued function.** If more than two images existed for some members of the domain, the function was called a **multivalued function.** The restriction on a function that it be single-valued is a rather recent development—say, after 1950.

So we see that *function* is the word we use now to describe what used to be called a single-valued equation. But because of the way the word is now defined, it covers any relationship that specifies for every value of the domain one and only one image in the range. For instance, a set of ordered pairs such as (5, 4), (6, 2), and (3, −7) can be called a function even though there is no equation. The members of the domain are specified and each member of the domain is paired with exactly one image in the range. Many equations and other mathematical relationships are single-valued relationships, and it is convenient to have a word to describe them collectively. This word is function.

example 114.1 Does the diagram designate a function?

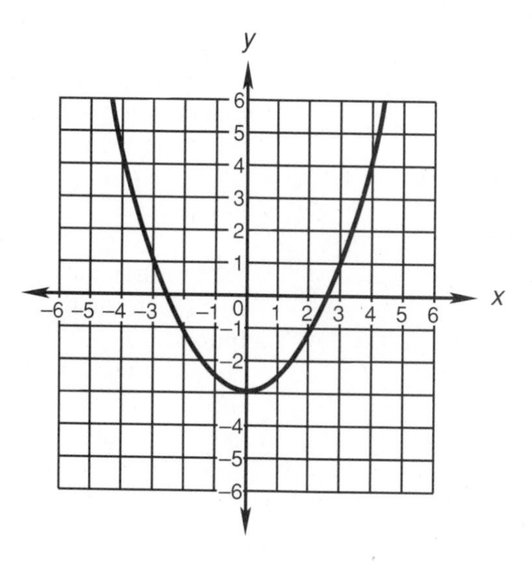

solution **Yes.** We assume that the figure is a portion of the entire graph and that the domain is all real numbers. If this is true, we see that the graph will match exactly one real value of y for each value of x.

example 114.2 Does the equation $x = y^2$ describe a function?

solution **No.** For example, if x is 4, then y could be either positive 2 or negative 2 because both

$$(2)^2 = 4 \qquad \text{and} \qquad (-2)^2 = 4$$

114.B vertical line test

As we noted in the preceding lesson, it is customary to use the letter x to designate the independent variable in a two-variable equation, and it is customary to use the letter y to designate the dependent variable. There is no requirement that this convention be followed, but everyone does so. It is also customary to graph the independent variable on the horizontal axis and the dependent variable on the vertical axis. If this convention is followed, the so-called **vertical line test** can be used on the graph of a relation to see if the relation is also a function. Remember that a relation has one or more images for each element of the domain and a function is a relation that has exactly one image for every member of the domain.

Below we see graphs of three different relations. For each value of x, the graph designates at least one value of y. **If every possible vertical line that we can draw touches the graph of the relation at only one point, then the graph is the graph of a function.** Every point on a vertical line has the same x coordinate, and if the vertical line touches the graph in only one point, the y coordinate of this point is the only paired y value for the x value of the vertical line.

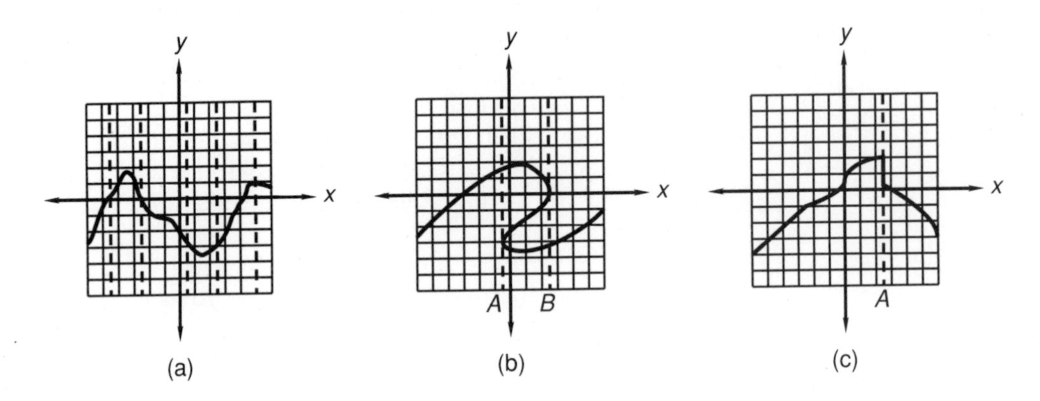

Look at Figure (a). We have drawn several vertical lines, and each of these touches the graph at only one point. If we look carefully at the figure, we can tell that no matter where we draw the vertical line, it will touch the graph of the relation at only one point. Thus by using the vertical line test, we have shown that graph (a) is the graph of a function.

In Figure (b) we drew two vertical lines at values of x arbitrarily called A and B. We see that each of these lines or any vertical line drawn anywhere between them will touch the given graph in at least two points, so the graph is not the graph of a function.

In Figure (c) the graph is vertical at one value of x that we have designated by the letter A. The vertical line drawn at this value of x touches the graph for its entire vertical length. Thus this graph is not the graph of a function.

The vertical line test can be used only if the convention of graphing the independent variable on the horizontal axis is followed. If the independent variable were graphed on the vertical axis, then we would have to use a horizontal line test. We will always follow the convention and thus will not dwell on other possibilities.

example 114.3 Is the relation graphed in the accompanying figure also a function?

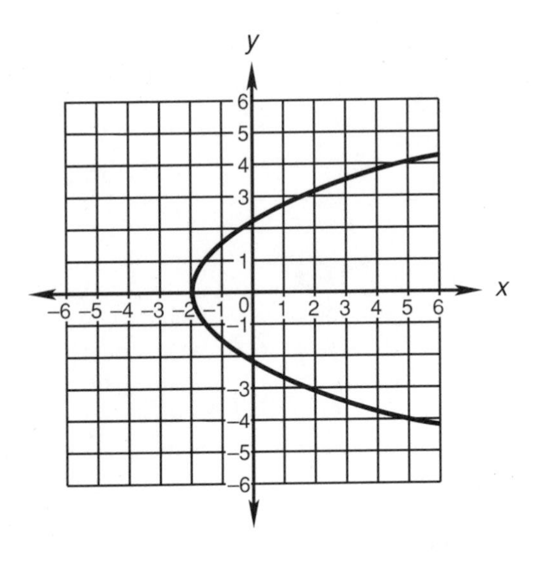

solution (The domain is assumed to be the values of x greater than -2.) **No.** Any vertical line drawn at a value of x such that $x > -2$ will touch the line in at least two points. This graph is the graph of a relation, not a function.

114.C functional notation

In functional notation we use symbols such as $f(x)$, $g(x)$, $h(x)$ to designate the dependent variable instead of using the letter y. If we use functional notation, the equation $y = 2x + 4$ can be written as

$$f(x) = 2x + 4 \qquad \text{or} \qquad g(x) = 2x + 4 \qquad \text{or} \qquad h(x) = 2x + 4$$

We do not restrict ourselves to the letters f, g, and h in designating functions since any letter or symbol may be used.

One advantage of the functional notation is that the functional notation always indicates the element of the domain that is paired with a given image in the range. To demonstrate this, on the left below we write the equation $y = 2x + 4$, and on the right we use functional notation for the same equation and write it as $f(x) = 2x + 4$. Then we replace x in each equation with the number 5 and solve to find the value of the dependent variable.

$$y = 2x + 4 \qquad f(x) = 2x + 4$$
$$y = 2(5) + 4 \qquad f(5) = 2(5) + 4$$
$$y = 10 + 4 \qquad f(5) = 10 + 4$$
- -
$$\mathbf{y = 14} \qquad \mathbf{f(5) = 14}$$

Of course, we get the same value, 14, for the dependent variable in both cases. If we cover up everything above the dashed line, however, we see in the last line on the left only the value for y is shown, but in functional notation on the right we see the value of y is shown, and the value of x that is paired with this value of y is also shown.

example 114.4 If $\phi(x) = 3x + 5$, $D = \{\text{Reals}\}$, find $\phi(-2)$.

solution In nonfunctional notation the problem would have read "given the equation $y = 3x + 5$, find y if x equals -2," and we would proceed as follows to find y.

$$y = 3x + 5 \longrightarrow y = 3(-2) + 5 \longrightarrow \mathbf{y = -1}$$

Now we will work the same problem again using functional notation as was used in the statement of the problem.

$$\phi(x) = 3x + 5 \longrightarrow \phi(-2) = 3(-2) + 5 \longrightarrow \boldsymbol{\phi(-2) = -1}$$

example 114.5 If $g(x) = x^2 + 4$, $D = \{\text{Reals}\}$, find $g(4)$.

solution In nonfunctional notation the problem would have read "given the equation $y = x^2 + 4$, find y if x equals 4."

$$y = x^2 + 4 \longrightarrow y = (4)^2 + 4 \longrightarrow \mathbf{y = 20}$$

Now we will work the same problem, but this time we will use functional notation.

$$g(x) = x^2 + 4 \longrightarrow g(4) = (4)^2 + 4 \longrightarrow \mathbf{g(4) = 20}$$

example 114.6 Given $f(x) = x + 2$, $D = \{\text{Reals}\}$, find $f(\frac{1}{2})$.

solution

$$f\left(\frac{1}{2}\right) = \left(\frac{1}{2}\right) + 2 \longrightarrow \mathbf{f\left(\frac{1}{2}\right) = \frac{5}{2}}$$

example 114.7 Given $f(x) = x + 2$, $D = \{\text{Integers}\}$, find $f(\frac{1}{2})$.

solution The domain of the function is the set of integers. The problem asks that we find the image of $\frac{1}{2}$. One-half does not have an image under this function because the replacement values of x are restricted to the set of integers. Thus we say that there is no real number that satisfies both conditions so **the solution is the empty set { }, or the null set ∅.**

practice Given that $\phi(x) = 3x + 5$, $D = \{\text{Reals}\}$; $g(x) = x^2 + 4$, $D = \{\text{Reals}\}$; $f(x) = x + 2$, $D = \{\text{Integers}\}$, find:

a. $f(4)$ **b.** $g(-2)$ **c.** $f\left(\frac{1}{5}\right)$ **d.** $\phi(-4)$

problem set 114

1. The brigantine was sailing at full speed and was 30 miles at sea when Lord Nelson began to give chase at twice the speed of the brigantine. If Nelson caught up in 6 hours, how fast was the brigantine traveling?

2. The freight train headed north at 9 a.m. at 40 miles per hour. Two hours later the express train headed north at 60 miles per hour. What time was it when the express was 20 miles farther from town than the freight?

3. When the announcer asked the question in the shopping mall, 60 percent of the responses were fatuous. If 3000 answers were not fatuous, how many answers were fatuous? How many people answered the question?

4. The sable was marked down 23 percent for the sale, yet its sale price was still \$15,400. What was the original price of the sable?

5. Find three consecutive integers such that −7 times the sum of the first and the third is 12 greater than the product of 10 and the opposite of the second.

6. Find x if $x = \dfrac{-b \pm \sqrt{b^2 - 4ac}}{2a}$ and if $b + 4 = 1$, $a = -2$, and $c - 7 = -5$.

7. Given that $\phi(x) = 3x + 5$, $D = \{\text{Reals}\}$; $g(x) = x^2 + 4$, $D = \{\text{Reals}\}$; $f(x) = x + 2$, $D = \{\text{Integers}\}$, find:

   (a) $f(2)$ (b) $g(0.6)$ (c) $\phi\left(\frac{1}{4}\right)$

8. Add: $\dfrac{7x + 2}{x + 3} - \dfrac{x}{x^2 - 9}$

9. What is the product of −105 and the multiplicative identity?

Find the equation of the lines through the following pairs of points:

10. (5, −2) and (−4, 3) **11.** (−2, −2) and (5, 5) **12.** (3, −2) and (7, −3)

13. Which of the following diagrams depict functions?

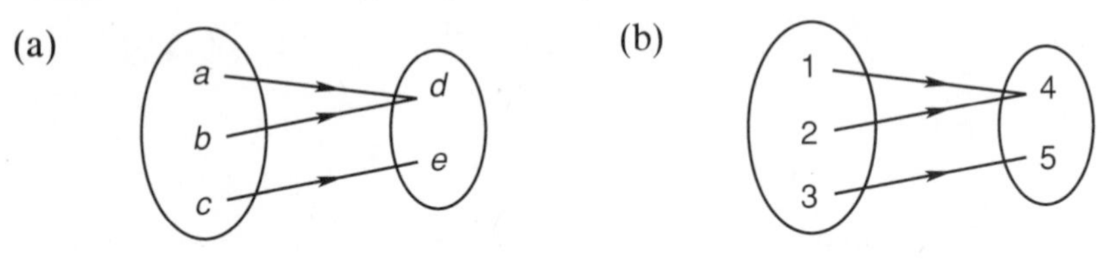

(c) (d), (e)

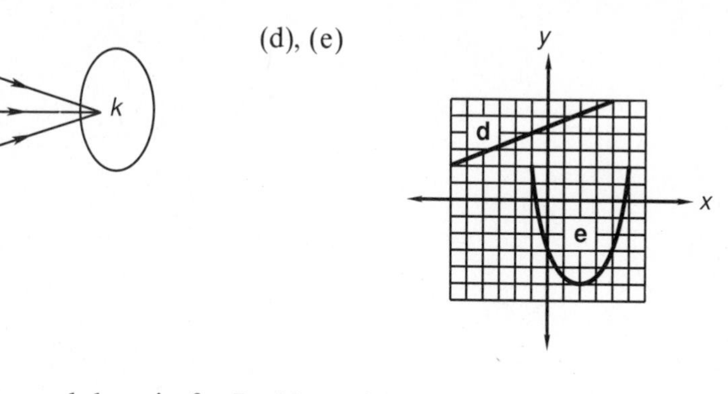

14. What are the range and domain for Problem 13(a)?

15. Which of the following sets of ordered pairs are functions?
(a) $(-3, 2)$ $(3, 2)$, $(5, 2)$
(b) $(-3, -2)$, $(5, -2)$, $(7, -2)$
(c) $(-3, -2)$, $(-3, -5)$, $(-3, -6)$

Solve:

16. $\frac{y}{3} + \frac{1}{4} = 2y$

17. $\frac{4}{p} - \frac{3}{p - 4} = 0$

18. Find x: $\frac{a}{c} + \frac{1}{x} = k$

19. Find f.

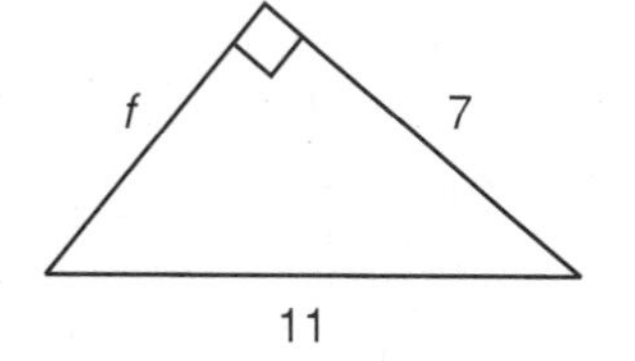

20. Simplify: $\frac{x^2 + 5x + 6}{x^3 + 7x^2 + 10x} \div \frac{x^3 + 11x^2 + 24x}{x^2 + 2x - 15}$

21. Solve by graphing: $\begin{cases} y = 2x + 2 \\ y = -x - 1 \end{cases}$

22. Find the equations of lines (a) and (b).

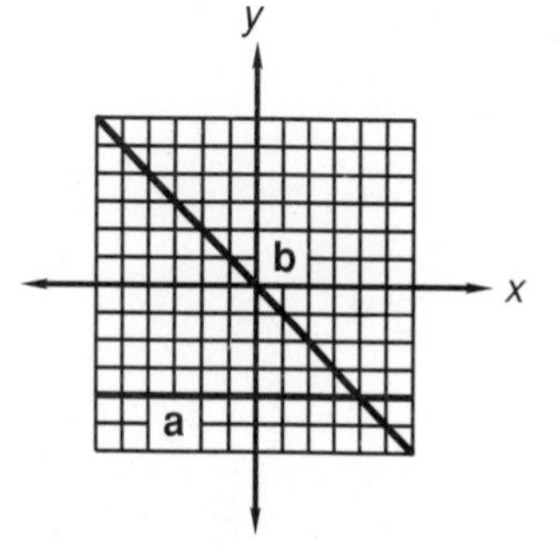

23. Simplify: $\dfrac{\frac{a}{x^2} - \frac{x}{a}}{\frac{x}{a} - \frac{1}{x^2}}$

24. Divide: $(x^3 + 5) \div (x - 2)$

25. Simplify: $-3[(-2^0 - 3) - (-5 + 7)(-2^2 + 3)] - [(-6^0 - 2) + \sqrt[3]{-64}]$

26. What fraction of $\frac{3}{5}$ is $1\frac{1}{3}$?

27. Evaluate: $-x^0 - x^2(x - xy)$ if $x = -3$ and $y = 2$

28. Add like terms: $\frac{x^2yya}{y^{-2}x^4} + \frac{3x^{-2}yy^{-5}y^9}{a^{-1}yxx^{-1}} - \frac{3x^2yyy^3a}{x^2yay^{-4}}$

29. $0.002\sqrt{3} \in$ {What sets of numbers}?

30. Graph: $4 - |x| < 3$; $D =$ {Integers}

Additional

18 Lessons

Topics

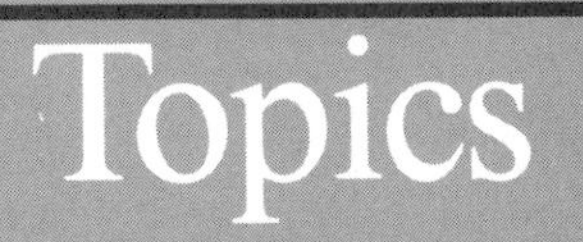

The topics in the following lessons will be considered again in depth in the next book in this series, *Algebra 2*.

LESSON 115 Line parallel to a given line

To find the slope of a line through two specified points, we have to graph the points, draw the line, draw the slope triangle, and compute the slope. To find the slope of a line that is to be parallel to a given line, all that we need to do is realize that two parallel lines have the same slope.

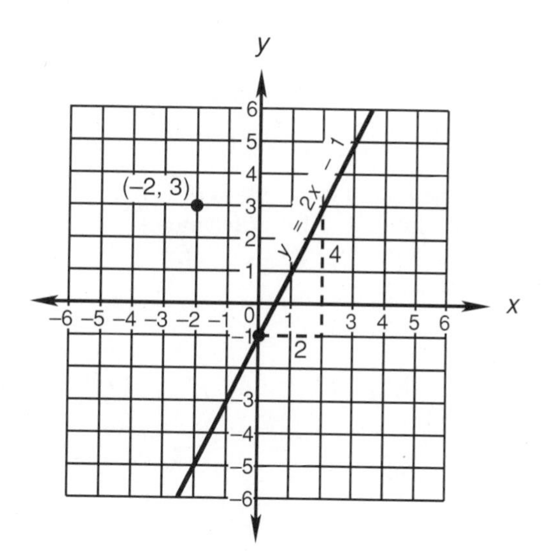

In the figure we have graphed the line whose equation is $y = 2x - 1$. We see from the equation and from the triangle drawn in the graph that the slope of this line is 2. **Any line that is parallel to this line must also have a slope of 2.** Thus if we are asked to find the equation of the line that is parallel to this line and that passes through $(-2, 3)$ we are already halfway home, for the slope of the new line has to be 2.

$$y = 2x + b$$

Now we can use the coordinates $(-2, 3)$ for x and y and find b algebraically.

$3 = 2(-2) + b$	replaced x and y with -2 and 3
$3 = -4 + b$	multiplied
$7 = b$	added 4 to both sides

So the equation of the line is

$$\mathbf{y = 2x + 7}$$

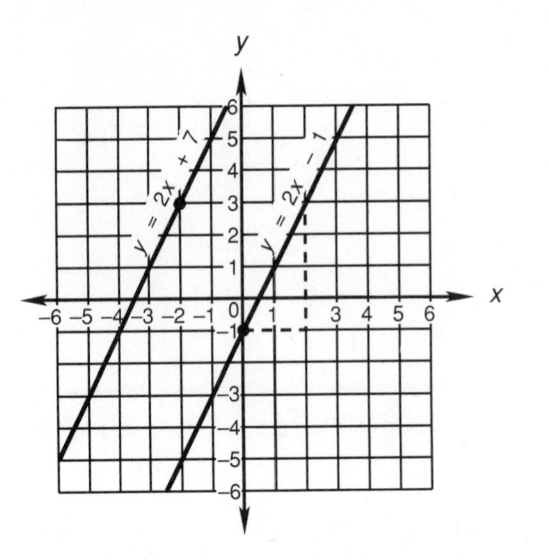

We have found the exact slope and the exact intercept. Estimated values for the slope and intercept will not be acceptable for this kind of problem.

example 115.1 Find the equation of the line through $(-1, -3)$ that is parallel to $4x + 3y = 7$.

solution The equation of the new line is

$$y = mx + b$$

and we have to find m and b. **If the new line is to be parallel to the line $4x + 3y = 7$, the new line must have the same slope as the line $4x + 3y = 7$.** If we write $4x + 3y = 7$ in slope-intercept form, we find

$$y = -\frac{4}{3}x + \frac{7}{3}$$

and the slope of this line is $-\frac{4}{3}$. So the slope of the new line must be $-\frac{4}{3}$.

$$y = -\frac{4}{3}x + b$$

We find the value of b by using coordinates $(-1, -3)$ for x and y and solving algebraically for b.

$$-3 = -\frac{4}{3}(-1) + b \qquad \text{substituted}$$

$$-\frac{9}{3} = \frac{4}{3} + b \qquad \text{multiplied}$$

$$-\frac{13}{3} = b \qquad \text{added } -\frac{4}{3} \text{ to both sides}$$

Thus the desired equation is

$$\mathbf{y = -\frac{4}{3}x - \frac{13}{3}}$$

and again we have found the exact value of the slope and the exact value of the intercept.

practice Find the equation of the line:

a. That goes through the point $(-1, 2)$ and is parallel to $y = -3x + 1$.

b. That goes through the point $(-2, -3)$ and is parallel to $3x + 2y = 5$.

problem set 115

1. Birthe rode to town in the bus at 20 miles per hour. Then she trotted back home at 8 miles per hour. If her total traveling time was 14 hours, how far was it to town?

2. Soren had a 36-mile head start. If Erik caught him in 3 hours, how fast was Erik traveling if his speed was twice that of Soren?

3. Flexner marked the bench down \$20 and sold it for 60 percent of the original price. What was the original price of the bench?

4. Weir and Max put \$75 in quarters and dimes in the box. There were 400 more dimes than quarters. How many coins of each type were there?

5. Find three consecutive odd integers such that the product of -7 and the sum of the first and third is 27 greater than the product of 11 and the opposite of the second.

6. Find x if $x = \dfrac{-b \pm \sqrt{b^2 - 4ac}}{2a}$ and $\dfrac{a}{3} = \dfrac{16}{12}$, $b = -6$, and $c - 9 = -13$.

7. Find the lateral surface area of a right circular cylinder 10 meters high whose base is also 10 meters.

Find the equation of the line:

8. That goes through the points $(-3, -2)$ and $(5, -3)$.

9. That goes through the points $(5, -1)$ and $(0, 0)$.

Given that $p(x) = x^2 + 5$, $D = \{\text{Negative integers}\}$, and $k(x) = x + 4$, $D = \{\text{Integers}\}$, find:

10. $k(4)$ **11.** $p(4)$

12. Which of the following diagrams depict functions?

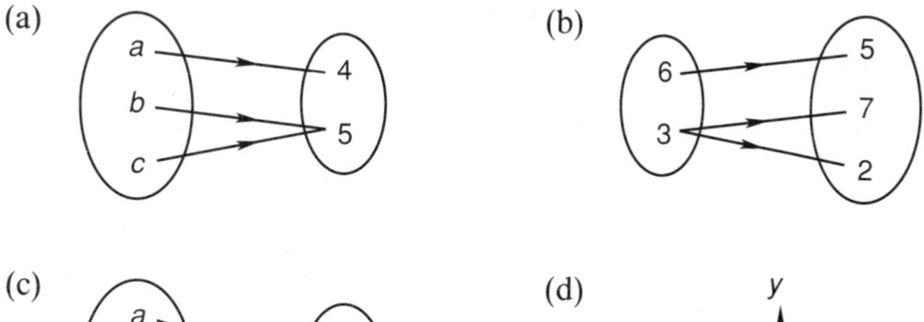

(c) a b c 4

(d)

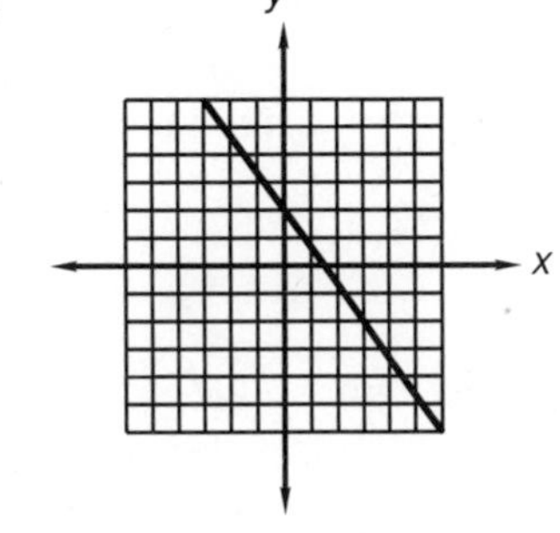

13. What are the domain and range for Problem 12(a)?

14. Which of the following sets of ordered pairs are functions?
(a) $(4, -7), (5, -7), (3, -2)$ (b) $(5, -3), (-5, 3), (7, -2)$
(c) $(-7, 4), (5, -3), (-7, 6)$

Solve:

15. $\frac{7}{y} + \frac{3}{y - 2} = 0$

16. $\frac{y}{7} - \frac{3}{4} = \frac{2y}{5}$

17. Find c: $\frac{a}{c} - \frac{1}{x} = b$

18. Graph: $+2 - |x| \geq -2$; $D = \{\text{Negative integers}\}$

19. Find the distance between $(-3, -2)$ and $(5, -3)$.

20. Solve: $x^0 - 3x(2 - 4^0) - (-3) - 2(x - 3) = 3x - (-4)$

Simplify:

21. $\frac{(30{,}000 \times 10^{-42})(7000 \times 10^{15})}{(0.00021 \times 10^{14})(1000 \times 10^{-23})}$

22. $\frac{(2x)^{-2}y^2x^2y^4y}{y^0(x^{-4})y^2y^{-2}y(x^{-4})^{-2}}$

23. $4\sqrt{50{,}000} - 3\sqrt{125}$

24. $\frac{-3 - 3x}{-3}$

25. $\frac{-2^{-2}}{(-2^0)^{-2}}$

26. Solve by factoring: $80 = -x^2 + 18x$

27. Evaluate: $-x^3 - x^2 - x(x^0 - yx)$ if $x = -4$ and $y = 3$

28. Graph on a number line: $-4 - |x| \leq -4$

29. Multiply: $\frac{x^{-3}}{y^2}\left(\frac{y^2}{x^3} - \frac{3y^{-3}}{x^{-2}}\right)$

30. Round $0.00\overline{318}$ to the nearest thousandth.

LESSON 116 Equation of a line with a given slope

Finding the equation of a line with a given slope and through a given point is the easiest problem type of all. The slope of the desired equation is *given* in the statement of the problem. All we have to do is find the value of the intercept.

example 116.1 Find the equation of the line that goes through the point (3, 4) and has a slope of $-\frac{3}{4}$.

solution The equation of the line in question is $y = mx + b$, and we need to determine the proper values for m and b. The statement of the problem tells us that the slope is $-\frac{3}{4}$ and if we use this value for m in the equation, we find

$$y = -\frac{3}{4}x + b$$

If we use (3, 4) for x and y in this equation, we can solve algebraically for b.

$$4 = -\frac{3}{4}(3) + b$$

$$4 = -\frac{9}{4} + b \longrightarrow \frac{16}{4} = -\frac{9}{4} + b \longrightarrow b = \frac{25}{4}$$

So the equation is

$$\mathbf{y = -\frac{3}{4}x + \frac{25}{4}}$$

and the numbers $-\frac{3}{4}$ and $\frac{25}{4}$ are the exact values of the slope and the intercept.

example 116.2 Find the equation of the line that goes through the point $(-5, 11)$ and has a slope of $\frac{1}{7}$.

solution The statement of the problem gives us the slope, so we can write

$$y = \frac{1}{7}x + b$$

Now we use the values -5 and 11 for x and y and solve for b.

$$11 = \frac{1}{7}(-5) + b \longrightarrow \frac{77}{7} = -\frac{5}{7} + b \longrightarrow b = \frac{82}{7}$$

So the equation is

$$\mathbf{y = \frac{1}{7}x + \frac{82}{7}}$$

and the numbers $\frac{1}{7}$ and $\frac{82}{7}$ are the exact values of the slope and the intercept.

example 116.3 Find the equation of the line that goes through the point $(-50, 40)$ and has a slope of $-\frac{1}{2}$.

solution Again we have been given the slope, so we can write

$$y = -\frac{1}{2}x + b$$

Now to find b, we replace y with 40 and x with -50 and solve.

$$40 = -\frac{1}{2}(-50) + b$$

$$40 = 25 + b$$

$$15 = b$$

Thus the desired equation is

$$\mathbf{y = -\frac{1}{2}x + 15}$$

practice Find the equation of the line that goes through:

a. The point (2, 3) and has a slope of $-\frac{2}{5}$.

b. The point $(-3, 7)$ and has a slope of $\frac{1}{4}$.

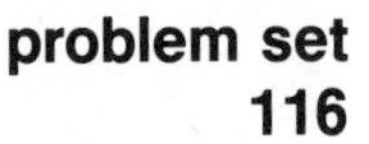

problem set 116

1. The track team ran to the cemetery at 8 miles per hour and trotted back home at 6 miles per hour. If the total trip took 7 hours, how far was it to the cemetery?

2. At 4 a.m. the northbound train left at 40 miles per hour. At 6 a.m. the southbound train left the same station at 60 miles per hour. At what time will the trains be 880 miles apart?

3. Small pizzas were \$3 and large pizzas were \$5. To feed the throng, it was necessary to spend \$475 for 125 pizzas. How many pizzas of each type were purchased?

4. Only 300 of the larvae metamorphosed into butterflies. If there were 2500 larvae at the outset, what percent became butterflies?

5. Mark thought of three consecutive odd integers. He added the first to twice the third and multiplied this sum by -3. The result was 3 less than the product of 8 and the opposite of the second. What were the numbers?

6. The ratio of bees to moths was 13 to 5. If there were a total of 2610 bees and moths in the bar, how many were moths and how many were bees?

Find the equation of the line that goes through:

7. The points $(-2, 3)$ and $(4, 5)$.

8. The point $(-2, 3)$ and is parallel to $2x + 3y = 7$.

9. Find x if $x = \dfrac{-b \pm \sqrt{b^2 - 4ac}}{2a}$ and $\dfrac{a}{3} - 3 = 9$, $b = 7$, and $4c + 11 = 9$.

10. Use the letters a, b, and c as necessary to state (a) the associative property of addition, and (b) the commutative property of multiplication.

11. Given $f(x) = 2x + 4$, $D = \{\text{Negative integers}\}$, and $p(x) = x + 5$, $D = \{\text{Positive integers}\}$, find: (a) $f(2)$ (b) $p(2)$

12. Which of the following diagrams or sets of ordered pairs represent functions?

(a)

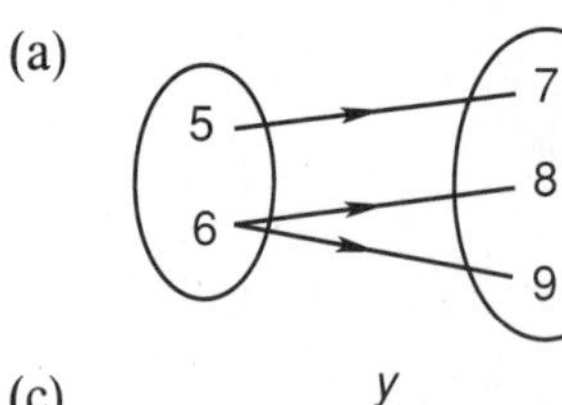

(b)

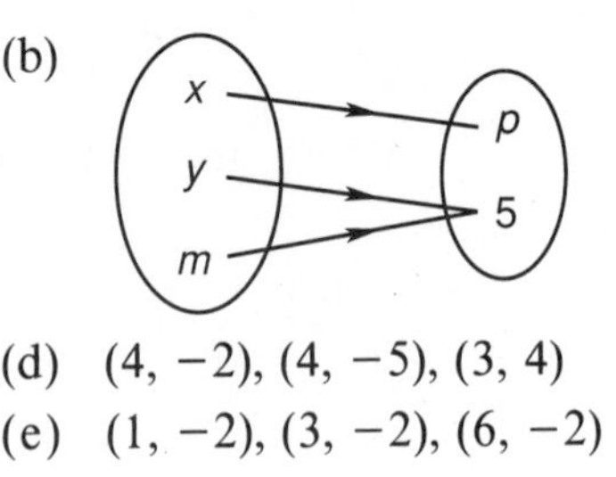

(c)

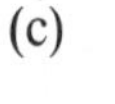

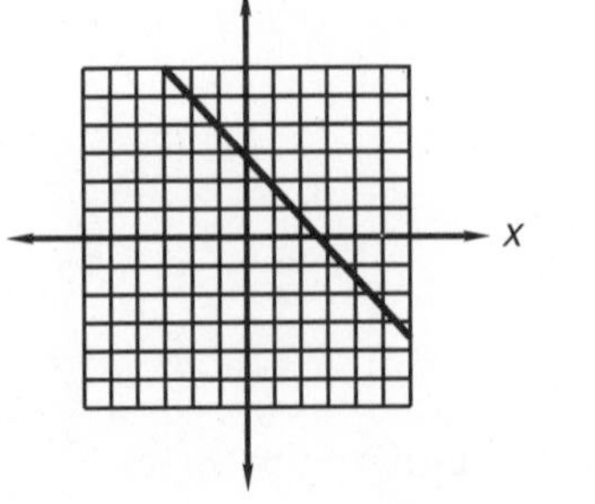

(d) $(4, -2), (4, -5), (3, 4)$

(e) $(1, -2), (3, -2), (6, -2)$

13. What is the range for Problem 12(b) above?

Solve:

14. $\dfrac{5x}{3} - \dfrac{1}{3} = \dfrac{2x}{5}$

15. $\dfrac{x - 2}{3x} = \dfrac{4}{x} - \dfrac{1}{5}$

16. Find c: $\frac{k}{m} - \frac{1}{c} + \frac{x}{y} = p$

17. Estimate. Then use the square root table to simplify:

$$\sqrt{0.000178563 \times 10^{-13}}$$

18. Find the distance between $(-2, 3)$ and $(4, 5)$.

19. Solve: $-x^0 - (2x - 5) + x - (-3^2) - 2 = 3x(4^0 - 2) - 2^2$

Simplify:

20. $\frac{(21{,}000 \times 10^{-42})(7{,}000{,}000)}{(0.0003 \times 10^{-21})(700 \times 10^{15})}$

21. $\frac{x(x^{-2}y)^{-2}(x^{-2}y)x^{-2}ya^2x}{(xy^{-2})^{-2}x^{-2}y^{-4}yy^3x^2}$

22. $\sqrt{150{,}000} + 2\sqrt{3} \cdot 5\sqrt{5} + 2\sqrt{15}(\sqrt{15} - 3)$

23. $\frac{6xy + 6xy^2}{6xy}$

24. $\frac{-3^{-2}}{-(-3)^{-3}}$

25. Solve by factoring: $120 = -22x - x^2$

26. Evaluate: $-x^0 - x^2 - xy(x - y)$ if $x = -3$ and $y = 4$

27. Graph on a number line: $-3 - |x| \le -3$; $D = \{\text{Integers}\}$

28. Multiply: $\frac{4x^{-2}}{yx}\left(\frac{x^3}{y} - \frac{3y^3}{x^3}\right)$

29. Find the volume in cubic meters of a cylinder whose base is shown and whose sides are 4 centimeters high. If the cylinder is a right cylinder, find the surface area in square centimeters. What is the volume of a cone that has the same base and the same altitude? Dimensions are in centimeters. The long lines are parallel.

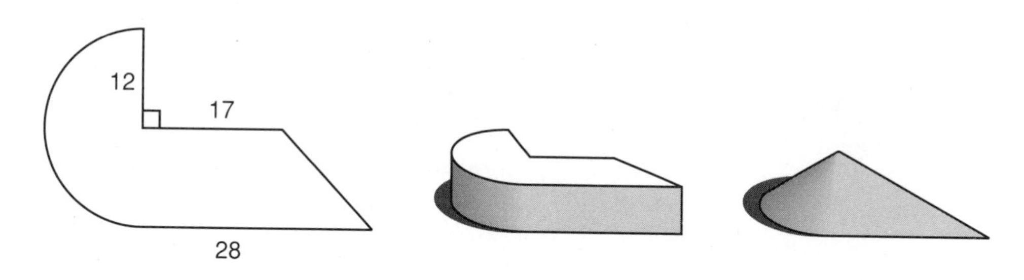

30. Use 10 unit multipliers to convert 26,000 square miles to square kilometers.

LESSON 117 ***Radical equations***

117.A square roots revisited

Both -2 and $+2$ are square roots of 4 because

$$(-2)^2 = 4 \qquad \text{and} \qquad (+2)^2 = 4$$

But when we write $\sqrt{4}$, we are indicating the positive or principal square root of 4, which, of course, is the number 2.

DEFINITION OF SQUARE ROOT

If x is greater than zero, then $\sqrt{x}$ is the unique positive real number such that

$$(\sqrt{x})^2 = x$$

We can state this definition in words by saying that the principal square root of a given positive number is that positive number which, multiplied by itself, yields the given number. Thus

$$\sqrt{2}\sqrt{2} = 2 \qquad \sqrt{7}\sqrt{7} = 7 \qquad \text{and} \qquad \sqrt{3.14}\sqrt{3.14} = 3.14$$

and also

$$(\sqrt{2})^2 = 2 \qquad (\sqrt{7})^2 = 7 \qquad \text{and} \qquad (\sqrt{3.14})^2 = 3.14$$

It is necessary to remember that algebraic expressions represent particular real numbers that are determined by the values assigned to the variables. Thus, if a particular algebraic expression represents a positive real number, the definition of the square root given in the box above applies to the expression. For example,

$$(\sqrt{x^2 + 4})^2 = x^2 + 4 \qquad \left(\sqrt{\frac{amx^2}{p}}\right)^2 = \frac{amx^2}{p} \qquad (\sqrt{x + 6})^2 = x + 6$$

$$(\sqrt{x^4 + 3x^2 + 5}) = x^4 + 3x^2 + 5$$

117.B radical equations

There is only one number that will satisfy the equation $x = 2$, and that number is 2. If we replace x with 2 in this equation, we find

$$2 = 2$$

which is a true statement. If we square both sides of the original equation,

$$(x)^2 = (2)^2 \longrightarrow x^2 = 4$$

the result is the equation $x^2 = 4$. While the equation $x = 2$ had only one solution, the equation $x^2 = 4$ has two numbers that satisfy it, the numbers +2 and −2.

REPLACING x WITH +2		REPLACING x WITH −2	
$(+2)^2 = 4$		$(-2)^2 = 4$	
$4 = 4$	True	$4 = 4$	True

We began with the equation $x = 2$, whose only solution is 2. We squared both sides and got the equation $x^2 = 4$, which also has the number 2 as a solution but has another solution, which is the number −2. **It can be shown that if both sides of an equation are squared, all of the solutions to the original equation (if any exist) are also solutions to the resulting equation, but the reverse is not true, for all of the solutions of the resulting equation are not necessarily solutions of the original equation.**

example 117.1 Solve: $\sqrt{x - 2} + 3 = 0$

solution We wish to **isolate the radical** on one side of the equation so that we may square both sides of the equation and use the fact that $(\sqrt{x-2})^2 = x - 2$, so we add −3 to both sides of the equation and get

$$\sqrt{x - 2} = -3$$

Now square both sides and get

$$(\sqrt{x-2})^2 = (-3)^2 \longrightarrow x - 2 = 9 \longrightarrow \mathbf{x = 11}$$

Now we must check our solution in the original equation.

$$\sqrt{11-2} + 3 = 0 \longrightarrow \sqrt{9} + 3 = 0 \longrightarrow 3 + 3 = 0 \longrightarrow 6 = 0 \qquad \text{False}$$

We see that while 11 is a solution of the second equation, $x - 2 = 9$, it is not a solution to the original equation $\sqrt{x-2} + 3 = 0$. Thus we see that there is no real number replacement for x that will satisfy the first equation, and we say that **the solution set of this equation is the empty set.**

example 117.2 Solve: $\sqrt{x-2} - 6 = 0$

solution We first **isolate the radical** on one side by adding +6 to both sides of the equation.

$$\sqrt{x-2} = 6$$

Now we square both sides to eliminate the radical and then solve the resulting equation.

$$(\sqrt{x-2})^2 = (6)^2 \longrightarrow x - 2 = 36 \longrightarrow \mathbf{x = 38}$$

Now we will check this solution in the original equation.

$$\sqrt{(38)-2} - 6 = 0 \longrightarrow \sqrt{36} - 6 = 0 \longrightarrow 6 - 6 = 0 \longrightarrow 0 = 0 \quad \text{Check}$$

Thus **38** is a solution to the original equation.

example 117.3 Solve: $\sqrt{x^2+9} - 5 = 0$

solution First **isolate the radical** and get

$$\sqrt{x^2+9} = 5$$

Now eliminate the radical by squaring both sides of the equation.

$$(\sqrt{x^2+9})^2 = (5)^2 \longrightarrow x^2 + 9 = 25$$

Now we will simplify, factor, and use the zero theorem to solve.

$$x^2 - 16 = 0 \longrightarrow (x+4)(x-4) = 0$$

$$\longrightarrow \mathbf{x = 4} \quad \text{or} \quad \mathbf{x = -4}$$

Since neither of these solutions to the second equation is guaranteed to be a solution of the original equation, both solutions must be checked in the original equation.

CHECK +4		CHECK −4	
$\sqrt{(4)^2+9} = 5$		$\sqrt{(-4)^2+9} = 5$	
$\sqrt{25} = 5$		$\sqrt{25} = 5$	
$5 = 5$	Check	$5 = 5$	Check

Thus both **+4** and **−4** are solutions.

example 117.4 Solve: $\sqrt{x-1} - 3 + x = 0$

solution We begin by adding $+3 - x$ to both sides of the equation to **isolate the radical** and get

$$\sqrt{x-1} = 3 - x$$

Now we square both sides to eliminate the radical.

$$(\sqrt{x-1})^2 = (3-x)^2 \longrightarrow x - 1 = 9 - 6x + x^2$$

Next we simplify, factor, and use the zero factor theorem to solve.

$$x^2 - 7x + 10 = 0 \longrightarrow (x-2)(x-5) = 0$$

$$\longrightarrow \mathbf{x = 2} \quad \text{or} \quad \mathbf{x = 5}$$

Now we must check both 2 and 5 in the original equation.

CHECK $x = 2$		CHECK $x = 5$	
$\sqrt{x-1} = 3 - x$		$\sqrt{x-1} = x$	
$\sqrt{(2)-1} = 3 - (2)$		$\sqrt{(5)-1} = 3 - (5)$	
$\sqrt{1} = 1$		$\sqrt{4} = -2$	
$1 = 1$	Check	$2 \neq -2$	Does not check

Thus $\mathbf{x = 2}$ is a solution of the original equation, and $\mathbf{x = 5}$ **is not a solution** of the original equation.

example 117.5 Solve: $\sqrt{2x-3} = \sqrt{x+2}$

solution One radical expression is isolated on each side of the equation, so we begin by squaring both sides of the equation.

$$(\sqrt{2x-3})^2 = (\sqrt{x+2})^2 \longrightarrow 2x - 3 = x + 2 \longrightarrow \mathbf{x = 5}$$

Now we will check $x = 5$ in the original equation.

$$\sqrt{2(5)-3} = \sqrt{5+2} \longrightarrow \sqrt{7} = \sqrt{7} \qquad \text{Check}$$

practice Solve:

a. $\sqrt{x-6} - 3 = 0$ **b.** $\sqrt{x-6} - 6 + x = 0$

problem set 117

1. The goliards sang songs before the banquet. If the ratio of ribald songs to scandalous songs in their repertoire of 3102 songs was 7 to 4, how many ribald songs did they know?

2. The doyenne walked to the meeting at 2 miles per hour and caught a ride home in an old truck at 10 miles per hour. How far was it to the meeting place if her total traveling time was 18 hours?

3. The gun sounded, and 3 hours later Bill was 6 miles ahead of Rose. How fast did Rose run if Bill's speed was 10 miles per hour?

4. Roger and Gwenn picked 178 quarts of berries. If Roger picked 8 more quarts than Gwenn picked, how many quarts did Gwenn pick?

5. Gold bricks sold for \$400 each, while pyrite bricks were only \$3 each. To stock his booth, Grimsby bought 123 bricks for \$21,013. How many of each kind did he buy?

6. Find the equation of the line that passes through (4, 2) and (−5, −7).

7. Given $f(x) = x + 3$, $D = \{\text{Reals}\}$; $g(x) = x - 4$, $D = \{\text{Integers}\}$. Find:
(a) $f(-2)$ (b) $g(-2)$

8. Use the letters a, b, and c as necessary to state the distributive property.

9. The final exam grade of 94 was given the weight of 10 weekly grades. If the average of the 10 weekly grades was 76, what was the weighted average grade for the entire course?

Solve:

10. $\sqrt{x^2 + 11} - 9 = 0$

11. $\sqrt{x} = 5\sqrt{2}$

12. $\frac{x}{3} - \frac{2 + x}{5} = -3$

13. $\frac{4}{x + 3} - \frac{2}{2x} = 0$

14. Find y: $\frac{x}{y} - \frac{1}{c} - d = k$

15. Find c.

c 5 9

16. Simplify: $\frac{x^2 - 25}{x^2 - 12x + 35} \div \frac{x^2 + x - 6}{x^2 - 4x - 21}$

17. Solve by graphing: $\begin{cases} y = 3x \\ y = -x + 4 \end{cases}$

18. Find the equations of lines (a) and (b).

y x b a

19. Simplify: $\dfrac{\frac{x}{yz} - \frac{1}{z^2}}{\frac{a}{z} - \frac{3}{yz^2}}$

20. Divide: $(2x^3 + 3x^2 + 5x + 4) \div (x - 1)$

21. Simplify: $-[(-2^0)(-3^2) - (-7 - 2) - \sqrt[5]{-32}] - [-3(-5 + 7)]$

22. What fraction of $2\frac{1}{8}$ is $3\frac{4}{5}$?

23. Evaluate: $-x^3 + (-x)^2 - x^2 - x(x - xy^2)$ if $x = -3$ and $y = -2$

24. Add like terms: $-3x^2yy^{-2} + \frac{2x^2}{y} - \frac{3xy^{-1}}{x^{-1}} + \frac{4x^3x^{-1}}{y}$

25. $\frac{3\sqrt{2}}{5} \in$ {What sets of numbers}?

26. Solve: $2\frac{1}{8}x + \frac{1}{4} = \frac{7}{8}$

Add:

27. $\frac{x}{y^{-2}x} - \frac{3}{y^3x^2} - \frac{2}{x + y}$

28. $\frac{3x + 2}{x - 4} - \frac{2x}{x^2 - 16}$

29. Round $0.03\overline{74}$ to the nearest ten-millionth.

30. Find x if $x = \frac{-b \pm \sqrt{b^2 - 4ac}}{2a}$ and $\frac{a}{7} - 3 = 0$, $b = -9$, and $8c + 13 = 7$.

LESSON 118 Slope formula

In Lesson 85 the slope of a straight line was defined to be the ratio of the change in the y coordinate to change in the x coordinate as we move from one point on the line to another point on the line. To demonstrate, we will find the slope of the line through the points (4, 3) and (−2, −2) by first graphing the points and drawing the line as shown in the figure on the left. The sign of the slope of this line is positive because the line segment graphed points toward the upper right. On the right we draw the triangle to determine the magnitude (absolute value) of the slope. From the triangle we see that the difference in the x coordinates of the two points is 6 and the difference in the y coordinates is 5. Thus the slope of this line is $+\frac{5}{6}$.

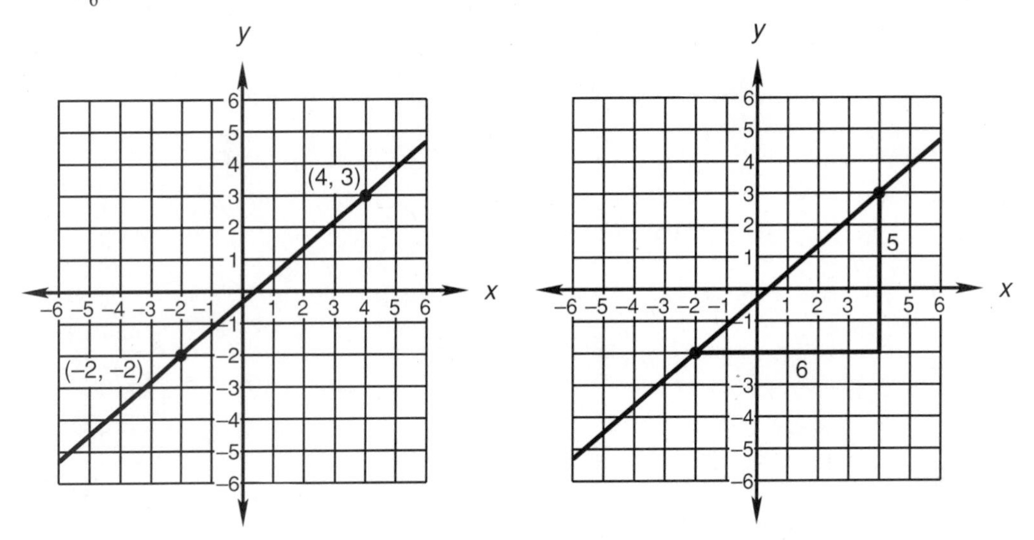

It is not necessary to graph the points to find the slope of the line. If we call the two points point 1 and point 2 and give them to coordinates (x_1, y_1) and (x_2, y_2), respectively, as shown below, we can derive a relationship from which the slope for the line through any two points can be determined algebraically.

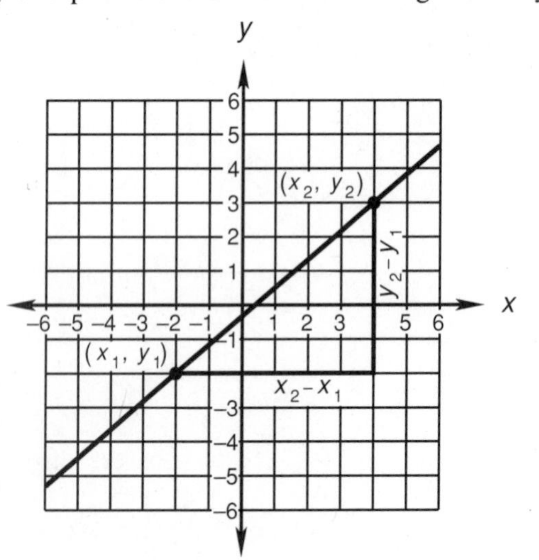

We begin at point 1 and move along the line to point 2. The change in the value of the x coordinate is $x_2 - x_1$, and the change in the value of the y coordinate is

$y_2 - y_1$. The slope m is defined to be the ratio of the change in the y coordinate to the change in the x coordinate, so

$$m = \frac{y_2 - y_1}{x_2 - x_1}$$

We can use this relationship or formula to find the sign and the magnitude of the slope of the line through any two given points.

example 118.1 Find the slope of the line that passes through the points $(-3, 4)$ and $(5, -2)$.

solution Either point can be designated as point 1. We will use $(-3, 4)$ as point 1 and $(5, -2)$ as point 2.

$$m = \frac{y_2 - y_1}{x_2 - x_1} \longrightarrow m = \frac{-2 - (4)}{5 - (-3)} \longrightarrow m = \frac{-6}{8} \longrightarrow \mathbf{m = -\frac{3}{4}}$$

example 118.2 Work Example 118.1 again, but this time use $(5, -2)$ as point 1 and $(-3, 4)$ as point 2.

solution

$$m = \frac{y_2 - y_1}{x_2 - x_1} \longrightarrow m = \frac{4 - (-2)}{-3 - (5)} \longrightarrow m = \frac{6}{-8} \longrightarrow \mathbf{m = -\frac{3}{4}}$$

The slope of the line was found to be $-\frac{3}{4}$ no matter which point was designated as point 1. **We see from these two examples that the slope can be determined by using this formula, and we also see that care must be exercised to prevent mistakes in handling the positive and negative signs of the numbers.** In the figure below, we use the familiar graphical method. We see from this figure that the sign of the slope is negative and the magnitude is $\frac{6}{8}$, so the slope is $-\frac{6}{8}$, or $-\frac{3}{4}$, the same slope found by using the formula.

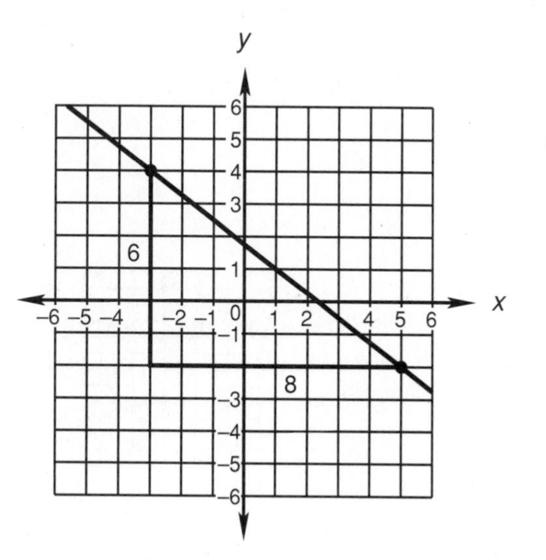

practice Use the slope formula to find the slope of the line that passes through:

a. $(-4, -7)$ and $(-8, -3)$ **b.** $(4, 7)$ and $(-14, -2)$

problem set 118

1. The ratio of Dobermans to terriers in the county was 2 to 7. If there were 774 Dobermans and terriers in the county, how many were Dobermans?

2. Paul and Kit had 3 hours to wait until the game began. They used the time to ride a bus into the country at 14 miles per hour and then to jog back at 7 miles per hour. How far did they go into the country?

3. Find four consecutive even integers such that -5 times the sum of the first and fourth is 10 greater than the opposite of the sum of the third and fourth multiplied by 6.

4. Smiley and Linda cut a 40-foot rope into two lengths. The ratio of the lengths was 3 to 1. How long was each length?

5. Nineteen percent of the bats had returned to the belfrey by 8 p.m. If 855 bats had returned, how many bats lived in the belfrey?

6. Find the equation of the line that passes through $(5, -2)$ and is parallel to the line $y = -\frac{1}{2}x - \frac{2}{3}$.

7. Find the equations of lines (a) and (b).

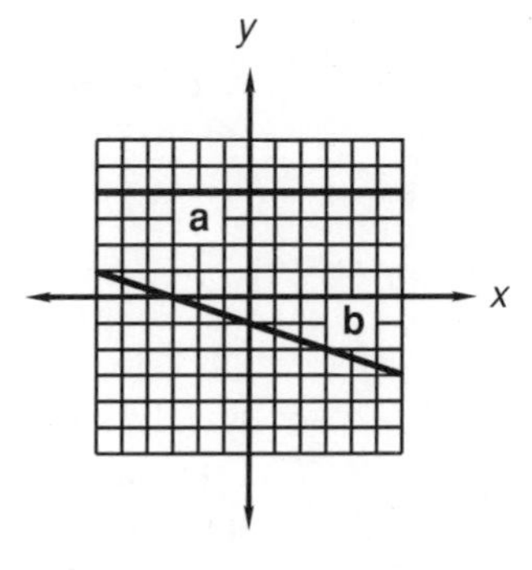

8. Which of the following diagrams or sets of ordered pairs depict functions?

(a)

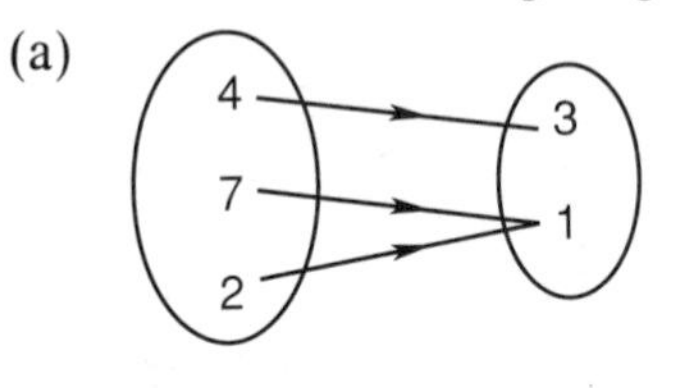

(b)

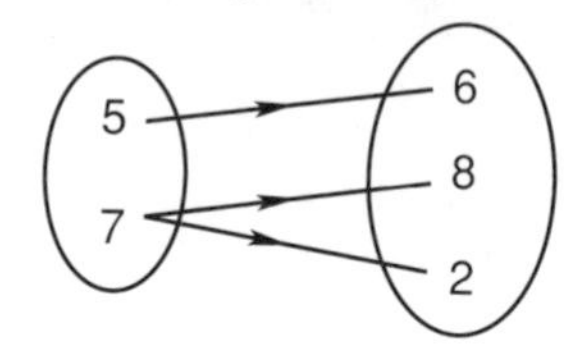

(c), (d)

(e) $(4, -2), (3, 5), (7, 6)$

(f) $(4, -2), (5, -2), (3, 5)$

(g) $(4, -2), (4, 7), (3, 5)$

Solve:

9. $\sqrt{x - 1} - 4 = 0$ 

10. $\sqrt{3x} + 4 = 7$

11. $5\sqrt{2x} = 4$

12. Find the slope of the line that passes through $(-2, 4)$ and $(5, -3)$ by using the slope formula and letting $(-2, 4)$ be point 1 and $(5, -3)$ be point 2. Repeat, and this time use $(5, -3)$ as point 1 and $(-2, 4)$ as point 2. The answers should be the same.

Solve:

13. $\frac{2x}{3} - \frac{x-2}{5} + x = 7$ **14.** $\frac{x-2}{2x} - \frac{3}{x} = -\frac{1}{5}$

15. Find y: $\frac{x}{y} + \frac{m}{n} - \frac{1}{c} = k$

16. Estimate. Then use the square root table to help simplify:

$$\sqrt{0.00052843 \times 10^{40}}$$

17. Find the distance between $(-3, 2)$ and $(5, -7)$.

18. Solve: $-3(x - 4^0) - (-2) - 3(-x - y^0) = 3(x - 2^2)$

Simplify:

19. $\frac{(5000 \times 10^{-15})(30{,}000 \times 10^{41})}{(6000 \times 10^{-14})(0.000025 \times 10^{-50})}$ **20.** $\frac{x^{-2}(x^{-4}x)^3}{x^{-3}x^0xx^{-2}}$

21. $\sqrt{2} \cdot 3\sqrt{12} + 2\sqrt{3} \cdot \sqrt{2} - 2\sqrt{6}(4\sqrt{6} - \sqrt{24})$

22. $\frac{4x+4}{4}$ **23.** $\left(\frac{-3^{-2}}{3^{-3}}\right)^{-2}$

24. Solve by factoring: $84 = -19x - x^2$

25. Evaluate: $-xy(y - x^0) - x^2 - x^0$ if $x = -3$ and $y = -5$

26. Graph on a number line: $+4 - |x| - 2 \geq 4$; $D = \{\text{Reals}\}$

27. Multiply: $xy^2\left[\frac{y^{-2}}{x} - \frac{3x^0x}{(y^{-3})^2}\right]$

28. True or false? $\{\text{Reals}\} \subset \{\text{Integers}\}$

29. Simplify: $\frac{\frac{a}{x^2y} - \frac{1}{y^2}}{\frac{bc}{x^2y^2} - 1}$ **30.** What fraction of $3\frac{1}{8}$ is $\frac{2}{3}$?

31. If the letters represent unspecified real numbers, use the associative and commutative properties to prove that $abcd = bdac$.

LESSON 119 *Consistent, inconsistent, and dependent equations*

119.A solutions of linear systems: consistent equations

We have considered three methods of finding the ordered pair that is the solution to a system of two first-degree equations in two unknowns. We call the methods the **substitution method**, the **elimination method**, and the **graphical method.** We will review these procedures by solving a system of equations by each of the three methods.

The system is

$$\begin{aligned} &(a) \quad \begin{cases} -x + y = 1 \\ x + y = 5 \end{cases} \\ &(b) \end{aligned}$$

SUBSTITUTION		ELIMINATION	
$x + (1 + x) = 5$	substituted $1 + x$ for y	$-x + y = 1$ $x + y = 5$	equation (a) equation (b)
$2x + 1 = 5$	simplified	$2y = 6$	added
$2x = 4$	added -1		

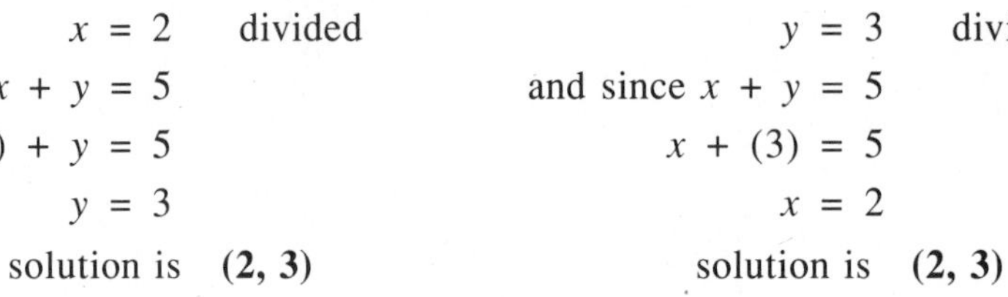

$x = 2$	divided	$y = 3$	divided
and since $x + y = 5$		and since $x + y = 5$	
$(2) + y = 5$		$x + (3) = 5$	
$y = 3$		$x = 2$	
solution is **(2, 3)**		solution is **(2, 3)**	

On the left we substituted for y in equation (b) the equivalent expression for y from equation (a). On the right above we added equations (a) and (b) algebraically and eliminated the variable x. The figure shows the graphs of the two linear equations. Note the coordinates of the intersection of the two lines. In each case, we find that the ordered pair (2, 3) is the solution to the system of equations. If a solution exists, the solution to such a system can always be found using any one of the three methods. **Equations that have a single solution are called *consistent equations.***

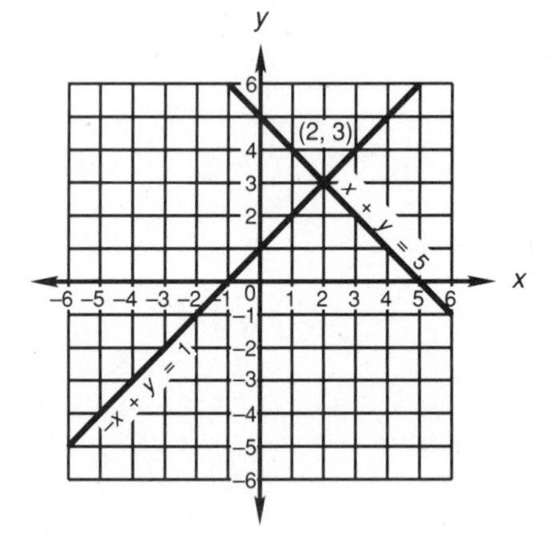

119.B inconsistent equations

Inconsistent equations have no common solution. On the left is a system of two linear equations, and the figure shows the graph of the two equations.

$$\begin{cases} x + y = 2 \\ x + y = -3 \end{cases}$$

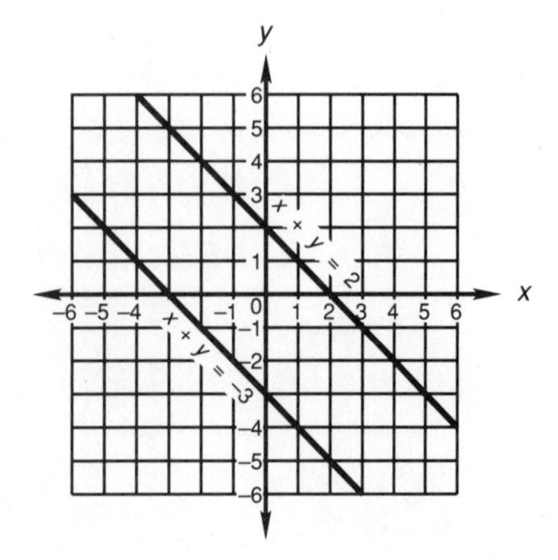

From the figure we see that the lines appear to be parallel and do not appear to intersect. If they do not intersect, there is no one point that lies on both of the lines, and hence no ordered pair of x and y will satisfy both equations. If we try to use either the substitution method or the elimination method, we will end up with a false numerical statement. To demonstrate this result, we will attempt to solve the given system by both methods.

SUBSTITUTION	ELIMINATION
$(-3 - y) + y = 2$	$-x - y = -2$
$-3 - y + y = 2$	$x + y = -3$
$-3 = 2$ False	$0 = -5$ False

In both cases the final result is a false numerical statement.

In Lessons 58 and 71 we discussed the substitution and elimination methods in some detail. We saw that before either method was used, an assumption was necessary. We assumed that a value of x and y existed that would simultaneously satisfy both equations and that the symbols x and y in the equations represented these numbers. From the graph of the parallel lines shown previously, it is obvious that in this problem our assumption was invalid, for in the graph we see that there is no point that lies on both lines and whose coordinates therefore satisfy both equations. Since our assumption was invalid, the substitution and elimination methods do not produce a solution.

119.C dependent equations

We have defined equivalent equations to be equations that have the same solution set. If we multiply every term on both sides of an equation by the same nonzero quantity, the resulting equation is an equivalent equation to the original equation. In the figure on the left, we have graphed the equation

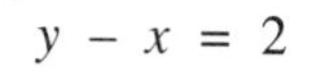

$$y - x = 2$$

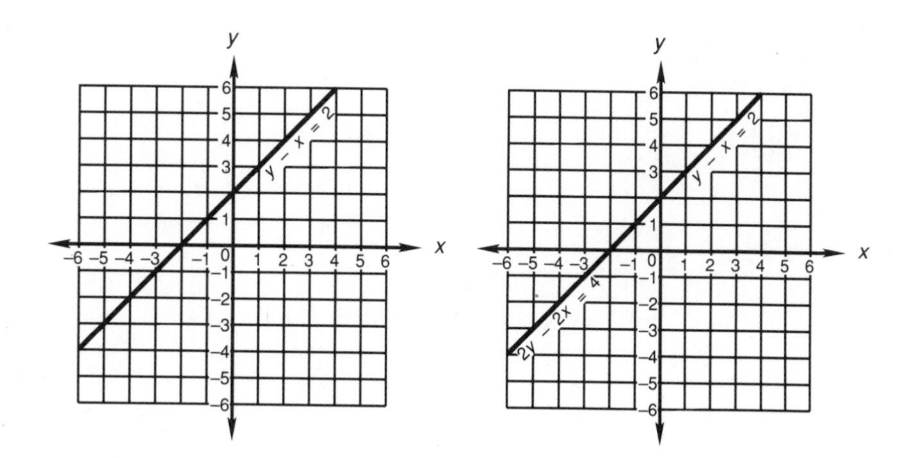

There is an infinite number of ordered pairs of x and y whose coordinates satisfy the equation, and the graph of these points is the line in the figure. If we multiply every term in the given equation by some quantity, say the number 2,

$$y - x = 2 \qquad \text{multiplied by (2) yields} \qquad 2y - 2x = 4$$

we get the equation $2y - 2x = 4$, which is an equivalent equation to the original equation. Thus all ordered pairs of x and y that satisfy the original equation also satisfy the new equation, and the graph of the new equation is the same as the graph of the original equation. If we are asked to find a graphical solution to a system that consists of a pair of equivalent equations such as the pair being discussed,

$$\begin{aligned} &\text{(a)} \quad \begin{cases} y - x = 2 \\ 2y - 2x = 4 \end{cases} \\ &\text{(b)} \end{aligned}$$

we find that the graph of both equations is the single line of either figure above. Thus any ordered pair that is a solution to one of the equations is also a solution to the other equation. **Equivalent linear equations are called *dependent equations.*** If we try to find the solution to our pair of dependent equations, we find that the result is a true numerical statement. We will attempt to solve the system above by using both substitution and elimination.

SUBSTITUTION

$$\begin{aligned} \text{(a)} \qquad y - x &= 2 \\ \text{(b)} \qquad 2y - 2x &= 4 \\ 2(x + 2) - 2x &= 4 \\ 2x + 4 - 2x &= 4 \\ 4 &= 4 \qquad \text{True} \end{aligned}$$

ELIMINATION

$$\begin{aligned} \text{(a)} \qquad y - x &= 2 \\ \text{(b)} \qquad 2y - 2x &= 4 \\ -2y + 2x &= -4 \\ 2y - 2x &= 4 \\ \hline 0 &= 0 \qquad \text{True} \end{aligned}$$

On the left we solved equation (a) for y and substituted this expression for y in equation (b). The result reduced to the true statement that 4 equals 4. On the right we multiplied each term in equation (a) by -2 and then added the resulting equation to equation (b). The result was the true statement that 0 equals 0. This indicates that any ordered pair of x and y that satisfies one of the equations will satisfy the other equation. Thus there is an infinite number of ordered pairs that will satisfy two dependent equations.

119.D summary

Two linear equations in two unknowns fall into one of three categories.

1. **Consistent equations,** which are equations that have a single ordered pair as a common solution. The graphs of consistent equations intersect, as shown in Figure (a).
2. **Inconsistent equations,** which are equations that have no common solution. The graphs of inconsistent equations are parallel lines, as shown in Figure (b).
3. **Dependent equations,** which are equivalent equations, or, if you will, the same equation in two different forms. The graph of two dependent equations is a single line, as shown in Figure (c).

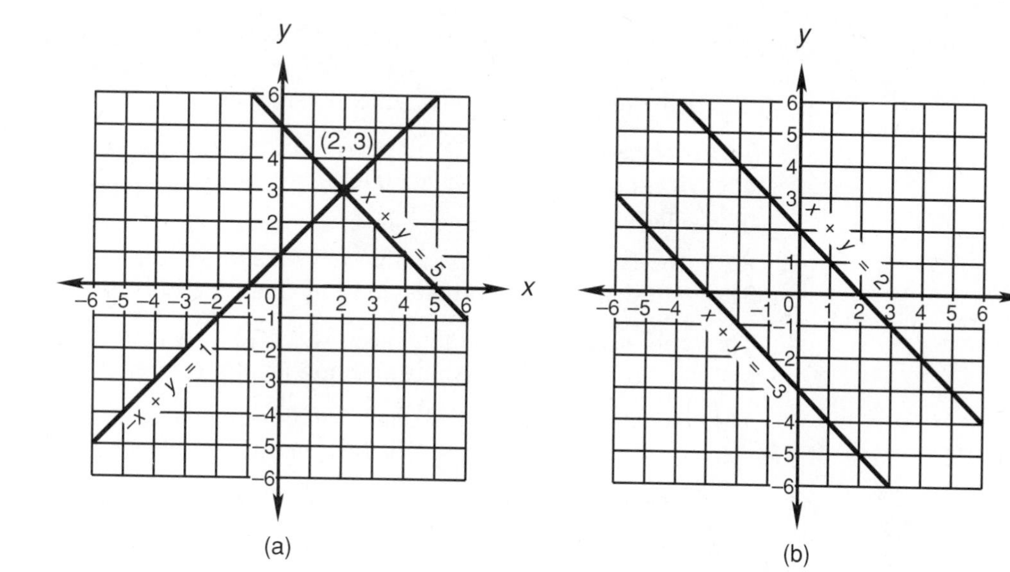

(a) (b)

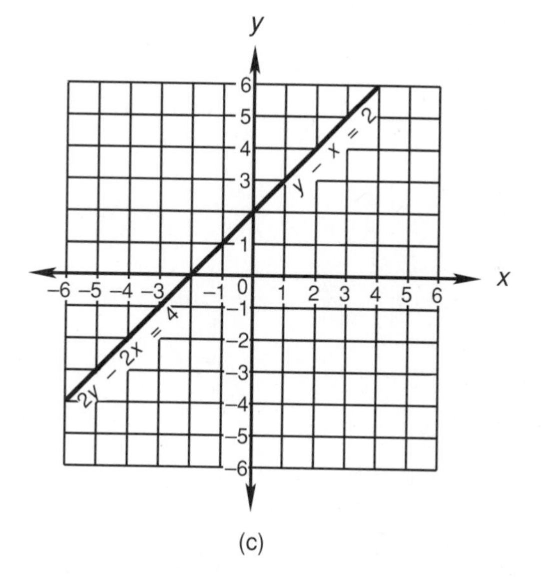

(c)

example 119.1 Consider the following equations:

$$y = 2x + 5 \qquad \text{and} \qquad y = x$$

Is this pair of equations consistent, inconsistent, or dependent?

solution Let's see if these equations have one point in common. Since $y = x$, we substitute x for y in the final equation:

$$x = 2x + 5 \qquad \text{substituted}$$

$$x = -5 \qquad \text{solved}$$

Since $y = x$, y is also equal to -5. The graph's equations have only one point in common, so the equations are **consistent.**

practice Are the equations $x + y = 7$ and $x + y = 9$ consistent, inconsistent, or dependent?

problem set 119

1. Poltavia walked for a while at 3 miles per hour and then rode a bus at 15 miles per hour. Her time on the bus was twice her time walking. How long did she ride if the total distance covered was 66 miles?

2. Marlow headed south at 20 miles per hour at noon. At 4 p.m. Mangum headed north from the same station at 60 miles per hour. At what time will they be 880 miles apart?

3. Rabbits were \$2 each and cats were \$7 each. Marcia bought 52 animals for \$259. How many were rabbits and how many were cats?

4. The ratio of raccoons to mongooses was 5 to 3. If there were 968 total in the pen, how many were raccoons and how many were mongooses?

5. Thirty-seven percent of the women had flowers in their hair. If 4788 women in the parade did not wear flowers, how many women were in the parade?

6. Find the equation of the line that passes through the points (4, 2) and (7, −3).

7. Given that $f(x) = x^2 + 4$; $D = \{\text{Reals}\}$. Find $f(\frac{1}{2})$.

Solve:

8. $7\sqrt{x} = 14\sqrt{2}$ 9. $\sqrt{x - 3} = 5$ 10. $\sqrt{x} + 2 = 7$

11. Are the equations of lines that are parallel called inconsistent or consistent?

Solve:

12. $\dfrac{4x}{3} - \dfrac{x + 2}{5} = 7$ 13. $\dfrac{p - 3}{p} = \dfrac{5}{3p} - \dfrac{1}{4}$

14. Find x: $\dfrac{k}{x} - \dfrac{1}{c} + p = m$

15. Find b.

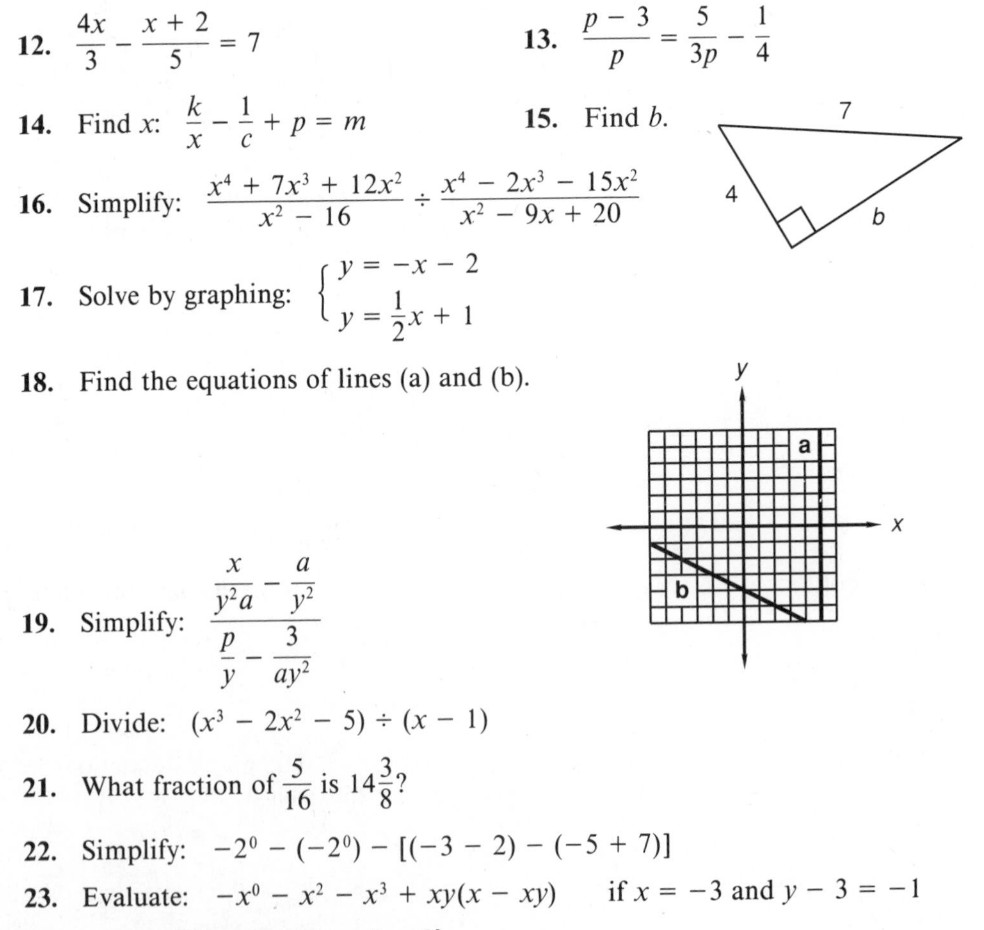

16. Simplify: $\dfrac{x^4 + 7x^3 + 12x^2}{x^2 - 16} \div \dfrac{x^4 - 2x^3 - 15x^2}{x^2 - 9x + 20}$

17. Solve by graphing: $\begin{cases} y = -x - 2 \\ y = \frac{1}{2}x + 1 \end{cases}$

18. Find the equations of lines (a) and (b).

19. Simplify: $\dfrac{\dfrac{x}{y^2a} - \dfrac{a}{y^2}}{\dfrac{p}{y} - \dfrac{3}{ay^2}}$

20. Divide: $(x^3 - 2x^2 - 5) \div (x - 1)$

21. What fraction of $\frac{5}{16}$ is $14\frac{3}{8}$?

22. Simplify: $-2^0 - (-2^0) - [(-3 - 2) - (-5 + 7)]$

23. Evaluate: $-x^0 - x^2 - x^3 + xy(x - xy)$ if $x = -3$ and $y - 3 = -1$

24. $-3 \in \{\text{What sets of numbers?}\}$

25. Solve by factoring: $x^2 + 40 = -22x$

26. Simplify by adding like terms: $x^2y - \frac{3x^3y}{x} + \frac{2x^4y^2}{yx^{-2}} + \frac{5x^2}{y^{-1}}$

Add:

27. $\frac{x}{y^2} + \frac{2x^2}{xy^2} - \frac{x^2}{y - 1}$

28. $\frac{4x - 2}{x - 3} - \frac{x + 3}{x^2 - 9}$

29. Round $32.07\overline{581}$ to the nearest hundred-millionth.

30. If the reciprocal of a number is $-\frac{1}{5}$, what is the product of the number and the multiplicative identity?

LESSON 120 More on conjunctions · Disjunctions

120.A more on conjunctions

We remember that a conjunction is a statement of two conditions, both of which must be met. The conjunction

$$4 < x < 14$$

tells us that x must be greater than 4 **and** that x must also be less than 14. Another way to state these conditions is to write

$$x > 4 \quad \text{and} \quad x < 14$$

The statement

$$4 < x + 3 < 10$$

is also a conjunction. It says that $x + 3$ is greater than 4 and that $x + 3$ is also less than 10. A good way to read these is to cover up the end numbers and the signs of inequality one at a time and read

$$4 < x + 3 \qquad \text{and} \qquad x + 3 < 10$$

We solve the inequality on the left and then solve the inequality on the right.

$$\begin{array}{rl} 4 < x + 3 \\ \underline{-3 \phantom{< x + 3}} \\ 1 < x \end{array} \qquad \begin{array}{rr} x + 3 < & 10 \\ \underline{-3 \quad} & \underline{-3} \\ x < & 7 \end{array} \qquad \text{subtracted 3 from both sides}$$

We could have simplified the conjunction in one step by adding -3 twice.

$$\begin{array}{c} 4 < x + 3 < 10 \\ \underline{-3 \qquad -3 \quad -3} \\ 1 < x < 7 \end{array}$$

example 120.1 Graph: $-5 \le x - 4 < 2$; $D = \{\text{Integers}\}$

solution We will begin by simplifying the conjunction by adding +4 in three places.

$$\begin{array}{c} -5 \le x - 4 < 2 \\ \underline{+4 \qquad +4 \quad +4} \\ -1 \le x \qquad < 6 \end{array} \qquad D = \{\text{Integers}\}$$

The following graph shows all integers that are greater than or equal to -1 and are also less than 6.

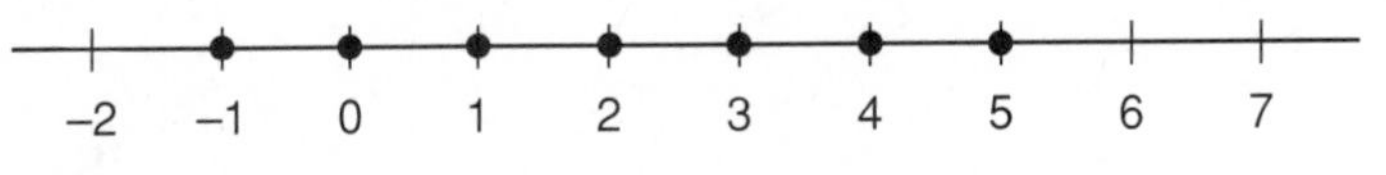

120.B disjunctions

The conjunction is a statement of two conditions, both of which must be met. The conjunction uses or implies the use of the word **and.**

$$x > -2 \quad \textbf{and} \quad x \leq 4 \quad \text{can be written} \quad -2 < x \leq 4$$

A **disjunction** is also a statement of two conditions. A disjunction is satisfied if either of the conditions is met. A disjunction uses the word **or.**

$$x \leq -2 \quad \textbf{or} \quad x > 1;\ D = \{\text{Reals}\}$$

This disjunction is satisfied by any real number that is less than or equal to -2. It is also satisfied by any real number that is greater than 1. This is a graph of the disjunction.

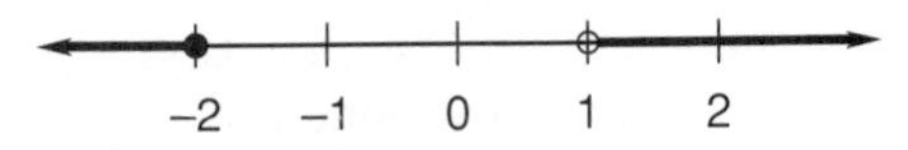

example 120.2 Graph $x \geq 3$ or $x < -2$; $D = \{\text{Integers}\}$.

solution This disjunction is satisfied by any integer that is greater than or equal to 3. It is also satisfied by any integer that is less than -2.

example 120.3 Graph $x > 15$ *or* $x \leq 10$; $D = \{\text{Reals}\}$.

solution The word *or* tells us that there are two sets of numbers that satisfy this dual condition.

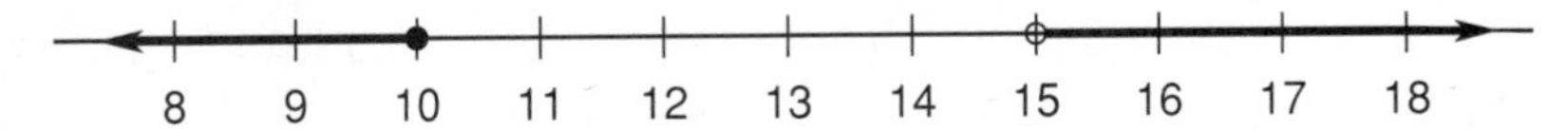

We have indicated all real numbers that are less than or equal to 10 *or* are greater than 15.

practice Graph:

a. $6 \leq x - 2 < 7$; $D = \{\text{Reals}\}$

b. $-x > -3$ or $x \geq 7$; $D = \{\text{Integers}\}$

problem set 120

1. Homer drove as fast as he could to get 60 miles ahead of Mae. If Homer drove at 17 miles per hour and the race lasted 20 hours, how fast did Mae drive?

2. Wanatobe walked to the moot at 4 miles per hour and then rode back home in a bus at 24 miles per hour. If her total traveling time was 14 hours, how far was it to the moot?

3. The sum of 13 and the opposite of a number was multiplied by 3. This result was 11 less than twice the number. What was the number?

4. The dress was marked down 20 percent for the sale, and it still sold for \$120. What would it sell for if the markdown was only 10 percent? (*Hint*: Find the original price and take off 10 percent of this price.)

5. Peaches were \$7 a bushel and apples were \$6 a bushel. Harry sold \$346 worth and sold 29 more bushels of peaches than apples. How many bushels of each did he sell?

6. Find the equation of the line that passes through the point $(-3, 2)$ and has a slope of $-\frac{1}{7}$.

7. Which of the following diagrams or sets of ordered pairs depict functions?

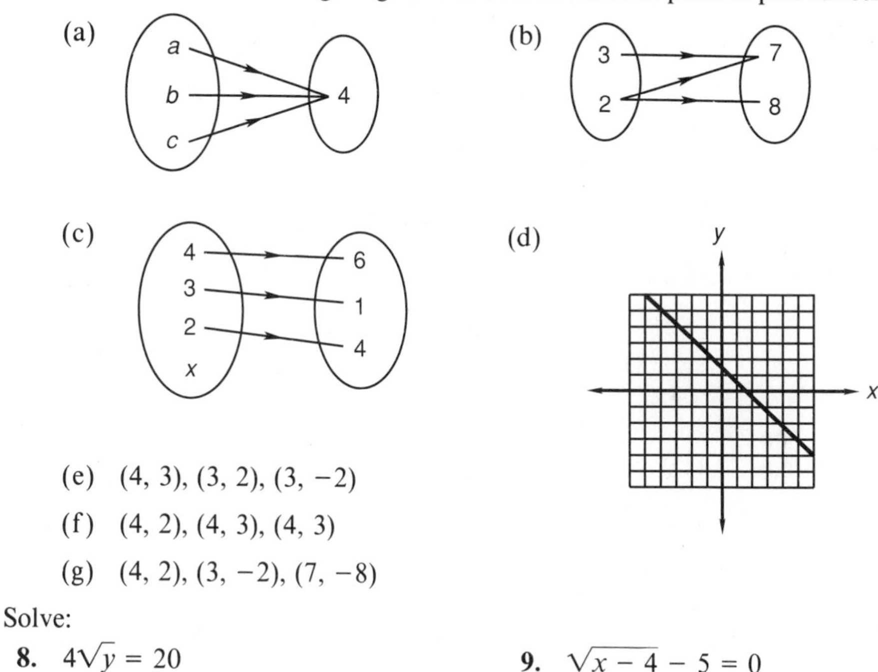

(e) $(4, 3), (3, 2), (3, -2)$

(f) $(4, 2), (4, 3), (4, 3)$

(g) $(4, 2), (3, -2), (7, -8)$

Solve:

8. $4\sqrt{y} = 20$

9. $\sqrt{x - 4} - 5 = 0$

Graph the following conjunctions and disjunctions on a number line:

10. $0 \leq x + 6 < 11$; $D = \{\text{Reals}\}$

11. $x < -1$ or $x \geq 5$; $D = \{\text{Integers}\}$

12. The set of negative odd integers is closed for what operations?

13. $\frac{x}{4} - \frac{x - 2}{3} = 7$

14. $\frac{p - 5}{p} = \frac{5}{3p} - \frac{1}{5}$

15. Find x: $\frac{a}{x} - \frac{1}{c} = \frac{1}{d}$

16. Estimate. Then use the square root table to simplify: $\sqrt{0.0004168521 \times 10^{-30}}$

17. Find the distance between $(-5, 2)$ and $(3, -7)$.

18. Solve: $p - 3p^0 - 2(p - 4^0) - (-3) - 2 = -3^0(2 - p)$

Simplify:

19. $\frac{(2000 \times 10^{15})(0.0004 \times 10^{21})}{(4000 \times 10^{-23})(1000 \times 10^{14})}$

20. $\frac{3(x^{-2}y)^{-3}x^{-3}y^0y^2}{(x^{-3}y^{-2})^{-2}xx^0}$

21. $2\sqrt{800} - 3\sqrt{18}$

22. $4\sqrt{2}(5\sqrt{2} - 2\sqrt{12})$

23. $\dfrac{4x^2 - 4x}{4x}$

24. $\dfrac{-3^{-3}(-3)^{-2}}{3^{-2}}$

25. Solve by factoring: $45 = x^2 + 4x$

26. Evaluate: $-x - ab(x - b^0)$ if $x = -3$, $a = 4$, $b = -5$

27. Graph on a number line: $-4 - |x| \le -4$; $D = \{\text{Reals}\}$

28. Multiply: $\dfrac{x^{-2}a^2}{y}\left(\dfrac{ya^{-2}}{x^{-2}} - \dfrac{3x^{-2}a^2}{y}\right)$

29. Round $478{,}325.0\overline{63}$ to the nearest thousand.

30. Find x if $x = \dfrac{-b \pm \sqrt{b^2 - 4ac}}{2a}$ and $a = -\frac{1}{2}$, $b = -1$, and $4c + 7 = 37$.

LESSON 121 More on multiplication of radical expressions

Thus far, our most advanced radical multiplication problem has been of the form

$$\sqrt{3}(5 + \sqrt{12})$$

This notation indicates that $\sqrt{3}$ is to be multiplied by both of the terms inside the parentheses and that the products are to be added.

$$\sqrt{3}(5 + \sqrt{12}) = 5\sqrt{3} + \sqrt{36} = \mathbf{5\sqrt{3} + 6}$$

We have a problem of the form

$$(4 + \sqrt{3})(5 + \sqrt{12})$$

We have four multiplications indicated. Both 4 and $\sqrt{3}$ must be multiplied by each term in the second parentheses and the four products simplified as shown here.

$$\begin{aligned}(4 + \sqrt{3})(5 + \sqrt{12}) &= 4 \cdot 5 + 4\sqrt{12} + 5\sqrt{3} + \sqrt{3} \cdot \sqrt{12}\\ &= 20 + 8\sqrt{3} + 5\sqrt{3} + 6\\ &= \mathbf{26 + 13\sqrt{3}}\end{aligned}$$

example 121.1 Multiply: $(2 + \sqrt{2})(3 + \sqrt{8})$

solution We will multiply 2 and $\sqrt{2}$ by both numbers in the second parentheses and simplify the result.

$$\begin{aligned}2 \cdot 3 + 2\sqrt{8} + 3\sqrt{2} + \sqrt{2} \cdot \sqrt{8} &= 6 + 4\sqrt{2} + 3\sqrt{2} + 4\\ &= \mathbf{10 + 7\sqrt{2}}\end{aligned}$$

example 121.2 Multiply: $(4 + \sqrt{5})(2 - 2\sqrt{5})$

solution We have four multiplications to perform, which are

$$4(2) + 4(-2\sqrt{5}) + \sqrt{5}(2) + \sqrt{5}(-2\sqrt{5})$$

Now we multiply and simplify the results.

$$8 - 8\sqrt{5} + 2\sqrt{5} - 10 = \mathbf{-2 - 6\sqrt{5}}$$

example 121.3 Multiply: $(2 + \sqrt{2})(3 + 2\sqrt{2})$

solution We perform the multiplications and then simplify by adding like terms.

$$6 + 4\sqrt{2} + 3\sqrt{2} + 4 = \mathbf{10 + 7\sqrt{2}}$$

example 121.4 Simplify: $(\sqrt{2}\,x + \sqrt{3}\,y)^2$

solution We rewrite the problem as $(\sqrt{2}\,x + \sqrt{3}\,y)(\sqrt{2}\,x + \sqrt{3}\,y)$. Now we perform the four multiplications and get

$$2x^2 + \sqrt{6}\,xy + \sqrt{6}\,xy + 3y^2$$

Now we add like terms and get

$$\mathbf{2x^2 + 2\sqrt{6}\,xy + 3y^2}$$

practice Multiply:

a. $(5 + \sqrt{2})(3 + \sqrt{8})$

b. $(2 + \sqrt{5})(4 - 3\sqrt{5})$

c. $(\sqrt{2} + \sqrt{5})^2$

d. $(\sqrt{2}x + \sqrt{7}y)^2$

problem set 121

1. Of the 1200 displays at the flower show, 300 had at least 1 rose. What percent of the displays did not contain any roses?

2. The northbound bus had been on the road at 50 miles per hour for 4 hours before the southbound bus left the same station at 45 miles per hour. How long was the southbound bus on the road before the buses were 580 miles apart?

3. Judy walked into the country at 5 kilometers per hour and then rode a bus back home at 30 kilometers per hour. It was a long trip—she was gone 21 hours. How far did she walk?

4. The ratio of haves to have-nots was 11 to 2. If the ruler heard requests for boons from 195 subjects, how many of these subjects were haves?

5. Maxine bought 176 stamps for \$10.75. She bought some 5-cent stamps and some 20-cent stamps. How many of each kind did she buy?

6. Is the following set of ordered pairs a function: $(4, -2)$, $(3, -5)$, $(8, -11)$, $(-2, 4)$, $(-5, 3)$, $(-11, -5)$?

7. Find the equation of the line that passes through the point $(-2, -3)$ that is parallel to the line $y = -\frac{2}{3}x + 5$.

8. Find the equations of lines (a) and (b).

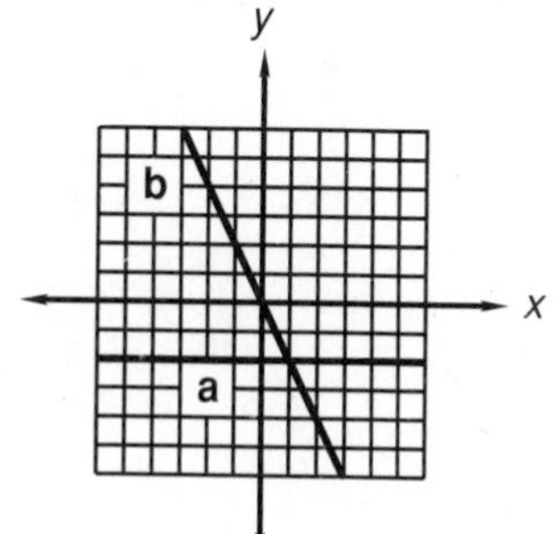

9. Given that $g(x) = x + 4$ and $D = \{\text{Integers}\}$, find $g\left(\frac{1}{2}\right)$.

Solve:

10. $2\sqrt{x} - 4 = 3$

11. $\sqrt{x + 5} - 3 = 2$

Graph:

12. $4 \le x - 3 < 6$; $D = \{\text{Integers}\}$

13. $x + 2 < 5$ or $x + 2 \ge 6$; $D = \{\text{Reals}\}$

14. Find the volume in cubic feet of a right prism whose base is shown and whose sides are 2 yards high. What is the surface area? Dimensions are in feet. Corners that look square are square.

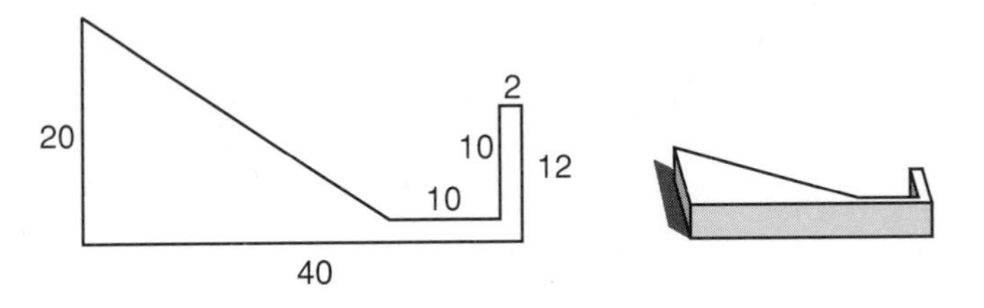

15. Use 12 unit multipliers to convert 1,000,000 cubic meters to cubic miles.

16. If the letters represent unspecified real numbers, use the associative and commutative properties to prove that $a + x + y + m = y + a + m + x$.

17. Multiply: (a) $(2 - 3\sqrt{12})(3 + 2\sqrt{12})$ (b) $(\sqrt{2}\,a - \sqrt{3}\,p)^2$

18. Solve: $\dfrac{k - 5}{k} = \dfrac{1}{5k} - \dfrac{1}{5}$

19. Find m: $\dfrac{a}{b} + \dfrac{x}{m} - \dfrac{1}{c} = p$

20. Solve by graphing: $\begin{cases} y = -x + 3 \\ y = -2 \end{cases}$

21. Simplify: $\dfrac{\dfrac{xy}{a^2} - \dfrac{1}{a}}{\dfrac{1}{a} + \dfrac{x^2y}{a^2}}$

22. Divide: $(x^3 - 2x - 4) \div (x + 2)$

23. What fraction of $3\frac{1}{7}$ is $\frac{3}{8}$?

24. $-3\sqrt{2} \in \{\text{What sets of numbers}\}$?

25. Simplify: $-[(-3 - 2)(-2) - 3^0(-3)] - [(-2^0 - 3^2) - (-2 - 4)]$

26. Evaluate: $-y^2 - y^0 - y(xy - y)$ if $x = 4$ and $y = -3$

27. Solve: $3\frac{1}{8}m + 2\frac{1}{4} = \frac{3}{8}$

28. Add: $\dfrac{5x + 2}{x - 3} - \dfrac{2x + 2}{x^2 - 9}$

Graph on a number line:

29. $-2 < x \le 2$; $D = \{\text{Reals}\}$

30. $-|x| + 2 \ge -1$; $D = \{\text{Positive integers}\}$

LESSON 122 Direct variation

In every classroom of a particular school, there are twice as many boys as there are girls. We can express this relationship mathematically by writing the equation

$$B = 2G$$

where B stands for the number of boys and G stands for the number of girls. Then we can use the equation to find the number of boys in any classroom if we are told the number of girls in that room, or to find the number of girls in a room if we are told the number of boys in that room.

In another school there are three times as many boys in each room as there are girls. For this school, the relationship can be expressed mathematically by writing

$$B = 3G$$

In the equation for the first school we had the variables B and G, and the constant was the number 2. In the equation for the other school the equation is the same equation except that the constant is the number 3. The general form of the relationship is

$$B = kG$$

where k represents the constant for any particular school.

We call a relationship such as this, where one variable is expressed as a constant times another variable, a *direct variation* or a *direct proportion,* and the constant in the equation is called the *constant of proportionality.*

If we consider a third school and are told

1. That the number of boys in any room **varies directly** as the number of girls in the room or
2. That the number of boys in any room **is directly proportional to** the number of girls in the room,

we have been told that the relationship between the number of boys and girls in any room in the school may be stated mathematically as

$$B = kG$$

Before we can solve any problem, we need to know the constant of proportionality for this school. If we are told the number of boys and girls in any room in the school, we can use these values in the equation to solve for k. If there are 30 boys and 5 girls in one room and we use these values for B and G in the equation, we find

$$30 = k(5) \qquad \text{and thus} \qquad k = 6$$

Since we have found the constant for this school, we may write the relationship for this school as

$$B = 6G$$

Now, if we are given the number of girls in any room in this school, we can solve for the number of boys, and, conversely, we can find the number of girls if we are given the number of boys.

The key to this type of problem is recognizing that the verbal statement denotes a direct variation or a direct proportion and realizing the equation that is implied by this statement. We will give some examples here. On the left is the key verbal statement, and on the right is the equation that is implied by the statement.

STATEMENT	EQUATION
The weight of a substance **varies directly** as the volume of the substance.	$W = kV$
Force **varies directly** as the current.	$F = kC$
The distance traveled **varies directly** as the time.	$D = kT$

The words **directly proportional** imply the same equation as do the words **varies directly.**

STATEMENT	EQUATION
The circumference of a circle is **directly proportional** to the length of the radius.	$C = kR$
The volume of a right circular cylinder of fixed radius is **directly proportional** to its height.	$V = kH$

In each of the relationships just discussed, the equation contained an unknown **constant of proportionality *k*. We begin the solution of any direct variation problem by finding the constant of proportionality for that problem.** Then we can solve for the value of one unknown if we are given the value of the other unknown.

example 122.1 The mass of a substance varies directly as the volume of the substance. If the mass of 2 liters of the substance is 10 kilograms, what will be the volume of 35 kilograms of the substance?

solution We will solve the problem in *four steps.*

Step 1: Recognize that the words **varies directly** imply the relationship

$$M = kV$$

Step 2: Reread the problem to find the values of M and V that can be used to find the value of k. Use these values to solve for k.

$$(10) = k(2) \longrightarrow k = 5$$

Step 3: Replace k in the equation with the value we have found

$$M = 5V$$

Step 4: Replace the problem to find that we are asked to find the value of V if M is 35. We replace M in the equation with 35 and solve for V by dividing by 5.

$$35 = 5V \longrightarrow \frac{35}{5} = \frac{\cancel{5}V}{\cancel{5}} \longrightarrow \mathbf{V = 7\ liters}$$

example 122.2 The distance traversed by a car traveling at a constant speed is directly proportional to the time spent traveling. If the car goes 75 kilometers in 5 hours, how far will it go in 7 hours?

solution We will use the same four steps.

Step 1: $D = kT$	write the equation
Step 2: $75 = k(5) \longrightarrow k = 15$	solve for k
Step 3: $D = 15T$	put k in the equation
Step 4: $D = (15)(7) \longrightarrow$ **D = 105 kilometers**	solve for D

example 122.3 Under certain conditions the pressure of a gas varies directly as the temperature. When the pressure is 800 pascals, the temperature is 400 K. What is the temperature when the pressure is 400 pascals?

solution Again we will use four steps in our solution.

Step 1:	$P = kT$	write the equation
Step 2:	$800 = k(400) \longrightarrow k = 2$	solve for k
Step 3:	$P = 2T$	put k in the equation
Step 4:	$400 = 2T \longrightarrow \boldsymbol{T = 200\ K}$	solve for T

practice

a. The mass of a substance varies directly as the volume of the substance. If a mass of 30 kg of a substance has a volume of 6 liters, what is the volume of 65 kg of the substance?

b. Under certain conditions the pressure of a gas varies directly as the temperature. When the pressure is 1200 pascals, the temperature is 300 K. What is the temperature when the pressure is 300 pascals?

problem set 122

1. Wormley ran to the store at 8 kilometers per hour and rode a bus home at 20 kilometers per hour. If his round trip traveling time was 7 hours, how far was it to the store?

2. The distance traversed by a car traveling at a constant speed is directly proportional to the time spent traveling. If the car goes 90 kilometers in 9 hours, how far will it go in 5 hours?

3. The mass of a substance varies directly as the volume of the substance. If the mass of 7 liters of the substance is 42 kilograms, what will be the volume of 63 kilograms of the substance?

4. The red carnations sold for 50 cents a bunch and the white carnations sold for 40 cents a bunch. Mike bought 27 bunches for $12.30. How many of each kind did he get?

5. Find x if $x = \dfrac{-b \pm \sqrt{b^2 - 4ac}}{2a}$ and $a = 3$, $b^2 = 3$, and $4c = 1$.

6. Find the equation of the line that passes through $(-2, -5)$ that is parallel to $y = 2x + 4$.

7. Which of the following sets of ordered pairs are functions?
(a) $(4, -3), (5, -3), (6, -3)$ (b) $(4, -3), (-3, 4), (-2, 4)$
(c) $(4, -3), (4, 3), (-4, 6)$ (d) $(4, -3), (-2, -4), (8, 3)$

Solve:

8. $2\sqrt{x} + 2 = 5$ **9.** $\sqrt{x - 4} - 2 = 6$

10. Graph: $-2 \le x + 5 < 3$; $D = \{\text{Reals}\}$

Multiply:

11. $(2 + \sqrt{3})(4 - 5\sqrt{12})$ **12.** $(2 + \sqrt{2})(4 - 3\sqrt{8})$ **13.** $(5 + \sqrt{6})(2 - 3\sqrt{24})$

Solve:

14. $\dfrac{x}{4} - \dfrac{x - 3}{2} = \dfrac{1}{5}$ **15.** $\dfrac{m - 2}{m} = \dfrac{1}{2m} - \dfrac{1}{3}$

16. Find a: $\frac{x}{a} - \frac{1}{k} = \frac{m}{c}$

17. Estimate. Then use the square root tables to simplify:

$$\sqrt{0.000416852 \times 10^{-13}}$$

18. Find the distance between $(-5, -2)$ and $(3, 7)$.

19. Solve: $3x^0 - 2(-x - 5) - (-3) + 2(x - 5) = -2(x + 3^0)$

Simplify:

20. $\frac{(35{,}000 \times 10^{-41})(700 \times 10^{14})}{(7000 \times 10^{21})(0.00005 \times 10^{15})}$

21. $\frac{(x^{-2})^{-3}(x^{-2}y^2)}{x^2yy^0(x^0y)^{-2}}$

22. $(3 + 3\sqrt{2})(4 - \sqrt{2})$

23. $2\sqrt{5}(\sqrt{5} - 2\sqrt{75})$

24. $\frac{2x^2yz - 2x^2yz^2}{2x^2yz}$

25. $\frac{3^{-2}}{-2^{-3}}$

26. Solve by factoring: $-8 = -x^2 + 7x$

27. Evaluate: $-x^2 - x^0 - x^3 + xy(y - xy)$ if $x = -3$ and $y = \sqrt[5]{-1024}$

28. Graph on a number line: $-3 - |-x| \geq -2$; $D = \{\text{Integers}\}$

29. Multiply: $\frac{a^{-2}x}{y}\left(\frac{ya^2}{x} - \frac{4x^2y}{a^2}\right)$

30. Round $4.0\overline{60}$ to the nearest ten-thousandth.

LESSON 123 Inverse variation

When a problem states that one variable **varies inversely** as the other variable or that the value of one variable is **inversely proportional** to the value of the other variable, an equation of the form

$$V = \frac{k}{W}$$

is implied, where k is the constant of proportionality and V and W are the two variables. If we look at the equations for direct variation and inverse variation

DIRECT VARIATION EQUATION	INVERSE VARIATION EQUATION
$V = kW$	$V = \frac{k}{W}$

we see that each equation contains two variables and one constant of proportionality *k*. In both equations, the constant *k* is the numerator! In a direct variation equation, both variables are in the numerator; and in an inverse variation equation, one variable is in the numerator and the other variable is in the denominator!

STATEMENT	EQUATION
The pressure of a perfect gas **varies inversely as the volume.**	$P = \frac{k}{V}$
The current **varies inversely** as the resistance.	$C = \frac{k}{R}$
The velocity is **inversely proportional** to the time.	$V = \frac{k}{T}$

Inverse variation problems are solved in the same way as the direct variation problems. First we recognize the equation implied by the statement of inverse variation. Then we find the constant of proportionality for this problem and use this constant in the equation to find the final solution.

example 123.1 Under certain conditions, the pressure of a perfect gas varies inversely as the volume. When the pressure of a quantity of gas is 7 pascals, the volume is 75 liters. What would be the volume if the pressure is increased to 15 pascals?

solution We will use the same four steps that we used for direct variation problems.

Step 1: $P = \frac{k}{V}$ — write the equation

Step 2: $7 = \frac{k}{75} \longrightarrow k = 525$ — solve for k

Step 3: $P = \frac{525}{V}$ — substitute 525 for k

Step 4: $15 = \frac{525}{V} \longrightarrow V = \frac{525}{15}$ — substitute 15 for P and solve

$\longrightarrow$ **V = 35 liters**

example 123.2 To travel a fixed distance, the rate is inversely proportional to the time required. When the rate is 60 kilometers per hour, the time required is 4 hours. What would be the time required for the same distance if the rate were increased to 80 kilometers per hour?

solution We will use the same four steps.

Step 1: $R = \frac{k}{T}$ — write the equation

Step 2: $60 = \frac{k}{4} \longrightarrow k = 240$ — solve for k

Step 3: $R = \frac{240}{T}$ — substitute 240 for k

Step 4: $80 = \frac{240}{T} \longrightarrow T = \frac{240}{80} \longrightarrow$ **T = 3 hours**

practice Under certain conditions, the pressure of a perfect gas varies inversely as the volume. When the pressure of a quantity of gas is 9 pascals, the volume is 100 liters. What would be the volume if the pressure is increased to 20 pascals?

problem set 123

1. To travel a given distance, the rate is inversely proportional to the time required. Tom noted that if he drove home at 100 kilometers per hour, the trip would take 5 hours. How long would it take him to drive home if he drove at 125 kilometers per hour?

2. Use 15 unit multipliers to convert 1,000,000 cubic yards to cubic kilometers.

3. The number of girls in a class varied directly as the number of boys. One class had 3 boys and 21 girls. If another class had 5 boys, how many girls were in this class?

4. Peaches varied directly as apples. When there were 40 peaches, there were 120 apples. How many apples went with 500 peaches?

5. Silent Steve planted his farm acreage in cotton and peanuts in the ratio of 4 to 5. If his farm had 1278 acres, how many acres were in cotton?

6. Find the equation of the line that passes through $(-2, -3)$ and has a slope of $-\frac{1}{5}$.

7. Which of the following diagrams depict functions?

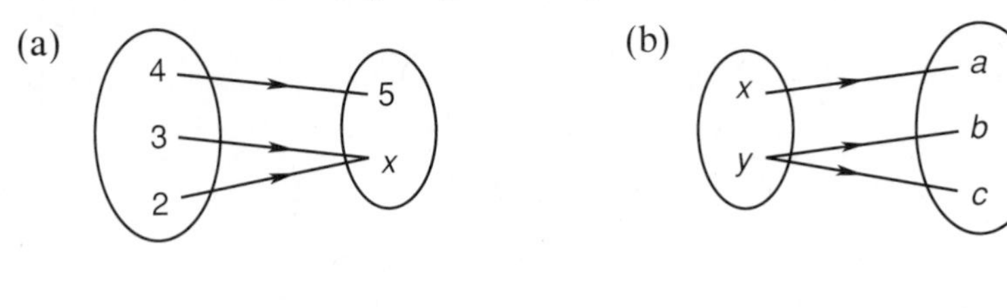

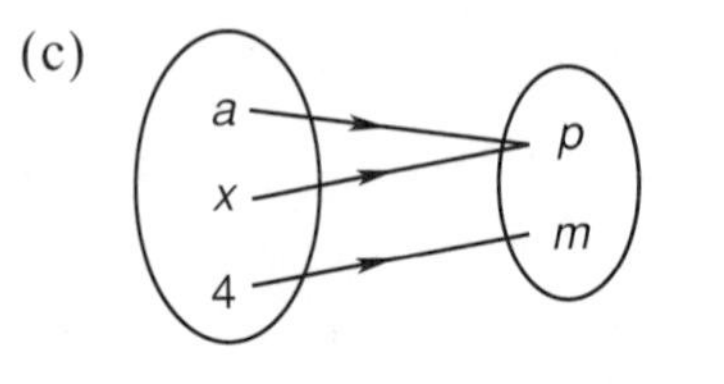

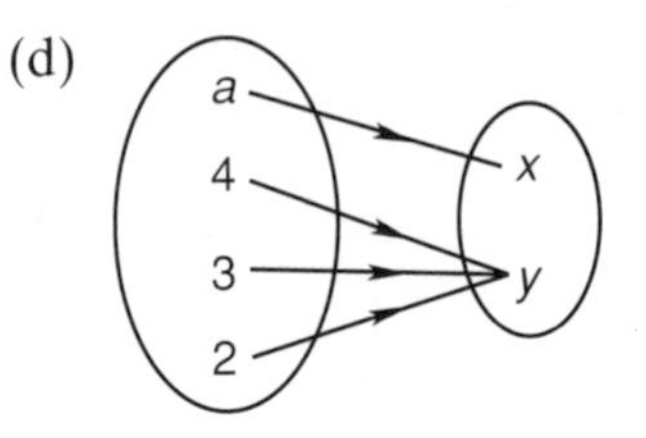

Solve:

8. $\sqrt{x} - 4 = 2$

9. $\sqrt{x - 5} - 3 = 2$

10. Graph: $x + 2 \geq 5$ or $x + 3 \leq 0$; $D = \{\text{Integers}\}$

Multiply:

11. $(4 - 3\sqrt{2})(2 + 6\sqrt{2})$

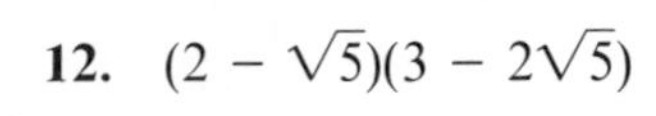

12. $(2 - \sqrt{5})(3 - 2\sqrt{5})$

13. $(3 - \sqrt{3})(2 - 2\sqrt{3})$

Solve:

14. $\frac{x}{4} - \frac{x - 5}{7} = 2$

15. $\frac{m + 5}{m} = \frac{3}{2m} - \frac{2}{5}$

16. Find y: $\frac{x}{y} - m + \frac{1}{c} = k$

17. Find k.

5

k

8

18. Simplify: $\dfrac{x^5 - 5x^4}{x^2 - 25} \div \dfrac{x^4 + 4x^3 - 32x^2}{x^2 + x - 20}$

19. Solve by graphing: $\begin{cases} y = \frac{1}{2}x - 2 \\ x = -4 \end{cases}$

20. Find the equations of lines (a) and (b).

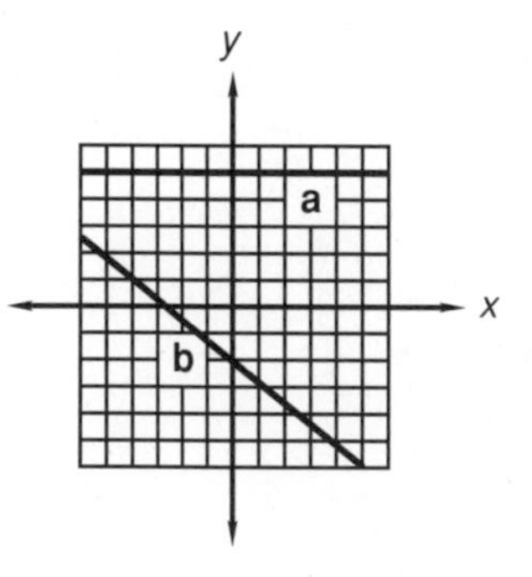

21. Simplify: $\dfrac{a + \frac{a}{x}}{\frac{1}{x} + a^2}$

22. Divide: $(x^3 - 2) \div (x^2 - 1)$

23. Simplify: $-2^0(-3^0 - 5)[(7 - 2^2)(-3 - 4^0) - (-2)][-(-2)]$

Graph on a number line:

24. $+1 - |x| \leq -1$; $D = \{\text{Positive integers}\}$

25. $-1 < x + 2 \leq 1$; $D = \{\text{Reals}\}$

26. Evaluate: $-p^0 - (p^0)^2 - p^2 - p^3 - p(p - y)$ if $p = -3$ and $y = 2$

27. Simplify by adding like terms: $x^2yz^{-1} + \dfrac{3y}{x^{-2}z} - \dfrac{4yx^2}{z} + 2y^2x^2z^{-1}$

28. Solve: $3\frac{1}{6}p + \frac{1}{4} = \frac{7}{8}$

29. Add: $\dfrac{3x - 2}{x - 3} - \dfrac{2x + 5}{x^2 - 9}$

LESSON 124 y^x *key · Exponential increases*

124.A y^x key

The y^x key on a calculator will let us raise numbers to powers quickly and accurately. The value of y is entered. Then the y^x key is depressed. Then the value of x is entered. Then we depress the $=$ key. Let's begin by finding 2^4, which we know equals 16.

	KEY	DISPLAY
(a)	2	2
(b)	y^x	2
(c)	4	4
(d)	$=$	16

example 124.1 Evaluate: $(1.06)^8$

solution The four steps are the same.

	KEY	DISPLAY
(a)	1.06	1.06
(b)	y^x	1.06
(c)	8	8
(d)	=	1.593848

124.B exponential growth

Jimmy had 150 rabbits. If the number of rabbits doubled every year, he would have 300 rabbits at the end of 1 year. We use R_0 to indicate the number of rabbits in the beginning, we use R_1 to indicate the number of rabbits at the end of the first year, etc.

In the beginning: $R_0 = 150 = 150 \cdot 2^0$

At the end of 1 year: $R_1 = 150 \cdot 2 = 150 \cdot 2$

At the end of 2 years: $R_2 = 150 \cdot 2 \cdot 2 = 150 \cdot 2^2$

At the end of 3 years: $R_3 = 150 \cdot 2 \cdot 2 \cdot 2 = 150 \cdot 2^3$

At the end of x years: $R_x = 150 \cdot 2^x$

The graph of the number of rabbits as time goes on looks like this:

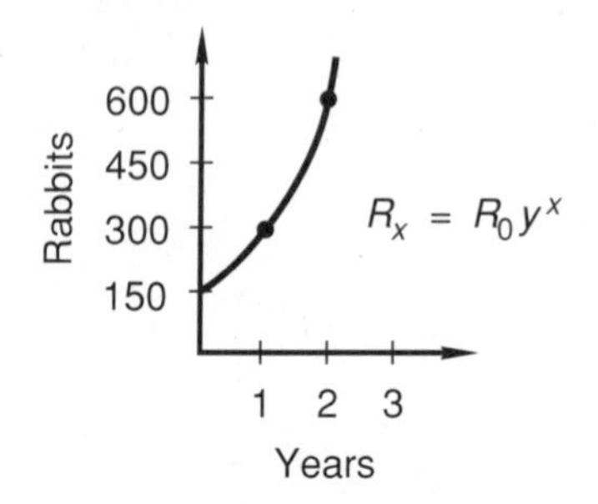

This type of growth is called **exponential growth.** It shows that the number of rabbits at the end of x years has the equation

$$R_x = R_0 y^x$$

We call the number represented by y the **growth multiplier.**

A bank will pay you to let the bank use your money. Each year the bank will pay you a percentage of the amount you deposit. The amount you deposit is called the **principal.** The amount they pay you is called the **interest.** If you deposit \$100 at 4 percent, at the end of 1 year the bank will give you back your \$100 plus 4 percent of \$100, which is \$4. This equals 1.04 times your original deposit.

Principal: \$100

At the end of 1st year: (\$100)(1.04)

If you leave your money in the bank for 1 more year, you will have 1.04 times the money you left in.

At the end of 2d year:

$$[(\$100)(1.04)](1.04) = \$100(1.04)^2$$

If you leave your money in the bank for 1 more year, you will have 1.04 times the amount you left in.

At the end of 3d year:

$$(\$100)(1.04)(1.04)(1.04) = (\$100)(1.04)^3$$

When the bank pays you interest on the interest you have earned, we say that the bank is paying **compound interest.** The amount you have in the bank in x years is

$$A_x = A_0(1 + r)^x$$

where A_0 is the principal and the rate r is percent divided by 100. We see that the equation has the form

$$A_x = A_0 y^x$$

example 124.2 The number of bacteria in the dish tripled every month. If there were 2500 bacteria at first, how many bacteria were there in 10 months?

solution The form of the equation is

$$A_x = A_0 y^x$$

where y is the multiplier, which is 3, x is the number of months (10), and A_0 is the number of bacteria (2500). Now we substitute.

$$A_{10} = 2500(3)^{10}$$

We use the y^x key to evaluate 3^{10}.

ENTER	DISPLAY
3	3
y^x	3
10	10
=	59049

Thus our answer is

$$2500(59{,}049) = \mathbf{1.4762 \times 10^8}$$

example 124.3 James deposited \$500 at 7 percent compound interest. How much money did he have in 14 years? How much interest did he earn?

solution The equation is

$$A_0 = A_0(1 + r)^x$$

Now we substitute and get

$$A_{14} = \$500(1 + 0.07)^{14}$$

First we evaluate $(1.07)^{14}$.

ENTER	DISPLAY
1.07	1.07
y^x	1.07
14	14
=	2.578342

We multiply this by \$500 and get

$$\$1289.27$$

In 14 years the amount on deposit went from \$500 to \$1289.27! Since James began with \$500 and ended with \$1287.27, his interest was

$$\$1287.27 - \$500 = \$787.27$$

practice Harriet deposited \$900 at 8 percent compound interest. How much money did she have in 12 years? How much interest did she earn?

problem set 124

1. Productivity varied inversely as the number of distractions. When there were 6 distractions, 500 items were produced. What would be the number of items produced if there were 10 distractions?
2. Howell hopped to the hostel at 5 miles per hour and hobbled back to the helicopter at 2 miles per hour. How far was it to the hostel if the round trip took 28 hours?
3. Allegra deposited \$700 at 9 percent compound interest. How much money will she have in 11 years? How much interest will she earn?
4. The number of bacteria in the petri dish quadrupled every hour. If there were 1000 bacteria at first, how many were there 24 hours later?
5. The ratio of the number of nuances to the number of blatancies was 3 to 17. If there were 90 nuances, how many blatancies were there?
6. Find the equations of lines (a) and (b).

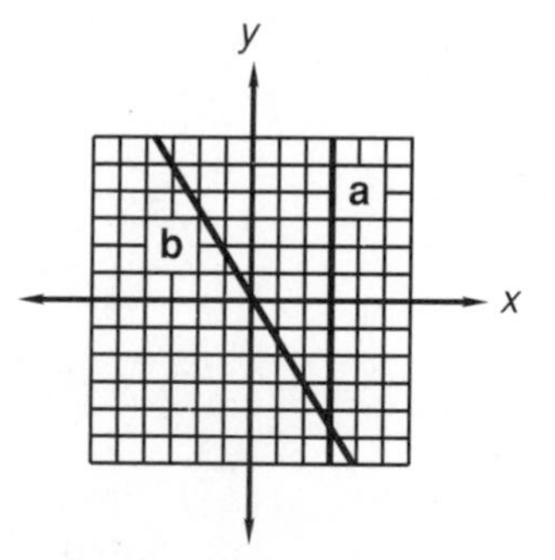

7. Which of the following sets of ordered pairs are functions?
   (a) $(-2, -1), (2, -1), (7, -1)$ (b) $(-2, -1), (-2, 1), (-2, 3)$
   (c) $(-2, -1), (2, 1), (-2, 7)$ (d) $(1, 3), (-1, -3), (-3, 1)$
8. Solve: $\sqrt{x - 7} + 4 = 9$
9. Graph on a number line: $x + 4 > 7$ or $x - 2 \leq 0$; $D = \{\text{Reals}\}$

Multiply:

10. $(4 + 3\sqrt{5})(1 - \sqrt{5})$ 11. $(3 + 2\sqrt{2})(3 - \sqrt{2})$ 12. $(5 + \sqrt{2})(2 - 4\sqrt{2})$

Simplify:

13. $\dfrac{3 + 2\sqrt{5}}{\sqrt{2}}$ 14. $\sqrt{\dfrac{3}{8}}$ 15. $\sqrt{\dfrac{2}{7}}$ 16. $\dfrac{3 + 3\sqrt{8}}{\sqrt{3}}$

17. Find the area of the shaded portion of this figure. Dimensions are in feet.

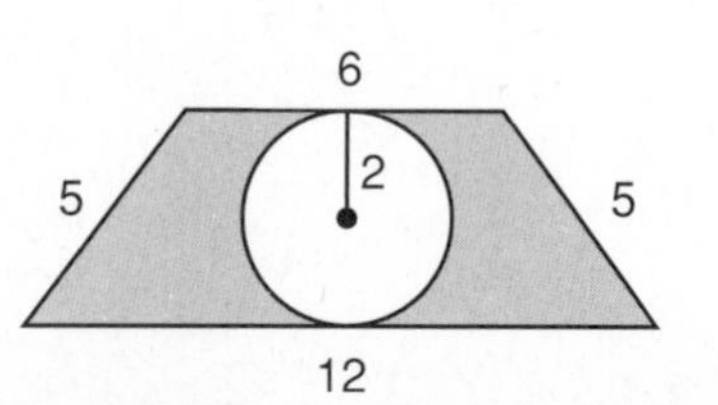

18. Use six unit multipliers to convert 160,000 cubic meters to cubic inches.

Solve:

19. $\frac{p + 8}{3p} = \frac{5}{2p} + \frac{1}{4}$

20. Find z: $\frac{p}{m} - \frac{x}{z} + a = k$

21. Simplify: $\dfrac{\dfrac{mp^2}{x} - \dfrac{z}{x^2}}{\dfrac{y}{x^2} - \dfrac{5a}{x}}$

22. Divide: $(7x^3 - 2x - 2) \div (x + 2)$

23. Simplify: $-3[(-3^0 - 3)^2(-3^3 - 3) - (-3)] - \sqrt[3]{-27}$

24. Evaluate: $-xy - y^x - x\left(\frac{y}{x}\right)$ if $y = -3$ and $x = -2$

25. Solve: $4\frac{1}{2}x + \frac{3}{5} = \frac{1}{4}$

26. Find the equation of the line through $(4, -1)$ that is parallel to $y = -\frac{1}{4}x - 2$.

27. Find the equation of the line that goes through $(4, -1)$, and $(-1, 4)$.

28. Find the equation of the line whose slope is -3 that passes through the point $(-1, -1)$.

29. Graph $y = -3x - 1$ on a rectangular coordinate system.

30. Do x values of either -1 or $\frac{1}{4}$ make $4x^2 + 3x - 1 = 0$ a true statement? (Show all computations.)

LESSON 125 Linear inequalities

Below we have graphed the line whose equation is $y = \frac{1}{2}x + 1$. We see that this line divides the set of all the points of the plane into three mutually exclusive subsets:

1. The set of points that lie on the line.
2. The set of points that lie above the line (region A).
3. The set of points that lie below the line (region B).

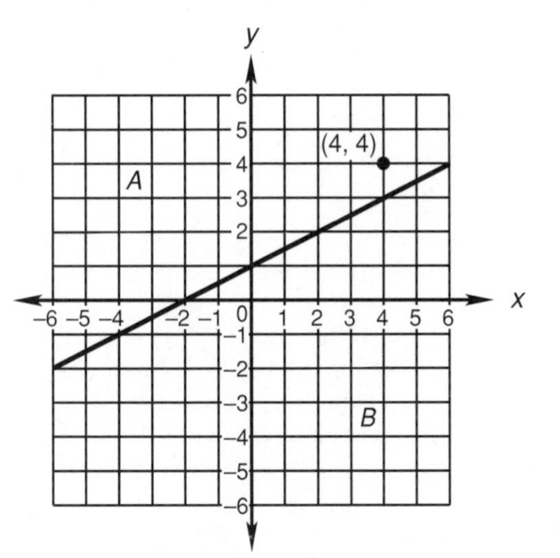

It can be shown that the coordinates of any point in the plane either will satisfy the equation of a given line or will satisfy one of the two linear inequalities that define the regions on either side of the line. For this particular line, we can say that the coordinates of any point in the plane will satisfy one and only one of the following:

$$y > \frac{1}{2}x + 1 \qquad y = \frac{1}{2}x + 1 \qquad y < \frac{1}{2}x + 1$$

The points in the regions denoted by A and B above do not lie on the line, and the coordinates of any point in region A or region B will satisfy one and only one of the inequalities. To see which of these inequalities defines the region above the line, we will choose a test point that clearly lies on one side of the line and test the coordinates of this point in both inequalities. We choose the point (4, 4).

$$y > \frac{1}{2}x + 1 \qquad\qquad y < \frac{1}{2}x + 1$$

$$4 > \frac{1}{2}(4) + 1 \qquad\qquad 4 < \frac{1}{2}(4) + 1$$

$$4 > 2 + 1 \qquad\qquad 4 < 2 + 1$$

$$4 > 3 \quad \text{True} \qquad\qquad 4 < 3 \quad \text{False}$$

Thus the coordinates of all points above the line satisfy the inequality $y > \frac{1}{2}x + 1$. If we choose a point below the line in region B, we can show that all points in this region satisfy the other inequality, $y < \frac{1}{2}x + 1$. We choose the point (0, 0).

$$y < \frac{1}{2}x + 1 \longrightarrow 0 < \frac{1}{2}(0) + 1 \longrightarrow 0 < 1 \quad \text{True}$$

We indicate in the next figure that the coordinates of all points above the line satisfy the inequality $y > \frac{1}{2}x + 1$ and that the coordinates of all points below the line satisfy the inequality $y < \frac{1}{2}x + 1$.

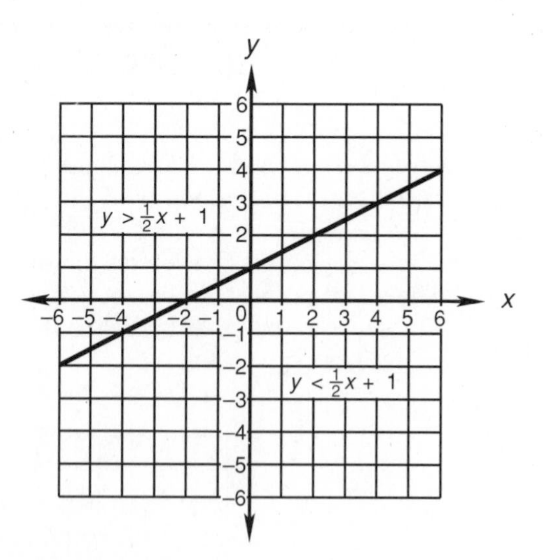

example 125.1 Graph the inequality: $y > -x + 2$

solution We first graph the line $y = -x + 2$ in the figure on the left at the top of the next page. We show the line as a dashed line to indicate that the points on the line do not satisfy the stated inequality, which uses a *greater than* symbol. Had the inequality used an *equal to or greater than* symbol, the line would have been drawn as a solid line. We choose the point (0, 0) as a test point because it clearly lies on one side of the line and also because it is the easiest test point to use.

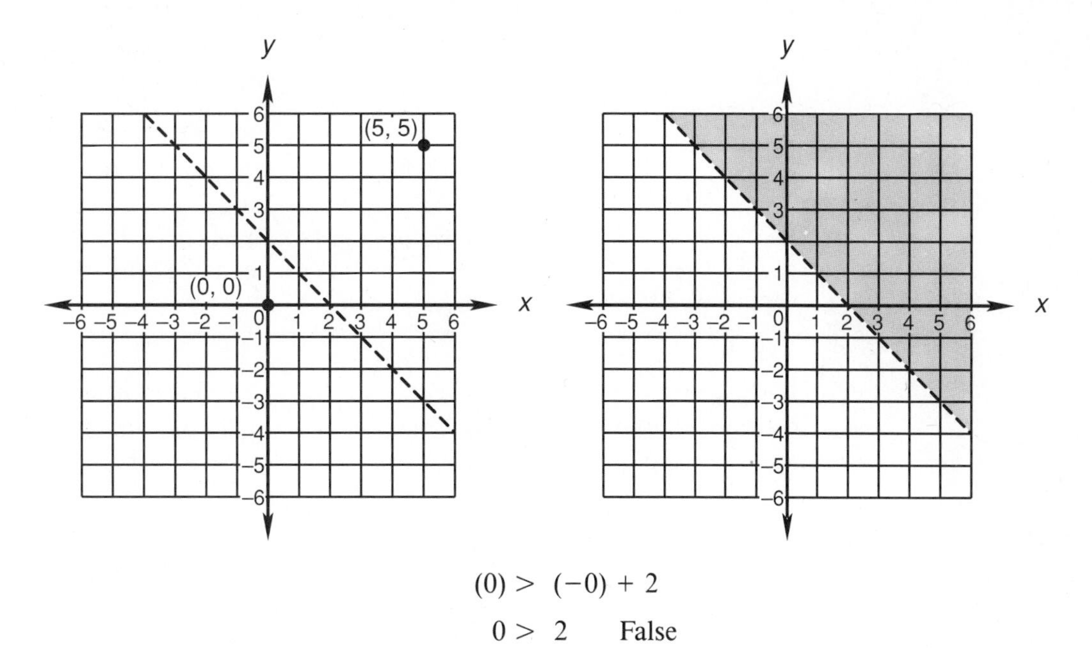

$$(0) > (-0) + 2$$

$$0 > 2 \qquad \text{False}$$

The coordinates of this point do not satisfy the inequality, so the coordinates of any point above the line must satisfy the inequality. We will use the point (5, 5).

$$(5) > (-5) + 2$$

$$5 > -3 \qquad \text{True}$$

We indicate that the coordinates of all points above the line satisfy the given condition by shading the region above the line in the figure on the right.

example 125.2 Graph: $y \le 2x - 1$

solution In the figure on the left we have graphed the line $y = 2x - 1$. We show the line as a solid line because we wish to indicate that the points on the line satisfy the stated inequality, which is read from left to right as "less than or equal to."

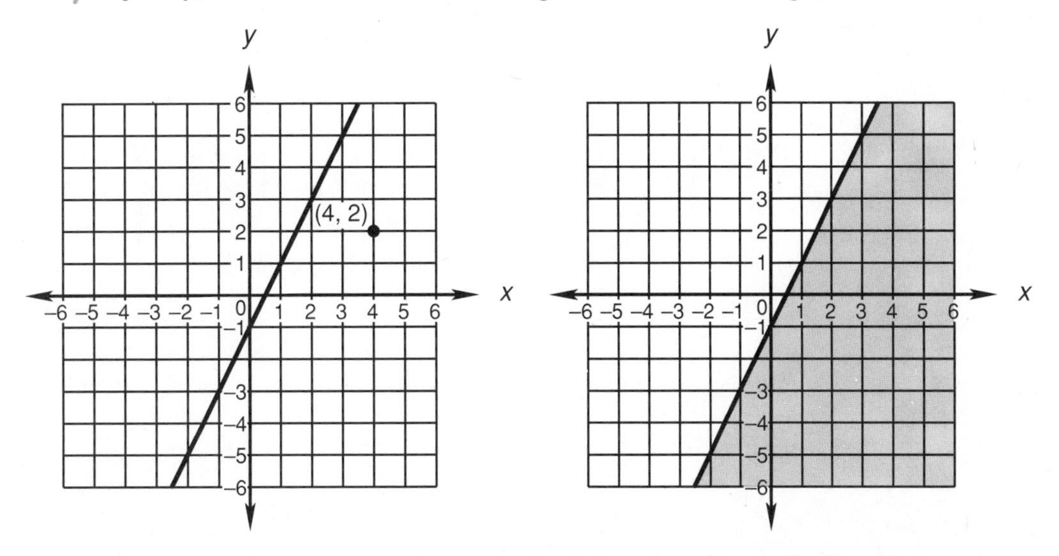

We could use the test point (1, 0), but it is rather close to the line. To be sure we have a point that is well on one side of the line, we choose the point (4, 2).

$$2 < 2(4) - 1$$

$$2 < 7 \qquad \text{True}$$

Thus the coordinates of the points on the line or in the region to the right of the line satisfy the stated inequality. We indicate the solution by shading this region in the figure on the right.

example 125.3 Graph: $y > -2$

solution In the figure on the left we have graphed the line $y = -2$. We show the line as a dashed line because the points on the line do not satisfy the inequality $y > -2$. We choose (0, 0) as the test point.

$$0 > -2 \qquad \text{True}$$

We shade the region above the line in the figure on the right to indicate that the coordinates of any point above the line will satisfy the stated inequality.

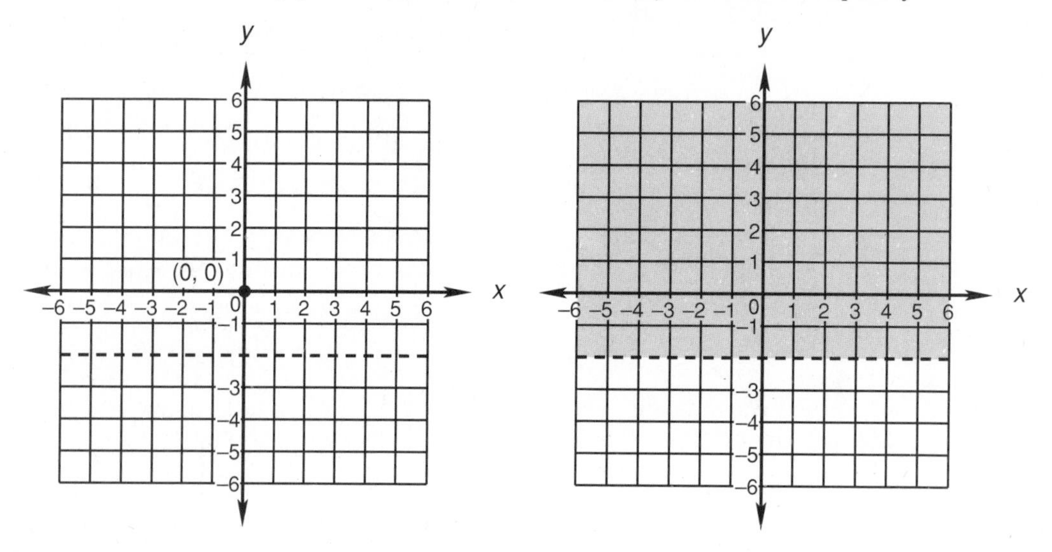

example 125.4 Graph: $x < 3$

solution In the figure on the left we graph the equation $x = 3$. We draw the line as a dashed line because points on the line do not satisfy $x < 3$, the given inequality. We will use the point (0, 0) as our test point.

$$0 < 3 \qquad \text{True}$$

We shade in the region to the left of the dashed line in the figure on the right to indicate that all points in this region will satisfy the inequality $x < 3$.

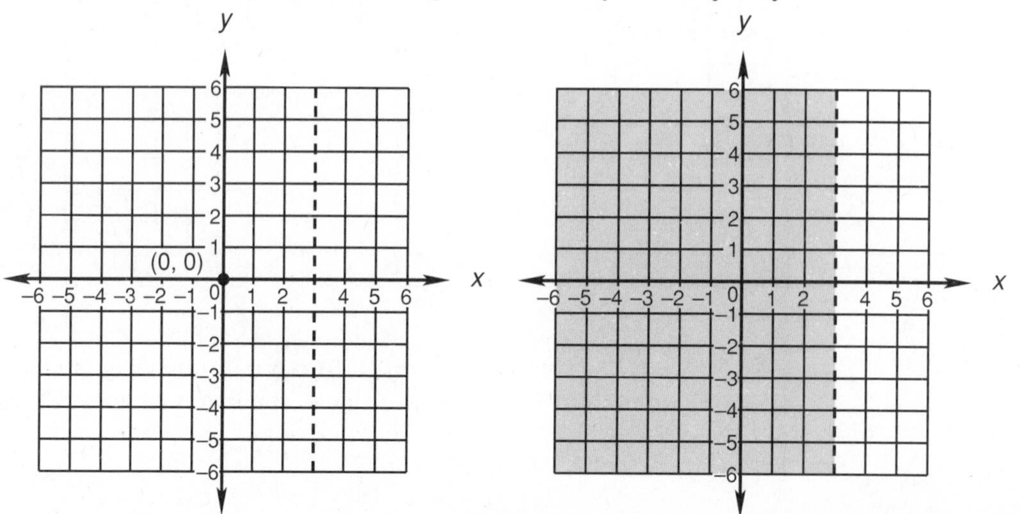

practice Graph:

a. $x < -2$

b. $y \le 3x - 2$

problem set 125

1. Pressure varies inversely as the volume. When the pressure is 10 pascals, the volume is 150 liters. What would the volume be if the pressure is reduced to 3 pascals?
2. Attractiveness varies inversely with the wiggles. At 300 wiggles the attractiveness is 10. What would the attractiveness be at 150 wiggles?
3. Mickey and Yarberry saved nickels and dimes. They had a total of 34 coins whose value was \$2.70. How many of each kind of coin did they have?
4. Rosie ran to the park at 7 miles per hour and walked back home at 3 miles per hour. How far was it to the park if the round trip took 20 hours?
5. At first there were 17 boring beetles. If the number of boring beetles doubled every day, how many boring beetles would there be in 30 days? ($N_x = N_0 \cdot 2^x$)
6. Graph $x < -5$ on a rectangular coordinate system.
7. Given $p(x) = x^2 + 2x + 5$; $D = \{\text{Reals}\}$. Find $p(-2)$.

Solve:

8. $-2\sqrt{x} + 4 = -1$
9. $2\sqrt{p + 2} - 4 = 3$
10. Graph on a number line: $4 \le x + 3 < 7$; $D = \{\text{Integers}\}$

Multiply:

11. $(3 + 2\sqrt{2})(5 - 3\sqrt{2})$
12. $(4 + \sqrt{3})(2 - 4\sqrt{3})$
13. $(2 + \sqrt{8})(3 - 2\sqrt{2})$

Solve:

14. $\dfrac{2x}{5} - \dfrac{x + 4}{2} = 3$
15. $\dfrac{k - 3}{2k} = \dfrac{3}{6k} - \dfrac{1}{4}$
16. Find m: $\dfrac{x}{m} - \dfrac{c}{d} = d$
17. Estimate. Then use the square root tables to simplify:

$$\sqrt{0.0001234567 \times 10^{-15}}$$

18. Find the distance between $(-4, -4)$ and $(3, 0)$.
19. Solve: $-(-3)x^0 - (-2)(x - 4) = -3(-x^0)$

Simplify:

20. $\dfrac{(21{,}000 \times 10^{-40})(5000 \times 10^{-20})}{(0.00003 \times 10^{15})(0.0007 \times 10^{28})}$
21. $\dfrac{4x^2y^{-2}(x^2)^{-2}y^2xy}{(2x^0)^2x^2y^{-2}(xy)}$
22. $3\sqrt{2} \cdot \sqrt{3} - 5\sqrt{24} + 3\sqrt{54}$
23. $3\sqrt{2}(\sqrt{2} - 4\sqrt{8})$
24. $\dfrac{xy - 4xy^2}{xy}$
25. $\dfrac{-2^{-4}}{(-2)^{-3}}$
26. Solve by factoring: $45 = x^2 - 4x$
27. Evaluate: $xy - a - ya(y - a)$ if $x = -2$, $y = \sqrt[5]{-3125}$, and $a = -1$
28. Graph on a number line: $4 \ge |x|$; $D = \{\text{Positive integers}\}$
29. Multiply: $\dfrac{3ax}{y}\left(\dfrac{y}{3ax} - \dfrac{a^{-1}x}{3y}\right)$
30. True or false? $4 \in \{\text{Naturals}\}$

LESSON 126 Quotient rule for square roots

The product rule for square roots tells us that the square root of a product equals the product of the square roots.

$$\sqrt{3 \cdot 2} = \sqrt{3}\sqrt{2}$$

In a similar fashion, the quotient rule for square roots tells us that the square root of a quotient (fraction) equals the quotient of the square roots.

$$\sqrt{\frac{3}{2}} = \frac{\sqrt{3}}{\sqrt{2}}$$

The expression on the right has the irrational number $\sqrt{2}$ in the denominator. We can change the denominator to the rational number 2 by multiplying both the numerator and the denominator by $\sqrt{2}$.

$$\frac{\sqrt{3}}{\sqrt{2}} = \frac{\sqrt{3}}{\sqrt{2}} \cdot \frac{\sqrt{2}}{\sqrt{2}} = \frac{\sqrt{6}}{2}$$

This process of changing a denominator to a rational number is called *rationalizing the denominator.* Many people prefer fractions with rational denominators. If we go along with their preference, we can say that: **An expression containing square roots is in the simplified form when no square roots are in the denominator and no radicand has a factor that is a perfect square.**

example 126.1 Write $\sqrt{\frac{5}{3}}$ in simplified form.

solution We begin by writing the radical as a fraction of radicals.

$$\sqrt{\frac{5}{3}} = \frac{\sqrt{5}}{\sqrt{3}}$$

We finish by multiplying both top and bottom by $\sqrt{3}$.

$$\frac{\sqrt{5}}{\sqrt{3}} = \frac{\sqrt{5}}{\sqrt{3}} \cdot \frac{\sqrt{3}}{\sqrt{3}} = \frac{\sqrt{15}}{3}$$

example 126.2 Simplify: $\frac{4 + \sqrt{3}}{\sqrt{2}}$

solution To simplify, we must change the denominator to a rational number. Thus, we multiply by $\sqrt{2}$ over $\sqrt{2}$.

$$\frac{4 + \sqrt{3}}{\sqrt{2}} \cdot \frac{\sqrt{2}}{\sqrt{2}} = \frac{4\sqrt{2} + \sqrt{6}}{2}$$

This may appear to some as more complicated than the original expression, but this form is preferred by many people because no radical appears in the denominator.

example 126.3 Simplify: $\frac{2 + \sqrt{15}}{\sqrt{5}}$

solution We will multiply top and bottom by $\sqrt{5}$.

$$\frac{2 + \sqrt{15}}{\sqrt{5}} \cdot \frac{\sqrt{5}}{\sqrt{5}} = \frac{2\sqrt{5} + \sqrt{75}}{5} = \frac{2\sqrt{5} + 5\sqrt{3}}{5}$$

practice Simplify:

a. $\sqrt{\frac{6}{23}}$ b. $\frac{4 + \sqrt{5}}{\sqrt{3}}$

problem set 126

1. For a meshed gear of a fixed radius, the number of revolutions per minute varies inversely as the number of gear teeth. If a particular gear had 100 teeth, it would revolve at 10 revolutions per minute (rpm). What rpm would result if the number of teeth were reduced to 25?

2. Hannibal rode the elephant to the outskirts of Rome at 2 kilometers per hour and then took a chariot back to camp at 10 kilometers per hour. If the total trip took 18 hours, how far was it from camp to the outskirts of Rome?

3. Bustles just weren't moving so Julie marked them down 40 percent so that the sale price would be \$3.60 each. What was the price before the sale?

4. The cheap dresses were \$15 and the more expensive ones were \$50. On Saturday the store took in \$550 and sold 15 more cheap dresses than expensive dresses. How many dresses of each type did the store sell?

5. Anastasia deposited \$1100 in the bank and received 6 percent interest compounded annually. How much money did she have in 20 years? How much interest did she receive?

6. Find the equations of lines (a) and (b).

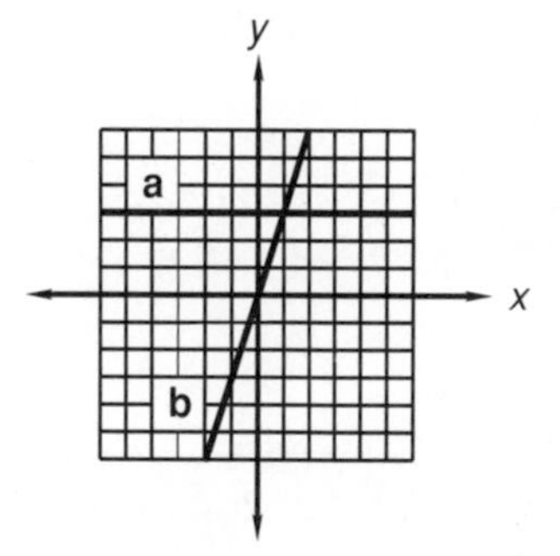

7. Which of the following sets of ordered pairs are functions?
(a) $(-5, -3), (5, 3), (-3, 5)$ (b) $(-5, -3), (-5, 3), (-3, 5)$
(c) $(-5, -3), (5, -3), (6, -3)$ (d) $(4, -2), (-4, 2), (-2, 4)$

8. Solve: $\sqrt{x - 3} - 2 = 5$

9. Graph on a number line: $x + 2 > 6$ or $x - 3 \leq -6$; $D = \{\text{Reals}\}$

Multiply:

10. $(3 + 2\sqrt{2})(2 - 4\sqrt{2})$ 11. $(2 + 3\sqrt{3})(2 - \sqrt{3})$ 12. $(3 + \sqrt{5})(2 - 4\sqrt{5})$

Simplify:

13. $\frac{2 + 3\sqrt{6}}{\sqrt{2}}$ 14. $\sqrt{\frac{2}{5}}$ 15. $\sqrt{\frac{3}{7}}$ 16. $\frac{4 + 2\sqrt{10}}{\sqrt{5}}$

17. Find the area of the shaded portion of this figure. Dimensions are in meters. Corners that look square are square.

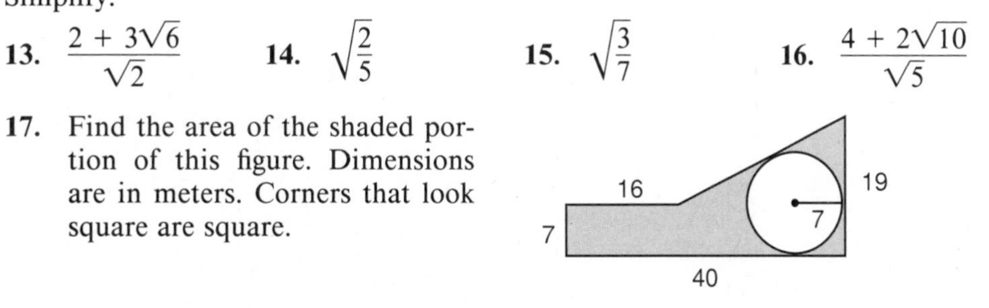

18. Use six unit multipliers to convert 145,000 cubic inches to cubic meters.

19. Solve: $\frac{m+2}{2m} = \frac{7}{3m} + \frac{2}{5}$

20. Find c: $\frac{a}{x} - \frac{m}{c} + b = k$

21. Simplify: $\dfrac{\dfrac{ax^2}{y} - \dfrac{x}{y^2}}{\dfrac{p}{y^2} - \dfrac{3k}{y}}$

22. Divide: $(4x^3 - 2x^2 + 4) \div (x + 2)$

23. Simplify: $-2[(-2^0 - 2^2)(-2^3 - 2) + (-2)][-(-3)(-2^2)]$

24. Evaluate: $-xy - y^2 - y^0(x - y)$ if $x = 3$ and $y = -2$

25. Solve: $3\frac{1}{5}k + \frac{2}{3} = \frac{1}{9}$

26. Find the equation of the line that goes through $(5, -2)$ and is parallel to $y = -3x + 2$.

27. Find the equation of the line that goes through $(4, -3)$ and $(-5, 2)$.

28. Find the equation of the line whose slope is -5 that passes through the point $(-4, -3)$.

29. Graph $y \leq -x - 3$ on a rectangular coordinate system.

30. Does $x = \frac{1}{4}$ or $x = -\frac{1}{2}$ make $8x^2 + 2x - 1 = 0$ a true equation? (Show computation.)

LESSON 127 Advanced trinomial factoring

Thus far, we have restricted our trinomial factoring to trinomials such as

$$x^2 - x - 6$$

in which the coefficient of the x^2 term is 1 and to trinomials that can be reduced to this form by factoring a common factor. Trinomials whose leading coefficient is not 1 can be formed by multiplying binomials, as we see here.

$$\begin{array}{l} 3x + 2 \\ \underline{2x - 3} \\ 6x^2 + 4x \\ \underline{\qquad - 9x - 6} \\ 6x^2 - 5x - 6 \end{array}$$

We note that the first term of the trinomial is the product of the first two terms of the binomials and that the last term of the trinomial is the product of the last terms of the binomials, **but, alas, the coefficient of the middle term of the trinomial is not the sum of the last two terms of the binomials.** The middle term is the sum of the product of the first term of the first binomial and the last term of the second binomial and the product of the last term of the first binomial and the first term of the second binomial. It is easier to see this if we write the original indicated multiplication in horizontal form and note that the middle term is the sum of the products of the means and the extremes.[†]

$$\overbrace{(3x + \underbrace{2)(2x}_{\text{means}} - 3)}^{\text{extremes}} = 6x^2 - 5x - 6$$

Product of means	=	$4x$
Product of extremes	=	$-9x$
Sum	=	$-5x$

example 127.1 Factor: $-7x - 15 + 2x^2$

solution We begin by writing the trinomial in descending powers of the variable.

$$2x^2 - 7x - 15$$

Now, to factor $2x^2 - 7x - 15$, we remember that the product of the first terms of the binomials is $2x^2$; the product of the last terms of the binomials is -15; and the middle term is the sum of the products of the means and extremes. Since the term $2x^2$ is the product of the first terms of the binomial we write

$$(2x \qquad)(x \qquad)$$

Now, the four pairs of integral factors of -15 are (a) $+15$ and -1, (b) -15 and $+1$, (c) $+5$ and -3, and (d) -5 and $+3$. Now we must try each pair **twice** and see what middle term will result in each case.

For $(15, -1)$	$(2x + 15)(x - 1)$	middle term is $13x$
	$(2x - 1)(x + 15)$	middle term is $29x$
For $(5, -3)$	$(2x + 5)(x - 3)$	middle term is $-x$
	$(2x - 3)(x + 5)$	middle term is $7x$
For $(-15, 1)$	$(2x - 15)(x + 1)$	middle term is $-13x$
	$(2x + 1)(x - 15)$	middle term is $-29x$
For $(-5, 3)$	$(2x - 5)(x + 3)$	middle term is x
	$(2x + 3)(x - 5)$	middle term is $-7x$

We will use the last entry because the sum of the products of the means and the extremes is $-7x$. Thus we see that $2x^2 - 7x - 15$ can be factored over the integers as $\mathbf{(2x + 3)(x - 5)}$.

[†] Mathematicians sometimes use the word *mean* to mean middle and the word *extreme* to mean end. Thus the mean terms in the multiplication shown are the middle terms, and the extreme terms are the end terms.

example 127.2 Factor: $3x^2 - x - 2$

solution We begin by writing as follows.

$$(3x \quad)(x \quad)$$

The second terms of the binomials must have a product of -2. The two pairs of integral factors of -2 are (a) -2 and $+1$, and (b) $+2$ and -1. We will try each pair **twice** and check to see what middle term results.

$(3x + 1)(x - 2)$ middle term is $-5x$

$(3x - 2)(x + 1)$ middle term is $+x$

$(3x - 1)(x + 2)$ middle term is $+5x$

$(3x + 2)(x - 1)$ middle term is $-x$

The sum of the products of the means and extremes of the last multiplication gives us a middle term of $-x$. Thus, these are the desired factors.

$$3x^2 - x - 2 = \mathbf{(3x + 2)(x - 1)}$$

example 127.3 Factor: $5x^2 - 13x - 6$

solution To begin we write as follows:

$$(5x \quad)(x \quad)$$

The pairs of integral factors of -6 are (a) $+1$ and -6, (b) -1 and $+6$, (c) $+3$ and -2, and (d) $+2$ and -3. We try each pair twice if necessary to find that the pair we need is $+2$ and -3 because

$$\mathbf{(5x + 2)(x - 3)} = 5x^2 - 13x - 6$$

practice

Factor:

a. $-11x - 21 + 2x^2$ **b.** $3x^2 - 6x + 3$

problem set 127

1. Up to a point, the yield varied directly as the amount of fertilizer used. If 500 pounds of fertilizer resulted in 2000 tons of produce, how much produce would be harvested if only 400 pounds of fertilizer were used?

2. The ratio of boys to girls in every class in the school was 7 to 5. If there were 2160 students in the school, how many were boys and how many were girls?

3. At first there were 10,000 rabbits in Australia. Their number doubled every year for 5 years. How many rabbits were there at the end of 5 years?

Graph the solution to the following linear inequalities:

4. $y > x$ 5. $y < -1$

6. Find m if $m = 5x^2 + 2x - 3$ and $x = -1$.

7. Find m if $m = 5x^2 + 2x - 3$ and $x = \frac{3}{5}$.

8. Find x if $x = \dfrac{-b \pm \sqrt{b^2 - 4ac}}{2a}$ and $a = 5$, $b = 2$, and $c = -3$.

Factor the following trinomials. Always begin by writing the trinomial in descending powers of the variable. Factor like terms first, if possible.

9. $3x^2 - 14x - 5$ **10.** $2x^2 + 8 + 10x$ **11.** $18 - 15x + 2x^2$

12. $-15 + 7x + 2x^2$ **13.** $8x - 24 + 2x^2$ **14.** $2x^2 - 24 - 8x$

15. $2x^2 - 6x + 4$ **16.** $2x^2 - 18 + 9x$ **17.** $2x^2 + 4 + 6x$

18. $3x^2 - 7 - 20x$ **19.** $3x^2 - 8 - 23x$ **20.** $4 - 7x + 3x^2$

Multiply:

21. $(4\sqrt{2} + 2)(3\sqrt{2} - 4)$ **22.** $(2\sqrt{3} + 4)(5\sqrt{6} - 2)$

Simplify:

23. $\sqrt{\frac{3}{7}}$ **24.** $\sqrt{\frac{5}{8}}$ **25.** $\frac{2 + \sqrt{3}}{\sqrt{5}}$ **26.** $\frac{4 + \sqrt{2}}{\sqrt{3}}$

27. Graph on a number line: $3 - |x| \geq 1$; $D = \{\text{Reals}\}$

28. Find the volume in cubic inches of the rectangular solid from which a right prism has been removed. The base of the solid is the shaded area in the figure. The solid is 2 feet long. Dimensions are in inches.

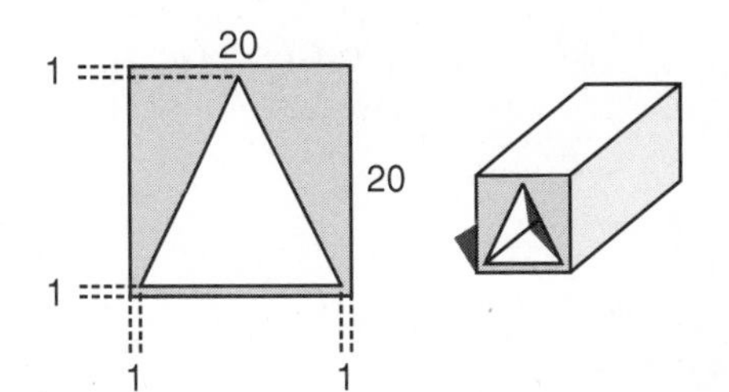

29. Mark Antony traveled the Appian Way from Brundisium to the rest stop at 3 miles per hour. He traveled back to Brundisium in his chariot at 9 miles per hour. If the round trip took 16 hours, how far is it from Brundisium to the rest stop?

30. Find the equation of the line through $(4, -3)$ that is parallel to $y = -2x + 4$.

LESSON 128 Factoring by grouping

Some expressions can be simplified if we note that two terms have common factors. If we consider the expression

$$xy + 3ay + bx + 3ba$$

we note that the first two terms have y as a factor. If we factor y out of the first two terms, we get

$$y(x + 3a) + bx + 3ba$$

Now we note that the last two terms have b as a factor. If we factor b out of the last two terms, we get

$$y(x + 3a) + b(x + 3a)$$

Now we can factor $(x + 3a)$ and get

$$(x + 3a)(y + b)$$

Many books call this type of factoring **factoring by grouping** because we grouped the terms that had y as a factor and those that had b as a factor.

example 128.1 Factor: $xya - 4a + xyb - 4b$

solution We note that the first two terms have a as a factor and that the last two terms have b as a factor. We begin by using parentheses to **group** these terms.

$$(xya - 4a) + (xyb - 4b)$$

Now we factor a from the first group and b from the second.

$$a(xy - 4) + b(xy - 4)$$

Lastly, we recognize that both of these terms have $xy - 4$ as a factor, so we factor this expression.

$$\mathbf{(xy - 4)(a + b)}$$

We can get the same result via a different route. We note that the first and third terms have xy as a factor and the second and fourth terms have 4 as a factor. We group these terms and get

$$(xya + xyb) + (-4a - 4b)$$

We factor xy from the first group and -4 from the second group and get

$$xy(a + b) + (-4)(a + b)$$

Now we factor $(a + b)$ from both terms and get

$$\mathbf{(a + b)(xy - 4)}$$

example 128.2 Factor: $ac + 2ad + 2bc + 4bd$

solution We recognize the form and note that the first and third terms have a common factor of c. We note that the second and fourth terms have a common factor of $2d$, so we rearrange the terms and use parentheses as follows:

$$(ac + 2bc) + (4bd + 2ad)$$

Now we factor these terms as

$$c(a + 2b) + 2d(2b + a)$$

and complete the problem by factoring $a + 2b$. The final result is

$$\mathbf{(a + 2b)(c + 2d)}$$

Now we try another way. We group the terms that have a as a factor and the terms that have b as a factor.

$$(ac + 2ad) + (2bc + 4bd)$$

From the first group we factor out a, and from the second group we factor out $2b$.

$$a(c + 2d) + 2b(c + 2d)$$

Now we note the common factor of $c + 2d$. So we factor one more time.

$$\mathbf{(c + 2d)(a + 2b)}$$

We always begin the factoring process by searching for common monomial factors. These problems are just problems in which two terms have one common factor and two other terms have another common factor.

practice Factor:

a. $mba - 7a + mbn - 7n$

b. $ns + 3nx + 2cs + 6cx$

problem set 128

1. In a particular experiment, the pressure varied inversely as the volume. When the pressure was 15 pounds per square inch, the volume was 20 liters. What was the pressure when the volume was reduced to 10 liters?

2. In another experiment, the pressure varied directly as the temperature. When the pressure was 1000 pounds per square inch, the temperature was 250°. What was the pressure when the temperature was 1000°?

3. The ratio of the pigs to chickens in the barnyard was 2 to 11. If there were 169 chickens and pigs in the barnyard, how many were chickens and how many were pigs?

4. David and Wade had a race. David started out at 9 a.m. and drove at 40 miles per hour. Wade waited until 11 a.m. to start, and he drove at 70 miles per hour. What time was it when Wade was 10 miles in front of David?

5. Calvin and Tooley hoarded nickels and pennies. They had 280 coins whose value was \$6.80. How many nickels did they have?

6. Graph $y < 2x + 2$ on a rectangular coordinate system.

Factor the following trinomials. Begin by writing the terms in descending order of the variable.

7. $-14 + 3x^2 - 19x$
8. $-14 + 19x + 3x^2$
9. $2x^2 - 15 + 7x$
10. $2x^2 - 18 + 9x$
11. $3x^2 + 14 + 23x$
12. $2x^2 - 17x + 21$
13. $3x^2 + 16 - 26x$
14. $18 + 15x + 2x^2$
15. $3x^2 + 13x + 14$
16. Find the equation of the line that goes through $(2, -3)$ and $(-3, -6)$.
17. Use 10 unit multipliers to convert 4,000,000 square miles to square kilometers.

Factor by grouping:

18. $ac - ad + bc - bd$
19. $ab + 4a + 2b + 8$
20. $ab + ac + xb + xc$
21. $2mx - 3m + 2pcx - 3pc$
22. $4k - kxy + 4pc - pcxy$
23. $ac - axy + dc - dxy$

Simplify:

24. $\sqrt{\dfrac{3}{11}}$
25. $\dfrac{2\sqrt{3} + 2}{\sqrt{5}}$
26. Graph on a number line: $-2 < x + 2 \leq 3$; $D = \{\text{Positive integers}\}$
27. Find m if $m = 3x^2 + 20x + 12$ and $x = -\dfrac{2}{3}$.
28. Find x if $x = \dfrac{-b \pm \sqrt{b^2 - 4ac}}{2a}$ and $a = 3$, $b = 20$, and $c = 12$.

29. Find the volume in cubic inches of a right prism whose base is the figure shown and whose height is 1 foot. What is the surface area? Dimensions are in inches. Corners that look square are square.

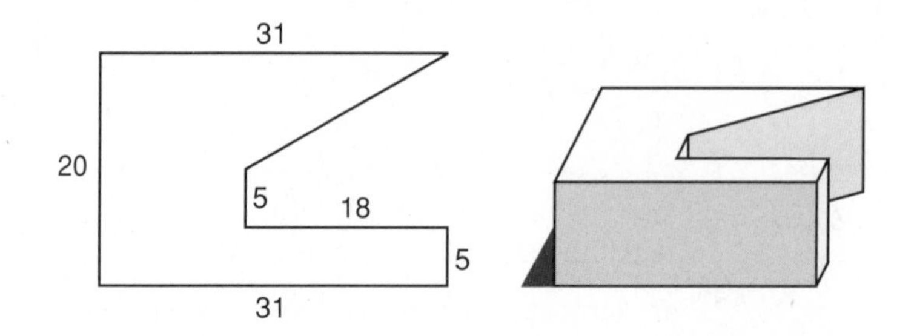

30. Use nine unit multipliers to convert 12,000 cubic feet to cubic meters.

LESSON 129 Direct and inverse variation squared

Direct and inverse variation statements are not always simple statements of direct variation or inverse variation. Often one variable will vary as the other variable squared or the other variable cubed.

STATEMENT	IMPLIED EQUATION
The weight of a body varies inversely with the square of the distance to the center of the earth.	$W = \frac{k}{D^2}$
The distance required to stop is directly proportional to the square of the velocity.	$D = kV^2$
The price of a diamond varies directly as the square of its weight.	$P = kW^2$
The strength of the field is inversely proportional to the cube of the radius.	$S = \frac{k}{R^3}$

These problems are solved in the same way that simple variation problems are solved. First we recognize the statement of the implied variation and write down the indicated equation. Then we find k, insert its value in the equation, and solve for the required unknown.

example 129.1 The distance required for an automobile to stop is directly proportional to the square of its velocity. If a car can stop in 200 meters at 20 kilometers per hour, what will be the required distance at 28 kilometers per hour?

solution

Step 1:	$D = kV^2$	write equation
Step 2:	$200 = k(20)^2 \longrightarrow k = 0.5$	find k
Step 3:	$D = 0.5V^2$	put k in equation
Step 4:	$D = 0.5(28)^2 \longrightarrow$ **D = 392 meters**	solve for D

example 129.2 The distance a body falls varies directly as the square of the time that it falls. If it falls 144 feet in 3 seconds, how far will it fall in 10 seconds?

solution

Step 1:	$D = kt^2$	write equation
Step 2:	$144 = k(3)^2 \longrightarrow k = 16$	find k
Step 3:	$D = 16t^2$	put k in equation
Step 4:	$D = 16(10)^2 \longrightarrow$ **D = 1600 feet**	solve for D

example 129.3 The weight of a body on or above the surface of the earth varies inversely with the square of the distance from the body to the center of the earth. If a body weighs 10,000 pounds at a distance of 5000 miles from the center of the earth, how much would it weigh 50,000 miles from the center of the earth?

solution

Step 1: $W = \dfrac{k}{D^2}$ write equation

Step 2: $10{,}000 = \dfrac{k}{(5000)^2} \longrightarrow 10{,}000(5000)^2 = k$ find k

$\longrightarrow 25 \times 10^{10} = k$

Step 3: $W = \dfrac{25 \times 10^{10}}{D^2}$ put k in equation

Step 4: $W = \dfrac{25 \times 10^{10}}{(50{,}000)^2} \longrightarrow W = \dfrac{25 \times 10^{10}}{25 \times 10^8}$ solve for W

$\longrightarrow$ **W = 100 pounds**

practice The weight of a body on or above the surface of the earth varies inversely with the square of the distance from the body to the center of the earth. If a body weighs 50,000 pounds at a distance of 2500 miles from the center of the earth, how much would it weigh 25,000 miles from the center of the earth?

problem set 129

1. The distance required for an automobile to stop is directly proportional to the square of its velocity. If a car can stop in 1800 meters from a velocity of 30 kilometers per hour, what will be the required distance at 28 kilometers per hour?

2. The distance a body falls varies directly as the square of the time that it falls. If it falls 256 feet in 3 seconds, how far will it fall in 8 seconds?

3. Greens vary directly as purples squared. When there were 4 greens, there were 2 purples. How many greens would be present if there were 6 purples?

4. The number of red marbles varied inversely as the square of number of blue marbles. When there were 4 reds, there were 20 blues. How many reds would there be if there were only 4 blues?

5. The ratio of rabbits to squirrels in the forest was 7 to 5. If there were 16,800 animals total, how many were rabbits and how many were squirrels?

6. Graph $y \geq \frac{1}{2}x - 2$ on a rectangular coordinate system.

7. Find x: $\frac{a}{b} + \frac{c}{x} = m$

Factor the trinomials. Begin by writing the terms in descending order of the variable.

8. $3x^2 + 25x - 18$
9. $3x^2 - 4 - x$
10. $2x^2 - 6 - 4x$
11. $3x^2 + 28x - 20$
12. $2x^2 + 15x + 25$
13. $2x^2 - 5x - 25$

Factor by grouping:

14. $ab + 15 + 5a + 3b$
15. $ay + xy + ac + xc$
16. $3mx - 2p + 3px - 2m$
17. $kx - 15 - 5k + 3x$
18. $xpc + pc^2 + 4x + 4c$
19. $acb - ack + 2b - 2k$

Graph on a number line:

20. $-2 - |x| > -4$; $D = \{\text{Reals}\}$
21. $4 \le x + 2 < 7$; $D = \{\text{Integers}\}$
22. $x \le 2$ or $x > 5$; $D = \{\text{Reals}\}$

Solve:

23. $\sqrt{x + 2} - 4 = 1$
24. $\sqrt{x - 3} - 5 = 3$
25. Find the equation of the line through $(-2, 5)$ and $(3, -2)$.
26. Find the equation of the line through $(-2, 5)$ that has a slope of $-\frac{1}{4}$.
27. Find the equation of the line through $(-2, 5)$ that is parallel to $y = -\frac{1}{3}x + 2$.

Simplify:

28. $\sqrt{\frac{3}{5}}$
29. $\sqrt{\frac{7}{3}}$
30. $\frac{2\sqrt{2} + \sqrt{2}}{\sqrt{2}}$

LESSON 130 Completing the square

We can solve a quadratic equation by taking the square root of both sides. We must remember not to forget the ± sign.

$x^2 = 3$	equation
$\sqrt{x^2} = \sqrt{3}$	square root of both sides
$x = \pm\sqrt{3}$	simplified

If the squared term is a sum, the procedure is the same.

$(x + 7)^2 = 4$	equation
$\sqrt{(x + 7)^2} = \sqrt{4}$	square root of both sides
$x + 7 = \pm 2$	simplified
$x = -7 \pm 2$	
$x = -5, -9$	solved

If the constant term is not a perfect square, the answer will contain a radical.

$$(x + 2)^2 = 3 \qquad \text{equation}$$
$$\sqrt{(x + 2)^2} = \sqrt{3} \qquad \text{square root of both sides}$$
$$x + 2 = \pm\sqrt{3} \qquad \text{simplified}$$
$$x = -2 \pm \sqrt{3} \qquad \text{solved}$$

Every quadratic equation can be written in the same form that these two equations have. For example,

$$x^2 + 4x - 5 = 0 \quad \text{can be written as} \quad (x + 2)^2 = 9$$

We can solve this equation by taking the square root of both sides.

$$(x + 2)^2 = 9 \longrightarrow x + 2 = \pm 3 \longrightarrow x = \pm 3 - 2 = \mathbf{-5, 1}$$

The process of writing the equation in the form

$$(x + 9)^2 = b$$

is called **completing the square.** We will consider quadratic equations in which the coefficient of the x term is 1. In the next book we will consider coefficients of x that are not 1. We can complete the square by using six steps. We will use parentheses to help us with the concept.

1. Write the equation with descending powers of the variable on the left-hand side of the equals sign. The right-hand side of the equation is the number zero.
2. Put parentheses around the x^2 term and the x term.
3. Move the constant term to the right-hand side of the equals sign.
4. Divide the coefficient of the x term by 2. Square the result and add this new term to both sides of the equation.
5. The term inside the parentheses is now a perfect square. Write the left-hand side as a perfect square. Simplify the right-hand side.
6. Complete the solution by taking the square root of both sides and simplifying.

example 130.1 Solve $x^2 + 6 = -10x$ by completing the square.

solution

1. The first step is to put all three terms on the left-hand side of the equals sign.

$$x^2 + 10x + 6 = 0$$

2. Next we use parentheses.

$$(x^2 + 10x \qquad) + 6 = 0$$

Note how we left a space inside the parentheses for another term.

3. Now we move the 6 to the right-hand side.

$$(x^2 + 10x \qquad) = -6$$

4. The coefficient of x is 10. We divide 10 by 2 and square the result.

$$\left(\frac{10}{2}\right)^2 = 25$$

We now add 25 to both sides of the equation and get

$$(x^2 + 10x + 25) = -6 + 25$$

5. The term **inside the parentheses** is a perfect square. Next we simplify both sides of the equation.

$$(x + 5)^2 = 19$$

6. Now we take the square root of both sides and solve for x.

$$x + 5 = \pm\sqrt{19} \qquad \text{square root of both sides}$$

$$x = -5 \pm \sqrt{19} \qquad \text{solved}$$

Many people have difficulty writing

$$(x^2 + 10x + 25) \qquad \text{as} \qquad (x + 5)^2$$

There are two ways we can remember. The first way is to remember that the number in the binomial is half the coefficient of x in the trinomial.

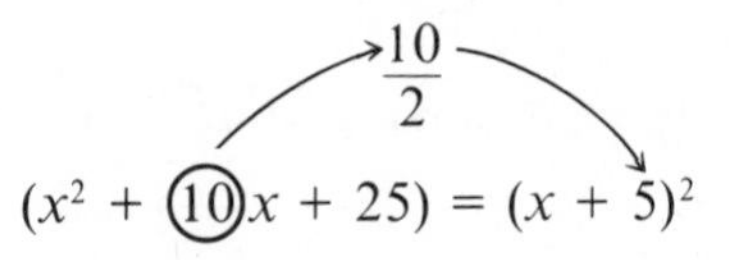

The second way is to remember that the number in the binomial is also the square root of the constant term in the trinomial.

$$\sqrt{25}$$

$$(x^2 + 10x + 25) = (x + 5)^2$$

Either of these memory aids will suffice. If the trinomial had been a perfect square,

$$\left(x^2 + 9x + \frac{81}{4}\right)$$

we would write this as $(x + \frac{9}{2})^2$. Note that $\frac{9}{2}$ is half of 9 and is also the square root of $\frac{81}{4}$.

example 130.2 Complete the square to solve $x^2 + 3 = -5x$.

solution First we rewrite the equation in standard form.

$$x^2 + 5x + 3 = 0$$

To complete the square, we will have to do three things that often cause difficulty. Let's look at these three things first. Then we will complete the square.

1. First we will have to find the square of one-half the coefficient of x.

$$\left(\frac{5}{2}\right)^2 = \frac{25}{4}$$

2. Then after we add $\frac{25}{4}$ to both sides, we will have to combine $\frac{25}{4}$ and -3. To combine these numbers, we will write -3 so that it has a denominator of 4.

$$\frac{25}{4} - 3 \qquad \text{expression}$$

$$\frac{25}{4} - \frac{12}{4} \qquad 3 \text{ equals } \frac{12}{4}$$

$$\frac{13}{4} \qquad \text{added}$$

3. Then we will have to write $(x^2 + 5x + \frac{25}{4})$ as a perfect square. The constant term of the binomial will be $\frac{5}{2}$ because this is half of 5 and is also $\sqrt{\frac{25}{4}}$.

$$\left(x^2 + 5x + \frac{25}{4}\right) = \left(x + \frac{5}{2}\right)^2$$

Now we will use the first five steps in completing the square.

Step 1:	$x^2 + 5x + 3 = 0$	rearranged
Step 2:	$(x^2 + 5x \qquad) + 3 = 0$	parentheses
Step 3:	$(x^2 + 5x \qquad) = -3$	moved constant term
Step 4:	$\left(x^2 + 5x + \frac{25}{4}\right) = -3 + \frac{25}{4}$	added $\frac{25}{4}$ to both sides
Step 5:	$\left(x + \frac{5}{2}\right)^2 = \frac{13}{4}$	simplified

Finally we take the square root of both sides and solve for x.

Step 6:	$x + \frac{5}{2} = \pm\sqrt{\frac{13}{4}}$	square root of both sides
	$x = -\frac{5}{2} \pm \frac{\sqrt{13}}{2}$	solved

example 130.3 Complete the square to solve $x^2 + 1 = 3x$.

solution We will use the same six steps.

Step 1:	$x^2 - 3x + 1 = 0$	rearranged
Step 2:	$(x^2 - 3x \qquad) + 1 = 0$	parentheses
Step 3:	$(x^2 - 3x \qquad) = -1$	moved constant

Now we will take one-half of -3, square it, and add to both sides.

Step 4:	$\left(x^2 - 3x + \frac{9}{4}\right) = -1 + \frac{9}{4}$	added $\frac{9}{4}$ to both sides
Step 5:	$\left(x - \frac{3}{2}\right)^2 = \frac{5}{4}$	simplified
Step 6:	$x - \frac{3}{2} = \pm\sqrt{\frac{5}{4}}$	square root of both sides
	$x = +\frac{3}{2} \pm \frac{\sqrt{5}}{2}$	added $\frac{3}{2}$ to both sides

practice Solve $x^2 - 9 = -7x$ by completing the square.

problem set 130

1. The freight train headed south at 9 a.m., and the express train headed north from the same station at noon. At 3 p.m., the trains were 420 miles apart. What was the speed of each if the speed of the express train was 20 miles per hour greater than the speed of the freight?

2. The ratio of greens to whites was 2 to 11. If the total was 2340, how many were greens and how many were whites?

3. Ninety-eight percent of the citizens favored the resolution. If 480 were against it or did not care, how many citizens lived in the town?

4. There was \$2900 in the pot. If there were 293 more \$1 bills than \$10 bills, how many bills of each kind were there?

5. Reds varied inversely as yellows squared. When there were 10 reds, there were 100 yellows. How many reds were present when the yellows were reduced to 5?

6. Find four consecutive even integers such that the product of -12 and the sum of the first and fourth is 6 less than the product of 19 and the opposite of the third.

7. Find m if $m = 2x^2 - 2x - 4$ if (a) $x = 2$ and (b) $x = -1$.

8. Find the volume in cubic feet of a prism whose base is shown and whose height is 2 yards. Dimensions are in feet.

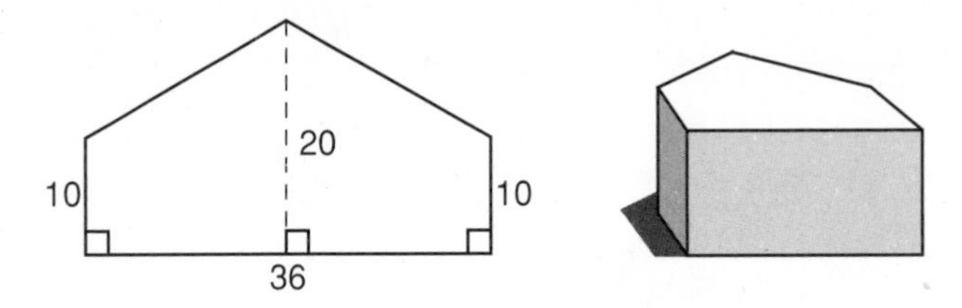

Solve the following quadratic equations by completing the square.

9. $x^2 + 2x - 4 = 0$

10. $x^2 + 3x - 8 = 0$

11. $x^2 + 2x - 5 = 0$

12. $x^2 + 4x - 7 = 0$

Factor the trinomials. Begin by writing the terms in descending order of the variables.

13. $3x^2 - 35 - 16x$

14. $-2x + 3x^2 - 5$

15. $2x^2 - 5x - 12$

Factor by grouping:

16. $p^2c - ab + p^2b - ac$

17. $ax^2 - ca + cx^2 - c^2$

18. $2y + mx^3 + my + 2x^3$

19. $4ab + 4x + abc + cx$

Graph on a number line:

20. $4 \le x - 2 \le 8$; $D = \{\text{Reals}\}$

21. $3 - |x| > 2$; $D = \{\text{Integers}\}$

22. $x < -2$ or $x \ge 4$; $D = \{\text{Integers}\}$

23. Find the equation of the line through $(-3, 2)$ and $(5, -3)$.

Solve:

24. $\sqrt{4x + 1} - 1 = 2$

25. $\sqrt{5m - 5} + 6 = 7$

26. Find the equation of the line through $(2, 4)$ that is parallel to $y = \frac{1}{5}x - 6$.

27. Graph: $\begin{cases} y \le -x + 2 \\ y \ge x \end{cases}$

Simplify:

28. $\sqrt{\frac{2}{7}}$

29. $\sqrt{\frac{5}{12}}$

30. $\frac{4 + \sqrt{3}}{\sqrt{6}}$

LESSON 131 The quadratic formula

131.A the quadratic formula

In the preceding lesson we learned how to complete the square to solve a quadratic equation. This method requires several steps and is time-consuming. **If we write a quadratic equation in standard form, using *a, b,* and *c* to represent the constants, and complete the square, we can develop a formula that can be used to find the values of *x* that will satisfy any quadratic equation.**

$ax^2 + bx + c = 0$	general form of a quadratic equation
$x^2 + \frac{b}{a}x + \frac{c}{a} = 0$	divide by a
$x^2 + \frac{b}{a}x = -\frac{c}{a}$	add $-\frac{c}{a}$ to both sides (additive property of equality)
$x^2 + \frac{b}{a}x + \frac{b^2}{4a^2} = \frac{b^2}{4a^2} - \frac{c}{a}$	add $\left(\frac{b}{2a}\right)^2 = \frac{b^2}{4a^2}$ to both sides
$\left(x + \frac{b}{2a}\right)^2 = \frac{b^2 - 4ac}{4a^2}$	simplification of both sides
$x + \frac{b}{2a} = \pm\sqrt{\frac{b^2 - 4ac}{4a^2}}$	square root of both sides
$x + \frac{b}{2a} = \pm\frac{\sqrt{b^2 - 4ac}}{\sqrt{4a^2}}$	quotient of square roots rule
$x = -\frac{b}{2a} \pm \frac{\sqrt{b^2 - 4ac}}{2a}$	added $-\frac{b}{2a}$ to both sides
$x = \frac{-b \pm \sqrt{b^2 - 4ac}}{2a}$	addition of rational expressions

131.B use of the quadratic formula

By completing the square on $ax^2 + bx + c = 0$, we have derived the **quadratic formula**

$$x = \frac{-b \pm \sqrt{b^2 - 4ac}}{2a} \qquad (a \neq 0)$$

This formula expresses x in terms of the constants a, b, and c of the equation. If we wish to use this formula to find the values of x that satisfy a particular quadratic equation, it is necessary to compare the particular quadratic equation with the equation $ax^2 + bx + c$ to determine the values of a, b, and c.

example 131.1 Use the quadratic formula to determine the roots of the equation $3x^2 + 2x - 7 = 0$.

solution The general form of the quadratic equation is $ax^2 + bx + c = 0$. If we write our equation and the general equation one over the other

$ax^2 + bx + c = 0$	general equation
$3x^2 + 2x - 7 = 0$	our equation

we see that 3 corresponds to a, 2 corresponds to b, and -7 corresponds to c. If we use these numbers as replacements for a, b, and c in the quadratic formula, we can find the roots of the equation.

$$x = \frac{-b \pm \sqrt{b^2 - 4ac}}{2a} \qquad \text{quadratic formula}$$

$$x = \frac{-(2) \pm \sqrt{(2)^2 - 4(3)(-7)}}{2(3)} \qquad \text{substituted}$$

$$x = \frac{-2 \pm \sqrt{4 + 84}}{6} \qquad \text{simplified under the radical}$$

$$x = \frac{-2 \pm \sqrt{88}}{6} \qquad \text{added 4 and 84}$$

$$x = \frac{-2 \pm 2\sqrt{22}}{6} \qquad \sqrt{88} \text{ equals } 2\sqrt{22}$$

$$x = \frac{-1 \pm \sqrt{22}}{3} \qquad \text{simplified}$$

Thus the two real numbers that will satisfy the given equation are

$$\mathbf{x = \frac{-1 + \sqrt{22}}{3}} \quad \text{and} \quad \mathbf{x = \frac{-1 - \sqrt{22}}{3}}$$

example 131.2 Use the quadratic formula to determine the roots of $-6 = -x^2 + x$.

solution We begin by writing the given equation in standard form as

$$x^2 - x - 6 = 0$$

Now we compare this equation to the equation $ax^2 + bx + c = 0$ to determine the numbers that correspond to a, b, and c.

$$ax^2 + bx + c = 0 \qquad \text{general equation}$$

$$x^2 - x - 6 = 0 \qquad \text{our equation}$$

We see that $a = 1$, $b = -1$, and $c = -6$.

$$x = \frac{-b \pm \sqrt{b^2 - 4ac}}{2a} \qquad \text{quadratic formula}$$

$$x = \frac{-(-1) \pm \sqrt{(-1)^2 - 4(1)(-6)}}{2(1)} \qquad \text{substituted}$$

$$x = \frac{1 \pm \sqrt{25}}{2} \qquad \text{simplified}$$

$$x = \frac{1 \pm 5}{2} \qquad \sqrt{25} \text{ equals } 5$$

$$\mathbf{x = 3, -2} \qquad \text{solved}$$

example 131.3 Use the quadratic formula to find the values of x that satisfy the equation $-x = 7 - x^2$.

solution We begin by writing the given equation in standard form and comparing it to the equation $ax^2 + bx + c = 0$.

$$ax^2 + bx + c = 0 \qquad \text{general equation}$$

$$x^2 - x - 7 = 0 \qquad \text{our equation}$$

We see that $a = 1$, $b = -1$, and $c = -7$.

$$x = \frac{-b \pm \sqrt{b^2 - 4ac}}{2a} \longrightarrow x = \frac{-(-1) \pm \sqrt{(-1)^2 - 4(1)(-7)}}{2(1)}$$

$$\longrightarrow x = \frac{1 \pm \sqrt{1 + 28}}{2} \longrightarrow \boldsymbol{x = \frac{1 \pm \sqrt{29}}{2}}$$

practice Use the quadratic formula to solve $3x - 1 = -2x^2$.

problem set 131

1. The express train made the trip in 20 hours. The freight train took 25 hours because it was 10 miles per hour slower than the express. What was the speed of each train?
2. Inflation took its toll, and the merchant had to mark the suit up 30 percent so it would sell for \$156. What was the original price of the suit?
3. Greens varied inversely as blues squared. When there were 5 greens, there were 50 blues. How many greens were there when there were only 10 blues?
4. The ratio of friendlies to enemies was 11 to 5. If there were 800 people hiding in the ruins, how many were friendlies?
5. Good ones were \$7 each and sorry ones were only \$3 each. If Fiona spent \$414 and bought 2 more good ones than sorry ones, how many of each kind did she buy?
6. Find three consecutive odd integers such that -3 times the sum of the first and third is 50 greater than 8 times the opposite of the second.
7. Simplify: $\dfrac{x + \dfrac{4x}{3y}}{\dfrac{2ax}{y} + 4}$
8. $R_E T_E = R_W T_W$, $R_E = 145$, $R_W = 200$, $T_E + T_W = 6$. Find T_E and T_W.

Use the quadratic formula to solve the following quadratic equations. Begin by writing each equation in standard form so that it can be compared to $ax^2 + bx + c = 0$ in order to determine the values of a, b, and c.

9. $-3x = -2x^2 + 10$
10. $-2x = 5 - x^2$
11. $x^2 + 2x - 11 = 0$
12. $5x^2 - 6x - 4 = 0$

Solve the equations by completing the square:

13. $-3x = -x^2 + 10$
14. $-2x = 5 - x^2$
15. $x^2 + 2x - 11 = 0$

Factor. Begin by writing terms in descending order of the variable.

16. $3x^2 - 5 + 14x$
17. $-27 + 24x + 3x^2$
18. $9x - 5 + 2x^2$

Factor by grouping:

19. $km^2 + 2c - 2m^2 - kc$
20. $6a - xya - xyb + 6b$
21. $abx - 2yc + xc - 2yab$
22. $4xn + abn - abm - 4xm$
23. Find the equation of the line through (2, 5) and $(-3, -4)$.
24. Find the equation of the line through $(-2, 5)$ that is parallel to $y = \frac{2}{5}x - 3$.

25. Graph: $\begin{cases} y \geq x \\ y \leq -x + 2 \end{cases}$

Simplify:

26. $(4 + 2\sqrt{2})(\sqrt{2} + 2)$ **27.** $\sqrt{\frac{3}{8}}$ **28.** $\frac{\sqrt{2} + 1}{\sqrt{2}}$

29. Find the volume in cubic inches of a cylinder whose base is shown and whose height is 8 feet. If the cylinder is a right cylinder, what is the surface area? Dimensions are in feet. The triangle is a right triangle.

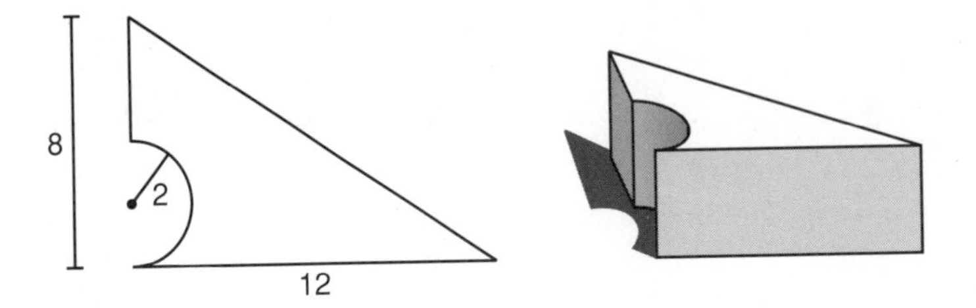

30. Four times the sum of twice a number and 7 exceeded the value of the number by 69. What is the number?

LESSON 132 Probability

132.A probability

On the left we show a pair of dice. The singular of **dice** is **die,**

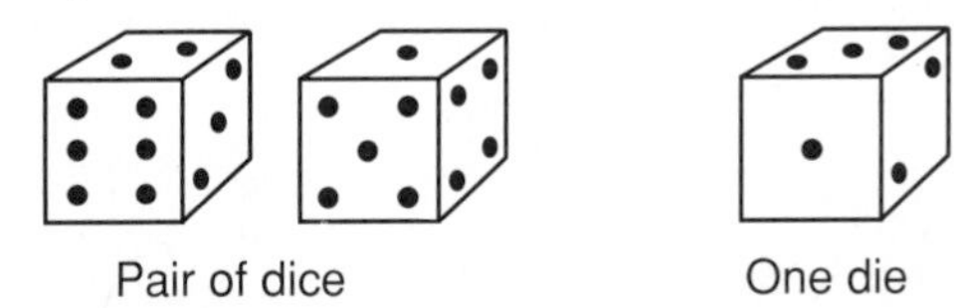

so we call one of these a die. A die has six faces, each of which has one to six dots.

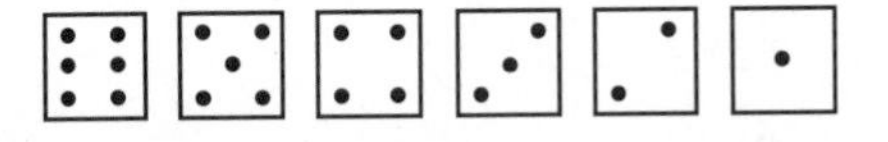

If we roll a die that has no flaws, it should come up 3 just as often as it comes up 5. Thus, we say that the probability of rolling a 3 is the same as the probability of rolling a 5. Since the die has six faces, if we rolled it a great number of times, we would expect each face to come up one-sixth of the time. We will use the symbol

$$P(\ \ )$$

to designate the word **probability.** Thus, we would write that the probability of 3 equals $\frac{1}{6}$ by writing.

$$P(3) = \frac{1}{6}$$

We will restrict our investigations to outcomes that are equally likely to happen. The probability P of an event E is the number of ways the event can happen divided by the number of possible outcomes.

$$P(E) = \frac{\text{number of ways event can happen}}{\text{total number of possible outcomes}}$$

example 132.1 A single die is rolled. What is the probability that it will come up either 5 or 6?

solution There are 6 possible outcomes and 2 of them give us the desired result. Thus, the probability of rolling a 5 or 6 is

$$P(5 \text{ or } 6) = \frac{2}{6} = \mathbf{\frac{1}{3}}$$

example 132.2 A deck of cards contains 52 cards: 13 are spades, 13 are hearts, 13 are clubs, and 13 are diamonds. Jimmy tears up the ace of spades. Then he shuffles the deck and draws 1 card. What is the probability that the card is a spade?

solution There are 51 cards and 12 are spades. So

$$P(S) = \frac{12}{51} = \mathbf{\frac{4}{17}}$$

132.B independent events

Events that do not affect one another are called **independent events.** If a fair coin is flipped once, the probability of getting heads is $\frac{1}{2}$. Whether or not the first toss comes up heads or tails, the probability of getting heads on the second toss is $\frac{1}{2}$, the probability of getting heads on the third toss is $\frac{1}{2}$, and the probability of getting heads on any toss is $\frac{1}{2}$. Each flip of the coin is an independent event.

The probability of two independent events happening in a designated order is the product of the probabilities of the individual events. If we look at what could happen on two tosses of a coin, we get the following diagram:

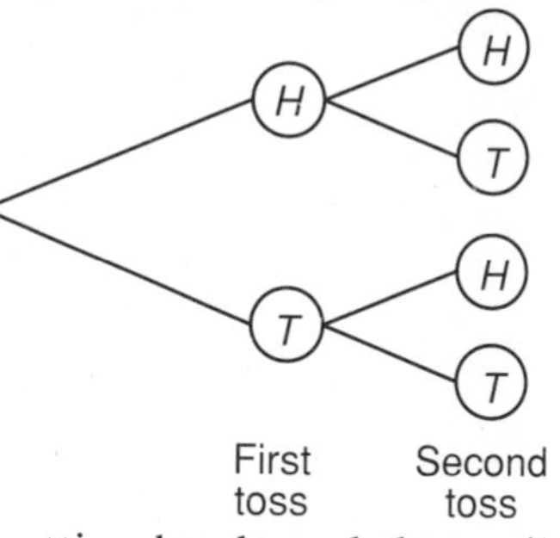

What is the probability of getting heads and then tails? There are four possible outcomes: (H, H), (H, T), (T, H), and (T, T). The probability of (H, T) is the probability of heads on the first toss times the probability of tails on the second toss, or $\frac{1}{2}$ times $\frac{1}{2}$.

$$P(H, T) = P(H) \times P(T) = \frac{1}{2} \cdot \frac{1}{2} = \mathbf{\frac{1}{4}}$$

example 132.3 A fair coin is tossed four times. What is the probability that the first two times it comes up heads and the last two times it comes up tails?

solution The probability of independent events happening in a designated order is the product of the probability of the individual events.

$$P(H, H, T, T) = \frac{1}{2} \cdot \frac{1}{2} \cdot \frac{1}{2} \cdot \frac{1}{2} = \mathbf{\frac{1}{16}}$$

example 132.4 The spinner on this card is spun twice. What is the probability that the spinner will stop on a 4 and then on a 3?

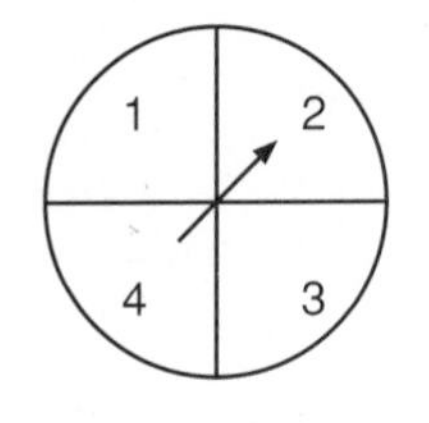

solution The probability of getting a 4 and then getting a 3 is the probability of getting a 4 times the probability of getting a 3.

$$P(4, 3) = P(4) \times P(3) = \frac{1}{4} \times \frac{1}{4} = \mathbf{\frac{1}{16}}$$

practice

a. In a standard deck of 52 cards, what is the probability of pulling a 10 from a deck of cards? (Remember that there are four 10s.)

b. A coin is flipped three times. What is the probability that all three flips come up heads?

problem set 132

1. Paula ran to Hugo in 6 hours, while Busking walked the same distance in 72 hours. How fast did Paula run if her speed was 11 kilometers per hour faster than Busking's? How far was it to Hugo?

2. The weight of a body on or above the surface of the earth varies inversely with the square of the distance from the body to the center of the earth. If a body weighs 25,000 pounds at a distance of 100,000 miles from the center of the earth, how much would it weigh 5000 miles from the center of the earth?

3. Horses sold for \$400 each and ponies for only \$100. Weir spent \$4500 and bought 5 more horses than ponies. How many horses did she buy?

4. Beaty and George judged the pigs at the fair. Beaty gave the black pigs only 35 percent of the points that George gave the pink pigs. If Beaty gave the black pigs 1351 points, how many points did the pink pigs get?

5. The ratio of dullards to scholars was 2 to 17. If there were 38,000 people at the university, how many were scholars?

6. (a) The spinner is spun once. What is the probability that the spinner will stop on a 2?
(b) In two spins what is the probability that the spinner will stop on a 3 and then on a 2?
(c) In three spins what is the probability that the spinner will stop on a 4 and then on a 3 and then on a 1?

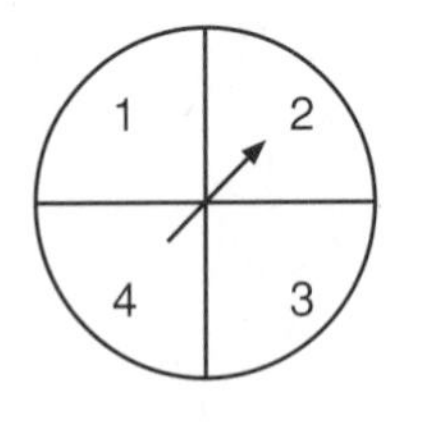

7. Graph on a number line: $4 \leq x < 10$; $D = \{\text{Integers}\}$

Solve by graphing:

8. $\begin{cases} y = -3x + 2 \\ x = 3 \end{cases}$

Use elimination to solve:

9. $\begin{cases} 4x - 5y = -3 \\ 2x + y = 9 \end{cases}$

10. Simplify: $\dfrac{(0.0005)(0.08 \times 10^{14})}{(40{,}000)(200 \times 10^{-5})}$

11. Find the equations of lines (a) and (b).

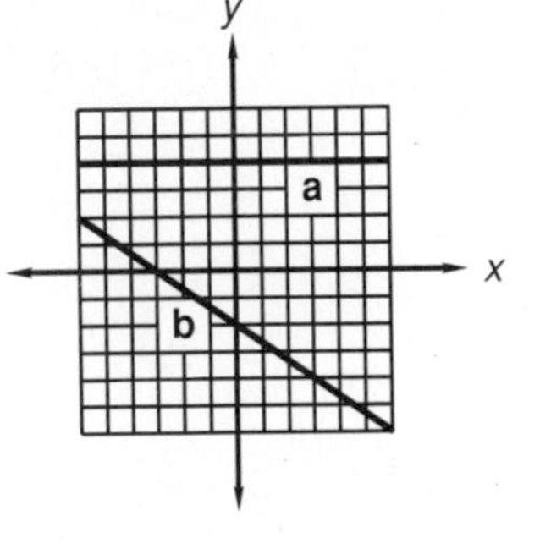

12. Divide: $(3x^3 - 5) \div (x + 3)$

13. Simplify: $-3\sqrt{12}(2\sqrt{6} - 5\sqrt{8})$

Use the quadratic formula to solve the following quadratic equations:

14. $-2 = x^2 + 6x$

15. $-7x = 4 - 2x^2$

Solve by factoring:

16. $-4 + 9x^2 = 0$

17. $5 = -x^2 - 6x$

18. Solve: $\dfrac{4x}{3} - \dfrac{2x + 4}{2} = 5$

19. $4 \in$ {What sets of numbers}?

20. Solve: $\dfrac{1}{2}x - 2\dfrac{3}{5} = \dfrac{1}{10}$

21. $R_H T_H = R_R T_R$, $R_H = 2$, $R_R = 12$, $T_H = T_R + 5$. Find T_H and T_R.

Solve the following quadratic equations by completing the square:

22. $-5 = x^2 - 7x$

23. $-3x = 4 - x^2$

24. Graph on a number line: $0 \not\leq -x - 3 \not\leq 2$; $D = \{\text{Integers}\}$

25. Find the equation of a line through $(-3, 4)$ and parallel to $y = -\dfrac{2}{5}x - 1$.

26. Demonstrate the additive property of zero.

Simplify:

27. (a) $\sqrt{\dfrac{5}{8}}$ (b) $\dfrac{\sqrt{5} + 3}{\sqrt{5}}$

28. $\dfrac{(x^0)^{-2}(xx^2)^3(y^{-2})^4y}{(xyy)^{-2}(yx^{-2})^4}$

29. Simplify by adding like terms: $x^2y^{-2} - \dfrac{3x^2}{y^2} + \dfrac{12x^4xy^{-2}}{x^3} - \dfrac{3x^2y^2}{x^{-4}}$

30. A cylinder whose radius is 1 m is removed from the right prism as shown. The ends of the prism have the shape of an equilateral triangle whose sides are 6 meters long. Find the volume of the remaining solid in cubic meters. Dimensions are in meters.

APPENDIX A Additional practice sets

practice set 1

1. The express train made the trip in 25 hours. The freight train took 30 hours because it was 15 kilometers per hour slower than the express. What was the speed of each train?
2. If a merchant had to reduce prices by 45 percent, what was the original price of a floor lamp which is now priced at \$275?
3. Reds varied inversely as blues squared. When there were 7 reds, there were 60 blues. How many reds were there when there were only 20 blues?
4. The ratio of harriers to mollifiers was 13 to 8. If there were 420 people at the meeting, how many were harriers?
5. Rebecca deposited \$22,000 at 11 percent compound interest. How much money did she have in 5 years? How much interest did she earn in 5 years?
6. Find four consecutive even integers such that -5 times the sum of the first and third is 40 greater than 6 times the opposite of the fourth.
7. Simplify: $\dfrac{3a + \dfrac{x}{2z}}{\dfrac{4am}{y} + 2}$
8. $R_E T_E = R_W T_W$, $R_E = 16$, $R_W = 60$, $T_E + T_W = 5$. Find T_E and T_W.

Use the quadratic formula to solve:

9. $-5 = -3x^2 - x$
10. $4x = 2x^2 - 2$
11. $3x^2 + 3x - 8 = 0$
12. $4x^2 - 5x - 3 = 0$

Solve by completing the square:

13. $4x = x^2 - 2$
14. $x^2 + 3x - 8 = 0$
15. $-5 = -x^2 - x$

Factor. Begin by writing terms in descending order of the variable.

16. $2x^2 - 6 + 11x$
17. $-33 + 30x + 3x^2$
18. $-20x - 7 + 3x^2$

Factor by grouping:

19. $ax^3 + 5a - 4x^3 - 20$
20. $7z + cmz - cmx - 7x$
21. $5a - mna - mnb + 5b$
22. $7ys + zfs - zfx - 7yx$
23. Find the equation of the line through (3, 6) and $(-2, -5)$.
24. Find the equation of the line through $(-3, 4)$ that is parallel to $y = \frac{1}{5}x - 1$.

25. Graph: $\begin{cases} y \leq x \\ y \geq -x - 1 \end{cases}$

Simplify:

26. $(5 + 3\sqrt{3})(\sqrt{3} + 4)$ 27. $\sqrt{\dfrac{5}{7}}$ 28. $\dfrac{2\sqrt{3} + 2}{\sqrt{3}}$

29. Expand: $(9x^2 - 4z^2)^2$

30. Find the volume in cubic inches of a right circular cylinder with a diameter of $6\sqrt{2}$ inches and a height of 5 inches.

practice set 2

1. Raoul ran to Bakersfield in 7 hours, while Gena walked the same distance in 63 hours. How fast did Raoul run if his speed was 16 kilometers per hour faster than Gena's? How far was it to Bakersfield?

2. The weight of a body on or above the surface of the earth varies inversely with the square of the distance from the body to the center of the earth. If a body weighs 6000 pounds at a distance of 1000 miles from the center of the earth, how much would it weigh 7000 miles from the center of the earth?

3. Box seats sold for \$125 per seat, and general seats sold for only \$25 per seat. Becky Jo spent \$3500 and bought 8 more general seats than box seats. How many box seats did she buy?

4. Rhonda and Mabel rated the commercial presentations. Rhonda gave the marketing teams only 30 percent of the points that Mabel gave the individual representatives. If Rhonda gave the marketing teams 1653 points, how many points did the individual representatives get?

5. The ratio of the exuberant to the depressed was 7 to 3. If there were 26,000 people at the fair, how many were exuberant?

6. On two rolls of a die, what is the probability of rolling a 2 and then a 5?

7. Graph on a number line: $-6 \leq x < 4$; $D = \{\text{Reals}\}$

Solve by graphing:

8. $\begin{cases} y = -2x + 3 \\ x = 1 \end{cases}$

Use elimination to solve:

9. $\begin{cases} 2x - 3y = -1 \\ -3x + y = 5 \end{cases}$

10. Simplify: $\dfrac{(0.000006)(0.03 \times 10^{16})}{(30{,}000)(600 \times 10^{-11})}$

11. Find the equations of lines (a) and (b).

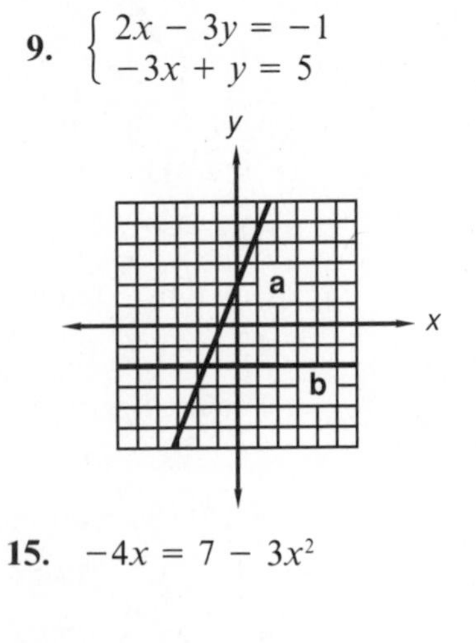

12. Divide: $(5x^3 - 2) \div (x + 4)$

13. Simplify: $-3\sqrt{18}(5\sqrt{16} - \sqrt{27})$

Solve:

14. $-3 = x^2 + 5x$

15. $-4x = 7 - 3x^2$

Solve by factoring:

16. $-9 + 25z^2 = 0$

17. $5 = -2x^2 - 11x$

18. Solve: $\dfrac{4x}{5} - \dfrac{3x + 3}{4} = 6$

19. $\dfrac{\sqrt{2}}{3} \in \{\text{What sets of numbers}\}$

20. Simplify: $\dfrac{b^2 + \dfrac{3}{b^3}}{4 - \dfrac{1}{b^2}}$

21. $R_H T_H = R_X T_X$, $R_H = 3$, $R_X = 7$, $T_X = T_H - 3$. Find T_X and T_H.

Solve by completing the square:

22. $-4 = -x^2 - 5x$

23. $-4x = 5 - x^2$

24. Graph on a number line: $-4 \not> x + 4 \not> 5$; $D = \{\text{Reals}\}$

25. Find the equation of a line through $(-1, -1)$ and parallel to $y = -\frac{3}{4}x + 3$.

26. Use the numbers 2, 3, and 4 in that order to demonstrate the distributive property of real numbers.

Simplify:

27. $\sqrt{\dfrac{3}{2}}$

28. $\dfrac{\sqrt{2} + 7}{\sqrt{2}}$

29. $\dfrac{(k^2)^{-3}(kk^{-1})^2(z^{-3})^3 z}{(kzz^2)^{-3}(zk^{-1})^4}$

30. Find the volume in cubic meters of a cylinder whose base is shown and whose sides are 3 yards high. If the cylinder is a right cylinder, what is the surface area? What is the volume of a cone 3 yards high that has the same base? Dimensions are in feet.

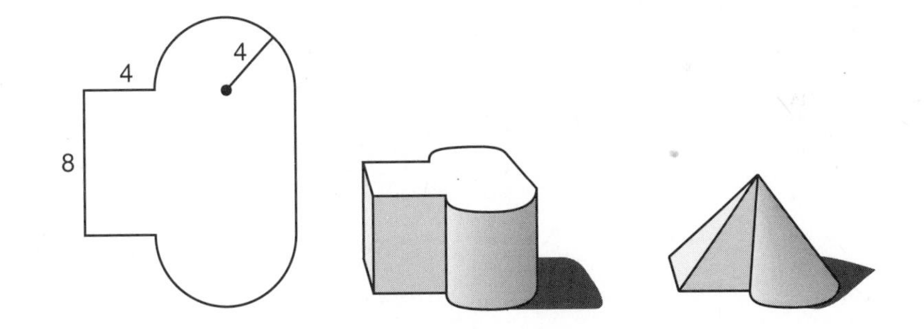

practice set 3

1. The charter flight headed east at 7 p.m., and the public flight headed west at midnight. At 4 a.m. the planes were 3500 miles apart. What was the speed of each if the speed of the charter flight was 100 miles per hour greater than the speed of the public flight?

2. The ratio of polymorphous to uniform was 11 to 5. If the total was 16,800, how many were polymorphous and how many were uniform?

3. Eighty-six percent of the populace voted in the election. If 1197 abstained from voting, how many citizens voted?

4. There was \$3600 in the pot. If there were 354 more \$1 bills than \$5 bills, how many bills of each kind were there?

5. Maureen deposited \$19,000 at 12 percent compound interest. How much money did she have in 10 years? How much interest did she earn?

6. Find four consecutive odd integers such that the product of -8 and the sum of the second and third is 6 less than the product of 14 and the opposite of the fourth.

Use the quadratic formula to solve:

7. $5 = 3x^2 + 5x$

8. $-5x = 8 - 3x^2$

Solve by completing the square:

9. $x^2 + 8x - 2 = 0$

10. $x^2 + 6x - 3 = 0$

Factor:

11. $-36y^6 = -121x^2$

12. $-41x = 6 - 7x^2$

13. $a^2b - 2c^3m + mb - 2a^2c^3$

14. $6ax + 7am^3 + 12d^3x + 14m^3d^3$

15. In a standard deck of 52 cards what is the probability of pulling a black jack from a deck of cards and then pulling another jack? The first card is replaced before the second card is drawn.

Graph on a number line:

16. $5 \geq x \geq -9$; $D = \{\text{Reals}\}$

17. $-2 - |x| > 2$; $D = \{\text{Integers}\}$

18. Find the equation of the line through $(-1, 3)$ and $(3, -1)$.

19. Find the equation of the line through $(3, 3)$ that is parallel to $y = \frac{2}{3}x + 7$.

20. Simplify: $\dfrac{2m + \dfrac{x}{3z}}{\dfrac{3mc}{z} + 1}$

21. $R_ET_E + R_WT_W = 688$, $R_E = 44$, $R_W = 100$, $T_E + T_W = 8$. Find T_E and T_W.

22. $-3\sqrt[3]{-27} \in \{\text{What sets of numbers}\}$?

Solve:

23. $\sqrt{5x + 2} - 5 = 2$

24. $\sqrt{4c - 7} + 6 = -3$

Simplify:

25. $\sqrt{\dfrac{3}{8}}$

26. $\dfrac{5 + \sqrt{10}}{\sqrt{5}}$

27. $(5 + 3\sqrt{3})(\sqrt{3} + 3)$

28. $\left(y - \frac{1}{2}p\right)^2$

29. $4z^{-4}x^2 - \dfrac{1}{x^{-2}z^4} + \dfrac{14z^5x^6}{yx^{-8}z^2z^7}$

30. Find the volume in cubic meters of a right solid whose base is shown and whose sides are 10 feet high. Dimensions are in feet.

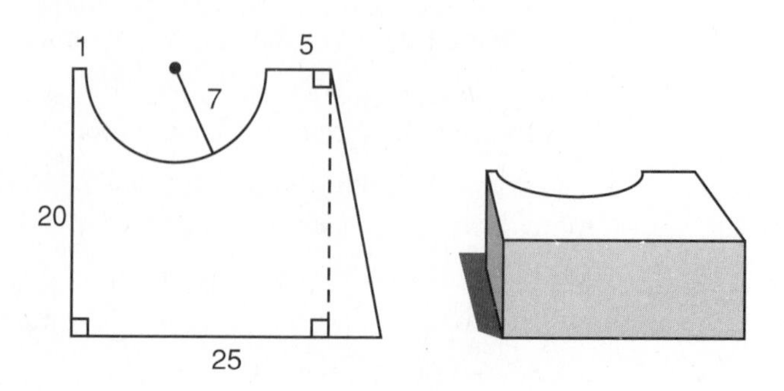

practice set 4

1. In a particular experiment, the pressure varied inversely as the volume. When the pressure was 20 pounds per square inch, the volume was 30 liters. What was the pressure when the volume was reduced to 5 liters?

2. In another experiment, the pressure varied directly as the temperature. When the pressure was 4000 pounds per square inch, the temperature was 125°. What was the pressure when the temperature was 1000°?

3. Chad had 200 mice. If the number of mice tripled every year, how many mice would Chad have in 30 years?

4. Dieta and Rhonda had a race. Dieta started out at 6 a.m. and drove at 55 miles per hour. Rhonda waited until 9 a.m. to start and drove at 65 miles per hour. What time was it when Rhonda was 15 miles in front of Dieta?

5. Sarah and Selby collected dimes and quarters. They collected 330 coins whose value was \$38.85. How many quarters did they have?

6. Graph $y \geq 3x + 4$ on a rectangular coordinate system.

Factor:

7. $2x^2 + 13x + 21$

8. $5x^2 - 34x - 7$

9. $4m - sam + 4xy - saxy$

10. $5cx + 3x - 5czx - 3xz$

Expand:

11. $(7x - \sqrt{13})^2$

12. $(\sqrt{3}\,x + \sqrt{3})^2$

Simplify:

13. $\sqrt{\frac{7}{12}}$

14. $\frac{7\sqrt{11} + 3}{\sqrt{11}}$

15. $\frac{6p + \frac{3m}{p}}{\frac{3mp}{z} + 5}$

16. $(4 + 3\sqrt{2})(\sqrt{2} + 3)$

17. $0.00361 \in$ {What sets of numbers}?

Use the quadratic formula to solve:

18. $-3 = x^2 + 7x$

19. $-5x + 3 = 4x^2$

Solve by completing the square:

20. $-3 = -x^2 - 9x$

21. $-5x = 1 - x^2$

22. Find the equation of the line through (1, 4) and (−2, 3).

23. Find the equation of the line through (−3, 3) that is parallel to $y = -\frac{1}{4}x + 1$.

24. $R_ET_E + R_XT_X = 1100$, $R_E = 50$, $R_X = 200$, $T_E + T_X = 10$. Find T_E and T_X.

Solve by graphing:

25. $\begin{cases} y = -\frac{1}{2}x + 1 \\ x = 1 \end{cases}$

Use elimination to solve:

26. $\begin{cases} x - y = -5 \\ 3x + y = 25 \end{cases}$

Simplify:

27. $\frac{(0.00006)(0.09 \times 10^{11})}{(70{,}000)(500 \times 10^{-9})}$

28. $\frac{x^{-4}(xx^{-1})^3(y^{-5})^4y}{x^2y^{-10}x^{-5}x^{-1}}$

29. $3x^4y^{-6} - \dfrac{x^6}{3^{-1}y^6x^2} - \dfrac{9y^{-9}x^{-3}x^{12}}{x^5y^{-3}}$

30. Find the volume in cubic centimeters of a right cylinder whose base is shown and whose sides are 3 feet high. Dimensions are in inches.

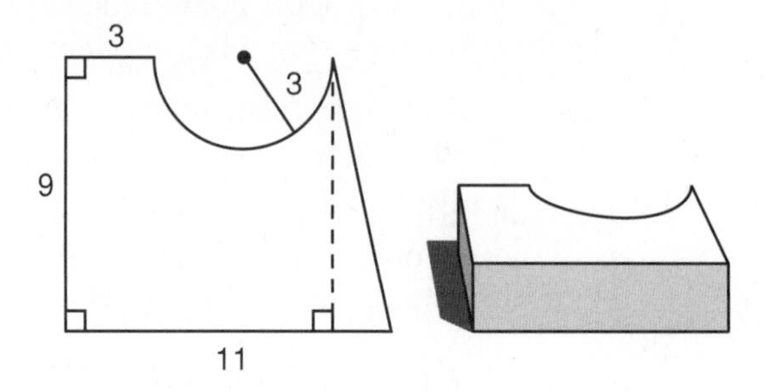

practice set 5

1. The distance required for an automobile to stop is directly proportional to the square of its velocity. If a car can stop in 1600 meters from a velocity of 40 kilometers per hour, what will be the required stopping distance at 32 kilometers per hour?

2. The distance a body falls in a vacuum varies directly as the square of the time that it falls. If an object falls 432 feet in 6 seconds, how far will it fall in 11 seconds?

3. The number of philanthropists varied inversely as the square of the number of misanthropes. When there were 15 philanthropists, there were 6 misanthropes. How many philanthropists would there be if there were only 2 misanthropes?

4. The ratio of bovine to porcine on the farm was 9 to 7. If there were 33,920 animals total, how many were bovine and how many were porcine?

5. Kahil rode the camel to the market in Cairo at 3 kilometers per hour and then took a horse back to the oasis at 8 kilometers per hour. If the total trip took 11 hours, how far was it from the oasis to the market?

6. Find the equation of lines (a) and (b).

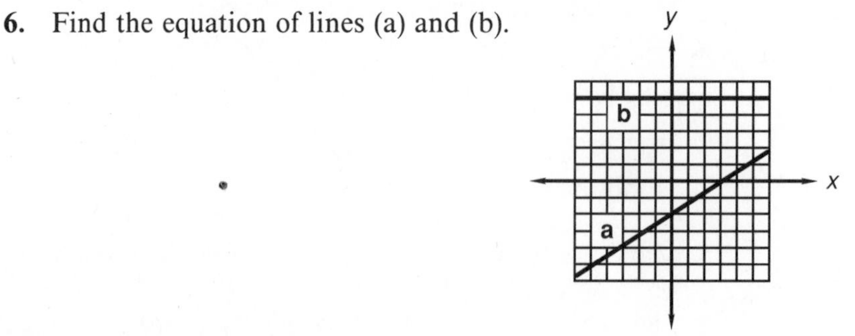

7. Which of the following sets of ordered pairs are functions?
(a) (4, 1), (4, 2), (4, 3) (b) (−1, −2), (1, −2), (4, −2)
(c) (−4, −1), (−4, 1), (−1, 4) (d) (−6, 2), (6, −2), (2, 6)

8. Solve: $\sqrt{x + 4} - 4 = 6$

9. Graph on a number line: $x + 3 > 5$ or $x \leq -4$; $D = \{\text{Reals}\}$

Multiply:

10. $(3\sqrt{5} + \sqrt{3})(\sqrt{5} + \sqrt{3})$

11. $(4 + 2\sqrt{11})(4 - \sqrt{11})$

Expand:

12. $(2a^2 + 4b^2)^2$

13. $(b^2 - \sqrt{2})^2$

Simplify:

14. $\sqrt{\frac{7}{8}}$

15. $\frac{4 + 2\sqrt{3}}{3\sqrt{5}}$

16. $\frac{5 + 2\sqrt{8}}{\sqrt{3}}$

17. Find the volume in cubic feet of a right circular cylinder whose diameter measures $2\sqrt{5}$ meters and whose sides are 3 meters high.

18. Use six unit multipliers to convert 10,000,000 cubic feet to cubic centimeters.

19. Solve: $\frac{x + 3}{3x} = \frac{5}{x} + \frac{1}{4}$

20. Find y: $\frac{2b}{y} - \frac{x}{4z} + m = s$

21. Simplify: $\dfrac{\frac{my^2}{x^2} - \frac{y}{x^3}}{\frac{z}{2x} - \frac{3m}{x^2}}$

22. Divide: $(6x^4 - 6) \div (x + 1)$

23. Simplify: $-4[(-1^2 - 2^0)(-3^3 - 2^2) + (-5)][-(-1)(-7^0)]$

24. Find the equation of the line through $(-2, 4)$ that is parallel to $y = -2x + \frac{1}{2}$.

25. Find the equation of the line that goes through $(-2, 1)$ and $(1, -2)$.

26. Find the equation of the line whose slope is $-\frac{1}{4}$ that passes through the point $(-1, -2)$.

27. Graph $y \leq -x + 5$ on a rectangular coordinate system.

28. Simplify: $\frac{(8 \times 10^{-17})(9{,}000{,}000)}{(3 \times 10^{-5})(4 \times 10^6)}$

29. Use the quadratic formula to solve: $-3x^2 + 12x = 3$

30. Solve by completing the square: $-9x + 2 = x^2$

practice set 6

1. Credenzas sold for \$350 each and hutches sold for \$500. Horatio spent \$6400 and bought 6 more hutches than credenzas. How many hutches did he buy?

2. Kitty flew to Sparks in 3 hours while Roland drove the same distance in 23 hours. How fast did Kitty fly if her speed was 200 miles per hour faster than Roland's? How far was it to Sparks?

3. The ratio of the suspenseful to the anticlimactic was 3 to 11. If 8800 events were anticlimactic, how many were suspenseful?

4. The year's yield varied inversely as the square of the amount of fertilizer used. If 8 tons of fertilizer resulted in 4000 tons of produce, how much produce would have been harvested if only 10 tons of fertilizer had been used?

5. Find four consecutive odd integers such that 6 times the sum of the first and the second is 9 greater than 9 times the fourth.

6. The number of bacteria in the petri dish quintupled every day. If there were 50,000 bacteria at first, how many bacteria were there in 6 days?

Use the quadratic formula to solve:

7. $4 = -5x - x^2$

8. $3x^2 - 2 = 5x$

Solve by completing the square:

9. $-3x = x^2 + 2$ 10. $-5x = 4 - x^2$ 11. $x^2 = -5x - 6$

12. Graph on a rectangular coordinate system: $y \nleq \frac{1}{5}x - 3$

13. Find p: $\frac{m}{p} + \frac{x}{z} = y$

Factor the trinomials:

14. $5x^2 + 6x - 11$ 15. $6x^2 + 25x + 25$ 16. $6x^2 - 149x - 25$

Factor by grouping:

17. $4ma + 6 + 24m + a$ 18. $xyz + xy^2 + 6z + 6y$

Graph on a number line:

19. $-5 + |-x| < -3$; $D = \{\text{Reals}\}$

20. $6 \le x + 3 < 8$; $D = \{\text{Integers}\}$

21. $x \le -3$ or $x > 0$; $D = \{\text{Reals}\}$

Solve:

22. $\sqrt{x + 5} - 6 = -1$ 23. $\sqrt{2x - 3} - 2 = 4$

24. Find the equation of the line through $(-4, -2)$ and $(0, 3)$.

25. Find the equation of the line through $(-3, 2)$ that has a slope of $-\frac{2}{3}$.

26. Find the equation of the line through $(3, -1)$ that is parallel to $y = -\frac{3}{5}x + 2$.

Simplify:

27. $\sqrt{\frac{4}{7}}$ 28. $\frac{3\sqrt{5} + \sqrt{5}}{\sqrt{5}}$

29. $\frac{(0.003 \times 10^{-11})(700 \times 10^{20})}{(0.00009 \times 10^{-10})(8 \times 10^{25})}$

30. Find the volume in cubic centimeters of a right solid whose base is the figure shown and whose height is 8 inches. Dimensions are in inches.

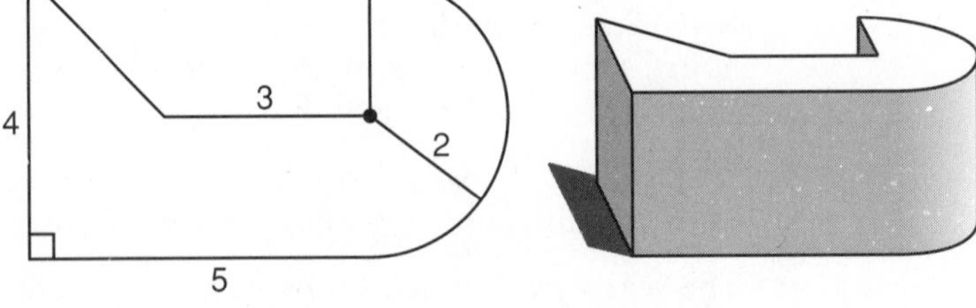

practice set 7

1. The ratio of boisterous to quiescent was 5 to 13. If 810 attended the melee, how many were boisterous and how many were quiescent?

2. Eloise rode the train to Cloverdale for 3 hours while Eric drove to Cloverdale. If Eloise arrived 1 hour before Eric and her speed was 20 kilometers per hour faster than Eric's speed, what was the distance to Cloverdale?

3. A fair coin is tossed four times. What is the probability that the first toss will be heads, the next two tosses will be tails, and the last toss will be heads?

4. Reds varied inversely as blues squared. When there were 2 reds, there were 40 blues. How many reds were there when there were only 10 blues?

5. Find four consecutive odd integers such that 2 times the sum of the second and the third is 20 greater than the sum of 10 times the first and -2.

6. Between the shoe factory and the retail shoe store, the price of a pair of shoes increases 225 percent. If the retail price at the shoe store is \$78, what was the wholesale cost at the factory?

7. Simplify: $\dfrac{5a + \dfrac{3m}{x}}{\dfrac{4ma}{x} + 7}$

8. $R_E T_E + 8 = R_W T_W$, $R_E = 7$, $R_W = 23$, $T_E + T_W = 60$. Find T_E and T_W.

Use the quadratic formula to solve:

9. $4 = 11x^2 + x$

10. $3x = 4x^2 - 5$

Solve by completing the square:

11. $-9 = -x^2 - 13x$

12. $x^2 + 3 = 7x$

Factor:

13. $-42x + 49 + 8x^2$

14. $-8x - 13 + 5x^2$

15. $mnf + p^2af - 4mn - 4p^2a$

16. $5pa^4 - 7a + 25pa^3 - 35$

17. Graph on a rectangular coordinate system: $y > 4x + 2$

18. Find the volume in cubic feet and the surface area of a right circular cylinder with a radius of $\sqrt{2}$ centimeters and height of 10 centimeters.

19. Solve: $\dfrac{p + 4}{3p} = \dfrac{5}{p} + \dfrac{1}{4}$

Simplify:

20. $\dfrac{(0.00014)(0.07 \times 10^{-12})}{(5000)(300 \times 10^{8})}$

21. $\dfrac{3\sqrt{7} + 7}{\sqrt{7}}$

22. $(5 + 3\sqrt{5})(\sqrt{5} + 2)$

23. $\dfrac{3mp^{-2}}{m^{-5}} - \dfrac{5m^4}{p^2} + \dfrac{m^5p^{-4}}{mp^{-2}}$

24. Solve: $\sqrt{3x + 4} - 5 = 3$

25. Find p: $\dfrac{2x + 3}{my} + \dfrac{y}{p} = x + 4$

26. Graph on a number line: $8 \le x + 5 \le 10$; $D = \{\text{Integers}\}$

27. Find the equation of the line through $(-3, -1)$ and $(1, 1)$.

28. Find the equation of the line through $(-2, 2)$ that has a slope of $\dfrac{3}{4}$.

29. Find the equation of the line through $(2, -1)$ that is parallel to $y = -\frac{3}{8}x - 4$.

30. Find the volume in cubic meters of a right solid whose base is shown and whose sides are 16 inches high. Dimensions are in feet.

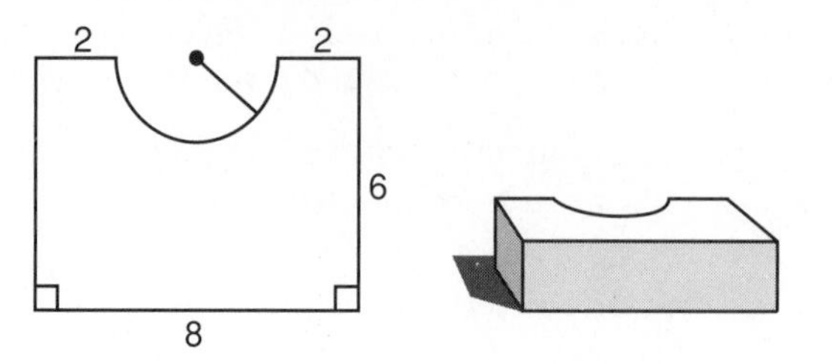

practice set 8

1. The ratio of equine to piscine was 2 to 61. If there was a total of 7686 in the pond and field, how many were equine and how many were piscine?

2. The number of opportunities varied inversely as the number of applicants squared. When there were 5 opportunities, there were 1000 applicants. How many opportunities were there when there were 20 applicants?

3. Chad walked to a meeting in Alvin at 4 miles per hour and caught a ride home at 10 miles per hour. How far was it to the meeting if his total traveling time was 14 hours?

4. Find four consecutive odd integers such that the product of -5 and the sum of the second and the third is 45 greater than the product of -7 and the fourth.

Solve by completing the square:

5. $-12x = -x^2 - 4$

6. $x^2 = 11x - 2$

Use the quadratic formula to solve:

7. $-9x = 7x^2 + 2$

8. $-3 - 12x = 4x^2$

9. Find the equations of lines (a) and (b).

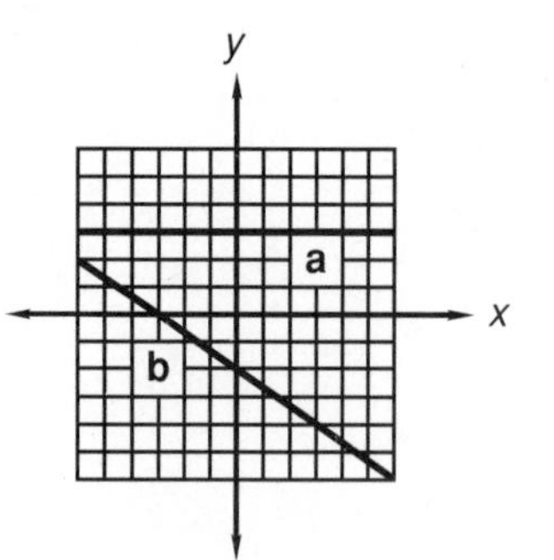

10. In three spins what is the probability that the spinner will stop on a 2 and then on a 5 and finally on a 3?

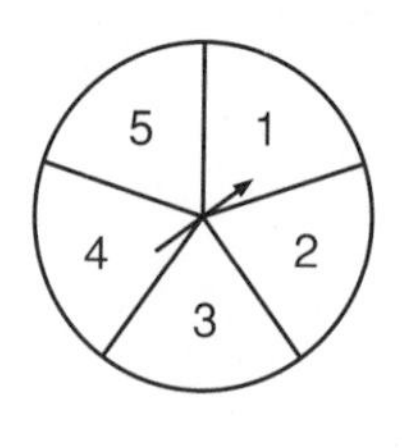

Graph on a rectangular coordinate system:

11. $y \le -4x - 5$

12. $x - 3 \le y \le -x$

13. Graph on a number line: $x - 3 > 7$ or $x + 2 \le 8$; $D = \{\text{Reals}\}$

14. $R_S T_S + 4 = R_B T_B$, $R_S = 4$, $R_B = 32$, $T_S + T_B = 200$. Find T_B and T_S.

Simplify:

15. $$\frac{m + \frac{x}{y}}{3mx + \frac{7}{4}}$$

16. $$\frac{(7 \times 10^{-11})(5000)(3 \times 10^4)}{(2 \times 10^{-8})(100{,}000)}$$

17. $(6 + 3\sqrt{2})(2\sqrt{2} + 2)$

18. $\dfrac{\sqrt{5} + 7}{3\sqrt{5}}$

19. $\sqrt{\dfrac{3}{7}}$

20. $-3[(-4^2 - 4^0)(-3^2 - 3) + (-104)]$

21. Find a: $\dfrac{m}{3a} + \dfrac{p}{x} + k = z$

22. Expand: $(5a^2 + 2b^2)^2$

23. Simplify: $(5a^2 + 2b^2)(5a^2 - 2b^2)$

Solve:

24. $\sqrt{4x + 4} + 6 = 12$

25. $3\frac{3}{8}m + \frac{1}{2} = 2\frac{3}{7}$

26. Use 10 unit multipliers to convert 10 square miles to square kilometers.

27. Find the equation of the line through $(3, -3)$ that is parallel to $y = \frac{1}{3}x + 1$.

28. Find the equation of the line that goes through $(2, 4)$ and $(-1, -1)$.

29. Find the equation of the line whose slope is -3 that passes through $(2, -3)$.

30. Find the volume in cubic meters of a right solid whose base is shown and whose sides are 2 feet high. Dimensions are in inches. Angles that look square are square.

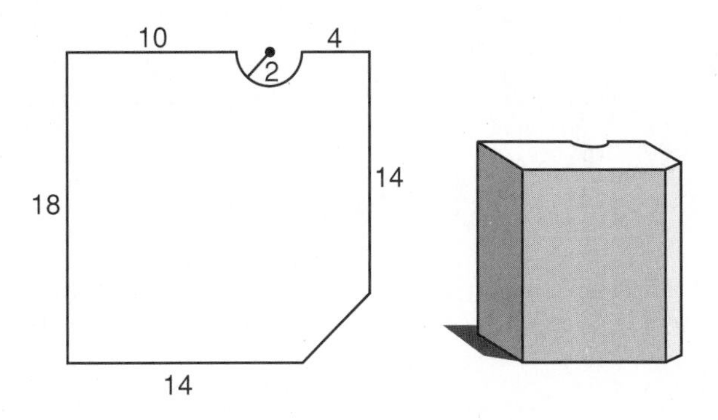

practice set 9

1. Inflation exacts its toll in the appreciation of land values. The average cost of houses in the area rose 28 percent. If a house now cost \$265,000, what was its price before the impact of inflation?

2. Jill deposited \$1500 at 9 percent compound interest. How much money did she have in 20 years? How much interest did she earn?

3. The blues varied inversely as the square of the greens. When there were 12 blues, there were 5 greens. How many blues would there be if there were 10 greens?

4. The ratio of eloquent to inarticulate was 3 to 14. If there were 71,400 involved in dialogue, how many spoke eloquently?

5. The express train made the trip in 24 hours. The freight train took 30 hours, as it was 17 miles per hour slower than the express. What was the speed of each train?

6. One card was drawn from a standard deck of 52 cards and the card was replaced. Then another card was drawn. What is the probability that the first card was an 8 *and* the second card was a 4?

7. Graph on a number line: $-2 \leq x + 1 \leq 4$; $D = \{\text{Reals}\}$

8. Find the equations of lines (a) and (b).

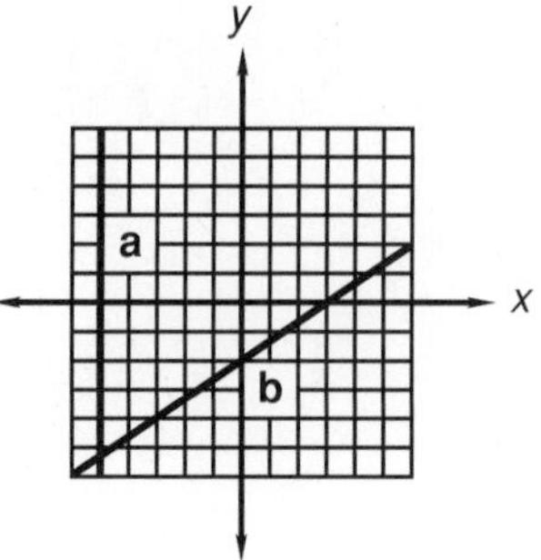

9. Solve by graphing: $\begin{cases} y = 2x - 3 \\ x = 1 \end{cases}$

10. Graph on a rectangular coordinate system: $y < -2x + 2$

11. Multiply: $(4 + 3\sqrt{5})(2 + \sqrt{5})$

12. Expand: $(4a^2 + 4b^2)^2$

Simplify:

13. $\dfrac{(36{,}000 \times 10^{-22})(4000 \times 10^{-4})}{(0.000009 \times 10^{6})(0.0007 \times 10^{-3})}$

14. $\dfrac{6x^5y^{-3}(x^6)^{-2}y^3x^2y}{(3x^0)^2x^2y^{-2}(xy)^2}$

15. $\dfrac{3 + \sqrt{5}}{\sqrt{3}}$

16. $-2[(-2^3 - 2^0)(2^3 - 3^2)][-(-2)(1^0)]$

Solve:

17. $3\sqrt{m + 6} - 5 = 2$

18. $\dfrac{x + 6}{4x} = \dfrac{4}{3x} + \dfrac{3}{4}$

19. Find the volume in cubic meters of a right solid whose base is shown and whose sides are 3 yards high. Dimensions are in feet.

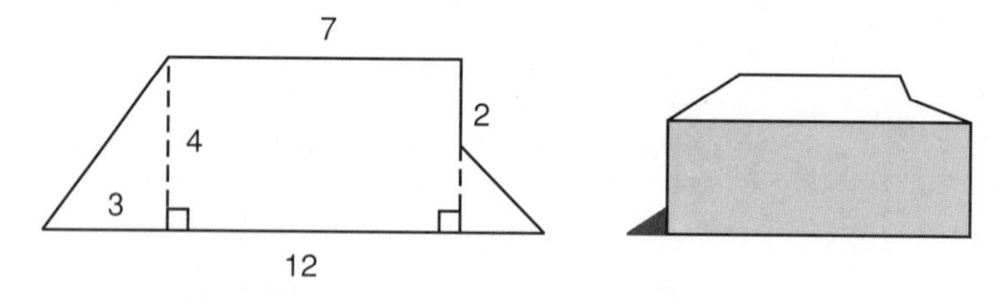

Use the quadratic formula to solve:

20. $5 = 4x^2 + 3x$

21. $-3x = 9 - 2x^2$

Solve by completing the square:

22. $-3 = -x^2 + 5x$

23. $-4x = 3 - x^2$

24. $R_CT_C = R_BT_B$, $R_C = 12$, $R_B = 4$, $T_C + T_B = 60$. Find T_C and T_B.

25. Which of the following sets are functions?
(a) $(-4, -2), (4, 2), (-2, 4)$
(b) $(-4, -2), (-4, 2), (-2, 4)$
(c) $(-4, -2), (4, -2), (5, -2)$
(d) $(3, -1), (-3, 1), (-1, 3)$

26. Divide: $(7x^4 - x^2 + 1) \div (x - 1)$

27. Find the total surface area of a right circular cylinder with a radius of $\sqrt{3}$ inches and a height of 13 inches.

28. Use 15 unit multipliers to convert 1000 cubic miles to cubic kilometers.

29. Find the equation of the line through $(-1, -4)$ and $(3, 3)$.

30. Find the equation of the line through $(3, 3)$ that is parallel to $y = -4x - 4$.

practice set 10

1. Find four consecutive even integers such that 6 times the sum of the first and the fourth is 42 greater than the product of 9 times the second.

2. Roland ran to Jackson in 9 hours while Raoul walked the same distance in 81 hours. How fast did Roland run if his speed was 16 kilometers per hour faster than Raoul's? How far was it to Jackson?

3. Martine had 10 rabbits. If the number of rabbits tripled every year, she would have 30 rabbits at the end of 1 year. How many rabbits would she have in 7 years?

4. On three rolls of a die, what is the probability of getting a 5, a 3, and a 6 in that order?

5. Zinnias varied inversely as the square of the number of poppies. When there were 200 zinnias, there were 30 poppies. How many zinnias were there if there were 50 poppies?

6. Johnny and Amy collected nickels and dimes. They collected 500 coins whose value was $28. How many nickels did they have?

Factor:

7. $6x^2 + 15x - 9$

8. $3mn + 4yn - 3mo - 4yo$

Simplify:

9. $\sqrt{\dfrac{3}{11}}$

10. $\dfrac{5\sqrt{11} + 3}{\sqrt{11}}$

11. $\dfrac{8x + \dfrac{2p}{c}}{\dfrac{3cx}{z} - 4}$

12. $(5\sqrt{2} + 3)(2\sqrt{2} - \sqrt{2})$

13. $\dfrac{(0.0006 \times 10^{-4})(6 \times 10^{20})}{(1{,}000{,}000)(300 \times 10^{-22})}$

14. $11x^{10}y^{-8} - \dfrac{xy^5x}{y^{10}y^3x^{-12}} - \dfrac{8x^{14}y^{10}}{3x^{-3}y^7xx^{-2}y^{11}}$

15. $\sqrt{2} \in$ {What sets of numbers}?

16. Expand: $(8x^4 + 3p^2)^2$

17. Graph on a number line: $x - 11 > -16$ or $x + 5 \leq 8$; $D = \{\text{Integers}\}$

18. Find the equations of lines (a) and (b).

y
x
a
b

19. Solve: $\dfrac{z + 4}{z} = \dfrac{2}{z} + \dfrac{1}{9}$

20. Find p: $\dfrac{3c}{m} - \dfrac{y}{4p} + z = 5$

21. Solve: $3\sqrt{m + 5} - 2 = 9$

22. Graph on a rectangular coordinate system: $y \leq -3x + \frac{1}{2}$

Solve by completing the square:

23. $-5x = -x^2 - 4$

24. $-x^2 = -9x + 2$

Use the quadratic formula to solve:

25. $-6x = -3x^2 + 2$

26. $5x^2 - 3 = 11x$

27. Find the volume in cubic inches of a right solid whose base is the figure shown and whose sides are 1 meter high. Dimensions are in centimeters.

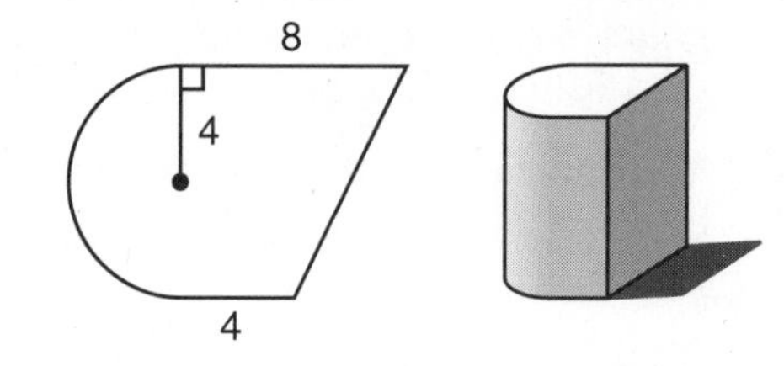

28. Find the equation of the line through $(-2, -2)$ and $(1, 4)$.

29. Find the equation of the line through $(1, 4)$ that is parallel to $y = -\frac{2}{5}x - 2$.

30. Find the equation of the line whose slope is $-\frac{2}{3}$ and that passes through $(3, 3)$.

APPENDIX B *Table of squares and square roots*

n	n^2	$\sqrt{n}$	n	n^2	$\sqrt{n}$	n	n^2	$\sqrt{n}$	n	n^2	$\sqrt{n}$	n	n^2	$\sqrt{n}$	n	n^2	$\sqrt{n}$
1.00	1.0000	1.00000	1.50	2.2500	1.22474	2.00	4.0000	1.41421	2.50	6.2500	1.58114	3.00	9.0000	1.73205	3.50	12.2500	1.87083
1.01	1.0201	1.00499	1.51	2.2801	1.22882	2.01	4.0401	1.41774	2.51	6.3001	1.58430	3.01	9.0601	1.73494	3.51	12.3201	1.87350
1.02	1.0404	1.00995	1.52	2.3104	1.23288	2.02	4.0804	1.42127	2.52	6.3504	1.58745	3.02	9.1204	1.73781	3.52	12.3904	1.87617
1.03	1.0609	1.01489	1.53	2.3409	1.23693	2.03	4.1209	1.42478	2.53	6.4009	1.59060	3.03	9.1809	1.74069	3.53	12.4609	1.87883
1.04	1.0816	1.01980	1.54	2.3716	1.24097	2.04	4.1616	1.42829	2.54	6.4516	1.59374	3.04	9.2416	1.74356	3.54	12.5316	1.88149
1.05	1.1025	1.02470	1.55	2.4025	1.24499	2.05	4.2025	1.43178	2.55	6.5025	1.59687	3.05	9.3025	1.74642	3.55	12.6025	1.88414
1.06	1.1236	1.02956	1.56	2.4336	1.24900	2.06	4.2436	1.43527	2.56	6.5536	1.60000	3.06	9.3636	1.74929	3.56	12.6736	1.88680
1.07	1.1449	1.03441	1.57	2.4649	1.25300	2.07	4.2849	1.43875	2.57	6.6049	1.60312	3.07	9.4249	1.75214	3.57	12.7449	1.88944
1.08	1.1664	1.03923	1.58	2.4964	1.25698	2.08	4.3264	1.44222	2.58	6.6564	1.60624	3.08	9.4864	1.75499	3.58	12.8164	1.89209
1.09	1.1881	1.04403	1.59	2.5281	1.26095	2.09	4.3681	1.44568	2.59	6.7081	1.60935	3.09	9.5481	1.75784	3.59	12.8881	1.89473
1.10	1.2100	1.04881	1.60	2.5600	1.26491	2.10	4.4100	1.44914	2.60	6.7600	1.61245	3.10	9.6100	1.76068	3.60	12.9600	1.89737
1.11	1.2321	1.05357	1.61	2.5921	1.26886	2.11	4.4521	1.45258	2.61	6.8121	1.61555	3.11	9.6721	1.76352	3.61	13.0321	1.90000
1.12	1.2544	1.05830	1.62	2.6244	1.27279	2.12	4.4944	1.45602	2.62	6.8644	1.61864	3.12	9.7344	1.76635	3.62	13.1044	1.90263
1.13	1.2769	1.06301	1.63	2.6569	1.27671	2.13	4.5369	1.45945	2.63	6.9169	1.62173	3.13	9.7969	1.76918	3.63	13.1769	1.90526
1.14	1.2996	1.06771	1.64	2.6896	1.28062	2.14	4.5796	1.46287	2.64	6.9696	1.62481	3.14	9.8596	1.77200	3.64	13.2496	1.90788
1.15	1.3225	1.07238	1.65	2.7225	1.28452	2.15	4.6225	1.46629	2.65	7.0225	1.62788	3.15	9.9225	1.77482	3.65	13.3225	1.91050
1.16	1.3456	1.07703	1.66	2.7556	1.28841	2.16	4.6656	1.46969	2.66	7.0756	1.63095	3.16	9.9856	1.77764	3.66	13.3956	1.91311
1.17	1.3689	1.08167	1.67	2.7889	1.29228	2.17	4.7089	1.47309	2.67	7.1289	1.63401	3.17	10.0489	1.78045	3.67	13.4689	1.91572
1.18	1.3924	1.08628	1.68	2.8224	1.29615	2.18	4.7524	1.47648	2.68	7.1824	1.63707	3.18	10.1124	1.78326	3.68	13.5424	1.91833
1.19	1.4161	1.09087	1.69	2.8561	1.30000	2.19	4.7961	1.47986	2.69	7.2361	1.64012	3.19	10.1761	1.78606	3.69	13.6161	1.92094
1.20	1.4400	1.09545	1.70	2.8900	1.30384	2.20	4.8400	1.48324	2.70	7.2900	1.64317	3.20	10.2400	1.78885	3.70	13.6900	1.92354
1.21	1.4641	1.10000	1.71	2.9241	1.30767	2.21	4.8841	1.48661	2.71	7.3441	1.64621	3.21	10.3041	1.79165	3.71	13.7641	1.92614
1.22	1.4884	1.10454	1.72	2.9584	1.31149	2.22	4.9284	1.48997	2.72	7.3984	1.64924	3.22	10.3684	1.79444	3.72	13.8384	1.92873
1.23	1.5129	1.10905	1.73	2.9929	1.31529	2.23	4.9729	1.49332	2.73	7.4529	1.65227	3.23	10.4329	1.79772	3.73	13.9129	1.93132
1.24	1.5376	1.11355	1.74	3.0276	1.31909	2.24	5.0176	1.49666	2.74	7.5076	1.65529	3.24	10.4976	1.80000	3.74	13.9876	1.93391
1.25	1.5625	1.11803	1.75	3.0625	1.32288	2.25	5.0625	1.50000	2.75	7.5625	1.65831	3.25	10.5625	1.80278	3.75	14.0625	1.93649
1.26	1.5876	1.12250	1.76	3.0976	1.32665	2.26	5.1076	1.50333	2.76	7.6176	1.66132	3.26	10.6276	1.80555	3.76	14.1376	1.93907
1.27	1.6129	1.12694	1.77	3.1329	1.33041	2.27	5.1529	1.50665	2.77	7.6729	1.66433	3.27	10.6929	1.80831	3.77	14.2129	1.94165
1.28	1.6384	1.13137	1.78	3.1684	1.33417	2.28	5.1984	1.50997	2.78	7.7284	1.66733	3.28	10.7584	1.81108	3.78	14.2884	1.94422
1.29	1.6641	1.13578	1.79	3.2041	1.33791	2.29	5.2441	1.51327	2.79	7.7841	1.67033	3.29	10.8241	1.81384	3.79	14.3641	1.94679
1.30	1.6900	1.14018	1.80	3.2400	1.34164	2.30	5.2900	1.51658	2.80	7.8400	1.67332	3.30	10.8900	1.81659	3.80	14.4400	1.94936
1.31	1.7161	1.14455	1.81	3.2761	1.34536	2.31	5.3361	1.51987	2.81	7.8961	1.67631	3.31	10.9561	1.81934	3.81	14.5161	1.95192
1.32	1.7424	1.14891	1.82	3.3124	1.34907	2.32	5.3824	1.52315	2.82	7.9524	1.67929	3.32	11.0224	1.82209	3.82	14.5924	1.95448
1.33	1.7689	1.15326	1.83	3.3489	1.35277	2.33	5.4289	1.52643	2.83	8.0089	1.68226	3.33	11.0889	1.82483	3.83	14.6689	1.95704
1.34	1.7956	1.15758	1.84	3.3856	1.35647	2.34	5.4756	1.52971	2.84	8.0656	1.68523	3.34	11.1556	1.82757	3.84	14.7456	1.95959
1.35	1.8225	1.16190	1.85	3.4225	1.36015	2.35	5.5225	1.53297	2.85	8.1225	1.68819	3.35	11.2225	1.83030	3.85	14.8225	1.96214
1.36	1.8496	1.16619	1.86	3.4596	1.36382	2.36	5.5696	1.53623	2.86	8.1796	1.69115	3.36	11.2896	1.83303	3.86	14.8996	1.96469
1.37	1.8769	1.17047	1.87	3.4969	1.36748	2.37	5.6169	1.53948	2.87	8.2369	1.69411	3.37	11.3569	1.83576	3.87	14.9769	1.96723
1.38	1.9044	1.17473	1.88	3.5344	1.37113	2.38	5.6644	1.54272	2.88	8.2944	1.69706	3.38	11.4244	1.83848	3.88	15.0544	1.96977
1.39	1.9321	1.17898	1.89	3.5721	1.37477	2.39	5.7121	1.54596	2.89	8.3521	1.70000	3.39	11.4921	1.84120	3.89	15.1321	1.97231
1.40	1.9600	1.18322	1.90	3.6100	1.37840	2.40	5.7600	1.54919	2.90	8.4100	1.70294	3.40	11.5600	1.84391	3.90	15.2100	1.97484
1.41	1.9881	1.18743	1.91	3.6481	1.38203	2.41	5.8081	1.55242	2.91	8.4681	1.70587	3.41	11.6281	1.84662	3.91	15.2881	1.97737
1.42	2.0164	1.19164	1.92	3.6864	1.38564	2.42	5.8564	1.55563	2.92	8.5264	1.70880	3.42	11.6964	1.84932	3.92	15.3664	1.97990
1.43	2.0449	1.19583	1.93	3.7249	1.38924	2.43	5.9049	1.55885	2.93	8.5849	1.71172	3.43	11.7649	1.85203	3.93	15.4449	1.98242
1.44	2.0736	1.20000	1.94	3.7636	1.39284	2.44	5.9536	1.56205	2.94	8.6436	1.71464	3.44	11.8336	1.85472	3.94	15.5236	1.98494
1.45	2.1025	1.20416	1.95	3.8025	1.39642	2.45	6.0025	1.56525	2.95	8.7025	1.71756	3.45	11.9025	1.85742	3.95	15.6025	1.98746
1.46	2.1316	1.20830	1.96	3.8416	1.40000	2.46	6.0516	1.56844	2.96	8.7616	1.72047	3.46	11.9716	1.86011	3.96	15.6816	1.98997
1.47	2.1609	1.21244	1.97	3.8809	1.40357	2.47	6.1009	1.57162	2.97	8.8209	1.72337	3.47	12.0409	1.86279	3.97	15.7609	1.99249
1.48	2.1904	1.21655	1.98	3.9204	1.40712	2.48	6.1504	1.57480	2.98	8.8804	1.72627	3.48	12.1104	1.86548	3.98	15.8404	1.99499
1.49	2.2201	1.22066	1.99	3.9601	1.41067	2.49	6.2001	1.57797	2.99	8.9401	1.72916	3.49	12.1801	1.86815	3.99	15.9201	1.99750
1.50	2.2500	1.22474	2.00	4.0000	1.41421	2.50	6.2500	1.58114	3.00	9.0000	1.73205	3.50	12.2500	1.87083	4.00	16.0000	2.00000
n	n^2	$\sqrt{n}$	n	n^2	$\sqrt{n}$	n	n^2	$\sqrt{n}$	n	n^2	$\sqrt{n}$	n	n^2	$\sqrt{n}$	n	n^2	$\sqrt{n}$

Table of squares and square roots (*Continued*)

n	n^2	$\sqrt{n}$	n	n^2	$\sqrt{n}$	n	n^2	$\sqrt{n}$	n	n^2	$\sqrt{n}$	n	n^2	$\sqrt{n}$	n	n^2	$\sqrt{n}$
4.00	16.0000	2.00000	4.50	20.2500	2.12132	5.00	25.0000	2.23607	5.50	30.2500	2.34521	6.00	36.0000	2.44949	6.50	42.2500	2.54951
4.01	16.0801	2.00250	4.51	20.3401	2.12368	5.01	25.1001	2.23830	5.51	30.3601	2.34734	6.01	36.1201	2.45153	6.51	42.3801	2.55147
4.02	16.1604	2.00499	4.52	20.4303	2.12603	5.02	25.2004	2.24054	5.52	30.4704	2.34947	6.02	36.2404	2.45357	6.52	42.5104	2.55343
4.03	16.2409	2.00749	4.53	20.5209	2.12838	5.03	25.3009	2.24277	5.53	30.5809	2.35160	6.03	36.3609	2.45561	6.53	42.6409	2.55539
4.04	16.3216	2.00998	4.54	20.6116	2.13073	5.04	25.4016	2.24499	5.54	30.6916	2.35372	6.04	36.4816	2.45764	6.54	42.7716	2.55734
4.05	16.4025	2.01246	4.55	20.7025	2.13307	5.05	25.5025	2.24722	5.55	30.8025	2.35584	6.05	36.6025	2.45967	6.55	42.9025	2.55930
4.06	16.4836	2.01494	4.56	20.7936	2.13542	5.06	25.6036	2.24944	5.56	30.9136	2.35797	6.06	36.7236	2.46171	6.56	43.0336	2.56125
4.07	16.5649	2.01742	4.57	20.8849	2.13776	5.07	25.7049	2.25167	5.57	31.0249	2.36008	6.07	36.8449	2.46374	6.57	43.1649	2.56320
4.08	16.6464	2.01990	4.58	20.9764	2.14009	5.08	25.8064	2.25389	5.58	31.1364	2.36220	6.08	36.9664	2.46577	6.58	43.2964	2.56515
4.09	16.7281	2.02237	4.59	21.0681	2.14243	5.09	25.9081	2.25610	5.59	31.2481	2.36432	6.09	37.0881	2.46779	6.59	43.4281	2.56710
4.10	16.8100	2.02485	4.60	21.1600	2.14476	5.10	26.0100	2.25832	5.60	31.3600	2.36643	6.10	37.2100	2.46982	6.60	43.5600	2.56905
4.11	16.8921	2.02731	4.61	21.2521	2.14709	5.11	26.1121	2.26053	5.61	31.4721	2.36854	6.11	37.3321	2.47184	6.61	43.6921	2.57099
4.12	16.9744	2.02978	4.62	21.3444	2.14942	5.12	26.2144	2.26274	5.62	31.5844	2.37065	6.12	37.4544	2.47386	6.62	43.8244	2.57294
4.13	17.0569	2.03224	4.63	21.4369	2.15174	5.13	26.3169	2.26495	5.63	31.6969	2.37276	6.13	37.5769	2.47588	6.63	43.9569	2.57488
4.14	17.1396	2.03470	4.64	21.5296	2.15407	5.14	26.4196	2.26716	5.64	31.8096	2.27487	6.14	37.6996	2.47790	6.64	44.0896	2.57682
4.15	17.2225	2.03715	4.65	21.6225	2.15639	5.15	26.5225	2.26936	5.65	31.9225	2.37697	6.15	37.8225	2.47992	6.65	44.2225	2.57876
4.16	17.3056	2.03961	4.66	21.7156	2.15870	5.16	26.6256	2.27156	5.66	32.0356	2.37908	6.16	37.9456	2.48193	6.66	44.3556	2.58070
4.17	17.3889	2.04206	4.67	21.8089	2.16102	5.17	26.7289	2.27376	5.67	32.1489	2.38118	6.17	38.0689	2.48395	6.67	44.4889	2.58263
4.18	17.4724	2.04450	4.68	21.9024	2.16333	5.18	26.8324	2.27596	5.68	32.2624	2.38328	6.18	38.1924	2.48596	6.68	44.6224	2.58457
4.19	17.5561	2.04695	4.69	21.9961	2.16564	5.19	26.9361	2.27816	5.69	32.3761	2.38537	6.19	38.3161	2.48797	6.69	44.7561	2.58650
4.20	17.6400	2.04939	4.70	22.0900	2.16795	5.20	27.0400	2.28035	5.70	32.4900	2.38747	6.20	38.4400	2.48998	6.70	44.8900	2.58844
4.21	17.7241	2.05183	4.71	22.1841	2.17025	5.21	27.1441	2.28254	5.71	32.6041	2.38956	6.21	38.5641	2.49199	6.71	45.0241	2.59037
4.22	17.8084	2.05426	4.72	22.2784	2.17256	5.22	27.2484	2.28473	5.72	32.7184	2.39165	6.22	38.6884	2.49399	6.72	45.1584	2.59230
4.23	17.8929	2.05670	4.73	22.3729	2.17486	5.23	27.3529	2.28692	5.73	32.8329	2.39374	6.23	38.8129	2.49600	6.73	45.2929	2.59422
4.24	17.9776	2.05913	4.74	22.4676	2.17715	5.24	27.4576	2.28910	5.74	32.9476	2.39583	6.24	38.9376	2.49800	6.74	45.4276	2.59615
4.25	18.0625	2.06155	4.75	22.5625	2.17945	5.25	27.5625	2.29129	5.75	33.0625	2.39792	6.25	39.0625	2.50000	6.75	45.5625	2.59808
4.26	18.1476	2.06398	4.76	22.6576	2.18174	5.26	27.6676	2.29347	5.76	33.1776	2.40000	6.26	39.1876	2.50200	6.76	45.6976	2.60000
4.27	18.2329	2.06640	4.77	22.7529	2.18403	5.27	27.7729	2.29565	5.77	33.2929	2.40208	6.27	39.3129	2.50400	6.77	45.8329	2.60192
4.28	18.3184	2.06882	4.78	22.8484	2.18632	5.28	27.8784	2.29783	5.78	33.4084	2.40416	6.28	39.4384	2.50599	6.78	45.9684	2.60384
4.29	18.4041	2.07123	4.79	22.9441	2.18861	5.29	27.9841	2.30000	5.79	33.5241	2.40624	6.29	39.5641	2.50799	6.79	46.1041	2.60576
4.30	18.4900	2.07364	4.80	23.0400	2.19089	5.30	28.0900	2.30217	5.80	33.6400	2.40832	6.30	39.6900	2.50998	6.80	46.2400	2.60768
4.31	18.5761	2.07605	4.81	23.1361	2.19317	5.31	28.1961	2.30434	5.81	33.7561	2.41039	6.31	39.8161	2.51197	6.81	46.3761	2.60960
4.32	18.6624	2.07846	4.82	23.2324	2.19545	5.32	28.3024	2.30651	5.82	33.8724	2.41247	6.32	39.9424	2.51396	6.82	46.5124	2.61151
4.33	18.7489	2.08087	4.83	23.3289	2.19773	5.33	28.4089	2.30868	5.83	33.9889	2.41454	6.33	40.0689	2.51595	6.83	46.6489	2.61343
4.34	18.8356	2.08327	4.84	23.4256	2.20000	5.34	28.5156	2.31084	5.84	34.1056	2.41661	6.34	40.1956	2.51794	6.84	46.7856	2.61534
4.35	18.9225	2.08567	4.85	23.5225	2.20227	5.35	28.6225	2.31301	5.85	34.2225	2.41868	6.35	40.3225	2.51992	6.85	46.9225	2.61725
4.36	19.0096	2.08806	4.86	23.6196	2.20454	5.36	28.7296	2.31517	5.86	34.3396	2.42074	6.36	40.4496	2.52190	6.86	47.0596	2.61916
4.37	19.0969	2.09045	4.87	23.7169	2.20681	5.37	28.8369	2.31733	5.87	34.4569	2.42281	6.37	40.5769	2.52389	6.87	47.1969	2.62107
4.38	19.1844	2.09284	4.88	23.8144	2.20907	5.38	28.9444	2.31948	5.88	34.5744	2.42487	6.38	40.7044	2.52587	6.88	47.3344	2.62298
4.39	19.2721	2.09523	4.89	23.9121	2.21133	5.39	29.0521	2.32164	5.89	34.6921	2.42693	6.39	40.8321	2.52784	6.89	47.4721	2.62488
4.40	19.3600	2.09762	4.90	24.0100	2.21359	5.40	29.1600	2.32379	5.90	34.8100	2.42899	6.40	40.9600	2.52982	6.90	47.6100	2.62679
4.41	19.4481	2.10000	4.91	24.1081	2.21585	5.41	29.2681	2.32594	5.91	34.9281	2.43105	6.41	41.0881	2.53180	6.91	47.7481	2.62869
4.42	19.5364	2.10238	4.92	24.2064	2.21811	5.42	29.3764	2.32809	5.92	35.0464	2.43311	6.42	41.2164	2.53377	6.92	47.8864	2.63059
4.43	19.6249	2.10476	4.93	24.3049	2.22036	5.43	29.4849	2.33024	5.93	35.1649	2.43516	6.43	41.3449	2.53574	6.93	48.0249	2.63249
4.44	19.7136	2.10713	4.94	24.4036	2.22261	5.44	29.5936	2.33238	5.94	35.2836	2.43721	6.44	41.4736	2.53772	6.94	48.1636	2.63439
4.45	19.8025	2.10950	4.95	24.5025	2.22486	5.45	29.7025	2.33452	5.95	35.4025	2.43926	6.45	41.6025	2.53969	6.95	48.3025	2.63629
4.46	19.8916	2.11187	4.96	24.6016	2.22711	5.46	29.8116	2.33666	5.96	34.5216	2.44131	6.46	41.7316	2.54165	6.96	48.4416	2.63818
4.47	19.9809	2.11424	4.97	24.7009	2.22935	5.47	29.9209	2.33880	5.97	35.6409	2.44336	6.47	41.8609	2.54362	6.97	48.5809	2.64008
4.48	20.0704	2.11660	4.98	24.8004	2.23159	5.48	30.0304	2.34094	5.98	35.7604	2.44540	6.48	41.9904	2.54558	6.98	48.7204	2.64197
4.49	20.1601	2.11896	4.99	24.9001	2.23383	5.49	30.1401	2.34307	5.99	35.8801	2.44745	6.49	42.1201	2.54755	6.99	48.8601	2.64386
4.50	20.2500	2.12132	5.00	25.0000	2.23607	5.50	30.2500	2.34521	6.00	36.0000	2.44949	6.50	42.2500	2.54951	7.00	49.0000	2.64575
n	n^2	$\sqrt{n}$	n	n^2	$\sqrt{n}$	n	n^2	$\sqrt{n}$	n	n^2	$\sqrt{n}$	n	n^2	$\sqrt{n}$	n	n^2	$\sqrt{n}$

Table of squares and square roots (*Continued*)

n	n^2	$\sqrt{n}$	n	n^2	$\sqrt{n}$	n	n^2	$\sqrt{n}$	n	n^2	$\sqrt{n}$	n	n^2	$\sqrt{n}$	n	n^2	$\sqrt{n}$
7.00	49.0000	2.64575	7.50	56.2500	2.73861	8.00	64.0000	2.82843	8.50	72.2500	2.91548	9.00	81.0000	3.00000	9.50	90.2500	3.08221
7.01	49.1401	2.64764	7.51	56.4001	2.74044	8.01	64.1601	2.83019	8.51	72.4201	2.91719	9.01	81.1801	3.00167	9.51	90.4401	3.08383
7.02	49.2804	2.64953	7.52	56.5504	2.74226	8.02	64.3204	2.83196	8.52	72.5904	2.91890	9.02	81.3604	3.00333	9.52	90.6304	3.08545
7.03	49.4209	2.65141	7.53	56.7009	2.74408	8.03	64.4809	2.83373	8.53	72.7609	2.92062	9.03	81.5409	3.00500	9.53	90.8209	3.08707
7.04	49.5616	2.65330	7.54	56.8516	2.74591	8.04	64.6416	2.83549	8.54	72.9316	2.92233	9.04	81.7216	3.00666	9.54	91.0116	3.08869
7.05	49.7025	2.65518	7.55	57.0025	2.74773	8.05	64.8025	2.83725	8.55	73.1025	2.92404	9.05	81.9025	3.00832	9.55	91.2025	3.09031
7.06	49.8436	2.65707	7.56	57.1536	2.74955	8.06	64.9636	2.83901	8.56	73.2736	2.92575	9.06	82.0836	3.00998	9.56	91.3936	3.09192
7.07	49.9849	2.65895	7.57	57.3049	2.75136	8.07	65.1249	2.84077	8.57	73.4449	2.92746	9.07	82.2649	3.01164	9.57	91.5849	3.09354
7.08	50.1264	2.66083	7.58	57.4564	2.75318	8.08	65.2864	2.84253	8.58	73.6164	2.92916	9.08	82.4464	3.01330	9.58	91.7764	3.09516
7.09	50.2681	2.66271	7.59	57.6081	2.75500	8.09	65.4481	2.84429	8.59	73.7881	2.93087	9.09	82.6281	3.01496	9.59	91.9681	3.09677
7.10	50.4100	2.66458	7.60	57.7600	2.75681	8.10	65.6100	2.84605	8.60	73.9600	2.93258	9.10	82.8100	3.01662	9.60	92.1600	3.09839
7.11	50.5521	2.66646	7.61	57.9121	2.75862	8.11	65.7721	2.84781	8.61	74.1321	2.93428	9.11	82.9921	3.01828	9.61	92.3521	3.10000
7.12	50.6944	2.66833	7.62	58.0644	2.76043	8.12	65.9344	2.84956	8.62	75.3044	2.93598	9.12	83.1744	3.01993	9.62	92.5444	3.10161
7.13	50.8369	2.67021	7.63	58.2169	2.76225	8.13	66.0969	2.85132	8.63	74.4769	2.93769	9.13	83.3569	3.02159	9.63	92.7369	3.10322
7.14	50.9796	2.67208	7.64	58.3696	2.76405	8.14	66.2596	2.85307	8.64	74.6496	2.93939	9.14	83.5396	3.02324	9.64	92.9296	3.10483
7.15	51.1225	2.67395	7.65	58.5225	2.76586	8.15	66.4225	2.85482	8.65	74.8225	2.94109	9.15	83.7225	3.02490	9.65	93.1225	3.10644
7.16	51.2656	2.67582	7.66	58.6756	2.76767	8.16	66.5856	2.85657	8.66	74.9956	2.94279	9.16	83.9056	3.02655	9.66	93.3156	3.10805
7.17	51.4089	2.67769	7.67	58.8289	2.76948	8.17	66.7489	2.85832	8.67	75.1689	2.94449	9.17	84.0889	3.02820	9.67	93.5089	3.10966
7.18	51.5524	2.67955	7.68	58.9824	2.77128	8.18	66.9124	2.86007	8.68	75.3424	2.94618	9.18	84.2724	3.02985	9.68	93.7024	3.11127
7.19	51.6961	2.68142	7.69	59.1361	2.77308	8.19	67.0761	2.86182	8.69	75.5161	2.94788	9.19	84.4561	3.03150	9.69	93.8961	3.11288
7.20	51.8400	2.68328	7.70	59.2900	2.77489	8.20	67.2400	2.86356	8.70	75.6900	2.94958	9.20	84.6400	3.03315	9.70	94.0900	3.11448
7.21	51.9841	2.68514	7.71	59.4441	2.77669	8.21	67.4041	2.86531	8.71	75.8641	2.95127	9.21	84.8241	3.03480	9.71	94.2841	3.11609
7.22	52.1284	2.68701	7.72	59.5984	2.77849	8.22	67.5684	2.86705	8.72	76.0384	2.95296	9.22	85.0084	3.03645	9.72	94.4784	3.11769
7.23	52.2729	2.68887	7.73	59.7529	2.78029	8.23	67.7329	2.86880	8.73	76.2129	2.95466	9.23	85.1929	3.03809	9.73	94.6729	3.11929
7.24	52.4176	2.69072	7.74	59.9076	2.78209	8.24	67.8976	2.87054	8.74	76.3876	2.95635	9.24	85.3776	3.03974	9.74	94.8676	3.12090
7.25	52.5625	2.69258	7.75	60.0625	2.78388	8.25	68.0625	2.87228	8.75	76.5625	2.95804	9.25	85.5625	3.04138	9.75	95.0625	3.12250
7.26	52.7076	2.69444	7.76	60.2176	2.78568	8.26	68.2276	2.87402	8.76	76.7376	2.95973	9.26	85.7476	3.04302	9.76	95.2576	3.12410
7.27	52.8529	2.69629	7.77	60.3729	2.78747	8.27	68.3929	2.87576	8.77	76.9129	2.96142	9.27	85.9329	3.04467	9.77	95.4529	3.12570
7.28	52.9984	2.69815	7.78	60.5284	2.78927	8.28	68.5584	2.87750	8.78	77.0884	2.96311	9.28	86.1184	3.04631	9.78	95.6484	3.12730
7.29	53.1441	2.70000	7.79	60.6841	2.79106	8.29	68.7241	2.87924	8.79	77.2641	2.96479	9.29	86.3041	3.04795	9.79	95.8441	3.12890
7.30	53.2900	2.70185	7.80	60.8400	2.79285	8.30	68.8900	2.88097	8.80	77.4400	2.96648	9.30	86.4900	3.04959	9.80	96.0400	3.13050
7.31	53.4361	2.70370	7.81	60.9961	2.79464	8.31	69.0561	2.88271	8.81	77.6161	2.96816	9.31	86.6761	3.05123	9.81	96.2361	3.13209
7.32	53.5824	2.70555	7.82	61.1524	2.79643	8.32	69.2224	2.88444	8.82	77.7924	2.96985	9.32	86.8624	3.05287	9.82	96.4324	3.13369
7.33	53.7289	2.70740	7.83	61.3089	2.79821	8.33	69.3889	2.88617	8.83	77.9689	2.97153	9.33	87.0489	3.05450	9.83	96.6289	3.13528
7.34	53.8756	2.70924	7.84	61.4656	2.80000	8.34	69.5556	2.88791	8.84	78.1456	2.97321	9.34	87.2356	3.05614	9.84	96.8256	3.13688
7.35	54.0225	2.71109	7.85	61.6225	2.80179	8.35	69.7225	2.88964	8.85	78.3225	2.97489	9.35	87.4225	3.05778	9.85	97.0225	3.13847
7.36	54.1696	2.71293	7.86	61.7796	2.80357	8.36	69.8896	2.89137	8.86	78.4996	2.97658	9.36	87.6096	3.05941	9.86	97.2196	3.14006
7.37	54.3169	2.71477	7.87	61.9369	2.80535	8.37	70.0569	2.89310	8.87	78.6769	2.97825	9.37	87.7969	3.06105	9.87	97.4169	3.14166
7.38	54.4644	2.71662	7.88	62.0944	2.80713	8.38	70.2244	2.89482	8.88	78.8544	2.97993	9.38	87.9844	3.06268	9.88	97.6144	3.14325
7.39	54.6121	2.71846	7.89	62.2521	2.80891	8.39	70.3921	2.89655	8.89	79.0321	2.98161	9.39	88.1721	3.06431	9.89	97.8121	3.14484
7.40	54.7600	2.72029	7.90	62.4100	2.81069	8.40	70.5600	2.89828	8.90	79.2100	2.98329	9.40	88.3600	3.06594	9.90	98.0100	3.14643
7.41	54.9081	2.72213	7.91	62.5681	2.81247	8.41	70.7281	2.90000	8.91	79.3881	2.98496	9.41	88.5481	3.06757	9.91	98.2081	3.14802
7.42	55.0564	2.72397	7.92	62.7264	2.81425	8.42	70.8964	2.90172	8.92	79.5664	2.98664	9.42	88.7364	3.06920	9.92	98.4064	3.14960
7.43	55.2049	2.72580	7.93	62.8849	2.81603	8.43	71.0649	2.90345	8.93	79.7449	2.98831	9.43	88.9249	3.07083	9.93	98.6049	3.15119
7.44	55.3536	2.72764	7.94	63.0436	2.81780	8.44	71.2336	2.90517	8.94	79.9236	2.98998	9.44	89.1136	3.07246	9.94	98.8036	3.15278
7.45	55.5025	2.72947	7.95	63.2025	2.81957	8.45	71.4025	2.90689	8.95	80.1025	2.99166	9.45	89.3025	3.07409	9.95	99.0025	3.15436
7.46	55.6516	2.73130	7.96	63.3616	2.82135	8.46	71.5716	2.90861	8.96	80.2816	2.99333	9.46	89.4916	3.07571	9.96	99.2016	3.15595
7.47	55.8009	2.73313	7.97	63.5209	2.82312	8.47	71.7409	2.91033	8.97	80.4609	2.99500	9.47	89.6809	3.07734	9.97	99.4009	3.15753
7.48	55.9504	2.73496	7.98	63.6804	2.82489	8.48	71.9104	2.91204	8.98	80.6404	2.99666	9.48	89.8704	3.07896	9.98	99.6004	3.15911
7.49	56.1001	2.73679	7.99	63.8401	2.82666	8.49	72.0801	2.91376	8.99	80.8201	2.99833	9.49	90.0601	3.08058	9.99	99.8001	3.16070
7.50	56.2500	2.73861	8.00	64.0000	2.82843	8.50	72.2500	2.91548	9.00	81.0000	3.00000	9.50	90.2500	3.08221	10.00	100.0000	3.16228
n	n^2	$\sqrt{n}$	n	n^2	$\sqrt{n}$	n	n^2	$\sqrt{n}$	n	n^2	$\sqrt{n}$	n	n^2	$\sqrt{n}$	n	n^2	$\sqrt{n}$

APPENDIX C Glossary

abscissa The x coordinate of a point in a Cartesian (rectangular) coordinate system.

absolute value In reference to a number, the positive number that describes the distance on a number line of the graph of the number from the origin. The absolute value of zero is zero.

abstract fraction A fraction that contains one or more variables.

acute angle An angle whose degree measure is between 0° and 90°.

acute triangle A triangle in which all the angles are acute.

addend A number that is to be added to one or more other numbers to form a sum.

addition The operation of combining two numbers to form a sum.

additive inverse In reference to a given number, the opposite of the number. For example, the additive inverse of –2 is 2, and the additive inverse of 2 is –2.

additive property of inequality A property of real numbers such that, for any real numbers a, b, and c, if $a > b$, then $a + c > b + c$ and also $c + a > c + b$.

algebraic addition A thought process that can be used to combine positive and negative numbers. For example, the expression $6 - 4$ is considered to indicate the addition of +6 and –4 as $(6) + (-4)$.

algebraic expression An expression that contains a meaningful arrangement of digits and/or variables.

algebraic phrase A meaningful arrangement of numbers and variables.

algebraic proof Use of definitions, axioms, and deductive reasoning to prove algebraic assertions.

altitude In reference to a triangle, the perpendicular distance from one side of the triangle to the opposite vertex.

angle The geometric figure formed by two rays that have a common endpoint; also the measure of the rotation of a ray about its endpoint.

area The number that tells how many squares of a certain size are needed to cover the surface of a figure.

average Of a group of numbers, the sum of the numbers divided by the number of numbers in the group.

axiom A statement that is assumed to be true without proof.

basic arithmetic operations Addition, subtraction, multiplication, and division.

binomial A polynomial of two terms.

Cartesian coordinates A standard method of locating points in the plane by pairs of numbers denoting distances along two fixed intersecting number lines, called the *axes*. The axes are perpendicular to each other and intersect at the origin of both axes. The system is named for the famous French mathematician René Descartes.

centimeter Metric unit of measurement: 1 centimeter = 10 millimeters; 100 centimeters = 1 meter.

circle A planar geometric figure in which every point on the figure is the same distance from a point called the *center* of the circle.

circumference The distance around a circle. Also called the *perimeter* of the circle.

closure A property of some sets of numbers. If every pair of numbers that can be related from a set is used in an operation and the result is always a member of the set, the set is said to be *closed* under that operation.

coefficient Any factor or any product of factors of a product.

commutative property A property of real numbers which notes that the order in which two real numbers are added or multiplied does not affect the sum or product, respectively.

complex fraction A fraction whose numerator or denominator (or both) contains a fraction.

concave polygon A polygon in which at least one interior angle has a measure greater than 180°.

conditional equation An equation whose truth or falsity depends on the numbers used to replace the variables in the equation.

conjunction A statement of two conditions which must both be true in order for the statement to be true.

consecutive integers Integers that are 1 unit apart.

constant A quantity whose value does not change.

constant of proportionality A constant in an equation that defines the relationship of two or more variables. For example, in the equation $y = 4m^2x$, the number 4 is the constant of proportionality.

convex polygon A polygon in which no interior angle has a measure greater than 180°.

coordinate A number that is associated with a point on a graph.

coordinate plane A plane with a coordinate system that can be used to designate the position of any point in the plane.

counting numbers *See* natural numbers.

cubic measure A unit or system of units for measuring volume.

curve The path "traced" by a moving point.

decimal number A number designated by a linear arrangement of one or more of the 10 digits and that uses a decimal point to define the place value of the digits.

decimal system The system of numeration that uses decimal numbers.

deductive reasoning A method of reasoning from a premise to a conclusion.

degree A unit of measure for angles. A right angle is a 90° angle and a straight angle is a 180° angle.

degree of a polynomial The degree of the highest-degree term in the polynomial, calculated as follows: The degree of a term in a polynomial is the sum of the exponents in the term. For example, the terms x^5, y^2x^3, and xy^2mp are all fifth-degree terms.

denominator The number under the fraction bar in a fraction.

dependent equations Equations whose solution sets are equal.

dependent variable A variable whose value depends on the value assigned to another variable, called the *independent variable.*

diameter The length of a chord of a circle that passes through the center of the circle.

digit Any of the 10 symbols of the decimal system: 0, 1, 2, 3, 4, 5, 6, 7, 8, 9.

dimension A measure of spatial extent, especially length, height, and width.

direct variation A relationship between two variables in which the value of one variable is the product of a constant and the value of the other variable. For example, the equation $y = 4x$ defines a direct variation between x and y.

disjunction A statement of two conditions of which only one condition must be true in order for the statement to be true.

distributive property A property of real numbers that notes that, for any real numbers a, b, and c, $a(b + c) = ab + ac$.

dividend The number that is to be divided.

division The inverse operation of multiplication. If one number is divided by another number, the result is called the *quotient.*

divisor The quantity by which another quantity, the dividend, is to be divided.

domain The set of numbers which are permissible replacement values for the independent variable in a particular equation or inequality.

elements The individual objects or members that make up a set.

empty set The set which has no members, denoted by the symbol Ø or the notation { }. Also called the *null set.*

equality The state of being equal.

equals sign Symbol of equality (=).

equation An algebraic statement consisting of two algebraic expressions connected by an equals sign.

equiangular In reference to a geometric figure, used if all the angles of the figure have the same measure.

equilateral triangle A triangle whose three sides all have equal length. All the angles in an equilateral triangle have a measure of 60°.

equivalent equations Equations that have the same solution set.

even integer Any member of the set $\{\ldots, -4, -2, 0, 2, 4, \ldots\}$.

event In statistics, an outcome of an experiment.

exponent The number that indicates the number of times the base of a power is to be used as a factor.

exponential function A function of the form $y = kb^x$, where k is constant and b is a nonunity constant greater than zero.

exponential increase In reference to an exponential function $y = kb^x$, where b is greater than 1; if the domain of x is the set of positive numbers, the value of the function will increase exponentially as x increases.

factor Any one of several quantities that are multiplied to form a product.

fraction A quotient indicated by writing two numbers vertically and separating them with a short line segment called the *fraction bar.*

function A mapping that pairs each member of a set called the *domain* with exactly one member of another set called the *range*.

functional notation The use of letters and parentheses to indicate a functional relationship. For example, $f(x) = x^2 + 2x + 2$.

geometric figure A figure made up of straight lines or curved lines or both.

geometric solid A geometric figure that has three dimensions.

graph The mark(s) made on a coordinate system that indicates the location of a point or a set of points.

graphical solution A solution that is displayed pictorially on a coordinate system.

greatest common factor Of two or more terms, the product of all prime factors common to every term, each to the highest power that it occurs in any of the terms.

horizontal line A line that is parallel to the plane of the horizon.

hypotenuse The side opposite the right angle in a right triangle.

image In a function, the element of the range that is paired with a particular element of the domain.

improper fraction A fraction whose numerator is equal to or greater than the denominator.

improper subset Designation of a set, defined as follows: If two sets have the same members, each set is an improper subset of the other set.

inconsistent equations Equations that have no common solution.

independent equations Equations such that no one of them is necessarily satisfied by the set of values of the variables that satisfy all the others.

independent events In statistics, events such that the outcome of one event does not affect the probability of the occurrence of another event.

independent variable A variable in an equation whose value can be chosen.

index In a radical expression, the number that indicates what root is to be taken.

inequality A mathematical statement comparing quantities that are not equal.

integers All numbers in the set $\{\ldots, -4, -3, -2, -1, 0, 1, 2, 3, 4, \ldots\}$.

intercept In reference to a graph on a coordinate plane, the y intercept is the point where the graph crosses the y axis and the x intercept is the point where the graph crosses the x axis.

inverse operation An operation which "undoes" another operation. For example, addition and subtraction are inverse operations. Also, multiplication and division are inverse operations.

inverse variation A comparison of two quantities (they are said to *vary inversely*) such that their product is a constant. For example, the equation $xy = 10$ expresses an inverse relationship between x and y.

irrational numbers A number that cannot be expressed as a quotient of integers. For example, the numbers π, $\sqrt{2}$, $\sqrt[4]{17}$, and e are irrational numbers.

isosceles triangle A triangle that has at least two sides of equal length.

lateral surface area The total area of the "sides" of a geometric solid.

lead coefficient Of a polynomial, the constant factor of the first term of the polynomial when the polynomial is arranged in descending powers of the variable.

least common denominator The least common multiple of the denominators of a set of fractions.

least common multiple The smallest number that can be divided by each of a group of specified numbers and have no remainder.

length The distance between two designated points in a geometric figure.

less than Designation in a comparison of two numbers, stated as follows: One number is less than another number if its graph is farther to the left on a number line than the graph of the other number.

like terms Terms that have the same variables in the same form or in equivalent forms so that the terms (excluding numerical coefficients) represent the same number regardless of the nonzero values assigned the variables.

line In mathematics, a straight curve that has no width and no end.

line segment A part of a line that consists of two points and all points between them.

linear equation A first-degree polynomial equation in one or more variables.

linear inequality A first-degree polynomial inequality in one or more variables.

literal coefficient A coefficient containing only letters.

literal factor A factor that is a letter.

members of a set Elements of a set.

meter Metric unit of measurement; 100 centimeters = 1 meter.

millimeter Metric unit of measurement; 10 millimeters = 1 centimeter.

minuend The number that is subtracted from in a subtraction problem.

mixed number The sum of a whole number and a fraction, written without a plus sign (e.g., $2\frac{1}{3}$).

monomial A polynomial that has only one term.

multiplication Operation of repeated addition indicated by the times sign × or the center dot · (e.g., $4 \times 3 = 4 + 4 + 4 = 12$).

multiplicative inverse The reciprocal of a number.

multiplicative property of equality A property of real numbers stated as follows: If every term on both sides of an equation is multiplied or divided by the same nonzero quantity, the resulting equation will be an equivalent equation to the original equation, and thus every solution of either equation will be a solution of the other equation.

multivariable equation An equation that contains more than one variable.

natural numbers The set of numbers that we use to count objects or things; also called the *positive integers*.

negated inequality An inequality that has been negated by a slash through the symbol (e.g., $x \not> 10$, read as "x is not greater than 10").

negative exponent An exponent preceded by a minus sign, defined as follows: If n is any real number and x is any real number that is not zero, then $1/x^n = x^{-n}$.

null set The set which has no members, denoted by the symbol Ø or the notation { }. Also called the *empty set*.

number An idea that is designated by a numeral.

number line A line divided into units of equal length with one point chosen as the origin, or zero point. The numbers to the right of zero are the positive real numbers, and those to the left of zero are the negative real numbers.

numeral A single symbol or a collection of symbols that is used to express the idea of a particular number.

numerator The expression above the fraction bar in a fraction.

numerical coefficient A coefficient that is a number.

numerical expression A meaningful arrangement of numerals that has a single value.

numerical factor A factor that is a number.

obtuse angle An angle whose measure is between 90° and 180°.

odd integer Any member of the set $\{\ldots, -3, -1, 1, 3, \ldots\}$.

operation The process of carrying out a rule or procedure such as adding, subtracting, or taking a root of.

opposites Two numbers with the same absolute value but with different signs.

order of operations The order in which operations are performed on an algebraic expression.

ordered pair A pair of numbers in a designated order that are enclosed in parentheses. For example, if the notation (5, 4) is an ordered pair of x and y, the value of x is 5 and the value of y is 4.

ordinate The y coordinate of a point in rectangular coordinates.

origin A beginning point. On a number line or coordinate plane, the number zero is associated with the origin.

outcome A possible result in a probability problem.

overall average Of a group of numbers, the sum of all the numbers divided by the number of numbers.

parallel lines Co-planar lines that do not intersect.

parallelogram A quadrilateral that has two pairs of parallel sides.

percent One part in 100. For example, 60 percent means sixty-hundredths.

perimeter The measure around the outside of a planar geometric figure.

perpendicular At right angles; two lines are perpendicular if their intersection forms "square corners."

pi (π) The ratio of the circumference of a circle to the diameter of that circle; $\pi \approx 3.14$.

planar geometric figure Any figure that is drawn on a flat surface (i.e., any figure that has two dimensions).

point of intersection In reference to two lines, the point where the lines cross.

polygon Any closed, planar geometric figure whose sides are straight lines.

polynomial An algebraic expression with one or more variables having only terms with real number coefficients and whole number powers of the variables.

positive real number Any number that can be used to describe a physical distance greater than zero.

postulate A statement that is assumed true without proof; axiom.

power rule for exponents A rule for exponents: If m, n, and x are real numbers and $x \neq 0$, then $(x^m)^n = x^{mn}$.

prime factor A factor that is a prime number.

prime number A natural number greater than 1 whose only integer factors are 1 and the number itself.

primitive term A basic mathematical term that cannot be defined exactly. The term is defined as best as is possible and then used to define other terms.

product of square roots rule A rule for evaluating products of radical expressions: If m and n are nonnegative real numbers, then $\sqrt{m}\sqrt{n} = \sqrt{mn}$ and $\sqrt{mn} = \sqrt{m}\sqrt{n}$.

proportion The equality of two ratios.

quadratic equation A polynomial equation in which the highest power of the variable is 2. General form: $ax^2 + bx + c = 0$.

quadratic trinomial A trinomial in which the highest power of the variable is 2.

quotient The answer obtained when one number is divided by another number.

quotient rule for exponents A rule for exponents: If m and n are real numbers and $x \neq 0$, then $x^m/x^n = x^{m-n} = 1/x^{n-m}$.

quotient rule for square roots A rule for square roots: The square root of a quotient (fraction) equals the quotient of the square roots; for example, $\sqrt{3/2} = \sqrt{3}/\sqrt{2}$.

rational expression An algebraic expression that is written in fractional form.

rational equation An equation in which at least one term is a rational expression.

real numbers Zero is a real number. Any number that can be used to describe a distance greater than zero is a real number. The negatives of these numbers are real numbers.

reciprocal Of two fractions, one is the inverted form of the other (e.g., $\frac{4}{3}$ is the reciprocal of $\frac{3}{4}$ and $\frac{3}{4}$ is the reciprocal of $\frac{4}{3}$).

relation A pairing that matches each element of the domain with one or more images in the range.

rhombus An equilateral parallelogram.

right angle An angle that measures 90°.

scientific notation A method for expressing numbers whereby a number between 1 and 10 is multiplied by a power of 10 (e.g., 7×10^{-7}).

set notation The method of designating a set by enclosing the numbers of the set within braces.

simplify To break down into the simplest, most easily understood form.

square root Of a number x, the number that, when multiplied by itself, equals x.

standard form In reference to a polynomial equation, one in which the terms are in descending powers of the variable with all nonzero terms to the left of the equals sign.

subscripted variable A variable with a subscript, i.e., a little letter, set slightly below and to the right of the variable (e.g., N_d).

subtraction The arithmetic operation of reducing the value of an expression by a designated amount.

trinomial A polynomial of three terms.

unit multiplier A fraction that has units and has a value of 1, used to change the units of a number.

variable A quantity that may take on any one of a designated set of values; also, the symbol, usually a letter, representing that quantity.

vertex In reference to a polygon, a "corner" of the polygon.

weighted average The sum of the products of values and their weights, divided by the sum of the weights.

whole numbers All numbers in the set: $\{1, 2, 3, \ldots\}$

zero exponent An exponent of zero, which is evaluated as follows: If x is any real number that is not zero, then $x^0 = 1$.

zero factor theorem A theorem involving a factor or factors of zero, stated as follows: If p and q are any real numbers and if $pq = 0$, then either $p = 0$ or $q = 0$, or both p and $q = 0$.

Answers

problem set A

1. $\frac{3}{5}$ **2.** $\frac{1}{8}$ **3.** $3\frac{2}{3}$ **4.** $\frac{8}{15}$ **5.** $\frac{7}{40}$ **6.** $\frac{13}{24}$ **7.** $\frac{18}{65}$ **8.** $\frac{7}{15}$

9. $\frac{43}{45}$ **10.** $\frac{11}{17}$ **11.** $\frac{11}{26}$ **12.** $\frac{6}{35}$ **13.** $1\frac{11}{56}$ **14.** $\frac{17}{20}$ **15.** $\frac{21}{22}$ **16.** $5\frac{7}{10}$

17. $13\frac{17}{24}$ **18.** $8\frac{21}{40}$ **19.** $7\frac{8}{15}$ **20.** $20\frac{5}{8}$ **21.** $8\frac{14}{15}$ **22.** $402\frac{25}{33}$

23. $63\frac{7}{10}$ **24.** $36\frac{47}{104}$ **25.** $5\frac{43}{65}$ **26.** $13\frac{77}{190}$ **27.** $21\frac{47}{170}$ **28.** $12\frac{23}{56}$

29. $17\frac{74}{77}$ **30.** $7\frac{1}{11}$

practice

a. $(16 + 2\pi)$ cm $\approx$ 22.28 cm **b.** $(22 + 2\pi)$ cm^2 $\approx$ 28.28 cm^2 **c.** 7 m

d. 2.23 in. **e.** 26 ft **f.** 40 m^2 **g.** 100 ft^2

problem set B

1. $\frac{31}{35}$ **2.** $6\frac{3}{4}$ **3.** $2\frac{11}{15}$ **4.** $13\frac{1}{2}$ **5.** $12\frac{11}{15}$ **6.** $17\frac{4}{15}$ **7.** $8\frac{59}{70}$

8. $27\frac{5}{8}$ **9.** $7\frac{13}{15}$ **10.** $408\frac{14}{33}$ **11.** $3\frac{1}{15}$ **12.** $36\frac{54}{65}$ **13.** $7\frac{8}{35}$

14. 12.56 cm^2 **15.** 6 in. **16.** 6.69 m **17.** 40 cm **18.** 44 cm

19. $\left(36 + \frac{9\pi}{2}\right)$ m^2 $\approx$ 50.13 m^2 **20.** 75 m^2 **21.** $(192 - 4\pi)$ cm^2 $\approx$ 179.44 cm^2

22. 1 **23.** $2\frac{3}{4}$ **24.** $5\frac{1}{6}$ **25.** $108\frac{5}{8}$ **26.** $21\frac{4}{35}$ **27.** $78\frac{17}{35}$ **28.** $6\frac{8}{15}$

29. $6\frac{15}{22}$ **30.** $84\frac{8}{15}$

practice

a. $(40 + 2\pi)$ ft^2 $\approx$ 46.28 ft^2 **b.** $(400 + 20\pi)$ ft^3 $\approx$ 406.28 ft^3

c. An *equilateral* triangle has three sides whose lengths are equal.

d. An *isosceles* triangle has two sides of equal length.

e. A quadrilateral with exactly two parallel sides is called a *trapezoid.*

f. Straight angle, 180°; right angle, 90°

problem set C

1. $9\frac{2}{35}$ **2.** $6\frac{31}{33}$ **3.** $13\frac{31}{56}$ **4.** $1\frac{17}{88}$ **5.** $2\frac{1}{40}$ **6.** $12\frac{1}{33}$ **7.** $24\frac{19}{40}$

8. $88\frac{22}{65}$ **9.** $3\frac{87}{104}$ **10.** 785 ft^3 **11.** 280 in.^{3}

12. A *rectangle* is a parallelogram with four right angles. A *square* is a rhombus with four right angles. Yes.

13. $26\frac{11}{15}$ cm **14.** 3.18 cm **15.** 60 ft **16.** 224 miles **17.** 2240 miles **18.** 92 m^2

19. $(16 + 2\pi)\ m^2 \approx 22.28\ m^2$ **20.** $(60 - 9\pi)\ km^2 \approx 88.26\ km^2$ **21.** $\frac{23}{24}$ **22.** $22\frac{1}{2}$

23. $26\frac{9}{40}$ **24.** $\frac{155}{156}$ **25.** $9\frac{35}{88}$ **26.** $91\frac{29}{60}$ **27.** $2\frac{86}{143}$ **28.** $2\frac{11}{40}$ **29.** $28\frac{10}{21}$ **30.** 395.942

practice

a. 760.939 **b.** 724.74 **c.** 302.061 **d.** 125.1875 **e.** 44

f. 75(12)(2.54) cm **g.** $\frac{450}{12(5280)}$ miles

problem set 1

1. A number is an idea. A numeral is a symbol or a group of symbols used to express a number.

2. The decimal system **3.** The Hindus of India **4.** 0, 1, 2, 3, 4, 5, 6, 7, 8, 9

5. 1, 2, 3, . . . **6.** 1, 2, 3, . . .

7. Those numbers which can describe a physical distance greater than zero

8. The origin **9.** (a) 8π cm $\approx$ 25.12 cm (b) $16\pi\ cm^2 \approx 50.24\ cm^2$

10. 8 in. **11.** $\frac{314}{\pi}$ cm $\approx$ 100 cm **12.** $\frac{157}{\pi}$ ft $\approx$ 50 ft **13.** 48 in.

14. $(35 - 4\pi)\ cm^2 \approx 22.44\ cm^2$ **15.** $(32 + 2\pi)\ in.^2 \approx 38.28\ in.^2$ **16.** 30

17. $\frac{27}{40}$ **18.** $4\frac{20}{39}$ **19.** 20 **20.** $\frac{12}{35}$ **21.** $36\frac{9}{128}$ **22.** $3\frac{3}{35}$ **23.** $8\frac{2}{5}$

24. $22\frac{25}{32}$ **25.** $132\frac{5}{6}$ **26.** $28\frac{7}{15}$ **27.** $12\frac{7}{45}$ **28.** $5\frac{8}{15}$

29. 800(12)(2.54) cm **30.** 4260 cm^3

practice

a. {. . . , −3, −2, −1, 0, 1, 2, 3, . . .} **b.** {0, 1, 2, 3, 4, . . .} **c.** {1, 2, 3, 4, . . .}

d. 52 ft^2 **e.** $8800\pi\ cm^2 \approx 27{,}632\ cm^2$ **f.** $(240 + 50\pi)\ cm^2 \approx 397\ cm^2$

problem set 2

1. The surface area of the solid **2.** 75.45 **3.** 83.317 **4.** {1, 2, 3, 4, . . .}

5. 90° **6.** Members of the set

7. The numbers are addends; the answer is the sum.

8. Any number that can be used to describe a distance greater than zero **9.** 180°

10. If its graph is farther to the right on the number line

11. 4000(12)(2.54) cm **12.** The quotient

13. $4 + 0 = 4$ The sum of zero and any number is the number itself.
$0 \times 4 = 0$ The product of zero and any number is zero.

14. (a) $\frac{20}{\pi}$ cm $\approx$ 6.37 cm (b) $\frac{40}{\pi}$ cm $\approx$ 12.74 cm **15.** $(14 + 3\pi)$ ft $\approx$ 23.42 ft

16. $\left(60 + \frac{9}{2}\pi\right)\ cm^2 \approx 74.13\ cm^2$ **17.** $(16\pi - 8)\ in.^2 \approx 42.24\ in.^2$

18. (42.6)(2.54) cm **19.** (5.6)(5280)(12) in. **20.** $648\pi\ ft^3 \approx 200.96\ ft^3$

21. 36 **22.** $\frac{27}{46}$ **23.** $1\frac{109}{176}$ **24.** $32\frac{7}{8}$ **25.** $90\frac{5}{16}$ **26.** $86\frac{9}{10}$ **27.** $3\frac{1}{6}$

28. 0.19642 ft^2 **29.** $13\frac{13}{56}$ **30.** $32\frac{53}{56}$ **31.** $7\frac{7}{40}$ **32.** 208 m^2 **33.** 108 m^2

practice

a. 4 **b.** 4.2 **c.** −4

d. The sum of the numbers is −3.

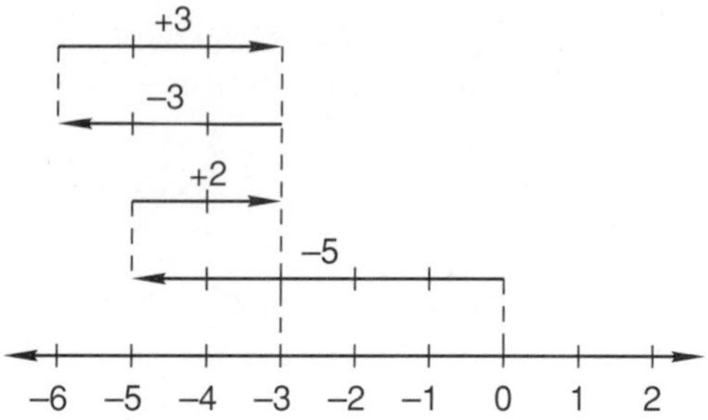

+3
−3
+2
−5
−6 −5 −4 −3 −2 −1 0 1 2

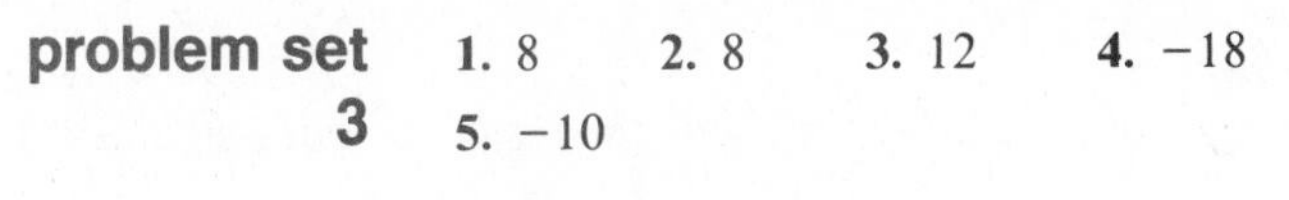

problem set 3

1. 8 **2.** 8 **3.** 12 **4.** −18
5. −10

6. The sum is −5.

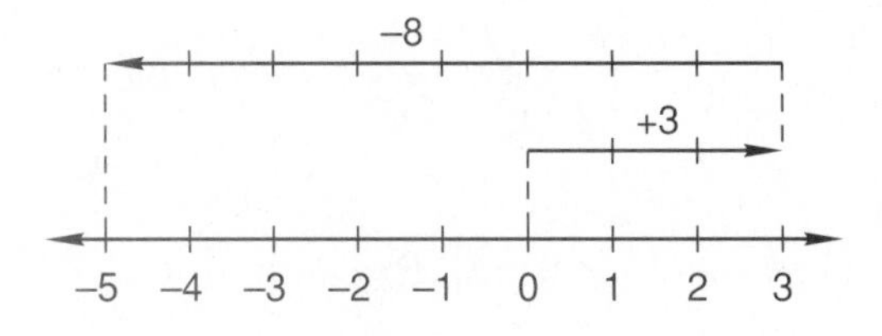

7. The sum is 1.

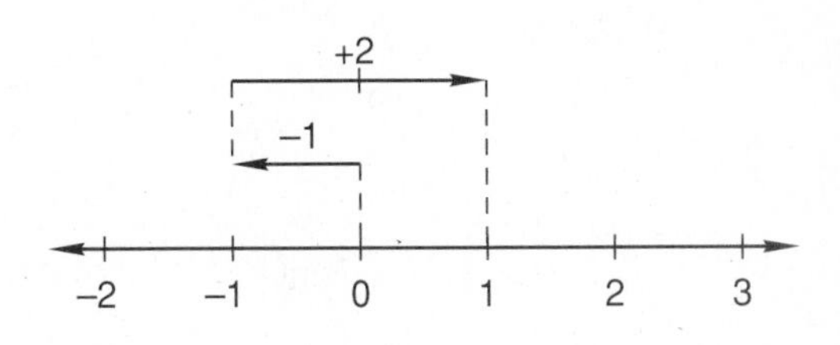

8. The sum is 7.

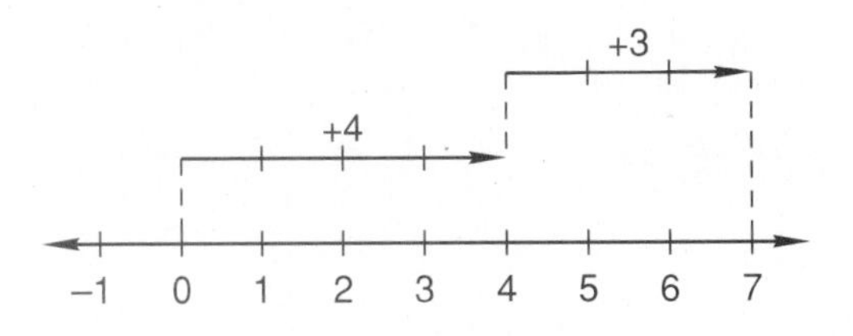

9. The sum is 2.

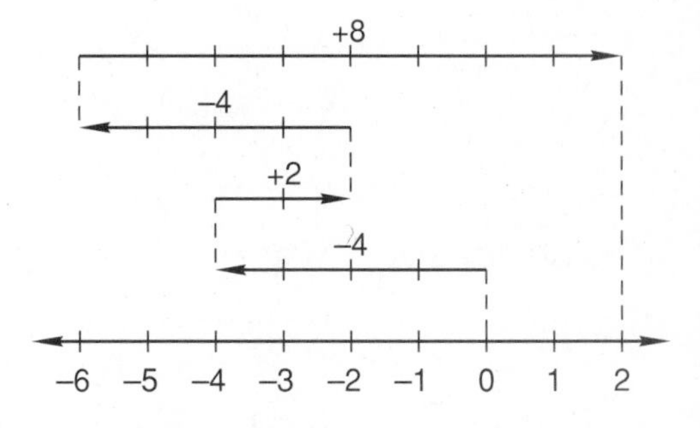

10. The sum is 4.

11. The sum is −5.

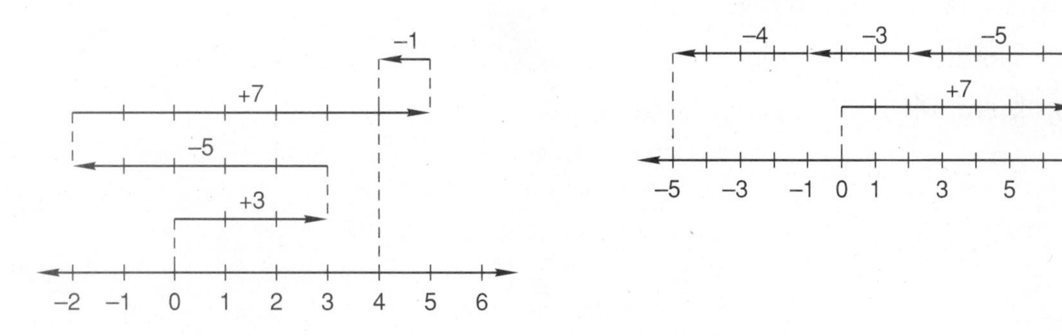

12. {1, 2, 3, 4, . . . } **13.** { . . . , −3, −2, −1, 0, 1, 2, 3, . . . } **14.** Quotient
15. The number that is associated with the point
16. The mark made on the number line to indicate the numbers' locations
17. Product **18.** One of two or more numbers to be multiplied
19. 10 is the dividend; 5 is the divisor
20. All real numbers can be graphed on the number line.
21. $39\frac{9}{16}$ **22.** $3\frac{5}{16}$ **23.** $\frac{1}{6}$ **24.** $11\frac{11}{24}$ **25.** 4.003
26. 150 in. **27.** Volume = 240 cm^3; surface area = 296 cm^2
28. $\frac{18}{25}$ **29.** $27\frac{17}{35}$ **30.** $11\frac{27}{40}$ **31.** 4.002 **32.** 443.61413 **33.** 0.1465712
34. $1\frac{1}{4}$ **35.** $\frac{51}{104}$ **36.** 100(12)(2.54) cm

practice a. -6 b. -17 c. -17 d. -14

problem set 4

1. We add the absolute value of both numbers and give the result the same sign as the numbers' signs.
2. We take the difference in the absolute values of the numbers and give the result the sign of the greater.
3. (a) A number that is to be multiplied (b) Answer to a division problem (c) Answer to a multiplication problem
4. 1660 m^2 5. 150 km
6. 4200 m^3 7. $56\pi\ cm^2 \approx 175.84\ cm^2$ 8. $\frac{3000}{12(5280)}$ miles 9. -11
10. -17 11. -35 12. -28 13. -31 14. 4 15. -1 16. -9
17. -13 18. 1 19. -38 20. -20 21. 0 22. 0 23. -5
24. -11 25. -7 26. -15 27. 0 28. -12 29. -3 30. 0
31. -11 32. -8 33. -19 34. 3 35. -9 36. -7 37. -2
38. $7\frac{23}{40}$ 39. 92.5

practice a. 7 b. 1 c. 7 d. 5

problem set 5

1. 45,654 2. 0 3. -4 4. 4 5. -4 6. 4 7. 142 ft
8. $(72\pi + 150)\ m^2 \approx 376.08\ m^2$ 9. 444(12)(2.54) cm 10. 8 11. -2
12. -2 13. 1 14. 8 15. -4 16. 8 17. 3 18. 5 19. 9 20. 0
21. -3 22. -19 23. 10 24. 7 25. 20 26. -11 27. -14 28. 0
29. -9 30. 4 31. $26\frac{109}{120}$ 32. $1\frac{3}{65}$ 33. $1\frac{32}{45}$ 34. 5.423 35. 0.402
36. 6.314408 37. 0.06

practice a. 20 b. -40 c. -18 d. -12 e. 40 f. 40

problem set 6

1. The number associated with the point 2. $+2$
3. If its graph is farther right on the number line 4. The product
5. A quotient 6. $\{\ldots, -2, -1, 0, 1, 2, \ldots\}$ 7. 172 m 8. 54 cm^2
9. 1350 m^2 10. Volume = 276 ft^3; surface area = 392 ft^2 11. 15 12. -10
13. -10 14. -4 15. 0 16. -2 17. 5 18. -3 19. -13 20. 2
21. -17 22. -14 23. -7 24. 1 25. -5 26. 15 27. 6 28. -2
29. 7 30. -7 31. 1 32. -10 33. 41.265 34. $13\frac{21}{40}$ 35. 8100
36. $\frac{33}{40}$ 37. $-1\frac{3}{40}$ 38. $\frac{18}{49}$ 39. $\frac{620}{12(2.54)}$ ft

practice a. 2 b. -8 c. -2 d. 8 e. 8 f. -2 g. -8 h. 2

problem set 7

1. Division 2. Multiplication 3. $\frac{4000}{(2.54)(12)}$ ft 4. 12 5. -48
6. -24 7. 8 8. 6000 ft^3 9. 120 m 10. 1660 cm^2 11. 4
12. -2 13. -4 14. -11 15. -17 16. -3 17. -13 18. -5
19. 1 20. -8 21. -8 22. -9 23. 3 24. 2 25. -7 26. 4
27. -4 28. 4 29. -7 30. -4 31. -15 32. $1\frac{1}{2}$ 33. $\frac{17}{37}$ 34. 0.02

practice a. Undefined b. Undefined c. 16 d. 240 e. 44(5280)(5280) ft^2
f. $\frac{4400}{3(3)(3)}$ yd

problem set 8

1. Yes **2.** No

3. Because we are trying to undo a multiplication that was never performed

4. 24 **5.** 48 **6.** -24 **7.** 5 **8.** -2 **9.** -2 **10.** 18 **11.** -24

12. -24 **13.** 48 **14.** -5 **15.** 2 **16.** 13 **17.** -6 **18.** -13

19. -4 **20.** -4 **21.** -4 **22.** -10 **23.** 2 **24.** 1 **25.** 1 **26.** -13

27. 5 **28.** 8 **29.** -5 **30.** $147(100)(100)(100)\text{ cm}^3$

31. $112\pi\text{ cm}^2 \approx 351.68\text{ cm}^2$

32. 4100 one-centimeter cubes; surface area = 2176.2 cm^2 **33.** $\frac{51}{110}$ **34.** $2\frac{8}{51}$

35. 3.00093 **36.** 3.03 **37.** $-\frac{14}{15}$ **38.** $-4\frac{43}{88}$

practice

a. -102 **b.** 3 **c.** Undefined **d.** $-1\frac{1}{5}$

problem set 9

1. Zero; if we try to write the reciprocal, we get a meaningless statement.

2. (a) 1 (b) No

3. (a) Answer to a division problem (b) Answer to a multiplication problem

4. 8 **5.** $420(12)(12)(12)\text{ in.}^3$ **6.** $\frac{465(100)}{2.54}$ in.

7. $768\pi\text{ m}^3 \approx 2411.52\text{ m}^3$ **8.** 24 m^2

9. 4 **10.** 1 **11.** -18 **12.** 32 **13.** 0 **14.** -20 **15.** 1 **16.** 18

17. 22 **18.** -20 **19.** 2 **20.** 2 **21.** -9 **22.** 6 **23.** -16 **24.** -12

25. -2 **26.** 0 **27.** 0 **28.** 8 **29.** 18 **30.** 4 **31.** $-4\frac{4}{35}$ **32.** $\frac{16}{37}$

33. 8 **34.** $13\frac{3}{10}$

practice

a. $3\frac{1}{2}$ **b.** Undefined

problem set 10

1. $\{0, 1, 2, 3, 4, 5, \ldots\}$ **2.** $\{\ldots, -3, -2, -1, 0, 1, 2, 3, \ldots\}$

3. (a) One of the quantities multiplied to form a product
(b) Answer to a division problem (c) Answer to an addition problem

4. 28 **5.** 33 **6.** -1 **7.** -2 **8.** 79 **9.** -40 **10.** $\frac{420}{(2.54)(12)}$ ft

11. $\frac{420}{12(12)(12)}\text{ ft}^3$ **12.** $(24 - 2\pi)\text{ m}^2 \approx 17.72\text{ m}^2$ **13.** 24 in.^2 **14.** 6

15. -25 **16.** 42 **17.** 0 **18.** 8 **19.** 4 **20.** -31 **21.** -27 **22.** -12

23. -74 **24.** 62 **25.** -4 **26.** 118 **27.** 35 **28.** -17 **29.** -19

30. 145 **31.** -6 **32.** -22 **33.** -4.02 **34.** $-\frac{22}{31}$ **35.** $-1\frac{13}{15}$

practice

a. 396 **b.** -10

problem set 11

1. $\{0, 1, 2, 3, \ldots\}$ **2.** Any number that can be graphed on the number line

3. (a) One of the quantities multiplied to form a product
(b) Answer to a multiplication problem (c) Answer to a division problem

4. 18 **5.** -17 **6.** 10 **7.** 6 **8.** -12 **9.** -2 **10.** -20

11. -6 **12.** -4 **13.** 15 **14.** 36 **15.** 13 **16.** -17 **17.** -21

18. (a) $(80 + 10\pi)\text{ m} \approx 111.4\text{ m}$ (b) $(600 + 50\pi)\text{ m}^2 \approx 757\text{ m}^2$

19. $(792 - 25\pi)\text{ in.}^2 \approx 713.5\text{ in.}^2$

20. $-\frac{1}{13}$ 21. $\frac{1}{2}$ 22. $\frac{5}{7}$ 23. -2 24. $-\frac{3}{4}$ 25. -20 26. -9 27. 3

28. -16 29. $-\frac{9}{10}$ 30. 1.05 31. $\frac{485(2.54)}{100}$ m 32. $476(5280)(5280)$ ft^2

practice

a. 36 b. 48

c. The number produced is negative because there is an odd number of negative factors.

d. -3 e. 2

problem set 12

1. Negative 2. Negative 3. $-\frac{1}{5}$ 4. Multiplicative inverse

5. $(27{,}000 + 5000\pi)$ cm^3 $\approx$ 42,700 cm^3 6. 88 in.^{2} 7. $\frac{60(100)}{2.54}$ in.

8. $\frac{48{,}700}{(5280)(5280)(5280)}$ mi^3 9. 10 10. 30 11. 84 12. 30 13. 3

14. 2 15. $-\frac{5}{4}$ 16. $-\frac{18}{11}$ 17. -13 18. 3 19. $-\frac{10}{7}$ 20. 16

21. 13 22. 13 23. -1 24. 2 25. -6 26. -16 27. -23

28. -12 29. 61 30. -15 31. $-1\frac{1}{30}$ 32. -2.2 33. $-1\frac{31}{65}$

practice

a. 4 b. -16 c. -19 d. 44

problem set 13

1. A numerical expression contains only numbers, and an algebraic expression contains only numbers or only letters or both.
2. The value of an expression is the one number being represented.
3. (a) A letter that represents an unspecified number (b) A variable
4. 75(12)(2.54) cm 5. 7000(5280)(5280) ft^2 6. 3872π m^3 $\approx$ 12,158.08 m^3
7. 800 cm^2 8. 12 9. 14 10. -30 11. -4 12. 32 13. 13
14. -31 15. -70 16. -12 17. -2 18. 1 19. -11 20. 5
21. -16 22. -126 23. -4 24. 6 25. 1 26. -32 27. -7
28. -40 29. -20 30. 12 31. -3 32. $-\frac{43}{44}$ 33. $-\frac{4}{5}$

practice

a. -24 b. 64 c. 18

problem set 14

1. (a) One of the quantities multiplied to form a product (b) A quotient is the answer to a division problem (c) A sum is the answer to an addition problem
2. -8 3. -2 4. -25 5. 0 6. -1 7. 48 8. 18
9. -30 10. 70 11. 30 12. 9 13. 1 14. -30
15. Volume = 144 m^3; surface area = 230.4 m^2 16. 56 ft 17. $\frac{4700}{3(3)(3)}$ yd^3
18. -54 19. -6 20. -14 21. 5 22. 15 23. $\frac{29}{3}$ 24. $-\frac{9}{14}$
25. $-\frac{11}{6}$ 26. 14 27. -13 28. 30 29. -34 30. 7 31. 0 32. 15

practice

a. $a(b + c) = ab + ac$ b. 8 c. 8 d. 8 e. $axy + bxy - 2cxy$

f. $4mnx - 18x + 12$

problem set 15

1. A coefficient is a factor. Usually this word is reserved for the numerical factor of a term.

2. A factor made up of letters **3.** 250 ft^2 **4.** 224π $cm^3 \approx 703.36$ cm^3

5. $\frac{280}{12(12)(12)}$ ft^3 **6.** 35 **7.** −45 **8.** $abmx - bmx$ **9.** $-4dy - 4cxy$

10. $2ax + 2bcx$ **11.** $3ax + 6ay$ **12.** 10 **13.** 9 **14.** 35 **15.** 4 **16.** 22

17. −60 **18.** 0 **19.** −2 **20.** −20 **21.** −40 **22.** −15 **23.** 4

24. −48 **25.** −2 **26.** 36 **27.** $\frac{11}{8}$ **28.** −7 **29.** −9 **30.** 20

31. −0.0162 **32.** $-1\frac{17}{33}$ **33.** $-1\frac{33}{40}$

practice a. $x - 6xy + 4$ b. $3xy - 3xyz$ c. $9acy - 2ac$ d. $2x - 9xy + 4$

problem set 16

1. A single symbol, a product, or a quotient **2.** Like terms

3. (a) One of the things that is multiplied to form a product (b) Answer to a multiplication problem (c) Answer to a division problem

4. 108π $ft^2 \approx 339.12$ ft^2 **5.** 66 $in.^2$

6. $\frac{6(12)(2.54)}{100}$ m **7.** $\frac{42{,}000}{12(12)(5280)(5280)}$ mi^2 **8.** $-2xyz + 2xy$

9. $2yx - x - 4$ **10.** $4x + 2xy$ **11.** $3xy - 6xm$ **12.** $2pxy - 6pk$ **13.** 27

14. 26 **15.** −10 **16.** 0 **17.** −2 **18.** 0 **19.** −180 **20.** −162

21. −21 **22.** −7 **23.** −7 **24.** −7 **25.** 8 **26.** −49 **27.** 10

28. $-\frac{4}{3}$ **29.** $-\frac{7}{6}$ **30.** $-\frac{32}{57}$ **31.** $2\frac{7}{15}$ **32.** −2.03

practice a. 4 b. −4 c. −30 d. −35 e. −11

problem set 17

1. One solution is: $2(3 + 4) = 2(3) + 2(4)$ **2.** All integers are real numbers.

3. Area = $(150 - \frac{25}{2}\pi)$ $in.^2$; so it would take approximately 111 one-inch-square floor tiles to cover the figure.

4. Volume = 648 ft^3; surface area = 540 ft^2 **5.** $\frac{42(12)(2.54)}{100}$ m

6. $\frac{170}{12(12)(3)(3)}$ yd^2 **7.** 42(100)(100)(100) cm^3 **8.** $-6xmy - my$

9. $-5pxk - kp - 3kx$ **10.** $-3mc + 2m - 2$ **11.** $-2ax - 4a - 8$

12. $2ax - 6p - 4$ **13.** $4x - apx$ **14.** $20xyp - 8xyc$ **15.** $8kc - 4ka + 12km$

16. −5 **17.** 4 **18.** 4 **19.** 64 **20.** 31 **21.** 36 **22.** 9 **23.** −32

24. −9 **25.** −10 **26.** −17 **27.** 140 **28.** −11 **29.** 1 **30.** −4 **31.** 3

32. 6

practice a. 144 b. −44 c. −32

problem set 18

1. (a) {0, 1, 2, 3, . . .} (b) {. . . , −3, −2, −1, 0, 1, 2, 3, . . .} **2.** Reciprocal

3. Positive **4.** 18 **5.** −16 **6.** 12 **7.** −1 **8.** −38 **9.** 18

10. 768 1-ft sugar cubes; surface area = 608 ft^2 **11.** 51 1-in.-square tiles

12. $\frac{49(100)}{12(2.54)}$ ft **13.** −1 **14.** 22 **15.** 5 **16.** 8 **17.** −20 **18.** −10

19. 3 **20.** 36 **21.** $-2xy + x + 3$ **22.** $-3bpx + 2pb - 3$ **23.** $15 - 5k + kx$

24. $-4 + py - y$ **25.** $16x - 8xp$ **26.** $3x + 12$ **27.** $-2xa + 6px$ **28.** 40

29. −27 **30.** 39 **31.** $-\frac{8}{15}$ **32.** $1\frac{5}{16}$ **33.** −0.004

practice a. $x^{11}y^7m^2$ b. x^8y^6 c. $-6x^2y^3 - 4xy$ d. $2x^6y - 5xy^6 + 3xy$

problem set 19

1. A letter that stands for an unspecified number
2. Terms whose letters represent the same number no matter what numbers are used to replace the letters
3. 24 1-in.-square floor tiles 4. $(48 - 4\pi)$ cm^2 $\approx$ 35.44 cm^2 5. x^6y^4
6. x^5m^6 7. k^6y^7 8. a^5b^8 9. $\frac{40(100)}{(2.54)(12)}$ ft 10. $\frac{400}{1000(1000)(1000)}$ km^3
11. $8b^2a - 3ab$ 12. $2x^2y - 4xy + x^2$ 13. $5m^2pxy - 3y^2pxm$ 14. $3px - 2xy$
15. $3a^2 - 2a^2b$ 16. $10 - 20p$ 17. 5 18. 15 19. -35 20. 124
21. -49 22. -72 23. -1 24. 2 25. -4 26. -3 27. 5 28. 16
29. -35 30. 26 31. $-\frac{3}{4}$ 32. $-2\frac{7}{30}$ 33. 0.000048

practice a. Neither -2 nor -5 is a root of $x - 2 = 0$.
b. Both -2 and -5 are roots of $x^2 + 7x = -10$.

problem set 20

1. (a) $6 = 4 + 1 + 1$ (b) $x + 2 = x$ (c) $x + 2 = 4$ 2. 142 ft
3. 2112π in.^{2} $\approx$ 6631.68 in.^{2} 4. $\frac{12}{3(3)(3)}$ yd^3 5. x^5y^5 6. $p^6y^3m^6$ 7. $8k^{12}n^6$
8. $a^{11}b^3$ 9. $m^3p^3a^7$ 10. $4p^3x^5k^4$ 11. $x^2 + 3x - 2$ 12. $-3xy + 2xy^2$
13. $-8ymx^2 + 23x$ 14. 0 15. $-4x^2 + 6x - 13$ 16. $-8m^2y + 6y$
17. $3x^2 - 5x + 8$ 18. $3ax - 2a$ 19. $20xy - 8axy$ 20. $18ax - 24x$
21. 1 22. -16 23. -37 24. 45 25. -10 26. -35 27. -5
28. -18 29. -60 30. -17 31. $-1\frac{15}{76}$ 32. $-\frac{15}{28}$ 33. 0.12

practice a. $x = \frac{7}{8}$ b. $k = -11$ c. $d = -\frac{41}{42}$ d. $p = 12$

problem set 21

1. Find the value(s) of the variable that will make the equation a true equation.
2. Equations that have the same answer(s)
3. $\frac{30(100)(100)}{(2.54)(2.54)}$ in.^{2} 4. 45 5. 36 6. $-2\frac{3}{20}$ 7. 86 8. 19
9. -11 10. $(1176 - 50\pi)$ cm^2 $\approx$ 1019 cm^2 11. $(20 + 4\pi)$ yd $\approx$ 32.56 yd
12. $x^4y^5m^5$ 13. $x^4y^3m^2$ 14. x^5y^6 15. $-3x^2y + 3yx - 2y^2x$
16. $-2x^3 + 5x^2 + x + 4$ 17. -5 18. $-4x^2 + 6x - 9$ 19. $4ax + 8bx$
20. $6x + 12$ 21. $4pxmy - 12pxab^2$ 22. -35 23. -30 24. -26 25. 7
26. -16 27. -16 28. 22 29. -5 30. $-\frac{57}{10}$

practice a. $x = 5$ b. $x = \frac{45}{4}$ c. $x = 45$ d. $z = 10$

problem set 22

1. Reciprocal 2. 1 3. 16 4. -10 5. 7 6. 35 7. 10
8. 15 9. $\frac{16,000}{(2.54)(12)(5280)}$ miles 10. $(220 - 25\pi)$ m^2 $\approx$ 141.5 m^2
11. 210π cm^2 $\approx$ 659.4 cm^2 12. 2 13. $\frac{17}{10}$ 14. 5 15. -2 16. 57 17. 2
18. $\frac{3}{2}$ 19. $\frac{9}{8}$ 20. 4 21. $\frac{5}{34}$ 22. $m^2x^4y^6p^2$ 23. $3x^3p^6y^3$
24. $-5a^2 - 2a + 8$ 25. $2x^2ym - 2my^2x$ 26. $3axy - 5pxy$ 27. -11
28. 4 29. -21 30. 24

practice a. $x = 2$ b. $x = -\frac{2}{63}$ c. $x = 0.13$ d. $x = 0.8$

problem set 23

1. $\frac{3}{2}$ 2. $\frac{1}{6}$ 3. $\frac{14}{3}$ 4. $-\frac{26}{5}$ 5. 16 6. $\frac{33}{4}$ 7. $-\frac{7}{2}$ 8. $\frac{35}{2}$
9. $\frac{26}{3}$ 10. $\frac{5}{3}$ 11. $\frac{27}{28}$ 12. $-\frac{1}{14}$ 13. $\frac{300(1000)(100)}{2.54}$ in.
14. Volume = 4800 in.^{3}; surface area = 1846 in.^{2} 15. x^7k^4y 16. $a^8b^5x^5$
17. $6a^2xy$ 18. $c - 4$ 19. $-x^2a^2$ 20. $8xy - 12x + 8xa$ 21. $3x - 12$
22. 19 23. −14 24. −3 25. −10 26. −24 27. −43 28. $\frac{39}{8}$
29. −44

practice a. $m = \frac{1}{2}$ b. $p = \frac{3}{5}$ c. $x = 2$ d. $p = -6$

problem set 24

1. Equations that have the same solution sets
2. Yes; the number 4 satisfies both equations
3. 120(5280)(5280)(12)(12) in.^{2} 4. 200(100)(100)(100) cm^3
5. Area = $(96 + 8\pi)$ yd^2; approx. 121.13 1-yd-square floor tiles 6. 48π cm^3 ≈ 150.72 cm^3
7. $\frac{6}{35}$ 8. −3 9. $\frac{11}{3}$ 10. 120.3 11. −3 12. 3 13. 1 14. $\frac{9}{5}$
15. −1 16. m^6y^9 17. $k^{10}m^6a^3$ 18. $-5m^2xyz$ 19. $5a^2bc - bc$
20. $xa - 2a - 3$ 21. $28 - 12x$ 22. $3amxy + 6apxy$ 23. 0 24. 24
25. 136 26. −9 27. 21 28. 4 29. 27 30. 0

practice a. $xy^4p - xy^2p$ b. $2x^2y^2 - 2x^2y$ c. $3xp^8 - 3x^3p^{11}$ d. $2x^2m^4 - 8x^2m^3$
e. $m = \frac{34}{13}$ f. $x = 210$

problem set 25

1. $a(b + c) = ab + ac$ 2. −4 3. $\frac{430(12)(2.54)}{100}$ m 4. $\frac{6500}{3(3)}$ yd^2 5. $\frac{5}{3}$
6. $-\frac{33}{14}$ 7. $\frac{33}{20}$ 8. $\frac{9}{35}$ 9. $-\frac{1}{5}$ 10. $\frac{1}{2}$ 11. $-\frac{6}{5}$ 12. 1 13. $-\frac{3}{4}$
14. 1 15. $p^2x^6y^4$ 16. $3p^7x^6y^4$ 17. $-4x^2y + xy - 8$ 18. $8x^2 - 7x - 5$
19. $-5p^2xy$ 20. $x^5y - x^3y^2z^3$ 21. $4ax^3 - 8x^2$
22. Area = 192π cm^2; approximately 602.88 one-cm-square floor tiles
23. Volume = $(264 + 24\pi)$ in.^{3}; approximately 339.36 one-inch sugar cubes; surface area = $(272 + 28\pi)$ in.^{2} ≈ 359.92 in.^{2}
24. −48 25. −8 26. 32 27. −11 28. 0 29. −31 30. $-\frac{1}{7}$

practice a. $WN = \frac{93}{2}$ b. $WF = \frac{7}{15}$ c. $WN = 3708$ d. $WN = \frac{25}{2}$

problem set 26

1. $\frac{50(12)(2.54)}{100}$ m 2. 50(100)(100)(100) cm^3 3. 2070 cm^2 4. 425 ft^2
5. $\frac{5}{6}$ 6. 98 7. $\frac{17}{5}$ 8. $\frac{13}{4}$ 9. −279 10. $-\frac{5}{4}$ 11. $\frac{17}{6}$ 12. $-\frac{5}{2}$
13. $-\frac{16}{3}$ 14. $\frac{10}{3}$ 15. 4 16. x^3y^{11} 17. m^7y^4 18. $-3m^2xy + 8mxy^2$
19. $4pc - p - 6c$ 20. $4a^2xy + 3ax^2y$ 21. $3x^3y^4 - 5x^2y^4$ 22. $3x^7a - 6x^8a^4$
23. $p^3x^2y^2 - 3p^3x^2y$ 24. 108 25. 59 26. −43 27. −30 28. −20

29. -23 30. $\frac{20}{7}$

practice a. $\frac{1}{16}$ b. 8 c. $\frac{1}{16}$ d. 9 e. x^3m^{-5} f. $x^3y^{-3}z^6$

problem set 27 1. (a) $\{\ldots, -3, -2, -1, 0, 1, 2, 3, \ldots\}$ (b) $\{0, 1, 2, 3, \ldots\}$ 2. $\frac{1000}{3}$

3. $\frac{7}{8}$ 4. 225 5. $\frac{44}{5}$ 6. 2 7. $\frac{3}{2}$ 8. 0 9. 4 10. $\frac{9}{2}$

11. 1577.2 12. -5 13. $-\frac{3}{13}$ 14. $\frac{7200(2.54)}{100}$ m

15. Volume = $(48 + 6\pi)(4)(12)$ in.^{3}; $\approx$ 3208.32 in.^{3} surface area = $(1171.2 + 300\pi)$ in.^{2} $\approx$ 2113.2 in.^{2}

16. 912 ft^2 17. 4 18. $\frac{1}{16}$ 19. 1 20. $y^2a^{-3}x^{-4}$ 21. p^2m^{-1}

22. $6px^2y - 4$ 23. $3m^2x^2y + 8m^2xy^2$ 24. $6x^4y^4p^2 - 4x^2y^3p^2 + 2x^2y^3p$

25. $4cm^4z^3 - 20m^2z^6$ 26. 18 27. 85 28. 36 29. -5 30. 6

practice a. $WD = 1.33$ b. $WN = 301$ c. $WN = 3.84$ d. 1 e. -1 f. 1

g. 1 h. $\frac{42}{(2.54)(2.54)(2.54)(12)(12)(12)}$ ft^3

problem set 28 1. $80(3)(3)(3)(12)(12)(12)$ in.^{3} 2. $\frac{42}{(2.54)(2.54)(2.54)}$ in.^{3} 3. $WN = 9.6$

4. $\frac{49}{3}$ 5. $\frac{44}{5}$ 6. $-\frac{11}{2}$ 7. 5.05 8. -2 9. $-\frac{7}{5}$ 10. $-\frac{5}{6}$

11. $\frac{3}{2}$ 12. $-\frac{5}{6}$ 13. $\frac{1}{9}$ 14. 19 15. -64 16. 8 17. $-\frac{1}{125}$ 18. 32 cm^2

19. Volume = $(480 + 48\pi)$ ft^3 $\approx$ 630.72 ft^3; surface area = $(328 + 40\pi)$ ft^2 $\approx$ 453.6 ft^2

20. $3 - 12x^3y^{-5}$ 21. $2 - 8x^{-8}y^4$ 22. $1 - 3y^2x$ 23. $1 - y^2x$

24. $7m^3x^2y - 2x^2y$ 25. $7abc^2 - 6ab^2c$ 26. -24 27. -12 28. 132

29. -43 30. $\frac{16}{21}$

practice a. $5(3N - 5)$ b. $3(N - 50)$ c. $5N - 13$ d. $3(-N - 7)$

problem set 29 1. 2000 2. 105 3. 22,214 4. $\frac{4200(2.54)(2.54)(2.54)}{100(100)(100)}$ m^3

5. $(96 - 8\pi)$ m^2 $\approx$ 70.88 m^2 6. $5N - 8$ 7. $3(-N - 7)$ 8. $-\frac{11}{2}$

9. 4 10. -45 11. 15 12. $\frac{-11}{4}$ 13. $\frac{6}{5}$ 14. $\frac{1}{36}$ 15. $\frac{1}{27}$ 16. 16

17. 130 18. $1 - \frac{2}{x}$ 19. $12x^3p^5 - 8$ 20. $1 + 3xp^2$ 21. $8x^{-10} - 12x^{-5}$

22. $2x^4 - 6p^{-2}x^3y^{-2}$ 23. $4y^5p^3 - 20x^6y^2p^{-6}$ 24. $3xmp^{-2} - 5mx$

25. $-7k^2yp^{-4}$ 26. -31 27. -21 28. -16 29. 27 30. -35

practice a. $c = 2$ b. $c = 1$ c. $z = 8$

problem set 30 1. $7(n - 5)$ 2. $-2n - 7$ 3. $7n - 51$ 4. $4n - 15$ 5. 180 6. $\frac{2}{55}$

7. $\frac{441}{8}$ 8. $\frac{70(12)(12)(2.54)(2.54)}{100(100)}$ m^2 9. $\frac{10{,}000(1000)(100)}{(2.54)(12)}$ ft

10. 4800π cm^2 $\approx$ 15,072 cm^2 11. $(110 + 18\pi)$ ft^2 $\approx$ 166.52 ft^2 12. 66

13. $-\frac{1}{36}$ 14. $\frac{22}{3}$

15. $\frac{69}{20}$ **16.** -66 **17.** $\frac{1}{64}$ **18.** $1 - 5y^4$ **19.** $1 + 3x^5y^7$ **20.** $-6x^6 + 12p$
21. $5x^6y^6 - 25y^{-2}$ **22.** $3m^6n^3 - 3$ **23.** $2p^5x^5 - 6x^5p^{-5}$ **24.** $3xym^2z^3 + x^3ym^2z^2$
25. $11xy$ **26.** -10 **27.** 52 **28.** -56 **29.** 3 **30.** -57

practice a. $N = 25$ b. $N = 2$

problem set 31

1. $\frac{800}{(2.54)(2.54)(2.54)(12)(12)(12)}$ ft^3 **2.** 7 **3.** 8

4. Volume = 720 in.^{3}; surface area = 522 in.^{2} **5.** $\frac{1}{2}(4)(12)(20)$ in.^{2} = 480 in.^{2}

6. $\frac{9}{5}$ **7.** 25 **8.** 49 **9.** $-\frac{1}{18}$ **10.** 2 **11.** $\frac{13}{2}$ **12.** -12 **13.** $\frac{4}{3}$

14. $\frac{28}{3}$ **15.** 4 **16.** $\frac{1}{36}$ **17.** -27 **18.** $-\frac{15}{4}$ **19.** -64 **20.** $15xp + 6$

21. $2x^{-3}p^{-3} - 6x^3p^{-6}$ **22.** $2p^9xym^4 - 10p^4x^{-2}y^{-3}$ **23.** $3x^4p^{-4} - 6p^{-2}$
24. $5x^{-9}y^{-2} - x^3y^2$ **25.** $2xym^2 - x^2ym$ **26.** 4 **27.** -327 **28.** -14
29. -129 **30.** 5

practice a. $N = 4$ b. $N = -2$ c. $2 \cdot 2 \cdot 2 \cdot 2 \cdot 5 \cdot 5$ d. $2 \cdot 2 \cdot 3 \cdot 3 \cdot 3$

problem set 32

1. 120(12)(12)(12)(2.54)(2.54)(2.54) cm^3 **2.** 72 cm^2

3. $\frac{1}{2}(80)(5)(12)$ in.^{2} = 2400 in.^{2} **4.** $\frac{77}{15}$ **5.** $\frac{9}{4}$ **6.** 51.25 **7.** $\frac{28}{5}$

8. $\frac{119}{4}$ **9.** 92 **10.** 4 **11.** $-\frac{4}{7}$ **12.** $\frac{9}{2}$ **13.** $-\frac{7}{2}$

14. $2 \cdot 2 \cdot 2 \cdot 2 \cdot 2 \cdot 5$ **15.** $2 \cdot 3 \cdot 7 \cdot 7$ **16.** $2 \cdot 5 \cdot 5 \cdot 5$ **17.** $2 \cdot 3 \cdot 3 \cdot 5 \cdot 5$

18. $-\frac{1}{27}$ **19.** $\frac{1}{8}$ **20.** $2x^{-4} + 2y^5$ **21.** $x^3p^5 - 3$ **22.** $4x^2y^2 - 12x^4y^{-2}$

23. $4 - 3p^7x^{-3}$ **24.** $5x^2y^3$ **25.** xyz **26.** 2 **27.** -72 **28.** 62

29. -101 **30.** $\frac{5}{4}$

practice a. $2xy^3m$ b. $5a^2b^2c^2$ c. $4xyp^3$

problem set 33

1. $-\frac{110}{3}$ **2.** 5 **3.** 42 **4.** $\frac{1}{16}$ **5.** 4 **6.** -1 **7.** 2 **8.** -460

9. $-\frac{9}{5}$ **10.** 2 **11.** 500(2.54)(2.54)(2.54) cm^3 **12.** $\frac{500(2.54)(2.54)(2.54)}{100(100)(100)}$ m^3

13. 1(12)(32) in.^{2} = 384 in.^{2} **14.** $2ab^2c$ **15.** $5xy^2m^2$ **16.** $2 \cdot 3 \cdot 3 \cdot 5 \cdot 7$

17. $2 \cdot 2 \cdot 2 \cdot 3 \cdot 5 \cdot 5$ **18.** $\frac{1}{16}$ **19.** 66 **20.** $3 - 9x^6y^2$ **21.** $2p^{-3}x^5 - 6$

22. $4x^{-6} - 8xy^6$ **23.** $1 - 2y^{12}x^5$ **24.** $2x^2yz - xyz^2$ **25.** $7x^2 + x^3y^3$

26. 32 **27.** -37 **28.** -4 **29.** -49 **30.** $\frac{40}{13}$

practice a. $2a^2b^2(1 + ab^2 - b^4)$ b. $5az^5(3az^3 - 7)$ c. $7xmz^4(4z^6 - xm^2)$

problem set 34

1. $-\frac{49}{5}$ **2.** $-\frac{5}{8}$ **3.** $\frac{87}{104}$ **4.** 0.012 **5.** $\frac{85}{8}$ **6.** $\frac{11}{21}$ **7.** $\frac{13}{7}$

8. 2.55 **9.** $-\frac{13}{4}$ **10.** 721.17 **11.** 20(5280)(5280)(12)(12)(2.54)(2.54) cm^2

12. $\left[0.2 + \frac{2(15)(0.1)}{10} + \frac{2(50)(0.1)}{10}\right]$ m = 1.5 m **13.** 1750 m^2

14. $3x^2y^2p(x^2 - 2y^3p^3)$ **15.** $2a^2x^2m(3am^4 + a^2x^3m^4 + 2)$ **16.** $2 \cdot 5 \cdot 5 \cdot 5$

17. $2 \cdot 2 \cdot 2 \cdot 3 \cdot 3 \cdot 5$ **18.** $\frac{1}{81}$ **19.** 64 **20.** $1 - 4x^2y^6$ **21.** $p^5y^{10} - 1$
22. $x^5 - 3x^7y^{-1}$ **23.** $2 - 4x^2p^5y^5$ **24.** $x^2y^{-2}p^4 + 7x^2y^2p^4$ **25.** $6y^2 - 10x^2$
26. 48 **27.** -16 **28.** -12 **29.** -18 **30.** 3

practice

a. $1 - x$ **b.** $x + 1$ **c.** $1 - 7x$

problem set 35

1. 11 **2.** -1 **3.** $\frac{5}{3}$ **4.** $\frac{19}{6}$ **5.** 0.04515 **6.** $\frac{235}{48}$ **7.** 8 **8.** $\frac{3}{4}$
9. $\frac{10}{3}$ **10.** $2a^2x(2y^4p - 3x^3)$ **11.** $3ax^2y^4(ax^2y^2 + 3 - 2ax^2yz)$
12. $\frac{3000(100)(100)(100)}{(2.54)(2.54)(2.54)}$ in.^{3} **13.** $\left[15(0.4)(100) + \frac{(0.4)(100)(7)}{2}\right]$ cm^2 = 740 cm^2
14. Volume = 756 ft^3; surface area = 1011.6 ft^2 **15.** $x + 3$ **16.** $\frac{1 - k}{y}$
17. 729.72 **18.** $2 \cdot 3 \cdot 3 \cdot 3 \cdot 5$ **19.** 251 **20.** $1 - x^{-1}y^4$
21. $12z - 21x^2y^{-3}$ **22.** $3x^5y^{-2} - 9y^7$ **23.** $8xy - 6x^5y^{-1}$
24. $7x^2ym^5 - 8xym^5$ **25.** $6x^4y^{-4} + x^2y^{-3} - x^4y^{-3}$ **26.** -3 **27.** 125
28. 24 **29.** -10 **30.** $-\frac{34}{7}$

practice

a. $\frac{x^4}{y^2} - \frac{3x^2y^2}{m}$ **b.** $\frac{15y}{z} - \frac{4x^2}{yz}$ **c.** $\frac{1}{9}$ **d.** $\frac{1}{9}$ **e.** 9 **f.** $-\frac{1}{9}$

problem set 36

1. 5 **2.** 20 **3.** $\frac{16}{3}$ **4.** 6 **5.** 3.05 **6.** $\frac{68}{65}$ **7.** $\frac{11}{7}$ **8.** $\frac{1}{2}$
9. 3.15 **10.** $3ax^2y^2(4x^3ay^5 - 1)$ **11.** $3a^2x^3y(5a^3xy^5 + a^2y^6 - 3x^3)$
12. $x - 1$ **13.** $1 - 3x$ **14.** $x + 1$ **15.** $\frac{x - 1}{m}$ **16.** $2 \cdot 3 \cdot 5 \cdot 5 \cdot 5$
17. $-\frac{1}{9}$ **18.** $\frac{1}{9}$ **19.** -11 **20.** $\frac{1}{9}$ **21.** $\frac{500(2.54)(2.54)(2.54)}{100(100)(100)}$ m^3
22. $\left[\frac{2(3)(48)}{12} + 2(2)(3) + \frac{2(2)(48)}{12}\right]$ ft^2 = 52 ft^2 **23.** $\frac{ab^4}{c^2k} - \frac{2axb^2}{c^2}$ **24.** $1 - 4p^{-6}x^{-1}y^5$
25. $5m^2xy - 12x^2ym^2$ **26.** $-3x$ **27.** -34 **28.** -15 **29.** -2 **30.** -1

practice

a. $x \leq 5$ (number line: 3 5 7) **b.** $x > 4$

problem set 37

1. 15 **2.** 4 **3.** $\frac{3}{2}$ **4.** 0.06 **5.** $\frac{43}{18}$ **6.** $\frac{22}{25}$ **7.** 7 **8.** -1
9. -2 **10.** $4a^2x^2y^4(xy - 2a^2)$ **11.** $2axm^5(3a + x^3m - 9a^4x^2)$ **12.** $x - 3$
13. $\frac{4x - 8}{x}$ **14.** $\frac{100(1000)(100)}{(2.54)(12)(5280)}$ miles
15. Volume = $(37{,}500 + 2500\pi)$ cm^3 $\approx$ 45,350 cm^3; surface area = $(6700 + 600\pi)$ cm^2 $\approx$ 8584 cm^2 **16.** (number line: 0 2 4)
17. $x \geq 2$ **18.** $-\frac{1}{27}$ **19.** $-\frac{1}{27}$ **20.** -30 **21.** $\frac{p^3k^2}{x^3} - \frac{p^4}{x^2}$ **22.** $4ax - \frac{8x^3}{a^2}$
23. $\frac{a^2x^3}{c^3} - \frac{3a^2}{x^2}$ **24.** $1 - 5x^2p^7$ **25.** $-2x^3y^3p$ **26.** $13a^2x - 2ax^2$ **27.** 0
28. -23 **29.** 9 **30.** $\frac{5}{9}$

practice a. $M = 204$ b. $C = 10{,}000$

problem set 38

1. $17(5280)(12)(2.54)$ cm 2. $216\pi \text{ cm}^2 \approx 678.24 \text{ cm}^2$ 3. 1560 4. 5

5. 0.06 6. $\frac{216}{77}$ 7. -1.7 8. $-\frac{13}{28}$ 9. 0 10. -5 11. -5

12. $3xy^3z^5(x - 3y^3z)$ 13. $4xy(x - 3y + 6x^2y^2)$ 14. $x + 1$ 15. $1 - 2kp$

16. −4 −2 0 17. $x > 4$ 18. $\frac{1}{4}$ 19. $-\frac{1}{4}$ 20. $-\frac{1}{8}$

21. $4 - \frac{p^4}{x^2}$ 22. $\frac{4a^3}{x} - 12a$ 23. $\frac{m^2p^2}{k^2} - 1$ 24. $x^{-3}y^{-2} - 3x^{-2}k^5$

25. $-\frac{2p^6}{x^3} - \frac{p^7}{x^3}$ 26. $3xy$ 27. 11 28. 301 29. 322 30. -6

practice a. $W = 39$ b. $E = 217$ c. $x \not< -5 \rightarrow x \geq -5$ −7 −5 −3

d. $x \not\leq -2 \rightarrow x > -2$ −4 −2 0

problem set 39

1. -20 2. $-\frac{12}{5}$ 3. $\frac{28{,}000}{12(12)(5280)(5280)} \text{ mi}^2$ 4. 784 kg 5. $\frac{28}{3}$

6. $\frac{33}{7}$ 7. $\frac{5}{4}$ 8. $\frac{3}{2}$ 9. -30 10. $4ax(3a + x^3)$

11. $ay(2x^2a^2 - x + 4y)$ 12. $x - 3$ 13. $1 + 4y$

14. Volume = $(200 + 100\pi)(5)(12) \text{ in.}^3$; approximately 30,840 one-inch sugar cubes; surface area = $[400 + 200\pi + (20\pi + 28.3)(5)(12)] \text{ in.}^2 \approx 6494 \text{ in.}^2$

15. $2(100) + 2(300)$ cm = 800 cm 16. $x \not> 3$

17. 1 3 5 18. $-\frac{1}{4}$ 19. $\frac{1}{4}$ 20. $-\frac{1}{4}$ 21. $\frac{p^3}{a^2} - \frac{p^3}{xa^2}$

22. $\frac{4x^5y}{m^3} - \frac{4x^2y^2}{m^2}$ 23. $\frac{mpa}{x^2} - \frac{mp}{x^3}$ 24. $1 - 3p^3x^5$ 25. $3xyp^2$

26. $-8ay - 4a^4y$ 27. 1 28. $\frac{35}{8}$ 29. -3 30. $-\frac{7}{4}$

practice a. $p^{-6}y^2z^6$ b. $y^{-1}z^3$ c. $m^6p^{-1}z^{16}d^3$

problem set 40

1. -3 2. $-\frac{5}{2}$ 3. 980 4. $\frac{48}{17}$ 5. 3.856 6. $\frac{21}{2}$ 7. 0 8. $\frac{8}{5}$

9. -2.1 10. $3x^2yp^3(y^4p^3 - 3y^3 + 4p)$ 11. $2xy^2(x - 3x^3 - 6y^3)$ 12. $x - 5$

13. $1 + 4xy$ 14. −4 −2 0 15. $2 \cdot 2 \cdot 2 \cdot 3 \cdot 3 \cdot 5$ 16. $-\frac{1}{16}$

17. 30 18. $200(5280)(5280)(12)(12) \text{ in.}^2$

19. Volume = $(92 - 8\pi)(1)(12) \text{ in.}^3$; approximately 802.56 one-inch sugar cubes; surface area = $[184 - 16\pi + (33.6 + 4\pi)(1)(12)] \text{ in.}^2 \approx 687.68 \text{ in.}^2$

20. $m^5z^{-1}x^5y^{-9}$ 21. $\frac{p^2a}{xbc} - \frac{p^2}{xc}$ 22. $\frac{3x^3y}{m^3} - \frac{6x^3y}{m^2}$ 23. $1 - 4p^2k^5$

24. $3x^6y^{-6} - 6y^{-2}$ 25. 149.9125 26. $\frac{1}{9}$ 27. -12 28. -19 29. $\frac{10}{3}$

practice **a.** $z^{-3}x^4m^{-3} - 3z^{-2}m^{-2}$ **b.** $\frac{1}{3}m^{-8}w^{-4}c^{-1}x^{-1} - \frac{1}{3}m^{-4}w^2x^{-3}$

problem set 41 **1.** -6 **2.** $-\frac{1}{3}$ **3.** 918 **4.** $\frac{2}{9}$ **5.** $2 \cdot 5 \cdot 13$ **6.** 25 **7.** $\frac{1}{2}$

8. $\frac{1}{3}$ **9.** 667 **10.** $2x^2m^3y(2m^2 - x^2y^2)$ **11.** $2m^2x^2(2x^3 - 1 + 3m^3)$

12. $1 - x$ **13.** $3 - x$ **14.** $x \leq -5$ [number line: −7, −5, −3]

15. $x > -2, x \neq -2$ **16.** $-\frac{1}{9}$ **17.** $\frac{1}{9}$ **18.** $\frac{x^2}{y^2}$ **19.** $\frac{x}{y^4z^4}$ **20.** $\frac{x}{y^4p^2}$

21. $\frac{390(5280)(2.54)(12)}{100(1000)}$ km **22.** 1014 cm² **23.** $\frac{1}{a^6x^5} - \frac{1}{x^2}$ **24.** $\frac{b}{m^5} - \frac{4a}{b^5}$

25. $-6x^2yp$ **26.** $2m^2x^2y - 2x^2my$ **27.** $-\frac{26}{9}$ **28.** 48 **29.** -8 **30.** -6

practice **a.** $4x^{-2}y - 5xy$ **b.** $-3a^{-8}b^{11} + 6a^{-8}b^6$ **c.** 16 **d.** 19

problem set 42 **1.** -4 **2.** 8 **3.** 1848 **4.** 2.02 **5.** $\frac{261}{32}$ **6.** 120 **7.** 2

8. $-\frac{8}{3}$ **9.** 3.5 **10.** $mk(6k^4m - 2k^2 - 1)$ **11.** $x^3y^2m(x - ym + 5x^3m)$

12. $1 - 3x$ **13.** $\frac{y-1}{x}$ **14.** [number line: 3, 5, 7] **15.** $x > 2$

16. $2 \cdot 2 \cdot 2 \cdot 5 \cdot 7$ **17.** $x^{-2}y^8m^{-2}$ **18.** xy^8m^{-5} **19.** $x^3y^5p^2$ **20.** $\frac{z}{xy^2} - \frac{y^3}{x^2}$

21. $\frac{b}{a} - \frac{2b^3}{a^3}$ **22.** $\frac{80(1000)(100)}{2.54}$ in. **23.** -6

24. Volume = 700 m³; surface area = 614 m²

25. $-2m^2y^{-2}$ **26.** $-\frac{9}{4}$ **27.** -1 **28.** -2 **29.** $\frac{1}{27}$ **30.** -2

practice **a.** $y = \frac{13}{8}x + \frac{3}{2}$ **b.** $p = -\frac{1}{5}W - \frac{3}{2}$

problem set 43 **1.** -3 **2.** 8000 farthings **3.** $\frac{3}{5}$ **4.** 6 **5.** $\frac{3}{8}$ **6.** $\frac{9}{4}$ **7.** 1.45

8. $\frac{400(1000)(100)}{(2.54)(12)(5280)}$ miles

9. Volume = $(\frac{125}{2}\pi)(2)(3)$ ft³ ≈ 1177.5 ft³; surface area = $[125\pi + (10 + 15\pi)(2)(3)]$ ft² ≈ 735.1 ft²

10. 27.6 **11.** 13 **12.** $y = \frac{5}{3} - x$ **13.** $y = \frac{1}{4}x + 1$ **14.** $2x^2y^2(3x^2 - 2z)$

15. $xyz(8x^4y - 16xyz - 1)$ **16.** $1 - 2y$ **17.** $5 - 25xy$ **18.** [number line: 0, 2, 4]

19. $5 \cdot 5 \cdot 5 \cdot 3 \cdot 3$ **20.** $\frac{1}{x^9y}$ **21.** $\frac{4m^5}{x^7y^{11}}$ **22.** $x^5z^{-6} - 3$ **23.** xy^5 **24.** 0

25. $-2x^2y^{-2}$ 26. 18 27. 120 28. $\frac{17}{36}$ 29. -12 30. -2

practice a. 420 b. 216 c. 600

problem set 44 1. 7 2. 600 3. $\frac{8}{85}$ 4. $\frac{54}{25}$ 5. 39 6. $\frac{24}{11}$ 7. $-\frac{4}{3}$

8. $y = \frac{4}{3} - \frac{1}{3}x$ 9. $y = \frac{1}{4}x + \frac{7}{4}$ 10. $y = 2 - k - 2x$ 11. $y = \frac{2}{3}x + \frac{7}{3}$

12. $3a^2b^4c^5(1 - 2b^2c)$ 13. $2xa(4x - 2xa + a)$ 14. $z - 1$ 15. $2x$

16. $x \geq -2, x \not< -2$ 17. 10.7 18. 80(3)(12)(2.54) cm 19. 1200

20. $p^{-2}z^6$ 21. $a^{-1}pk^{-3}$ 22. p^{-1} 23. $1 - 3m^4z$ 24. $3a^2x^2y^{-3} - 4ax^3y^{-3}$

25. $2m^2xy^{-2} - 3m^2x^{-1}y^2$ 26. $\frac{71}{8}$ 27. 9 28. 3 29. $-\frac{95}{8}$ 30. $\frac{2}{3}$

practice a. $12y^3w^2$ b. $15x^6y^2m^3$ c. $60a^{10}b^4$

problem set 45 1. $\frac{38}{7}$ 2. 450 3. 0.622 4. $\frac{100}{13}$ 5. 8 6. $-\frac{5}{2}$ 7. -0.5

8. $y = \frac{2}{5}x + \frac{4}{5}$ 9. $2 - x = y$ 10. $5x^2y^2m^2(y^3 - 2x^2m)$ 11. $xyz(-x + 2z)$

12. $x + 1$ 13. $y + 1$ 14. [number line: open circle at 2, arrow to the right; marks 0, 2, 4] 15. 600 16. 270

17. Volume = 6200 in.^{3}; surface area = 2310 in.^{2} 18. $\frac{200}{(2.54)(2.54)(12)(12)(3)(3)}$ yd^2

19. -18 20. $\frac{p^4}{x^4y^2}$ 21. $\frac{m^7}{k^2}$ 22. $\frac{b^4}{a^2}$ 23. $y^2 + 4m^3y^7$ 24. $5ay^2x^{-1}$

25. $5m^2k^5 - 3mk^5$ 26. 12 27. -18 28. $\frac{3}{4}$ 29. $-\frac{1}{27}$ 30. $-\frac{2}{3}$

practice a. $\frac{6}{11}$ b. $\frac{10m - 6}{3m + 2}$ c. $\frac{4x + 2m + 1}{3x^2m}$ d. $\frac{9 - 7ap}{xy^3 + m}$

problem set 46 1. $\frac{40}{3}$ 2. 21,530 3. $\frac{1}{3}$ 4. 14 5. $-\frac{13}{11}$ 6. $y = \frac{5}{2} - \frac{3}{2}x$

7. $y = \frac{1}{3}x + \frac{7}{3}$ 8. $\frac{50.8}{(2.54)(2.54)(12)(12)}$ ft^2 9. -32 10. 95.2

11. Surface area = 344(100)(100) cm^2 = 3,440,000 cm^2; volume = 240(100)(100)(100) cm^3 = 240,000,000 cm^3

12. $\frac{c^2 + 4}{b}$ 13. $x \leq -4, x \not> -4$ 14. 1125 15. $2c^3$ 16. $12c^4$ 17. b^3c^2

18. $x^2y^4p^2(4y - 3x^3)$ 19. $x - 1$ 20. $\frac{1}{x^{-2}y^3}$ 21. $\frac{1}{m^2p^{-8}}$ 22. $\frac{1}{x^{-8}y^{-8}}$

23. $y^{-4}p^{-4} - x^2p^6y^{-4}$ 24. $-6x^2y^{-2}$ 25. $3a^2xy - 6xy^{-1}$ 26. 1 27. 5

28. $-\frac{9}{2}$ 29. -13 30. $-\frac{15}{2}$

practice a. $\frac{4x + 20y + xy}{4xy}$ b. $\frac{4x + c + 10}{4}$ c. $\frac{ac^3 + m^4 + xmc^3}{c^3m^4}$ d. $\frac{mp + 3 - 4p^3}{p^3}$

problem set 47 1. -2 2. 1278 3. 0.023 4. $-\frac{100}{11}$ 5. $\frac{7}{8}$ 6. $\frac{5}{4}$ 7. 0.9

8. $y = \frac{5}{3}x + \frac{4}{3}$ 9. $y = \frac{1}{2}x + \frac{5}{2}$ 10. $\frac{50(5280)(5280)(12)(12)(2.54)(2.54)}{100(100)}$ m^2

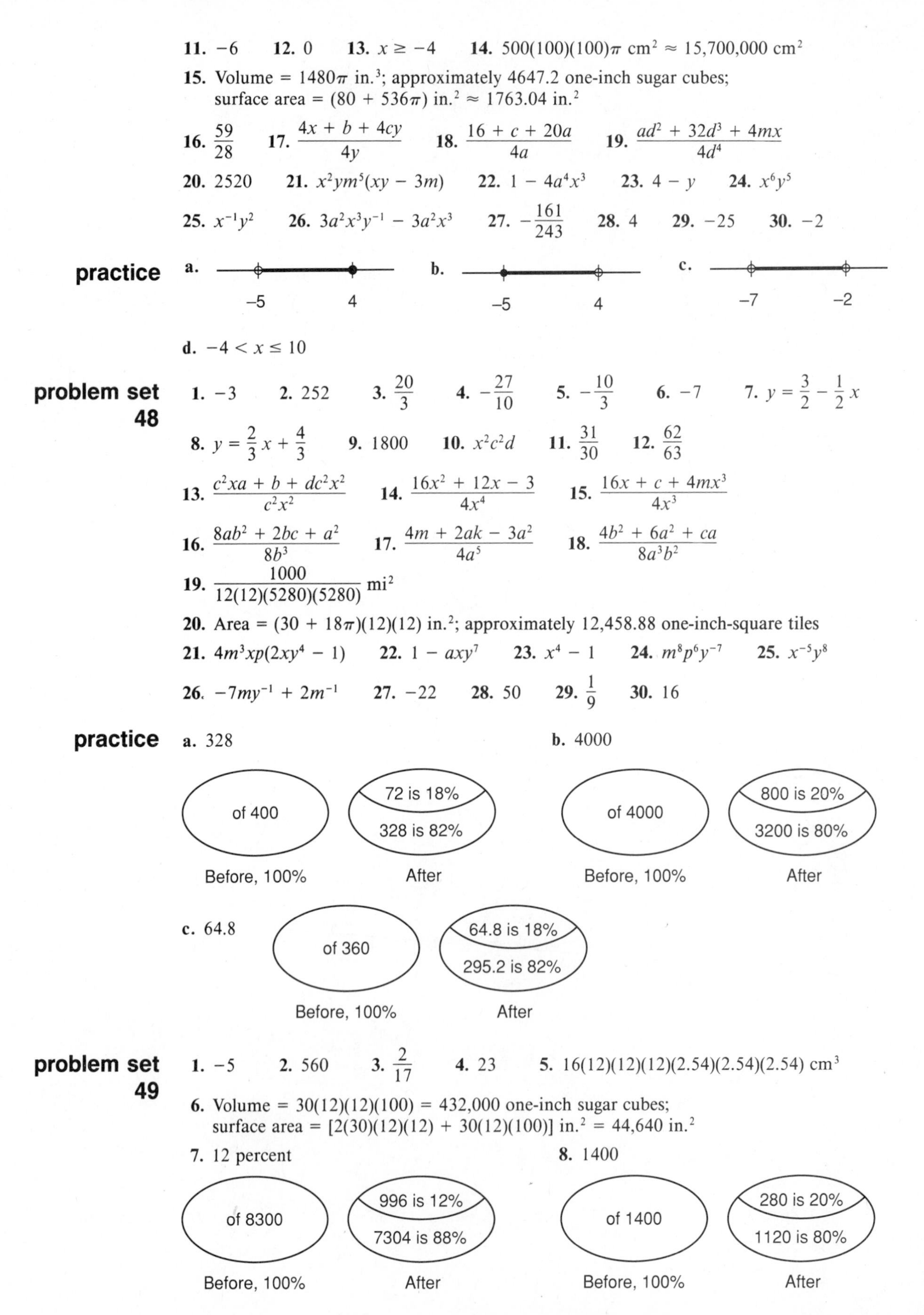

11. -6 **12.** 0 **13.** $x \geq -4$ **14.** $500(100)(100)\pi \text{ cm}^2 \approx 15{,}700{,}000 \text{ cm}^2$

15. Volume = 1480π in.^{3}; approximately 4647.2 one-inch sugar cubes; surface area = $(80 + 536\pi)$ in.$^2 \approx 1763.04$ in.^{2}

16. $\frac{59}{28}$ **17.** $\frac{4x + b + 4cy}{4y}$ **18.** $\frac{16 + c + 20a}{4a}$ **19.** $\frac{ad^2 + 32d^3 + 4mx}{4d^4}$

20. 2520 **21.** $x^2ym^5(xy - 3m)$ **22.** $1 - 4a^4x^3$ **23.** $4 - y$ **24.** x^6y^5

25. $x^{-1}y^2$ **26.** $3a^2x^3y^{-1} - 3a^2x^3$ **27.** $-\frac{161}{243}$ **28.** 4 **29.** -25 **30.** -2

practice

a.

b.

c.

d. $-4 < x \leq 10$

problem set 48

1. -3 **2.** 252 **3.** $\frac{20}{3}$ **4.** $-\frac{27}{10}$ **5.** $-\frac{10}{3}$ **6.** -7 **7.** $y = \frac{3}{2} - \frac{1}{2}x$

8. $y = \frac{2}{3}x + \frac{4}{3}$ **9.** 1800 **10.** x^2c^2d **11.** $\frac{31}{30}$ **12.** $\frac{62}{63}$

13. $\frac{c^2xa + b + dc^2x^2}{c^2x^2}$ **14.** $\frac{16x^2 + 12x - 3}{4x^4}$ **15.** $\frac{16x + c + 4mx^3}{4x^3}$

16. $\frac{8ab^2 + 2bc + a^2}{8b^3}$ **17.** $\frac{4m + 2ak - 3a^2}{4a^5}$ **18.** $\frac{4b^2 + 6a^2 + ca}{8a^3b^2}$

19. $\frac{1000}{12(12)(5280)(5280)}$ mi^2

20. Area = $(30 + 18\pi)(12)(12)$ in.^{2}; approximately 12,458.88 one-inch-square tiles

21. $4m^3xp(2xy^4 - 1)$ **22.** $1 - axy^7$ **23.** $x^4 - 1$ **24.** $m^8p^6y^{-7}$ **25.** $x^{-5}y^8$

26. $-7my^{-1} + 2m^{-1}$ **27.** -22 **28.** 50 **29.** $\frac{1}{9}$ **30.** 16

practice

a. 328

b. 4000

c. 64.8

problem set 49

1. -5 **2.** 560 **3.** $\frac{2}{17}$ **4.** 23 **5.** $16(12)(12)(12)(2.54)(2.54)(2.54) \text{ cm}^3$

6. Volume = $30(12)(12)(100) = 432{,}000$ one-inch sugar cubes; surface area = $[2(30)(12)(12) + 30(12)(100)]$ in.$^2 = 44{,}640$ in.^{2}

7. 12 percent

8. 1400

9. 14 **10.** −9.2 **11.** $y = \frac{3}{4}x - \frac{7}{4}$ **12.** $y = \frac{3}{2}x + \frac{5}{2}$ **13.** 840

14. $8x^2ym^3$ **15.** 1 **16.** $\frac{ac + xb + dxc}{xc}$ **17.** $\frac{k^2c^2m + kd^2 - 3p}{xk^2c^3}$

18. $\frac{4ad - a^2b^2c - bdm}{a^3db^2}$ **19.**

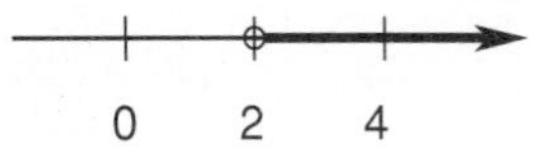
0
2
4

20.

2
5

21. $9x^3ym(2x^2y - m^4)$ **22.** $x^{-2} - 3y^{-6}p^{-6}$ **23.** $x - 1$ **24.** k^3p^{-8}

25. $m^8x^{-6}y^4$ **26.** $m^2x^2 - 3m^{-6}x^2 - 3mx^{-1}$ **27.** $\frac{-37}{9}$ **28.** −13 **29.** −5

30. 40

practice

a. $-2x^5 - 7x^4 + 7x^3 + 3x^2 - 2x - 3$ **b.** $-14x^4 - 2x^3 + 9x^2 - 11x + 5$

c. (2), (3), and (4) are polynomials

problem set 50

1. −8 **2.** 294 **3.** $\frac{2}{9}$

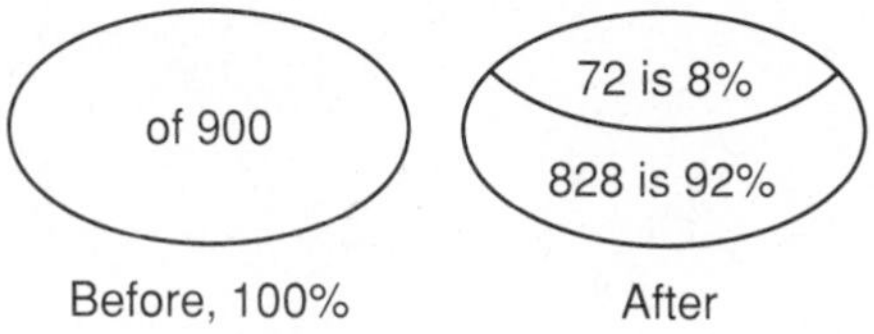
of 900
72 is 8%
828 is 92%
Before, 100%
After

5. 5 percent

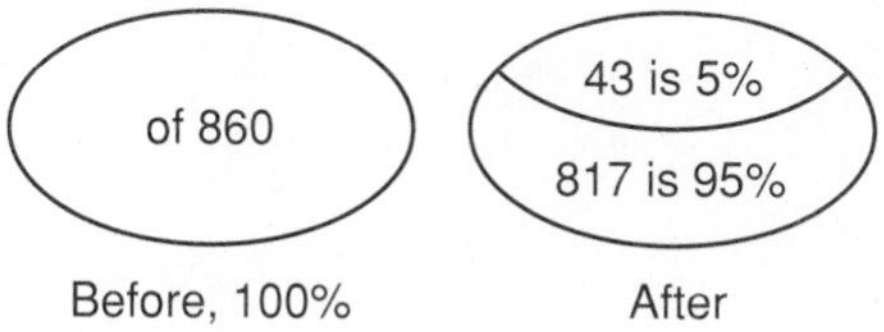
of 860
43 is 5%
817 is 95%
Before, 100%
After

6. 672

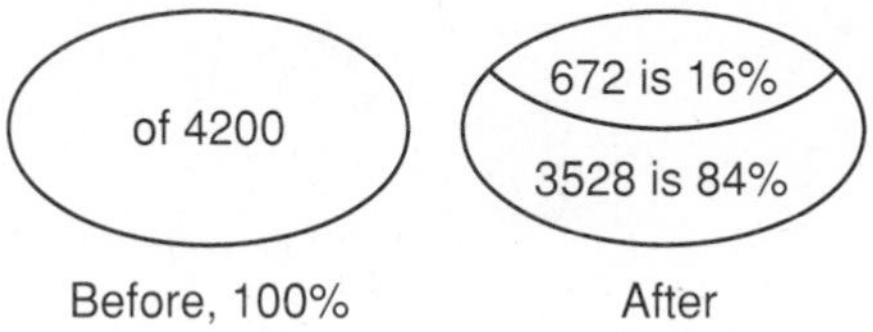
of 4200
672 is 16%
3528 is 84%
Before, 100%
After

7. 5000

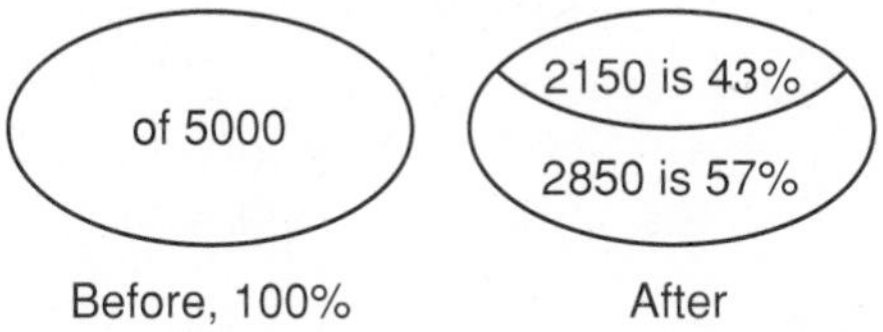
of 5000
2150 is 43%
2850 is 57%
Before, 100%
After

8. 2 percent

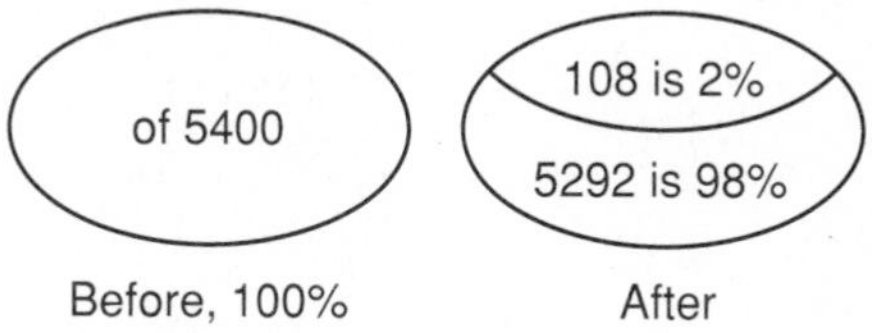
of 5400
108 is 2%
5292 is 98%
Before, 100%
After

9. $-\frac{19}{7}$ **10.** −9 **11.** $y = \frac{3}{4}x + \frac{7}{4}$ **12.** 1500 **13.** $\frac{103}{105}$

14. $\frac{xym^2 + xmb + ac}{x^2ym}$ **15.** $\frac{24cd + 10d + 7c}{8c^2d^2}$ **16.** $\frac{pm + 5 + x^2am}{xam}$

17. $\frac{65{,}000}{5280(5280)}$ mi^2 **18.** $\frac{960}{100}$ m = 9.6 m **19.** −16 **20.** $-3x^3 - 7x^2 + 11$

21.

−4
1

22. $x^2ym(1 - 4m^2 + 2x^2y^2m^5)$ **23.** $1 - x^2$

24. x^3ym^{-5} **25.** $y^{-5}p^3 - 3y^{-3}p^3$ **26.** $4x^2y^{-2}p + 2xy^2$ **27.** −120 **28.** $-\frac{9}{2}$

29. $-\frac{1}{27}$ **30.** 23

practice a. $a^2 - 36$ b. $10x^2 - 14x - 12$ c. $25x^2 - 60x + 36$

problem set 51

1. -8 2. 1377 3. $\frac{184}{399}$ 4. 5600

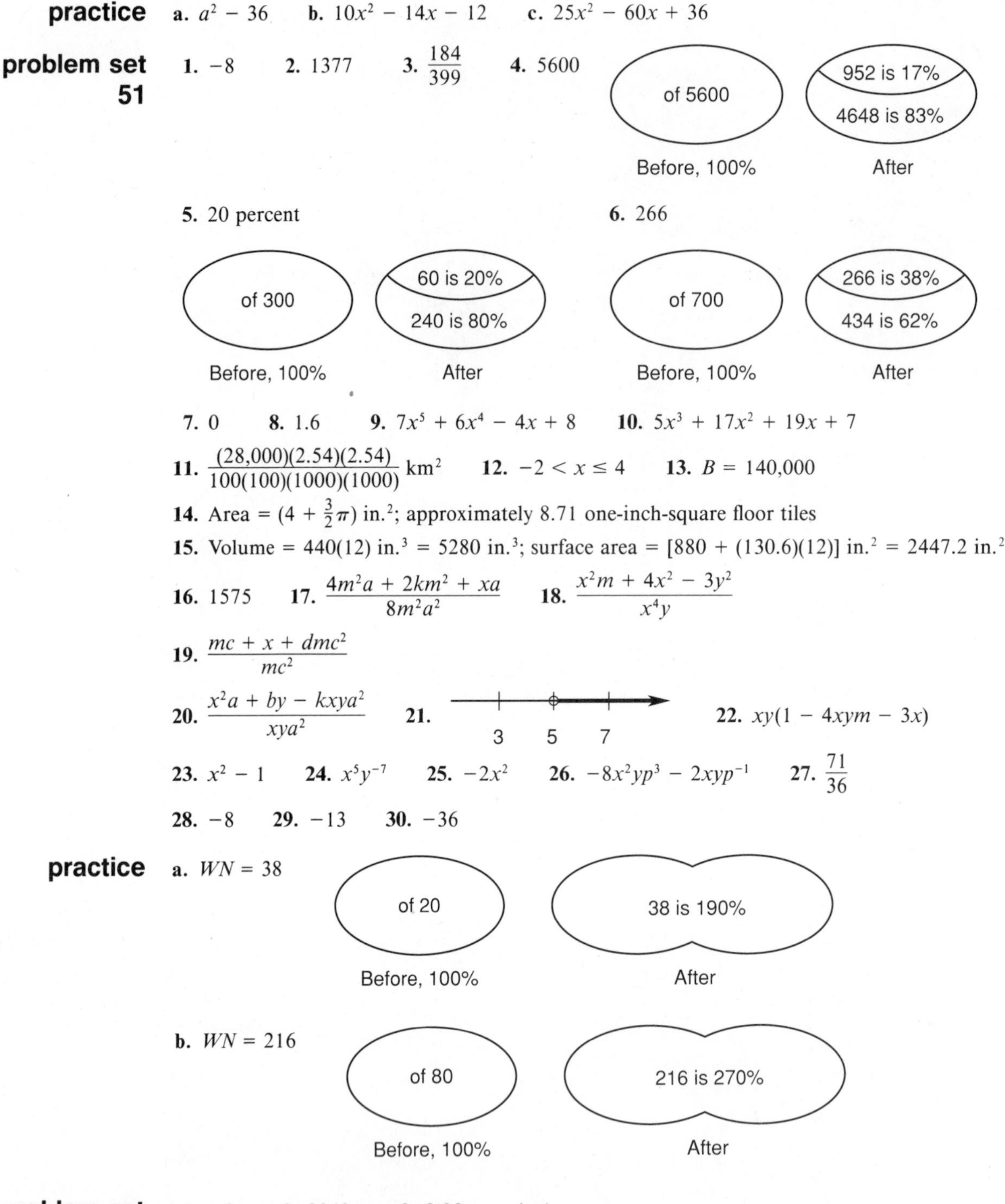

5. 20 percent

6. 266

7. 0 8. 1.6 9. $7x^5 + 6x^4 - 4x + 8$ 10. $5x^3 + 17x^2 + 19x + 7$

11. $\frac{(28{,}000)(2.54)(2.54)}{100(100)(1000)(1000)}$ km^2 12. $-2 < x \leq 4$ 13. $B = 140{,}000$

14. Area $= (4 + \frac{3}{2}\pi)$ in.^{2}; approximately 8.71 one-inch-square floor tiles

15. Volume $= 440(12)$ in.^{3} $= 5280$ in.^{3}; surface area $= [880 + (130.6)(12)]$ in.^{2} $= 2447.2$ in.^{2}

16. 1575 17. $\frac{4m^2a + 2km^2 + xa}{8m^2a^2}$ 18. $\frac{x^2m + 4x^2 - 3y^2}{x^4y}$

19. $\frac{mc + x + dmc^2}{mc^2}$

20. $\frac{x^2a + by - kxya^2}{xya^2}$ 21.

22. $xy(1 - 4xym - 3x)$

23. $x^2 - 1$ 24. x^5y^{-7} 25. $-2x^2$ 26. $-8x^2yp^3 - 2xyp^{-1}$ 27. $\frac{71}{36}$

28. -8 29. -13 30. -36

practice a. $WN = 38$

b. $WN = 216$

problem set 52

1. -6 2. 2940 3. 2.03 4. 4

5. 202 cm^2 6. 1400 percent

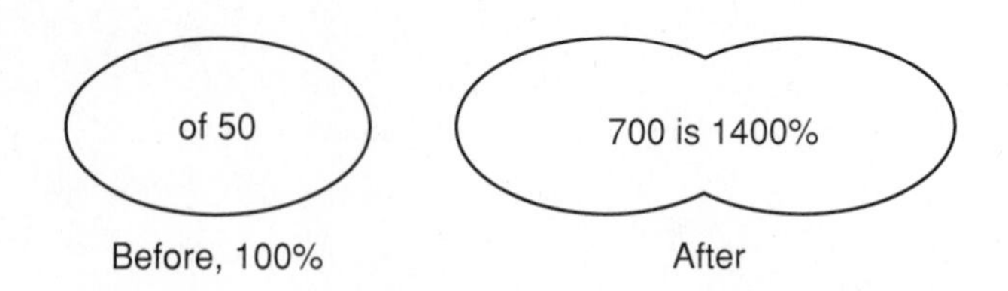

7. $-\frac{11}{14}$ **8.** $\frac{13}{4}$ **9.** $-\frac{3}{4}$ **10.** $-2x^4 + x^3 - x^2 - 12x + 25$

11. $10x^2 + 14x - 12$ **12.** $x^2 + 6x + 9$ **13.** $3x^2 + 5x - 12$ **14.** $4x^2 - 49$

15. 1800 **16.** $\frac{cxa + b + dcx^2}{cx^2}$ **17.** $\frac{15km - 20apk + 18p}{15p^2k^2}$

18. $\frac{amx^2 + bcx^2 - 4m^2c}{mx^2}$ **19.** $\frac{2kxc + 2bkmc - 4m}{2kmc^2}$ **20.** $4 \leq x < 9$

21. $2xym(2x - 3 + xym)$ **22.** $x - 1$ **23.** $x^5p^{-2}y^{-2}$ **24.** $1 + 4x^8a^{-2}y^{-4}$

25. $-xym^2$ **26.** $\frac{8}{3}$ **27.** 126 **28.** $-\frac{1}{81}$ **29.** -21 **30.** -21

practice

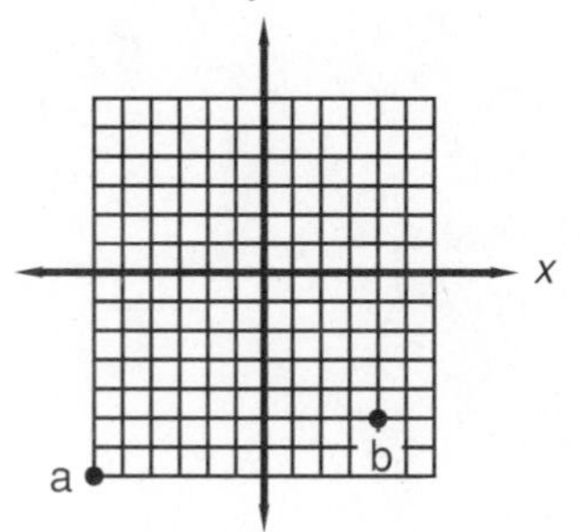

problem set 53

1. -6 **2.** 9380 **3.** $\frac{2}{17}$

4. 60

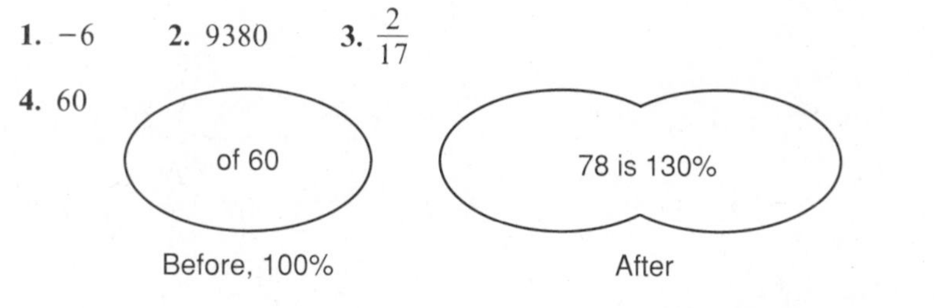

5. 70 **6.** 64 percent

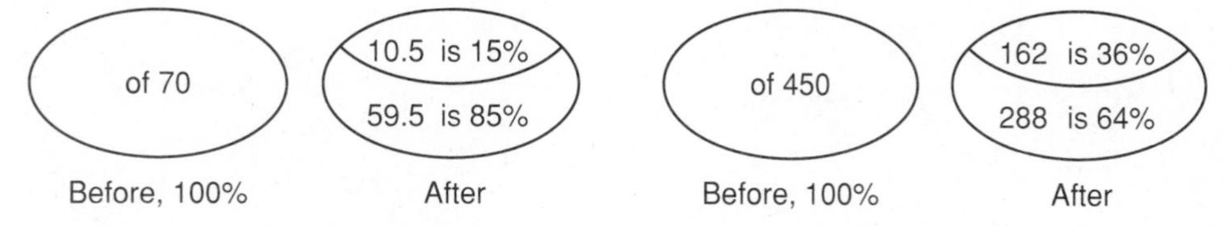

7. $\frac{285}{136}$ **8.** -4.25 **9.** $3x^2 + 10x - 8$ **10.** $25x^2 - 30x + 9$

11. $6x^2 - 19x + 10$ **12.** $30x^2 - 47x + 7$

13. Volume = 1264 yd^3; surface area = 1579.6 yd^2

14. $\frac{400(2.54)(2.54)(12)(12)}{100(100)(1000)(1000)}$ km^2 **15, 16.**

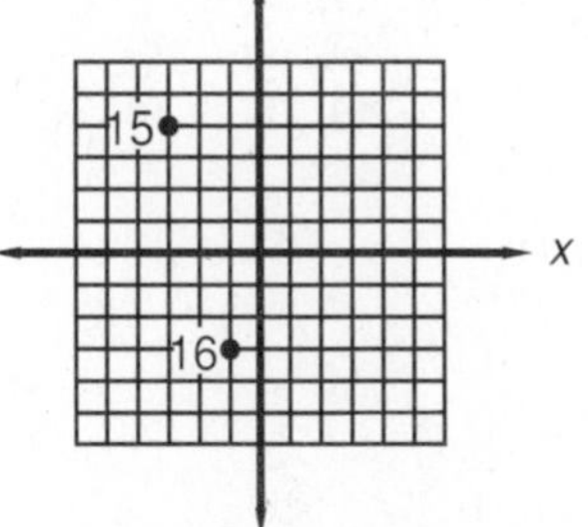

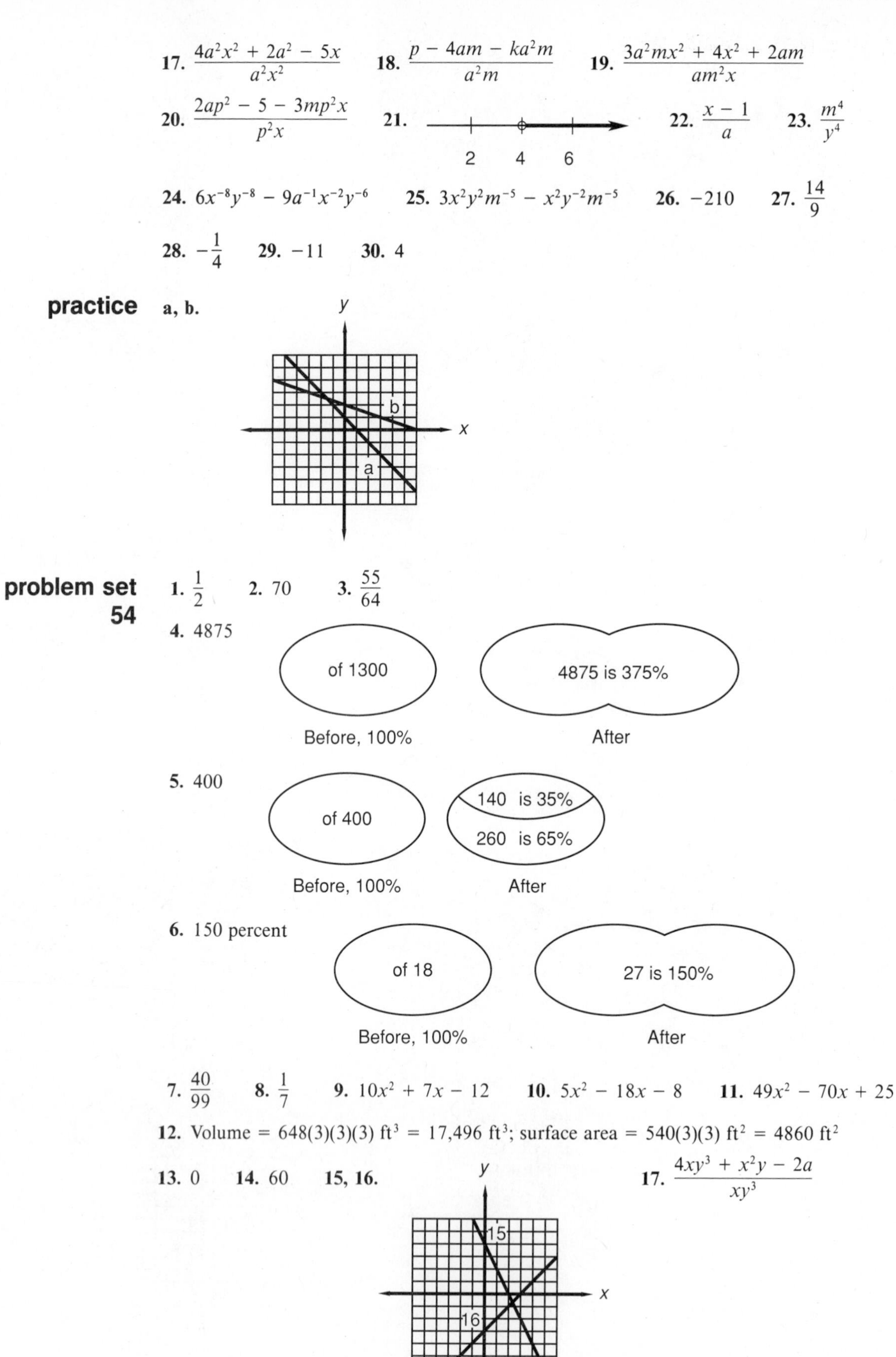

17. $\dfrac{4a^2x^2 + 2a^2 - 5x}{a^2x^2}$ 18. $\dfrac{p - 4am - ka^2m}{a^2m}$ 19. $\dfrac{3a^2mx^2 + 4x^2 + 2am}{am^2x}$

20. $\dfrac{2ap^2 - 5 - 3mp^2x}{p^2x}$ 21. (number line: 2 4 6) 22. $\dfrac{x - 1}{a}$ 23. $\dfrac{m^4}{y^4}$

24. $6x^{-8}y^{-8} - 9a^{-1}x^{-2}y^{-6}$ 25. $3x^2y^2m^{-5} - x^2y^{-2}m^{-5}$ 26. -210 27. $\dfrac{14}{9}$

28. $-\dfrac{1}{4}$ 29. -11 30. 4

practice a, b.

problem set 54

1. $\dfrac{1}{2}$ 2. 70 3. $\dfrac{55}{64}$

4. 4875

5. 400

6. 150 percent

7. $\dfrac{40}{99}$ 8. $\dfrac{1}{7}$ 9. $10x^2 + 7x - 12$ 10. $5x^2 - 18x - 8$ 11. $49x^2 - 70x + 25$

12. Volume = 648(3)(3)(3) ft^3 = 17,496 ft^3; surface area = 540(3)(3) ft^2 = 4860 ft^2

13. 0 14. 60 15, 16. 17. $\dfrac{4xy^3 + x^2y - 2a}{xy^3}$

18. $\frac{xca - 3a - 4x^2c + 3ac + 2c}{x^3c^2}$ **19.** $\frac{mc^2b - p + 4m^3pc^2}{m^3pc^2}$

20. $\frac{4x^3p^5 + 8 + x^2yp^4}{4x^2p^5}$ **21.** $-2 < x \leq 4$ **22.** $3y$ **23.** $p^{14}x^{-2}$ **24.** 1

25. $x^{-2} - 12y^{-9}p^{-1}$ **26.** $-2xy^{-1} - 2xy^{-2}$ **27.** $\frac{7}{4}$ **28.** 28 **29.** -39 **30.** -24

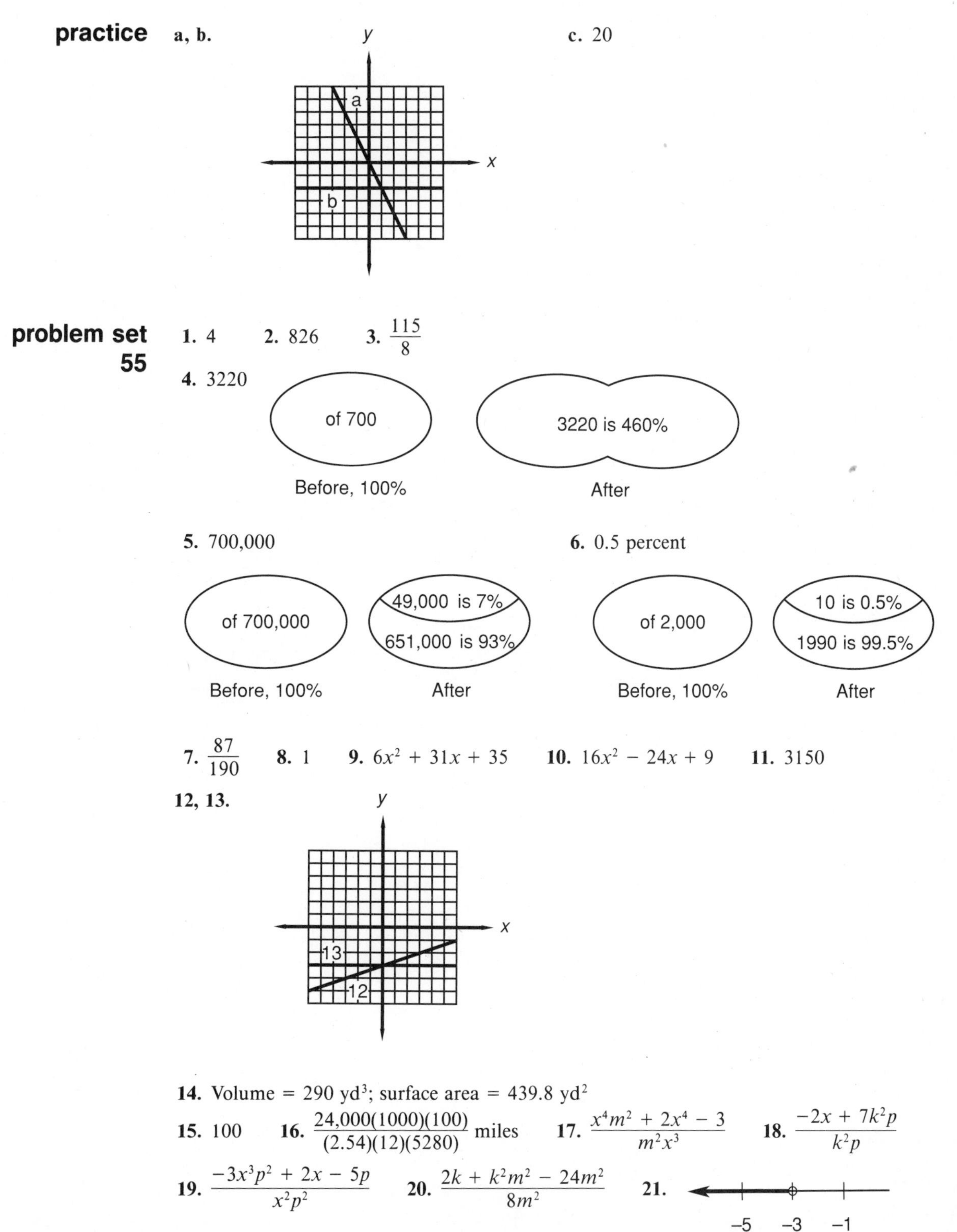

practice **a, b.** (graph) **c.** 20

problem set 55

1. 4 **2.** 826 **3.** $\frac{115}{8}$

4. 3220

5. 700,000 **6.** 0.5 percent

7. $\frac{87}{190}$ **8.** 1 **9.** $6x^2 + 31x + 35$ **10.** $16x^2 - 24x + 9$ **11.** 3150

12, 13. (graph)

14. Volume = 290 yd^3; surface area = 439.8 yd^2

15. 100 **16.** $\frac{24{,}000(1000)(100)}{(2.54)(12)(5280)}$ miles **17.** $\frac{x^4m^2 + 2x^4 - 3}{m^2x^3}$ **18.** $\frac{-2x + 7k^2p}{k^2p}$

19. $\frac{-3x^3p^2 + 2x - 5p}{x^2p^2}$ **20.** $\frac{2k + k^2m^2 - 24m^2}{8m^2}$ **21.** (number line)

22. $2kp(2kz - 3k^2pz^5 - kpz^2 - 2)$ **23.** $1 - 4y$ **24.** $k^{11}m^{-6}p^{-2}$
25. $1 - 15x^3m^2$ **26.** $-a^2k^2y^{-1} - 6k^{-6}y^{-1}$ **27.** 18 **28.** 52 **29.** -21
30. -47

practice a. $\dfrac{mx - bmy + y(c + d)}{my}$ b. $\dfrac{b(5b + c) - x(a + b) + bc(a + b)}{b(a + b)}$

problem set 56 **1.** -5 **2.** 280 **3.** 0.003
4. 700 **5.** 261

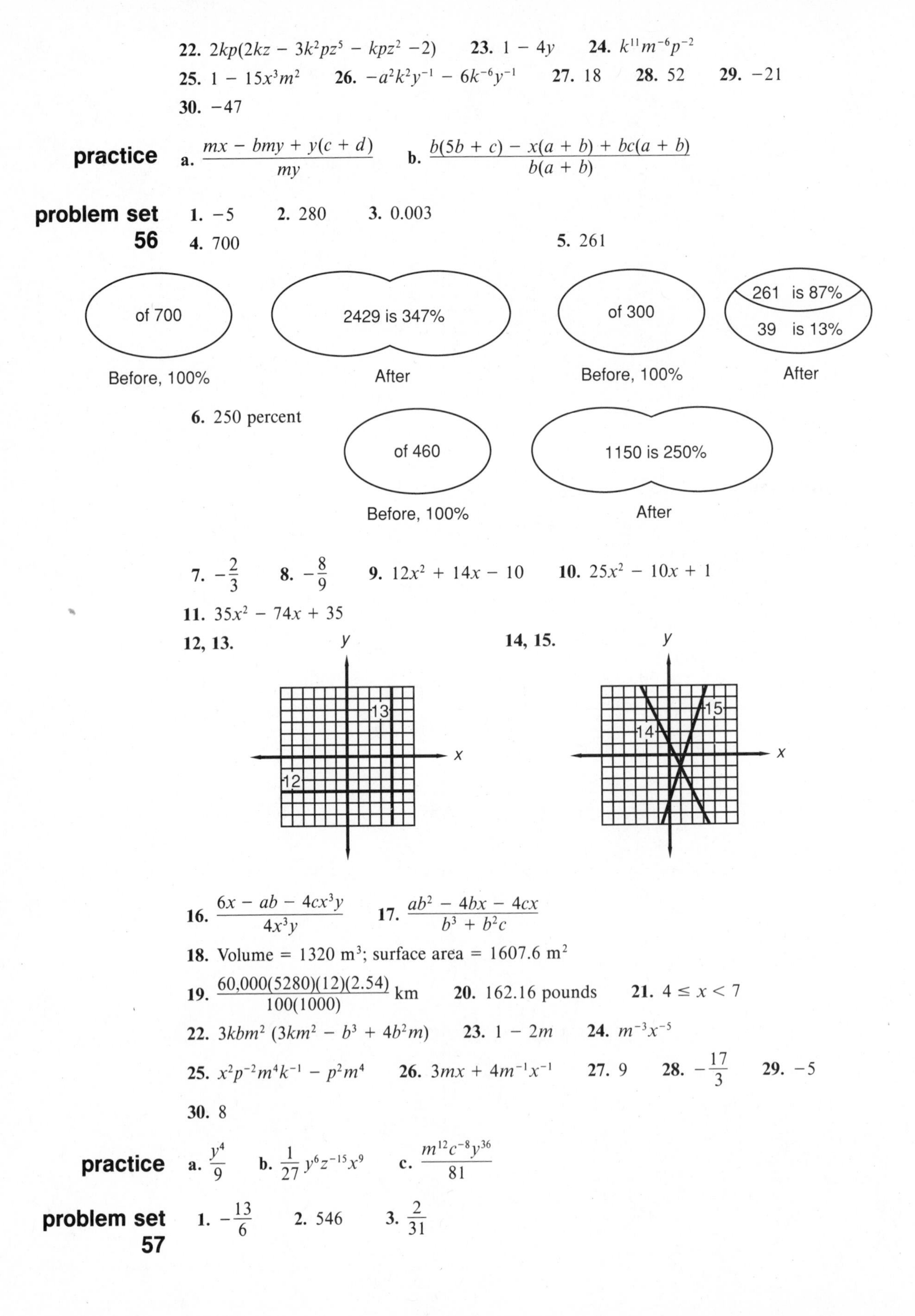

6. 250 percent

7. $-\dfrac{2}{3}$ **8.** $-\dfrac{8}{9}$ **9.** $12x^2 + 14x - 10$ **10.** $25x^2 - 10x + 1$
11. $35x^2 - 74x + 35$
12, 13. **14, 15.**

16. $\dfrac{6x - ab - 4cx^3y}{4x^3y}$ **17.** $\dfrac{ab^2 - 4bx - 4cx}{b^3 + b^2c}$
18. Volume = 1320 m^3; surface area = 1607.6 m^2
19. $\dfrac{60{,}000(5280)(12)(2.54)}{100(1000)}$ km **20.** 162.16 pounds **21.** $4 \le x < 7$
22. $3kbm^2(3km^2 - b^3 + 4b^2m)$ **23.** $1 - 2m$ **24.** $m^{-3}x^{-5}$
25. $x^2p^{-2}m^4k^{-1} - p^2m^4$ **26.** $3mx + 4m^{-1}x^{-1}$ **27.** 9 **28.** $-\dfrac{17}{3}$ **29.** -5
30. 8

practice a. $\dfrac{y^4}{9}$ b. $\dfrac{1}{27}y^6z^{-15}x^9$ c. $\dfrac{m^{12}c^{-8}y^{36}}{81}$

problem set 57 **1.** $-\dfrac{13}{6}$ **2.** 546 **3.** $\dfrac{2}{31}$

4. 60

of 60 — Before, 100%

192 is 320% — After

5. 4 percent

of 98 — Before, 100%

3.92 is 4% / 94.08 is 96% — After

6. 165.6

of 72 — Before, 100%

165.6 is 230% — After

7. $-\frac{625}{396}$ **8.** $\frac{1}{6}$ **9.** $2x^2 - 10x + 12$ **10.** $9x^2 + 30x + 25$

11, 12, 13.

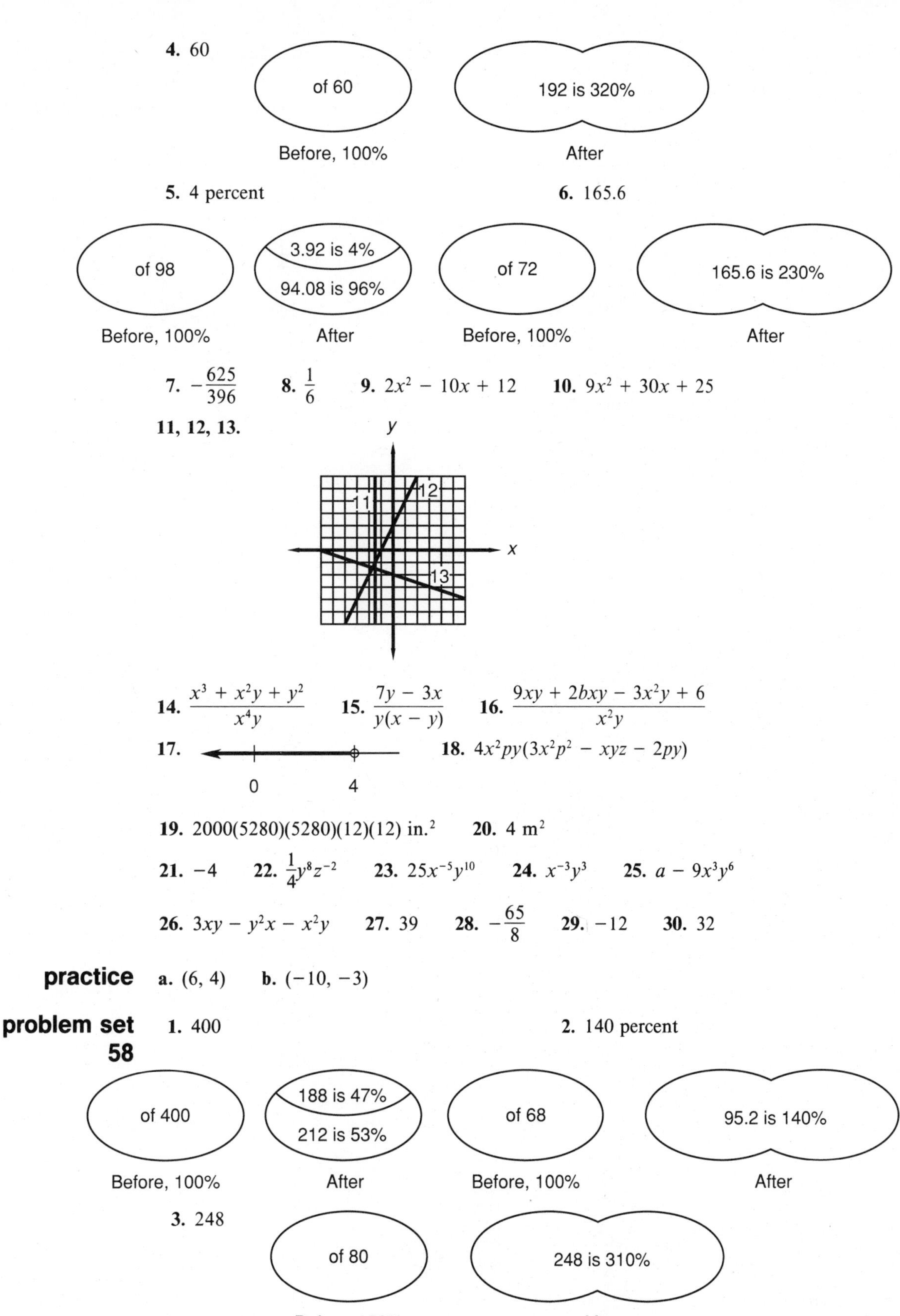

14. $\frac{x^3 + x^2y + y^2}{x^4y}$ **15.** $\frac{7y - 3x}{y(x - y)}$ **16.** $\frac{9xy + 2bxy - 3x^2y + 6}{x^2y}$

17. (number line: 0, 4) **18.** $4x^2py(3x^2p^2 - xyz - 2py)$

19. $2000(5280)(5280)(12)(12)\text{ in.}^2$ **20.** 4 m^2

21. -4 **22.** $\frac{1}{4}y^8z^{-2}$ **23.** $25x^{-5}y^{10}$ **24.** $x^{-3}y^3$ **25.** $a - 9x^3y^6$

26. $3xy - y^2x - x^2y$ **27.** 39 **28.** $-\frac{65}{8}$ **29.** -12 **30.** 32

practice **a.** (6, 4) **b.** (−10, −3)

problem set 58

1. 400

of 400 — Before, 100%

188 is 47% / 212 is 53% — After

2. 140 percent

of 68 — Before, 100%

95.2 is 140% — After

3. 248

of 80 — Before, 100%

248 is 310% — After

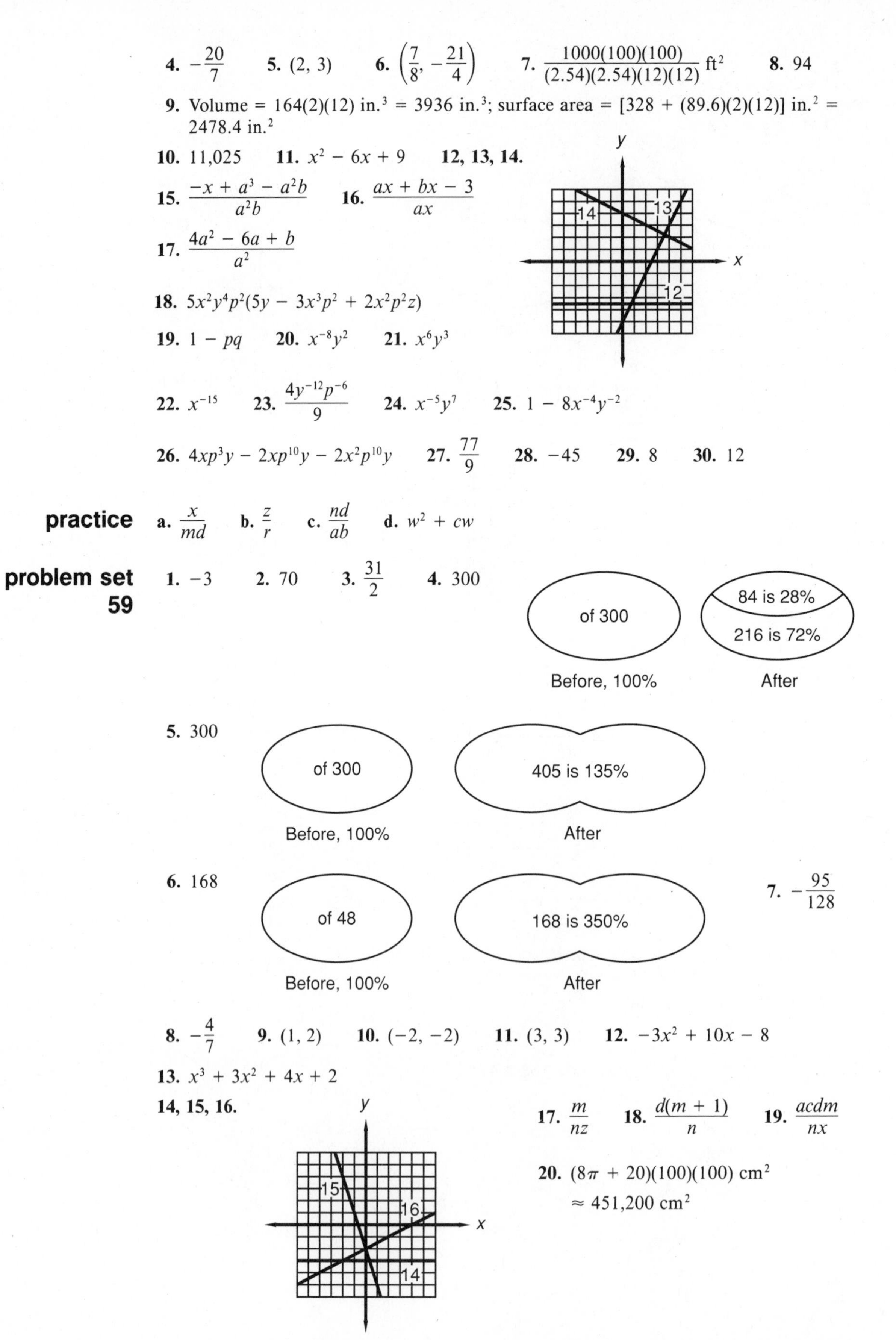

4. $-\frac{20}{7}$ 5. (2, 3) 6. $\left(\frac{7}{8}, -\frac{21}{4}\right)$ 7. $\frac{1000(100)(100)}{(2.54)(2.54)(12)(12)}$ ft^2 8. 94

9. Volume = 164(2)(12) in.^{3} = 3936 in.^{3}; surface area = [328 + (89.6)(2)(12)] in.^{2} = 2478.4 in.^{2}

10. 11,025 11. $x^2 - 6x + 9$ 12, 13, 14.

15. $\frac{-x + a^3 - a^2b}{a^2b}$ 16. $\frac{ax + bx - 3}{ax}$

17. $\frac{4a^2 - 6a + b}{a^2}$

18. $5x^2y^4p^2(5y - 3x^3p^2 + 2x^2p^2z)$

19. $1 - pq$ 20. $x^{-8}y^2$ 21. x^6y^3

22. x^{-15} 23. $\frac{4y^{-12}p^{-6}}{9}$ 24. $x^{-5}y^7$ 25. $1 - 8x^{-4}y^{-2}$

26. $4xp^3y - 2xp^{10}y - 2x^2p^{10}y$ 27. $\frac{77}{9}$ 28. -45 29. 8 30. 12

practice a. $\frac{x}{md}$ b. $\frac{z}{r}$ c. $\frac{nd}{ab}$ d. $w^2 + cw$

problem set 59 1. -3 2. 70 3. $\frac{31}{2}$ 4. 300

5. 300

6. 168

7. $-\frac{95}{128}$

8. $-\frac{4}{7}$ 9. (1, 2) 10. $(-2, -2)$ 11. (3, 3) 12. $-3x^2 + 10x - 8$

13. $x^3 + 3x^2 + 4x + 2$

14, 15, 16.

17. $\frac{m}{nz}$ 18. $\frac{d(m + 1)}{n}$ 19. $\frac{acdm}{nx}$

20. $(8\pi + 20)(100)(100)$ cm^2 $\approx 451{,}200$ cm^2

21. $\dfrac{4y^2 - 3x - 3y}{xy^2 + y^3}$ **22.** $\dfrac{ax + ay + 4ax^3y + 4ax^2y^2 - mx^2y}{x^3y + x^2y^2}$ **23.** $\dfrac{x + y + ay^2}{a^2}$

24. $x^{-9}y^{10}$ **25.** $x^{-3}y^{-9}$ **26.** $9p^9x^5y^7$ **27.** $x^2 - 1$ **28.** $x^{-8}p^{-2} + 2x^{-8}p^{-4}y^2$

29. $5xy^3z^{-1}$ **30.** 26

practice **a.** $K = \{0, 1, 3, 5, 9\}$ **b.** (a) True (b) True (c) True (d) False

c. $y = \dfrac{2}{3}x - 2$

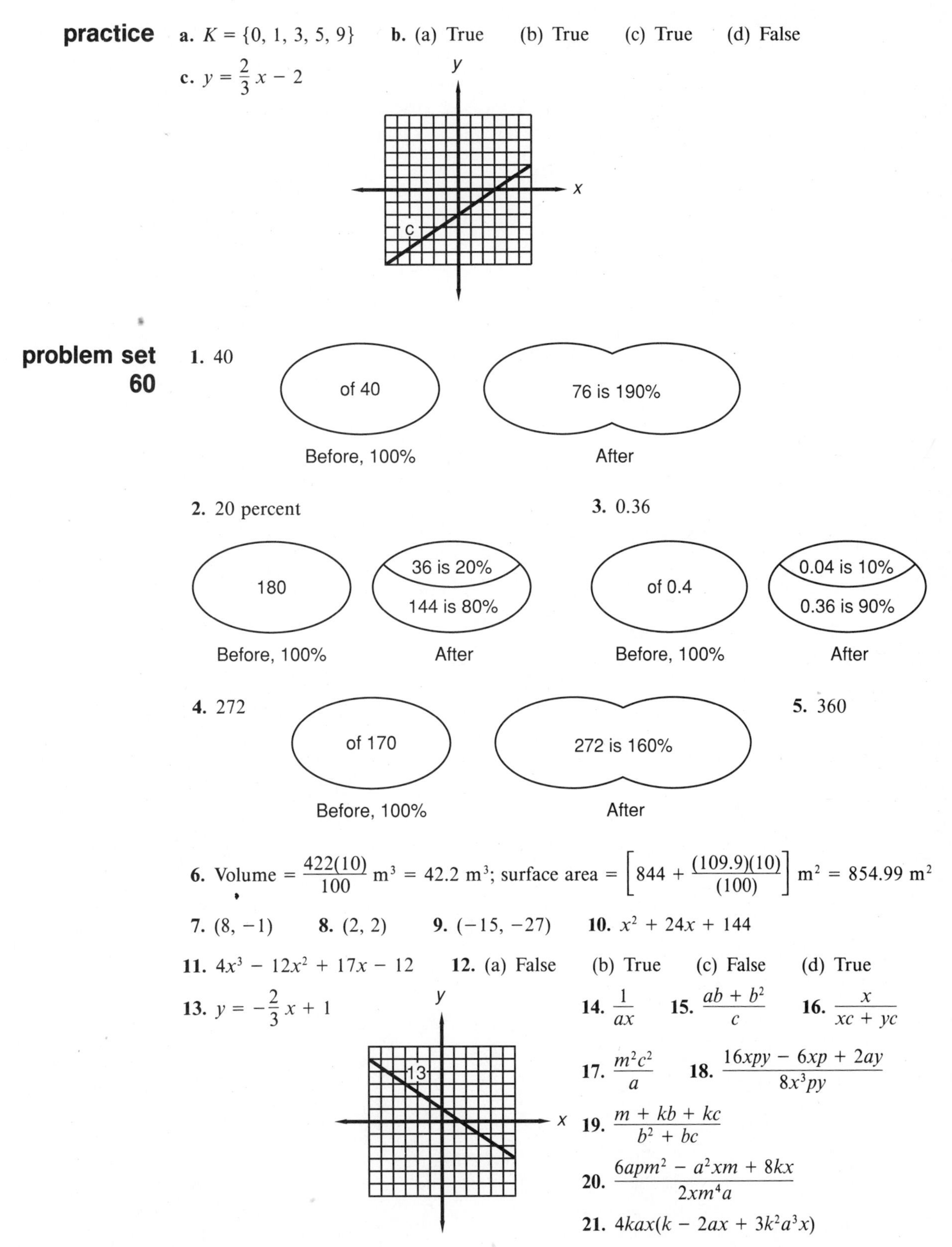

problem set 60

1. 40

2. 20 percent

3. 0.36

4. 272

5. 360

6. Volume $= \dfrac{422(10)}{100}$ m^3 $= 42.2$ m^3; surface area $= \left[844 + \dfrac{(109.9)(10)}{(100)}\right]$ m^2 $= 854.99$ m^2

7. $(8, -1)$ **8.** $(2, 2)$ **9.** $(-15, -27)$ **10.** $x^2 + 24x + 144$

11. $4x^3 - 12x^2 + 17x - 12$ **12.** (a) False (b) True (c) False (d) True

13. $y = -\dfrac{2}{3}x + 1$

14. $\dfrac{1}{ax}$ **15.** $\dfrac{ab + b^2}{c}$ **16.** $\dfrac{x}{xc + yc}$

17. $\dfrac{m^2c^2}{a}$ **18.** $\dfrac{16xpy - 6xp + 2ay}{8x^3py}$

19. $\dfrac{m + kb + kc}{b^2 + bc}$

20. $\dfrac{6apm^2 - a^2xm + 8kx}{2xm^4a}$

21. $4kax(k - 2ax + 3k^2a^3x)$

22. $\frac{x - xz}{z}$ **23.** $x^4y^{-20}m^{-2}$ **24.** $k^{-3}p^{10}$ **25.** $x^{-5}y^2p^8$ **26.** $1 - 2m^{-4}k$

27. $6x^2y^2m - 3x^3y^3$ **28.** $\frac{95}{8}$ **29.** 8 **30.** 1

practice

a. $(240 + 20\pi)\text{ m}^3 \approx 302.8\text{ m}^3$ **b.** $(48 + 4\pi)\text{ m}^3 \approx 60.56\text{ m}^3$
c. $\frac{4000}{3}\pi\text{ ft}^3 \approx 4186.67\text{ ft}^3$

problem set 61

1. 400 **2.** 25 percent **3.** 7000 **4.** 288 **5.** 18
6. Volume of cylinder $= 2000\pi\text{ cm}^3 \approx 6280\text{ cm}^3$; volume of sphere $= \frac{4000}{3}\pi\text{ cm}^3 \approx 4186.67\text{ cm}^3$

7. $\left(\frac{4}{5}, \frac{9}{5}\right)$ **8.** (2, 3) **9.** (2, 1) **10.** $x^2 + 10x + 25$

11. $6x^3 - 13x^2 + 10x - 6$ **12.** (a) False (b) True (c) True

13. y, x, 13

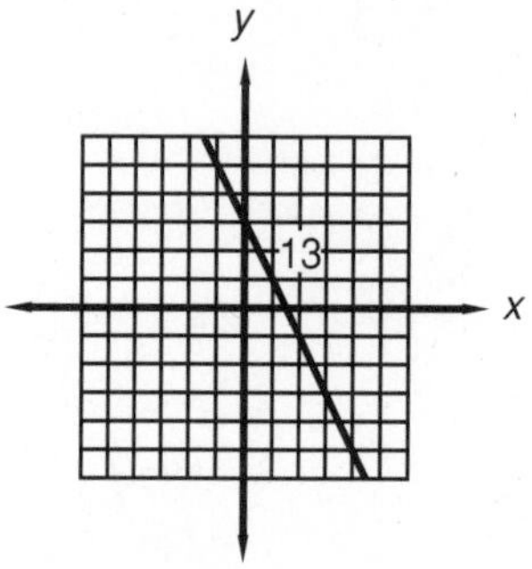

14. $\frac{1}{ax}$ **15.** $\frac{b}{ax + ay}$ **16.** $c(x + y)$

17. $\frac{bx}{a}$ **18.** $\frac{4py - 2px^2 + 3axy}{6px^4y}$

19. $\frac{km + cm + m}{k(c + k)}$

20. $\frac{3a^2p^2 - 2a^2mp + 8z}{2a^2p^4}$

21. $2bc(p^2 - 4bc + 6b^3cp^3)$

22. $x^2 + 1$ **23.** $4m^4x^4y^4$ **24.** $m^{10}x^{-3}$ **25.** $p^8x^2m^{-5}$

26. $-2a^{-2}b^{-3}m^{-4}p^{-4} + \frac{1}{2}p^{-1}$ **27.** $\frac{m^4x^2(2 + 7x)}{y}$ **28.** $\frac{33}{4}$ **29.** $\frac{4}{5}$ **30.** $3 + x$

practice

a. $NT = 100$

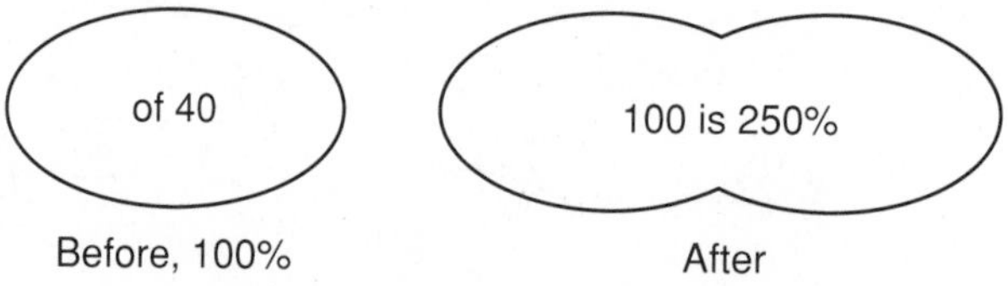

of 40
Before, 100%

100 is 250%
After

b. $WP = 20\%$

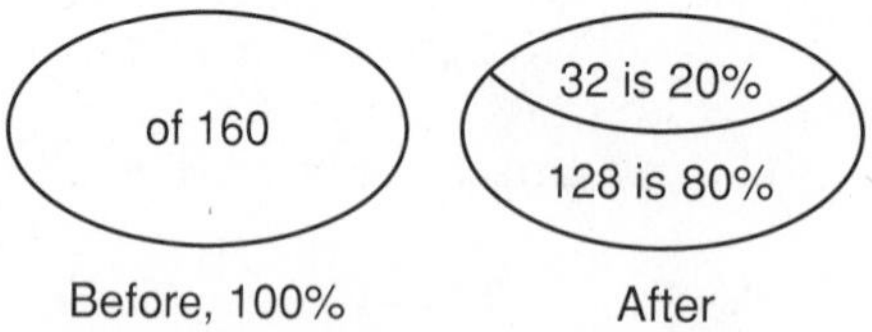

of 160
Before, 100%

32 is 20%
128 is 80%
After

c. $NR = 1680$

of 2400
Before, 100%

720 is 30%
1680 is 70%
After

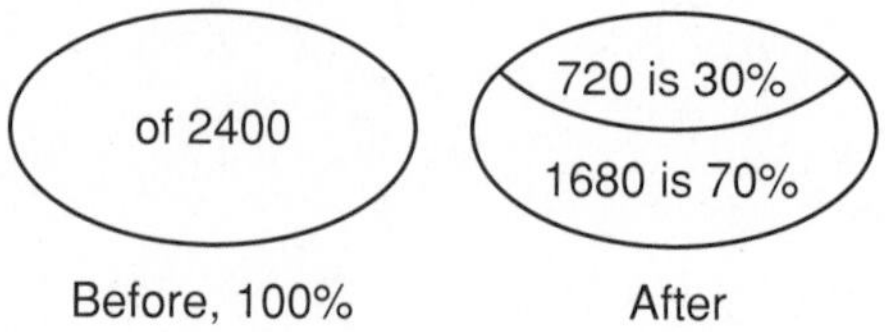

problem set 62

1. $-\frac{1}{3}$ **2.** 5

3. Volume of prism $= 3015\text{ m}^3$; volume of pyramid $= 1105.5\text{ m}^3$

4. $\frac{420(1000)(100)}{(2.54)(12)}$ ft **5.** $\frac{3}{82}$ **6.** 0.02 **7.** $-\frac{155}{136}$ **8.** $-\frac{1}{7}$

9. 2 4 **10.** $K = \{0, 1, 2, 5, 7\}$

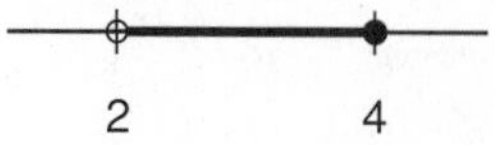

11. (a) False (b) False (c) True **12.** $(-1, -3)$ **13.** $(-2, -3)$

14. $(-3, -3)$ **15.** $-6x^2 + 14x + 40$ **16.** $16x^2 + 16x + 4$

17, 18.

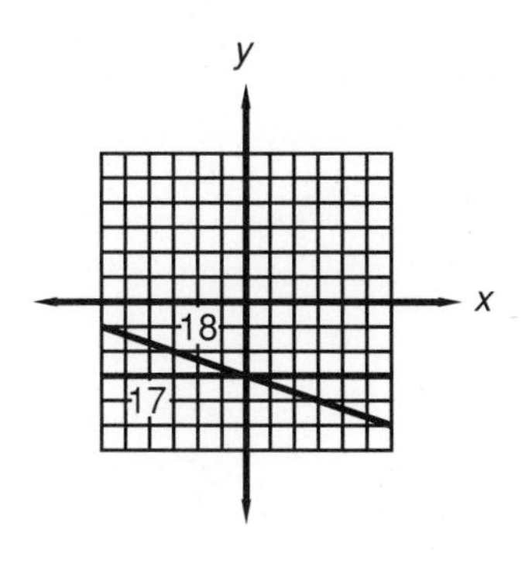

19. $xa + xb$

20. $\frac{1}{xa + xb}$ **21.** $\frac{ax}{b}$

22. $a + b$ **23.** $\frac{xy + x + y}{y^2 + y}$

24. $\frac{y + x}{y}$ **25.** $\frac{y^2 - 1}{y}$

26. $5x^2y^2z(2y^3 - x^3z^4 - 2x^2y^2z^3)$

27. $x^2m^{-4}y^3$ **28.** x^4y^3 **29.** $8x^3y^6m^6$

30. 10

practice

a. (11, 6) **b.** (11, 3)

problem set 63

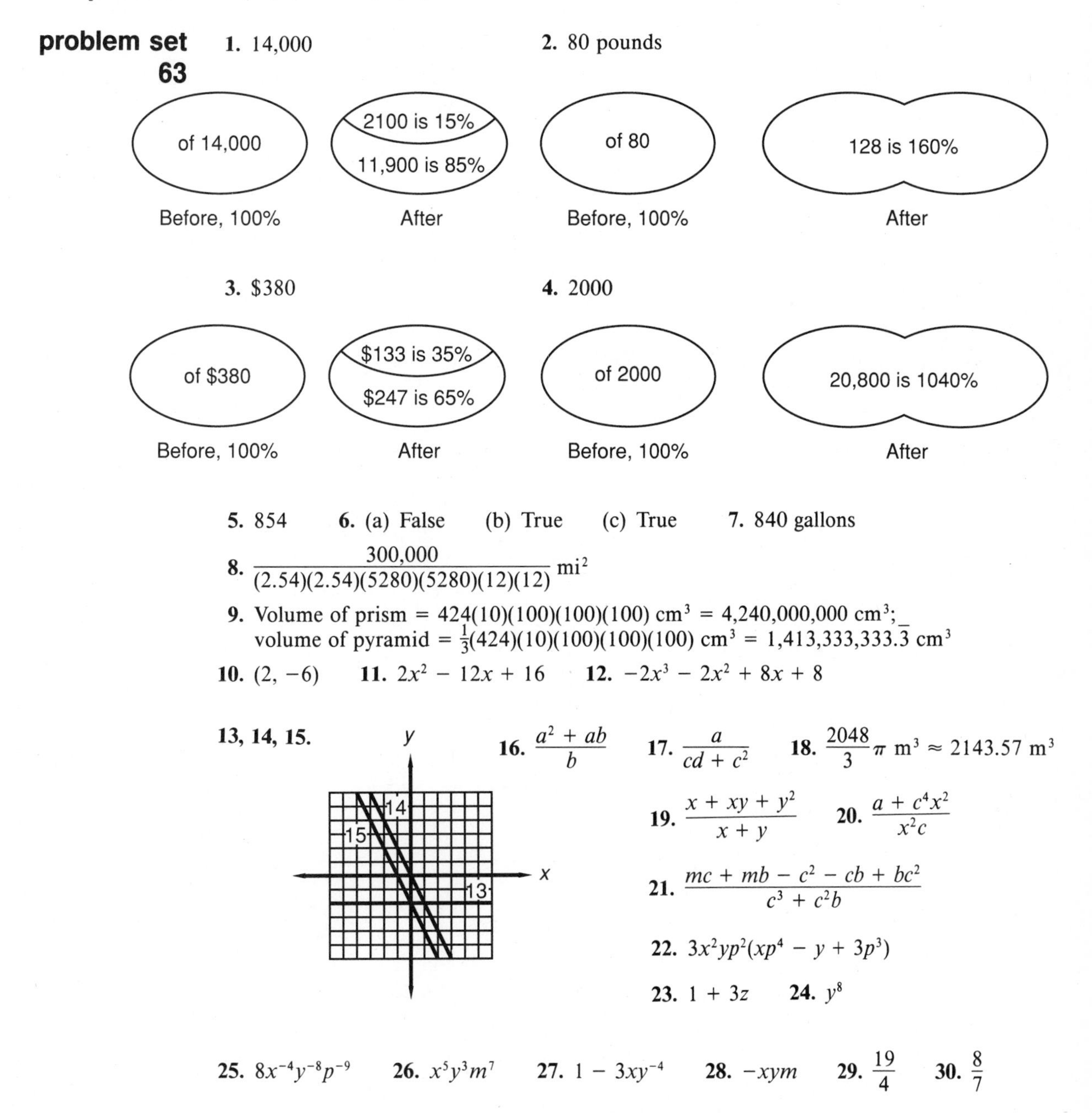

1. 14,000 **2.** 80 pounds

3. \$380 **4.** 2000

5. 854 **6.** (a) False (b) True (c) True **7.** 840 gallons

8. $\frac{300{,}000}{(2.54)(2.54)(5280)(5280)(12)(12)}$ mi^2

9. Volume of prism = 424(10)(100)(100)(100) cm^3 = 4,240,000,000 cm^3; volume of pyramid = $\frac{1}{3}$(424)(10)(100)(100)(100) cm^3 = 1,413,333,333.$\overline{3}$ cm^3

10. (2, −6) **11.** $2x^2 - 12x + 16$ **12.** $-2x^3 - 2x^2 + 8x + 8$

13, 14, 15.

16. $\frac{a^2 + ab}{b}$ **17.** $\frac{a}{cd + c^2}$ **18.** $\frac{2048}{3}\pi$ m^3 ≈ 2143.57 m^3

19. $\frac{x + xy + y^2}{x + y}$ **20.** $\frac{a + c^4x^2}{x^2c}$

21. $\frac{mc + mb - c^2 - cb + bc^2}{c^3 + c^2b}$

22. $3x^2yp^2(xp^4 - y + 3p^3)$

23. $1 + 3z$ **24.** y^8

25. $8x^{-4}y^{-8}p^{-9}$ **26.** $x^5y^3m^7$ **27.** $1 - 3xy^{-4}$ **28.** $-xym$ **29.** $\frac{19}{4}$ **30.** $\frac{8}{7}$

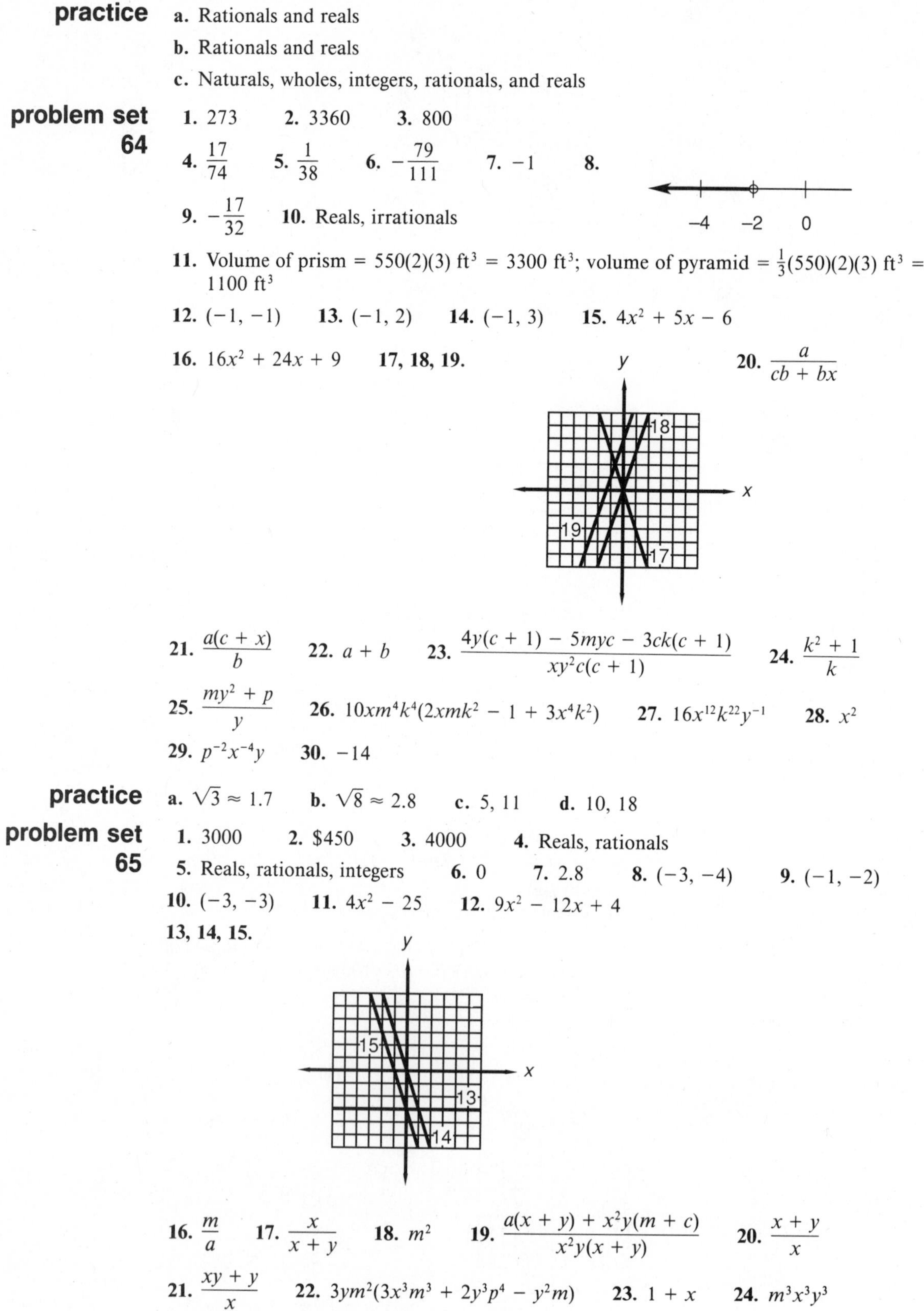

practice

a. Rationals and reals

b. Rationals and reals

c. Naturals, wholes, integers, rationals, and reals

problem set 64

1. 273 **2.** 3360 **3.** 800

4. $\frac{17}{74}$ **5.** $\frac{1}{38}$ **6.** $-\frac{79}{111}$ **7.** -1 **8.**

9. $-\frac{17}{32}$ **10.** Reals, irrationals

11. Volume of prism = 550(2)(3) ft^3 = 3300 ft^3; volume of pyramid = $\frac{1}{3}$(550)(2)(3) ft^3 = 1100 ft^3

12. $(-1, -1)$ **13.** $(-1, 2)$ **14.** $(-1, 3)$ **15.** $4x^2 + 5x - 6$

16. $16x^2 + 24x + 9$ **17, 18, 19.**

20. $\frac{a}{cb + bx}$

21. $\frac{a(c + x)}{b}$ **22.** $a + b$ **23.** $\frac{4y(c + 1) - 5myc - 3ck(c + 1)}{xy^2c(c + 1)}$ **24.** $\frac{k^2 + 1}{k}$

25. $\frac{my^2 + p}{y}$ **26.** $10xm^4k^4(2xmk^2 - 1 + 3x^4k^2)$ **27.** $16x^{12}k^{22}y^{-1}$ **28.** x^2

29. $p^{-2}x^{-4}y$ **30.** -14

practice

a. $\sqrt{3} \approx 1.7$ b. $\sqrt{8} \approx 2.8$ c. 5, 11 d. 10, 18

problem set 65

1. 3000 **2.** \$450 **3.** 4000 **4.** Reals, rationals

5. Reals, rationals, integers **6.** 0 **7.** 2.8 **8.** $(-3, -4)$ **9.** $(-1, -2)$

10. $(-3, -3)$ **11.** $4x^2 - 25$ **12.** $9x^2 - 12x + 4$

13, 14, 15.

16. $\frac{m}{a}$ **17.** $\frac{x}{x + y}$ **18.** m^2 **19.** $\frac{a(x + y) + x^2y(m + c)}{x^2y(x + y)}$ **20.** $\frac{x + y}{x}$

21. $\frac{xy + y}{x}$ **22.** $3ym^2(3x^3m^3 + 2y^3p^4 - y^2m)$ **23.** $1 + x$ **24.** $m^3x^3y^3$

25. a^5p^{-2} **26.** $9mx^3y^{-2}$ **27.** $-1 + 3x^{-6}y^{-3}$ **28.** $4y^2 - y^2x^{-2}$ **29.** 8

30. $k = 1$

practice a. $5\sqrt{3}$ b. $10\sqrt{2}$ c. $3\sqrt{21}$

problem set 66

1. 4
2. 6000
3. 36

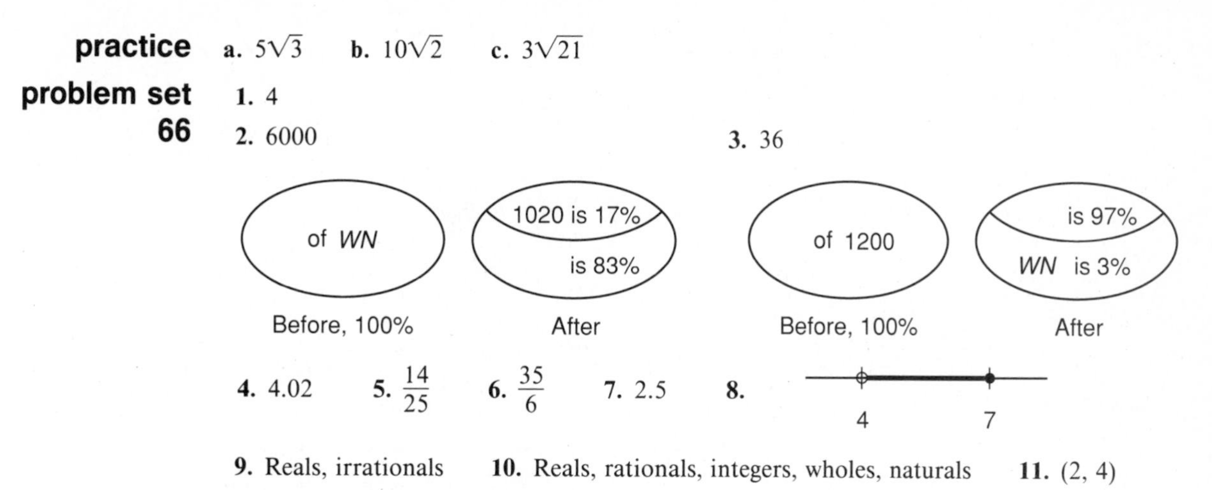

4. 4.02 5. $\frac{14}{25}$ 6. $\frac{35}{6}$ 7. 2.5 8.

4 7

9. Reals, irrationals 10. Reals, rationals, integers, wholes, naturals 11. (2, 4)
12. $(-29, -18)$ 13. $(5, -6)$ 14. 20
15. Volume of prism = 540(2)(3) ft^3 = 3240 ft^3; volume of pyramid = $\frac{1}{3}$(540)(2)(3) ft^3 = 1080 ft^3
16. $\frac{8(3)(3)(3)(12)(12)(12)(2.54)(2.54)(2.54)}{100(100)(100)}$ m^3 17. 93 lb
18. Volume of cylinder = $16{,}000\pi$ m^3 $\approx$ 50,240 m^3; volume of sphere = $\frac{32{,}000}{3}\pi$ m^3 $\approx$ $33{,}493.\overline{3}$ m^3

19, 20, 21.

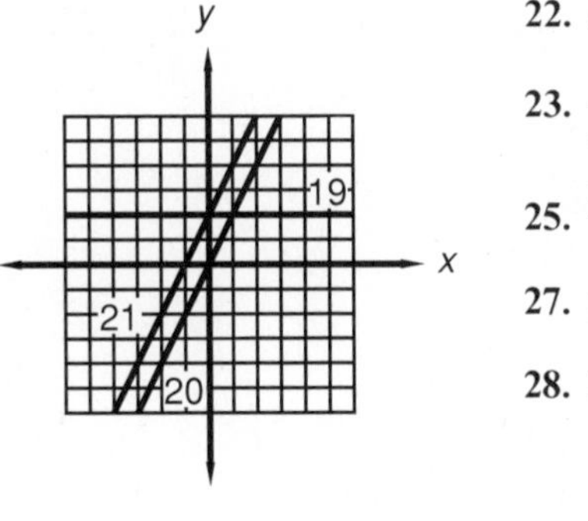

22. $a(a + b)$
23. $\frac{x^2}{x + y}$ 24. $\frac{4x + y}{x^2 + xy}$
25. $\frac{xy + 1}{y}$ 26. $\frac{y + 1}{y}$
27. $5m^2xk^4(3x^4 - m^4x^5k^2 + 4m^2k)$
28. $2x^4m^6$ 29. $16p^{-1}x^5$ 30. $\frac{x^{12}}{9y^{18}}$
31. -3

practice

a. $x \not< -2 \longrightarrow x > -2$; $D = \{\text{positive integers}\}$

−1 0 1 2 3 4

b. $x \not< 4 \longrightarrow x \geq 4$; $D = \{\text{Reals}\}$

2 3 4 5 6 7

problem set 67

1. 114,000 2. 6000 3. 460 4. $6\sqrt{2}$ 5. $15\sqrt{3}$ 6. 72
7. $48x^2 - 28x - 6$ 8, 9, 10.

y
x
9
10
8

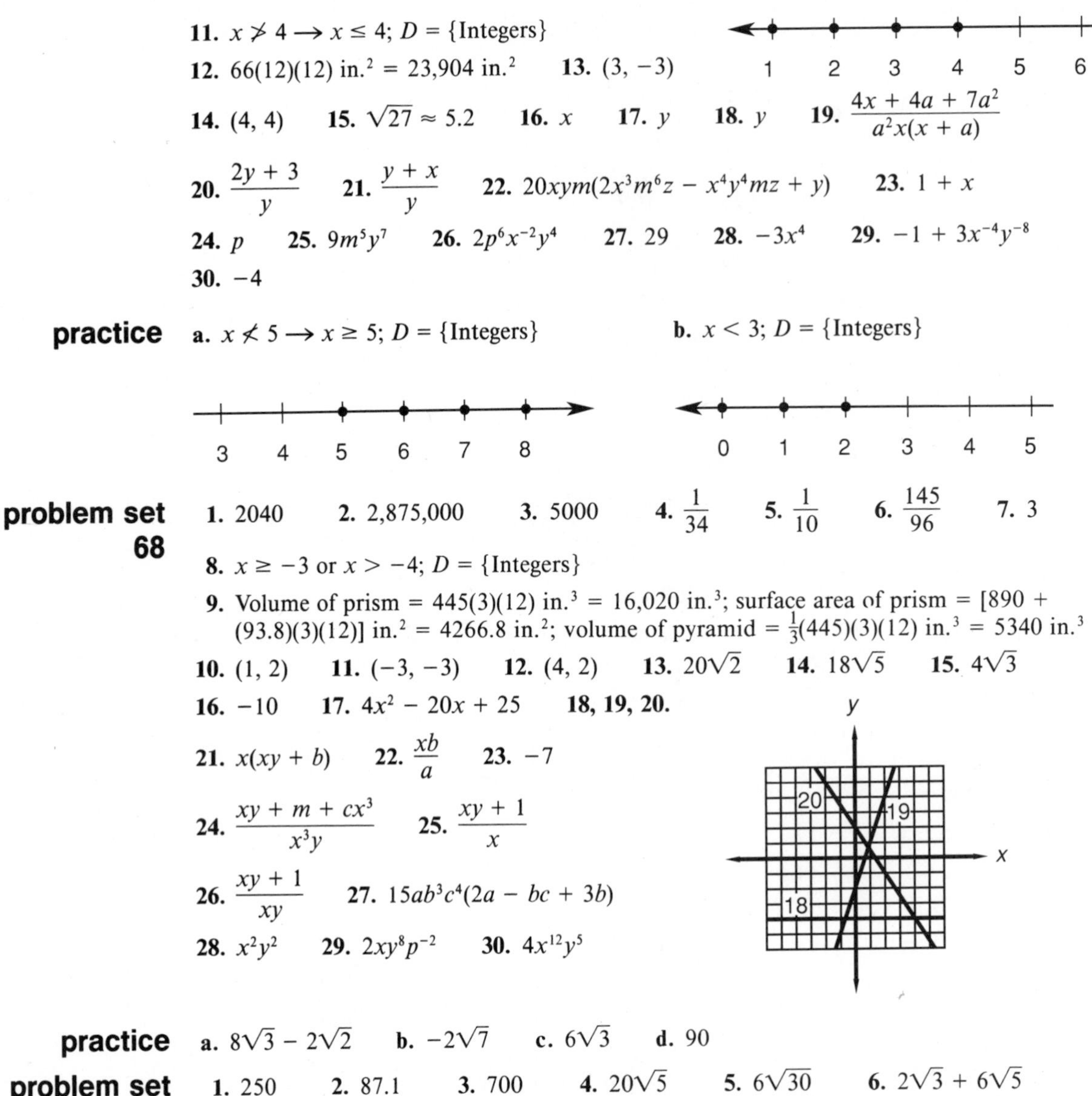

11. $x \not> 4 \longrightarrow x \leq 4$; $D = \{\text{Integers}\}$

12. $66(12)(12) \text{ in.}^2 = 23{,}904 \text{ in.}^2$ **13.** $(3, -3)$

14. $(4, 4)$ **15.** $\sqrt{27} \approx 5.2$ **16.** x **17.** y **18.** y **19.** $\dfrac{4x + 4a + 7a^2}{a^2x(x + a)}$

20. $\dfrac{2y + 3}{y}$ **21.** $\dfrac{y + x}{y}$ **22.** $20xym(2x^3m^6z - x^4y^4mz + y)$ **23.** $1 + x$

24. p **25.** $9m^5y^7$ **26.** $2p^6x^{-2}y^4$ **27.** 29 **28.** $-3x^4$ **29.** $-1 + 3x^{-4}y^{-8}$

30. -4

practice **a.** $x \not< 5 \longrightarrow x \geq 5$; $D = \{\text{Integers}\}$ **b.** $x < 3$; $D = \{\text{Integers}\}$

problem set 68 **1.** 2040 **2.** 2,875,000 **3.** 5000 **4.** $\dfrac{1}{34}$ **5.** $\dfrac{1}{10}$ **6.** $\dfrac{145}{96}$ **7.** 3

8. $x \geq -3$ or $x > -4$; $D = \{\text{Integers}\}$

9. Volume of prism $= 445(3)(12) \text{ in.}^3 = 16{,}020 \text{ in.}^3$; surface area of prism $= [890 + (93.8)(3)(12)] \text{ in.}^2 = 4266.8 \text{ in.}^2$; volume of pyramid $= \frac{1}{3}(445)(3)(12) \text{ in.}^3 = 5340 \text{ in.}^3$

10. $(1, 2)$ **11.** $(-3, -3)$ **12.** $(4, 2)$ **13.** $20\sqrt{2}$ **14.** $18\sqrt{5}$ **15.** $4\sqrt{3}$

16. -10 **17.** $4x^2 - 20x + 25$ **18, 19, 20.**

21. $x(xy + b)$ **22.** $\dfrac{xb}{a}$ **23.** -7

24. $\dfrac{xy + m + cx^3}{x^3y}$ **25.** $\dfrac{xy + 1}{x}$

26. $\dfrac{xy + 1}{xy}$ **27.** $15ab^3c^4(2a - bc + 3b)$

28. x^2y^2 **29.** $2xy^8p^{-2}$ **30.** $4x^{12}y^5$

practice **a.** $8\sqrt{3} - 2\sqrt{2}$ **b.** $-2\sqrt{7}$ **c.** $6\sqrt{3}$ **d.** 90

problem set 69 **1.** 250 **2.** 87.1 **3.** 700 **4.** $20\sqrt{5}$ **5.** $6\sqrt{30}$ **6.** $2\sqrt{3} + 6\sqrt{5}$

7. Volume of cylinder $= (450 + 50\pi)(2)(12) \text{ in.}^3 \approx 14{,}568 \text{ in.}^3$; surface area of cylinder $= [900 + 100\pi + (71.6 + 10\pi)(2)(12)] \text{ in.}^2 \approx 55{,}171.76 \text{ in.}^2$; volume of cone $= \frac{1}{3}(450 + 50\pi)(2)(12) \text{ in.}^3 \approx 4856 \text{ in.}^3$

8. $6p^2 + 7p - 20$

9, 10, 11.

y
x
9
10
11

12.

−10 −9 0 1 2 3

13. $(5, 5)$ **14.** $(2, 1)$ **15.** $(4, 4)$

16. $\dfrac{a^3}{x}$ **17.** $\dfrac{1}{a + b}$ **18.** x

19. $\dfrac{ax^2 + 5x + 5y}{x^2(x + y)}$ **20.** $\dfrac{b + a}{a}$

21. $\dfrac{x^2 + 1}{x}$ **22.** $4x^2y^2z(1 - 2z^4)$

23. $-3 - 12y^3$ **24.** $1 - x$

25. $p^{-4}x^4y^7$ **26.** $16x^{-6}p^{-6}y^2$ **27.** x^8 **28.** $4a^2xm^{-1}$ **29.** $\frac{319}{16}$ **30.** -53

practice **a.** $9\sqrt{2}$ **b.** $-9\sqrt{3}$

problem set 70 **1.** -2 **2.** 75 **3.** 111 **4.** $\frac{287}{60}$ **5.** 0.05 **6.** $\frac{133}{85}$ **7.** 1

8. –2 –1 0 1 2 3 **9.** (4, 2) **10.** $(-4, 2)$ **11.** $(-4, -2)$

12. Volume of prism = 980(100) cm^3 = 98,000 cm^3; surface area of prism = [1960 + (100)(182.8)] cm^2 = 20,240 cm^2; volume of pyramid = $\frac{1}{3}$(980)(100) cm^3 = 32,666.$\bar{6}$ cm^3

13. 30 **14.** $10\sqrt{5} - 24\sqrt{2}$ **15.** 0 **16.** $16x^2 + 40x + 25$

17, 18. y x 17 18

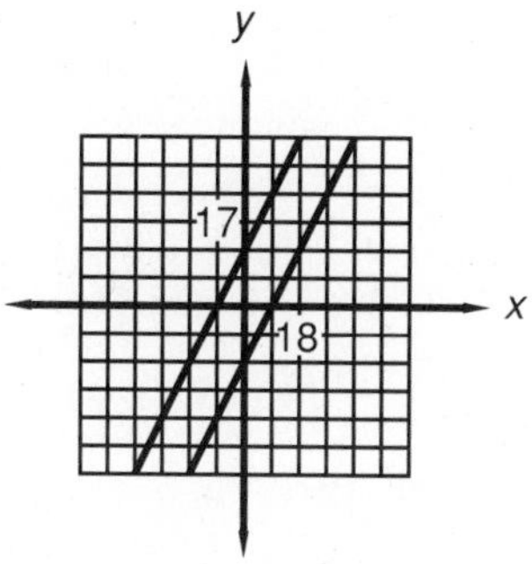

19. Reals, irrationals **20.** ax **21.** $\frac{ac}{b}$

22. $\frac{a}{bc}$ **23.** $\frac{a(x + y) + bx^2y}{x^2y(x + y)}$ **24.** $\frac{ky^2 + 1}{y^2}$

25. $\frac{m^3 + 1}{m^2}$ **26.** $4x^2y^2p^4(3y - xp^2 + 4x^2y^2)$

27. $9y^8m^4$ **28.** $8x^{-1}y^5$

29. $16x^{-12}y^{-3}$ **30.** -4

practice **a.** $(3, -4)$ **b.** $(1, -4)$

problem set 71 **1.** 220 **2.** 2610 **3.** 5000 tons **4.** $24\sqrt{5}$ **5.** $-8\sqrt{2}$

6. $4\sqrt{3} - 9\sqrt{2}$ **7, 8.** y x 7 8 **9.** (5, 5)

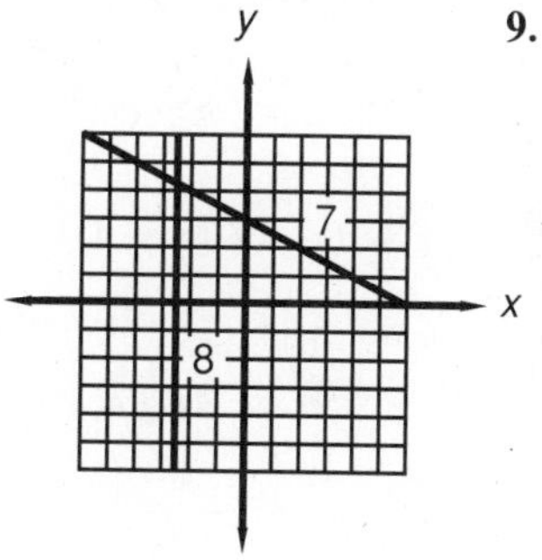

10. 0 1 2 3 4 5

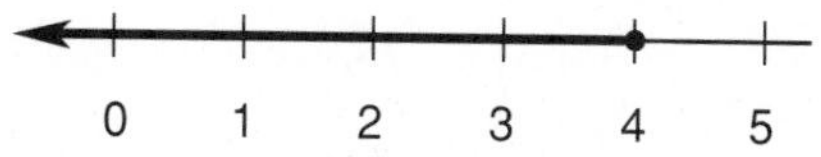

11. Volume of cylinder = $(72 + \frac{43}{2}\pi)(5)(12)$ in.^{3} $\approx$ 8370.6 in.^{3}; volume of cone = $\frac{1}{3}(72 + \frac{43}{2}\pi)(5)(12)$ in.^{3} $\approx$ 2790.2 in.^{3}

12. $\frac{200(100)(100)(100)}{(2.54)(2.54)(2.54)}$ in.^{3} **13.** $91\frac{1}{3}$ **14.** (4, 3) **15.** (3, 2) **16.** x^2a^2

17. $\frac{a + b}{a}$ **18.** $\frac{ma + mx + 3x^2}{x^2a(a + x)}$ **19.** $\frac{4xy + 1}{y}$ **20.** $\frac{y + x}{y}$ **21.** $1 + y$

22. $x^{13}y^{-1}$ **23.** $64x^6y^9p^{12}$ **24.** $x^{-4}y^{-10} - 3x^{-4}p^{-1}$ **25.** $\frac{151}{8}$ **26.** 12

27. Undefined **28.** -96 **29.** 1

practice a. $\frac{c^2 + c}{w}$ b. $\frac{5m + 1}{2 - x}$ c. $\frac{8xy + x}{xy + y}$

problem set 72 1. 5 2. 150 3. 3000 4. $\frac{240}{49}$ 5. $\frac{16}{315}$ 6. $\frac{25}{324}$ 7. $\frac{5}{2}$

8. [number line from −1 to 4 with dots at 1, 2, 3, 4 and arrow to the right] 9. True 10. (5, 5) 11. (2, −2)

12, 13. [graph: lines labeled 12 and 13 on x–y grid]

14. (4, 5) 15. (2, −3) 16. (1, 1)

17. $\frac{ay}{bx}$ 18. $\frac{3b - a}{1 + b^2}$

19. Volume of cylinder $= \frac{(270 + 72\pi)(4)}{100}\text{ m}^3 \approx 11.74\text{ m}^3$; surface area of cylinder $= \left[540 + 144\pi + \frac{(73.3 + 12\pi)(4)}{100}\right]\text{ m}^2 \approx 996.60\text{ m}^2$; volume of cone $= \left(\frac{1}{3}\right)\frac{(270 + 72\pi)(4)}{100}\text{ m}^3 \approx 3.91\text{ m}^3$

20. $-\frac{607}{32}$ 21. $\frac{12{,}000(100)(100)(100)}{(2.54)(2.54)(2.54)(12)(12)(12)(3)(3)(3)}\text{ yd}^3$ 22. 7.6 jousts

23. $8\sqrt{2} - 6\sqrt{3}$ 24. $-26\sqrt{3}$ 25. $-9x^2y$ 26. $\frac{x^4}{16m^2y^3}$ 27. $x^{-14}y^{-2}p^{-6}$

28. $x^{10}y^{-10}$ 29. $x^2y^{-6}p^6$ 30. 19

practice a. $(x - 7)(x + 6)$ b. $(x + 7)(x - 6)$ c. $(x - 8)(x + 2)$

problem set 73 1. $\frac{600(3)(3)(3)(12)(12)(12)(2.54)(2.54)(2.54)}{100(100)(100)}\text{ m}^3$

2. Volume of right solid $= (288 + 36\pi)\text{ ft}^3 \approx 401.04\text{ ft}^3$; surface area of right solid $= (624 + 84\pi)\text{ ft}^2 \approx 887.76\text{ ft}^2$; volume of cone $= (96 + 12\pi)\text{ ft}^3 \approx 133.68\text{ ft}^3$

3. $\frac{320}{27}$ 4. $(x + 8)(x - 2)$ 5. $(x - 3)(x - 3)$ 6. $(x - 9)(x + 3)$

7. $(p - 5)(p + 4)$ 8. $(x - 5)(x + 3)$ 9. $(p - 7)(p + 3)$

10. $(p + 5)(p - 4)$ 11. $(k - 8)(k + 5)$ 12. $(m + 4)(m + 5)$

13. $(x + 3)(x + 11)$ 14. $(p - 4)(p - 9)$ 15. $(m - 6)(m + 5)$

16. $(n + 9)(n + 2)$ 17. $(x + 3)(x + 9)$ 18. $(x - 9)(x - 10)$

19. $(x + 12)(x - 11)$ 20. $(a - 45)(a - 2)$ 21. $(m + 8)(m + 2)$ 22. (2, 3)

23. (1, 1) 24. (3, −4) 25. (4, 5) 26. $14\sqrt{5} - 20\sqrt{2}$ 27. $16\sqrt{2}$

28. $y + 1$ 29. $\frac{a - 4b}{x - b^2}$ 30. $\frac{a - ax}{x^2 + y}$

practice a. $-4x(x + 3)(x + 4)$ b. $-(x - 6)(x + 4)$ c. $N_Q = 13, N_N = 8$

d. $N_D = 18, N_N = -6$

problem set 74 1. $(x - 5)(x + 2)$ 2. $(x + 4)(x + 3)$ 3. $(x - 6)(x + 5)$

4. $(x + 2)(x + 5)$ 5. $(x + 2)(x + 6)$ 6. $(x - 2)(x - 2)$ 7. $(x + 2)(x + 7)$

8. $(x - 7)(x + 2)$ 9. $(x - 6)(x + 3)$ 10. $(x + 2)(x + 4)$

11. $(x + 4)(x - 2)$ 12. $(x - 4)(x + 2)$ 13. $2(x + 2)(x + 3)$

14. $5(x + 2)(x + 4)$ 15. $x(x + 4)(x - 5)$ 16. $a(x + 3)(x + 3)$

17. $ab(x + 3)(x - 2)$ **18.** $x(x + 4)(x + 5)$ **19.** $-b(b - 8)(b + 3)$
20. $-3(m + 2)(m + 8)$ **21.** $-2(p - 11)(p + 5)$
22. $\dfrac{10{,}000(1000)(1000)(100)(100)}{(2.54)(2.54)(12)(12)(5280)(5280)}$ mi^2 **23.** $(20 + 10\pi)$ ft ≈ 51.4 ft **24.** $\frac{43}{9}$
25. Undefined **26.** $\frac{1}{1 - x}$ **27.** $-60\sqrt{2}$ **28.** $\frac{a + x^2}{1 - x}$ **29.** $\frac{1}{a^2 + 1}$ **30.** $\frac{x + 4}{7x}$

practice **a.** $(a + b)(x + 5)(x + 3)$ **b.** $c(m - b)(x - 6)(x + 4)$

problem set 75
1. $(m + 8)(m + 2)$ **2.** $(n - 12)(n + 4)$ **3.** $(y - 8)(y - 7)$
4. $(p - 11)(p + 5)$ **5.** $(t + 7)(t + 5)$ **6.** $(y + 50)(y + 1)$
7. $(r - 7)(r - 11)$ **8.** $(m + 15)(m + 6)$ **9.** $(v + 5)(v + 11)$
10. $(h - 9)(h + 7)$ **11.** $(x - 15)(x + 2)$ **12.** $(w - 2)(w - 11)$
13. $-(x - 3)(x - 4)$ **14.** $-(s + 3)(s + 5)$ **15.** $-(a + 8)(a - 5)$
16. $2x(x + 3)(x + 5)$ **17.** $4(a + 8)(a - 5)$ **18.** $ab(x - 8)(x + 3)$
19. $(x - 1)(x + 2)(x + 5)$ **20.** $-5 < x \le 2$; $D = \{\text{Reals}\}$
21. $(360 - 36\pi)(100)(100)$ cm^2 $\approx 4{,}730{,}400$ cm^2 **22.** (5, 5) **23.** $N_D = 7$, $N_Q = 17$
24. (3, 3) **25.** (2, 1) **26.** $\sqrt{2}$ **27.** $-6\sqrt{5}$ **28.** $\frac{a^2 + b}{1 - 4a}$ **29.** $\frac{m + p^2}{1 - xp}$
30. -2000

practice **a.** $(8x + 9y)(8x - 9y)$ **b.** $(10m - 5)(10m + 5)$ **c.** $(y^2x + 13z^5)(y^2x - 13z^5)$

problem set 76
1. 360 **2.** 100 **3.** 176 pounds **4.** [number line from −7 to 4, closed dots at −6 and 3, segment between]

−7 −6 −5 −4 −3 −2 −1 0 1 2 3 4

5. $\dfrac{25{,}000(5280)(5280)(12)(12)(2.54)(2.54)}{100(100)(1000)(1000)}$ km^2
6. Surface area of cylinder = $(3450\pi)(100)(100)$ cm^2 $\approx 108{,}330{,}000$ cm^2; volume of cylinder = $(22{,}500\pi)(100)(100)(100)$ cm^3 $\approx 70{,}600{,}000{,}000$ cm^3; volume of sphere = $(4500\pi)(100)(100)(100)$ cm^3 $\approx 14{,}130{,}000{,}000$ cm^3
7. $(2px - k)(2px + k)$ **8.** $(5px + 2m)(5px - 2m)$ **9.** $(2y - 3x)(2y + 3x)$
10. $(3ka - 7)(3ka + 7)$ **11.** $(p - 2k)(p + 2k)$ **12.** $(6ax - k)(6ax + k)$
13. $(x - 5)(x + 4)$ **14.** $4(x - 5)(x + 4)$ **15.** $2(b - 8)(b + 3)$
16. $3(x - 15)(x + 2)$ **17.** $(a + b)(x + 2)(x + 5)$ **18.** $p(m + 4)(m + 5)$
19. $5(k + 2)(k + 3)$ **20.** $-(x + 7)(x + 1)$ **21.** $5(m - 1)(m - 1)$ **22.** (4, 4)
23. (3, −3) **24.** (1, 1) **25.** $N_P = 100$, $N_N = 75$ **26.** $21\sqrt{5}$ **27.** $4\sqrt{3}$
28. $\frac{1 + x}{y + x^2}$ **29.** $\frac{2x + y}{x(x + y)}$ **30.** $\frac{6}{5}$

practice **a.** 4.99×10^4 **b.** 4.99×10^{-7} **c.** 4.99×10^{-4}

problem set 77
1. 1200 **2.** \$88,800 **3.** \$7000 **4.** 117
5. [number line from 3 to 8, arrow pointing left from closed dot at 7]

3 4 5 6 7 8

6. $x^3 + 6x^2 + 11x + 12$
7. $\dfrac{xc + x^2 + bxc^2 + 5c + 5x}{x^2c^2(c + x)}$ **8.** $\frac{xb + a}{b - 1}$ **9.** 18 **10.** 81
11. Surface area of right prism = $344(100)(100)$ cm^2 = $3{,}440{,}000$ cm^2; volume of right prism = $240(100)(100)(100)$ cm^3 = $240{,}000{,}000$ cm^3; volume of pyramid = $80(100)(100)(100)$ cm^3 = $80{,}000{,}000$ cm^3

12. 9.16 points **13.** 4.78×10^{-4} **14.**

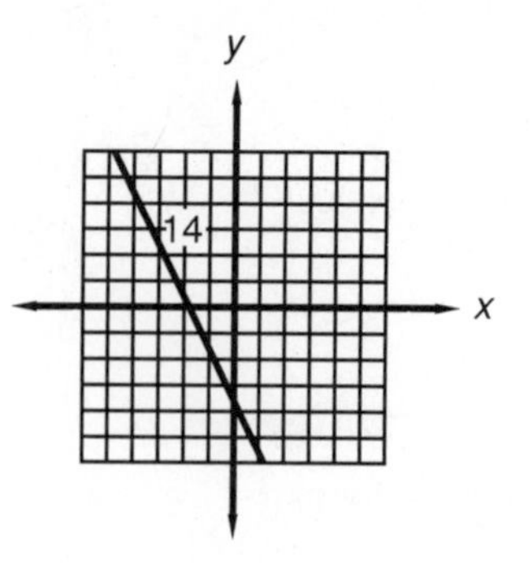

15. $(-2, -4)$ **16.** $N_D = 35, N_Q = 5$
17. $-x^4 + 32x^4y^{-2}a^6$ **18.** $x^{-2}y^{-4}$
19. $-13\sqrt{15}$ **20.** $40\sqrt{5}$ **21.** $(x + 4)(x + 5)$
22. $(x + 3)(x + 5)$ **23.** $(x + 7)(x + 4)$
24. $x(x + 6)(x + 4)$ **25.** $a(x - 5)(x + 3)$
26. $5(x + 7)(x - 4)$ **27.** $5(x + y)(x - y)$
28. $5(3x - 2m)(3x + 2m)$
29. $(2a - 3b)(2a + 3b)$ **30.** $(7ap - k)(7ap + k)$

practice
a. Addition, subtraction, and multiplication
b. None
c. Addition and multiplication

problem set 78
1. 4000 **2.** 1632 **3.** 56 percent patched, 44 percent not patched **4.** $\frac{1}{40}$
5. Reals, rationals, integers **6.**

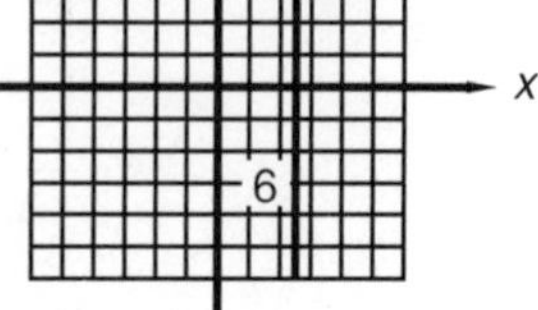

7. $\frac{4c + 4x + 11xc + 5x^2}{x^2c(c + x)}$
8. x^8y^{-5} **9.** -27 **10.** -43
11. $6ax^{-1} - ax$
12. 1.23×10^{-8} **13.** 1.23×10^5
14. Addition
15. None
16.

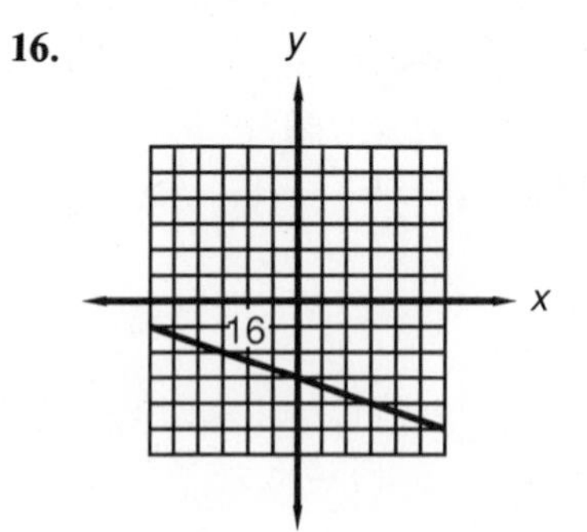

17. $(1, -5)$ **18.** $(1, 1)$ **19.** $y^{-4} - 4a^{-2}xy^{-1}$ **20.** $-40\sqrt{2}$
21. $(x + 5)(x - 2)$ **22.** $(x + 7)(x - 3)$ **23.** $(x + 3)(x + 6)$
24. $5(x - 5)(x + 2)$ **25.** $x(x - 2)(x - 1)$ **26.** $-x(x - 6)(x + 3)$
27. $b^3(x + 2)(x - 2)$ **28.** $(4x - a)(4x + a)$ **29.** $(3p - m)(3p + m)$

30. Surface area of cylinder = $(2200\pi)(100)(100)$ cm^2 $\approx$ 69,080,000 cm^2; volume of cylinder = $(10{,}000\pi)(100)(100)(100)$ cm^3 $\approx$ 31,400,000,000 cm^3; volume of cone = $\frac{1}{3}(10{,}000\pi)(100)(100)(100)$ cm^3 $\approx 10{,}466{,}666{,}666.\overline{6}$ cm^3; volume of sphere = $\frac{4}{3}(1000\pi)(100)(100)(100)$ cm^3 $\approx 4{,}186{,}666{,}666.\overline{6}$ cm^3

practice **a.** 70, 71, 72 **b.** 9, 10, 11, 12

problem set 79

1. Addition **2.** Addition and multiplication **3.** 10, 11, 12, 13 **4.** 81

5. $\frac{83}{156}$ **6.** $-\frac{2}{5}$ **7.** False **8.** (graph) **9.** 13

10. 4.3×10^3 **11.** 4.3×10^{10} **12.** 14,352 **13.** $\dfrac{a^2cx + 2cx + ab}{acx^3}$

14. $\dfrac{ax + 6a + bx}{x(x + 6)}$ **15.** x^4a^{-4} **16.** $2a^2x^5y^{-1} - 3a^2x^3y^{-3}$ **17.** $N_N = 30$, $N_Q = 15$

18. (2, 5) **19.** $3x^6a^{-1}y^{-4} - 6xy^{-4}$ **20.** $8\sqrt{5}$ **21.** $(x - 7)(x + 2)$

22. $-x(x - 6)(x + 2)$ **23.** $a(x + 2)(x + 5)$ **24.** $2(x + 7)(x + 5)$

25. $3(x + 8)(x + 1)$ **26.** $p(x - 2)(x + 1)$ **27.** $(m - 3x)(m + 3x)$

28. $(2x - 3m)(2x + 3m)$ **29.** $5(5m - x)(5m + x)$ **30.** $2(x - 6k)(x + 6k)$

practice **a.** −16, −14, −12 **b.** 216 yd **c.** 4480 tickets

problem set 80

1. 19, 21, 23, 25 **2.** 7.2 **3.** None **4.** 165

5. Volume of cylinder = $(552 - 48\pi)(12)(12)(12) \approx$ 693,411.84 one-inch sugar cubes; surface area of cylinder = $(385.6 + 8\pi)(12)(12)$ in.^{2} $\approx$ 59,143.68 in.^{2}; volume of cone = $\frac{1}{3}(552 - 48\pi)(12)(12)(12)$ in.^{3} $\approx$ 231,137.28 in.^{3}

6. (number line with points at 4, 5, 6, 7, 8, 9) **7.** $\dfrac{x + c + ac(x + c) + mxc}{xc^2(x + c)}$ **8.** $\dfrac{xy + 1}{x^2 - 5y}$

9. $\frac{12}{5}$ **10.** 7×10^{-12} **11.** 7×10^{-21}

12. Addition, multiplication, and division

13, 14. (graph)

15. (4, 4) **16.** $N_N = 400$, $N_D = 100$

17. $3x^2y^{-4} - 12a^3x^2y^{-3}$ **18.** $\dfrac{x^4y^7}{4}$

19. $-2\sqrt{15}$ **20.** $-8\sqrt{3}$

21. $(x - 3)(x - 3)$ **22.** $2(x - 2)(x - 2)$

23. $2(x + 2)(x + 2)$ **24.** $2(x + 5)(x + 5)$

25. $3(x - 5)(x - 5)$ **26.** $a(x - 6)(x - 6)$

27. $(2x - 7)(2x + 7)$ **28.** $(k - 3xy)(k + 3xy)$ **29.** $3(p - 2k)(p + 2k)$

30. $(k - 2m)(k + 2m)$

practice a. $z = -\frac{4}{3}$ b. $y = 16$

problem set 81

1. $-10, -9, -8$ 2. 3, 5, 7, 9 3. 4000

4. 6000 5. $\frac{1}{4}$ 6. Reals, irrationals 7. $\frac{ay^2 + by + c}{x^2y^3}$ 8. x^8y^4

9. $-\frac{1}{27}$ 10. -7 11. $-5m^2y$ 12. 3×10^{-19} 13. 4×10^{-37}

14. (graph)

15. $N_D = 20, N_Q = 320$ 16. $(-3, -3)$

17. $2ax^{-1}y^{-4} - 6x^{-3}y^{-4}$ 18. $12\sqrt{7} - 10\sqrt{14}$ 19. 3953

20. $(30 + 18\pi)(12)(12) \approx 12{,}458.88$ one-inch-square tiles 21. $\frac{1}{4}$ 22. $\frac{161}{11}$

23. $(x - 4)(x - 5)$ 24. $2a(x - 7)(x - 3)$ 25. $m(x + 6)(x + 7)$

26. $(4x - 3a)(4x + 3a)$ 27. $(5m - 2)(5m + 2)$ 28. $9(my - 2k)(my + 2k)$

29. Addition, subtraction, multiplication, and division

practice a. $R_A = 130$ b. $T_T = 10, T_R = 20$

problem set 82

1. $-8, -6, -4, -2$ 2. 2 3. 145 pounds 4. 400 acres

5. 3200 6. Reals, rationals, integers, naturals, wholes 7. $\frac{476}{125}$ 8. $-\frac{3}{4}$

9. 1.35×10^{-21} 10. 1.35×10^{-12} 11. $\frac{ax + ay + 4y}{xy(x + y)}$ 12. $\frac{x^{-2}a^{-2}}{64}$

13. $4x^2y - 3x^2y^4$ 14. (4, 4) 15. $N_N = 17, N_D = 5$ 16. $1 - 3x^2y^{-6}$

17. $\sqrt{5}$ 18. -152 19. $\frac{21}{2}$ 20. $\frac{2}{3}$ 21. $R_A = 55$ 22. -1

23. $\frac{23{,}000(100)(100)(100)}{(2.54)(2.54)(2.54)(12)(12)(12)}$ ft^3 24. 64 in.^{2} 25. $(p - 11)(p + 5)$

26. $(x - 15)(x + 2)$ 27. $2(m - 7)(m - 5)$ 28. $-x(x - 10)(x - 4)$

29. $(2m - 7xp)(2m + 7xp)$ 30. Addition, subtraction, and multiplication

practice a. 1.4×10^9 b. 1×10^8

problem set 83

1. 34, 36, 38 2. $-3, -1, 1, 3$ 3. 200 4. 50 5. 10,725 pounds

6. (number line with points at 5, 6, 7; labels 4 5 6 7) 7. $\frac{3a^2 + 3ax + 4a + 4x + 7}{a^2(a + x)}$ 8. -2

9. $\frac{4y^2 + 1}{xy + m}$ 10. $\frac{55}{18}$

11. Volume of cylinder $= (480 + 170\pi)(12)(12)(12)$ in.$^3 \approx 1{,}751{,}846.4$ in.^{3}; volume of cone $= \frac{1}{3}(480 + 170\pi)(12)(12)(12)$ in.$^3 \approx 583{,}948.8$ in.^{3}; surface area $= (256 + 114\pi)(12)(12)$ in.$^2 \approx 88{,}410.24$ in.^{2}

12. 3400 **13.**

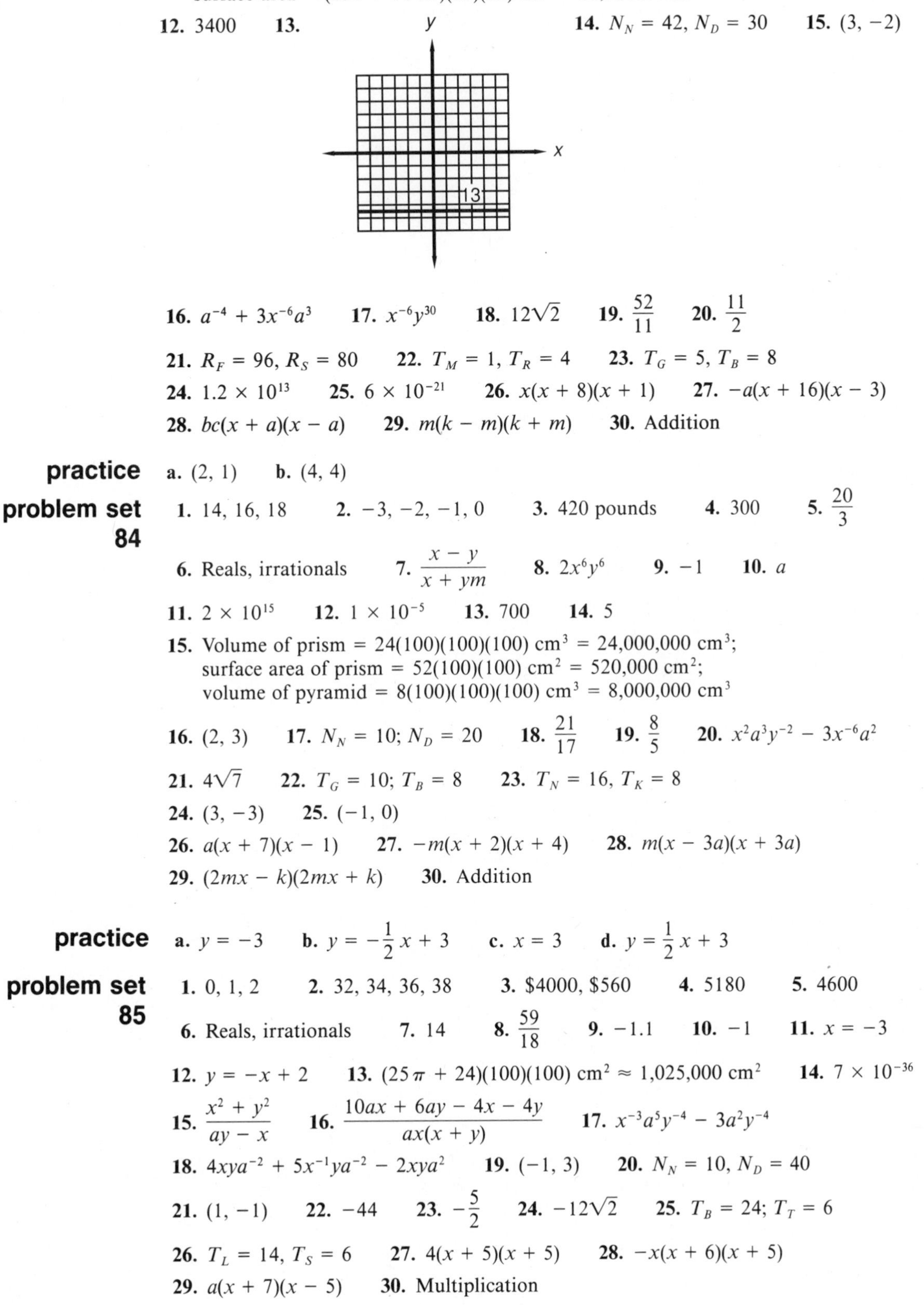

14. $N_N = 42, N_D = 30$ **15.** $(3, -2)$

16. $a^{-4} + 3x^{-6}a^3$ **17.** $x^{-6}y^{30}$ **18.** $12\sqrt{2}$ **19.** $\frac{52}{11}$ **20.** $\frac{11}{2}$

21. $R_F = 96, R_S = 80$ **22.** $T_M = 1, T_R = 4$ **23.** $T_G = 5, T_B = 8$

24. 1.2×10^{13} **25.** 6×10^{-21} **26.** $x(x + 8)(x + 1)$ **27.** $-a(x + 16)(x - 3)$

28. $bc(x + a)(x - a)$ **29.** $m(k - m)(k + m)$ **30.** Addition

practice

a. $(2, 1)$ **b.** $(4, 4)$

problem set 84

1. 14, 16, 18 **2.** $-3, -2, -1, 0$ **3.** 420 pounds **4.** 300 **5.** $\frac{20}{3}$

6. Reals, irrationals **7.** $\frac{x - y}{x + ym}$ **8.** $2x^6y^6$ **9.** -1 **10.** a

11. 2×10^{15} **12.** 1×10^{-5} **13.** 700 **14.** 5

15. Volume of prism $= 24(100)(100)(100)$ cm$^3 = 24{,}000{,}000$ cm^3; surface area of prism $= 52(100)(100)$ cm$^2 = 520{,}000$ cm^2; volume of pyramid $= 8(100)(100)(100)$ cm$^3 = 8{,}000{,}000$ cm^3

16. $(2, 3)$ **17.** $N_N = 10; N_D = 20$ **18.** $\frac{21}{17}$ **19.** $\frac{8}{5}$ **20.** $x^2a^3y^{-2} - 3x^{-6}a^2$

21. $4\sqrt{7}$ **22.** $T_G = 10; T_B = 8$ **23.** $T_N = 16, T_K = 8$

24. $(3, -3)$ **25.** $(-1, 0)$

26. $a(x + 7)(x - 1)$ **27.** $-m(x + 2)(x + 4)$ **28.** $m(x - 3a)(x + 3a)$

29. $(2mx - k)(2mx + k)$ **30.** Addition

practice

a. $y = -3$ **b.** $y = -\frac{1}{2}x + 3$ **c.** $x = 3$ **d.** $y = \frac{1}{2}x + 3$

problem set 85

1. 0, 1, 2 **2.** 32, 34, 36, 38 **3.** \$4000, \$560 **4.** 5180 **5.** 4600

6. Reals, irrationals **7.** 14 **8.** $\frac{59}{18}$ **9.** -1.1 **10.** -1 **11.** $x = -3$

12. $y = -x + 2$ **13.** $(25\pi + 24)(100)(100)$ cm$^2 \approx 1{,}025{,}000$ cm^2 **14.** 7×10^{-36}

15. $\frac{x^2 + y^2}{ay - x}$ **16.** $\frac{10ax + 6ay - 4x - 4y}{ax(x + y)}$ **17.** $x^{-3}a^5y^{-4} - 3a^2y^{-4}$

18. $4xya^{-2} + 5x^{-1}ya^{-2} - 2xya^2$ **19.** $(-1, 3)$ **20.** $N_N = 10, N_D = 40$

21. $(1, -1)$ **22.** -44 **23.** $-\frac{5}{2}$ **24.** $-12\sqrt{2}$ **25.** $T_B = 24; T_T = 6$

26. $T_L = 14, T_S = 6$ **27.** $4(x + 5)(x + 5)$ **28.** $-x(x + 6)(x + 5)$

29. $a(x + 7)(x - 5)$ **30.** Multiplication

practice a. $N_D = 22, N_N = 14$ b. $N_D = 12, N_Q = 21$

problem set 86

1. $N_D = 31, N_N = 20$
2. Volume of cylinder = $(72 - 4\pi)(12)(12)(12)$ in.$^3 \approx 102{,}712.32$ in.^{3}; volume of cone = $\left(\frac{72 - 4\pi}{3}\right)(12)(12)(12)$ in.$^3 \approx 34{,}237.44$ in.^{3}; surface area = $(118.8 + 7\pi)(12)(12)$ in.$^2 \approx 20{,}272.32$ in.^{2}
3. \$800 4. 6800 5. 4×10^{-8} 6. 1×10^2
7. (a) $x = 4$ (b) $y = 2x$ 8. Addition and multiplication 9. $(-1, 1)$
10. $(5, -2)$ 11. $6\sqrt{2}$ 12. $\frac{62}{13}$ 13. [number line from −3 to 2 with a dot at 1]
14. $T_S = 3; T_M = 4$ 15. $T_P = -4; T_M = -12$
16. $R_G = 40; R_P = 85$ 17. $\frac{21}{4}$ 18. 2 19. $-\frac{122}{13}$ 20. $-\frac{38}{3}$ 21. $\frac{4a + 5b}{4a^3}$
22. $\frac{4a^2 + 6a + 6b}{a^2(a + b)}$ 23. $\frac{33}{16}$ 24. $\frac{xy + x}{ax + y}$ 25. $\frac{abc - 1}{4c^2 - a}$ 26. $-\frac{1}{16}$
27. $x(x + 7)(x + 4)$ 28. $x(2a - y)(2a + y)$ 29. $1 - 3y^{-4}$ 30. $x^{-10}y^4$ 31. x^2y^5

practice a. 120 b. $12\sqrt{6} + 30\sqrt{3}$

problem set 87

1. $N_P = 100, N_N = 75$ 2. $N_N = 42, N_D = 30$
3. 160 4. $-6, -4, -2$ 5. 4×10^1 6. $y = 4$ 7. $y = -2x - 2$
8. $(136)(100)(100) = 1{,}360{,}000$ one-cm-square tiles 9. $x \not< -3$, $D = \{\text{Integers}\}$
10. $10\sqrt{6} + 10\sqrt{3}$ 11. $-6\sqrt{2}$ 12. $(1, -4)$ 13. $(-2, -2)$ 14. $\frac{9}{25}$
15. Reals, integers, rationals 16. -51 17. $-\frac{65}{4}$ 18. $T_K = 12; T_M = 4$
19. $\frac{x - a - 3ax^2 + 2x^2 - 2ax}{x^2(x - a)}$ 20. $\frac{bcx - x^2c^2a - d}{x^3c^2}$ 21. $\frac{a^3 + a}{3 - ba^2}$
22. $\frac{abx - 4c}{1 + ac}$ 23. $-\frac{353}{16}$ 24. -5 25. 9 26. 1.6
27. $3y^{-10} - 3x^{-5}y^{-4}a^{-3}$ 28. x^{-13} 29. xy^{-3}
30. Addition, subtraction, and multiplication

practice a. $5x^2 - 9x + 1$ b. $6x^2 + 9x + 19 - \frac{2}{x - 2}$

problem set 88

1. $N_D = 35, N_Q = 5$ 2. $N_N = 30, N_Q = 15$
3. $-13, -11, -9, -7$ 4. 350 5. 6×10^{-5} 6. $y = -2x + 4$
7. $x = -3$ 8. $72\sqrt{2}$ 9. $42 - 6\sqrt{3}$ 10. $9\sqrt{5}$ 11. $-\frac{3}{2}$
12. Volume of cylinder = $(1000 + 500\pi)(12)(12)(12) \approx 4{,}440{,}960$ one-inch sugar cubes; surface area of cylinder = $(541.5 + 300\pi)(12)(12)$ in.$^2 \approx 213{,}624$ in.^{2}; volume of cone = $\frac{1}{3}(1000 + 500\pi)(12)(12)(12)$ in.$^3 \approx 1{,}480{,}320$ in.^{3}
13. $-5x^2 + 24x - 49 + \frac{108}{x + 2}$ 14. $-\frac{8}{3}$ 15. $\frac{8x^2 + 3y + 2x^3 + 2xy}{x^2(x^2 + y)}$
16. $\frac{x^2 + 4y^2}{x + 2y}$ 17. $1 - 3a^{-4}x^{-8}$ 18. $3x - 4xy^{-1}$ 19. $(1, 1)$ 20. $(-4, 2)$
21. $(5, -5)$ 22. $\frac{11}{3}$ 23. 6 24. $T_D = 2, T_M = 6$ 25. $8x^3y^2$

26. $a(x + 2)(x + 2)$ **27.** $(x - 5)(x + 2)$ **28.** $a(3 - 2x)(3 + 2x)$
29. $x(x + 2)(x + 10)$ **30.** Addition

practice

a. $T_1 = 40$; $T_2 = 40$ **b.** $R_S = -2$, $R_O = 1$

problem set 89

1. $N_N = 400$, $N_D = 100$ **2.** $N_D = 20$, $N_Q = 320$
3. 11, 12, 13, 14 **4.** 1737 **5.** 1.2×10^{19} **6.** $y = 3$ **7.** $y = -\frac{1}{3}x - 4$
8. $24 + 36\sqrt{3}$ **9.** -150 **10.** $x^2 + 4 + \frac{8}{x - 2}$ **11.** $2x^2 - 9x + 29 - \frac{91}{x + 3}$
12. 0.018 **13.** $\frac{35ax + 12y}{840a^2}$ **14.** $\frac{k - 7x}{42}$ **15.** $\frac{x - y}{a + by}$ **16.** $2x^2y^2 - 5xy^{-1}$
17. $-\frac{17}{4}$ **18.** $(0, -2)$ **19.** $(1, 1)$ **20.** $(4, 4)$
21. $1 - 3x^{-6}ay^{-2}$ **22.** -3 **23.** $0, -98$ **24.** $R_T = 10$, $R_J = 10$ **25.** y^6x^{-5}
26. $ma(x + 7)(x + 2)$ **27.** $-x(x + 5)(x + 7)$ **28.** None
29. Surface area $= 768\pi$ in.$^2 \approx 2411.52$ in.^{2}; volume $= \frac{2048}{3}\pi$ in.$^3 \approx 2143.57$ in.^{3}
30. $a^2 + 2\sqrt{2}ab + 2b^2$

practice

a. $x^2 + 2x + 4 + \frac{3}{x - 2}$ **b.** $3x^2 + 12x + 43 + \frac{176}{x - 4}$

problem set 90

1. 151 **2.** 20 **3.** 228 **4.** 12, 13, 14 **5.** 5×10^{-21} **6.** $y = -x$
7. $y = x$ **8.** $30 - 20\sqrt{6}$ **9.** $30 - 90\sqrt{3}$ **10.** $-2, -6$
11. $3x^2 + 15x + 74 + \frac{363}{x - 5}$ **12.** $(2, 0)$ **13.** $(1, -1)$ **14.** $(1, -2)$
15. $24 - 6\sqrt{6}$ **16.** $20 - 12\sqrt{10}$ **17.** $6 - 12\sqrt{5}$ **18.** $\frac{133}{90}$
19. [number line with points at 3, 4, 5, 6] **20.** -88 **21.** 66 **22.** $\frac{6a + 6b - 4ab - b^2}{b^2(a + b)}$
23. $\frac{3x + 3a + 2bax}{a^2x(x + a)}$ **24.** $\frac{xy - 4}{a + 3y}$ **25.** 50 **26.** $\frac{1}{2}$ **27.** $a(x + 8)(x + 7)$
28. $T_P = 3$, $T_D = 6$ **29.** $a^{-2}y^{-4}x^{-5} - 3y^{-3}x^{-3}$ **30.** $\frac{yx^8}{12}$

practice

a. $x = -8$ or $x = 3$ **b.** $x = -12$ or $x = 4$

problem set 91

1. $N_P = 120$, $N_N = 30$ **2.** $N_P = 90$, $N_N = 36$ **3.** 31,200 **4.** 7, 9, 11, 13
5. 6×10^{-2} **6.** $y = -2$ **7.** $y = -\frac{1}{2}x + 3$ **8.** $50\sqrt{2} - 75$
9. $56 - 84\sqrt{2}$ **10.** $2x^2 - 4x + 3 - \frac{2}{x + 2}$ **11.** $3x^2 + 15x + 75 + \frac{371}{x - 5}$
12. $350(10)(12) = 42{,}000$ one-inch sugar cubes; volume of pyramid $= \frac{1}{3}(350)(10)(12)$ in.^{3} $= 14{,}000$ in.^{3}
13. Addition, multiplication, and division **14.** $0, -2$ **15.** 99.2
16. $x^2 - 2\sqrt{3}xy + 3y^2$ **17.** $7, -4$ **18.** ± 5 **19.** $3, -2$ **20.** -4
21. $(-4, -11)$ **22.** $(2, 2)$ **23.** [number line with points at 3, 4, 5; labels 2, 3, 4, 5]

24. $\frac{7}{92}$ **25.** -1 **26.** $R_K = 30, R_P = 40$ **27.** $\frac{93}{4}$ **28.** $\frac{4a^2 + ax + 2a + 2x}{a^2(a + x)}$

29. -4 **30.** $a^{-2}x^{-13}y^{-2}$

practice $S_G = 13$

problem set 92

1. Surface area = $(2112\pi)(100)(100)$ cm^2 $\approx$ 66,316,800 cm^2;
volume of cylinder = $(11{,}520\pi)(100)(100)(100)$ cm^3 $\approx$ 36,172,800,000 cm^3;
volume of cone = $(3840\pi)(100)(100)(100)$ cm^3 $\approx$ 12,057,600,000 cm^3;
volume of sphere = $(18{,}432\pi)(100)(100)(100)$ cm^3 $\approx$ 57,876,480,000 cm^3

2. 7 **3.** 7000 **4.** $-5, -3, -1$ **5.** 2×10^{-40}

6. (a) $y = 4$ (b) $y = -\frac{1}{2}x - 2$ **7.** $75\sqrt{2} - 30$ **8.** $42\sqrt{2} - 20\sqrt{7}$

9. $x - 3$ **10.** $3x^2 - 12x + 48 - \frac{193}{x + 4}$ **11.** 5, 7 **12.** $-5, -7$ **13.** 8, 4

14. $-5, -12$ **15.** $\pm\frac{3}{2}$ **16.** $\pm\frac{7}{3}$ **17.** -5 **18.** 8, 3 **19.** $\pm\frac{2}{3}$ **20.** (2, 2)

21. (4, 4) **22.** $\frac{45}{28}$ **23.** 3400 **24.** $T_T = 2, T_M = 5$ **25.** -9 **26.** -2

27. -3 **28.** (a) -9 (b) 9 (c) $-\frac{1}{9}$ **29.** $\frac{xy^2 + a}{ax - y}$ **30.** $36a^6b^2$

practice **a.** $y = -\frac{2}{3}x + 2$ **b.** $y = \frac{1}{3}x - \frac{5}{3}$

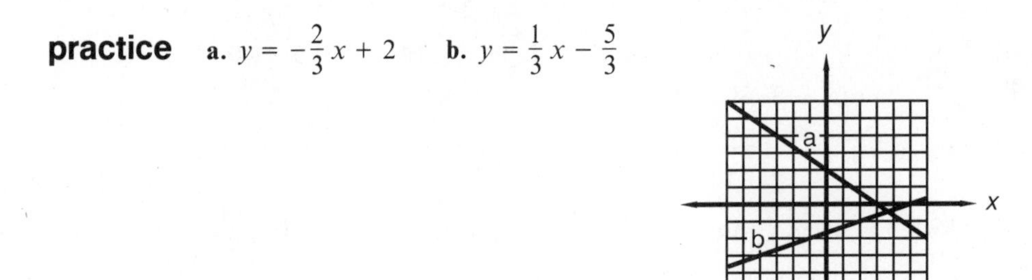

problem set 93

1. $N_P = 200, N_N = 250$ **2.** $N_I = 20, N_W = 15$ **3.** 2160

4. $-9, -7, -5, -3$

5. Volume of cylinder = $1250 + 250\pi$ in.^{3}; $\approx 203\overline{5}$ in.^{3}
volume of cone = $\frac{1}{3}(1250 + 250\pi)$ in.^{3}; $\approx 678.\overline{3}$ in.^{3}
surface area = $785 + 150\pi$ in.^{2} $\approx$ 1256 in.^{2}

6. Addition **7, 8.**

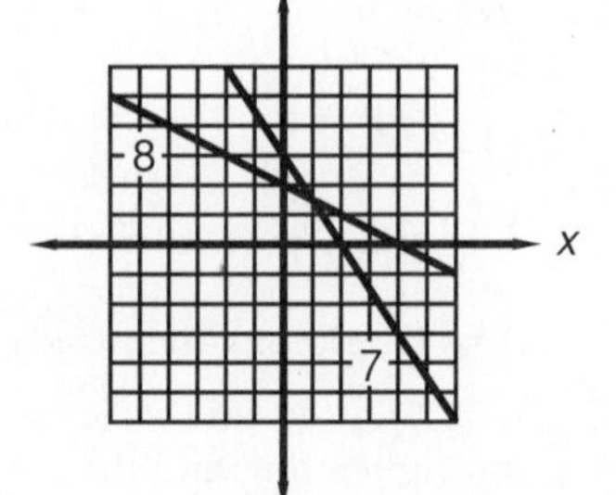

9. 2.8×10^{-17}

10. (a) $x = 5$ (b) $y = -x$ **11.** $6\sqrt{6}$ **12.** $30\sqrt{6} - 96$ **13.** $x^2 + 2 + \frac{11}{x - 3}$

14. $x^2 - 3x + 9 - \frac{28}{x + 3}$ **15.** 4, 5 **16.** 6, 7 **17.** $\pm\frac{3}{2}$ **18.** $\pm\frac{2}{3}$ **19.** (2, 2)

20. (4, 5) **21.** $\frac{29}{35}$ **22.** Reals, irrationals **23.** $4 - y$ **24.** $R_K = 38, R_M = 87$

25. $\frac{9}{7}$ **26.** $\frac{4x^2y + 16x^2 - 2xy}{y(y + 4)}$ **27.** -58 **28.** 18

29. (a) -27 (b) $-\frac{1}{27}$ (c) $\frac{1}{27}$ **30.** $x^{-3} - 4y^5$

practice **a.** $N_1 = 69, N_2 = 29$ **b.** $G = 24$ or $B = 13$

problem set 94

1. 90 **2.** $\frac{1000(3)(3)(3)(12)(12)(12)(2.54)(2.54)(2.54)}{100(100)(100)}$ m^3

3. $9x^2 + 12xy + 4y^2$ **4.** $L = 44; S = 32$ **5.** 80 **6.** $(3, -3)$ **7.** $(3, -1)$

8. 3×10^{-5} **9.** (a) $y = -2x$ (b) $y = -3$ **10.** $9\sqrt{3} + 36$ **11.** $180\sqrt{2}$

12. $3x^2 - 3x + 1 - \frac{5}{x + 1}$ **13.** $2x^2 - 4x + 4 - \frac{8}{x + 1}$ **14.** -5 **15.** $-6, -5$

16. ± 3 **17.** ± 3 **18.** (3, 1) **19.** 46 **20.** $-\frac{19}{22}$

21. Reals, rationals, integers **22.** $\frac{5m^3y}{x^2}$ **23.** $T_B = 8, T_M = 10$ **24.** $\frac{2x}{x + 1}$

25. $\frac{ay + x}{a - xy}$ **26.** [number line: closed dot at 2, arrow to the right; labels 0, 2, 4] **27.** -7 **28.** 3

29. (a) $-\frac{1}{9}$ (b) $\frac{1}{9}$ (c) $-\frac{1}{9}$ **30.** x^{-7}

practice

a. $x \leq -5, D = \{\text{Reals}\}$ [number line: closed dot at −5, arrow to the left; labels −7, −5, −3]

b. $x < 1, D = \{\text{Integers}\}$ [number line: dots at −2, −1, 0, arrow to the left; labels −2, −1, 0, 1]

c. $x < \frac{4}{3}, D = \{\text{Integers}\}$ [number line: dots at −1, 0, 1, arrow to the left; labels −1, 0, 1, 2]

problem set 95

1. $N_D = 15, N_Q = 20$ **2.** 30 **3.** $N_F = 360, N_S = 270$ **4.** 2000

5. [number line: closed dot at −3, arrow to the left; labels −7, −5, −3] **6.** Addition, multiplication **7.** 8, −6

8. $(-2, 3)$ **9.** $(2, -2)$ **10.** 2×10^{25} **11.** (a) $y = -3$ (b) $y = \frac{1}{2}x + 3$

12. $23\sqrt{3}$ **13.** $36 - 24\sqrt{2}$ **14.** $x^3 - 2x^2 + 2x - 4 + \frac{4}{x + 2}$

15. 3, 7 **16.** $\pm\frac{7}{2}$ **17.** $-4, -8$ **18.** 9 **19.** $\frac{187}{78}$

20. [number line: dots at 4, 5, 6; labels 4, 5, 6, 7] **21.** Reals, rationals **22.** $\frac{x^2 + 4x + 4}{x(x + 1)}$

23. $5x^2ym^2 - 3x^2ym + 2xmy^{-3}$ **24.** $R_H = 76, R_O = 9$ **25.** $\frac{1}{4}$ **26.** $\frac{3x - 2y}{ay - 4}$

27. $1 - 3y$ **28.** (a) $-\frac{1}{8}$ (b) $-\frac{1}{8}$ (c) 8 **29.** -25 **30.** $2x^3$

practice

a. $R_W T_W = R_J T_J$, $T_J = 9$, $T_J = 3$, $R_J = R_W + 4$; 18 miles

b. $R_M T_M = R_T T_T$, $R_M = 60$, $R_T = 40$, $T_M + T_T = 15$; 360 miles

problem set 96

1. 3000.04 **2.** $\frac{10{,}000(3)(3)(3)(12)(12)(12)(2.54)(2.54)(2.54)}{100(100)(100)}$ m^3

3. $R_F T_F = R_P T_P$, $T_F = 7$, $T_P = 3$, $R_F = R_P - 40$; $R_P = 70$ mph, $R_F = 30$ mph

4. $R_R T_R = R_W T_W$, $R_R = 10$, $R_W = 4$, $T_R + T_W = 14$; 40 kilometers

5. 19 **6.** $\frac{10}{7}$

7. [number line: arrow to the left from a closed dot at −1; marks −2, −1, 0] **8.** (3, −2) **9.** (5, 5) **10.** 2×10^{-20}

11. (a) $x = -3$ (b) $y = \frac{1}{3}x + 2$ **12.** $5\sqrt{5}$ **13.** $24\sqrt{10} - 18$

14. $x^2 + 5x + 25 + \frac{121}{x - 5}$ **15.** −2, −4 **16.** $\pm\frac{3}{2}$ **17.** −4, −8 **18.** $\frac{195}{2}$

19. $\frac{5}{2}$ **20.** $\frac{46}{5}$ **21.** False **22.** $T_W = 4$, $T_R = 6$ **23.** $\frac{12x^2 + 15x + 25}{30x}$

24. $\frac{3a - 2b^2}{b^2 - 4}$ **25.** $1 + k$ **26.** 18 **27.** −10

28. (a) $-\frac{1}{4}$ (b) −4 (c) $-\frac{1}{4}$ (d) −2 **29.** $y^2 - x^{-7}y^2a^6$

30. Volume of prism = 41,000 cm^3; surface area = 10,260 cm^2; volume of pyramid = $13{,}666.\overline{6}$ cm^3

practice

a. +6 **b.** 1 **c.** $4 \cdot 3 = 3 \cdot 4$

d. $(3 + 4) + 5 = 3 + (4 + 5)$, $(3 \cdot 4) \cdot 5 = 3 \cdot (4 \cdot 5)$ **e.** $4(5 + 6) = 4 \cdot 5 + 4 \cdot 6$

problem set 97

1. $R_G T_G = R_S T_S$, $R_G = 12$, $R_S = 8$, $T_G = T_S - 5$; $T_S = 15$ hours, $T_G = 10$ hours

2. $R_L T_L = R_H T_H$, $R_L = 30$, $R_H = 4$, $T_L + T_H = 17$; 15 hours, 60 miles

3. 22 **4.** 1200

5. $a(b + c) = ab + ac$ **6.** $(a + b) + c = a + (b + c)$ **7.** Reciprocal

8. $-m(m + 8)(m + 3)$ **9.** $\frac{28}{5}$ **10.** [number line: arrow to the right from a closed dot at 2; marks 0, 2, 4] **11.** (−1, 1)

12. (6, 6) **13.** 1×10^{15} **14.** (a) $y = -2$ (b) $y = -\frac{1}{4}x + \frac{3}{2}$

15. $288 - 3\sqrt{2}$ **16.** $30\sqrt{6} - 108\sqrt{2}$ **17.** $x^3 + x^2 + x - \frac{4}{x - 1}$ **18.** −4, −6

19. $\pm\frac{2}{3}$ **20.** $x = \frac{13}{8}$ **21.** $\frac{3}{68}$ **22.** $\frac{8 - 2p}{p(p + 4)}$ **23.** $\frac{xy^2 - 1}{x - 4y}$ **24.** $\frac{231}{305}$

25. Reals, irrationals **26.** 0 **27.** 4 **28.** (a) -9 (b) -9 (c) $-\frac{1}{4}$

29. $y^4 + 4x^6$ **30.** (a) $a(b + c) = ab + ac$ (b) $+2$ (c) $-\frac{1}{b}$

practice **a.** $\frac{x + 5}{x}$ **b.** $\frac{x - 3}{x - 4}$

problem set 98

1. $R_C T_C = R_H T_H$, $R_C = 300$, $R_H = 400$, $T_C + T_H = 7$; $T_C = 4$ min; 1200 cm

2. $R_G T_G = R_B T_B$, $T_G = 4$, $T_B = 3$, $R_B = R_G + 11$; $R_G = 33$ mph, $R_B = 44$ mph

3. 20 percent

4. 83 girls, 96 boys **5.** (a) $(a \cdot b) \cdot c = a \cdot (b \cdot c)$ (b) $a + b = b + a$

6. $+9$ **7.** Associative property of addition **8.** 650 **9.** -4

10. (number line: −6 −5 −4 −3 −2 −1 0) **11.** (2, 0) **12.** $(-3, -1)$

13. 3×10^{-12} **14.** (a) $x = -5$ (b) $y = -2x$ **15.** $4\sqrt{30}$ **16.** $24\sqrt{6} - 96$

17. $3x^2 - 9x + 27 - \frac{85}{x + 3}$ **18.** $-4, -10$ **19.** (number line: −3 2)

20. $\frac{34}{11}$ **21.** $\frac{21}{25}$ **22.** $9x^2 + 18xy + 9y^2$ **23.** False **24.** $\frac{xy + xa}{a^2y^2}$

25. $\frac{ky + k}{y^2 + a}$ **26.** -14 **27.** 5 **28.** (a) -27 (b) 27 (c) $\frac{1}{27}$ (d) -4

29. $4y^4x^{-5}$ **30.** $x - 5$

practice **a.** $R_E T_E + R_W T_W = 500$, $T_E = 6$, $T_W = 3$; $R_E = R_W = \frac{500}{9}$ km/hr

b. $R_P T_P + R_R T_R = 666$, $T_P = 3$, $T_R = 3$, $R_R = 85$; $R_P = 137$ km/hr

problem set 99

1. $R_1 T_1 + R_2 T_2 = 700$, $T_1 = 10$, $T_2 = 10$, $R_1 = R_2 + 30$; $R_2 =$ 20 mph, $R_1 = 50$ mph

2. (a) 0 (b) 1 (c) -9 **3.** $97\frac{2}{3}$

4. 90 buckets, 102 pots **5.** 4000 **6.** $\frac{16}{(x + 6)(x - 7)}$ **7.** 4 **8.** 900

9. $-\frac{3}{2}$ **10.** (number line: −4 −3 −2 −1 0 1) **11.** $(-3, -3)$ **12.** $(-3, -2)$

13. 1×10^{63} **14.** (a) $y = -2$ (b) $y = \frac{3}{2}x + 3$ **15.** $60\sqrt{2} + 2\sqrt{5}$

16. $24\sqrt{6} - 144$ **17.** $x^2 + 2x - 1$ **18.** $10, -3$ **19.** $\pm\frac{10}{3}$ **20.** $-\frac{15}{2}$

21. $-\frac{45}{124}$ **22.** $\frac{axy(x + y) + a^2(x + y) + a^3x^2y}{x^3y(x + y)}$ **23.** $\frac{1}{y - 1}$ **24.** $\frac{26}{7}$

25. Reals, rationals **26.** -24 **27.** 24 **28.** (a) $x - 1$ (b) $\frac{1}{8}$ **29.** $4 - 12y^3$

30. Volume of cylinder = $(110 + \frac{125}{2}\pi)(2)(12)$ in.$^3 \approx 7350$ in.^{3}; surface area of cylinder = $[220 + 125\pi + (32 + 15\pi)(2)(12)]$ in.$^2 \approx 2510.9$ in.^{2}; volume of cone = $\frac{1}{3}(110 + \frac{125}{2}\pi)(2)(12)$ in.$^3 \approx 2450$ in.^{3}

practice

a. $p = \pm 13$ **b.** $q = \pm\sqrt{23}$ **c.** $w = \pm\sqrt{14}$

problem set 100

1.  $R_BT_B + R_LT_L = 340$, $R_B = 30$, $R_L = 40$, $T_B = T_L + 2$; 9 p.m.

2. 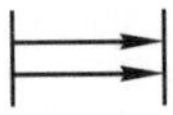 $R_AT_A = R_BT_B$, $R_A = 30$, $R_B = 60$, $T_A = T_B + 2$; 1 p.m.

3.  $R_RT_R = R_WT_W$, $R_R = 8$, $R_W = 3$, $T_R + T_W = 11$; 24 km

4. $N_Q = 90$, $N_H = 100$

5. 340 **6.** 1 **7.** (a) $p = \pm 7$ (b) $p = \pm\sqrt{39}$ (c) $k = \pm\sqrt{11}$

8. 2800 **9.** 12 **10.**

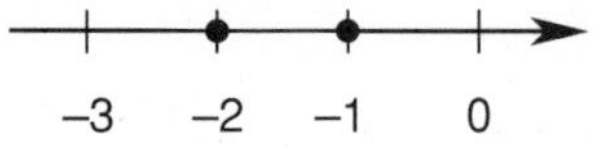

−3 −2 −1 0

11. $(-1, -2)$ **12.** $(4, -5)$ **13.** 3×10^{-17}

14. (a) $x = -5$ (b) $y = -\frac{3}{2}x + 3$ **15.** 360 **16.** -24

17. $x^2 - 2x + 2 - \frac{4}{x + 1}$ **18.** $-7, -8$ **19.** $\pm\frac{9}{2}$ **20.** 54

21. $\frac{7}{40}$ **22.** $\frac{x^2 + 8x - 12}{x(x - 3)}$ **23.** 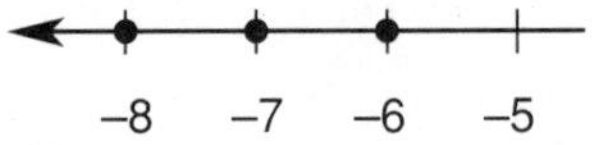 **24.** $\frac{14}{19}$

−8 −7 −6 −5

25. Reals, irrationals **26.** -1 **27.** -20 **28.** (a) $1 + ax$ (b) $-\frac{1}{9}$

29. $7xy^3$ **30.** $-\frac{1}{7}$

practice

a. $p = \sqrt{61}$ **b.** $f = 10\sqrt{2}$

problem set 101

1. -18 **2.** $-5, -4, -3, -2$ **3.** 160

4. 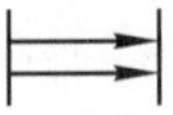 $R_WT_W = R_DT_D$, $R_W = 60$, $R_D = 50$, $T_W = T_D - 1$; $T_W = 5$ hr, $T_D = 6$ hr, 300 miles

5. $N_F = 700$, $N_T = 550$ **6.** $\frac{x + 5}{x - 2}$ **7.** $R_2 = 132$

8. $\frac{5280(5280)(5280)(12)(12)(12)(2.54)(2.54)(2.54)}{100(100)(100)(1000)(1000)(1000)}$ km^3 **9.** $k = 2\sqrt{5}$ **10.** $-\frac{5}{4}$

11. $(1, 3)$ **12.** 4×10^4 **13.** (a) $x = -2$ (b) $y = \frac{1}{3}x - 2$ **14.** $30\sqrt{6}$

15. $30\sqrt{3} - 60\sqrt{2}$ **16.** 5, 20 **17.** $x^2 - 2x + 3 - \frac{6}{x + 2}$ **18.** $\frac{20}{3}$ **19.** $\frac{63}{52}$

20. $\frac{k^2 + 8k + 25}{k(k + 5)}$ **21.** $\frac{p - 4k}{k^2 - 1}$ **22.** $\frac{8}{75}$ **23.** False **24.** $-\frac{1}{2}, -2$ **25.** -9

26. (a) $-(1 + x)$ (b) 16 **27.** $1 - 4x^2y^2a^{-4}$ **28.** $4m^2 + 8mp + 4p^2$
29. Commutative property for addition
30. Volume of cylinder $= \frac{100 + 25\pi}{2}$ m$^3 \approx 89.25$ m^3;
volume of cone $= \frac{100 + 25\pi}{6}$ m$^3 \approx 29.75$ m^3;
surface area $= 210 + 55\pi$ m$^2 \approx 382.7$ m^2

practice **a.** $D = 2\sqrt{17}$

problem set 102
1. -7 **2.** 28, 30, 32, 34 **3.** 630
4. $R_ST_S + R_JT_J = 332$, $R_S = 60$, $R_J = 46$, $T_s = T_J + 2$; 4 p.m.
5. 40 hours **6.** 6 **7.** $D = 2\sqrt{10}$ **8.** 9 **9.** $\frac{1}{4}x$ **10.** 6×10^{-30}
11. -2 **12.** $(4, -4)$ **13.** (a) $y = 3$ (b) $y = -\frac{1}{3}x - 1$
14. $20\sqrt{2}$ **15.** $36\sqrt{3} - 24\sqrt{6}$ **16.** 2, -7
17. $x^2 + x + 1$ **18.** $\frac{39}{2}$ **19.** $-\frac{4}{195}$ **20.** $\frac{4m^2 + 6m + 30}{m(m + 5)}$ **21.** $\frac{5}{58}$
22. $\frac{3xy - 1}{2x - 4y}$ **23.** Reals, rationals **24.** -36 **25.** 0 **26.** (a) $x - 1$ (b) 9
27. Reciprocal **28.** $3x^{-7}$ **29.** (number line with points at −3, −2, −1; 0 marked)
30. Distributive property

practice
a.
$xyzo = (xy)zo$ associative
$= (yx)zo$ commutative
$= yx(zo)$ associative
$= yx(oz)$ commutative
$= y(xo)z$ associative
$= y(ox)z$ commutative
$= yoxz$ removed parentheses

b.
$s + r + p = (s + r) + p$ associative
$= (r + s) + p$ commutative
$= r + (s + p)$ associative
$= r + (p + s)$ commutative
$= r + p + s$ removed parentheses

problem set 103
1. $R_ST_S + R_PT_P = 490$, $R_S = 20$, $R_P = 35$, $T_S + T_P = 17$; 10 hr
2. $R_JT_j = R_RT_R$, $T_J = 8$, $T_R = 10$, $R_J = R_R + 10$; $R_R = 40$ km/hr, $R_J =$ 50 km/hr
3. 2, 4, 6, 8 **4.** $N_E = 300$, $N_H = 2700$ **5.** 4500
6.
$s + x + n = (s + x) + n$ associative
$= (x + s) + n$ commutative
$= x + (s + n)$ associative
$= x + (n + s)$ commutative
$= x + n + s$ removed parentheses

7. $A = \$690{,}000$ **8.** $\frac{10(1000)(1000)(1000)(100)(100)(100)}{(2.54)(2.54)(2.54)(12)(12)(12)(5280)(5280)(5280)}$ mi^3

9. $2\sqrt{6}$ **10.** $2\sqrt{10}$ **11.** $-x - 3$ **12.** $-\frac{19}{6}$ **13.** $(2, -7)$ **14.** 2×10^{-40}

15. (a) $y = -\frac{7}{2}$ (b) $y = -\frac{3}{5}x + 3$ **16.** $150 + 42\sqrt{2}$ **17.** $72 - 16\sqrt{3}$

18. $-10, 5$ **19.** $x^2 - x$ **20.** 11 **21.** $\frac{443}{180}$ **22.** $\frac{10x^2 + 15x + 2}{x^2(x + 1)}$

23. $\frac{x}{y - 1}$ **24.** $\frac{304}{565}$ **25.** Reals, irrationals **26.** 79

27. (a) $1 + xm$ (b) 16 **28.** $1 - 20x^5y^{-2}a^4$ **29.** -34 **30.** $x = \pm 4$

practice

a. 20 — $R_WT_W = R_NT_N + 20$, $R_N = 4$, $T_N = 9$, $T_W = 7$; $R_W = 8$ km/hr

b. 4 — $R_PT_P = R_H\,T_H + 4$, $R_H = 6$, $R_P = 8$, $T_H = T_P$; $T_P = 2$ hr

problem set 104

1. 500 — $R_JT_J = R_FT_F + 500$, $R_J = 250$, $R_F = 230$, $T_J = T_F$;

$T_J = 25$ min

2. $\frac{25{,}000(100)(100)(100)}{(2.54)(2.54)(2.54)(12)(12)(12)(5280)(5280)(5280)}$ mi^3

3. $x = \pm 20$ **4.** $N_N = 250$, $N_Q = 200$ **5.** 5000

6. (number line: −7, −5, −3; open circle at −5, arrow to the right) **7.** $\sqrt{65}$ **8.** $3\sqrt{7}$ **9.** $2\sqrt{10}$ **10.** $\sqrt{34}$

11. $-\frac{(x + 7)(x + 2)}{x - 3}$ **12.** -2 **13.** $(3, -1)$ **14.** 5×10^{-9}

15. (a) $y = 4$ (b) $y = -x - 3$ **16.** $364\sqrt{2}$ **17.** $30 - 24\sqrt{21}$ **18.** $\pm\frac{9}{2}$

19. $x^2 + 4x + 16 + \frac{60}{x - 4}$ **20.** -10 **21.** $-\frac{21}{128}$ **22.** $\frac{x^3 + 2x^2 + 6x - 8}{x^2(x + 4)}$

23. $\frac{1 + 4a}{a^3 + 4}$ **24.** $\frac{59}{168}$ **25.** False **26.** -31 **27.** (a) $2a + 1$ (b) 81

28. 31 **29.** $-x^2y^6$

30.

$a + c + x = (a + c) + x$	associative
$= (c + a) + x$	commutative
$= c + (a + x)$	associative
$= c + (x + a)$	commutative
$= (c + x) + a$	associative
$= (x + c) + a$	commutative
$= x + c + a$	removed parentheses

practice

a. $100\sqrt{15}$ **b.** $500\sqrt{2}$

problem set 105

1. 60 — $R_ET_E = R_AT_A + 60$, $R_E = 60$, $T_E = 6$, $T_A = 4$; $R_A = 75$ mph

2. 40 **3.** $R_WT_W = R_BT_B$, $T_W = 60$, T_B 100, $R_W = R_B + 2$; 300 miles

4. 7 **5.** $N_R = 10$, $N_W = 20$ **6.** $-7, -6, -5, -4$ **7.** $\sqrt{55}$ **8.** $4\sqrt{5}$

9. $\frac{x - 3}{x - 4}$ **10.** $\frac{15}{2}$ **11.** $(2, -4)$ **12.** 3×10^{-9}

13. (a) $x = 4$ (b) $y = \frac{5}{6}x + 1$ **14.** 8 **15.** Multiplication **16.** $3400\sqrt{5}$

17. [Number line from −5 to 3 with dots at −3, −2, −1, 0, 1, 2]

18. 8, −8 **19.** $x^3 + 2$ **20.** $-\frac{7}{2}$ **21.** 1

22. $\frac{-2x^3 + 3x^2 + 8x + 4}{x^2(x + 1)}$ **23.** $\frac{1 - 4x}{yx - 1}$ **24.** $\frac{22}{281}$ **25.** Reals, irrationals

26. 26 **27.** $1 - a$ **28.** $-\frac{1}{16}$ **29.** −20 **30.** $1 - x^3a^{-2}$

practice **a.** 59,740,000 **b.** 513.1293 **c.** 63.014915

problem set 106

1. [Diagram: 40, two arrows] $R_BT_B + 40 = R_LT_L$, $R_B = 6$, $R_L = 10$; $T_L = T_B = 10$ sec

2. 700

3. [Diagram: two arrows] $R_RT_R = R_WT_W$, $R_W = 8$, $R_R = 6$, $T_R = T_W + 2$; $T_W = 6$ hr, $T_R = 8$ hr

4. −11, −9, −7, −5 **5.** 10 feet, 50 feet

6. (a) 104.0625 (b) 400 **7.** $2\sqrt{5}$ **8.** $2\sqrt{2}$ **9.** $x - 4$ **10.** $\frac{4}{9}$

11. (2, 4) **12.** 8×10^{48} **13.** (a) $x = 5$ (b) $y = -\frac{4}{3}x - 2$ **14.** $240\sqrt{3}$

15. $48 - 18\sqrt{6}$ **16.** (a) 101 (b) Associative property

17. [Number line 0, 2, 4 with dots at 2 and 4, arrow to the right] **18.** −7, −5 **19.** $x^2 - 2x + 16 - \frac{27}{x + 2}$

20. 52 **21.** $-\frac{5}{88}$ **22.** $\frac{-9x^3 - 4x^2 + 13x + 16}{4x^2(x + 1)}$ **23.** $\frac{ay - 4x}{1 + 5x}$ **24.** $\frac{17}{121}$

25. Reals, irrationals **26.** −2 **27.** $1 - a$ **28.** $4xy^{-8}$ **29.** $-\frac{1}{27}$

30. −634

practice **a.** Estimate: $(2.3)(3.1) \times 10^2$; from table: $(2.31733)(3.16228) \times 10^2$
b. Estimate: $(2.6)(3.1) \times 10^{-3}$; from table: $(2.61534)(3.16228) \times 10^{-3}$

problem set 107

1. [Diagram: arrows in opposite directions] $R_RT_R = R_WT_W$, $R_R = 6$, $R_W = 3$, $T_R + T_W = 6$; 12 km **2.** 30

3. [Diagram: 500, two arrows] $R_RT_R + 500 = R_WT_W$, $R_W = 40$, $R_R = 20$; $T_W = T_R = 25$ sec

4. 600 **5.** 4, 6, 8, 10

6.

$abxy = ab(xy)$	associative
$= ab(yx)$	commutative
$= a(by)x$	associative
$= a(yb)x$	commutative
$= (ay)bx$	associative
$= (ya)bx$	commutative
$= yabx$	removed parentheses

7. $\frac{1}{5}$ **8.** $(1.80831)(3.16228) \times 10^{-11}$ **9.** $2\sqrt{10}$ **10.** $\sqrt{61}$ **11.** $\frac{x + 7}{x + 10}$

12. $-\frac{11}{8}$ **13.** $(-1, 2)$ **14.** 1×10^{-3} **15.** (a) $y = 3$ (b) $y = \frac{3}{2}x - 3$

16. $400\sqrt{2}$ **17.** $28\sqrt{6} - 12$ **18.** [number line with points at 2, 3, 4; labels 2 3 4 5] **19.** $-10, 7$

20. $\frac{54}{11}$ **21.** $\frac{31}{26}$ **22.** $\frac{-3x^3 + 3x^2 + 9x + 4}{x^2(x + 1)}$ **23.** $\frac{mx + 4}{1 + 4a}$ **24.** False

25. -70 **26.** $1 - x$ **27.** $\frac{1}{16}$ **28.** -11 **29.** $y^{-4} - 2a^{-2}y^{-1}x^{-2}$

30. 40.3737374

practice **a.** $\frac{2x - 21}{(x - 7)(x + 3)}$ **b.** $\frac{-x + 45}{x(x - 9)(x - 1)}$

problem set 108

1. [diagram: 40] $R_FT_F + 40 = R_MT_M$, $R_M = 54$, $R_F = 46$; $T_M = T_F = 5$ sec

2. $-8, -6, -4$

3. [diagram] $R_WT_W + R_BT_B = 20$, $R_W = 5$, $R_B = 15$, $T_W + T_B = 2$; 15 miles

4. 480 **5.** $N_O = 750$, $N_T = 600$ **6.** $(1.03923)(3.16228) \times 10^{-3}$

7. $a = 4\sqrt{5}$ **8.** $\sqrt{89}$ **9.** $\frac{x + 10}{x + 7}$ **10.** $\frac{3}{4}$ **11.** $(1, -2)$ **12.** 2×10^{-55}

13. (a) $y = -2$ (b) $y = -2x$ **14.** $200\sqrt{6}$ **15.** $18\sqrt{15} + 15$

16. [number line shaded left from closed point at −6; labels −8 −6 −4] **17.** $\pm\frac{9}{2}$ **18.** $\frac{69}{7}$ **19.** $\frac{8}{11}$

20. Volume of cylinder $= \frac{225\pi}{100(100)(100)}$ m$^3 \approx 7.07 \times 10^{-4}$ m^3;

volume of cone $= \frac{75\pi}{100(100)(100)}$ m$^3 \approx 2.36 \times 10^{-4}$ m^3;

surface area $= \frac{210\pi}{100(100)}$ m$^2 \approx 6.60 \times 10^{-2}$ m^2

21. None **22.** $\frac{2x + 6 + p}{x^2 - 9}$ **23.** Reals, rationals, integers, wholes, naturals

24. $1 + 2a$ **25.** $-\frac{1}{81}$ **26.** $\frac{5p - 4x}{3 - x^2}$ **27.** -176 **28.** $x^{-2}a^9$ **29.** 60

30. 7.18518519

practice **a.** $D = \{\text{Integers}\}$ [number line with points at −1, 0, 1; labels −2 −1 0 1 2]

b. $D = \{\text{Reals}\}$ [number line shaded in both directions with open point at 0; labels −1 0 1 2]

problem set 109

1. [diagram] $R_AT_A + R_RT_R = 38$, $R_A = 3$, $R_R = 5$, $T_A = T_R + 2$; 6 p.m.

2. 490 **3.** 4800 **4.** $-10, -9, -8, -7$ **5.** $N_W = 12, N_R = 40$

6. $(2.89310)(3.16228) \times 10^{-5}$ **7.** $\sqrt{146}$ **8.** $\sqrt{5}$ **9.** $x + 10$ **10.** $\frac{2}{3}$

11. $(2, 5)$ **12.** 2×10^{62} **13.** (a) $x = -4$ (b) $y = \frac{5}{2}x - 3$

14. $2506\sqrt{6} - 30$ **15.** $-8, -10$

16. Volume of cylinder = $7200 + 3750\pi$ cm^3 $\approx$ 18,975 cm^3; volume of cone = $2400 + 1250\pi$ cm^3 $\approx$ 6325 cm^3; surface area = $4248 + 1525\pi$ cm^2 $\approx$ 9036.5 cm^2

17. $\frac{3}{5}$ **18.** (number line: −3, −2, −1, 0, 1, 2, 3)

19. $\frac{4x + 21}{x^2 - 16}$ **20.** $\frac{3x^2 + 3x + 4}{x^2 + 2x + 1}$ **21.** $\frac{64}{23}$ **22.** $\frac{65}{56}$ **23.** -13 **24.** $1 + 3x$

25. 8 **26.** $\frac{4x^2 + y}{x^2y^2 + y}$ **27.** Reals, rationals **28.** $3x^4y^{-1}a - 12x^{-3}y^{-5}a^{-3}$

29. 3 **30.** 0

practice a. $t = 5$ b. $y = \frac{24}{13}$

problem set 110

1. 460

2. $R_ST_S + R_FT_F = 52, R_S = 3, R_F = 4, T_S + T_F = 15$; 8 hr

3. $R_WT_W = R_TT_T, R_W = 2, R_T = 4, T_W = T_T + 2$; 8 mi

4. -4 **5.** $N_D = 10, N_C = 15$

6. \$10 **7.** 1 **8.** 6 **9.** 1

10. (number line: −6, −4, −2, 0, 2, 4, 6) **11.** No real numbers

12. $2\sqrt{5}$ **13.** $x - 2$ **14.** $(2, 0)$ **15.** (a) $y = 4$ (b) $y = -\frac{1}{3}x - 3$

16. $\frac{1}{18}$ **17.** (a) $1 - x$ (b) $\frac{1}{4}$ **18.** $\frac{a^3 + 1}{a^2x + b}$ **19.** 0 **20.** $\frac{10a + 8}{a^2 - 4}$

21. $\frac{5x - 13}{x^2 + 2x - 8}$

22. $(2.04450)(3.16228) \times 10^{12}$ **23.** $(2.04450)(3.16228) \times 10^{-28}$

24. $-382\sqrt{6} - 6$ **25.** $x^2 - 7x + 49 - \frac{347}{x + 7}$ **26.** $\pm\frac{9}{2}$ **27.** $\frac{7}{18}$

28. Reals, irrationals **29.** 48 **30.** 0

practice a. $b = \frac{mn}{mf - 3z}$ b. $m = \frac{ky}{xy + sy - 3a}$

problem set 111

1. 5 **2.** $-1, 1, 3$ **3.** 270,500

4. $R_MT_M = R_HT_H, R_M = 8, R_H = 16, T_M = T_H + 4$; 4 hr

5. $R_TT_T = R_BT_B, T_T = 4, T_B = 48, R_T = R_B + 55$; 240 mi

6. 2 **7.** 3 **8.** $\frac{15}{2}$ **9.** -2 **10.** $n = \frac{ax}{xy + mx - 5k}$ **11.** $d = \frac{ab}{2c - ax}$

12. $abxy = (ab)xy$ associative
$= (ba)xy$ commutative
$= b(ax)y$ associative
$= b(xa)y$ commutative
$= (bx)ay$ associative
$= (xb)ay$ commutative
$= xb(ay)$ associative
$= xb(ya)$ commutative
$= xbya$ removed parentheses

13. $\frac{1}{2}, -2$ 14. $\frac{3x^2 + 6x + 4}{x^2 - 4}$ 15. $\frac{5x - 35}{x^2 - 25}$ 16. 2.67208×10^{-5}

17. $2\sqrt{13}$ 18. 7 19. 2×10^{84} 20. $-7, -9$ 21. $335\sqrt{2} + 24$

22. $\frac{5x^2 + 22x - 50}{x^3 - 25x}$ 23. $\frac{-x^2 - x}{x^2 + 3x - 10}$ 24. 44 25. y^{-4}

26. [number line: −4, −2, 0, 2, 4] 27. [number line: −7, −5, −3]

28. (a) $x - 1$ (b) $-\frac{1}{4}$ 29. $1 - 8x^{-4}a^{-6}$ 30. Volume $= 810\pi - 180 \text{ m}^3 \approx 2363.4 \text{ m}^3$

practice a. $y = -4x + 7$ b. $y = \frac{8}{9}x - \frac{1}{3}$

problem set 112

1. $R_PT_P + R_FT_F = 440, R_P = 70, R_F = 30, T_P = T_F + 2$; 11 a.m.

2. $R_ST_S = R_CT_C, R_S = 30, R_C = 50, T_S = T_C + 2$; 3 hr

3. $N_D = 43, N_Q = 21$ 4. $-20, -18, -16$

5. 22,500 6. $y = \frac{4}{3}x + \frac{7}{3}$ 7. $-\frac{3}{2}$

8. Volume $= 84\pi \text{ m}^3 \approx 263.76 \text{ m}^3$ 9. 6 10. 9 11. $\frac{7}{4}$ 12. $\frac{mx}{pm - 1}$

13. $\frac{m}{xm - k}$ 14. $\frac{x}{yx - k}$ 15. $\frac{yc}{xc - yb}$ 16. $\frac{-x^2 - 5x + 4}{x^2 - 25}$ 17. $\frac{-6}{x^2 - x - 6}$

18. $(1.80278)(3.16228) \times 10^{-23}$ 19. $\sqrt{106}$ 20. 2 21. 1×10^{-38}

22. $y^9x^2a^{-1}$ 23. $275\sqrt{3} + 30$ 24. (a) $x + 1$ (b) $-\frac{1}{36}$ 25. $-8, -7$

26. -57 27. No real number 28. $2x^2y^3a^4 - 4x^2y^7a^2$ 29. $4y^{-4} + 12x^3y^{-4}a^{-4}$

30. 50,000

practice a. Function b. Not a function c. $D = \{x, y\}$ d. $R = \{5, 11\}$

problem set 113

1. $R_MT_M = R_PT_P, R_M = 2, R_P = 11, T_M + T_P = 15$; 26 miles

2. $R_ST_S = R_TT_T, R_S = 2, R_T = 4, T_T + T_S = 120$; 160 miles

3. $N_{50} = 5000, N_{100} = 100$ 4. 20 percent

5. 9, 10, 11, 12 6. $y = -\frac{5}{7}x - \frac{22}{7}$ 7. $y = -2x - 1$ 8. $y = 2x - 17$

9. a, b, d, f 10. Multiplication 11. (a) $x = -3$ (b) $y = -x + 6$ 12. 8

13. 44 **14.** $\frac{ac}{dc - 1}$ **15.** $\frac{dx}{ad - bx}$ **16.** $\frac{x}{kx - p}$ **17.** $\frac{-3x + 13}{x^2 - 9}$

18. $\frac{2x + 15}{x^2 + 5x + 6}$ **19.** 2.28035×10^{-5} **20.** [number line: −9, −7, −5] **21.** $\frac{13}{5}$

22. 5×10^{34} **23.** $\frac{1}{x}$ **24.** $2\sqrt{5}$ **25.** $25 - 50\sqrt{5}$ **26.** $\frac{xa^2 + 1}{x + a^2}$ **27.** -1

28. Reals, irrationals **29.** $\frac{5}{168}$ **30.** (2, 0)

practice **a.** 6 **b.** 8 **c.** ∅ **d.** −7

problem set 114

1. [diagram: 30] $R_BT_B + 30 = R_NT_N$, $T_B = 6$, $T_N = 6$, $2R_B = R_N$; 5 mph

2. [diagram: 20] $R_FT_F + 20 = R_ET_E$, $R_F = 40$, $R_E = 60$, $T_F = T_E + 2$; 4 p.m.

3. 4500, 7500 **4.** $20,000 **5.** −4, −3, −2

6. $1, \frac{1}{2}$ **7.** (a) 4 (b) 4.36 (c) $\frac{23}{4}$ **8.** $\frac{7x^2 - 20x - 6}{x^2 - 9}$ **9.** -105

10. $y = -\frac{5}{9}x + \frac{7}{9}$ **11.** $y = x$ **12.** $y = -\frac{1}{4}x - \frac{5}{4}$ **13.** a, b, c, d, e

14. $D = \{a, b, c\}$; $R = \{d, e\}$ **15.** a, b **16.** $\frac{3}{20}$ **17.** 16 **18.** $\frac{c}{kc - a}$

19. $6\sqrt{2}$ **20.** $\frac{x - 3}{x^3 + 8x^2}$ **21.** (−1, 0) **22.** (a) $y = -4$ (b) $y = -x$

23. $\frac{a^2 - x^3}{x^3 - a}$ **24.** $x^2 + 2x + 4 + \frac{13}{x - 2}$ **25.** 13 **26.** $\frac{20}{9}$ **27.** −28

28. $4ax^{-2}y^4 - 3y^8$ **29.** Reals, irrationals

30. [number line: −3, −2, −1, 0, 1, 2, 3]

practice **a.** $y = -3x - 1$ **b.** $y = -\frac{3}{2}x - 6$

problem set 115

1. [diagram] $R_RT_R = R_TT_T$, $R_R = 20$, $R_T = 8$, $T_R + T_T = 14$; 80 miles

2. [diagram: 36] $R_ST_S + 36 = R_ET_E$, $T_E = 3$, $T_S = 3$, $R_E = 2R_S$; 24 mph

3. $50 **4.** $N_D = 500$, $N_Q = 100$ **5.** −11, −9, −7 **6.** $2, -\frac{1}{2}$

7. 314 m^2 **8.** $y = -\frac{1}{8}x - \frac{19}{8}$ **9.** $y = -\frac{1}{5}x$ **10.** 8 **11.** ∅ **12.** a, c, d

13. $D = \{a, b, c\}$; $R = \{4, 5\}$ **14.** a, b **15.** $\frac{7}{5}$ **16.** $-\frac{35}{12}$ **17.** $\frac{ax}{bx + 1}$

18. [number line: −5, −4, −3, −2, −1, 0] **19.** $\sqrt{65}$ **20.** $\frac{3}{4}$ **21.** 1×10^{-9}

22. $\frac{1}{4}x^{-4}y^6$ **23.** $385\sqrt{5}$ **24.** $1 + x$ **25.** $-\frac{1}{4}$ **26.** 8, 10 **27.** 100

28. All real numbers **29.** $x^{-6} - 3x^{-1}y^{-5}$ **30.** 0.003

practice a. $y = -\frac{2}{5}x + \frac{19}{5}$ b. $y = \frac{1}{4}x + \frac{31}{4}$

problem set 116

1. $R_RT_R = R_TT_T$, $R_R = 8$, $R_T = 6$, $T_R + T_T = 7$; 24 miles

2. $R_NT_N + R_ST_S = 880$, $T_N = T_S + 2$, $R_N = 40$, $R_S = 60$; 2 p.m.

3. $N_S = 75$, $N_L = 50$ 4. 12 percent

5. $-5, -3, -1$ 6. Bees, 1885; moths, 725 7. $y = \frac{1}{3}x + \frac{11}{3}$

8. $y = -\frac{2}{3}x + \frac{5}{3}$ 9. $\frac{1}{18}, -\frac{1}{4}$

10. (a) $a + (b + c) = (a + b) + c$; (b) $ab = ba$

11. (a) ∅ (b) 7 12. b, c, e 13. $R = \{p, 5\}$ 14. $\frac{5}{19}$

15. $\frac{35}{4}$ 16. $\frac{ym}{xm + ky - pym}$ 17. $(1.33791)(3.16228) \times 10^{-9}$ 18. $2\sqrt{10}$

19. $-\frac{15}{2}$ 20. 7×10^{-25} 21. $a^2x^2y^{-4}$ 22. $104\sqrt{15} + 30$ 23. $1 + y$

24. -3 25. $-12, -10$ 26. -94 27. All integers 28. $4y^{-2} - 12x^{-6}y^2$

29. Volume of cylinder = $\frac{288\pi + 1080}{100(100)(100)}$ m^3 $\approx 1.98 \times 10^{-3}$ m^3; volume of cone = $\frac{96\pi + 360}{100(100)(100)}$ m^3 $\approx 6.61 \times 10^{-4}$ m^3; surface area = $(768 + 4\sqrt{265} + 192\pi)$ cm^2 $\approx$ 1436.00 cm^2

30. $\frac{26{,}000(5280)(5280)(12)(12)(2.54)(2.54)}{100(100)(1000)(1000)}$ km^2

practice a. $x = 15$ b. $x = 6$

problem set 117

1. 1974

2. $R_RT_R = R_WT_W$, $R_W = 2$, $R_R = 10$, $T_W + T_R = 18$; 30 miles

3. 6 — $R_BT_B = R_RT_R + 6$, $T_B = 3$, $T_R = 3$, $R_B = 10$; 8 mph

4. 85 5. $N_G = 52$, $N_P = 71$

6. $y = x - 2$ 7. (a) 1 (b) -6 8. $a(b + c) = ab + ac$ 9. 85

10. $x = \pm\sqrt{70}$ 11. 50 12. $-\frac{39}{2}$ 13. 1 14. $\frac{xc}{1 + kc + dc}$ 15. $2\sqrt{14}$

16. $\frac{x + 5}{x - 2}$ 17. (1, 3) 18. (a) $y = -4$ (b) $y = \frac{3}{4}x + 2$ 19. $\frac{xz - y}{ayz - 3}$

20. $2x^2 + 5x + 10 + \frac{14}{x - 1}$ 21. -14 22. $\frac{152}{85}$ 23. 54 24. 0

25. Reals, irrationals 26. $\frac{5}{17}$ 27. $\frac{x^3y^5 + x^2y^6 - 3x - 3y - 2x^2y^3}{y^3x^2(x + y)}$

28. $\frac{3x^2 + 12x + 8}{x^2 - 16}$ 29. 0.0374747 30. $\frac{1}{2}, -\frac{1}{14}$

practice a. -1 b. $\frac{1}{2}$

problem set 118

1. 172 **2.** [diagram] $R_RT_R = R_JT_J$, $R_R = 14$, $R_J = 7$, $T_R + T_J = 3$; 14 miles

3. $-10, -8, -6, -4$ **4.** Short, 10 ft; long, 30 ft

5. 4500 **6.** $y = -\frac{1}{2}x + \frac{1}{2}$ **7.** (a) $y = 4$ (b) $y = -\frac{1}{3}x - 1$ **8.** a, c, e, f

9. 17 **10.** 3 **11.** $\frac{8}{25}$ **12.** -1 **13.** $\frac{9}{2}$ **14.** $\frac{40}{7}$ **15.** $\frac{ncx}{kcn - mc + n}$

16. 2.29783×10^{18} **17.** $\sqrt{145}$ **18.** $\frac{20}{3}$ **19.** 1×10^{99} **20.** x^{-7}

21. $8\sqrt{6} - 24$ **22.** $x + 1$ **23.** $\frac{1}{9}$ **24.** $-12, -7$ **25.** 80

26. No real number **27.** $1 - 3x^2y^8$ **28.** False **29.** $\frac{ay - x^2}{bc - x^2y^2}$ **30.** $\frac{16}{75}$

31.

$abcd = (ab)cd$	associative
$= (ba)cd$	commutative
$= ba(cd)$	associative
$= ba(dc)$	commutative
$= b(ad)c$	associative
$= b(da)c$	commutative
$= bdac$	removed parentheses

practice Inconsistent

problem set 119

1. [diagram] $R_WT_W + R_RT_R = 66$, $R_W = 3$, $R_R = 15$, $T_R = 2T_W$; 4 hours

2. [diagram] $R_NT_N + R_ST_S = 880$, $R_S = 20$, $R_N = 60$, $T_S = T_N + 4$; 2 a.m.

3. $N_R = 21$, $N_C = 31$ **4.** Raccoons, 605; mongooses, 363 **5.** 7600

6. $y = -\frac{5}{3}x + \frac{26}{3}$ **7.** $\frac{17}{4}$ **8.** 8 **9.** 28 **10.** 25 **11.** Inconsistent

12. $\frac{111}{17}$ **13.** $\frac{56}{15}$ **14.** $\frac{kc}{mc - pc + 1}$ **15.** $\sqrt{33}$ **16.** 1 **17.** $(-2, 0)$

18. (a) $x = 5$ (b) $y = -\frac{1}{2}x - 4$ **19.** $\frac{x - a^2}{pay - 3}$ **20.** $x^2 - x - 1 - \frac{6}{x - 1}$

21. 46 **22.** 7 **23.** -1 **24.** Reals, rationals, integers **25.** $-20, -2$

26. $3x^2y + 2x^6y$ **27.** $\frac{3xy - 3x - x^2y^2}{y^2(y - 1)}$ **28.** $\frac{4x - 3}{x - 3}$ **29.** 32.07581582

30. -5

practice

a. [number line: 3 4 5 6 7]

b. [number line: 1 2 3 4 5 6 7 8]

problem set 120

1. [diagram: 60] $R_HT_H = R_MT_M + 60$, $R_H = 17$, $T_H = 20$, $T_M = 20$; $R_M = 14$ mph

2. [diagram] $R_WT_W = R_RT_R$, $R_W = 4$, $R_R = 24$, $T_W + T_R = 14$; 48 miles

3. 10 **4.** \$135 **5.** $N_A = 11$, $N_P = 40$

6. $y = -\frac{1}{7}x + \frac{11}{7}$ **7.** a, d, g **8.** 25 **9.** 29

10. (number line) −8 −6 −4 −2 0 2 4 6

11. (number line) −2 −1 0 1 2 3 4 5 6

12. No operations **13.** −76 **14.** $\frac{50}{9}$ **15.** $\frac{acd}{c + d}$ **16.** 2.04206×10^{-17}

17. $\sqrt{145}$ **18.** 1 **19.** 2×10^{38} **20.** $3x^{-4}y^{-5}$ **21.** $31\sqrt{2}$

22. $40 - 16\sqrt{6}$ **23.** $x - 1$ **24.** $-\frac{1}{27}$ **25.** −9, 5 **26.** −77

27. All real numbers **28.** $1 - 3x^{-4}a^4y^{-2}$ **29.** 478,000 **30.** −5, 3

practice problem set 121

a. $19 + 13\sqrt{2}$ **b.** $-7 - 2\sqrt{5}$ **c.** $7 + 2\sqrt{10}$ **d.** $2x^2 + 2xy\sqrt{14} + 7y^2$

1. 75 percent

2. $R_NT_N + R_ST_S = 580, R_N = 50, R_S = 45, T_N = T_S + 4$; 4 hours

3. $R_WT_W = R_RT_R, R_W = 5, R_R = 30, T_W + T_R = 21$; 90 km

4. 165 **5.** $N_5 = 163, N_{20} = 13$ **6.** Yes **7.** $y = -\frac{2}{3}x - \frac{13}{3}$

8. (a) $y = -2$ (b) $y = -2x$ **9.** $\emptyset$ **10.** $\frac{49}{4}$ **11.** 20

12. (number line) 6 7 8 9

13. (number line) 2 3 4 5

14. Volume = $352(2)(3)$ ft^3 = 2112 ft^3; surface area = $[704 + (94 + 2\sqrt{277})(2)(3)]$ ft^2 ≈ 1467.72 ft^2

15. $\frac{(1{,}000{,}000)(100)(100)(100)}{(2.54)(2.54)(2.54)(12)(12)(12)(5280)(5280)(5280)}$ mi^3

16.

$a + x + y + m = a + (x + y) + m$	associative
$= a + (y + x) + m$	commutative
$= a + y + (x + m)$	associative
$= a + y + (m + x)$	commutative
$= (a + y) + m + x$	associative
$= (y + a) + m + x$	commutative
$= y + a + m + x$	removed parentheses

17. (a) $-66 - 10\sqrt{3}$ (b) $2a^2 - 2\sqrt{6}\,ap + 3p^2$ **18.** $\frac{13}{3}$ **19.** $\frac{cbx}{pbc + b - ac}$

20. (5, −2) **21.** $\frac{xy - a}{a + x^2y}$ **22.** $x^2 - 2x + 2 - \frac{8}{x + 2}$ **23.** $\frac{21}{176}$

24. Reals, irrationals **25.** −9 **26.** −37 **27.** $-\frac{3}{5}$ **28.** $\frac{5x^2 + 15x + 4}{x^2 - 9}$

29. (number line) −3 −2 −1 0 1 2 3

30. (number line) 0 1 2 3

practice a. $V = 13$ liters b. $T = 75$ K

problem set 122

1. $R_R T_R = R_B T_B$, $R_R = 8$, $R_B = 20$, $T_B + T_R = 7$; 40 km
2. 50 km 3. 10.5 liters 4. $N_W = 12$, $N_R = 15$ 5. $\pm\frac{\sqrt{3}}{6}$
6. $y = 2x - 1$ 7. a, b, d 8. $\frac{9}{4}$ 9. 68
10. −8 −7 −6 −5 −4 −3 −2 −1
11. $-22 - 16\sqrt{3}$ 12. $-4 - 8\sqrt{2}$
13. $-26 - 28\sqrt{6}$ 14. $\frac{26}{5}$ 15. $\frac{15}{8}$ 16. $\frac{ckx}{mk + c}$
17. $(2.04206)(3.16228) \times 10^{-9}$ 18. $\sqrt{145}$ 19. $-\frac{4}{3}$ 20. 7×10^{-56}
21. x^2y^3 22. $6 + 9\sqrt{2}$ 23. $10 - 20\sqrt{15}$ 24. $1 - z$ 25. $-\frac{8}{9}$ 26. $-1, 8$
27. -175 28. No integers 29. $1 - 4x^3a^{-4}$ 30. 4.0606

practice $V = 45$ liters

problem set 123

1. 4 hours 2. $\frac{(1,000,000)(3)(3)(3)(12)(12)(12)(2.54)(2.54)(2.54)}{100(100)(100)(1000)(1000)(1000)}$ km^3
3. 35 4. 1500 5. 568 acres 6. $y = -\frac{1}{5}x - \frac{17}{5}$ 7. a, c, d
8. 36 9. 30 10. −5 −3 −1 1 3 5 11. $-28 + 18\sqrt{2}$
12. $16 - 7\sqrt{5}$ 13. $12 - 8\sqrt{3}$ 14. 12 15. $-\frac{5}{2}$ 16. $\frac{cx}{kc + mc - 1}$
17. $\sqrt{39}$ 18. $\frac{x^2}{x + 8}$ 19. $(-4, -4)$ 20. (a) $y = 5$ (b) $y = -\frac{3}{4}x - 2$
21. $\frac{ax + a}{1 + a^2x}$ 22. $x + \frac{x - 2}{x^2 - 1}$ 23. -120 24. 1 2 3 4
25. −4 −3 −2 −1 0 26. 1 27. $2y^2x^2z^{-1}$ 28. $\frac{15}{76}$
29. $\frac{3x^2 + 5x - 11}{x^2 - 9}$

practice \$2266.35, \$1366.35

problem set 124

1. 300 2. $R_T T_T = R_B T_B$, $R_T = 5$, $R_B = 2$, $T_T + T_B = 28$; 40 miles
3. \$1806.30, \$1106.30 4. 2.81×10^{17} 5. 510
6. (a) $x = 3$ (b) $y = -\frac{3}{2}x$ 7. a, d 8. 32
9. 1 2 3 4 10. $-11 - \sqrt{5}$ 11. $5 + 3\sqrt{2}$ 12. $2 - 18\sqrt{2}$
13. $\frac{3\sqrt{2}}{2} + \sqrt{10}$ 14. $\frac{\sqrt{6}}{4}$ 15. $\frac{\sqrt{14}}{7}$(a) 16. $\sqrt{3} + 2\sqrt{6}$ 17. $(36 - 4\pi)$ ft^2 ≈ 23.44 ft^2

18. $\dfrac{(160{,}000)(100)(100)(100)}{(2.54)(2.54)(2.54)}$ in.^{3} **19.** -2 **20.** $\dfrac{mx}{am - km + p}$ **21.** $\dfrac{mp^2x - z}{y - 5ax}$

22. $7x^2 - 14x + 26 - \dfrac{54}{x + 2}$ **23.** 1434 **24.** $-\dfrac{28}{9}$ **25.** $-\dfrac{7}{90}$ **26.** $y = -\dfrac{1}{4}x$

27. $y = -x + 3$ **28.** $y = -3x - 4$ **29.** $y = -3x - 1$

30. Yes; both make the statement true

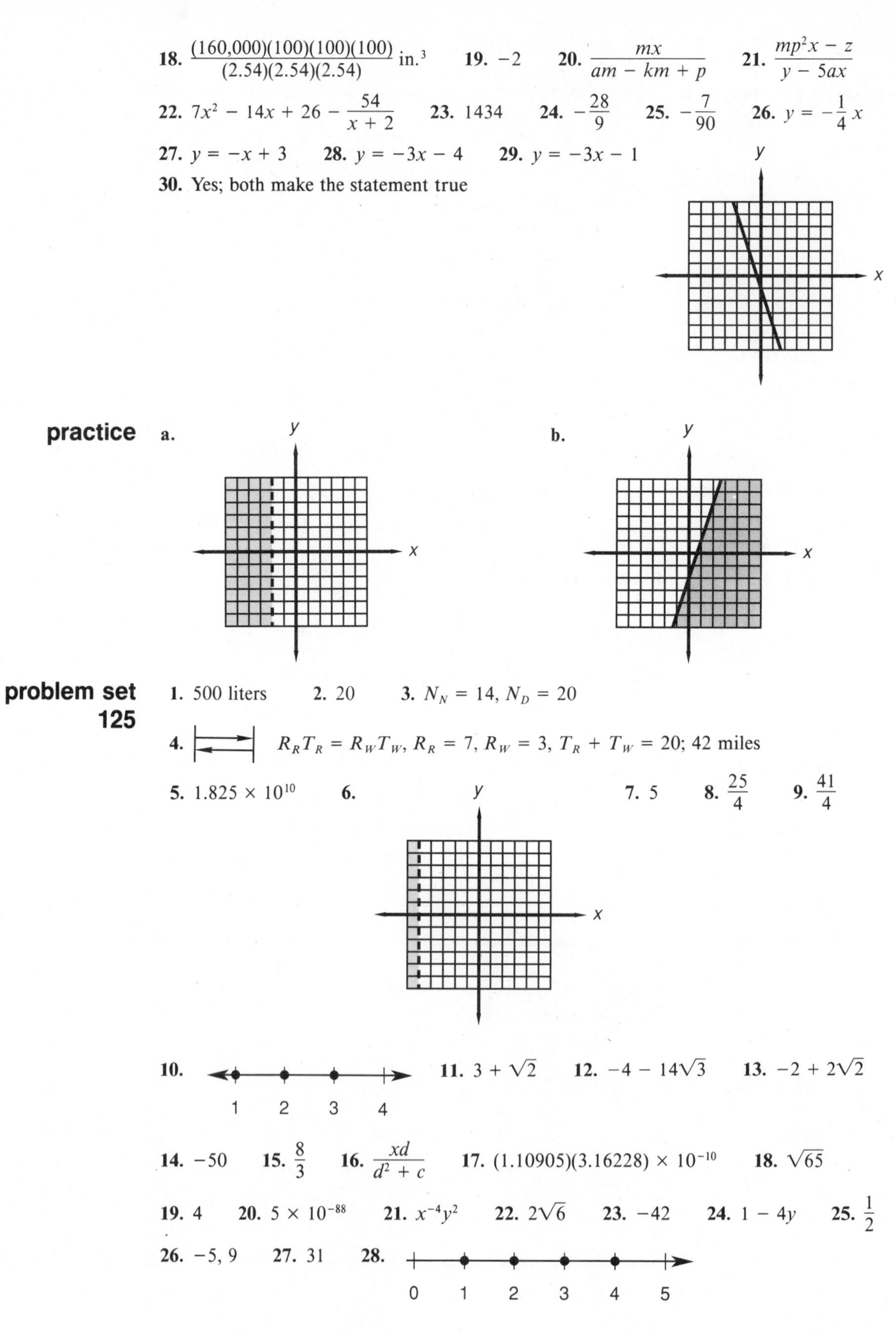

practice **a.** **b.**

problem set 125

1. 500 liters **2.** 20 **3.** $N_N = 14$, $N_D = 20$

4. $R_RT_R = R_WT_W$, $R_R = 7$, $R_W = 3$, $T_R + T_W = 20$; 42 miles

5. 1.825×10^{10} **6.** **7.** 5 **8.** $\dfrac{25}{4}$ **9.** $\dfrac{41}{4}$

10. **11.** $3 + \sqrt{2}$ **12.** $-4 - 14\sqrt{3}$ **13.** $-2 + 2\sqrt{2}$

14. -50 **15.** $\dfrac{8}{3}$ **16.** $\dfrac{xd}{d^2 + c}$ **17.** $(1.10905)(3.16228) \times 10^{-10}$ **18.** $\sqrt{65}$

19. 4 **20.** 5×10^{-88} **21.** $x^{-4}y^2$ **22.** $2\sqrt{6}$ **23.** -42 **24.** $1 - 4y$ **25.** $\dfrac{1}{2}$

26. $-5, 9$ **27.** 31 **28.**

29. $1 - x^2y^{-2}$ **30.** True

practice a. $\frac{\sqrt{138}}{23}$ b. $\frac{4\sqrt{3} + \sqrt{15}}{3}$

problem set 126

1. 40 rpm 2. 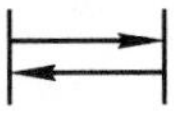 $R_E T_E = R_C T_C$, $R_E = 2$, $R_C = 10$, $T_E + T_C = 18$; 30 km

3. \$6 4. $N_C = 20$, $N_E = 5$

5. \$3527.85, \$2427.85 6. (a) $y = 3$ (b) $y = 3x$ 7. *a*, *c*, *d* 8. 52

9. 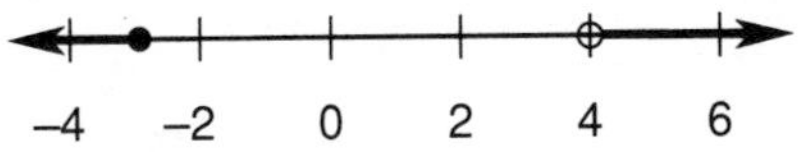 –4 –2 0 2 4 6

10. $-10 - 8\sqrt{2}$ 11. $-5 + 4\sqrt{3}$

12. $-14 - 10\sqrt{5}$ 13. $\sqrt{2} + 3\sqrt{3}$ 14. $\frac{\sqrt{10}}{5}$ 15. $\frac{\sqrt{21}}{7}$ 16. $\frac{4\sqrt{5} + 10\sqrt{2}}{5}$

17. $(424 - 49\pi)\text{ m}^2 \approx 270.14\text{ m}^2$ 18. $\frac{(145{,}000)(2.54)(2.54)(2.54)}{100(100)(100)}\text{ m}^3$ 19. $\frac{40}{3}$

20. $\frac{mx}{a + xb - xk}$ 21. $\frac{ax^2y - x}{p - 3ky}$ 22. $4x^2 - 10x + 20 - \frac{36}{x + 2}$ 23. 1152

24. -3 25. $-\frac{25}{144}$ 26. $y = -3x + 13$ 27. $y = -\frac{5}{9}x - \frac{7}{9}$

28. $y = -5x - 23$ 29. 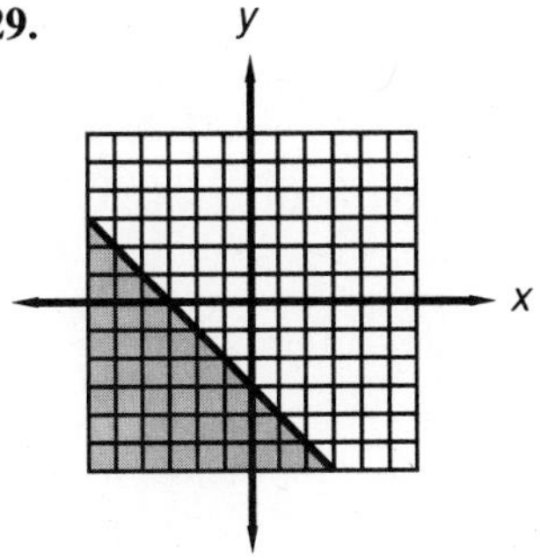 y x 30. $-\frac{1}{2}$ for x makes the equation true

practice a. $(2x + 3)(x - 7)$ b. $3(x - 1)(x - 1)$

problem set 127

1. 1600 tons 2. Boys, 1260; girls, 900 3. 320,000

4. 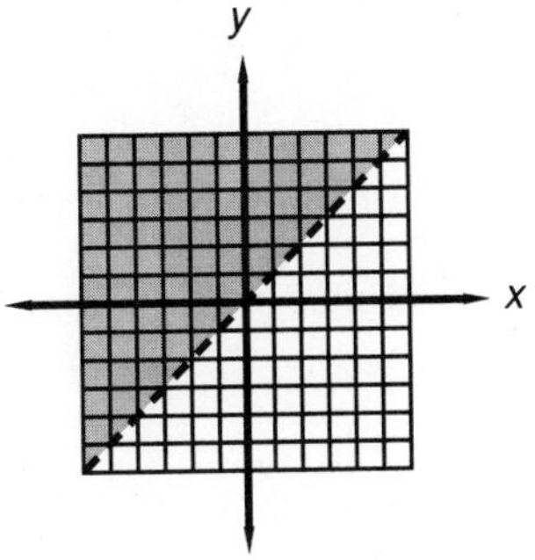 y x 5. 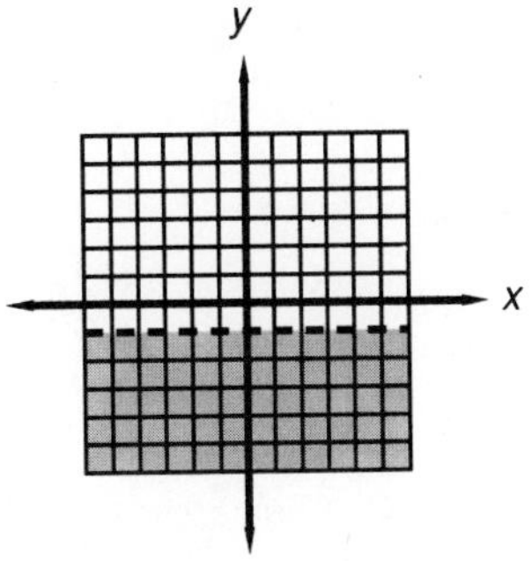 y x 6. 0 7. 0

8. $x = \frac{3}{5}$ or $x = -1$ 9. $(3x + 1)(x - 5)$ 10. $2(x + 4)(x + 1)$

11. $(2x - 3)(x - 6)$ 12. $(2x - 3)(x + 5)$ 13. $2(x + 6)(x - 2)$

14. $2(x - 6)(x + 2)$ 15. $2(x - 2)(x - 1)$ 16. $(2x - 3)(x + 6)$

17. $2(x + 2)(x + 1)$ 18. $(3x + 1)(x - 7)$ 19. $(3x + 1)(x - 8)$

20. $(3x - 4)(x - 1)$ 21. $16 - 10\sqrt{2}$ 22. $30\sqrt{2} - 4\sqrt{3} + 20\sqrt{6} - 8$

23. $\frac{\sqrt{21}}{7}$ 24. $\frac{\sqrt{10}}{4}$ 25. $\frac{2\sqrt{5} + \sqrt{15}}{5}$ 26. $\frac{4\sqrt{3} + \sqrt{6}}{3}$

27. 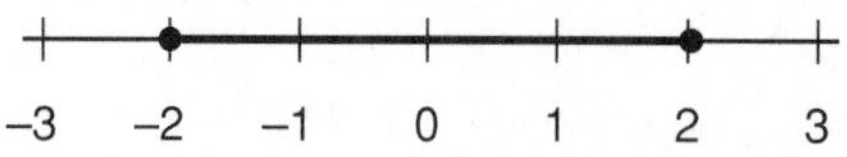 (number line from −3 to 3, closed dots at −2 and 2, segment between)

28. Volume = 5712 in.^3

29.  $R_1T_1 = R_2T_2$, $R_1 = 3$, $R_2 = 9$, $T_1 + T_2 = 16$; 36 miles

30. $y = -2x + 5$

practice

a. $(a + n)(mb - 7)$ **b.** $(n + 2c)(s + 3x)$

problem set 128

1. $30 \frac{\text{lb}}{\text{in.}^2}$ **2.** $4000 \frac{\text{lb}}{\text{in.}^2}$ **3.** Chickens, 143; pigs, 26

4. 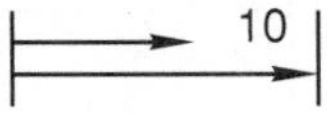 10 $R_DT_D = R_WT_W - 10$, $R_D = 40$, $R_W = 70$, $T_D = T_W + 2$; 2 p.m.

5. $N_N = 100$ **6.** 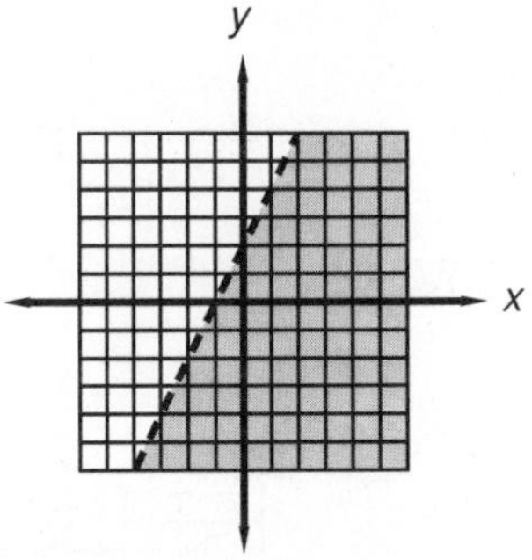 **7.** $(3x + 2)(x - 7)$

y x

8. $(3x - 2)(x + 7)$ **9.** $(2x - 3)(x + 5)$ **10.** $(2x - 3)(x + 6)$
11. $(3x + 2)(x + 7)$ **12.** $(2x - 3)(x - 7)$ **13.** $(3x - 2)(x - 8)$
14. $(2x + 3)(x + 6)$ **15.** $(3x + 7)(x + 2)$ **16.** $y = \frac{3}{5}x - \frac{21}{5}$

17. $\frac{(4,000,000)(5280)(5280)(12)(12)(2.54)(2.54)}{100(100)(1000)(1000)}\ \text{km}^2$ **18.** $(a + b)(c - d)$

19. $(2 + a)(b + 4)$ **20.** $(a + x)(b + c)$ **21.** $(m + pc)(2x - 3)$

22. $(k + pc)(4 - xy)$ **23.** $(a + d)(c - xy)$ **24.** $\frac{\sqrt{33}}{11}$ **25.** $\frac{2\sqrt{15} + 2\sqrt{5}}{5}$

26. 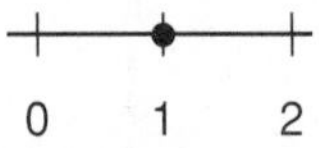 (number line 0 1 2, closed dot at 1) **27.** 0 **28.** $-\frac{2}{3}, -6$

29. Volume = 5280 in.^3; surface area = $(2200 + 24\sqrt{106})\ \text{in.}^2 \approx 2447.10\ \text{in.}^2$

30. $\frac{12,000(12)(12)(12)(2.54)(2.54)(2.54)}{100(100)(100)}\ \text{m}^3$

practice

$W = 500$ pounds

problem set 129

1. 1568 meters **2.** $1820\frac{4}{9}$ feet **3.** 36 **4.** 100

5. Rabbits, 9800; squirrels, 7000 **6.** 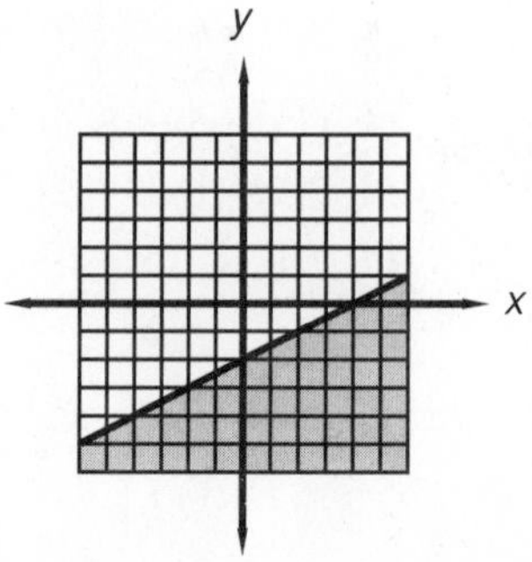 **7.** $\frac{bc}{mb - a}$

y x

8. $(3x-2)(x+9)$ **9.** $(3x-4)(x+1)$ **10.** $2(x+1)(x-3)$
11. $(3x-2)(x+10)$ **12.** $(2x+5)(x+5)$ **13.** $(2x+5)(x-5)$
14. $(5+b)(a+3)$ **15.** $(y+c)(a+x)$ **16.** $(3x-2)(m+p)$
17. $(k+3)(x-5)$ **18.** $(pc+4)(x+c)$ **19.** $(ac+2)(b-k)$

20. 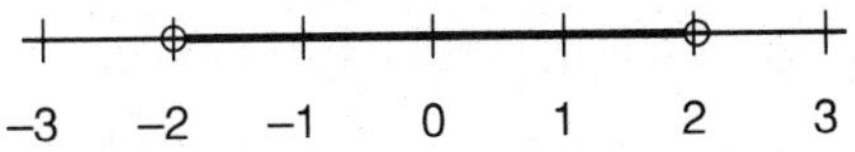 –3 –2 –1 0 1 2 3 **21.** 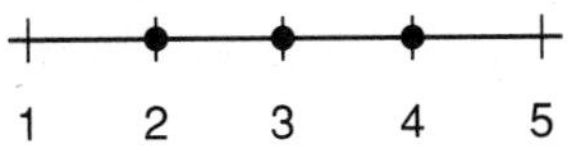 1 2 3 4 5

22. 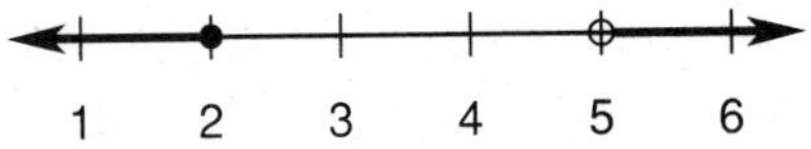 1 2 3 4 5 6 **23.** 23 **24.** 67

25. $y=-\frac{7}{5}x+\frac{11}{5}$ **26.** $y=-\frac{1}{4}x+\frac{9}{2}$ **27.** $y=-\frac{1}{3}x+\frac{13}{3}$ **28.** $\frac{\sqrt{15}}{5}$

29. $\frac{\sqrt{21}}{3}$ **30.** 3

practice $x=-\frac{7}{2}\pm\frac{\sqrt{85}}{2}$

problem set 130

1.  $R_FT_F+R_ET_E=420$, $T_F=6$, $T_E=3$, $R_E=R_F+20$;
$R_F=40$ mph, $R_E=60$ mph
2. Greens, 360; whites, 1980
3. 24,000 **4.** $N_O=530$, $N_T=237$ **5.** 4000 **6.** 2, 4, 6, 8
7. (a) 0 (b) 0 **8.** Volume = 3240 ft^3 **9.** $-1\pm\sqrt{5}$ **10.** $\frac{-3\pm\sqrt{41}}{2}$
11. $x=-1\pm\sqrt{6}$ **12.** $x=-2\pm\sqrt{11}$ **13.** $(3x+5)(x-7)$
14. $(3x-5)(x+1)$ **15.** $(2x+3)(x-4)$ **16.** $(p^2-a)(b+c)$
17. $(x^2-c)(a+c)$ **18.** $(x^3+y)(m+2)$ **19.** $(4+c)(ab+x)$

20. 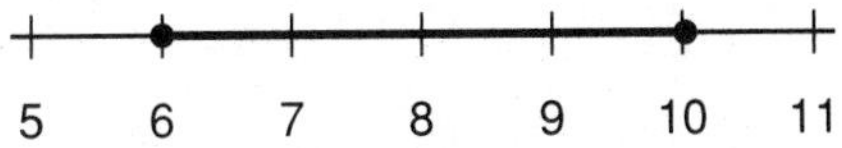 5 6 7 8 9 10 11 **21.** 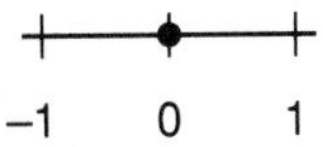 –1 0 1

22. 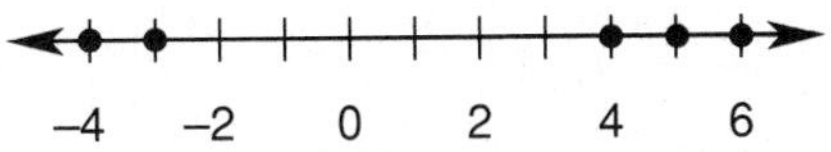 –4 –2 0 2 4 6

23. $y=-\frac{5}{8}x+\frac{1}{8}$ **24.** 2 **25.** $\frac{6}{5}$ **26.** $y=\frac{1}{5}x+\frac{18}{5}$

27. 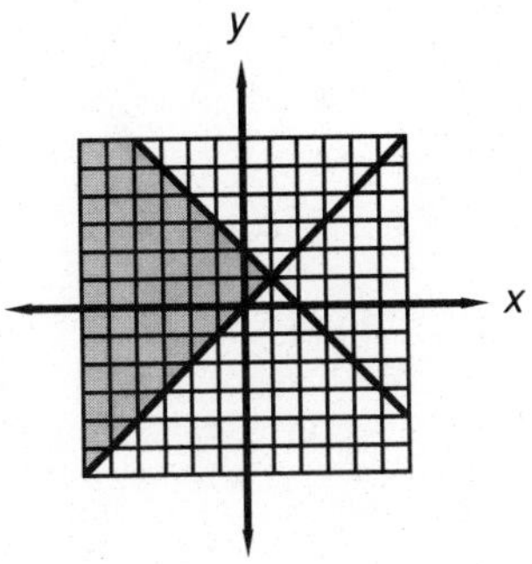 y x **28.** $\frac{\sqrt{14}}{7}$ **29.** $\frac{\sqrt{15}}{6}$ **30.** $\frac{4\sqrt{6}+3\sqrt{2}}{6}$

practice $\frac{-3\pm\sqrt{17}}{4}$

problem set 131

1. 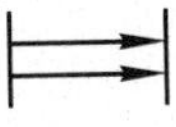 $R_FT_F=R_ET_E$, $T_E=20$, $T_F=25$, $R_F=R_E-10$;
$R_F=40$ mph, $R_E=50$ mph

2. \$120 **3.** 125 **4.** 550

5. $N_G = 42, N_S = 40$ **6.** 23, 25, 27 **7.** $\frac{3xy + 4x}{6ax + 12y}$ **8.** $T_E = \frac{80}{23}, T_W = \frac{58}{23}$

9. $\frac{3 \pm \sqrt{89}}{4}$ **10.** $1 \pm \sqrt{6}$ **11.** $-1 \pm 2\sqrt{3}$ **12.** $\frac{3 \pm \sqrt{29}}{5}$

13. 5, −2 **14.** $1 \pm \sqrt{6}$ **15.** $-1 \pm 2\sqrt{3}$ **16.** $(3x - 1)(x + 5)$

17. $3(x + 9)(x - 1)$ **18.** $(2x - 1)(x + 5)$ **19.** $(m^2 - c)(k - 2)$

20. $(6 - xy)(a + b)$ **21.** $(ab + c)(x - 2y)$ **22.** $(4x + ab)(n - m)$

23. $y = \frac{9}{5}x + \frac{7}{5}$ **24.** $y = \frac{2}{5}x + \frac{29}{5}$

25. [graph] **26.** $8\sqrt{2} + 12$ **27.** $\frac{\sqrt{6}}{4}$ **28.** $1 + \frac{\sqrt{2}}{2}$

29. Volume = $(384 - 16\pi)(12)(12)(12)$ in.$^3 \approx 576{,}737.28$ in.^{3};
surface area = $(224 + 32\sqrt{13} + 12\pi)(12)(12)$ in.$^2 \approx 54{,}296.30$ in.^{2}

30. $\frac{41}{7}$

practice

a. $\frac{1}{13}$ **b.** $\frac{1}{8}$

problem set 132

1. [diagram] $R_P T_P = R_B T_B$, $T_P = 6$, $T_B = 72$, $R_P = R_B + 11$;

$R_P = 12$ km/hr, $D = 72$ km

2. $W = 10{,}000{,}000$ pounds **3.** $N_H = 10$

4. 3860 **5.** $S = 34{,}000$ **6.** (a) $\frac{1}{4}$ (b) $\frac{1}{16}$ (c) $\frac{1}{64}$

7. [number line: 3 4 5 6 7 8 9 10, dots at 4, 5, 6, 7, 8, 9] **8.** (3, −7) **9.** (3, 3)

10. 5×10^7 **11.** (a) $y = 4$ (b) $y = -\frac{2}{3}x - 2$ **12.** $3x^2 - 9x + 27 - \frac{86}{x + 3}$

13. $-36\sqrt{2} + 60\sqrt{6}$ **14.** $-3 \pm \sqrt{7}$ **15.** $x = 4$ or $x = -\frac{1}{2}$ **16.** $x = -\frac{2}{3}$ or $x = \frac{2}{3}$

17. $x = -1$ or $x = -5$ **18.** $x = 21$ **19.** Reals, rationals, integers, wholes, naturals

20. $x = \frac{27}{5}$ **21.** $T_R = 1, T_H = 6$ **22.** $x = \frac{7}{2} \pm \frac{\sqrt{29}}{2}$ **23.** $x = 4$ or $x = -1$

24. [number line: −8 −7 −6 −5 −4 −3 −2 −1 0, dots at −8, −7, −6, −2, −1, 0] **25.** $y = -\frac{2}{5}x + \frac{14}{5}$

26. One solution is $5 + 0 = 5$. **27.** (a) $\frac{\sqrt{10}}{4}$ (b) $\frac{5 + 3\sqrt{5}}{5}$ **28.** $x^{19}y^{-7}$

29. $10x^2y^{-2} - 3x^6y^2$ **30.** $(90\sqrt{3} - 10\pi)$ m$^3 \approx 124.48$ m^3

practice set 1

1. $R_ET_E = R_FT_F$, $T_E = 25$, $T_F = 30$, $R_F = R_E - 15$; $R_E = 90$ km/hr, $R_F = 75$ km/hr

2. \$500 3. 63 4. 260

5. \$37,071.30; \$15,071.30 6. $-6, -4, -2, 0$ 7. $\frac{6azy + xy}{8amz + 4zy}$

8. $T_E = \frac{75}{19}$, $T_W = \frac{20}{19}$ 9. $x = \frac{-1 \pm \sqrt{61}}{6}$ 10. $x = 1 \pm \sqrt{2}$ 11. $x = -\frac{1}{2} \pm \frac{\sqrt{105}}{6}$

12. $x = \frac{5 \pm \sqrt{73}}{8}$ 13. $x = 2 \pm \sqrt{6}$ 14. $x = -\frac{3}{2} \pm \frac{\sqrt{41}}{2}$ 15. $x = -\frac{1}{2} \pm \frac{\sqrt{21}}{2}$

16. $(x + 6)(2x - 1)$ 17. $3(x - 1)(x + 11)$ 18. $(x - 7)(3x + 1)$ 19. $(a - 4)(x^3 + 5)$

20. $(7 + cm)(z - x)$ 21. $(5 - mn)(a + b)$ 22. $(s - x)(7y + fz)$ 23. $y = \frac{11}{5}x - \frac{3}{5}$

24. $y = \frac{1}{5}x + \frac{23}{5}$ 25. (graph) 26. $29 + 17\sqrt{3}$ 27. $\frac{\sqrt{35}}{7}$

28. $2 + \frac{2\sqrt{3}}{3}$ 29. $81x^4 - 72x^2z^2 + 16z^4$ 30. Volume $= 90\pi$ in.$^3 \approx 282.6$ in.^{3}

practice set 2

1. $R_RT_R = R_GT_G$, $T_R = 7$, $T_G = 63$, $R_R = R_G + 16$; $R_R = 18$ km/hr, 126 km

2. 122 pounds 3. 22 4. 5510 5. 18,200 6. $\frac{1}{36}$

7. (number line: −8 −6 −4 −2 0 2 4 6) 8. (1, 1) 9. $(-2, -1)$

10. 1×10^{13} 11. (a) $y = \frac{5}{2}x + 2$ (b) $y = -2$ 12. $5x^2 - 20x + 80 - \frac{322}{x + 4}$

13. $3\sqrt{3} - 180\sqrt{2}$ 14. $x = \frac{-5 \pm \sqrt{13}}{2}$ 15. $x = -1, x = \frac{7}{3}$ 16. $z = \pm\frac{3}{5}$

17. $x = -5, x = -\frac{1}{2}$ 18. $x = 135$ 19. Reals, irrationals 20. $\frac{b^5 + 3}{b(4b^2 - 1)}$

21. $T_X = \frac{9}{4}$, $T_H = \frac{21}{4}$ 22. $x = \frac{-5 \pm \sqrt{41}}{2}$ 23. $x = -1, x = 5$

24. (number line: −10 −8 −6 −4 −2 0 2) 25. $y = -\frac{3}{4}x - \frac{7}{4}$

26. $2(3 + 4) = 2(3) + 2(4)$ 27. $\frac{\sqrt{6}}{2}$ 28. $1 + \frac{7\sqrt{2}}{2}$ 29. $\frac{k}{z^3}$

30. Volume of cylinder $= \frac{(864 + 144\pi)(12)(12)(12)(2.54)(2.54)(2.54)}{100(100)(100)}$ m$^3 \approx 37.27$ m^3;
surface area $= \frac{(408 + 104\pi)(12)(12)(2.54)(2.54)}{100(100)}$ m$^2 \approx 68.24$ m^2;
volume of cone $= \frac{(288 + 48\pi)(12)(12)(12)(2.54)(2.54)(2.54)}{100(100)(100)}$ m$^3 \approx 12.42$ m^3

practice set 3

1.  $R_C T_C + R_P T_P = 3500$, $T_C = 9$, $T_P = 4$, $R_C = R_P + 100$; $R_P = 200$ mph, $R_C = 300$ mph

2. Polymorphous = 11,550; uniform = 5250

3. 7353 **4.** $N_O = 895$, $N_F = 541$ **5.** \$59,011.10; \$40,011.10

6. 21, 23, 25, 27 **7.** $\frac{-5 \pm \sqrt{85}}{6}$ **8.** $-1, \frac{8}{3}$ **9.** $-4 \pm 3\sqrt{2}$ **10.** $-3 \pm 2\sqrt{3}$

11. $(11x - 6y^3)(11x + 6y^3)$ **12.** $(x - 6)(7x + 1)$ **13.** $(b - 2c^3)(a^2 + m)$

14. $(a + 2d^3)(7m^3 + 6x)$ **15.** $\frac{1}{338}$ **16.** 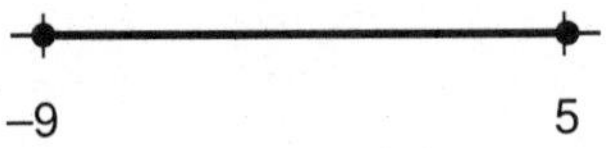 (number line with closed endpoints at −9 and 5) **17.** $\emptyset$

18. $y = -x + 2$ **19.** $x = \frac{2}{3}x + 1$ **20.** $\frac{x + 6mz}{9cm + 3z}$ **21.** $T_E = 2$, $T_W = 6$

22. Reals, rationals, integers, naturals, wholes **23.** $\frac{47}{5}$ **24.** 22 **25.** $\frac{\sqrt{6}}{4}$

26. $\sqrt{5} + \sqrt{2}$ **27.** $24 + 14\sqrt{3}$ **28.** $y^2 - py + \frac{1}{4}p^2$ **29.** $\frac{3x^2y + 14x^{14}}{yz^4}$

30. Volume $= \frac{(4500 - 245\pi)(12)(12)(12)(2.54)(2.54)(2.54)}{100(100)(100)}$ m$^3 \approx 105.64$ m^3

practice set 4

1. 120 lb/in^2 **2.** 32,000 lb/in.^{2} **3.** 4.118×10^{16}

4. 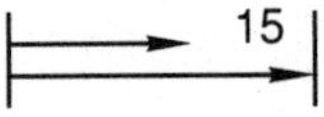 $R_D T_D = R_R T_R - 15$, $R_D = 55$, $R_R = 65$, $T_R = T_D - 3$; 3 a.m.

5. 39 **6.** 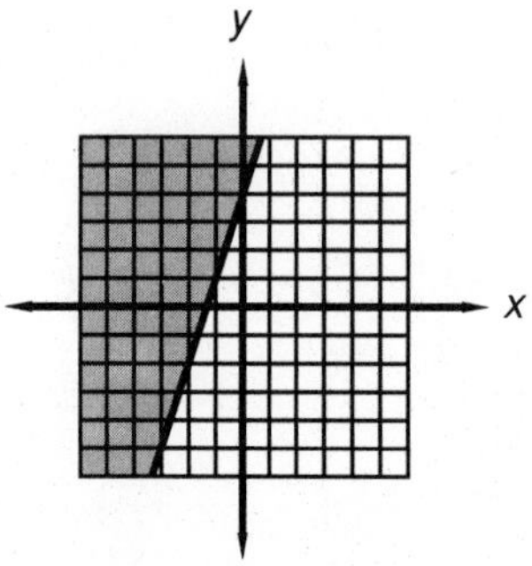 (graph) **7.** $(x + 3)(2x + 7)$ **8.** $(x - 7)(5x + 1)$

9. $(4 - as)(m + xy)$ **10.** $x(-5c - 3)(z - 1)$ **11.** $49x^2 - 14x\sqrt{13} + 13$

12. $3x^2 + 6x + 3$ **13.** $\frac{\sqrt{21}}{6}$ **14.** $\frac{77 + 3\sqrt{11}}{11}$ **15.** $\frac{3mz + 6p^2z}{3mp^2 + 5pz}$

16. $18 + 13\sqrt{2}$ **17.** Reals, rationals **18.** $\frac{-7 \pm \sqrt{37}}{2}$ **19.** $\frac{-5 \pm \sqrt{73}}{8}$

20. $\frac{-9 \pm \sqrt{93}}{2}$ **21.** $\frac{5 \pm \sqrt{29}}{2}$ **22.** $y = \frac{1}{3}x + \frac{11}{3}$ **23.** $y = -\frac{1}{4}x + \frac{9}{4}$

24. $T_E = 6$, $T_X = 4$ **25.** $\left(1, \frac{1}{2}\right)$ **26.** (5, 10) **27.** 1.54×10^7 **28.** y^{-9}

29. $-9x^4y^{-6}$ **30.** Volume $= (3240 - 162\pi)(2.54)(2.54)(2.54)$ cm$^3 \approx 44{,}758.32$ cm^3

practice set 5

1. 1024 m **2.** 1452 ft **3.** 135 **4.** Bovine = 19,080; porcine = 14,840

5. 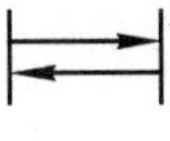 $R_C T_C = R_H T_H$, $R_C = 3$, $R_H = 8$, $T_C + T_H = 11$; 24 km

6. (a) $y = \frac{2}{3}x - 2$ (b) $y = 5$ **7.** b, d **8.** 96

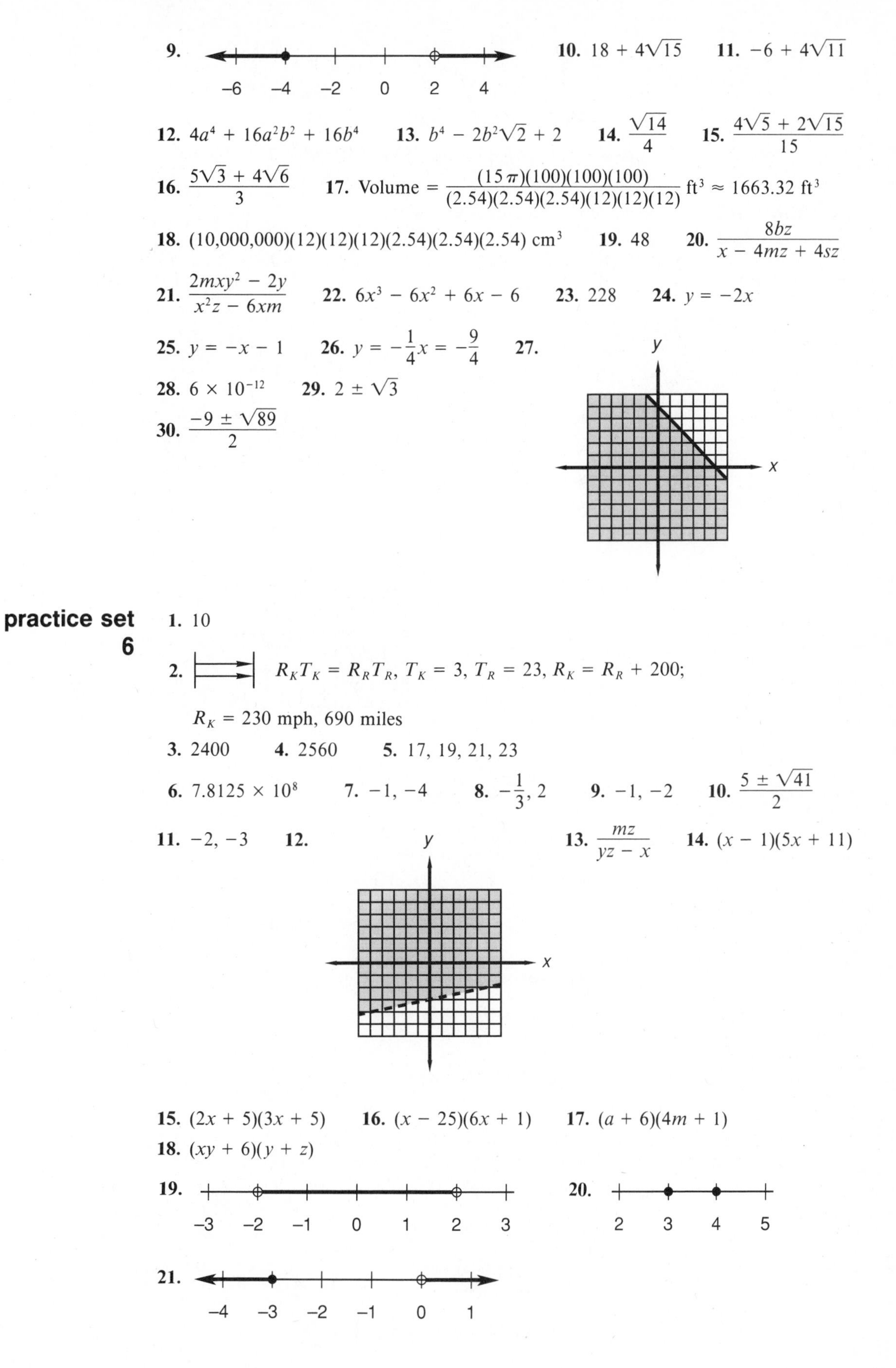

9. (number line: closed dot at −4 shaded left, open dot at 2 shaded right; labels −6, −4, −2, 0, 2, 4)   10. $18 + 4\sqrt{15}$ 11. $-6 + 4\sqrt{11}$

12. $4a^4 + 16a^2b^2 + 16b^4$ 13. $b^4 - 2b^2\sqrt{2} + 2$ 14. $\frac{\sqrt{14}}{4}$ 15. $\frac{4\sqrt{5} + 2\sqrt{15}}{15}$

16. $\frac{5\sqrt{3} + 4\sqrt{6}}{3}$ 17. Volume $= \frac{(15\pi)(100)(100)(100)}{(2.54)(2.54)(2.54)(12)(12)(12)}$ ft^3 $\approx$ 1663.32 ft^3

18. $(10{,}000{,}000)(12)(12)(12)(2.54)(2.54)(2.54)$ cm^3 19. 48   20. $\frac{8bz}{x - 4mz + 4sz}$

21. $\frac{2mxy^2 - 2y}{x^2z - 6xm}$ 22. $6x^3 - 6x^2 + 6x - 6$ 23. 228   24. $y = -2x$

25. $y = -x - 1$ 26. $y = -\frac{1}{4}x = -\frac{9}{4}$ 27.

28. 6×10^{-12} 29. $2 \pm \sqrt{3}$

30. $\frac{-9 \pm \sqrt{89}}{2}$

practice set 6

1. 10

2. $R_K T_K = R_R T_R$, $T_K = 3$, $T_R = 23$, $R_K = R_R + 200$;

$R_K = 230$ mph, 690 miles

3. 2400   4. 2560   5. 17, 19, 21, 23

6. 7.8125×10^8 7. −1, −4   8. $-\frac{1}{3}$, 2   9. −1, −2   10. $\frac{5 \pm \sqrt{41}}{2}$

11. −2, −3   12.

13. $\frac{mz}{yz - x}$ 14. $(x - 1)(5x + 11)$

15. $(2x + 5)(3x + 5)$ 16. $(x - 25)(6x + 1)$ 17. $(a + 6)(4m + 1)$

18. $(xy + 6)(y + z)$

19. (number line: open dots at −2 and 2, shaded between; labels −3, −2, −1, 0, 1, 2, 3)   20. (number line: closed dots at 3 and 4; labels 2, 3, 4, 5)

21. (number line: closed dot at −3 shaded left, open dot at 0 shaded right; labels −4, −3, −2, −1, 0, 1)

22. 20 **23.** $\frac{39}{2}$ **24.** $y = \frac{5}{4}x + 3$ **25.** $y = -\frac{2}{3}x$ **26.** $y = -\frac{3}{5}x + \frac{4}{5}$

27. $\frac{2\sqrt{7}}{7}$ **28.** 4 **29.** 2.9×10^{-3}

30. Volume = $(96 + 16\pi)(2.54)(2.54)(2.54)$ cm^3 $\approx$ 2396.44 cm^3

practice set 7

1. Boisterous = 225, quiescent = 585

2. $R_1T_1 = R_2T_2$, $T_1 = 3$, $T_2 = 4$, $R_1 = R_2 + 20$; 240 km

3. $\frac{1}{16}$ **4.** 32

5. $-1, 1, 3, 5$ **6.** \$24 **7.** $\frac{3m + 5ax}{4am + 7x}$ **8.** $T_E = \frac{686}{15}$, $T_W = \frac{214}{15}$

9. $\frac{-1 \pm \sqrt{177}}{22}$ **10.** $-\frac{1}{2}, \frac{5}{4}$ **11.** $\frac{-13 \pm \sqrt{205}}{2}$ **12.** $\frac{7 \pm \sqrt{37}}{2}$

13. $(2x - 7)(4x - 7)$ **14.** $(x + 1)(5x - 13)$ **15.** $(f - 4)(ap^2 + mn)$

16. $(a + 5)(5a^3p - 7)$ **17.**

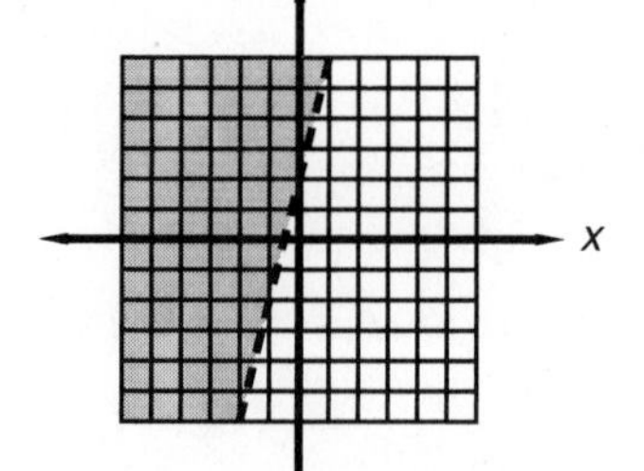

18. Volume = $\frac{20\pi}{(2.54)(2.54)(2.54)(12)(12)(12)}$ ft^3 $\approx 2.22 \times 10^{-3}$ ft^3;

surface area = $\frac{4\pi + 20\pi\sqrt{2}}{(2.54)(2.54)(12)(12)}$ ft^2 $\approx$ 0.11 ft^2

19. 44 **20.** 6.53×10^{-32} **21.** $3 + \sqrt{7}$

22. $25 + 11\sqrt{5}$ **23.** $3m^6p^{-2} - 4m^4p^{-2}$ **24.** 20 **25.** $\frac{my^2}{4my + mxy - 2x - 3}$

26. 2 3 4 5 6 **27.** $y = \frac{1}{2}x + \frac{1}{2}$ **28.** $y = \frac{3}{4}x + \frac{7}{2}$

29. $y = -\frac{3}{8}x - \frac{1}{4}$ **30.** Volume = $\frac{\left(64 - \frac{8}{3}\pi\right)(12)(12)(12)(2.54)(2.54)(2.54)}{100(100)(100)}$ m^3 $\approx$ 1.58 m^3

practice set 8

1. Equine = 244, piscine = 7442 **2.** 12,500

3. $R_WT_W = R_RT_R$, $R_W = 4$, $R_R = 10$, $T_W + T_R = 14$; 40 miles

4. $-11, -9, -7, -5$

5. $6 \pm 4\sqrt{2}$ **6.** $\frac{11 \pm \sqrt{113}}{2}$ **7.** $-1, -\frac{2}{7}$ **8.** $\frac{-3 \pm \sqrt{6}}{2}$

9. (a) $y = 3$ (b) $y = -\frac{2}{3}x - 2$ **10.** $\frac{1}{125}$

11. 12. 13.

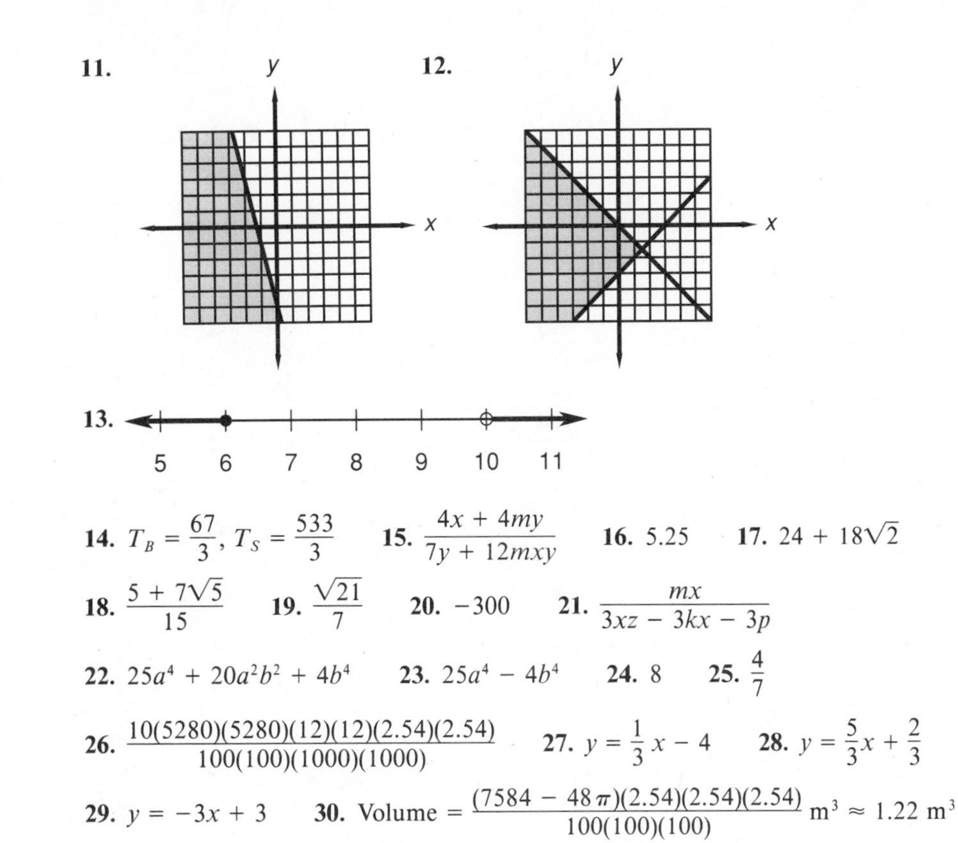

14. $T_B = \frac{67}{3}, T_S = \frac{533}{3}$ **15.** $\frac{4x + 4my}{7y + 12mxy}$ **16.** 5.25 **17.** $24 + 18\sqrt{2}$

18. $\frac{5 + 7\sqrt{5}}{15}$ **19.** $\frac{\sqrt{21}}{7}$ **20.** -300 **21.** $\frac{mx}{3xz - 3kx - 3p}$

22. $25a^4 + 20a^2b^2 + 4b^4$ **23.** $25a^4 - 4b^4$ **24.** 8 **25.** $\frac{4}{7}$

26. $\frac{10(5280)(5280)(12)(12)(2.54)(2.54)}{100(100)(1000)(1000)}$ **27.** $y = \frac{1}{3}x - 4$ **28.** $y = \frac{5}{3}x + \frac{2}{3}$

29. $y = -3x + 3$ **30.** Volume $= \frac{(7584 - 48\pi)(2.54)(2.54)(2.54)}{100(100)(100)}$ m$^3 \approx 1.22$ m^3

practice set 9

1. \$207,031 **2.** \$8406.62, \$6906.62 **3.** 3 **4.** 12,600

5. $R_E T_E = R_F T_F$, $T_E = 24$, $T_F = 30$, $R_F = R_E + 17$; $R_E = 85$ mph, $R_F = 68$ mph

6. $\frac{1}{169}$ **7.**

−3 −2 −1 0 1 2 3

8. (a) $x = -5$ (b) $y = \frac{2}{3}x - 2$ **9.** $(1, -1)$ **10.**

y

x

11. $23 + 10\sqrt{5}$ **12.** $16a^4 + 32a^2b^2 + 16b^4$

13. 2.29×10^{-13} **14.** $\frac{2}{3}x^{-9}y$

15. $\frac{3\sqrt{3} + \sqrt{15}}{3}$ **16.** -36 **17.** $-\frac{5}{9}$ **18.** $\frac{1}{3}$

19. Volume $= \frac{324(12)(12)(12)(2.54)(2.54)(2.54)}{100(100)(100)}$ m$^3 \approx 9.17$ m^3

20. $\frac{-3 \pm \sqrt{89}}{8}$ **21.** $-\frac{3}{2}, 3$ **22.** $\frac{5 \pm \sqrt{37}}{2}$ **23.** $2 \pm \sqrt{7}$ **24.** $T_B = 45, T_C = 15$

25. a, c, d **26.** $7x^3 + 7x^2 + 6x + 6 + \frac{7}{x - 1}$ **27.** $6\pi + 26\pi\sqrt{3}$ in.$^2 \approx 160.24$ in.^{2}

28. $\frac{1000(5280)(5280)(5280)(12)(12)(12)(2.54)(2.54)(2.54)}{100(100)(100)(1000)(1000)(1000)}$ km^3 **29.** $y = \frac{7}{4}x - \frac{9}{4}$

30. $y = -4x + 15$

practice set 10

1. 8, 10, 12, 14

2. $R_R T_R = R_W T_W$, $T_R = 9$, $T_W = 81$, $R_R = R_W + 16$;

$R_R = 18$ km/hr, 162 km

3. 21,870 **4.** $\frac{1}{216}$

5. 72 **6.** $N_N = 440$ **7.** $3(x + 3)(2x - 1)$ **8.** $(n - o)(3m + 4y)$

9. $\frac{\sqrt{33}}{11}$ **10.** $\frac{55 + 3\sqrt{11}}{11}$ **11.** $\frac{2pz + 8cxz}{3c^2x - 4cz}$ **12.** $10 + 3\sqrt{2}$ **13.** 1.2×10^{27}

14. $\frac{33x^{10} - 3x^{14} - 8x^{18}}{3y^8}$ **15.** Reals, irrationals **16.** $64x^8 + 48x^4p^2 + 9p^4$

17. All integers **18.** (a) $y = -\frac{4}{3}x - 4$ (b) $x = 5$ **19.** $-\frac{9}{4}$

20. $\frac{my}{12c - 20m + 4mz}$ **21.** $\frac{76}{9}$ **22.**

y

x

23. 1, 4 **24.** $\frac{9 \pm \sqrt{73}}{2}$ **25.** $1 \pm \frac{\sqrt{15}}{3}$

26. $\frac{11 \pm \sqrt{181}}{10}$

27. Volume $= \frac{4800 + 800\pi}{(2.54)(2.54)(2.54)}$ in.$^3 \approx 446.27$ in.^{3}

28. $y = 2x + 2$ **29.** $y = -\frac{2}{5}x + \frac{22}{5}$

30. $y = -\frac{2}{3}x + 5$

Index